Handbook of Biosensors and Biosensor Kinetics

Handbook of Biosensors and Biosensor Kinetics

Ajit Sadana
Neeti Sadana

ELSEVIER

Amsterdam • Boston • Heidelberg • London • New York • Oxford
Paris • San Diego • San Francisco • Singapore • Sydney • Tokyo

Elsevier
Radarweg 29, PO Box 211, 1000 AE Amsterdam, The Netherlands
Linacre House, Jordan Hill, Oxford OX2 8DP, UK

First edition 2011

British Library Cataloguing in Publication Data
A catalogue record for this book is available from the British Library

Library of Congress Cataloging-in-Publication Data
A catalog record for this book is available from the Library of Congress

ISBN: 978-0-444-53262-6

Printed and bound in Great Britain
11 12 10 9 8 7 6 5 4 3 2 1

Contents

Preface

This is the sixth book in the series of books written by one of the co-authors (A.S.), and the second by the other co-author (N.S.). In contrast to the previous five books that dealt with just the kinetics of binding and dissociation (if applicable) of analyte-receptors on biosensor surfaces, the present book analyzes a range of biosensor features, and is aptly titled a handbook. Some of the features analyzed besides kinetics include, in chronological order: fabrication of biosensors (Chapter 3), nanobiosensors (Chapter 5), binding of the same analyte on different biosensor surfaces (Chapter 6), binding of the same analyte (glucose) to different biosensor surfaces (Chapter 7), detection of gases on biosensor surfaces (Chapter 10), and detection of analytes on arrays/microarrays/DNA chips (Chapter 11).

It is hoped that the additional areas, besides the kinetics, covered in the present handbook will help provide a better perspective of the potential of biosensors and where they may be used more effectively. Biosensors are excellent medical devices, and it is anticipated that they will be used in more and more areas to advantage, especially for the detection of biomarkers for different diseases and their diagnosis. If this handbook along with others provides (or aids in the development of) ideas for the future development of biosensors in newer areas, then its purpose is well served, and the time and effort spent in writing it is well worth it.

The co-author, Ajit Sadana, expresses his appreciation to Dr. Alexander H.-D. Cheng and Kai-Fong Lee, Deans, School of Engineering at the University of Mississippi for their continued support and encouragement for research that facilitates the writing of treatises like this one. This co-author also wishes to thank Dr. Guang Shi for his help in putting this book together.

—Neeti Sadana, MD, Boston, Massachusetts
—Ajit Sadana, Oxford, Mississippi

CHAPTER 1

Introduction

Chapter Outline

1.1 Introduction

Biosensor applications have expanded considerably over the last few years after a modest beginning with the control application for the detection of glucose in the management of diabetes mellitus (DM). This is due to the ease of application of these biosensors to different areas of application. Some of the more recent areas of application include:

(a) Quantum dot biosensors for ultrasensitive multiplexed diagnostics (Berger, 2010).

(b) GE Healthcare have purchased Biacore™ and Microtal™ and combined the use of the SPR (surface plasmon resonance) biosensor with ITC (isothermal titration calorimetry) to provide orthogonal label-free determinations of affinity, binding kinetics, residence time, and enthalpic efficiency. They have also developed Bia2iTC software that compares outputs for both instruments (GE Healthcare, personal communication, 2010).

(c) Other interesting biosensor applications include (SPIE, Optics/Photonics in Security and Defence, 2006):
   (1) Unmanned/unattended sensors and sensor networks
   (2) Technologies for optical countermeasures
   (3) Optically based biological and chemical detection for defence
   (4) Femtosecond phenomena and nonlinear optics
   (5) Optics and photons for counter-terrorism and crime-fighting

(d) Different sensing platforms have been integrated to achieve results unattainable by a single sensor. For example, electrochemical, mechanical, electrical, and optical signal transduction have been integrated (Center for Biosensors and Bioelectronics, 2010).

(e) Recently, Silicon Kinetics, Inc. (2010) announced the development of new biosensor chips with different surface chemistries. These surface chemistries could accommodate

Handbook of Biosensors and Biosensor Kinetics. DOI: 10.1016/B978-0-444-53262-6.00001-2

carboxyl, streptavidin, benzaldehyde, and Ni-NTA for Histidine-tagged Protein A and Protein G for coupling antibodies. This was the first company that introduced the 3D biosensor surface for label-free biomolecular interaction analysis. The company indicates that their biosensor permits high throughput ranking in multiwall plates, and data-rich kinetic measurements in flow cells, using a single reader instrument and the same chemistry. The company emphasizes that its sensor permits the measurements of slow off-rates which are not measurable by traditional SPR biosensors.

(f) Uludag et al. (2010) have recently indicated that nucleic acid based recognition of viral sequences may be used for the rapid and accurate confirmation of viral infection. This is done using label-free biosensing. The authors indicate that gold nanoparticles may be used to enhance detection sensitivity. Quartz crystal microbalance biosensors may be used upon surfaces where nanoparticle oligonucleotides conjugates are complementary to surface-immobilized ss DNA probes. The authors indicate that their signal amplification assays may be used for the detection of specific DNA sequences of Herpes Simplex Virus (HSV) type 1.

More importantly, the authors developed the biosensor to understand the influence of mass transport in the flow cell (which incidentally is one of the major themes of this book), and the binding kinetics of targets to nanoparticles in solution. Other parameters analyzed include the binding geometries of the targets in the nanoparticle, and the packing of nanoparticles on the surface. All of this points to the influence of the biosensor surface characteristics, and how it may influence the binding kinetics. This is in fact one of the major themes of the book wherein the degree of heterogeneity on the biosensor surface is characterized by a fractal dimension, and how this fractal dimension influences the binding and dissociation kinetics.

The authors conclude by indicating that their analytical model permitted the determination of optimal nanoparticle diameter, concentration, and probe density. Their results were based on both numerical analysis as well as on subsequent associated experimental data. They further emphasize that their analysis suggests that the proximal contact area between the particle and the sensor surfaces, and the available capture area of the particle and the binding dynamics to this capture area very significantly influence the detection limit.

(g) Some of the more recent biosensor applications were presented at Biosensors 2010 held in Glasgow, Scotland from May 25-28, 2010 (http://www.biosensors=congress.elsevier.com). Some of the more interesting ones include:
   1. Direct detection of drugs in serum with an electrochemical aptamer-based biosensor (Rowe and Plaxco, 2010)
   2. The development of a novel theranostic platform for cytotoxicity evaluation of amyloid-forming neurodegenerative causative proteins (Kim et al., 2010)
   3. Theranostic biochips—from biosensors to personalized medicine (Bachmann, 2010)

4. Microfabricated electrochemical probe for the detection of signaling proteins released by single cells (Corgler et al., 2010)
5. Enhancing sensitivity of molecular biosensors through self-assembly (Wei et al., 2010)
6. Detection of amyloid precursor protein (APPTTO) using spectroscopic ellipsometry and QCM techniques (Mustafa et al., 2010)
7. Biomolecule/metal nanoparticle composites on electrodes for sensing biofuel cells and photoelectrochemical applications (Tel-Vered et al., 2010)
8. Single molecule applications for high-resolution AFM topography and recognition imaging (Ebner et al., 2010)
9. Biomarker pattern analysis as an analytical system in the evaluation of an atherosclerotic rich profile (Siegel et al., 2010)
10. Biosensors: a mixed market (Neuman and Turner, 2010)

A conference held in Baltimore, Maryland from May 5-7, 2010 on, "Sample preparation for virus toxin, and pathogen detection and identification" (Mello, 2010) discussed different sample preparation procedures to assist in the detection of harmful pathogens. McLaughlin (2010) of the U.S. Army indicates that the detection and identification of biological warfare agents rely heavily on the ability to purify, enrich, and concentrate molecular targets prior to analysis. Note that the use of analytical technologies in the field is complicated or limited by available methods for processing a wide variety of sample types into a form compatible with multiple analytical technologies, such as biosensors.

Some of the presentations scheduled are on:

(a) Sample preparation and microfluidic technologies (Maricella, 2010)
(b) Point-of-test sample prep and molecular analysis (Gau, 2010)
(c) Sample preparation as a part of an integrated fluidic process for rapid diagnostics (Clarkson, 2010)
(d) A fully integrated system for nucleic acid-based detection of bacteria and viruses in biological samples at POC (point-of-care) (Bau, 2010)
(e) A novel preanalysis system for rapid, quantitative diagnostics (Feaster, 2010)

At the recent 2nd European Congress on Immunology in Berlin, Germany (from September 13-16, 2009) there were quite a few research papers including poster presentations that analyzed different biosensor applications for use in clinical laboratories, not the ones for general home-use such as the glucose biosensor. Both researchers and vendors of different biosensor applications confirmed the above view that the biosensors were being developed for clinical laboratory usage only. Very little, if any, market research or survey has been done to help develop these newer biosensors for home use. Of course, economic feasibility will still have to be discussed with the administration to bring these biosensors into home use.

Nevertheless, some of the newer biosensor applications presented at the above-mentioned Immunology Congress include:

(a) Localization and fine mapping of an antigenic site on the nucleocapsid protein of human parainfluenza virus type 3 (Sezaile et al., 2009). The authors point out that human parainfluenza virus type 3 (hPIVN3) is a respiratory tract pathogen. The current study was performed by the authors to investigate immunodominant regions of hPIV3 nucleocapsid (N) protein by using monoclonal antibodies (mAbs) raised against recombinant N protein and human serum specimens from hPIV3 infected individuals. According to the authors the present study enhances the knowledge of the antigenic structure and should facilitate the development of better diagnostic methods for hPIV3 infection.

(b) Determination of myelin basic protein (MBP)-reactive antibodies in healthy individuals and patients with multiple sclerosis (MS) using a novel highly sensitive assay (Hedegard et al., 2009). The authors point out that autoantibodies to MBP are apparently absent in sera from healthy individuals but their presence has been reported in sera from some patients with MS. The authors have developed a novel assay for anti-MBP antibodies (MBPAbs) to analyze the influence of disease-associated MBPAbs and "natural" MBPAbs in MS patients and healthy individuals, respectively.

(c) Antibodies to Aquaporin-4 in neuromyletic optica (NMO): biological relevance and use as biomarkers (Mader et al., 2009). The authors point out that NMO is a devastating neurological disease, which is clinically characterized by optic neuritis and longitudinally extensive transverse myletis (LETM). Recently autoantibodies in the serum of NMO patients have been detected. These autoantibodies target aquaporin 4 (AQP4). AQP4 is a key constituent of the blood-brain CSF (cerebrospinal fluid). This is a membrane spanning water channel protein localized mainly in the brain and spinal cord. The authors show that various assays with different sensitivity and specificity have been developed. Note that in no other organ is consistency of the internal environment more important than in the brain. In the CNS (central nervous system) a change in the composition of the interstitial fluid could lead to uncontrolled brain activity.

(d) Systematic development of a novel biomarker for diagnostic protein biochips: the rheumatoid arthritis (RA) case study (Leuking et al., 2009). The authors point out that RA is a chronic, systemic, inflammatory disease. They emphasize that diagnosis and treatment of the disease is required (almost mandatory) to prevent extensive joint damage, deformity, and functional impairment. They indicate that the diagnosis of RA patients based on cyclic citrullinated peptides (CCP) is possible with high sensitivity and specificity. However, the authors emphasize that anti-CCP autoantibodies are preferentially detected in patients with severe RA, and less frequently in patients with mild RA. Thus, it may be stated that the present panel of anti-RF and anti-CCP markers may not effectively address the heterogeneity of the disease. The authors are currently developing a novel set of diagnostic markers based on their proprietary materials. In essence, stratified patient samples are incubated with the large collection of recombinant human proteins currently available for screening purposes to detect autoantibodies against specific targets.

(e) Acoustic biosensor for characterizing immune-cell receptor/ligand interactions (Saitakis and Giseli, 2009).

The authors point out that cells of the immune system come into contact with their environment through molecules on the cell membrane. They show that the interaction of these membrane molecules with their ligands is governed by two-dimensional (2D) kinetics and affinity. The intention of the authors was to develop a technique to analyze the binding of cell membrane molecules in their native state, that is, using whole cells. Acoustic measurements were performed by the authors to monitor in real time the binding of cell suspensions on the surface. The authors claim to have developed a simple approach to detect and to characterize whole-cell receptors interacting with surface immobilized ligands. Their analysis is label-free and noninvasive for investigating membrane interactions in the immune system.

(f) Biochemical characterization of soluble HLA-DR. A potential urinary biomarker for renal transplant rejection (Ding et al., 2009)

The authors have used a sandwich ELISA to determine the efficacy of monitoring soluble Major Histocompatibility Complex Class II (sHLA-DR) in urine for renal posttransplant patients. The authors provided biochemical characterization of the biomarker. The authors noted that a novel monoclonal antibody (mAb) generated in their laboratory was an epitope on the NH_2 domain of HLA-DR alpha chain. Finally, the authors claim that soluble HLA-DR excreted into urine is a useful indicator for kidney inflammation and their test should prove useful for posttransplant monitoring.

(g) Development of an *in vitro* sensitization assay based on monocyte-derived dendritic cells (Reuter et al., 2009)

The authors have developed an assay to assess the allergic potential of active components, and for the detection of surface marker expression changes. Their *in vitro* characterization assay is based on monocyte-derived dendritic cells. The authors state that the dendritic cells, including Langerhans cells, form a sentinel network for pathogen detection, and are the most abundant antigen presenting cells in the skin. The authors conclude by pointing out that their assay provides a basic application in assessing the allergic potential of active components.

(h) Optimization of diagnostic EILSA-based tests for the detection of autoantibodies against tumor antigens in the serum of patients with colorectal cancer (Stefatic et al., 2009).

The authors point out that colorectal cancer is one of the most common cancer types worldwide, and continues to be a serious public health problem. As can be expected, early detection and diagnosis are of great importance in cancer management. The authors emphasize that present day diagnostic tests are based on the detection of tumor-associated markers. However, the authors point out that the lack of specificity and sensitivity of these markers hinders their general use in cancer screening of an average risk population. Thus, newer cancer biomarkers or better screening methods are urgently required. The authors emphasize that they have successfully optimized their diagnostic ELISA-based kits for the detection of antibodies against tumor proteins.

Quite a few different types of vendors exhibited their latest detection devices for different analytes at the 2nd European Congress of Immunology held from September 13-16, 2009 in Berlin, Germany. Once again, as indicated above and as confirmed by the author's discussions with the different vendors, the devices were being primarily developed for use in clinical laboratories and not for home use. Of course, when these detection devices are used in clinical laboratories skilled technicians will be available for performing the tests. It would be ideal to be able to use these different types of detection devices to detect the different analytes (related to specific diseases) for home use. But, for this to become a reality, they would have to comply with the regulations in different countries, for example, the Federal Drug Administration (FDA) in the United States of America. There may also be some hesitancy in home use thanks to litigation possibilities. Some companies, which get these devices ready, may be extra careful in introducing their detection devices for home use to help minimize these litigation possibilities as they have to be absolutely sure that their device is safe to use at home even by an unskilled patient or individual.

(i) Dissociation of T-cell subsets by fluorophore spot assays (Mabtech AB, Sweden, 2009)
The FluoroSpot assay manufactured by Mabtech enables the rapid detection and enumeration of double-secreting cells making it an attractive device for more advanced analysis and screening of cellular immune responses. Mabtech indicates that their FluoroSpot is a good tool for the detection of dual cytokine secretion at the single cell level. Once again, the instrument is for use only in clinical laboratories. The detection kit consists of 10 plates, 96 wells, and costs between 1000 and 2000 Euros. Besides, it takes about 3 days to run an assay. The detection kit may be used for the possible diagnosis of TB.

(j) Cell-based ELISAs by R&D Systems (2009)

The company states that their R&D Systems Cell-based ELISAs are designed to measure phosphorylated and total proteins in the same microplate well without the lysis of cells. The company emphasizes that the fluorescence signal derived from the phosphor-protein in the same well allows for the accurate correction of well-to-well variabilities such as differences in cell numbers. Their cell-based ELISAs are for research purposes only and it takes about 5 h to get results.

(k) Hycult biotech products brochure (2009)

Hycult Biotech (2009) located in Udden, The Netherlands, makes and markets antibodies and immunoassays for innate immunity and related fields. The company specializes in complement, neutrophil proteins, and acute-phase proteins (APP). They indicate that they are focused on the progress of research in the area of inflammation and cell damage caused by pathogens and oxidation factors. They claim that one of their aims is to move their products from research to diagnostic testing, for example, APP and serum protein levels change in response to inflammation and are therefore useful indicators of stress and disease. However, the functions of APPs according to them remain largely unresolved. Nevertheless, some roles of the APPs have been postulated that include regulation of inflammation processes. For example, the elevation of C-reactive protein (CRP) relates to a number of diseases including atherosclerosis, RA, and sepsis. This makes APPs an important indicator of inflammation. Also mannose-binding lectin (MBL) is another well-characterized protein which activates the lectin complement pathway and is an important element in innate immunity.

Hycult Biotech also makes products related to oxidative stress. Oxidative stress is involved in many diseases, such as atherosclerosis, Parkinson disease, and Alzheimer disease. It is also important in the ageing process. However, reactive oxygen species (ROS) may also be beneficial, and may be used by the immune system to attack and kill pathogens. Hycult Biotech points out that detection or measurement of oxidation markers is helpful to assess oxidant activity and to monitor the effectiveness of the oxidant system. Hycult Biotech states that superoxide and other free radicals contribute along with inflammation, radiation, and carcinogen exposure to DNA damage. Antibodies and assays for the detection of DNA adducts are valuable markers for studies on DNA damage. Since nitrotyrosine is a stable end-product of peroxynitrite oxidation, Hycult Biotech indicates that its assessment in plasma concentration is a useful marker of NO-dependent damage *in vivo*.

(l) Enzo Life Sciences GmbH product brochure (2009)

Enzo Life Sciences GmbH located in Lorrach, Germany makes detection kits for ROS and for reactive nitrogen species (NOS). These kits are FDA approved, cost about 150-200 Euros, and the tests take about 15-30 min. A discussion with a sales director of Enzo indicated that the tests take about 1-1.5 years to develop, and the costs for development depend on where they are developed. For example, the tests are cheaper to develop in the United States. Once again, as indicated above, the tests are for research purposes only.

Enzo Life Sciences claims that their detection kit directly monitors ROS and NOS in live cells. Free radicals and other reactive species play important roles in many physiological and pathophysiological processes. These free radicals in cells can damage proteins, DNA, and lipids. However, at lower concentrations they also serve as second messengers in cellular signaling, and play a beneficial role. Thus, detection methods are required to determine the quantitative levels of these various reactive species. Enzo Life Sciences claims that these detection kits:

(1) can also discriminate among superoxides, nitric oxides, and peroxynitrites,
(2) are highly sensitive, specific, and accurate for cell studies,
(3) eliminate nonspecific assay artifacts by using stringent manufacturing conditions,
(4) finally, permits the detection of the different species simultaneously.

Examples of the detection of some of the different analytes in research and clinical laboratories have been presented above. They are primarily for the detection of pathogens and the stages of different diseases, besides attempting, wherever possible, a better understanding of the different pathophysiological processes involved in the progression of a disease.

However, in a recent editorial, "What's in a test?" in Nature Methods, Evanko (2009) cautions that customers of genetic and genomic services need better education even more than tighter regulation. He points out that in the last 3 years there has been a considerable increase in direct to customer (DTC) services. These provide single-gene tests to companies that screen customer's DNA for polymorphisms associated with certain diseases. Evanko (2009) points out that at present even the general public has access to information that was previously limited to professionals. Tighter regulation is required since concerns have been raised by public health and consumer advocates as well as by governmental institutions.

Specifically, Evanko (2009) raises the following three issues:

(a) Analytical validity: are the tests accurate?
(b) Clinical validity: are the genetic variants associated with increased disease being tested?
(c) Clinical utility: is the information of use to the customer?

Evanko (2009) calls attention to the fact that different countries have approached this problem differently. In the United States there is little Federal oversight and the responsibility rests with each individual state. In Germany on the other hand, DTC testing, in essence, is banned. In the United Kingdom, in a recent report on genomic medicine, self-policing by the industry is recommended. Evanko (2009) opines that the onus is on the customers to educate themselves about DTC genetic services as to what they offer. For example, he suggests, that in the case of tests for a single mutation known to be associated with a specific disease, the customer should examine the details of the science involved.

The Naval Research Laboratory (NRL) (2010) in Washington, DC has developed a sensor system with assistance from industrial partners for monitoring biomolecules in healthcare, veterinary diagnostics, food safety, environmental testing, and national security. This is a highly sensitive, portable biosensor system and is called the Bead Array Counter (BARC). The NRL system contains an embedded array of giant magnetoresistance (GMR) sensors.

Their biosensor is capable of detecting 64 different analytes. NRL has been working with this GMR concept for over 10 years. The NRL-based technology has been licensed to Seahawk Biosystem Corporation in Rockville, Maryland. NRL (2010) reports that GMRs are magnetic field dependent resistors, that is, their resistance changes when subjected to an externally applied magnetic field. These GMR devices are constructed of alternating magnetic and non-magnetic metal thin-film multilayers. These multilayers are nanometers in thickness.

1.2 Biosensor Markets and Economics

Global Industry Analysts, Inc. (2010) recently indicated that the global chemical biosensor market will touch $ 17.3 billion by the year 2015. They credit this to new product applications and to expanding applications of current biosensors. This, according to them, is in spite of slower growth in mature markets such as pH testing, industrial safety, and gas monitoring systems.

GIA (2010) points out that biosensors provide low cost, compact, and low power devices for environmental monitoring and for POC medical applications. Furthermore, they permit an ease of application in warfare threat detection and in security applications. GIA further adds that the demand for DO (dissolved oxygen) biosensors will continue to grow due to poor water quality control and the urgent need to preserve natural resources. Also, microfabrication and newer manufacturing techniques will continue to increase newer applications for current biosensors.

The GIA report provides a comprehensive industry overview, product overview, market trends, and profiles of major players in the biosensor arena. Finally, the GIA report indicates that, as expected, the United States and Europe account for a major share of the chemical sensor market, and that the medical diagnostic market continues to exhibit major growth.

Frost and Sullivan (2010) have recently come out with a report on "Developments in medical sensors – opportunities for biosensors in medical diagnosis and drug discovery/therapeutics/ (technical insights)." Over the past years Frost and Sullivan have come out almost yearly with fresh reports that provide a perspective on biosensor economics and markets. In their present effort Frost and Sullivan (2010) provide an overview of the industry with a complete analysis of the technology trends, key market drivers, constraints, and challenges faced by the biosensor industry. Frost and Sullivan (2010) critically examine and analyze medical devices and diagnostics, and drug discovery and therapeutics.

Some of the biosensor market aspects analyzed in detail include:

(a) noteworthy emerging technologies and applications,
(b) role in healthcare,
(c) analysis of applications,
(d) different technological developments from companies, and analysis of technology developments,

(e) technological developments from universities and national laboratories,
(f) analysis of funding sources and critical mergers and acquisitions,
(g) key patents.

The report is an excellent resource at present. However, as is the case with all other reports that emphasize economic analysis of biosensor markets, they will lose their full value in about a year or so due to the rapidly changing biosensor market and landscape. Nevertheless, these types of reports have some lasting value as they provide a perspective, based on which one can analyze further in particular to suit one's environment.

Eisner et al. (2009) of the Center for Healthcare Management at the Leipzig Graduate School of Management in Leipzig, Germany, have come up with an interesting economic model for biosensor applications in the medical industries. They point out that many medical decisions are initiated by technical feasibility rather than by economic feasibility. The authors emphasize that economic feasibility is gradually coming to the forefront since payoffs and health economic considerations are becoming more and more important. However, economic information in the literature, if available, targets only a single economic consideration, and even that for a specific aspect or a particular cohort of patients (such as patients who have suffered a stroke). In order that this economic aspect may be addressed in a more general sense for biosensor applications and corresponding planning decisions, as well as for presentations, and negotiations, the authors have developed a methodology to collect more relevant data for these economic feasibility studies. Considering the present emphasis on health care, and the urgent need to bring health care costs down, here in United States, as well as throughout the world, the analysis should be valuable. Also, the move to include more and more families and individuals under the health care umbrella (especially those that are under-or un-insured) is bound to place more pressure on finding ways and means to cut down delivery costs in health care.

Tae-Gyu (2010) points out that transistor-based biosensors will lead to cheap biosensors (about one percent of the present cost of biosensors), and this will lead to a medical revolution. A cell phone coupled with a biosensor used on a drop of blood should reveal the health of an individual. This according to the author should take place in the next 3-5 years. The author claims that since biosensors will be produced in a manner similar to transistors, the process will go into an economic free-fall. The authors emphasize that the demand for biosensors has increased at breakneck speed. Present-day electrochemical and optically based biosensors are expensive. The author opines that in the next 3-5 years tailor-made biosensors combined with chips will be developed.

Lee (2008) has described the past, present, and future state for OTC (over-the-counter) biosensors. The author opines that the demand for specific, low-cost, rapid and easy detection of different analytes is ever increasing. A good example is, of course, the use of biosensors for glucose detection in blood for DM management. These biosensors are based on electrochemical technology. The author feels that the inherent small size and simple construction is ideally suited for POC testing. The author notes that electrochemical biosensors have also been developed to detect some key metabolites, proteins, and nucleic acids. However, the primarily small market for the detection of these types of analytes restricts commercialization of protein and nucleic acid biosensors. They have met with limited success. The authors detail the electrochemical detection of metabolites, proteins, and DNA by biosensors particularly in its application to a home-use setting. The author points out that lab-on-a-chip microdevices as well as nanosensors (silicon and nanotube effect biosensors) offer the potential for the construction of next generation biosensors, which should exhibit better performance characteristics. It is expected that future improvements with respect to the different biosensor aspects (such as transducers, biorecognition molecules (receptors on the biosensor surface), immobilization and signal transduction) should pave the way for the development of the next-generation of biosensors for the detection of different types of analytes (especially for biomarker detection for certain diseases besides glucose level monitoring in blood for DM management). These biosensors should be available at reasonable prices as OTC biosensors in pharmacies in the not too distant future.

CleanFutures (2010) indicates that they have managed to secure funding to complete testing of its biosensor technology for detecting contaminants in water, wine, and food. The technology was developed at Monash University, Victoria, Australia, and started its commercialization process with assistance from the Victoria State Government body, Nanotechnology, Victoria. The author states that Bio Innovation SA recognized the potential for biosensor development and provided funding through its business development initiative (BDI). This funding will permit CleanFutures to design and manufacture the industrial prototypes of its biosensor as well as conduct customer trials. Their AquaSens biosensor is based on a rapid, highly sensitive sensor probe that detects nitrates and phosphates in water, and sulfites in wine and food.

CleanFutures (2010) points out that sulfite in wine is a major problem for the industry since it is difficult to detect. Also, it is estimated that 1 percent of the population is sulfite-resistant, creating the need for a suitable detection device. Besides, South Australia is a leader in wine production, and so the AquaSens biosensor is a natural fit.

1.3 Chapter Contents

In the following paragraphs we describe the chapter contents very briefly. Chapter 1 is introduction. Chapter 2 describes the modeling and theory for the binding and dissociation of the different analytes on biosensor surfaces. The fractal dimension provides a quantitative measure of the degree of heterogeneity on the biosensor surface. Chapter 3 describes the fabrication of biosensors. Different types of biosensor fabrication procedures that have recently appeared in the literature are presented.

Drug discovery is an important area of investigation for biosensor applications. Chapter 4 provides examples where biosensors have been used for drug discovery. Nanobiotechnology is currently an important area of investigation. Applications of nanobiotechnology to the area of biosensors are presented in Chapter 5. The same analyte may be detected by different biosensors. Chapter 6 provides examples wherein the same analyte has been detected by different biosensors.

The detection of glucose is another important area of application of biosensors. Chapter 7 analyzes the different types of biosensors that have been used to detect glucose in solution. The detection of analytes of medical relevance is yet another important area of biosensor applications. Chapter 8 analyzes examples of medical applications of biosensors. The detection of analytes involved in physiological cellular reactions is an important area of investigation of biosensor applications. Chapter 9 analyzes examples of physiological cellular reactions detection on biosensor surfaces.

Biosensors have been used to detect different gases in the environment and in the atmosphere. Chapter 10 analyzes the detection of different gases in the atmosphere. Microarrays/Arrays/DNA chips are frequently being used to detect different analytes. Chapter 11 analyzes the detection of different analytes on Microarrays/Arrays/DNA chips. Novel biosensors are continuously being developed to detect different analytes Chapter 12 provides examples of the detection of different analytes on novel biosensing surfaces.

Chapter 13 describes the binding and dissociation (if applicable) of different analytes on biosensor surfaces. The detection of toxins and pollutants in air and water environments is an important area of investigation for biosensors. Chapter 14 provides examples of the detection of different toxins and pollutants present in these environments. Protein biomarkers are important indicators of the onset (incidence) or presence of different diseases. Chapter 15 provides examples of the detection of different protein biomarkers and other medically related analytes that indicate the incidence of different diseases.

Hybridization is an important area of investigation for biosensors. Chapter 16 provides different examples where hybridization is involved in the binding and dissociation (if applicable) of different analytes on biosensor surfaces. Information on the economics of biosensors and what it takes to start a biosensor industry is very difficult to obtain in the open literature. Companies guard this type of information very carefully, and will not disclose this information. The last chapter, Chapter 17, attempts to provide a perspective of the markets for different types of biosensors, and what it takes to set up a biosensor industry. This is a fast changing market and the numbers may change very quickly (even yearly).

References

Bachmann TT, Theranostic biochips—From biosensors to personalized medicine. In *Parallel Session 3B, Immunosensors, BIOSENSORS 2010*, Glasgow, Scotland, May 25-28, 2010.

Bau HH, A fully integrated system for nucleic acid based detection of bacteria and viruses in biological samples at POC (point-of-care). In *Sampleprep2010*, Baltimore, MD, May 6-7, 2010, email personal communication from David Mello, February 24, 2010.

Berger M, Quantum dot biosensors for ultrasensitive multiplexed diagnostics, *Nanomagazine* (2010) http://www.nanowerk.com/spotlight/February17.

Clarkson J, Sample preparation as a part of integrated fluid process for rapid diagnostics. In *Sampleprep 2010*, Baltimore, MD, May 6-7, 2010, email personal communication from David Mello, February 24, 2010.

CleanFutures, *Clean Futures Secures Funding for Biosensor*, http://bridges8,wordpress.com/2009/09/16, downloaded, February 23, 2010.

Corgler BP, D Juncker, and CA Marquette, Microfabricated electrochemical probe for the detection of signaling proteins released by single cells. In *Parallel Session 3A, Nanobiosensors, nanomaterials, and nanoanalytical systems, BIOSENSORS 2010*, Glasgow, Scotland, May 25-28, 2010.

Ding JT, RT Walker, KC Dunn, RM Parker, RW Jack, HP Marti, PT Coates, and AD McCellan, In *Paper PC 20/3, Tuesday, 2nd Congress of Immunology, Immunity for Life, Immunology for Health*, Berlin, Germany, September 13-16, 2009.

Ebner A, B Mayer, L Wilding, H Gruber, P Hunterdorfer, and J Kepler, Single molecule biosensors for high resolution AFM topography and recognition imaging. In *Parallel Session 5D, Enzyme-based biosensors, BIOSENSORS 2010*, Glasgow, Scotland, May 25-28, 2010.

Eisner CH, D Hackl, and H Weismeth, Mivro- and nanosystems in medicine, active implants, biosensors. In *World Congress on Medical Physics, and Biomedical Engineering*, Munich, Germany, September 7-12, 2009.

Enzo Life Sciences GmbH, ROS/NOS Detection, product brochure. In *2nd European Congress of Immunology, Immunity for Life, Immunology for Health*, Berlin, Germany, September 13-16, 2009.

Evanko D, Editorial, What's in a test?, *Nature Methods*, **6**(11), 783 (2009).

Feaster SR, A novel pre-analysis system for rapid, quantitative diagnostics. In *Sampleprep2010*, Baltimore, MD, May 6-7, 2010, email personal communication from David Mello, February 24, 2010.

Frost and Sullivan, Developments in medical sensors—Opportunities for biosensors in medical diagnosis and drug discovery/therapeutics (technical insights), http://www.reseacrhandmarkets.com/reports/developments_medical_sensors.htm, February 23, 2010.

Gau V, Point-of-test sample prep and molecular analysis. In *Sampleprep2010*, Baltimore, MD, May 6-7, 2010, email private communication, February 24, 2010.

GE Healthcare, gelifesciences.com, personal communication (email), February 17, 2010.

Global Industry Analysts, Inc., http://www.prweb.com/releases/chemical/sensors, February 17, 2010.

Hedegard CJ, N Chen, K Bendtzen, and CH Nielsen, Demonstration of myelin basic protein–reactive antibodies in healthy individuals an patients with multiple sclerosis using a highly sensitive assay. In *Paper PC18/38, Tuesday, 2nd European Congress on Immunology, Immunity for Life, Immunology for Health*, Berlin, Germany, September 13-16, 2009.

Hycult biotech, In *2nd European Congress of Immunology, Immunity for Life, Immunology for Health*, Berlin, Germany, September 13-16, 2009, PO Box 30,5400 AA, UDDEN, The Netherlands.

Kim J, R Harada, N Kobayashi, K Ikebukuro, and K Sode, The development of a novel theranostic platform for cytotoxicity evaluation of amyloid-forming neurodegenerative disease causative proteins. In *Theranostics, Parallel Session 2C. BIOSENSORS 2010*, Glasgow, Scotland, May 25-28, 2010.

Lee JMH, Over-the-counter biosensors: Past, present, and future, *Sensors*, **81**(9), 5535–5559 (2008). http://www.mdpi.com/downloaded February 24, 2010.

Leuking A, A Koswald, HE Meyer, M Schneider, and S Mullner, Systematic development of novel antibody biomarkers for diagnostic protein biochips—The rheumatoid arthritis case study. In *Paper PC13/7, Tuesday, 2nd European Congress of Immunology, Immunity for Life, Immunology for Health*, Berlin, Germany, September 13-16, 2009.

MABTECH AB, 2009, Box 1233, SE-131-28, Nacka Strand, Sweden.

Mader S, C Rainer, B Keuntz, T Berger, W Kristo-feritsch, and M Reindl, In *Paper PC18/25, Tuesday, 2nd European Congress of Immunology, Immunity for Life, Immunology for Health*, Berlin, Germany, September 13-16, 2009.

Maricella RP, Sample preparation and microfluidic technologies. In *Sampleprep2010*, Baltimore, MD, May 6-7, 2010, email private communication from David Mello, February 24, 2010.

Mclaughlin J, US Army Chemical Biological Medical Systems Joint Project Management Office (CBMS-JPMO), In *Sampleprep2010*, Baltimore, MD, May 6-7, 2010, email personal communication, February 24, 2010.

Mello D, 2010 email personal communication, February 24.

Mustafa MK, AV Nabok, D Parkinson, A Tsargorodskaya, F Salaam, and IE Tothill, Detection of amyloidprecursor protein (APP770) using spectroscopic, ellipsometric, and QCM techniques. In *Poster Session 1, BIOSENSORS 2010*, Glasgow, Scotland, May 25-28, 2010.

Neuman JD and APF Turner, Parallel session 3D, BIOSENSORS 2010, Glasgow, Scotland, May 25-28, 2010.

NRL, NRL partners with industry to develop compact biosensor for wide-ranging applications, http://esciencenews.com/articles/2009/02/05, February 22, 2010.

Silicon Kinetics, Silicon kinetics announces new biosensor chips with wide palette of surface chemistries, http://www.pr.com/press-release/210990, February 17, 2010.

SPIE, *Optics/Photonics in Security and Defence*, Stockholm, Sweden, September 11-16, 2006.

Center for Biosensors and Bioelectronics, http://www.biodesign.asu.edu/research/research-centres/biosensoss-and-bioelecronics, February 15, 2010.

R&D Systems Ltd., UK and Europe, 1G Bartn Lane, Abingdon Science Park, Abingdon OX14 SNB, United Kingdom, 2009.

Reuter H, J Spieker, S Gerlach, L Kolbe, W Diembeck, H Wenck, KP Wittern, and A Schelpky, Development of an *in vitro* securitization assay based on monocyte-derived dendritic cells. In *PD 18/35, Tuesday, 2nd European Congress of Immunology, Immunity for Life, Immunology for Health*, Berlin, Germany, September 13-16, 2009.

Rowe AA and KW Plaxco, Direct detection of drugs in serum with an electrochemical aptamer-based biosensor. In *Theranostics, Parallel Session 2C, Biosensors 2010*, Glasgow, Scotland, 25-28 May, 2010.

Saitakis A and E Giseli, Acoustic biosensor for characterizing immune-cell receptor/ligand interactions. In *Paper PD18/6, Tuesday, 2nd European Congress of Immunology, Immunity for Life, Immunology for Health*, Berlin, Germany, September 13-16, 2009.

Sezaile J, M Pleckaiyte, I Kucinsbachte-Kodze, M Juozapartis, and K Sasnauskas, Localization and fine mapping of an antigenic site on the nucleocapsid protein of human parainfluenza virus type 3. In *Paper PC25/02, Tuesday, 2nd European Congress of Immunology, Immunity for Life, Immunology for Health*, Berlin, Germany, September 13-16, 2009.

Siegel G, M Rodriguez, K Winkler, AR Pries, and E Ermilov, Biomarker pattern analysis as nanoanalytical system in the evaluation of atherosclerotic risk profile. In *Parallel Session 7A, Nanobiosensors, nanomaterials, nanoanalytical systems, Biosensors*, Glasgow, Scotland, May 25-28, 2010.

Stefatic D, M Riederer, and T Bauernhofer, Optimization of diagnostic ELISA-based tests for the detection of auto-antibodies against tumor antigens in serum of patients with colorectal cancer. In *PD 18/38, Immunology for Life, Immunology for Health*, Berlin, Germany, September 13-16, 2009.

Tae-Gyu K, Cheap biosensors to bring medical revolution, http://www.koreatimes.co.kr/www/nes/tech/2009, February 22, 2010.

Tel-Vered R, O Yehezkell, I Baravik, and I Willner, Biomolecule nanoparticles composites electrodes for sensing biofuel cells and photoelectrochemical applications. In *Parallel Session 5D, Enzyme-based biosensors, Biosensors*, Glasgow, Scotland, May 25-28, 2010.

Uludag Y, R Hammond, and M Cooper, A signal amplification assay for HSV type 1 viral DNA detection using nanoparticles, *Journal of Nanobotechnology*, **8**(31), 201 (2010) http://7thspace.com/headlines/a_signal_amplification_assay_forhsv_type1-viral_DNA-detection_using-nanoparticles, February 17, 2010.

Wei HP, Y Leng, FY Li, ZP Zhang, Q Cui, XE Zhang, et al., Enhancing sensitivity of molecular biosensors through self assembly. In *Parallel Session 3D, Organism and Whole-Cell Based Biosensors, BIOSENSORS 2010*, Glasgow, Scotland, May 25-28, 2010.

CHAPTER 2

Modeling and Theory

Chapter Outline

2.1 Introduction

In a biosensor based assay the molecule to be detected (analyte) is present in solution and the appropriate receptor is immobilized on a solid surface. The interaction between the analyte and the receptor on the solid biosensor surface is detected either by a change in the refractive index (in SPR (surface plasmon resonance) instruments) or by changes in the fluorometric intensity, ultraviolet light intensity, etc. The SPR biosensor protocol analyzes the binding (and dissociation where applicable) kinetic curves using classical saturation models involving analyte-receptor binding using 1:1, 1:2, etc. ratios, generally under diffusion-free conditions and assuming that the receptors are homogeneously distributed over the sensor surface. Computer programs and software that come with the equipment provide values of the binding (and the dissociation) rate coefficients. Though a careful analysis and experimental protocol may eliminate or minimize the influence of diffusional limitations; realistically speaking, it is more appropriate to include a heterogeneous distribution on the sensing surface. Heterogeneity on the sensing surface and in the biosensor systems itself may be

Handbook of Biosensors and Biosensor Kinetics. DOI: 10.1016/B978-0-444-53262-6.00002-4

due to other reasons, such as nonspecific binding, inherent irregularities on the sensing surface, mixture of receptors on the surface, and mixture of analytes in solution which includes the analyte of interest.

Two factors need to be addressed while analyzing the analyte-receptor binding and dissociation kinetics. The system by its design is heterogeneous. For example, as indicated above, the receptors immobilized on the biosensor surface may exhibit some heterogeneity, that is, surface roughness. No matter how careful one is in immobilizing the receptors on the biosensor surface, there will be some degree of heterogeneity on the surface.

Henke et al. (2002) have used the atomic force microscopy (AFM) technique to determine the effects of cleaning fused silica and glass on surface roughness. This is for biosensor use. Note that prior to the immobilization of receptors on the surface, the surface needs to be cleaned to remove contaminants, and to create surface attachment sites for example, for hydroxyl groups. For the analyte-receptor binding (and dissociation) to take place the analyte, by the diffusion process, must come within the "proximity" of the active site on the receptor. Mass transport limitations may be minimized or eliminated if the system is either properly designed or properly operated or both. In most cases, however, both diffusional effects and heterogeneity aspects will be present in biosensor systems, and their influence on binding and dissociation kinetics need to be determined. Ideally, one would like to determine the influence of each of these separately on the binding and dissociation kinetics. In the theoretical analysis to be presented below (the Havlin, 1989, analysis) the effects of diffusion and of heterogeneity are presented coupled together. One possible way of accounting for the presence of diffusional limitations and the heterogeneity that exists on the surface is by using fractals. Ideally, and as indicated above, one would prefer to decouple the influence of diffusion and heterogeneity. Presumably, an approach other than fractal analysis is required to decouple these two effects.

A characteristic feature of fractals is self-similarity at different levels of the scale. Fractals exhibit dilatational symmetry. Fractals are disordered systems, and the disorder is described by nonintegral dimensions (Pfeifer and Obert, 1989). Fractals have nonintegral dimensions, and are smaller than the dimension they are embedded in. In other words, the highest value that a fractal can have is three. In our case, an increase in the degree of heterogeneity on the biosensor surface would lead to an increase in the value of the fractal dimension. Another way of looking at the fractal dimension is its "space filling" capacity. The more the space a surface fills, the higher is its fractal dimension. The fractal dimension cannot have a negative value, and very low values of the fractal dimension on the surface indicate that the surface exists as a Cantor like dust.

Kopelman (1988) points out that surface diffusion-controlled reactions that occur on clusters or islands are expected to exhibit anomalous and fractal-like kinetics. These kinetics exhibit anomalous reaction orders and time-dependent (e.g., binding) rate coefficients. As long as surface irregularities show scale invariance they can be characterized by a single number, the fractal dimension. Later on in the book we will characterize the surfaces of the biosensors used in the different examples by a fractal dimension. More specifically, we will characterize the heterogeneity present on these biosensor surfaces by a fractal dimension.

The fractal dimension is a global property, and it is insensitive to structural or morphological details (Pajkossy and Nyikos, 1989). Markel et al. (1991) point out that fractals are scale self-similar mathematical objects that possess nontrivial geometrical properties. Furthermore, these authors state that rough surfaces, disordered layers on surfaces, and porous objects all possess fractal structure. A consequence of the fractal nature is a power-law dependence of a correlation function (in our case the analyte-receptor on the biosensor surface) on a coordinate (e.g., time).

Pfeifer (1987) shows that fractals may be used to track topographical features of a surface at different levels of scale. Lee and Lee (1995) point out that the fractal approach permits a predictive approach for transport (diffusion-related) and reaction processes occurring on catalytic surfaces. This approach may presumably be extended to diffusion-limited analyte-receptor reactions occurring on biosensor surfaces.

The binding of an analyte in solution to a receptor attached to a solid (albeit flow cell or biosensor surface) is a good example of a low dimension reaction system in which the distribution tends to be "less random" (Kopelman, 1988), and a fractal analysis would provide novel physical insights into the diffusion-controlled reactions occurring at the surface. Also, when too many parameters are involved in a reaction, which is the case for these analyte-receptor reactions on a solid (e.g., biosensor surface), a fractal analysis provides a useful lumped parameter. It is appropriate to pay particular care to the design of such systems and to explore new avenues by which further insight or knowledge may be obtained on these biosensor systems. The fractal approach is not new and has been used previously in analyzing different phenomena on lipid membranes.

Fatin-Rouge et al. (2004) have recently presented a summary of cases where the analysis of diffusion properties in random media has provoked significant theoretical and experimental interest. These cases include soils (Sahimi, 1993), gels (Starchev et al., 1997; Pluen et al., 1999), bacteria cytoplasm (Berland et al., 1995; Schwille et al., 1999), membranes (Saffman and Delbruck, 1975; Peters and Cherry, 1982; Ghosh and Webb, 1988), and channels (Wei et al., 2000). Coppens and Froment (1995) have analyzed the geometrical aspects of diffusion and the reaction occurring in a fractal catalyst pore. In this chapter, and in this book as a whole, we are extending the analysis to analyte-receptor binding (and dissociation) on biosensor surfaces.

Fatin-Rouge et al. (2004) show that in most real systems disorder may exist over a finite range of distances. Harder et al. (1987) and Havlin (1989) point out that in this range the diffusion process cannot be characterized by the classical Fick's law. In this range, anomalous diffusion applies. Fatin-Rouge et al. (2004) emphasize that at larger distances than in the above window range, the effects of disorder on diffusion may be very small due to statistical effects, and may cancel each other.

Prior to presenting the Havlin (1989) analysis modified for the analyte-receptor binding occurring on biosensor surfaces, it is appropriate to discuss briefly the analysis presented by Fatin-Rouge et al. (2004) on size effects on diffusion processes within agarose gels, and apply it to analyte-receptor binding and dissociation for biosensor kinetics. This analysis provides some insights into general fractal-related processes. Fatin-Rogue et al. (2004) have considered diffusion within a fractal network of pores. They indicate that fractal networks such as percolating clusters may be characterized by a power law distribution (Havlin, 1989):

$$M(L)^{D_f} \tag{2.1}$$

Here M is the average number of empty holes in the (gel) space characterized by a linear size, L. The exponent, D_f is the mass fraction dimension. Fatin-Rogue et al. (2004) emphasize that in the general case of fractals, D_f is smaller than the dimension of space of interest. Furthermore, the independence of D_f on scale is also referred to as self-similarity, and is an important property of rigorous fractals.

Havlin and Ben-Avraham (1987) point out that the diffusion behavior of a particle within a medium can be characterized by its mean-square displacement, $r^2(t)$ versus time, Γt, which is written as:

$$r^2(t) = t^{(2/D_w)} \tag{2.2a}$$

Here Đ is the transport coefficient, and D_w is the fractal dimension for diffusion. Normal or regular diffusion occurs when D_w is equal to 2. In this case, $r^2(t)$ is equal to t. In other words,

$$r^2(t) = 2dDt \tag{2.2b}$$

Here d is the dimensionality of space, and D is the diffusion coefficient.

Harder et al. (1987) and Havlin (1989) describe *anomalous diffusion* where the particles sense obstructions to their movement. This is within the fractal matrix, or in our case due to heterogeneities on the biosensor surface, perhaps due to irregularities on the biosensor surface. Fatin-Rogue et al. (2004) are careful to point out that anomalous diffusion may also occur due to nonelastic interactions between the network and the diffusing particles in a gel matrix (Saxton, 2001). Furthermore, Fatin-Rouge et al. (2004) indicate that *anomalous diffusion* is different from *trapped diffusion* where the particles are permanently trapped in holes, and are unable to come out of these holes. When the particles (analytes in our case) are in these trapped holes, then as time $t \to \infty$, the mean-square displacement, $r^2(t)$ tends to a constant value. Fatin-Rouge et al. (2004) emphasize that in real heterogeneous porous media anomalous diffusion of particles occurs over a limited length- or time-scale since the structure is only fractal over a limited size scale. In other words, there is a lower bound and an upper bound beyond which the fractal structure applies. Similarly, in our case, the anomalous diffusion of the analyte on the biosensor surface occurs over a limited range of length- or time-scale.

For anomalous diffusion, one may combine the right-hand side of Equations (2.2a) and (2.2b). Then, the diffusion coefficient, D is given by (Fatin-Rouge et al., 2004):

$$D(t) = (1/4)t^{[(2/D_{\mathrm{w}})-1]} \tag{2.3}$$

Due to the temporal nature of $D(t)$, it is better to characterize the diffusion of the analyte in our case by D_{w}. If we were still talking about the medium and gels, then D_{w} would refer to the diffusing medium.

We will now develop the theory for the analyte-receptor binding and dissociation on biosensor surfaces. We will use the Havlin (1989) approach.

2.2 Theory

We present now a method of estimating fractal dimension values for analyte-receptor binding and dissociation kinetics observed in biosensor applications. The following chapters will present the different examples of data that have been modeled using the fractal analysis. The selection of the binding and dissociation data to be analyzed in the later chapters is constrained by whatever is available in the literature.

2.2.1 Variable Rate Coefficient

Kopelman (1988) points out that classical reaction kinetics are sometimes unsatisfactory when the reactants are spatially constrained at the microscopic level by either walls, phase boundaries, or force fields. Such heterogeneous reactions, for example, bioenzymatic reactions, that occur at interfaces of different phases, exhibit fractal orders for elementary reactions and rate coefficients with temporal memories. In such reactions, the rate coefficient exhibits a form given by:

$$k_1 = k' t^{-\mathrm{b}} \qquad 0 \le b \le 1 \; (t \ge 1) \tag{2.4}$$

In general, k_1 depends on time whereas $k' = k_1$ $(t = 1)$ does not. Kopelman (1988) points out that in three dimensions (homogeneous space) $b = 0$. This is in agreement with the results obtained in classical kinetics. Also, with vigorous stirring, the system is made homogeneous and b again equals zero. However, for diffusion-limited reactions occurring in fractal spaces, $\mathrm{b} > 0$; this yields a time-dependent rate coefficient.

Antibodies may form fractal clusters on biosensor surfaces. These antibodies or receptors on the biosensor surface may consist of islands of highly organized or disorganized antibodies. This is similar to the growth of crystalline structures. It is quite possible that a cooperative effect may arise due to this tightly organized fractal structures. This is one possibility that could lead to an increase in the binding rate coefficient with an increase in the fractal dimension or the degree of heterogeneity on the biosensor surface.

The diffusion-limited binding kinetics of antigen (or antibody or analyte or substrate) in solution to antibody (or antigen, or receptor, or enzyme) immobilized on a biosensor surface has been analyzed within a fractal framework (Sadana and Beelaram, 1994; Sadana et al., 1995). One of the findings, for example, is that an increase in the surface roughness or fractal dimension leads to an increase in the binding rate coefficient. Furthermore, experimental data presented for the binding of HIV virus (antigen) to the antibody immobilized on a surface displays characteristic ordered "disorder" (Anderson, 1993). This indicates the possibility of a fractal-like surface.

A biosensor system (wherein either the antigen, antibody, analyte, or substrate is attached to the surface), along with its different complexities, which include heterogeneities on the surface and in solution, diffusion-coupled reaction, time-varying adsorption, or binding rate coefficients, etc., can be characterized as a fractal system.

The diffusion of reactants toward fractal surfaces has been analyzed (De Gennes, 1982; Pfeifer et al., 1984a,b; Nyikos and Pajkossy, 1986). Havlin (1989) has briefly reviewed and discussed these results. The diffusion is in the Euclidean space surrounding the fractal surface (Giona, 1992). Havlin (1989) presents an equation that may be utilized to describe the build-up of the analyte-receptor on a biosensor surface during the binding reaction. The receptor is immobilized on the biosensor surface. This equation is given below. In all fairness, at the outset, it is appropriate to indicate that the biosensor surface is assumed to be fractal, or possibly so.

Ideally, it is advisable to provide independent proof or physical evidence for the existence of fractals in the analysis of analyte-receptor reactions occurring on biosensor surfaces. Also, and as indicated earlier, if the diffusion effects can be separated from the heterogeneity effects, then one may better understand the effects of each of these on analyte-receptors reactions occurring on biosensor surfaces. In general, diffusion effects may be minimized either by increasing flow rates or by immobilizing fewer receptors on the biosensor surface.

In general, to demonstrate fractal-like behavior log-log plots of distribution of molecules $M(r)$ as a function of the radial distance (r) from a given molecule are required. This plot should be close to a straight line. The slope of log $M(r)$ *versus* log(r) plot determines the fractal dimension. In our case, one could try to obtain a log-log plot of two variables, k and time, t and perform a least squares fit in this parameter space to find the slope of the curve.

A regression coefficient at this stage could be beneficial in understanding the efficacy of this metric. However, an easier method, without the use of the required log-log plots, is presented below.

This is the equation developed by Havlin (1989) for diffusion of analytes toward fractal surfaces.

2.2.2 Single-Fractal Analysis

In the literature some authors refer to binding as comprising of two phases, an association phase and a dissociation phase. In this chapter and in the book, we will refer to binding as just binding. The dissociation phase is separate.

Binding Rate Coefficient

Havlin (1989) indicates that the diffusion of a particle (analyte) from a homogeneous solution to a solid surface (e.g., receptor-coated surface) on which it reacts to form a product (analyte-receptor complex) is given by:

$$(\text{Analyte}\cdot\text{Receptor}) \sim \begin{cases} t^{(3-D_{\text{f, bind}})/2} = t^p, & t < t_c \\ t^{1/2}, & t > t_c \end{cases} \quad (2.5a)$$

where the analyte-receptor represents the association (or binding) complex formed on the surface. Here $p = -b$, and D_f is the fractal dimension of the surface. Havlin (1989) states that the crossover value may be determined by $r_c^2 \sim t_c$. Above the characteristic length, r_c, the self-similarity of the surface is lost and the surface may be considered homogeneous. Equation (2.5a) indicates that the concentration of the product [analyte-receptor] on a solid fractal surface scales at short and intermediate times as analyte-receptor $\sim t^p$ with the coefficient $p = (3\text{–}D_f)/2$ at short time scales and $p = 1/2$ at intermediate time scales. Note that D_f, $D_{f,assoc}$, and $D_{f,bind}$ are used interchangeably. This equation is associated with the short-term diffusional properties of a random walk on a fractal surface. Note that, in perfectly stirred kinetics on a regular (nonfractal) structure (or surface), the binding rate coefficient, k_1, is a constant, that is, it is independent of time. In other words, the limit of regular structures (or surfaces) and the absence of diffusion-limited kinetics leads to k_1 being independent of time. In all other situations, one would expect a scaling behavior given by $k_1 \sim k' t^{-b}$ with $-b = p < 0$. Also, the appearance of the coefficient, p different from $p = 0$ is the consequence of two different phenomena, that is, the heterogeneity (fractality) of the surface and the imperfect mixing (diffusion-limited) condition. Finally, for a homogeneous surface where D_f is equal to 2, and when only diffusional limitations are present, $p = ½$ as it should be. Another way of looking at the $p = 1/2$ case (where $D_{f,bind}$ is equal to 2) is that the analyte in solution views the fractal object, in our case the receptor-coated biosensor surface, from a "large distance." In essence, in the association process, the diffusion of the analyte from the solution to the receptor surface creates a depletion layer of width $(Ɖ^{t})^{1/2}$ where Ɖ is the diffusion constant. This gives rise to the fractal power law, (Analyte·Receptor) $\sim t^{(3-D_{f,\text{bind}})/2}$.

The values of the parameters k (binding rate coefficient), p, and D_f in Equation (2.5a) may be obtained for analyte-receptor association kinetics data. This may be done by a regression analysis using, for example, Corel Quattro Pro (1997) along with Equation (2.5a) where (analyte·receptor) $= kt^p$ (Sadana and Beelaram, 1994; Sadana et al., 1995). The fractal dimension may be obtained from the parameter p. Since $p = (3–D_{f,\text{bind}})/2$, $D_{f,\text{bind}}$ is equal to $(3–2p)$. In general, low values of p would lead to higher values of the fractal dimension, $D_{f,\text{bind}}$. Higher values of the fractal dimension would indicate higher degrees of "disorder" or heterogeneity or inhomogeneity on the surface.

Another way of looking at the diffusive process is that it inherently involves fluctuations at the molecular level that may be described by a random walk (Weiss, 1994). This author points out that the kinetics of transport on disordered (or heterogeneous) media needs to be described by a random-walk model. When both of these are present, that is the diffusion phenomena as well as a fractal surface, then one needs to analyze the interplay of both these fluctuations. In essence, the disorder on the surface (or a higher fractal dimension, D_f) tends to slow down the motion of a particle (analyte, in our case) moving in such a medium. Basically, according to Weiss (1994), the particle (random walker analyte) is trapped in regions in space as it oscillates for a long time before resuming its motion.

Havlin (1989) indicates that the crossover value may be determined by $r_c^2 \sim t_c$. Above the characteristic length, r_c, the self-similarity of the surface is lost. Above t_c, the surface may be considered homogeneous, and "regular" diffusion is now present. One may consider the analysis to be presented as an intermediate "heuristic" approach in that in the future one may also be able to develop an autonomous (and not time-dependent) model of diffusion-limited kinetics in disordered media.

It is worthwhile commenting on the units of the association and the dissociation rate coefficient(s) obtained for the fractal analysis. In general, for SPR biosensor analysis, the unit for the analyte·receptor complex on the biosensor surface is RU (resonance unit). One thousand resonance units is generally 1 ng/(mm)^2 (of surface), or 1 RU is 1 pg/(mm)^2. Here, ng and pg are nanogram and picogram, respectively. Then, to help determine the units for the binding coefficient, k, from Equation (2.5a):

$$(\text{analyte·receptor}), \text{pg}/(\text{mm})^2 = kt^p = kt^{(3-D_{f,\text{bind}})/2}$$

This yields a unit for the binding rate coefficient, k as (pg)(mm)$^{-2}$(sec)$^{(D_{f,\text{bind}}-3)/2}$. Note that the unit of dependence in time exhibited by the association (or binding) rate coefficient, k, changes slightly depending on the corresponding fractal dimension obtained in the binding phase, $D_{f,\text{bind}}$. The fractal dimension value is less than or equal to three. Three is the highest value of the fractal dimension, since the system is embedded in a three-dimensional system. k and k_{bind}, and D_{f}, and $D_{\text{f,bind}}$ are used interchangeably in this chapter and in the book.

It should be noted that different laboratories use different technologies or different experimental designs to analyze the binding affinity of ligands to target proteins (or analytes) of interest (or to determine the rate coefficients for association and dissociation kinetics for binding). The comparison of data between different technologies and experimental designs and conclusions thereof should be made with great caution. The fractal analysis is of value in that it provides the pros and cons of different *in vitro* technologies (or more precisely, in this case, analysis procedures). It makes the user of the technology aware of the quality of data generated and what can be done to improve the analysis.

One might very reasonably question the utility of the approach, considering the different dimensions, and subsequently the units one may obtain even for the same interactions. It would be difficult to compare this technique with other approaches for different interactions. Nevertheless, the inclusion of the surface effects is essential, albeit difficult. This is especially true, if the rate coefficients for association and dissociation for binding are very significantly dependent on the nature of the surface. Unless a simpler alternative approach that includes the surface effects is suggested, it is reasonable, for now, to follow this approach. Hopefully, modifications to this approach may be suggested that permit comparison for different interactions as well as with other approaches.

It would be useful to specify the carrier of fractal properties. It could either be the analyte surface, the receptor surface, or the immobilizing (in our case, the biosensor) surface. There is a considerable body of work on fractal surface properties of proteins (Li et al., 1990; Dewey and Bann, 1992; Le Brecque, 1992; Federov et al., 1999). Le Brecque (1992) points out that the active sites (in our case, the receptors on the biosensor surface) may themselves form a fractal surface. Furthermore, the inclusion of nonspecific association sites on the surface would increase the degree of heterogeneity on the surface, thereby leading to an increase in the fractal dimension of the surface. At present, we are unable to specify what the carrier of the fractal properties is. This is exacerbated by our reanalysis of kinetic data available in the literature. Presumably, it is due to a composite of some or all of the factors mentioned above. No evidence of fractality is presented.

Dissociation Rate Coefficient

The diffusion of the dissociated particle (receptor or analyte) from the solid surface (e.g., analyte-receptor complex coated surface) into the solution may be given as a first approximation by:

$$\begin{aligned}(\text{Analyte}\cdot\text{Receptor}) &\sim -t^{(3-D_{\text{f,diss}})/2} \\ &= -k_{\text{diss}}t^{(3-D_{\text{f,diss}})/2}, \quad t > t_{\text{diss}}\end{aligned} \tag{2.5b}$$

Here $D_{f,diss}$ is the fractal dimension of the surface for the dissociation step. t_{diss} represents the start of the dissociation step. This corresponds to the highest concentration of the analyte·receptor complex on the surface. Henceforth, its concentration only decreases. $D_{f,bind}$ may or may not be equal to $D_{f,diss}$. k_d and k_{diss}, and $D_{f,d}$ and $D_{f,diss}$ are used interchangeably in this chapter and in this book.

One may obtain a unit for the dissociation rate coefficient, k_d, in a similar manner as done for the binding rate coefficient. In this case, the units for the binding and the dissociation rate coefficient are the same. The unit for the dissociation rate coefficient, k_d is (pg)(mm)$^{-2}$ (sec)$^{(D_{f,diss}-3)/2}$. Once again, note that the unit dependence on time exhibited by k_d changes slightly due to the dependence on $D_{f,diss}$.

2.2.3 Dual-Fractal Analysis

Binding Rate Coefficient

The single-fractal analysis we have just presented is extended to include two fractal dimensions. At present, the time ($t = t_1$) at which the first fractal dimension "changes" to the second fractal dimension is arbitrary and empirical. For the most part it is dictated by the data analyzed and the experience gained by handling a single-fractal analysis. The r^2 (regression coefficient) value obtained is also used to determine if a single-fractal analysis is sufficient, or one needs to use a dual-fractal analysis to provide an adequate fit. Only if the r^2 value is less than 0.97 for a single-fractal analysis, do we use a dual-fractal model. In this case, the analyte-receptor complex is given by:

$$(\text{Analyte·Receptor}) \sim \begin{cases} t^{(3-D_{f1,\text{bind}})/2} = t^{p_1}, & t < t_1 \\ t^{(3-D_{f2,\text{bind}})/2} = t^{p_2}, & t_1 < t < t_2 = t_c \\ t^{1/2}, & t > t_c \end{cases} \tag{2.5c}$$

In analyte-receptor binding, the analyte-receptor binds with the active site on the surface and the product is released. In this sense the catalytic surface exhibits an unchanging fractal surface to the reactant in the absence of fouling and other complications. In the case of analyte-receptor association, the biosensor surface exhibits a changing fractal surface to the analyte in solution. This occurs because as each association reaction takes place, smaller and smaller amounts of "association" sites or receptors are available on the biosensor surface to which the analyte may bind. Furthermore, as the reaction proceeds, there is an increasing degree of heterogeneity on the biosensor surface for some reaction systems. This is manifested by two degrees of heterogeneity or two fractal dimensions on the biosensor surface. In the theoretical limit one might envisage a temporal fractal dimension wherein there is a continuous change in the degree of heterogeneity on the surface; though of course, such situations would be very rare, if at all.

Surfaces exhibit roughness, or a degree of heterogeneity at some scale. This degree of heterogeneity on the surface may be due to fracture or erosion. In our case of biosensors, this may arise due to (a) the inherent roughness of the biosensor surface, or (b) due to the immobilization or deposition of the receptors on the biosensor surface. The method of deposition of the receptors on the surface would also lead to different degrees of heterogeneity on the surface. The binding reaction takes place between the analyte in solution and the receptors on the surface through chemical bond formation and subsequent molecular association. The geometric nature (or parameter) of the surface will significantly influence these reactions. The influence of surface morphology and structure has been analyzed (Lee and Lee, 1994; Chaudhari et al., 2002, 2003). It would be of interest to determine the scale of these roughness heterogeneities. Are these at the Angstrom level or lower? With the current emphasis on nanotechnology and nanobiotechnology these types of questions are becoming more and more relevant and of significance. The nature of surfaces in general, and of biosensors in particular (our case), should exhibit a fractal nature at the molecular level. Furthermore, one of the reasons for the emphasis on nanotechnology is that as one goes down in scale, the properties of some substances change, sometimes for the better. It is these beneficial changes that one wishes to exploit in nanotechnology and nanobiotechnology. Hopefully, similar parallels can be drawn on analyzing the fractal nature of biosensor surfaces. Do they exhibit self-similarity; and if they do what are their limits? In other words, what are their lower and upper bounds?

Furthermore, each binding event need not result in the formation of an analyte-receptor complex on the biosensor surface. All of the receptors on the biosensor surface are presumably not the results of binding events, and do not exhibit the same activity. In other words, their active sites should comprise of presumably a probability distribution in "activity." *In lieu* of any prior information, it is reasonable to assume a bell-shaped Gaussian (or normal) distribution of active sites on the surface. A probabilistic approach is more realistic here. Analyses of this sort have presumably not been performed (at least this author is unaware of them) for analyte-receptor reactions occurring on biosensor surfaces. Thus the fractal analysis is a convenient method of providing a lumped parameter analysis of analyte-receptor reactions occurring on biosensor surfaces.

Note that, at present, the dual-fractal analysis does not have a basis at the molecular level. This represents two different levels of heterogeneity on the biosensor surface. But, in some of the examples presented, a single-fractal analysis is clearly inadequate to model the data. Only in these cases does one resort to a dual-fractal analysis. The binding rate coefficients, k_1 and k_2 in the dual-fractal analysis have the same units (pg)(mm)$^{-2}$(sec)$^{(D_{f1,bind}-3)/2}$ and (pg)(mm)$^{-2}$ (sec)$^{(D_{f2,bind}-3)/2}$, respectively, as the association rate coefficient, k, in the single-fractal analysis.

Dissociation Rate Coefficient

In this case the dissociation rate coefficient is given by:

$$(\text{Ab}\cdot\text{Ag}) \approx \begin{cases} -t^{(3-D_{f1,\,\text{diss}})/2}, & t_{\text{diss}} < t < t_{d1} \\ -t^{(3-D_{f2,\,\text{diss}})/2}, & t_{d1} < t < t_{d2} \end{cases} \tag{2.5d}$$

Here $D_{f,diss}$ is the fractal dimension of the surface for the dissociation step. t_{diss} represents the start of the dissociation step. This corresponds to the highest concentration of the analyte-receptor on the surface. Henceforth, its concentration only decreases. $D_{f,bind}$ or $D_{f,assoc}$ may or may not be equal to $D_{f,diss}$. The dissociation rate coefficients, k_{d1} and k_{d2} in the dual-fractal analysis have the same units (pg)(mm)$^{-2}$(sec)$^{(D_{fd1}-3)/2}$ and (pg)(mm)$^{-2}$ (sec)$^{(D_{fd2}-3)/2}$, respectively, as the dissociation rate coefficient, k_d, in the single-fractal analysis.

2.2.4 Triple-Fractal Analysis

As will be shown later in the book, one resorts to a triple-fractal analysis when the dual-fractal analysis does not provide an adequate fit. The equation for fractal analysis is generic in nature, and one may easily extend the single- and the dual-fractal analysis equations (Equations 2.5a and 2.5c) to describe the binding (and/or the dissociation) kinetics for a triple fractal analysis. In fact, in the extreme case, n fractal dimensions may be present. In this case, the degree of heterogeneity, D_f, or the fractal dimension is continuously changing on the biosensor surface, and the surface needs to be represented by D_{fi} where i goes from 1 to n. Similarly, we have n binding rate coefficients on the biosensor surface. A similar representation may also be made for the dissociation phase.

It is perhaps appropriate here to at least mention one more approach that has been used to model the binding kinetics on surfaces.

2.2.5 Pfeifer's Fractal Binding Rate Theory

Pfeifer and Obert (1989) have suggested an alternate form of the binding rate theory. In the equation given in this reference, N is the number of complexes, N_0 is the number of receptors on the solid surface, D is the diffusion coefficient of the analyte, L is the receptor diameter, and λ is the mean distance between two neighboring receptors. This equation may also be used to analyze the analyte-receptor binding kinetics. The problem, however, is that it may not be possible in all instances to estimate *a priori* all the parameters described in the equation (not given here). In that case, one may have to approximate or assume certain values, and this will affect the accuracy and reliability of the analysis. The suggested equation does have an advantage compared to the fractal analysis described above in that it does include a prefactor necessary to convert the time interval over which fractal scaling is observed into a length interval. It also provides an expression for t_c ($= L^2/D$), which separates the short-term regime from the long term regime. The short-term regime is the one in which the anomalous diffusion applies. At the end of the short term interval ($t = t_c$), the self-similarity of the system is lost, the surface is homogeneous, and regular diffusion applies.

Pfeifer and Obert (1989) state that the application of the above equation is contingent on the:

(a) analyte being uniformly distributed in the solution at time t equal to zero,
(b) binding being irreversible and first-order (N equals the number of analyte particles that have reached the receptors), and
(c) binding occurring whenever an incoming analyte particle hits a receptor surface for the first time. In other words, the "sticking" probability is one.

It is perhaps difficult to imagine any one or all of these conditions being satisfied for analyte-receptor binding interactions occurring in continuous-flow reactors.

Given the extremely small volume of the flow channels there is a high probability of the mixing of the analyte not being proper. This in turn may lead to analyte depletion in the flow channel. Also, the binding cannot be assumed to be irreversible in all instances. There may be cases of extremely fast binding and dissociation, especially for analytes with low affinity, which can dissociate in the continuously flowing buffer without any regeneration reagent.

Condition (c) may be satisfied. However, it does not include the "sticking" probability in that each collision leads to a binding event. Also, the presence of nonspecific binding, avidity effects, and binding with reactions or binding of dissociated analytes may interfere with condition (c) being satisfied. Furthermore, the equation makes assumptions about the number of active sites, and the immobilized receptors. For example it states that the analyte binds to one specific active site. The receptor cannot bind to more than one analyte molecule at a time (1:1 binding). The equilibrium dissociation rate coefficient, $K_D = k_{diss}/k_{assoc}$ can be calculated using the above models. The K_D value is frequently used in analyte-receptor reactions occurring on biosensor surfaces. The ratio, besides providing physical insights into the analyte·receptor system, is of practical importance as it may be used to help determine (and possibly enhance) the regenerability, reusability, stability, and other biosensor performance parameters. K_D has the unit (sec) $^{[D_{f,diss}-D_{f,assoc}]/2}$. This applies to both the single- as well as the dual-fractal analysis. For example, for a single-fractal analysis, K_D has the units (sec) $^{[D_{fd}-D_f]/2}$. Similarly, for a dual-fractal analysis, the affinity, K_{D1} has the units (sec) $^{[D_{fd1}-D_{fassoc1}]/2}$ and K_{D2} has the units (sec) $^{[D_{fd2}-D_{fassoc2}]1/2}$. Note the difference in the units of the equilibrium dissociation rate coefficient obtained for the classical as well as the fractal-type kinetics. Though the definition of the equilibrium dissociation rate coefficient is the same in both types of kinetics (ratio of the dissociation rate coefficient to the association rate coefficient), the difference(s) in the units of the different rate coefficients eventually leads to a different unit for the equilibrium dissociation rate coefficient in the two types of kinetics. This is not entirely unexpected as the classical kinetic analysis does not include the characteristics of the surface in the definition of the equilibrium dissociation rate coefficient whereas the present fractal analysis does. Thus, one may not be able to actually compare the equilibrium dissociation rate coefficient affinities in these two types of systems. This is a significant difference in the kinetic analysis of binding and dissociation reactions on biosensor surfaces from what is available in the literature.

It is perhaps appropriate to, at least, briefly present some of the other approaches that have recently appeared in the literature and to help model the binding and the dissociation kinetics of the different analytes (present in the liquid phase) on biosensor surfaces. The following kinetic modeling approaches will be presented and analyzed briefly:

(a) Application of the synthetic jet concept to low Reynolds number biosensor microfluidic flows for enhanced mixing (Mautner, 2004)
(b) Kinetics of analyte capture on nanoscale sensors (Solomon and Paul, 2006)
(c) Probing the functional heterogeneity of surface binding sites along with the effect of mass transport limitation and its influence on binding and dissociation of analytes on biosensor surfaces (Svitel et al., 2007)

2.2.6 The Mautner Model

Mautner (2004) points out that microfluidic components used in biosensors may be effectively used for the analysis of chemicals and biologicals. He emphasizes that themicrofluidic systems used in biosensors will operate at low Reynolds numbers. Reynolds number is a dimensionless number and is used to characterize flows in different types of systems. Reynolds number (Re) is equal to $(Dv(\rho)/\mu)$. ρ represents the density of the fluid, D is the diameter of the pipe or a characteristic dimension of the system, v is the velocity of the fluid in the system, and μ is the viscosity of the fluid. Low Reynolds number regime is characterized by Re less than 10. Furthermore, Mautner (2004) points out that these devices will have characteristic dimensions less than 100 μm. He also emphasizes that we are dealing here with small volumes of fluids (pico to microliters). This type of slow Re laminar flow is characterized by diffusion-only mixing. However, Mautner (2004) emphasizes that rapid mixing is essential in immunoassays. Thus, enhanced mixing is essential to overcome the slow fluid mixing in low Re number flow.

Mautner (2004) points out that the following techniques have been used to enhance mixing in microfluidic networks:

(i) Use of slanted wells to increase lateral flow transport (Johnson et al., 2002)
(ii) Flows over shallow grooves (Stroock et al., 2002)
(iii) Utilization of passive mixing in three-dimensional serpentine microchannels (Liu et al., 2000)

(iv) Application of surface layers creating hydrophobic or hydrophilic surface patterns to direct fluids (Zhao et al., 2002)

Mautner emphasizes that the application of readily available pumps can help the unsteady wall jets obtain the required time dependent wall jet conditions which would mix both existing and merging flow stream conditions.

Mautner (2004) has proposed the application of macroscale jets to be applied to the low Reynolds ($Re = 10$) two-dimensional channel flows that may be found in biosensor microfluiidc systems. The method includes a hybrid approach of the Lattice Boltzmann (LB) method for flow field computations and a finite difference, convection-diffusion equation for passive scalar transport. This author emphasizes that the forced jet imparts momentum to the channel flow, thereby enhancing fluid mixing.

2.2.7 Kinetics of Analyte Capture on Nanoscale Sensors (Solomon and Paul, 2006)

Solomon and Paul (2006) point out that nanoscale electromechanical systems have been used to detect biomolecular targets with increasing sensitivity (Roukes et al., 2000; Paul and Cross, 2004; Ekinci and Roukes, 2005). Solomon and Paul (2006) call these devices BioNems devices. These BioNems devices may be used to detect proteins, enzymes, viruses, and bacteria. They point out that the sensitivity of these devices is directly related to the binding kinetics of the analytes to the receptors immobilized on these BioNems devices. The BioNem device that these authors have developed and analyzed are different from the conventional devices that measure, for example, the binding rate coefficients, in that they identify the presence or absence of biological macromolecules such as large proteins, enzymes, viruses, and enzymes.

The Solomon and Paul (2006) system uses a single (or multiple) cantilever(s). The multiple cantilever configuration improves detection probabilities. The cantilever is located in a small channel (or "via"). Fluid containing the target analyte flows through these channels at some fixed velocity. The detection of the analyte by the cantilever is based on either (a) the mass-loading effects or (b) by a change in the effective damping constant of the cantilever that alters the mean-square displacement of the cantilever tip. Solomon and Paul (2006) point out that a small region of the cantilever tip is functionalized with immobilized receptors specific to the target analyte. SAM (self-assembling monolayers) constructed with alkanethiols permits the functionalization by a linkage to analyte specific receptors. These emphasize that their device is able to detect specific biomolecules down to concentration levels of 1 nM or less.

Solomon and Paul (2006) also attempted to analyze the influence of surface-diffusion enhancement on the analyte capturer. Their quantitative results indicate that under certain circumstances this mechanism might be advantageous. This will depend, the authors indicate, on the particular parameters that characterize a specific device implementation.

Solomon and Paul (2006) also attempted to analyze the influence of convection on these types of BioNems devices. Using simple fluid dynamics arguments these authors show that at the bulk flow velocities used in these devices and for Damkohler (Da) numbers less than unity, capture efficiencies would be dominated by reaction-diffusion mechanisms. In other words, convection will only play a minor role. They emphasize that for Damkohler numbers $\gg 1$, the reaction-diffusion-convection coupled equations need to be solved to obtain a better perspective of analyte-binding kinetics to receptors immobilized on these nanoscale sensors.

2.2.8 Probing the Functional Heterogeneity of Surface Binding Sites Along with the Effect of Mass Transport Limitation and Its Influence on Binding and Dissociation of Analytes on Biosensor Surfaces (Svitel et al., 2007)

Svitel et al. (2007) have recently probed the functional heterogeneity of surface binding sites under the influence of external mass transport limitations. They did this by analyzing experimental binding traces. These authors point out the need to analyze the binding of macromolecules to surfaces assuming that the surface binding sites are heterogeneous (Sips, 1948; Koopal and Vos, 1993; Vijayendran and Leckband, 2001; Lebedev et al., 2006). Svitel et al. (2007) suggest that that there may be two (or more) reasons for inhomogeneity on the surface during the interactions of analyte-receptors on the sensor surface: (a) the surface sites may be intrinsically inhomogeneous in their binding properties, and (b) the receptors may be rendered heterogeneous by attaching them to the surface. Yeung and Leckband (1997), Rabbany et al. (1997), and Kloss et al. (2000) have pointed out that the immobilization of chemically homogeneous species frequently results in functionally impaired subpopulations due to (a) constraints in orientation, (b) variable cross-linking, and (c) the influence of the microenvironment of the surface. This heterogeneity, Svitel et al. (2007) explain, will influence the application of antibody-based affinity biosensors (Wilson and Nock, 2002; Taitt et al., 2005), and the analysis of protein interactions by SPR biosensors (O'Shannessy, 1994; O'Shannessy and Winzor, 1996; Schuck, 1997).

Svitel et al. (2007) explain that evanescent field biosensors have been used to characterize protein-protein, protein-small molecule, protein-nucleic-acid interactions, and DNA hybridization reactions (Cooper, 2002; Livache et al., 2003). These biosensors have allowed the kinetic binding traces to be measured with high sensitivity. Svitel et al. (2007), however, also point out that these binding traces when analyzed kinetically are apparently not consistent with a simple 1:1 interaction (Karlsson et al., 1994; Glaser and Hausdorf, 1996; Schuck, 1997; Schuck et al., 1998). Svitel et al. (2003) attempt to interpret these deviations from a simple 1:1 interaction as a source of information on the homogeneity of the surface immobilized sites. They have proposed a computational model that assumes that the binding signal is a superposition of independent parallel reactions occurring on the biosensor surface. These parallel binding reactions result from a continuous distribution of thermodynamic and kinetic binding constants. In a more recent publication, Svitel et al. (2007) have expanded their computational model and their approach to include a compartment-like transport step, which describes the competitive binding to different surface sites in a zone of depleted analyte close to the biosensor surface. Just as in the fractal analysis approach presented in different chapters in the book to analyze the binding and the dissociation phase, the approach presented by Svitel et al. (2007) helps to analyze surface binding when both inhomogeneity on the biosensor surface and transport limitations are present simultaneously. Their approach, the authors claim, permits the evaluation of both the kinetic binding parameters as well as the effective transport rate coefficients.

References

Anderson J, *NIH Panel Review Meeting,* Case Western Reserve University, Cleveland, OH, July 1993.

Berland KM, PTC So, and E Gratton, Two-photon fluorescence correlation spectroscopy: Method and application to the intracellular environment, *Biophysical Journal*, **68**, 694–701 (1995).

Chaudhari A, CC Yan, and SL Lee, Multifractal analysis of diffusion-limited reactions over surfaces of diffusion-limited aggregates, *Chemical Physics Letters*, **207**, 220–226 (2002).

Chaudhari A, CC Yan, and SL Lee, Mathematical and general, Journal of Physics *A*, **36**, 3757 (2003).

Cooper MA, Optical biosensors in drug discovery, *National Reviews in Drug Discovery*, **1**, 515–528 (2002).

Coppens MO and GF Froment, Diffusion and reaction in a fractal catalyst pore. Geometrical aspects, *Chemical Engineering Science*, **50**(6), 1013–1026 (1995).

Corel Quattro Pro, 8.0, Corel Corporation Limited, Ottawa, Canada, 1997.

De Gennes PG, Diffusion-controlled reactions, *Polymer Melts Radiation and Physical Chemistry*, **22**, 193 (1982).

Dewey TG and JG Bann, Diffusion-controlled reaction in polymer melts, *Biophysical Journal*, **63**, 594 (1992).

Ekinci KL and MN Roukes, Nanoelectrical systems, *Reviews of Scientific Instrumentation*, **76**, 1–12 (2005).

Fatin-Rouge N, K Starchev, and J Buffle, Size effects on diffusion process with agarose gels, *Biophysical Journal*, **86**, 2710–2719 (2004).

Federov BA, BB Federov, and PW Schmidt, An analysis of the fractal properties of globular proteins, *Journal of Chemical Physics*, **99**, 4076–4083 (1999).

Giona M, First-order reaction-diffusion in complex fractal media, *Chemical Engineering Science*, **47**, 1503–1515 (1992).

Ghosh RN and WW Webb Results of automated tracking of low density lipoprotein receptors on cell surfaces, *Biophysical Journal*, **53**, A352 (1988).

Glaser RW and X Hausdorf Binding kinetics of an antibody against HIV p24 core protein measured with real-time biomolecular interaction analysis suggest a slow conformational exchange in antigen p24, *Journal of Immunological Methods*, **189**, 1–14 (1996).

Harder FH, S Havlin, and A Bunde, Diffusion in fractals with singular waiting-time distribution, *Physics Reviews B*, **36**, 3874–3879 (1987).

Havlin S, Molecular diffusion and reaction. In *The Fractal Approach to Heterogeneous Chemistry: Surfaces, Colloids, Polymers*, Avnir D (Ed.), Wiley, New York, pp. 251–269, 1989.

Havlin S and D Ben-Avraham, Diffusion in disordered media, *Advance in Physics*, **36**, 695–798 (1987).

Henke L, N Nagy, and UJ Krull, An AFM determination of the effects of surface roughness caused by cleaning of fused silica and glass substrates in the process of optical biosensor preparation, *Biosensors & Bioelectronics*, **17**, 547–555 (2002).

Johnson T, D Ross, and L Locascio, Rapid microfluidic mixing, *Analytical Chemistry*, **74**, 45–51 (2002).

Karlsson R, H Roos, L Fagerstam, and B Persson, Kinetic and concentration analysis using BIA technology, *Methods in Computation and Methods in Enzymology*, **6**, 99–110 (1994).

Kloss AA, N Lavrik, C Yeung, and DE Leckband, Effect of the microenvironment on the recognition of immobilized cytochrome by soluble redox proteins, *Langmuir*, **16**, 3414–3421 (2000).

Koopal LK and CHW Vos, Adsorption on heterogeneous surfaces. Calculation of the adsorption energy distribution function or the affinity spectrum, *Langmuir*, **9**, 2593–2605 (1993).

Kopelman R, Fractal reaction kinetics, *Science*, **241**, 1620–1624 (1988).

Lebedev K, S Mafe, and P Stroeve, Convection, diffusion and reaction in a surface-based biosensor: modeling of cooperativities and binding site competition on the surface and in the hydrogel, *Journal of Colloid and Interface Science*, **296**, 527–537 (2006).

Le Brecque M, *Mosaic*, **23**, 12–15 (1992).

Lee CK and SL Lee, Multifractal scaling analysis of autocatalytic and autopoisoning reactions over DLA surfaces, *Chemical Physics Letters*, **228**, 539–546 (1994).

Lee CK and SL Lee, Multifractal scaling analysis of reactions over fractal surfaces, *Surface Science*, **325**, 294–310 (1995).

Li HL, S Chen, and H Zhao, Fractal mechanisms for the allosteric effects of proteins and enzymes, *Biophysical Journal*, **58**, 1313–1320 (1990).

Liu R, M Stremler, K Sharp, M Olsen, J Santiago, R Adrian, H Aref, and D Beebe, Passive mixing in a three-dimensional serpentine microchannel, *Journal of MEMS*, **9**, 190–197 (2000).

Livache T, E Maillart, N Lasalle, P Mailley, B Corso, P Guiedon, A Roget, and Y Levy, Polypyrrole based DNA hybridization assays: study of label free detection processes versus fluorescence on microchips, *Journal of Pharmaceutical & Biomedical Analysis*, **32**, 687–696 (2003).

Markel VA, LS Muratov, MI Stockman, and TF George, Theory and numerical simulation of optical properties of fractal clusters, *Physics Reviews B*, **43**(10), 8183 (1991).

Mautner T, Application of the synthetic jet concept to low Reynolds number, biosensor mictofluidic flows for enhanced mixing: a numerical study using the lattice Boltzmann method, *Biosensors & Bioelectronics*, **19**, 14009–14019 (2004).

Nyikos L and T Pajkossy, Diffusion to fractal surfaces, *Electrochimica Acta*, **31**, 1347 (1986).

O'Shannessy DJ, Determination of kinetic rate and equilibrium binding constants for macromolecular interactions: a critique of the surface plasmon resonance literature, *Current Opinions in Biotechnology*, **5**, 65–71 (1994).

O'Shannessy DJ and DJ Winzor, Interpretations of deviations from pseudo-first-order kinetic behavior in the characterization of ligand binding by biosensor technology, *Analytical Biochemistry*, **236**, 275–283 (1996).

Pajkossy T and L Nyikos, Diffusion to fractal surfaces. II. Verification of theory, *Electrochimica Acta*, **34**, 171 (1989).

Paul MR and MC Cross, Stochastic dynamics of nanoscale mechanical oscillators immersed in a viscous fluid, *Physics Reviews Letters*, **92**, 235501–235504 (2004).

Peters R and RJ Cherry, Lateral and rotational diffusion of bacteriorhodopsin in lipid bilayers: experimental test of the Saffman-Delbruck equations, *Proceedings of the National Academy of Sciences, USA*, **79**, 4317–4321 (1982).

Pfeifer P and M Obert, In *The Fractal Approach to Heterogeneous Chemistry: Surfaces, Colloids, Polymers*, Avnir D (Ed.), Wiley, New York, pp. 251–269 (1989).

Pfeifer P, Characterization of surface irregularity, Chapter 2. In: *Preparative Chemistry Using Supported Reagents*, Academic Press, San Diego (1987).

Pfeifer P, D Avnir, and DJ Farin, Molecular fractal surfaces, *Nature* (London), **308**, 261 (1984a).

Pfeifer P, D Avnir, and DJ Farin, Surface geometric irregularity of particulate materials. The fractal approach, *Journal of Colloid and Interface Science*, **103**(1), 112 (1984b).

Pluen A, PA Netti, KJ Rakesh, and DA Berk, Diffusion of macromolecules in agarose gels: comparison of linear and globular configurations, *Biophysical Journal*, **77**, 542–552 (1999).

Rabbany SY, R Piervincenzi, L Judd, AW Kusterbeck, R Brederhorst, K Hakansson, and FS Ligler, Assessment of heterogeneity in antibody-antigen displacement reactions, *Analytical Chemistry*, **69**, 175–183 (1997).

Roukesm M, S Fraser, M Cross, and T Solomon, U.S Patent Application Number 60/224, vol. 109, 2000.

Sadana A and A Beelaram, Fractal analysis of antigen-antibody binding kinetics biosensor applications, *Biotechnology Progress*, **9**, 45 (1994).

Sadana A, JP Alarie, and T Vo-Dinh, A β-cyclodextrin based fiber-optic chemical sensor: a fractal analysis, *Talanta*, **42**, 1567 (1995).

Saffman PG and M Delbruck, Brownian motion in biological membranes, *Proceedings of the National Academy of Sciences, USA*, **72**, 3111–3113 (1975).

Sahimi M, Flow phenomena in rocks: From continuum models to fractals, percolation, cellular automata and simulated annealing, *Reviews in Modern Physics*, **65**, 1393–1534 (1993).

Saxton MJ, Anomalous diffusion due to binding: a Monte Carlo study, *Biophysical Journal*, **77**, 2251–2265 (2001).

Schwille P, U Haupts, S Maiti, and WW Webb, Molecular dynamics in living cells observed by fluorescence correlation spectroscopy with one- and two-photon excitation, *Biophysical Journal*, **77**, 2251–2265 (1999).

Schuck P, Use of surface plasmon resonance to probe the equilibrium and dynamic aspects of interactions between biological macromolecules, *Annual Reviews in Biophysical and Biomolecular Structures*, **26**, 541–566 (1997).

Schuck P, Reliable determination of binding affinity and kinetics using surface plasmon resonance biosensors, *Current Opinion in Biotechnology*, **8**, 498–502 (1997).

Schuck P, DB Millar, and AA Kortt, Determination of binding constants by equilibrium titration with circulating sample in a surface plasmon resonance biosensor, *Analytical Biochemistry*, **265**, 79–91 (1998).

Sips R, On the structure of a catalyst surface, *Journal of Chemical Physics*, **16**, 490–495 (1948).

Solomon JE and MR Paul, The kinetics of analyte capture on nanoscale sensors, *Biophysical Journal*, **90**(March), 1842–1852 (2006).

Stroock A, S Dertinger, G Whitesides, and A Ajdari, Patterning flows used grooved surfaces, *Analytical Chemistry*, **74**, 5306–5312 (2002).

Svitel J, A Balbo, RA Mariuzza, NR Gonzales, and P Schuck, Combined affinity and rate constant distributions of analyte or ligand populations from experimental surface binding and kinetics and equilibria, *Biophysics Journal*, **84**, 4062–4077 (2003).

Svitel J, H Boukari, D van Ryk, RC Willson, and P Schuck, Probing the functional heterogeneity of surface binding sites by analysis of experimental binding traces and the effect of mass transfer limitation, *Biophysical Journal*, **92**(March), 1742–1758 (2007).

Starchev K, J Sturm, G Weill, and CH Brogen, Brownian motion and electrophoretic transport in agarose gels studied by epifluorescence microscopy and simple particle tracking analysis, *Journal of Physical Chemistry*, **101**, 5659–5663 (1997).

Taitt CR, GP Anderson, and FS Ligler, Evanescent wave fluorescence biosensors, *Biosensors & Bioelectronics*, **20**, 2470–2487 (2005).

Vijayendran RA and DE Leckband, A quantitative assessment of heterogeneity for surface-immobilized proteins, *Analytical Chemistry*, **73**, 471–480 (2001).

Wei QH, C Bechinger, and P Leiderer, Single-file diffusion of colloids in one-dimensional channels, *Science*, **287**, 625–627 (2000).

Weiss GH, *Fractals in Science*, Springer-Verlag, Berlin, pp. 119, 1994.

Wilson DS and S Nock, Functional protein microarrays, *Current Opinion in Chemical Biology*, **6**, 81–85 (2002).

Yeung C and DE Leckband, Molecular level characterization of microenvironmental influences on the properties of immobilized proteins, *Langmuir*, **13**, 6746–6754 (1997).

Zhao B, J Moore, and D Beebe, Principles of surface-directed liquid flow in microfluidic channels, *Analytical Chemistry*, **74**, 4259–4268 (2002).

CHAPTER 3

Fabrication of Biosensors

Chapter Outline

Handbook of Biosensors and Biosensor Kinetics. DOI: 10.1016/B978-0-444-53262-6.00003-6

3.1 Introduction

In this chapter we analyze the different types of biosensor fabrication procedures that have recently appeared in the literature. Though fabrication of glucose biosensors still remains an important area of investigation (as a large share of the market is for biosensors in this area), the fabrication of biosensors for the detection of other analytes of interest will be presented in this chapter. Some of the recent methods of biosensor fabrication include:

(a) fabrication of a glucose biosensor based on a novel nanocomposite structure (Safavi et al., 2009),
(b) fabrication of a microbial glucose biosensor using a screen-oriented water-based carbon ink and its application in serum analysis (Pemberton et al., 2009),
(c) a potentiometric protein sensor built by the surface molecular imprinting method (Wang et al., 2008),
(d) fabrication of a molecularly imprinted polymer microarray on a chip by mid-infrared laser pulse initiated polymerization (Henry et al., 2008),
(e) fabrication and characterization of an indium tin oxide (ITO)-polyaniline biosensor (Tahir et al., 2007),
(f) a novel biosensor fabrication methodology based on processable conducting polyaniline nanoparticles (Morrin et al., 2005a,b),
(g) fabrication of a photometric dip-strip test system for the detection of β(1-3)-D-glucan using crude β(1-3)-D-glucanase from sprouts of *Vigna aconitifolia* (Bagal-Kestwal et al., 2009),
(h) screen-printing fabrication method for DNA chips with electrical readout for the detection of viral DNA (Schuler et al., 2009),
(i) characterization and fabrication of disposable screen-printed microelectrodes (Kadara et al., 2009),
(j) fabrication of optical fiber imaging sensors using inkjet printing technology (Carter et al., 2006),
(k) fabrication of a highly sensitive penicillin sensor based on a charge transfer technique (Lee et al., 2009),
(l) an overview of the biosensor manufacturing process (Biodot, biosensor introduction, 2009),
(m) manufacturing method for a biosensor and biosensing apparatus (Gu et al., 2009),
(n) introduction to biosensor technology (2008).
(o) the fabrication of a highly sensitive glucose biosensor using an osmium complex and glucose oxidase (GOD) immobilized onto carbon nanotube (CNT) modified electrodes (Salimi et al., 2009),
(p) fabrication of a porous silicon-based biosensor (Matthew and Alocilja, 2003),
(q) low cost silicon nanowire biosensors using recrystallization technologies (Ashburn and Sun, 2009),

(r) fabrication of screen-printed carbon electrodes (SPCEs) by a novel one-step process for manufacturing electrodes for injection molding (Wu et al., 2008), and
(s) a rapid and easy procedure for fabrication of biosensors by microencapsulation of an enzyme in hydrophilic synthetic latex films.

3.2 Different Methods of Biosensor Fabrication

We now examine some of the biosensor fabrication techniques that have appeared recently in the open literature. These include a novel nanocomposite electrode (Safavi et al., 2009), microbial biosensors using screen-printing water-based carbon ink (Pemberton et al., 2009), using surface molecular imprinting (Wang et al., 2008; Henry et al., 2008), using processable conducting polyaniline nanoparticles (Morrin et al., 2005a,b), using screen printing (Schuler et al., 2009; Kadara et al., 2009), using ink-jet printing (Carter et al., 2006), and using charge transfer techniques (Lee et al., 2009).

3.2.1 Fabrication of a Glucose Sensor Based on a Nanocomposite Electrode (Safavi et al., 2009)

Safavi et al. (2009) have very recently presented a novel technique for the fabrication of a glucose sensor based on a novel nanocomposite electrode. These authors point out that there has been considerable emphasis on the development of glucose sensors which involves the immobilization of glucose substrates (Lin et al., 2004; Maleki et al., 2007; Zhang et al., 2007; Musameh et al., 2008; Jeykumari and Narayanan, 2008; Deng et al., 2008).

As early as 1972, Wilson and Turner (1992) indicated that the loss of enzyme stability or insufficient stability as one of the major problems during the immobilization process which leads to a loss of sensitivity and affects reproducibility. Further problems arise due to the interference caused by endogeneous electroactive ascorbic acid (AA) and uric acid (UA) in blood samples. Safavi et al. (2009) emphasize that electrocatalytic activity is the key factor that affects sensitivity and selectivity during glucose detection.

Safavi et al. (2009) point out that nanomaterials have been introduced in electrochemical sensing. They exhibit increasing surface area, mass transfer, and catalysis (Katz et al., 2006). Some of the more recent applications for nanomaterials in biosensing include a Pt nanotubular array (Yuan et al., 2005), mesoporous Pt (Park et al., 2003), Ni nanoparticles (You et al., 2003), Au nanoparticles (Jena and Raj, 2006; Kurniawan et al., 2004), Pt nanoparticles (Rong et al., 2007), Pt/Pb nanoparticles (Cui et al., 2007; Wang et al., 2008), and CNTs (Ye et al., 2004; Tan et al., 2008).

Safavi et al. (2009) report on the fabrication of a novel nonenzymatic composite electrode based on powdered nanoscale nickel hydroxide, graphite powder, and ionic liquid (octylpyridium hexafluorophosphate, OPy^4PG_{6-}). They indicate that their nanosensor is stable and highly sensitive for the amperometric detection of glucose. They further emphasize that the electrode surface could be easily and reproducibly renewed by polishing lightly with a smooth paper.

The electrocatalyitc activity of their modified electrode is described by the following reaction (Zhao et al., 2007):

$$\begin{aligned} &Ni(OH)_2 + OH^- \rightarrow NiO(OH) + e \\ &NiO(OH) + \text{glucose} \rightarrow Ni(OH)_2 + \text{glucolactone} \end{aligned} \tag{3.1}$$

They indicate that Ni^{2+}/Ni^{3+} species on the electrode surface acts as a catalyst for the oxidation of glucose. Furthermore, Safavi et al. (2008) emphasize that the shape and morphology of the nanoscale materials significantly affects their catalytic behavior.

Safavi et al. (2009) report that their biosensor is stable as it exhibits 95% of its initial response value after 100 days. They also emphasize that their biosensor produces reproducible results in that a modified electrode yields results with an RSD (relative standard deviation) of 3.4% from six successive amperometric measurements with a 2 mM glucose solution. Finally, the use of their ionic binder not only increased the sensitivity of their biosensor, but also increased its resistance toward electrode fouling significantly (Zhao et al., 2007).

3.2.2 A Potentiometric Protein Sensor Using Surface Molecular Imprinting Method (Wang et al., 2008)

Wang et al. (2008) report that the molecular imprinting (MI) method is a fast developing technology, and has been used in biosensing applications (Viatakis et al., 1993; Shea, 1994; Wulff, 1995; Bowma et al., 1998; Yano and Karube, 1999; Sellergren, 2000; Haupt, 2003; Hayden et al., 2003; Guo et al., 2004). Wang et al. (2008) point out that the traditional MI recognition mechanism depends mainly on the precise spatial arrangement of functional groups in the matrix to ensure selective recognition of target molecules. These authors have fabricated a biosensor built with surface molecular imprinting of thiol SAMs (self-assembled monolayers). These SAMs have the capacity to detect complex molecules with parts per million accuracy. These authors demonstrate that their biosensor can detect globular proteins such as myoglobin and hemoglobin with good sensitivity and selectivity. Wang et al. (2008) report that the (a) attraction forces between the protein molecule and the gold surface, (b) the hydrogen bonds between the hydrophilic groups of the protein surfaces and the –OH groups at the thiol end, and (c) the specific arrangement of these interactions in shape and orientation is responsible for the recognition and selectivity of their biosensor (Shi et al., 1999; Kaufmann et al., 2007).

The Wang et al. (2008) fabrication technique is described briefly. A gold-coated silicon chip was used after cleaning with de-ionized water and dried with pure nitrogen gas. The authors report that proteins were dissolved in de-ionized water. The thiol was dissolved in acetic acid. A mixture solution was then made by blending them in 19:1 (water:acetic acid) ratio. Thus, a good solubility was obtained for thiol and there was no precipitation of the proteins and formation of supramolecular structure (Singh et al., 1999). The gold-coated plate was then immersed in the solution for at least 2 h. The thiol molecules are tightly attached to the electrode surface via the sulfur-metal bond. The proteins are adsorbed on the gold surface through hydrophilic interactions and electrostatic forces in the absence of strong chemical bindings (Kaufmann et al., 2007). Cavities that are complementary to the template proteins were created in the SAM matrix. This was done by repeated rinsing with deionized water to remove the protein molecules. This unique complementarity on the prepared electrode facilitated the high affinity for the adsorption of the same kind of template proteins. The authors then state that their prepared electrode was dried at room temperature overnight before electrochemical measurements were made.

Wang et al. (2008) report that proteins in aqueous solution act as polyelectrolytes. These have a net electrical charge. This net electrical charge depends on (a) the isoelectric point of the protein and (b) the ioinic composition of the solution. Janata (1975) points out that when charged proteins are trapped in a thin insulating layer deposited on a metallic conductor a change in surface potential occurs. This change in surface potential may be measured potentiometrically by using a reference electrode in the same solution.

Finally, Wang et al. (2008) report that their surface molecular imprinted biosensor is capable of detecting myoglobin and hemoglobin in solution. Their biosensor has the capability to recognize a specific protein in a solution of multiple proteins. Their results indicate that the size and shape match is critical for precise recognition.

3.2.3 Fabrication of Molecularly Imprinted Polymer Microarray on a Chip (Henry et al., 2008)

Henry et al. (2008) have recently fabricated a molecularly imprinted polymer microarray on a chip by mid-infrared laser pulse initiated polymerization. These authors indicate that MIPs have been employed in separation science (Wei and Mizaikoff, 2007) and catalysis (Tada and Iwasawa, 2005) as well as in chemical and biochemical sensing (Henry et al., 2005). Lotizero et al. (2004) have indicated that MIPs now compete with natural receptors in terms of selectivity and sensitivity. Henry et al. (2005) report that MIPs are now an attractive, cheap, and robust alternative for the detection of small (Jiang et al., 2007), and large molecules (Bossi et al., 2007).

Henry et al. (2008) point out that the scarcity of publications in the area of multi-MIP platforms wherein several MIPs prepared against different analytes have been integrated on a single chip vis-a-vis microarray technology used in genomics, proteomics, and drug screening.

The authors emphasize that the development of such multi-MIP platforms would not only extend their applications but also enhance their commercial aspects. Henry et al. (2008) have reported their initial results for the fabrication of a heterogeneous microarray of 14 different polymers on a single microfluidic chip. They determined the best polymers for MI by testing their polymer microarray against the fluorescent model template dansyl-L-phenylalanine.

The fabrication of the molecularly imprinted polymer consisted of the following steps, and is briefly described below:

(a) Derivatization of the glass substrates. The glass cover slips were cleaned by sonication in HCl, and rinsed and sonicated in deionized water.
(b) Assembly of the microfluidic device. Microfluidic channels 300 μm wide connected to a polymerization chamber 5 mm wide were laser cut into a silicon sheet 300 μm thick.
(c) Polymerization solutions. Fourteen monomers were used for the preparation of the polymer microarray.
(d) Polymerization set-up. A CFCD camera was used to visualize and localize the polymerization and the growth of the polymer dots. This procedure was repeated for each of the different polymer compositions. The polymerization was stopped once the desired diameter was obtained.

The binding assays were conducted in acetonitrile.

Henry et al. (2008) report that they have developed a new method for the fabrication of localized polymeric solutions. Their method also permitted the real-time monitoring of the growth of polymeric structures. They indicate that their technique helps overcome previous limitations observed in molecularly imprinted sensor microarrays. Finally, they emphasize that their technique should find application in HTS (high-throughput screening), chemical and biochemical sensing, and in biocompatability studies.

3.2.4 Fabrication of an Optical Fiber Imaging Sensor Using Inkjet Printing Technology (Carter et al., 2006): A pH Sensor

Carter et al. (2006) have fabricated an optical fiber imaging sensor using inkjet printing technology for pH sensing. These authors indicate that for fiber-based pH sensing an approach is to confine the pH sensitive indicators in substrates attached to the fiber surface (Peterson et al., 1980; Saari and Seitz, 1982; Gehrich et al., 1986; Munkholm et al., 1986; Jordan and Walt, 1987; Nivens et al., 1998, 2002). These authors further state that the traditional methods for fabricating fiber-based pH sensors include attachment of the substrate via mechanical (Peterson et al., 1980; Saari and Seitz, 1982; Gehrich et al., 1986, dip coating (Nivens et al., 1998, 2002), or photo-polymerization methods (Munkholm et al., 1986; Jordan and Walt, 1987).

Carter et al. (2006) report that mechanical methods generally use tubing (e.g., capillary) filled with an indicating reagent. The substrate may be directly bound (by epoxy) to the fiber surface (Fuh et al., 1987). Carter et al. (2006) state that dip-coating methods are generally used in many sol-gel sensor preparations. This technique produces micrometer thick sensing membranes per dip. The resulting membrane covers the entire fiber surface.

Carter et al. (2006) report that photopolymerization methods were among the earliest methods used for fiber-based sensor fabrication. Polymerized arrays of indicators may be produced by immersing the optical fiber tip in a polymerizable indicator chemistry. These indicator chemistries were selectively grown on the end of optical fiber strands by UV (ultra violet) radiation polymerization. Carter et al. (2006) have demonstrated the feasibility of using Drop-on-Demand printing technology for fabricating imaging sensors (Wallace et al., 2002). This was accomplished by printing an array of photo-polymerizable sensing elements on the surface of an optical fiber image guide. Carter et al. (2006) emphasize that the microjet procedure produces highly reproducible droplets by a piezoelectric-driven orifice. This results in a very uniform sensor array. They indicate that the reproducibility of the microjet printing process is excellent. The microdot sensor diameter is 92.2 ± 2.2 μm, the height is 35.0 ± 0 μm, and the roundness is 0.00072 ± 0.00023. This is calculated by dividing the difference between the maximum and minimum diameters by the average diameter. The diameter versus height (aspect ratio) is controlled by adjusting physical characteristics such as surface tension and viscosity of the polymer formulation.

Carter et al. (2006) point out that the inkjet printing chemistries on an optical substrate are very similar to those used to produce micro-optical components (Cox et al., 1995, 1996; Chen et al., 2002). This is the Drop-on-Demand microjet printing and provides for a highly reproducible droplet (that is indicator chemistry).

Carter et al. (2006) reiterate that microjet printing technology is a viable tool for fabricating fiber-based imaging sensors. They point out that though they have demonstrated pH sensing only so far, this microjet printing technique does have the potential for the fabrication of multianalyte sensors. This can then be used to detect and measure biologically important parameters such as blood/gas and other ions. Cross sensitivity is avoided in these types of sensors. Besides, these authors claim that there is excellent uniformity in the polymer sensor arrays on the fiber surface.

3.2.5 Fabrication and Characterization of an Indium Tin-Oxide-Polyaniline Biosensor (Tahir et al., 2007)

Tahir et al. (2007) recently reported that the conducting polymer polyaniline (Pani)-based biosensors have been used to detect different analytes. Pani exhibits excellent electrical and optical properties. Besides, antibodies may be immobilized on the polymer surface by entrapment in the polymer matrix (Sadik and Emon, 1996; Adeloju and Wallace, 1996). Tahir et al. (2007) indicate that this entrapment feature may be used for the direct measurement of antibody-antigen binding (Sergeyeva et al., 1996; Gerard et al., 2002). Besides, during the entrapment procedure, the biological activity is maintained.

Tahir et al. (2007) report that the electrical, optical, chemical, and electrochemical properties have been extensively investigated, including a wide range of applications (Winokur, 1998). Tahir et al. (2007) emphasize that the electrical property of Pani is pH dependent, with most analyses being performed at pH less than 4.0 (Shaolin and Jincui, 1999). Since most biological and immunological reactions occur optimally around a neutral pH of 7.0, Tahir et al. (2007) indicate that this poses a challenge to incorporate biological elements in pH-dependent Pani. The authors indicate that there have, however, been improvements to enhance the electrical and physiochemical properties of Pani (Cao et al., 1993; Stejskal et al., 1998; Lukachova et al., 2003).

Tahir et al. (2007) have developed a biosensor platform that uses ITO glass, spin-coating Pani, and antibodies on it. These authors initially synthesized the self-doped Pani, and evaluated its feasibility for incorporation in the biosensor design. They then evaluated the self-doped Pani and antibodies specific to bovine viral diarrhea virus (BVDV). The BVDV was selected as a model pathogen.

Initially, Tahir et al. (2007) attempted to characterize the Pani used. These authors state that the TEM (transmission electron microscope) showed that the higher the molecular weight of the self doped Pani the larger the polymer structure. Also, as the molecular weight increases, the length of the polymer backbone per unit area increases (Carbrey et al., 1971). This enhances the flow of electrons and thereby the conductivity. The Pani weight is also temperature dependent in the 22-75 °C range (Tahir et al., 2007). Thus, these authors indicate that a temperature-controlled mechanism is required in the biosensor design to minimize the temperature-dependent variations of the polymer properties.

Tahir et al. (2007) report that their ITO-Pani biosensor utilizes the antigen-antibody binding format with self-doped Pani as the transducer. Their biosensor design concept is based on the difference between the signal before (I^0), and the signal after (I^s) the antibody-antigen binding. The current drop is given by I^0–I^s, and the higher the current drop, the greater the amount of antibody-antigen binding on the biosensor surface. This binding of the surface blocks the transfer of electrons. In essence, Tahir et al. (2007) state that as the antibodies (with a molecular weight around 15-kDa) are immobilized within the polymer backbone, electron flows are reduced. As the antigen-antibody complex is formed (molecular weight of BVDV of at least 4 MDa), the electron flow is restricted even more. According to Tahir et al. (2007), the larger the antigen-antibody complex formed in the Pani backbone, the higher the restriction for the flow of electrons.

Finally, the fabrication of the ITO-Pani biosensor by Tahir et al. (2007) involves several steps. These include:

(a) surface treatment,
(b) Pani coating, and
(c) antibody functionalization.

A significant amount of antigen-antibody binding was demonstrated due to the significant drop in the current (amperometric response) before and after antigen-antibody binding. The authors also demonstrated the feasibility of their biosensor by detecting 10^6 CCID/ml of the BVDV in pure culture. The authors emphasize the need to further investigate the biosensor parameters such as antibody concentration, Pani size, immobilization method, and incubation time to enhance their biosensor performance parameters.

3.2.6 Novel Biosensor Fabrication Procedure Based on Processable Conducting Polyaniline Nanoparticles (Morrin et al., 2005a,b)

Morrin et al. (2005a,b) have recently analyzed polyaniline (Pani) nanoparticles using dodecylbenzenesulfonic acid (DBSA) as a dopant in an enzyme biosensing application. These authors indicate that this is a novel, highly processable, and nondiffusionable mediating species. These authors report that these nanoparticles are readily dispersable in aqueous media which facilitates their processability and overcomes the traditional issues associated with Pani. The authors emphasize that their screen-printed electrodes were readily modified by the aqueous nanoparticle dispersions. The nanoparticles, the authors claim, were simply cast by drop-coating them onto the sensing surface. They further emphasize that no electrochemical steps are involved, and hence their method is easily amenable to mass production.

Morrin et al. (2005a,b) report that polyaniline (Pani)-modified electrodes have been used in biosensing applications (Ramanathan et al., 1994; Shaolin and Jinqing, 1995; Killard et al., 2001; Tatsuma et al., 2001; Halliwell et al., 2002; Raitman et al., 2002; Shi et al., 2004). Here Pani acts as a nondiffusional mediating species. It couples electrons directly from the enzyme redox site to the electrode. This permits, the authors claim, a direct electrical communication between the biomolecule and the electrode surface. Morrin et al. (2005a,b) claim that Pani exhibits environmental stability, and its electrical properties may be modified by (a) the oxidation of the main chain, and (b) by the degree of protonation for different applications.

Morrin et al. (2005a,b) point out that the disadvantages in using Pani include its poor process ability and the fact that it is a carcinogenic monomer. Also, as indicated previously, acidic conditions are required for the formation of its highly conductive form. This, as mentioned earlier, limits the entrapment of biological molecules, such as proteins.

Morrin et al. (2005a,b) claim that electrodeposition is not a technique that leads to effective mass production of biosensors. They report that drop-coating is a simpler method of electrode modification. This, they indicate, combined with screen-printed electrodes could then be amenable to mass production. They further state that the enzyme could either be drop-coated in a simultaneous or sequential manner. This should lead to a simple method for biosensor fabrication.

Finally, Morrin et al. (2005a,b) report that the casting of conducting polyaniline nanoparticles and enzymes simultaneously by drop-coating on screen printed electrodes is an effective method of biosensor fabrication. The enzyme, horse-radish peroxidase (HRP) was incorporated before casting. The HRP was incorporated into the films so that a working biosensor could be constructed to detect the substrate H_2O_2. Different methods for immobilizing the HRP were examined. The pH of the nanoPani/DBSA (dodecylbenzenesulphonic acid) dispersion is adjusted to 7.0 prior to the addition of HRP. The authors report that the thickness of the polymer film deposited does limit its potential. A better technique is required for a thinner and more homogeneous film deposition. Morrin et al. (2005a,b) are examining ink-jet printing as an alternate casting method. The goal is to provide a single-step mass production method of fabricating enzyme biosensors. Perhaps, according to these authors, manipulation of the ink-jet printing technique may lead to a more sophisticated biosensor fabrication technique.

3.2.7 Screen-Printing for DNA Chips with an Electrical Readout for the Detection of Viral DNA (Schuler et al., 2009)

Schuler et al. (2009) have recently reported that screen printing is a cost-efficient fabrication technique for DNA chips with an electrical readout for the detection of viral DNA. These authors point out that the identification and analysis of biomolecules is important in the fields of life sciences, food technology, and in forensic and environmental research (Schena et al., 1998). Microarray technology provides for high throughput as well as high sensitivity (Wang, 2000; Schena, 2003). Biochips are used (as microarray-based approaches) to detect binding events (Choi et al., 2007; Marquette et al., 2008). Schuler et al. (2009) report that simple detection methods that require only simple instrumentation techniques are gaining popularity (Cheung et al., 1999; Heller, 2002).

Schuller et al. (2009) indicate that screen printing technology is one of the oldest forms of graphic art reproduction and its characteristic, as different from other forms of microstructuring, is its film deposition. It has been widely used for the large-scale fabrication of disposable biosensors. This is due to its low cost, versatility, and miniaturization. Tudorache and Bala (2007) report that modern sensors may be integrated into portable forms, thereby permitting analytical methods for on-site testing. Personal glucose biosensors used by diabetic patients is a good example of the use of screen printing technique for fabricating biosensors (Matthews et al., 1987; Nagata et al., 1995).

Schuler et al. (2008) analyzed and compared the applicability, sensitivity, and specificity of screen printed electrodes to electrodes produced by standard photolithography. These authors detected cytomegalovirus (CMV) DNA on their chip. The authors point out that CMV belongs to the family of human herpes virus. Their results demonstrated screen printing as a cost efficient and presumably large-scale production technique for microstructure chips. The electrical DNA detection was based on nanoparticle labeling and subsequent site-specific silver deposition.

For their investigation and analysis Schuler et al. (2008) used the following DNA-oligonucleotides (both capture and target probes):

(a) Positive biotin labeled control: (5′-NH_2-C6CAT AGA ATC AAG GAG CAC ATG CTG AAA AAA-biotin-3′)
(b) A complementary capture sequence: (5′-NH_2-C6-TTT TTT CAG CAT GT CTC CTT GAT TCT ATG-3′)
(c) A sequence containing 1 mismatch: (5′-NH_2-C6-TTT TTT CAG CAT **GGG** CTC CTT GAT TCT ATG-3′)
(d) A sequence containing 3 mismatches (5′-NH_2-C6-TTT TTT CAG CAT **TAT** CTC CTT GAT TCT ATG-3′)
(e) A negative control (5′-ACT-GAC TGA CTG ACT GAC TGA CTG GGC GGC GAC CT-NH_2-C7-3′)
(f) Biotin labeled target DNA (5′-biotin-CAT AGA ATC AAG GAG CAC ATG CTG AAA AAA-3′) had only a biotin modification

Schuler et al. (2009) report that the metallic microelectrode structures of the biochips (0.5 in. square in size) were printed on 50 × 50 mm glass slides. The screen-printed substrates were chemically modified by GOPS (3-glycidyloxypropyl-trimethoxysilane) for the binding of amino-modified ss (single-stranded) capture DNA molecules. Different 30 bp ss capture sequences were spotted by a noncontact Nanoplotter from Gessellschaft fuer Sclizium-Mikrosysteme mbH, Grosserkmannscdorf, Germany to detect CMV-DNA on the screen-printed chips. The concentration of the spotted capture DNA was 10 μM in the microarray printing buffer.

The principle of the electrical detection of DNA may be described in the following four steps:

(a) The ss capture molecules are spotted between two microelectrodes.
(b) The biotin-labeled target DNA hybridizes to the specific partner on the chip surface.
(c) The binding between the streptavidin-horseradish peroxidase-polymer occurs due to the binding molecule, biotin.
(d) A metallic layer between the electrodes is introduced due to a final enzyme-induced silver deposition. The readout is obtained by an electrical measurement. This occurs due to the conductivity of the gap filled by the interconnected silver nanoparticles.

Schuler et al. (2009) emphasize that their screen-printed chips may be produced at a cost that is an order of magnitude less than that of the chips produced by photolithography. The primary reasons for this, these authors indicate, are the expensive clean room technology required for photolithography, the equipment required, and the size of the chips. Photolithography becomes economical, according to these authors, if the size of the chips becomes very small. This, however, implies a larger number of chips will be required. Schuler et al. (2009) point out that the size reduction in chips is difficult, because the chip size is dependent on the number of measurement points, the size of the electrode gap, the distance between the conducting paths, and the contact areas for the electrical readout.

Finally, Schuler et al. (2009) conclude that screen printing technology is a cost-effective procedure to establish platforms for the chip-based analysis of biomolecules. They report that this method is an alternate technique which is especially suited for systems with sophisticated electrode layouts and structures. They point out that their screen printed substrates exhibited the same sensitivity and specificity of DNA chips when compared with standard photolithography on silicon techniques to produce these DNA chips. Their electrode structures did exhibit a high degree of stability. These authors intend to use their screen printed biochips to further detect proteins and RNA.

3.2.8 Fabrication of Microband Glucose Biosensors: Use of Screen-Printed Water-Based Carbon Ink and Use of the Biosensors in Serum Analysis (Pemberton et al., 2009)

Pemberton et al. (2009) have recently fabricated a microband glucose biosensor by screen-printing a water-based carbon ink formulation containing cobalt phthalocyanine redox mediator and GOD enzyme. They investigated the performance of these biosensors at 25°C. They optimized the working pH at 8.0. These authors were able to obtain steady-state responses under quiescent conditions. This suggested to the authors that a mixed mechanism is involved which is predominantly radial diffusion. This was indicative of microelectrode behavior.

Pemberton et al. (2009) report that electrochemical enzyme biosensors are commonly fabricated by screen printing of carbon inks as transducers along with inorganic mediators. These authors indicate that the screen-printing approach is adaptable to an industrial mass-production scale (Newman and Setford, 2006). Thus, it is suitable for producing low-cost disposable devices, and has been used to manufacture the amperometric test strips used to monitor sugar levels by diabetics (Matthews et al., 1987). Hart et al. (2004) indicate that SPCEs are used in combination with the enzyme GOD (Wilson and Turner, 1992).

Pemberton et al. (2009) point out that amperometric responses obtained at planar electrodes do not maintain steady-state conditions. Thus, this prevents their use in monitoring conditions. However, microelectrodes including microbands provide for efficient mass transfer via radial diffusion (Pletcher, 1991; Amatore, 1995). This is due to one dimension being less than 50 μm in size. This increase in mass transfer, the authors claim, permits steady-state current responses, and allows determination in quiescent sample conditions. Besides, continuous monitoring applications are now feasible.

Pemberton et al. (2009) report that ultramicroband electrodes using the SPCE method have been used by Chang and Zen (2006a). Because of the low ohmic drop for these electrodes, Chang and Zen (2006b) used these electrodes for the voltametric determination of trace nitrite ions in low ionic strength solutions.

Pemberton et al. (2009) wanted to analyze water-based COPG/GOD-containing carbon ink formulation to see the possibility of fabricating glucose microband biosensors possessing microelectrode-type behavior. They also wanted to optimize the performance of their resulting biosensors. The authors report that their devices operate by producing H_2O_2 (hydrogen peroxide) from glucose. This H_2O_2 is then electrocatalytically oxidized via the CoPC mediator. The authors report that they were careful to select and apply a potential of +0.4 V for the operation of their devices. They emphasize that the +0.4 V potential is sufficiently positive to permit the electrocatalytic oxidation of H_2O_2. It is also sufficiently low to avoid any loss of sensitivity that is observed at higher potentials. Also, the enhanced mass transport owing to the small size of the microbands (sufficiently small size in at least one dimension) yields an enhanced current density. This leads to better signal-to-noise ratios.

Pemberton et al. (2009) report that their biosensors may simply be dipped in solution, which leads them to believe that their biosensor may be used as a versatile portable device. Furthermore, owing to the ability of their biosensors to attain steady-state responses at low current, their microband biosensor may be used in continuous process-monitoring applications. These authors indicate that the single screen-printing step used in their biosensor fabrication process is relatively low cost, and could be used at a mass production scale. Finally, Pemberton et al. (2009) point out that SPCE-based biosensors may be used to detect analytes other than glucose (Hart et al., 2007).

3.2.9 *Fabrication of a Highly Sensitive Glucose Sensor Using Immobilized Enzymes (Lu et al., 2007)*

Lu et al. (2007) have recently fabricated a highly sensitive glucose biosensor using enzymes immobilized in exfoliated graphite nanoparticles Nafion membrane. These authors emphasize that nanotechnology provides the opportunity to increase the sensitivity, selectivity, and response speed of biosensors. Because of their unique chemical, physical, and optoelectronic properties nanomaterials have been used in different biosensor applications (Luo et al., 2006). Lu et al. (2007) report that the incorporation of CNTs and fullerenes in biosensor applications have substantially increased their sensitivity and response speed owing to their high chemical stability, high surface area, and unique electronic properties (Sherijara et al., 2003; Balasubramanian and Burghard, 2006; Merkoci, 2006). Lu et al. (2007) point out that CNTs have excellent electocatalytic activities (Lin et al., 2005). Furthermore, they promote electron transfer reactions involving hydrogen peroxide (Wang et al., 2003), NADH (Musameh et al., 2002; Wang and Musameh, 2003), cytochrome (Wang et al., 2002a), and AA (Wang et al., 2002b). Lu et al. (2007) also report that there is one drawback in the use of CNTs which is the high price of the CNTs. This can range from twenty to hundreds of dollars per gram.

Lu et al. (2007) report a more affordable alternative to CNTs which is exfoliated nanoplatelets (xGNP). The platelets, according to these authors consist of sp^2 hybridized carbon atoms which are arranged in a sheet-like structure unlike the cylindrical geometry found in CNTs. These authors have demonstrated the potential use of graphite nanoplatelets in biosensors. Glucose oxidase (GO_x) was used to develop a glucose biosensor. GO_x catalyzes the oxidation of glucose to gluconolactone (Wise, 1989):

$$\text{Glucose} + O_2 \rightarrow \text{gluconolactone} + H_2O_2 \tag{3.2}$$

Glucose is made quantitative by the electrochemical detection of H_2O_2.

Lu et al. (2007) point out that there are several methods to immobilize enzymes for glucose biosensor applications. A common method to prepare biosensors is to use Nafion encapsulation (Wise, 1989). Lu et al. (2007) indicate that Nafion is a sulfonated tetrafluoroethylene copolymer that has been used as a proton conductor for proton exchange membrane in fuel cells (Rizukawa and Sanui, 2000), and in biosensor applications (Fan and Harrison, 1992). Lu et al. (2007) point out that the main advantages for using Nafion in biosensor applications is its biocompatibility, excellent thermal and mechanical stability, mechanical strength, and antifouling properties. These authors have developed a highly sensitive and quick responding glucose biosensor using the combination of xGuP (graphite nanoplatelets), GO_x, and Nafion.

Drzal and Fukushima (U.S. patent application 20040127621) have previously demonstrated a method to produce exfoliated graphite nanoplatelets from natural crystalline graphite by microwave and milling. Lu et al. (2007) point out that a major challenge of using graphite nanoplatelets in biosensor applications is their insolubility in most enzyme compatible solvents owing to their large hydrophobic basal plane. These authors report that the polyelectrolytes used earlier to solubilize CNTs can also disperse graphite nanoplatelets. These polyelectrolytes are poly(diallydimethylammonium chloride) (PDAC), sulfonated poly(styrene) (SPS), and polyethyleneimine (PEI) (Lu et al., 2007). They indicate that Nafion is a negatively charged polyelectrolyte and can be used to suspend graphite nanoplatelets in water or alcohol.

Finally, Lu et al. (2007) have demonstrated that graphite nanoplatelets may be used as an inexpensive alternative to CNT for the fabrication of a highly sensitive and fast responding glucose biosensor. They report that a simple water-organic Nafion solution cast with a high (85 wt%) content of organic solvent may be used to prepare the glucose biosensor. The addition of the graphite nanoplatelets greatly enhanced the redox peak current of ferrocyanide solution. It also lowered the over potential for monitoring enzymatically produced hydrogen peroxide.

3.2.10 Fabricaton of a Zinc Oxide Nanoparticle/Glucose Oxidase Biosensor (Ren et al., 2009)

Ren et al. (2009) have recently fabricated a biosensor using zinc oxide (ZnO) nanoparticles/glucose oxidase. These authors report that enzymatic biosensors exhibit high sensitivity and quick response times to different substrates; thus, they have been the subject of much research. They further point out that different types of advanced materials have been used to increase the sensitivity of enzyme electrodes (Gerard et al., 2002; Xiao et al., 2003; Yang et al., 2004; Ren et al., 2005; Salimi et al., 2007). They concede, however, that there is a drawback in the sense that there is still inefficient electrical communication between the enzyme's redox center and the solid electrode surfaces (Zhou et al., 2005). This they opine is because the enzyme active site is deeply buried in the protein shell.

Ren et al. (2009) indicate that semiconductor colloids exhibit unique electrical and optical properties, and have been used in a variety of optoelectronic applications (Alivisatos, 2004). Ren et al. (2009) point out that ZnO is an n-type semiconductor, which exhibits both photoconductivity and photocatalytic activity and has been used in biosensor applications (Liu et al., 2001). In essence, Ren et al. (2009) report that the photophysical properties of semiconductor particles have been used to develop biosensor systems (Curri et al., 2002; Pardo-Yissar et al., 2003; Huang et al., 2005; Zhao et al., 2007; Umar et al., 2008). Ren et al. (2009) have developed and characterized the performance of a glucose biosensor based on ZnO particles.

Their results show that the ZnO nanoparticles significantly enhance the sensitivity of the GO_x enzyme electrode. They emphasize that their fabrication method of making a glucose biosensor is simple and effective.

Ren et al. (2009) have recently used glucose oxidase (GO_x)/ZnO as a model system to determine the photovoltaic effect of nanoparticles on the enzyme electrode. They noted that the structure of the GO_x was preserved after conjugation with ZnO particles. Also, the current response increased with ZnO particles. The current response increased by 30%, and the detecton limit was lowered by two orders of magnitude upon irradiating the enzyme electrode for 2 h with UV light.

These authors used glucose oxidase (130 U/mg) E.C.1.1.3.4 from *Aspergillus niger*, and β-D(+)-glucose oxidase. They synthesized the ZnO nanoparticles by precipitation from

2-propanol using the Pesika et al. (2002) method. They performed the amperometric measurements using a three electrode cell. They used an enzyme working electrode, a platinum wire counter electrode, and a Ag/AgCl reference electrode. Measurements were made at 35 °C and at a fixed potential of 0.4 V. The SEM (scanning electron microscope) indicated that the surface of their ZnO particles was rough, and consisted of smaller particles. These primary small nanoparticles were interconnected with each other and formed larger secondary particles. This indicates the fractal nature of these ZnO nanoparticles. An additional advantage of this rough or fractal nature of these ZnO nanoparticles provides for a larger ratio of surface to volume ratio, thereby making it more feasible to immobilize the GO_x.

Ren et al. (2009) also evaluated the influence of pH in the 4.0-9.0 range on the biosensor performance. They obtained the pH optimum at 6.8, which is also close to the pH optimum for free GO_x. Thus, a pH of 6.8 for the phosphate buffer was selected.

These authors also analyzed the influence of temperature (in the 10-70 °C range) on their biosensor performance for the 2.8 mmol/L concentration of glucose solution. An optimum temperature of 45 °C was noted; at higher temperatures GO_x was denatured. Ren et al. (2009) selected a temperature of 35 °C in order to keep it compatible with the human body temperature.

Ren et al. (2009) also noted an interesting photovoltaic effect of the ZnO nanoparticles on their biosensor performance. On irradiating their biosensor with UV light, a maximum current response was obtained after 2 h when compared with results with normal light. However, after 2 h the current declined presumably, according to these authors, due to enzyme (GO_x) denaturation.

Finally, the authors conclude that they have achieved a simple and effective method for biosensor fabrication. The structure of the glucose oxidase is maintained after conjugation with the ZnO nanoparticles. This is confirmed by the UV-spectrum and CD analysis. The enzyme electrode with the ZnO particles showed a significant improvement in their biosensor performance when compared with the enzyme electrode without the ZnO particles. Also, the photovoltaic effect of ZnO nanoparticles enhances the catalytic activity of the glucose oxidase enzyme, leading to an improvement of their enzyme electrode performance, and subsequently to an improvement in their biosensor performance.

3.2.11 Fabrication of a Highly Sensitive Glucose Biosensor Using an Osmium Complex and Glucose Oxidase (Salimi et al., 2009)

Salimi et al. (2009) have recently used a novel osmium complex as an electrocatalyst for the electroreduction of oxygen and H_2O_2 at relevant physiological conditions (pH). These authors used single-walled carbon nanotubes (SWCNTs), and reversibly adsorbed 1,4,8,12-tetraazacyclotetradecane osmium (III) chloride ($Os(III)Cl_2)ClO_4$ on the SWCNTs using electroless deposition. Their glucose biosensor was made by covering this CNTs/Osmium complex modified electrode with a thin film of glucose oxidase. Using a decreasing cathodic peak current of oxygen these authors were successful in detecting glucose. The performance characteristics of their biosensor was good exhibiting a high sensitivity of 826.3 nA/(μM cm^2), a low detection limit of 56 nM, a response time of less than 3 s, and a three-order calibration range of 1.0 μM-1.0 mM. The authors assert that the selectivity of their biosensor was very good as, owing to the relatively low potential applied, the interference from electroactive species was minimized. Furthermore, their glucose biosensor exhibits high stability and is technically simple.

3.2.12 Fabrication of a Porous Silicon-Based Biosensor (Matthew and Alocilja, 2003)

Matthew and Alocilja (2003) have fabricated a porous silicon-based biosensor to detect bacteria. The authors used a sponge like porous layer of silicon. For the fabrication of their biosensor the authors report that anodizing conditions of 5 mA/cm^2 were the best. Their porous silicon chips were functionalized using silanization and antibodies immobilized to the porous surface. They assessed the functionalization of their biosensor using a chemiluminescence-based assay. They report that light emission for the silanized porous silicon biosensor chip with *Salmonella* was an order of magnitude greater than that of the control and nonsilanized porous silicon with *Salmonella*. Also, these authors noted that a higher light emission was observed for the porous silanized biosensor with *Salmonella* compared with that observed with the nonporous chip. Thus, these authors fabricated the porous silicon-based biosensor and functionalized it to successfully detect *Salmonella*.

3.2.13 Recrystallization Technologies to Fabricate a Low Cost Si Nanowire Biosensor (Ashburn and Sun, 2009)

Ashburn and Sun (2009) report that approximately one billion diagnostic tests are performed in the United Kingdom every year to obtain an accurate diagnosis of a patient's condition. These authors point out that for the routine application of these diagnostics tests they need to be performed at a much larger scale, at a lower cost, and at POC (point-of-care) locations rather than at clinical laboratories. Thus, they note the need to develop newer fabrication technologies to yield a cost-effective biosensor. These authors emphasize that silicon nanowire biosensors exhibit the potential to provide real-time, high sensitivity, high selectivity, and label-free biosensing. They emphasize that the nanoscale diameter of the nanowires leads to high sensitivity. They further indicate that current fabrication techniques are expensive owing to the use of silicon-on-insulator wafers and electronic lithography. Ashburn and Sun (2009) have developed a low-cost fabrication process for silicon nanowire biosensors. These authors used thin film transistor technology. They used plastic substrates, which, according to them, require a low thermal budget process for their nanowire biosensor fabrication process.

Ashburn and Sun (2009) emphasize that amorphous silicon cannot be used because its mobility is low. They explored the possibility of using nickel-induced lateral crystallization to convert this amorphous silicon to polycrystalline silicon using a low temperature anneal. This, they felt, should lead to a better biosensor performance for their biosensor since polycrystalline silicon has better mobility than amorphous silicon. These authors looked at Si-on-oxide and Si-on-air. They noted that Si-on-oxide was easier to fabricate though the Si-on-air configurations permitted biomolecule attachment all around the nanowire. This, they report, should lead to a higher sensitivity. These authors are currently exploring other avenues such as fluorine implantation to enhance the crystallization process.

3.2.14 Clinical Evaluation of Bionime Rightest GM310 with a Simplified Electrode for Alternate Site Blood Glucose Tests (Wu et al., 2008)

Wu et al. (2008) have reported that most processes for fabricating biosensors by the SPCEs process are complex. These authors have recently developed a novel one-step process for manufacturing electrodes for injection molding. They inserted barrel-plated gold electrodes into an injection-molded base. Their electrode was in direct electrical contact with a meter. They measured glucose levels at the finger tip, palm, and arm, and compared the results with plasma values obtained by the hexokinase method on an Olympus AU640 instrument. They obtained good reproducible results. When their Rightest GM310 meter results were compared with those obtained by a laboratory method, their results complied with the ISO 15197:2003 criteria. These authors conclude that their Bionime Rightest GM310 meter used a simplified biosensor fabrication process, and exhibited reasonable performance for monitoring glucose at alternative test sites.

3.2.15 Biosensor Fabrication Method Using Microencapsulation of Enzyme in Hydrophobic Synthetic Latex Films: Amperometric Determination of Glucose (Cosnier et al., 2000a,b)

Cosnier et al. (2000a,b) have developed novel enzyme electrodes based on synthetic hydrophobic latex matrices which may be used for glucose detection. These authors used microencapsulation to immobilize glucose oxidase by using simple adsorption of enzyme-latex suspensions on the surface of a platinum electrode. They used two latex films functionalized by a hydroxyl or a glucoamide group. Their biosensors were able to provide a quantitative estimate of the glucose present by potentiostating the modified electrode at 0.6 V/SCE. This was done to oxidize the hydrogen peroxide generated by the enzymatic oxidation of glucose in the presence of dioxygen. The authors evaluated the response of their electrodes as a function of temperature and film thickness. They noted that their biosensor was stable; however, it is only at temperatures above 65 °C that the glucose oxidase started to denature and became inactive.

3.2.16 Fabrication of Dip-Strip Test Systems: Detection of β(1→3)-D-glucan (Bagal-Kestwal et al., 2009)

Bagal-Kestwal et al. (2009) have very recently fabricated photometric enzyme dip-strip test systems for the detection of β(1→3)-D-glucan using crude β(1→3)-D-glucanase from the sprouts of *V. aconitifolia* (moth bean). This moth bean (8-day old) was coentrapped with glucose oxidase in different combinations of composite polymer matrices of agarose, gelatin, polyvinyl alcohol and corn flour. The specific activity of the enzyme is 244 units/mg. The authors selected a 3% agarose, 2% corn flour, and 8% gelatin composite mixture for fabricating the enzyme dip-strip systems for the detection of β-glucan by a spectrometer using the DNSA (3,5-dinitrosalicyclic acid) method (system 1), and the AAP method (system 2). These authors report that the level of detection (LOD) for system 2 was lower than that observed for system 1, and thus system 2 was selected for analyzing and detecting β(1→3)-D-glucan content in different pharmaceutical samples.

Finally, these authors conclude that there is less than 5 percent error in detection for samples that were not pre-treated.

3.2.17 Fabrication of Disposable Screen Printed Electrodes (Kadara et al., 2009)

Kadara et al. (2009) have recently fabricated screen printed microelectrodes which are disposable and flexible, and may be characterized by microscopy and cyclic voltametry. These microelectrodes comprise microsized graphite typically with radii of 60-100 mm and are defined by an inert dielectric. The authors indicate that the advantage of this type of electrochemical sensing platform is that each of the microelectrodes is disposable. They are also cost effective and do not require either extensive cleaning or electrode pretreatment between measurements. The only requirement prior to measurements is that the microelectrode needs to be calibrated with a suitable redox probe. This is regular procedure with microelectrodes. The authors were able to detect lead via cathodic stripping volatmetry using their screen printed electrodes. The advantage of their screen printed electrodes is that they are able to obtain comparable detection limits to that obtained by insonated boron doped diamond electrodes. This is also without the need for power ultrasound. This is particularly advantageous, according to these authors, as this limits the widespread use and applicability of the detection procedure.

3.2.18 Fabrication of a Penicillin Biosensor Using Charge Transfer Techniques (Lee et al., 2009)

Lee et al. (2009) have recently fabricated and demonstrated a highly sensitive penicillin biosensor based on charge-transfer techniques (CTTPS). This CTTPS technique consists of a charge accumulation technique for penicilloic acid and H^+ ions. These authors report that there is no need to amplify the sensing signals using an external amplifier; use is made of the charge accumulation cycles. Their penicillin biosensor exhibits excellent performance characteristics: a high sensitivity (47.9 mV/mm), a high signal-to-noise ratio, a large span (1445 mV), a wide linear range (0-25 mM), and a quick response time (less than 3 s). Besides, their biosensor demonstrates good reproducibility. In addition, 0.01 mM was the lower detection limit exhibited by their biosensor. Finally, they assert that their biosensor clearly exhibits a better performance than the presently widely used ISFET penicillin biosensor. For example, their biosensor is about eight times more sensitive than the present ISFET biosensor. The authors have used their biosensor to measure penicillin concentrations in penicillin fermentation broths.

Lee et al. (2009) indicate that their CTTPS is constructed by immobilizing penicillinase onto the ion-sensitive membrane Si_3N_4 of a pH-CCD (pH sensor based on charge-coupled device). Their CTTPS device is able to detect variations in the hydrogen ion (H^+) concentration due to the catalyzed hydrolysis of penicillin by the enzymatic reaction. The authors indicate that charges corresponding to the penicillin concentration are repeatedly transferred from a sensing part to the floating diffusion region. The signal charges are accumulated here. The authors assert that their biosensor is stable, and could be a new device that may be used in the health care field.

3.2.19 *Fabrication of Biosensors Using Platinum Nanowire Nanoelectrode Array (Yang et al., 2006)*

Yang et al. (2006) have developed a glucose biosensor using a platinum nanowire nanoelectrode array (NEA). These authors used the platinum nanowires to fabricate a biosensor array. They indicate that platinum nanowire arrays can be grown by electrodeposition in a polycarbonate membrane. They report that the nanowire array so prepared may be considered as a NEA, which improves the signal-to-noise ratio and decreases the detection limit. They further assert that the high surface area of the platinum electrode helps minimize the problems associated with conventional platinum electrodes because of the limited electroactive sites. The sensitivity of their NEA for hydrogen peroxide is fifty times larger than that observed with conventional platinum electrodes. Their biosensor was able to determine glucose concentration in the 10^{-6} to 3×10^{-2} M range, interference free. They assert that their biosensor has a high efficiency of signal transduction, and is able to determine glucose concentrations in real blood samples. The authors report that their nanostructuring process not only increases surface area and the number of electroactive sites, but also expands the upper detection limit. They point out that their glucose oxidase is stable when adsorbed onto the electrode surface. Finally, their biosensor fabrication method should be readily applicable to other biosensor applications where other oxidases are used, for example, for the detection of choline, cholesterol, and alcohol. Thus, their biosensor fabrication method is perhaps generic in nature.

3.3 Conclusions

Different biosensor fabrication methods have been presented here. Some of the methods presented initially are dealt with in detail, whereas towards the end of the chapter others presented are only briefly alluded too. This is to prevent the biosensor fabrication chapter from becoming unmanageable in size. The examples of biosensor fabrication presented are selected at random from the literature. The only driving criterion, as is to be reasonably expected, is that these examples have recently appeared in the literature. Though glucose biosensor fabrication methods presented appear frequently, other examples of biosensor fabrication are also presented. Some of the biosensor fabrication presented include those for glucose (biosensor using a nanocomposite electrode, Safavi et al., 2009; screen-printed water-based carbon-ink microband biosensor, Pemberton et al., 2009; immobilized enzyme biosensor, Lu et al., 2007; ZnO nanoparticle/glucose oxidase biosensor, Ren et al., 2009; osmium complex and glucose oxidase biosensor, Salimi et al., 2009; microencapsulated enzyme in a hydrophobic synthetic latex film biosensor, Cosnier et al., 2000a,b; platinum nanowire array biosensor, Yang et al., 2006). Other fabrication techniques for biosensor fabrication mentioned include MI (for a potentiometric protein biosensor, Wang et al., 2008; polymer microarray on a chip, Henry et al., 2008), inkjet printing technology for an optical fiber imaging sensor (Carter et al., 2006), conducting polymer polyaniline (Pani)-based biosensors (using ITO, Tahir et al., 2007; biosensor fabrication procedure based on processable polyaniline nanoparticles, Morrin et al., 2005a,b; biosensor fabrication based on screen printing technology (DNA chips with an electrical readout for the detection of viral DNA), Schuller et al., 2009; microband glucose biosensor, Pemberton et al., 2009; disposable screen printed electrodes (Kadara et al., 2009) biosensor based on charge transfer technique (Lee et al., 2009), biosensor fabrication based on nanowire NEA (Yang et al., 2006), fabrication of dip-strip test systems (Bagal-kestwal et al., 2009), recrystallization technologies to fabricate a Si nanowire biosensor (Ashburn and Sun, 2009), and porous silicon-based biosensor (Matthew and Alocilja, 2003).

The presentation of different biosensor fabrication techniques together in one chapter quickly provides one with an overall perspective of the different fabrication techniques that have appeared in the recent literature, of the techniques that exhibit potential, and techniques that are being explored further research-wise. Needless to say, the fabrication techniques are application dependent, such as the determination of glucose, but as one might very reasonably expect, biosensors for the detection of other analytes is slowly but surely gaining importance and thus biosensor fabrication techniques other than for glucose detection need to be explored further. Surely, some of the biosensor fabrication techniques used for glucose determination may also be used for the fabrication of biosensors (with suitable modifications, if necessary) for the detection of other analytes of importance. It is very reasonable to expect that if the biosensor for the detection of some particular analyte is to be developed, then perhaps one may have to resort to a completely new process which is more suitable and amenable for that analyte's detection. This chapter provides one with an overall view of the processes that are being currently explored or researched. Surely, there will be other biosensor fabrication procedures that will be developed in future that will be more suitable for the detection of analytes in solution and in the gas phase, other than those that are mentioned in this chapter or those that have already appeared in the literature. Thus, in a sense the biosensor fabrication procedures mentioned here may be considered loosely as a starting point in an iterative solution, which continually improves (or converges to the optimal solution) with each iteration.

References

Adeloju SB and GG Wallace, *The Analyst*, **121**, 699–703 (1996).

Alivisatos P, The use of nanocrystals in biological detection, *Nature Biotechnology*, **22**, 47 (2004).

Amatore C, In *Physical Electrochemistry*, I Rubenstein (Ed.), Marcel Dekker, New York, 1995, pp. 136–208, Chapter 4.

Ashburn P and K Sun, Low cost Si nanowire biosensors by recrystallization technologies, http://www.ecs.soton.uc.uk/researchprojects/696, URI, http://id.ecs.soton.ac.uk/project/696/, 2009.

Balasubramanian K and M Burghard, *Analytical and Bioanalytical Chemistry*, **385**(3), 452 (2006).

Bagal-Kestwal D, RM Kestwal, and BH Chiang, Fabrication of photometric dip-strip test systems for detection of β (1-3)-D-glucan using crude β (1-3)-D-glucanase from sprouts of *Vigna aconitifolia, Biosensors and Bioelectronics*, 2566–2573 (2009).

Biodot, http://www.biodot.com/applications/biosensor/ps_main_page.htm,downloaded March 26, 2009.

Bossi A, F Bonini, APF Turner, and S Piletsky, Molecularly imprinted polymers for the recognition of proteins: The state of the art, *Biosensors & Bioelectronics*, **22**(6), 1131–1137 (2007).

Bowma MAE, CJ Allender, KR Brain, and CM Heard, Molecularly imprinted polymers as selective sorbents for the preliminary screening of combinatorial libraries, *Methodology Survival for Bioanalytical Drugs*, **25**, 37–43 (1998).

Cao Y, P Smith, and AJ Heeger, Counter-ion induced processibility of conducting polyaniline, *Synthetic Metals*, **57**, 3514–3519 (1993).

Carbrey EA, LN Brown, and TL Chow, Recommended standard laboratory techniques for diagnosing infectious bovine rhinotracheitis bovine virus diarrhea virus and shipping fever, *Proceedings of the US Animal Health Association*, , 629–701 (1971).

Carter JC, RM Alvis, SB Brown, KC Langry, TS Wilson, MT McBride, ML Myrick, WR Cox, ME Grove, and BW Colston, Fabricating optical fiber imaging sensors using inkjet printing technology: A pH proof-of-concept, *Biosensors and Bioelectronics*, 1355–1364 (2006).

Chang JL and JM Zen, Fabrication of disposable ultramicroelectrodes: Characterization and applications, *Electrochemistry Communications*, **8**, 571–576 (2006a).

Chang JL and JM Zen, Disposable screen-printed edge band ultramicroelectrodes for the determination of trace amounts of nitrite ion, *Electroanalysis*, **18**, 941–946 (2006b).

Chen T, WR Cox, D Lenhard, and DJ Hayes, Microjet printing of high precision microlens for packaging of fiber-optic components, *SPIE Proceedings*, **4652**, 136–141 (2002).

Cheung VG, M Morley, F Aguilar, A Missimi, R Kucherlapti, and G Childs, Making and reading microarrays, *Nature Genetics*, **21**(1 Suppl.), 15–19 (1999).

Choi JW, B Koh, YK Kim, and J Min, Nanotechnology in biodevices, *Microbiology and Biotechnology*, **17**(1), 5–14 (2007).

Cosnier S, Biomolecule immobilization on electrode surfaces by entrapment or attachment to electrochemically polymerized films. A review, *Biosensors & Bioelectronics*, **14**, 443–456 (2000a).

Cosnier S, S Szuneritis, RA Marks, A Novoa, L Puech, E Perez, and I Rico-lattes, A rapid and easy procedure of biosensor fabrication by micro-encapsulation of enzyme in hydrophobic synthetic latex films. Application to the amperometric determination of glucose, *Electrochemistry Communications*, **2**(12), 851–855 (2000b).

Cox WR, DJ Hayes, T Chen, and DW Ussery, Fabrication of micro-optics by microjet printing, *SPIE Proceedings*, **2383**, 110–115 (1995).

Cox WR, T Chen, D Ussery, DJ Hayes, JA Tatum, and DL Mcfarland, Microjet lenslet-tipped fibers, *Optical Communication*, **123**(4-6), 492–496 (1996).

Cui HF, JS Ye, WD Zhang, CM Li, JHT Luong, and FS Sheu, *Analytica Chimica Acta*, **594**, 175–183 (2007).

Curri ML, A Agostiano, G Leo, A Mallard, P Cosma, and MD Monica, *Materials Science and Engineering*, **22**, 449 (2002).

Deng C, J Chen, X Chen, C Xiao, L Nie, and S Yao, Direct electrochemistry of glucose oxidase and biosensing for glucose-based or boron-doped carbon nanotubes modified electrode, *Biosensors & Bioelectronics*, **23**, 1272–1277 (2008).

Drzal LT and H Fukushima, Expanded graphite and products produced therefrom, U.S. patent application, 20040127621.

Fan ZH and DJ Harrison, Permeability of glucose and other neutral species through recast perfluorosulfonated nanomer films, *Analytical Chemistry*, **64**(11), 1304 (1992).

Fuh MRS, LW Burgess, T Hirschfeld, GD Christian, and F Wang, Single fibre optic fluorescence pH probe, *Analyst*, **112**(8), 1159–1163 (1987).

Gehrich JL, DW Lubbers, N Opitz, DR Hansmann, WW Miller, JK Tussa, and M Yafuso, Optical fluorescence and its application to an intravascular blood-gas monitoring system, *IEEE Transactions of Biomedical Engineering*, **33**, 117–132 (1986).

Gerard M, A Chaubey, and BD Malhotra, *Biosensors & Bioelecronics*, **17**, 345–359 (2002).

Gilmartin MAT, JP Hart, and DT Patton, *Analyst*, **120**, 1973–1981 (1995).

Gu Y, HY Yu, AS Kim, CS Ah, JH Yang, IB Baek, CG Ahn, SJ Lee, and JH Yun, Biosensor manufacturing method thereof, and biosensing applications including the same, WO/2008/38906, publication date 03.04.08; http://wipo/worldintellectualpropertyorganization.int/pctdb/en/wo.jsp?WO=2008038906 downloaded March 26, 2009.

Guo TY, YQ Xia, GJ Hao, MD Song, and BH Zhang, Adsorptive separation of hemoglobin by molecularly imprinted chitosan beads, *Biomaterials*, **25**, 5905–5912 (2004).

Halliwell C, E Simon, C Toh, P Bartlett, and A Cass, Immobilization of lactate dehydrogenase on poly(aniline)-poly(acrylate) and poly(aniline)-poly(vinyl sulphonate) films for a lactate biosensor, *Analytica Chmica Acta*, **453**, 191–200 (2002).

Hart JP, A Crew, E Crouch, KC Honeychurch, and RM Pemberton, *Analytical Letters*, **37**, 789–830 (2004).

Hart JP, A crew, E Crouch, KC Honeychurch, and RM Pemberton, In S Alegret, A Merkoci, (Eds.), *Comprehensive Analytical Chemistry*, vol. **49**, Elsevier B.V., Amsterdam, 2007, pp 497–557, Chapter 23.

Haupt K, Imprinted polymers—Tailor made mimics of antibodies and receptors, *Chemical Communications*, 171–178 (2003).

Hayden O, R Bindeus, C Haderspock, KJ Mann, B Wirl, and FL Dickert, *Sensors & Actuators B*, **91**, 316–319 (2003).

Heller MJ, DNA microarray technology: Devices, systems, and applications, *Annual Reviews of Biomedical Engineering*, **4**, 129–153 (2002).

Henry OYF, DC Cullen, and SA Piletsky, *Analytical and Bioanalytical Chemistry*, **382**(4), 447–456 (2005).

Henry OYF, SA Piletsky, and DC Cullen, Fabrication of molecularly imprinted polymer microarray on a chip by mid-infrared laser pulse initiated polymerization, *Biosensors and Bioelectronics*, **23**, 1769–1775 (2008).

Huang YX, WJ Zhang, H Xiao, and GX Li, *Biosensors & Bioelectronics*, **21**, 817 (2005).

Introduction to biosensor technology, http://www.cranfield.ac.uk/health/shortcourses/page26524.jsp downloaded April 22, 2009.

Janata J, Immunoelectrode, *Journal of the American Chemical Society*, **97**, 2914–2916 (1975).

Jena BK and CR Raj, *European Journal of Chemistry*, **12**, 2702–2708 (2006).

Jeykumari DAS and AS Narayanan, A novel nanobiocomposite based glucose biosensor using neutral red functionalized nanotubes, *Biosensors & Bioelectronics*, **23**, 1404–1411 (2008).

Jiang XM, N Jiang, MX Xiang, and ML Liu, *Analytical and Bioanalytical Chemistry*, **389**(2), 355–368 (2007).

Jordan DM and DR Walt, Physiological pH fiber optic chemical sensor based on energy-transfer, *Analytical Chemistry*, **59**, 437–439 (1987).

Kadara RO, N Jenkinson, and CE Banks, Characterization and fabrication of disposable screen printed microelectrodes, *Electrochemistry Communications*, July, 1377–1389 (2009).

Katz E, L Willner, and J Wang, Gold nanoparticles in nonenzymatic electrochemical detection of sugars, *Electroanalysis*, **18**, 1937–1942 (2006).

Kaufmann ED, J Beylea, MC Johnson, ZM Nicholson, JL Richs, PK Shah, M Bayless, T Petterson, Z Feldoro, E Blomberg, P Claesson, and S Franzen, *Langmuir*, **22**, 6053–6062 (2007).

Killard AJ, L Micheli, K Grennan, M Franek, V Kolar, D Moscone, I Palchetti, and MR Smyth, Amperometric separation-free immunosensor for real-time monitoring, *Analytica Chimia Acta*, **427**, 173–180 (2001).

Kurniawan F, V Tsakova, and VM Mirskya, *Electroanalysis*, **16**, 19–44 (2004).

Lee SR, MM Rahman, K Sawada, and M Ishida, Fabrication of a highly sensitive penicillin sensor based on charge transfer technique, *Biosensors and Bioelectronics*, 15 March, 1877–1882 (2009).

Lin Y, F Lu, Y Tu, and Z Ren, *Nano Letters*, **4**, 191–195 (2004).

Lin YH, W Yantasee, and J Wang, *Frontiers in Bioscience*, **10**, 492 (2005).

Liu R, AA Vertegel, EW Bohannan, TA Sorenson, and JA Switzer, *Chemical Materials*, **13**, 508 (2001).

Lotizero M, OYF Henry, S Piletsky, I Tothill, D Cullen, M Kania, B Hock, and APF Turner, Surface plasmon resonance biosensor for domoic acid based on grafted imprinted polymer, *Biosensors & Bioelectonics*, **20**(2), 145–152 (2004).

Lu J, LT Drzal, RM Worden, and I Lee, Simple fabrication of a highly sensitive glucose biosensor using enzyme immobilized in exfoliated graphite nanoplatelets nafion membrane, *Chemical Materials*, **19**, 6240–6246 (2007).

Lukachova LV, et al., *Journal of Electroanalytical Chemistry*, **544**, 59–63 (2003).

Luo XL, A Morrin, AJ Killard, and MR Smyth, *Electroanalysis*, **18**, 319 (2006).

Maleki N, A Safavi, and F Tajabadi, Investigation of the role of ionic liquids in imparting electrocatalytic behavior to carbon paste electrode, *Electroanalysis*, **19**, 2247–2250 (2007).

Marquette LA, BP Corgier, KA Heyries, and LJ Blum, *Frontiers in Bioscience*, **13**, 382–400 (2008).

Matthew FP and EC Alocilja, Fabrication of porous silicon-based biosensor, *Sensors, Proceedings of IEEE*, **1**, 293–298 (2003), doi:10.1109/ICENS.2003.1278945, downloaded September 1, 2009.

Matthews DR, E Brown, A Watson, RR Holman, J Stemson, S Hughes, and D Scott, *Lancet*, **1**(8536), 778–779 (1987).

Merkoci A, *Microchimica Acta*, **152**(3-4), 157 (2006).

Morrin OYF, SA Piletsky, and DC Cullen, Fabrication of molecularly imprinted polymer microarray on a chip by mid-infrared laser pulse initiated polymerization, *Electrochemistry Communications*, March, 317–322 (2005a).

Morrin A, F Wilbeer, O Ngamna, SE Moulton, AJ Killard, GG Wallace, and MY Smyth, Novel biosensor fabrication methodology based on processable conducting polyaniline particles, *Electrochemistry Communications*, 317–325 (2005b).

Munkholm C, DR Walt, FP Milanovich, and SM Klainer, Polymer modification of fiber optic chemical sensors as a method of enhancing fluorescence signal for pH measurements, *Analytical Chemistry*, **58**, 1427–1430 (1986).

Musameh MM, RT Kachoosangi, L Xiao, A Russell, and RG Compton, Ionic liquid-carbon composite glucose biosensor, *Biosensors & Bioelectronics*, **24**, 87–92 (2008).

Musameh M, J Wang, A Merkoci, and YH Lin, *Electrochemical Communications*, **4**(1), 743 (2002).

Nagata R, K Yokoyama, SA Clark, and I Karube, *Biosensors & Bioelectronics*, **10**(3-4), 261–1167 (1995).

Newman JD and SJ Setford, *Molecular Biotechnology*, **32**(3), 249–268 (2006).

Nivens DA, MV Schiza, and SM Angel, Multilayer sol-gel membranes for optical sensing applications: Single pH and dual layer CO_2 and NH_3 sensors, *Talanta*, **58**(3), 543–550 (2002).

Nivens DA, Y Zhang, and SM Angel, A fiber-optic pH sensor using base-catalyzed organo-silica sol-gel, *Analytica Chimica Acta*, **376**, 235–245 (1998).

Pardo-Yissar V, E Katz, J wasserman, and I Willner, *Journal of the American Chemical Society*, **125**, 622 (2003).

Park S, TD Chung, and HC Kim, Nonenzymatic glucose detection using mesoporous platinum, *Analytical Chemistry*, **75**, 3046–3049 (2003).

Pemberton RM, R Pittson, N Biddle, and JP Hart, Fabrication of microband glucose biosensors using a screen-printing water-based carbon ink and their application in serum analysis, *Selected Papers from the Tenth World Congress on Biosensors*, Shanghai, China, May 14-16, 2008, *Biosensors and Bioelectronics*, 1 January, 1246–1252, 2009.

Pesika NS, Z Hu, KJ Stebe, and PC Sehrson, *Journal of Physical Chemistry B*, **106**, 6985 (2002).

Peterson JI, SR Goldstein, RV Fitzgerald, and DK Buckhold, Fiber optic pH probe for physiological use, *Analytical Chemistry*, **52**, 864–869 (1980).

Pletcher D, In *Microelectrodes Theory and Applications*, MI Montenegro, MA Queiros, JC Daschbach, (Eds.), Kluwer, Netherlands, 1991, pp. 3–16.

Raitman OA, E Katz, AF Buckman, and I Willner, Integration of polyaniline/poly(acrylic acid) films and redox enzyme electrode supports of the bioelectrocatalyzed oxidation of glucose or lactate in the integrated bioelectrocatalytic systems, *Journal of the American Chemical Society*, **124**, 6487 (2002).

Ramanathan K, S Annapoorni, and BD Malhotra, Application of poly(aniline) as a glucose biosensor, *Sensors & Actuators B*, **21**, 165–169 (1994).

Ren X, D Chen, X Meng, F Tang, X Hou, D Han, and L Zhang, Zinc oxide nanoparticles/glucose oxidase photoelectrochemical system for the fabrication of biosensor, *Journal of Colloid and Interface Science*, **334**, 183–187 (2009).

Ren XI, XW Meng, D Chen, FQ Tang, and J Jiao, *Biosensors & Bioelectronics*, **21**, 433 (2005).

Rizukawa M and K Sanui, *Progress in Polymer Science*, **25**(10), 1463 (2000).

Rong LQ, C Yang, QY Qian, and XH Xia, *Talanta*, **72**, 819–824 (2007).

Saari LA and WR Seitz, pH sensor based on immobilized fluoreseinamine, *Analytical Chemistry*, **54**, 821–823 (1982).

Sadik OA and JMV Emon, *Biosensors & Bioelectronics*, **11**, i–xi (1996).

Safavi A, G Absalan, and G Bamdad, *Analytica Chimica Acta*, **610**, 243–248 (2008).

Safavi A, N Maleki, and E Farjani, Fabrication of a glucose biosensor based on a novel nanocomposite electrode, *Biosensors and Bioelectronics*, **24**, 1655–1660 (2009).

Salimi A, B Kavosi, R Hallaj, and A Barbei, Fabrication of a highly sensitive glucose biosensor based on immobilization of osmium complex and glucose oxidase onto carbon nanotubes modified electrode, *Electroanalysis*, accepted December 24, 2008, DOI 10.1002/elan.200804486, downloaded September 1, 2009.

Salimi A, E Sharifi, A Noorbaksh, and S Soltanian, *Biosensors & Bioelectronics*, **22**, 3146 (2007).

Schena M, RA Heller, TP Theriault, K Konrad, E Lachenmeier, and RW Davis, *Trends in Biotechnology*, **16**(7), 301–306 (1998).

Schena M, *Microarray Analysis*, Hoboken, New Jersey, 648, 2003.

Schena M, RA Heller, TP Theriault, K Konrad, E Lachenmeier, and RW Davis, *Trends in Biotechnology*, **16**(7), 881–889 (2003).

Schuler T, T Asmus, W Fritzsche, and RM Moller, Screen printing as a cost-effective fabrication method for DNA-chips with electrical readout for detection of viral DNA, *Biosensors and Bioelectronics*, 15 March, 2077–2084 (2009).

Sellergren B, *Agnew Chemistry International Edition*, **39**, 1031–1037 (2000).

Sergeyeva TA, NV Lavrik, and AE Rachkov, Sensors & Actuators *B*, **34**, 283–288 (1996).

Shaolin M and K Jinqing, Bioelectrochemical activation and inhibition of polyaniline gkucose oxidase electrode by cations, *Electrochimica Acta*, **40**, 241–246 (1995).

Shaolin M and L Jincui, *Chemistry Journal on Internet*, 1 (1999).

Shea KJ, *Trends in Polymer Science*, **2**, 166–173 (1994).

Sherijara BS, W Kutner, and F D'Souza, *Electroanalysis*, **15**(9), 753 (2003).

Shi H, WB Tsai, MD Garrison, S Ferrari, and BD Ratner, *Nature*, **398**, 593–597 (1999).

Shi L, Y Xiao, and J Willner, Electrical contacting of glucose oxidase by DNA-templated polyaniline wires on surfaces, *Electrochemical Communications*, **6**(10), 1057–1060 (2004).

Singh S, DY Sasaki, J Cesarano III, and AJ Hurd, *Thin Solid Films*, **339**, 209–215 (1999).

Stejskal J et al., *Synthetic Metals*, **96**, 55–61 (1998).

Tada M and Y Iwasawa, *Annual Review of Materials Research*, **35**, 397–426 (2005).

Tahir ZM, EC Alocilja, and DL Grooms, Indium tin oxide-polyaniline biosensor: Fabrication and characterization, *Sensors*, **7**, 1123–1140 (2007).

Tan CK, KP Loh, and TTL John, Direct amperometric detection of glucose on a multiple- branching carbon nanotube forest, *Analyst*, **133**, 448–451 (2008).

Tatsuma T, T Ogawa, R Sato, and N Oyama, Peroxidase-incorporated sulfonated polyaniline-polycation complexes for electrochemical sensing of H_2O_2, *Journal of Electroanalytical Chemistry*, 180–185 (2001).

Tudorache M and C Bala, *Analytical and Bioanalytical Chemistry*, **388**(3), 565–587 (2007).

Umar A, MM Rahman, SH Kim, and YB Hahn, *Nanoscience and Nanotechnology*, **8**, 3216 (2008).

Viatakis G, LI Andersson, R Muller, and K Mosbach, *Nature*, **361**, 645–647 (1993).

Wallace DB, WR Cox, and DJ Hayes, In *Direct Write Using Ink-jet Techniques*, A Pique, DB Chrisey (Eds.), Academic Press, New York, 2002, Chapter 7.

Wang J, Survey and summary from DNA biosensors to gene chips, *Nucleic Acids Research*, **28**(16), 3011–3016 (2000).

Wang J and M Musameh, Carbon nanotube/Teflon composite electrochemical sensors and biosensors, *Analytical Chemistry*, **75**(9), 2075 (2003).

Wang J, M Musameh, and YH Lin, *Journal of the American Chemical Society*, **125**(9), 2408 (2003).

Wang JX, MX Li, ZJ Shi, NQ Li, and ZN Gu, Direct electrochemistry of cytochrome c at a glassy carbon electrode modified with single-wall carbon nanotubes, *Analytical Chemistry*, **74**(9), 1993 (2002a).

Wang Y, Y Zhou, J Sokolov, B Rigas, K Levon, and M Rafailovich, A potentiometric protein sensor built with surface molecular imprinting method, *Biosensors and Bioelectronics*, **24**, 162–166 (2008).

Wang ZH, J Liu, QL Liang, YM Wang, and G Luo, *Analyst*, **127**(5), 653 (2002b).

Wei S and B Mizaikoff, *Journal of Separation Science*, **30**(1), 1794–2805 (2007).

Wilson R and APF Turner, Glucose oxidase: An ideal enzyme, *Biosensors & Bioelectronics*, **7**, 165–185 (1992).

Winokur WJ, In *Handbook of Conducting Polymers*, TA Skotheim, RL Elsenbaumer, JR Reynolds, (Eds.), Marcel Dekker, New York, NY, 1998, pp. 707–726.

Wise DL, *Applied Biosensor*, Butterworths, Boston, 1989.

Wu MH, MY Fang, LN Jen, HC Hsiao, A Muller, and CT Hsu, Clinical evaluation of bionime rightest GM310 biosensors with a simplified electrode for alternative-site blood glucose tests, *Clinical Chemistry*, http://www.clinchem.org/cgi/content/abstract/clin.chem.2008106132801, accepted July 2, 2008, downloaded September 1, 2009.

Wulff G, *Agnew Chemistry International Edition, English*, **34**, 1612–1832 (1995).

Xiao Y, F Patolsky, E katz, JF Hainfield, and I Willner, *Science*, **299**, 1877 (2003).

Yang M, F Qu, Y Lu, Y He, G Shen, and R Yu, Platinum nanowire nanoelectrode array for the fabrication of biosensors, *Biomaterials*, 5944–5950 (2006).

Yang YH, HF Yang, MH Yang, YI Liu, GI Shen, and RQ Yu, *Analytica Chimica Acta*, **525**, 213 (2004).

Yano K and I Karube, *Trends in Analytical Science*, **18**, 199–204 (1999).

Ye JS, Y Wen, WD Zhang, LM Gan, GQ Xu, and FS Sheu, *Electrochemistry Communications*, **6**, 66–70 (2004).

Yuan J, K Wang, and X Xia, *Advances in Functional Materials*, **15**, 803–809 (2005).

Zhang N, T Wilkop, S Lee, and Q Cheng, *Analyst*, **132**, 164–172 (2007).

Zhao C, C Shao, M Li, and K Jiao, Flow injection analysis of glucose without enzyme based on electrocatalytic oxidation of glucose at a nickel electrode, *Talanta*, **71**, 1769–1773 (2007).

Zhou H, X Gan, J Wang, XI Zhu, and GX Li, *Analytical Chemistry*, **77**, 6102 (2005).

CHAPTER 4

Biosensors Involved in Drug Discovery

Chapter Outline

4.1 Introduction

Biosensors are finding increasing applications in drug discovery. Appropriate drugs have to be initially identified from a wide spectrum of possible drug candidates. Biosensors are an appropriate means to help identify these drugs. This can be a Herculean task at times, consuming tremendous amounts of time and resources. Biosensors can significantly shorten this time and also help minimize the resources required to help identify suitable drug candidates initially.

It must be recognized that identifying possible drug candidates is an important step, because in general, it takes about 10 years and around 600-700 million dollars to bring a drug from bench scale to the market. Surely, it would be tragic if during the initial identification of possible drug(s), one misses a good drug candidate either by using biosensors or by some other suitable identification technique. Thus, the importance of this initial identification step, wherein biosensors can be very useful, is quite clear. The biosensor technique is also a simple technique, which helps identify suitable drug candidates.

In this chapter we use fractal analysis to analyze (a) the binding and dissociation (if applicable) kinetics of the catalytic subunit (Cα) of CAPK in solution to adenosine-ologoarginine conjugates (ARC) using a surface plasmon resonance (SPR) biosensor (Viht et al., 2007), (b) the binding of phosphate ion (P_i) in solution to a rhodamine-PBP (phosphate binding protein) phosphate biosensor (Okoh et al., 2006), and (c) the binding of methionine-7-amido-4-

Handbook of Biosensors and Biosensor Kinetics. DOI: 10.1016/B978-0-444-53262-6.00004-8

methylcoumarin (MET-AMC) in solution in the competitive scintillation proximity aminoacyl-tRNA synthetase charging assay (cSPA) (Forbes et al., 2007). These authors report that protein kinases play a significant role in the regulation of protein function in living cells, and the above interactions help identify inhibitors for protein kinases. These authors point out that protein phosphatases are involved in the control of the phosphorylation state of many proteins. Furthermore, these authors add that the measurement of the phosphate ion, P_i is an important target for the understanding of cellular activities involving such proteins. They report that this is a good and high-throughput screening (HTS) compatible method for measuring the concentration of most naturally occurring amino acids. They further assert that the quantitative detection of amino acids is important in the areas of patient care and drug discovery.

The fractal analysis is just one method of providing values for the binding and the dissociation (if applicable) rate coefficient values. Other methods for obtaining values for the binding and the dissociation rate coefficient values on biosensor surfaces are also available. Needless to say each method of analysis has different assumptions involved. The present fractal analysis method also provides values for the fractal dimension that exists on the biosensor surface, or in other words, the degree of heterogeneity that exists on the biosensor surface.

4.2 Theory

4.2.1 Single-Fractal Analysis

Binding Rate Coefficient

Havlin (1989) reports that the diffusion of a particle (analyte [Ag]) from a homogeneous solution to a solid surface (e.g., receptor [Ab]-coated surface) on which it reacts to form a product (analyte-receptor complex; (Ab·Ag)) is given by:

$$(\mathrm{Ab\cdot Ag}) \approx \begin{cases} t^{(3-D_{\mathrm{f,bind}})/2} = t^p, & t < t_c \\ t^{1/2}, & t > t_c \end{cases} \tag{4.1}$$

Here $D_{\mathrm{f,bind}}$ or D_{f} is the fractal dimension of the surface during the binding step. t_c is the cross-over value. Havlin (1989) reports that the cross-over value may be determined by $r_c^2 \sim t_c$. Above the characteristic length, r_c, the self-similarity of the surface is lost and the surface may be considered homogeneous. Above time, t_c, the surface may be considered homogeneous, because the self-similarity property disappears, and "regular" diffusion is now present. For a homogeneous surface where D_{f} is equal to 2, and when only diffusional limitations are present, $p = ½$ as it should be. Another way of looking at the $p = ½$ case (where $D_{\mathrm{f,bind}}$ is equal to 2) is that the analyte in solution views the fractal object, in our case, the receptor-coated biosensor surface, from a "large distance." In essence, in the association process, the diffusion of the analyte from the solution to the receptor surface creates a depletion layer of width $(Đt)^{1/2}$ where $Đ$ is the diffusion constant. This gives rise to the fractal power law, (Analyte·Receptor) $\sim t^{(3-D_{f,bind})/2}$. For the present analysis, t_c is arbitrarily chosen and we assume that the value of the t_c is not reached. One may consider the approach as an intermediate "heuristic" approach that may be used in the future to develop an autonomous (and not time-dependent) model for diffusion-controlled kinetics.

Dissociation Rate Coefficient

The diffusion of the dissociated particle (receptor [Ab] or analyte [Ag]) from the solid surface (e.g., analyte [Ag]-receptor [Ab]) complex coated surface) into solution may be given, as a first approximation by:

$$(\mathrm{Ab{\cdot}Ag}) \approx -t^{(3-D_{f,\,diss})/2} = -t^{p}, \quad t > t_{diss} \tag{4.2}$$

Here $D_{f,diss}$ is the fractal dimension of the surface for the dissociation step. This corresponds to the highest concentration of the analyte-receptor complex on the surface. Henceforth, its concentration only decreases. The dissociation kinetics may be analyzed in a manner "similar" to the binding kinetics.

4.2.2 Dual-Fractal Analysis

Binding Rate Coefficient

Sometimes, the binding curve exhibits complexities and two parameters (k, D_f) are not sufficient to adequately describe the binding kinetics. This is further corroborated by low values of r^2 factor (goodness-of-fit). In that case, one resorts to a dual-fractal analysis (four parameters; k_1, k_2, D_{f1}, and D_{f2}) to adequately describe the binding kinetics. The single-fractal analysis presented above is thus extended to include two fractal dimensions. At present, the time ($t = t_1$) at which the "first" fractal dimension "changes" to the "second" fractal dimension is arbitrary and empirical. For the most part, it is dictated by the data analyzed and experience gained by handling a single-fractal analysis. A smoother curve is obtained in the "transition" region, if care is taken to select the correct number of points for the two regions. In this case, the product (antibody-antigen; or analyte-receptor complex, Ab·Ag or analyte·receptor) is given by:

$$(\mathrm{Ab{\cdot}Ag}) \approx \begin{cases} t^{(3-D_{f1,\,bind})/2} = t^{p1}, & t < t_1 \\ t^{(3-D_{f2,\,bind})/2} = t^{p2}, & t_1 < t < t_2 = t_c \\ t^{1/2}, \quad t > t_c & \end{cases} \tag{4.3}$$

In some cases, as mentioned above, a triple-fractal analysis with six parameters (k_1, k_2, k_3, D_{f1}, D_{f2}, and D_{f3}) may be required to adequately model the binding kinetics. This is when the binding curve exhibits convolutions and complexities in its shape due to perhaps to the very dilute nature of the analyte (in some of the cases to be presented) or for some other reasons. Also, in some cases, a dual-fractal analysis may be required to describe the dissociation kinetics.

4.3 Results

In this chapter we analyze the binding and dissociation (if applicable) kinetics of (a) different concentrations (in nM) of Cα in solution to ARC-704 immobilized on a SPR biosensor surface (Viht et al., 2007), (b) binding of different concentrations of inorganic phosphate (P_i) to a biosensor surface (Okoh et al., 2006), and (c) binding of different concentrations (in mU) of MET-AMC and methionine in solution in a cSPA (Forbes et al., 2007).

The fractal analysis will be applied to the different analyte-receptor reactions mentioned above with the intent of providing a better understanding of these types of reactions along with obtaining fresh physical insights into them. At the outset it should be indicated that alternative expressions for fitting the binding and dissociation data are available that include saturation, first-order reaction, and no diffusional limitations, but these expressions are deficient in describing the heterogeneity that inherently exists on the surface. It is this heterogeneity on the biosensor surface that one is attempting to relate to the different biosensor performance parameters. More specifically the question we wish to answer is how may one change the heterogeneity or the fractal dimension, D_f, on the biosensor chip surface in order that one may be able to enhance the different biosensor performance parameters.

Other modeling attempts also need to be mentioned. One might justifiably argue that appropriate modeling may be achieved by using a Langmuirian or other approach. The Langmuirian approach may be used to model the data presented if one assumes the presence of discrete classes of sites, for example double exponential analysis as compared with the single-fractal analysis. Lee and Lee (1995) indicate that the fractal approach has been applied to surface science, for example, adsorption and reaction processes. These authors report that the fractal approach provides a convenient means to represent the different structures and morphology at the reaction surface. These authors also suggest using the fractal approach to develop optimal structures and as a predictive approach. Another advantage of the fractal technique is that the analyte-receptor association is a complex reaction, and the fractal analysis via the fractal dimension and the rate coefficient provide a useful lumped parameter analysis of the diffusion-limited reaction occurring on a heterogeneous surface.

In a classical situation, to demonstrate fractality, one should make a log-log plot, and definitely have a large amount of data. It may be useful to compare the fit to some other forms, such as exponential, or involving saturation, etc. The authors do not present any independent proof or physical evidence of fractals in the examples presented here. Nevertheless, the authors still use fractals and the degree of heterogeneity on the biosensor surface to gain insights into enhancing the different biosensor performance parameters. The fractal approach is a convenient means (since it is a lumped parameter) to make the degree of heterogeneity that exists on the surface more quantitative. Thus, there is some arbitrariness in the fractal approach. The fractal approach provides additional information about interactions that may not be obtained by a conventional analysis of biosensor data. In this chapter as mentioned above, the authors are attempting to relate the fractal dimension, D_f, or the degree of heterogeneity on the biosensor surface to the different biosensor performance parameters. More specifically, the authors are interested in finding out how changes in the fractal dimension or the degree of heterogeneity on the biosensor chip surface affect the different biosensor parameters of interest.

Unless specifically mentioned there is no nonselective adsorption of the analyte. In other words, nonspecific binding is ignored. Nonselective adsorption would skew the results obtained very significantly. In these types of systems, it is imperative to minimize this nonselective adsorption. The authors also recognize that, in some cases, this nonselective adsorption may not be a significant component of the adsorbed material and that the rate of association, which is of a temporal nature would depend on surface availability.

Viht et al. (2007) point out that protein kinases play a significant role in the regulation of protein function in living cells. In fact, they report that more than 400 human diseases (including some forms of cancer) may be linked to incorrect protein kinase signaling. According to Cohen (2002) and Fischer (2004), protein kinases are the second largest drug targets after G-protein-coupled receptors. In fact, the protein kinases are the fastest growing category of drugs in development (Viht et al., 2007). Viht et al. (2007) further report that protein kinases follow a ternary complex mechanism wherein there is direct transfer of the phosphoryl group from the ATP to the protein substrate at the active site (Adams, 2001).

Viht et al. (2007) thus assert that three different types of inhibitors that target the active site are under development. Furthermore, new assays are under development for the evaluation of these protein kinase inhibitors. For example, these authors suggest that kinase inhibitors may be detected by the measurement of the binding potency of these compounds to kinase.

Viht et al. (2007) assert that SPR based biosensors have been used for a variety of biomolecular interactions (Karlsson, 2004; Boozer et al., 2006). Inhibitors of protein kinases have also been characterized by SPR biosensors. Viht et al. (2007) have used the SPR biosensor to help characterize the interactions between ARC and α isoforms of the catalytic subunit (Ca) of CAPK. These authors emphasize that the high affinity of the surface-immobilized ARC toward CAPK $C\alpha$ permits the affinity determination of the binding of the active site of the kinase.

Viht et al. (2007) report that it is possible to attach ARC-type inhibitors to a polymer matrix using a suitable tether. This tether is connected to the C terminus of the peptide moiety of the conjugate (Loog et al., 2000; Viht et al., 2003). Viht et al. (2007) point out that this type of attachment does not lead to a significant loss in inhibitor potency.

Figure 4.1a shows the binding and dissociation of 50 nM Cα in solution to ARC-704 immobilized on a SPR biosensor chip surface. A dual-fractal is required to adequately describe the binding kinetics. A single-fractal analysis is adequate to describe the dissociation kinetics. The values of (a) the binding rate coefficient, k and the fractal dimension, D_f for a single-fractal analysis, (b) the binding rate coefficients, $k_1$and k_2, and the fractal dimensions, D_{f1} and D_{f2}, and (c) the dissociation rate coefficient, k_d and the fractal dimension, D_{fd}, for a single-fractal analysis are given in Tables 4.1 and 4.2.

The values of the binding rate coefficients and the fractal dimensions presented in Table 4.1 were obtained from a regression analysis using Corel Quattro Pro 8.0 (Corel Quatro Pro 8.0, 1007) to model the data using Equations (4.1) and (4.3), wherein (Analyte·Receptor) $= kt^{(3-D_f)/2}$ for a single-fractal analysis, and (Analyte·Receptor) $= k_1t^{(3-D_{f1})/2}$ and (Analyte·Receptor) $=$ $k_2t^{(3-D_{f2})/2}$ for a dual-fractal analysis, and (Analyte·Receptor) $= -k_dt^{(3-D_{fd})/2}$ for the dissociation phase (eqn 4.2). The binding and the dissociation rate coefficient values presented in Table 4.1 are within 95% confidence limits. For example, for the binding of 50 nM Cα in solution to ARC-704 immobilized on a SPR biosensor chip surface, and for a dual-fractal analysis, the binding rate coefficient (k_1) value is 6.540 ± 0.876. This 95% confidence limit indicates that the k_1 value lies between 5.664 and 7.416. This indicates that the value is precise and significant.

Figure 4.1b shows the binding and dissociation of 25 nM Cα in solution to ARC-704 immobilized on a SPR biosensor chip surface. Once again, a dual-fractal is required to adequately describe the binding kinetics. A single-fractal analysis is adequate to describe the dissociation kinetics. The values of (a) the binding rate coefficient, k and the fractal dimension, D_f, for a single-fractal analysis, (b) the binding rate coefficients, k_1 and k_2, and the fractal dimensions, D_{f1} and D_{f2}, and (c) the dissociation rate coefficient, k_d and the fractal dimension, D_{fd} for a single-fractal analysis are given in Tables 4.1 and 4.2.

In this case it is of interest to note that as the fractal dimension or the degree of heterogeneity on the biosensor surface increases by a factor of 50.8% from a value of D_{f1} equal to 1.809 to D_{f2} equal to 2.729, the binding rate coefficient increases by a factor of 4.07 from a value of k_1 equal to 2.831 to k_2 equal to 11.53. Note that changes in the fractal dimension or the degree of heterogeneity on the biosensor chip surface and in the binding rate coefficient are in the same direction. In this case at least, an increase in the degree of heterogeneity or the fractal dimension on the biosensor chip surface leads to an increase in the binding rate coefficient.

Figure 4.1c shows the binding and dissociation of 10 nM Cα in solution to ARC-704 immobilized on a SPR biosensor chip surface. Once again, a dual-fractal analysis is required

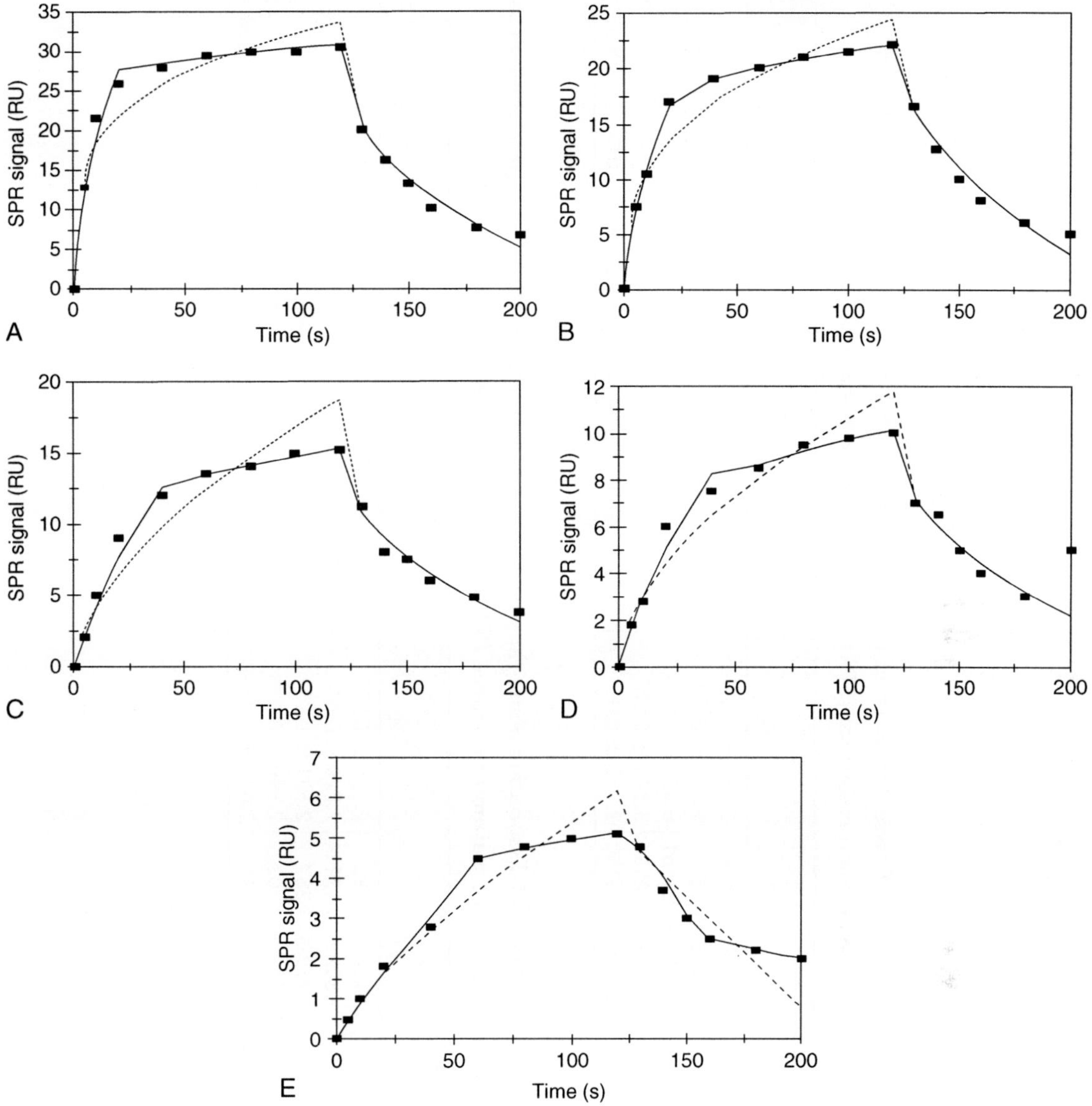

Figure 4.1

Binding and dissociation of different concentrations (in nM) of Cα in solution to ARC-704 immobilized on a surface plasmon resonance biosensor surface (Viht et al., 2007): (a) 50, (b) 25, (c) 10, (d) 5, (e) 2.5. If only a solid line is used (___) then a single-fractal analysis applies. If both a dashed (----) and a solid (___) line are used then the solid line is the best-fit line.

Table 4.1: Binding and dissociation rate coefficients for different concentrations (in nM) of Cα in solution to ARC-704 immobilized on a surface plasmon resonance (SPR) biosensor surface (Viht et al., 2007).

Cα concentration, nM	k	k_1	k_2	k_d	k_{d1}	k_{d2}
50	10.777 ± 1.680	6.540 ± 0.876	2.165 ± 0.256	3.993 ± 0.239	na	na
25	5.091 ± 0.699	2.831 ± 0.173	11.53 ± 0.050	1.721 ± 0.172	na	na
10	1.1174 ± 0.312	0.5869 ± 0.1461	6.321 ± 0.094	1.484 ± 0.158	na	na
5	0.8764 ± 0.165	0.5646 ± 0.0956	3.344 ± 0.087	0.926 ± 0.0798	na	na
2.5	0.1726 ± 0.026	0.1396 ± 0.0130	2.136 ± 0.0194	0.0579 ± 0.0186	0.00921 ± 0.00286	1.0175 ± 0.0053

Table 4.2: Fractal dimensions for the binding and dissociation phase rate for different concentrations (in nM) of Cα in solution to ARC-704 immobilized on a surface plasmon resonance (SPR) biosensor surface (Viht et al., 2007).

Cα concentration, nM	D_f	D_{f1}	D_{f2}	D_{fd}	D_{fd1}	D_{fd2}
50	2.5244 ± 0.0951	1.9454 ± 0.293	2.855 ± 0.027	2.152 ± 0.027	na	na
25	2.3454 ± 0.0844	1.809 ± 0.121	2.729 ± 0.010	1.907 ± 0.112	na	na
10	1.8220 ± 0.162	1.279 ± 0.287	2.631 ± 0.057	2.040 ± 0.119	na	na
5	1.914 ± 0.113	1.545 ± 0.202	2.536 ± 0.0994	2.024 ± 0.0977	na	na
2.5	1.504 ± 0.0921	1.339 ± 0.115	2.633 ± 0.0349	0.0849 ± 0.459	0.00921 ± 0.00286	2.4906 ± 0.0208

to adequately describe the binding kinetics. A single-fractal analysis is adequate to describe the dissociation kinetics. The values of (a) the binding rate coefficient, k and the fractal dimension, D_f, for a single-fractal analysis, (b) the binding rate coefficients, k_1 and k_2, and the fractal dimensions, D_{f1} and D_{f2}, and (c) the dissociation rate coefficient, k_d and the fractal dimension, D_{fd} for a single-fractal analysis are given in Tables 4.1 and 4.2.

Figure 4.1d shows the binding and dissociation of 5 nM Cα in solution to ARC-704 immobilized on a SPR biosensor chip surface. Once again, a dual-fractal is required to adequately describe the binding kinetics. A single-fractal analysis is adequate to describe the dissociation kinetics. The values of (a) the binding rate coefficient, k and the fractal dimension, D_f, for a single-fractal analysis, (b) the binding rate coefficients, $k_1$and k_2, and the fractal dimensions, D_{f1} and D_{f2}, and (c) the dissociation rate coefficient, k_d and the fractal dimension, D_{fd} for a single-fractal analysis are given in Tables 4.1 and 4.2.

Figure 4.1e shows the binding and dissociation of 2.5 nM Cα in solution to ARC-704 immobilized on a SPR biosensor chip surface. Once again, a dual-fractal is required to adequately describe the binding kinetics. A single-fractal analysis is adequate to describe the dissociation kinetics. The values of (a) the binding rate coefficient, k and the fractal dimension, D_f, for a single-fractal analysis, (b) the binding rate coefficients, $k_1$and k_2, and the fractal dimensions, D_{f1} and D_{f2}, and (c) the dissociation rate coefficient, k_d and the fractal dimension, D_{fd} for a single-fractal analysis are given in Tables 4.1 and 4.2.

Figure 4.2a and Table 4.1 show the increase in the binding rate coefficient, k_1, with an increase in the Cα concentration (in nM) in the 0-50 nM range in solution for a dual-fractal analysis. For the data shown in Figure 4.2a, the binding rate coefficient, k_1, is given by:

$$k_1 = (0.0513 \pm 0.0209)[\mathrm{C}\alpha, \mathrm{nM}]^{1.228 \pm 0.1422} \tag{4.4a}$$

The fit is very good. Only five data points are available. The availability of more data points would lead to a more reliable fit. The binding rate coefficient, k_1, exhibits an order of dependence between one and one and a half (equal to 1.228) on the Cα concentration in solution in the 0-50 nM concentration range.

Figure 4.2b and Table 4.1 show the increase in the binding rate coefficient, k_2, with an increase in the Cα concentration (in nM) in the 0-50 nM range in solution for a dual-fractal analysis. For the data shown in Figure 4.2b, the binding rate coefficient, k_2, is given by:

$$k_2 = (1.0212 \pm 0.0638)[\mathrm{C}\alpha, \mathrm{nM}]^{0.771 \pm 0.0252} \tag{4.4b}$$

The fit is very good. Only five data points are available. The availability of more data points would lead to a more reliable fit. The binding rate coefficient, k_1, exhibits an order of dependence between one-half and one (equal to 0.771) on the Cα concentration in solution in the 0-50 nM concentration range.

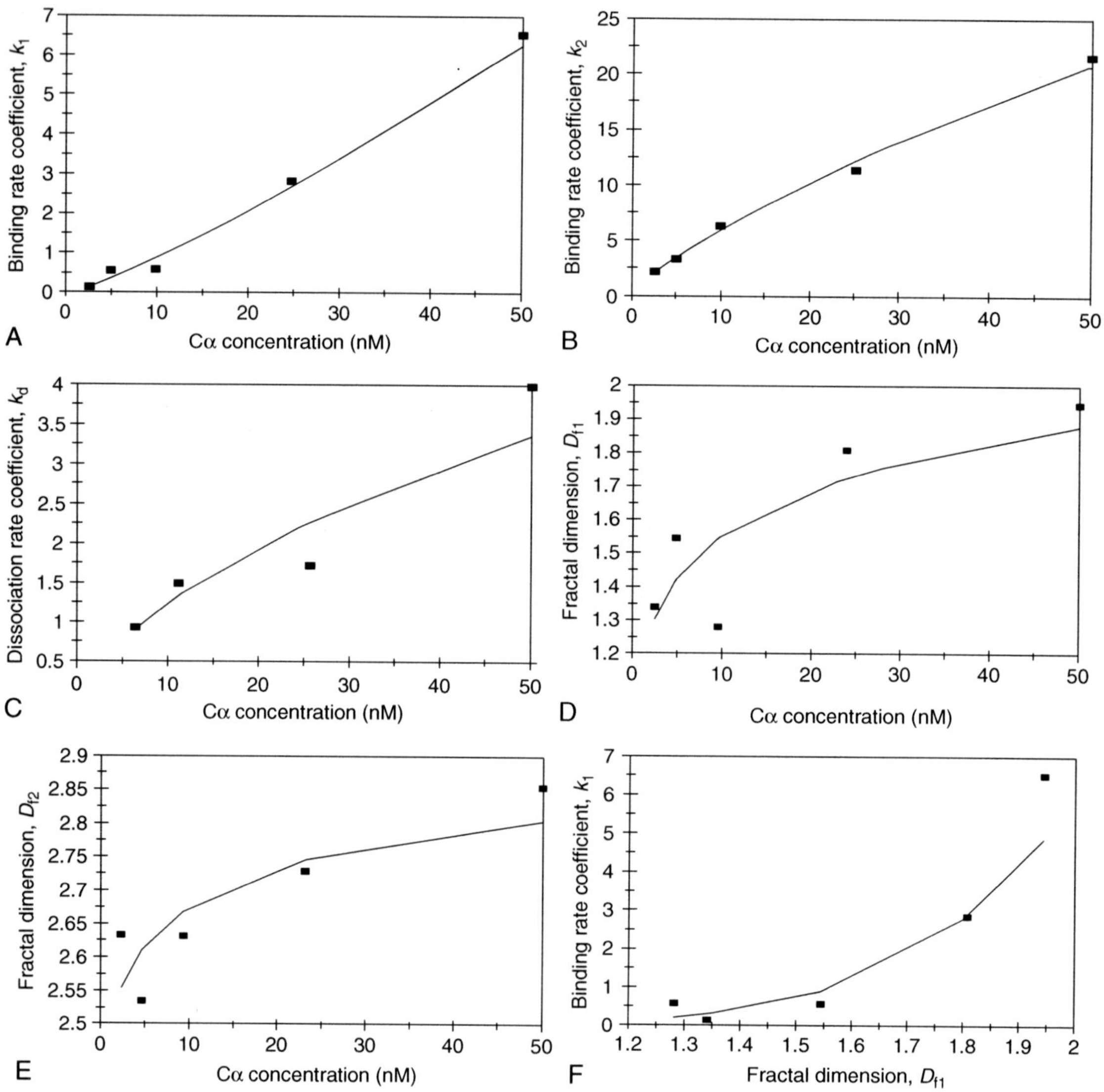

Figure 4.2

(a) Increase in the binding rate coefficient, k_1 with an increase in the Cα concentration (in nM) in solution. (b) Increase in the binding rate coefficient, k_2, with an increase in the Cα concentration (in nM) in solution. (c) Increase in the dissociation rate coefficient, k_2, with an increase in the Cα concentration (in nM) in solution. (d) Increase in the fractal dimension, D_{f1} with an increase in the Cα concentration (in nM) in solution. (e) Increase in the fractal dimension, D_{f2} with an increase in the Cα concentration (in nM) in solution. (f) Increase in the binding rate coefficient, k_1 with an increase in the fractal dimension, D_{f1}.

Continued

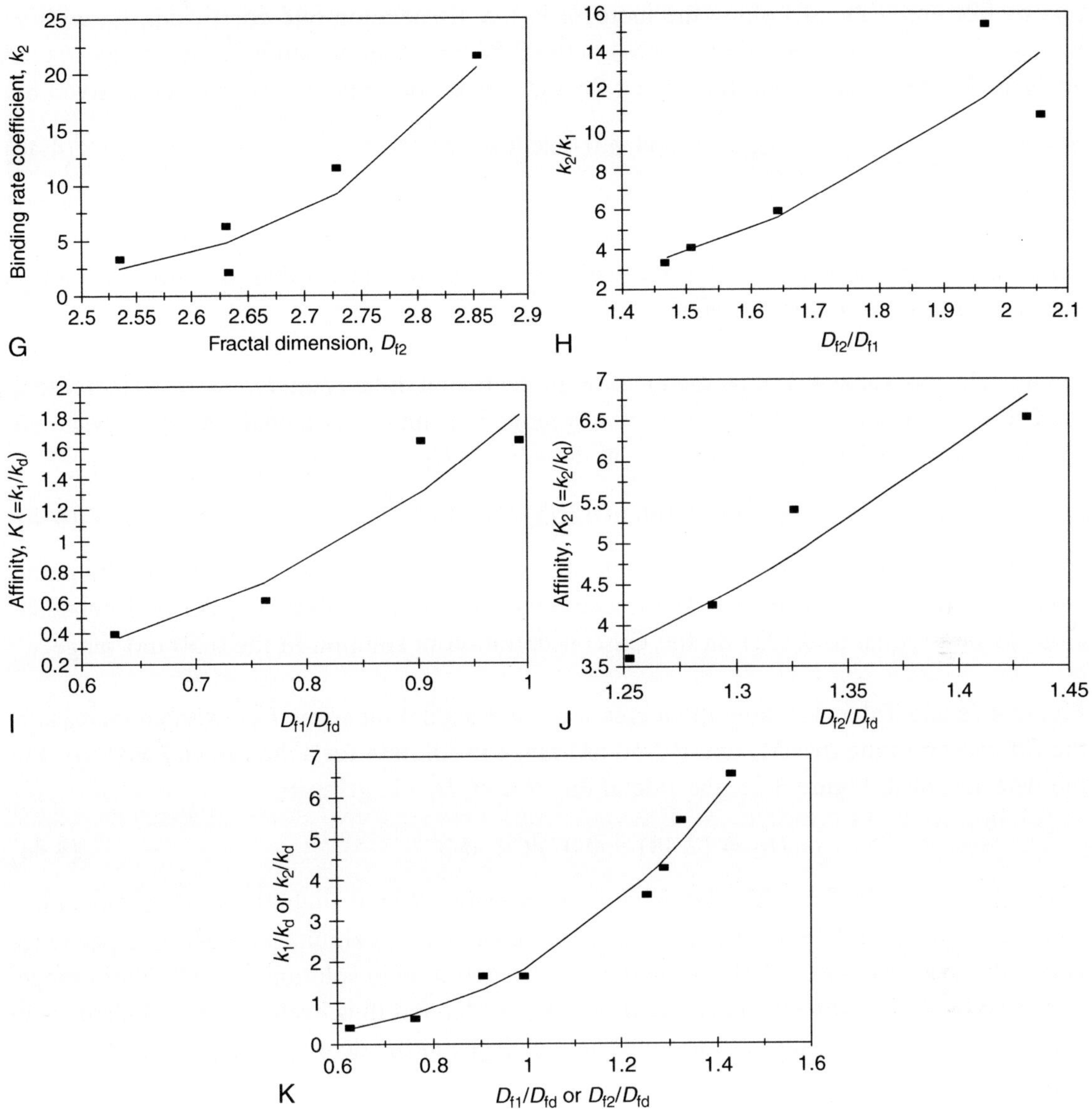

Figure 4.2—Cont'd

(g) Increase in the binding rate coefficient, k_2 with an increase in the fractal dimension, D_{f2}. (h) Increase in the binding rate coefficient ratio, $k_2/k_1$with an increase in the fractal dimension ratio, D_{f2}/D_{f1}. (i) Increase in the affinity, $K_1(=k_1/k_d)$with an n increase in the fractal dimension ratio, D_{f1}/D_{fd}. (j) Increase in the affinity, $K_2(=k_2/k_d)$with an n increase in the fractal dimension ratio, D_{f2}/D_{fd}. (k) Increase in the affinity, K_1 or K_2 with an increase in D_{f1}/D_{fd} or D_{f2}/D_{fd}.

Figure 4.2c and Table 4.1 show the increase in the dissociation rate coefficient, k_d, with an increase in the Cα concentration (in nM) in the 0-50 nM range in solution for a single-fractal analysis. For the data shown in Figure 4.2c, the dissociation rate coefficient, k_d, is given by:

$$k_d = (0.364 \pm 0.098)[\mathrm{C}\alpha, \mathrm{nM}]^{0.569 \pm 0.137} \qquad (4.4c)$$

The fit is very good. Only four data points are available. The availability of more data points would lead to a more reliable fit. The dissociation rate coefficient, k_d, exhibits an order of dependence between one-half and one (equal to 0.569) on the Cα concentration in solution in the 0-50 nM concentration range.

Figure 4.2d and Table 4.2 show the increase in the fractal dimension, D_{f1}, with an increase in the Cα concentration (in nM) in the 0-50 nM range in solution for a dual-fractal analysis. For the data shown in Figure 4.2d, the fractal dimension, D_{f1}, is given by:

$$D_{f1} = (1.166 \pm 0.155)[\mathrm{C}\alpha, \mathrm{nM}]^{0.122 \pm 0.0519} \qquad (4.4d)$$

The fit is reasonable. Only five data points are available. The availability of more data points would lead to a more reliable fit. The fractal dimension, D_{f1}, exhibits an order of dependence close to zero (equal to 0.122) on the Cα concentration in solution in the 0-50 nM range.

Figure 4.2e and Table 4.2 show the increase in the fractal dimension, D_{f2}, with an increase in the Cα concentration (in nM) in the 0-50 nM range in solution for a dual-fractal analysis. For the data shown in Figure 4.2e, the fractal dimension, D_{f2}, is given by:

$$D_{f2} = (2.481 \pm 0.07)[\mathrm{C}\alpha, \mathrm{nM}]^{0.0314 \pm 0.0115} \qquad (4.4e)$$

The fit is reasonable. Only five data points are available. The availability of more data points would lead to a more reliable fit. The fractal dimension, D_{f1}, exhibits an order of dependence very close to zero (equal to 0.0314) on the Cα concentration in solution in the 0-50 nM range. Note that the fractal dimension is based on a log scale, and thus even very small changes in the fractal dimension value lead to noticeable changes in the degree of heterogeneity on the biosensor chip surface.

Figure 4.2f and Tables 4.1 and 4.2 show the increase in the binding rate coefficient, k_1 with an increase in the fractal dimension, D_{f1} for the Cα concentration (in nM) in the 0-50 nM range in solution for a dual-fractal analysis. For the data shown in Figure 4.2f, the binding rate coefficient, k_1, is given by:

$$k_1 = (0.0364 \pm 0.044) D_{f1}^{7.351 \pm 2.167} \qquad (4.4f)$$

The fit is very good. Only five data points are available. The availability of more data points would lead to a more reliable fit. The binding rate coefficient, k_1, exhibits an order of dependence between seven and seven and a half (equal to 7.351) on the fractal dimension, D_{f1}, for the Cα concentration in solution in the 0-50 nM concentration range.

Figure 4.2g and Tables 4.1 and 4.2 show the increase in the binding rate coefficient, k_2, with an increase in the fractal dimension, D_{f1}, for the Cα concentration (in nM) in the 0-50 nM range in solution for a dual-fractal analysis. For the data shown in Figure 4.2g, the binding rate coefficient, k_2, is given by:

$$k_2 = (1.5 \times 10^{-7} \pm 1.1 \times 10^{-7}) D_{f2}^{17.86 \pm 6.105} \tag{4.4g}$$

The fit is very good. Only five data points are available. The availability of more data points would lead to a more reliable fit. The binding rate coefficient, k_2, exhibits an order of dependence between seventeen and a half and eighteen (equal to 17.86) on the fractal dimension, D_{f2}, for the Cα concentration in solution in the 0-50 nM concentration range.

Figure 4.2h and Tables 4.1 and 4.2 show the increase in the ratio of the binding rate coefficients, k_2/k_1, with an increase in the ratio of the fractal dimensions, D_{f2}/D_{f1}, for the Cα concentration (in nM) in the 0-50 nM range in solution for a dual fractal analysis. For the data shown in Figure 4.2h, the ratio of the binding rate coefficients, k_2/k_1, is given by:

$$k_2/k_1 = (0.772 \pm 0.193)(D_{f2}/D_{f1})^{4.01 \pm 0.728} \tag{4.4h}$$

The fit is good. Only five data points are available. The availability of more data points would lead to a more reliable fit. The ratio of the binding rate coefficients, k_2/k_1, exhibits an order of dependence very close to four (equal to 4.01) on the ratio of the fractal dimensions, D_{f2}/D_{f1}, for the Cα concentration in solution in the 0-50 nM concentration range.

Figure 4.2i and Tables 4.1 and 4.2 show the increase in the affinity, K_1 ($=k_1/k_d$) with an increase in the fractal dimension ratio, D_{f1}/D_{fd}. For the data shown in Figure 4.2i, the affinity, K_1, is given by:

$$K_1 = (1.856 \pm 0.458)(D_{f1}/D_{fd})^{3.441 \pm 0.629} \tag{4.4i}$$

The fit is good. Only four data points are available. The availability of more data points would lead to a more reliable fit. The affinity, K_1, is sensitive to the ratio of the fractal dimensions, D_{f1}/D_{fd}, as noted by the close to three and a half (equal to 3.441) order of dependence exhibited. This is one way of changing the affinity, K_1, by changing the degree of heterogeneity exhibited on the biosensor surface. Of course, some ingenuity is involved here in that one needs to change the degree of heterogeneity exhibited in the binding and in the dissociation phases differently so that one may change the ratio, D_{f1}/D_{fd}, in the required direction.

Figure 4.2j and Tables 4.1 and 4.2 show the increase in the affinity, K_2 ($=k_2/k_d$), with an increase in the fractal dimension ratio, D_{f2}/D_{fd}. For the data shown in Figure 4.2j, the affinity, K_2, is given by:

$$K_2 = (1.404 \pm 0.13)(D_{f2}/D_{fd})^{4.408 \pm 0.8925} \tag{4.4j}$$

The fit is good. Only four data points are available. The availability of more data points would lead to a more reliable fit. The affinity, K_2, is sensitive to the ratio of the fractal dimensions, D_{f2}/D_{fd}, as noted by the close to four and a half (equal to 4.408) order of dependence exhibited. Once again, this is one way of changing the affinity, K_2, by changing the degree of heterogeneity exhibited on the biosensor surface. Here again, one needs to change the degree of heterogeneity exhibited in the binding and in the dissociation phases differently so that one may change the ratio, D_{f2}/D_{fd}, in the required direction.

The data presented in Figures 4.2i and 4.2j are presented together in Figure 4.2k, since only four data points each were presented in Figures 4.2i and 4.2j, respectively. For the data shown in Figure 4.2k, the affinities, K_1 or K_2 are given by:

$$(K_1 \text{ or } K_2) = (1.849 \pm 0.283)[(D_{f2}/D_{fd}) \text{ or } (D_{f1}/D_{fd})]^{3.444 \pm 0.1818} \tag{4.4k}$$

The fit is very good considering that two different data sets are presented together. The affinities, K_1 and K_2, are quite sensitive to the fractal dimension ratios, (D_{f1}/D_{fd}) or (D_{f2}/D_{fd}) present on the biosensor chip surface as noted by the close to three and a half (equal to 3.444) order of dependence exhibited.

Figure 4.3a shows the binding of 50 nM ARC-704 in solution to Cα (950 RU) immobilized on a Biacore S51 sensor chip surface (Viht et al., 2007). A dual-fractal analysis is required to adequately describe the binding and the dissociation kinetics. The values of (a) the binding rate coefficient, k, and the fractal dimension, D_f, for a single-fractal analysis, (b) the binding rate coefficients, k_1 and k_2, and the fractal dimensions, D_{f1} and D_{f2}, for a dual-fractal analysis, (c) the dissociation rate coefficient, k_d, and the fractal dimension, D_{fd}, for a single-fractal analysis, and (d) the dissociation rate coefficients, k_{d1} and k_{d2}, and the fractal dimensions, D_{fd1} and D_{fd2}, for a dual-fractal analysis are given in Tables 4.3 and 4.4. It is of interest to note that for a dual-fractal analysis, as the fractal dimension increases by a factor of 1.50 from a value of D_{f1} equal to 1.9126 to D_{f2} equal to 2.8771, the binding rate coefficient increases by a factor of 3.92 from a value of k_1 equal to 4.492 to k_2 equal to 17.6.

Figure 4.3b shows the binding of 25 nM ARC-704 in solution to Cα (950 RU) immobilized on a Biacore S51 sensor chip surface (Viht et al., 2007). A dual-fractal analysis is required to adequately describe the binding kinetics. A single-fractal analysis is required to adequately describe the dissociation kinetics. The values of (a) the binding rate coefficient, k, and the fractal dimension, D_f, for a single-fractal analysis, (b) the binding rate coefficients, k_1 and k_2,

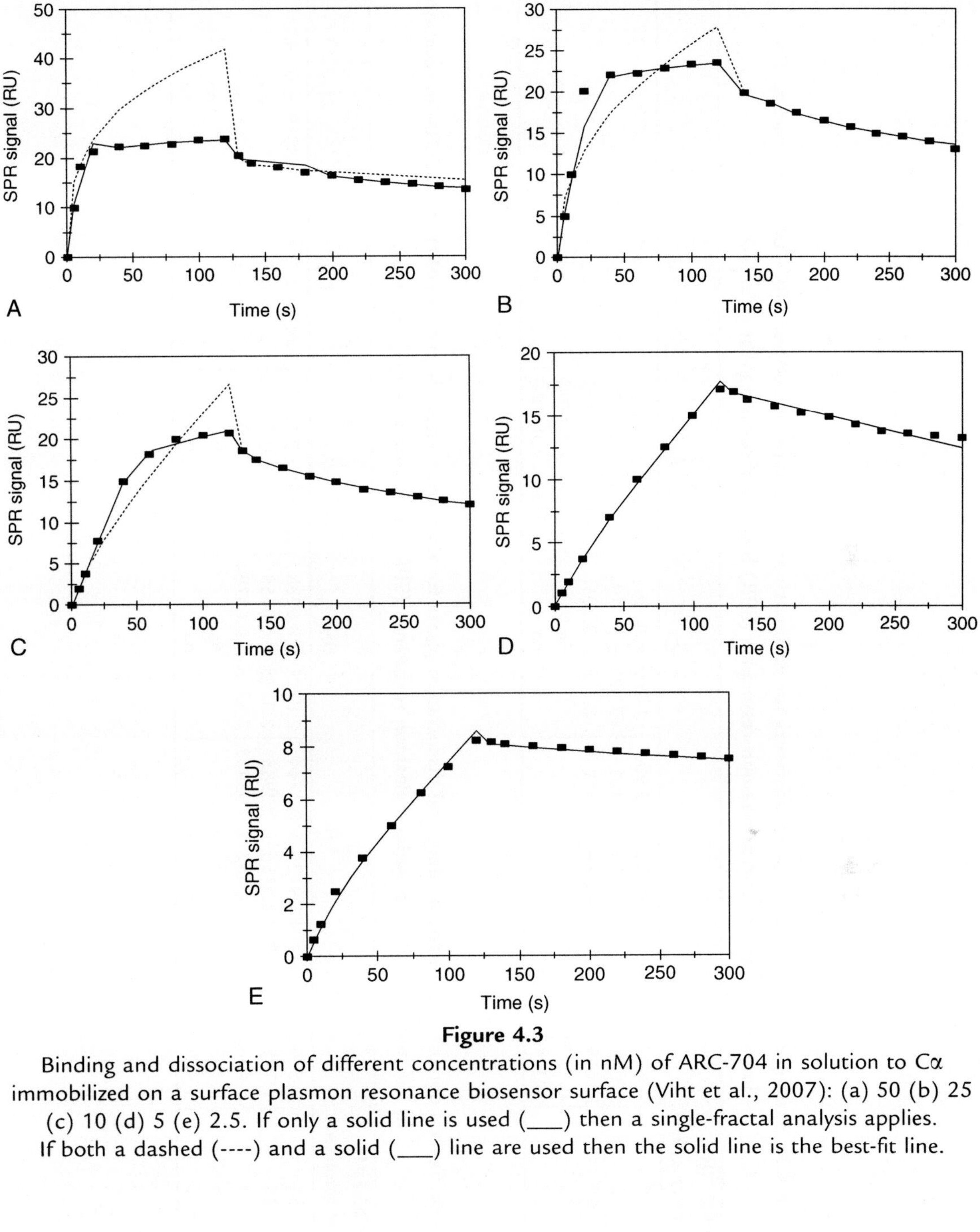

Figure 4.3

Binding and dissociation of different concentrations (in nM) of ARC-704 in solution to Cα immobilized on a surface plasmon resonance biosensor surface (Viht et al., 2007): (a) 50 (b) 25 (c) 10 (d) 5 (e) 2.5. If only a solid line is used (____) then a single-fractal analysis applies. If both a dashed (----) and a solid (____) line are used then the solid line is the best-fit line.

Table 4.3: Binding and dissociation rate coefficients for different concentrations (in nM) of ARC-704 in solution to Cα immobilized on a surface plasmon resonance (SPR) biosensor surface (Viht et al.,2007).

ARC-704 concentration, nM	k	k_1	k_2	k_d	k_{d1}	k_{d2}
50	9.264 ± 1.646	4.492 ± 0.0.904	17.60 ± 0.19	2.353 ± 0.661	2.632 ± 0.61	1.715 ± 0.019
25	3.465 ± 1.044	1.709 ± 0.445	17.322 ± 0.119	1.691 ± 0.032	na	na
10	0.6795 ± 0.1424	0.4089 ± 0.011	8.718 ± 0.201	0.7366 ± 0.0137	na	na
55	0.2726 ± 0.0068	na	na	0.03486 ± 0.0089	na	na
2.5	0.2046 ± 0.0174	na	na	0.01343 ± 0.0007	na	na

Table 4.4: Fractal dimensions for the binding and dissociation phase for different concentrations (in nM) of ARC-704 in solution to Cα immobilized on a surface plasmon resonance (SPR) biosensor surface (Viht et al., 2007).

ARC-704 concentration, nM	D_f	D_{f1}	D_{f2}	D_{fd}	D_{fd1}	D_{fd2}
50	2.5702 ± 0.1074	1.9126 ± 0.3744	2.8771 ± 0.0238	2.5134 ± 0.1314	2.6566 ± 0.1218	2.3510 ± 0.0478
25	2.1158 ± 0.1728	1.5156 ± 0.2988	2.8738 ± 0.0158	2.300 ± 0.013	na	na
10	1.4678 ± 0.1249	1.0482 ± 0.03408	2.6316 ± 0.08794	2.0442 ± 0.0129	na	na
5	1.2556 ± 0.0161	na	na	1.1130 ± 0.1592	na	na
2.5	1.4362 ± 0.0538	na	na	1.4642 ± 0.0352	na	na

and the fractal dimensions, D_{f1} and D_{f2}, for a dual-fractal analysis, and (c) the dissociation rate coefficient, k_d, and the fractal dimension in the dissociation phase, D_{fd}, for a single-fractal analysis are given in Tables 4.3 and 4.4. Note that an increase in the degree of heterogeneity on the biosensor chip surface leads to an increase in the binding rate coefficient. Similarly, a decrease in the fractal dimension in the dissociation phase by a factor of 1.13 from a value of D_{fd1} equal to 2.6566 to D_{fd2} equal to 2.3510 leads to a decrease in the dissociation rate coefficient by a factor of 1.53 from a value of k_{d1} equal to 2.632 to k_{d2} equal to 1.715. Once again, changes in the degree of heterogeneity present on the biosensor chip surface in the dissociation phase and in the dissociation rate coefficient are in the same direction.

Once again, note that for a dual-fractal analysis, as the fractal dimension increases by a factor of 1.896 from a value of D_{f1} equal to 1.5156 to D_{f2} equal to 2.8738, the binding rate coefficient increases by a factor of 10.14 from a value of k_1 equal to 1.709 to k_2 equal to 17.322. Once again, an increase in the degree of heterogeneity on the biosensor chip surface leads to an increase in the binding rate coefficient.

Figure 4.3c shows the binding of 10 nM ARC-704 in solution to Cα (950 RU) immobilized on a Biacore S51 sensor chip surface (Viht et al., 2007). A dual-fractal analysis is required to adequately describe the binding kinetics. A single-fractal analysis is required to adequately describe the dissociation kinetics. The values of (a) the binding rate coefficient, k, and the fractal dimension, D_f, for a single-fractal analysis, the binding rate coefficients, k_1 and k_2, and the fractal dimensions, D_{f1} and D_{f2}, (c) and the dissociation rate coefficient, k_d, and the fractal dimension, D_{fd}, for a single-fractal analysis, are given in Tables 4.3 and 4.4.

Figure 4.3d shows the binding of 5 nM ARC-704 in solution to Cα (950 RU) immobilized on a Biacore S51 sensor chip surface (Viht et al., 2007). A single-fractal analysis is required to adequately describe the binding and the dissociation kinetics. The values of (a) the binding rate coefficient, k, and the fractal dimension, D_f, for a single-fractal analysis, and (b) the dissociation rate coefficient, k_d, and the fractal dimension, D_{fd}, for a single-fractal analysis, are given in Tables 4.3 and 4.4.

Figure 4.3e shows the binding of 2.5 nM ARC-704 in solution to Cα (950 RU) immobilized on a Biacore S51 sensor chip surface (Viht et al., 2007). A single-fractal analysis is required to adequately describe the binding and the dissociation kinetics. The values of (a) the binding rate coefficient, k, and the fractal dimension, D_f, for a single-fractal analysis, and (b) the dissociation rate coefficient, k_d, and the fractal dimension, D_{fd}, for a single-fractal analysis, are given in Tables 4.3 and 4.4.

Figure 4.4a and Table 4.3 show the increase in the binding rate coefficient, k_1, with an increase in the ARC-704 concentration (in nM) in the 0-50 nM range in solution for a dual-fractal analysis. For the data shown in Figure 4.4a, the binding rate coefficient, k_1, is given by:

$$k_1 = (0.01339 \pm 0.00074)[\mathrm{ARC} - 704, \mathrm{nM}]^{1.4928 \pm 0.04688} \qquad (4.5a)$$

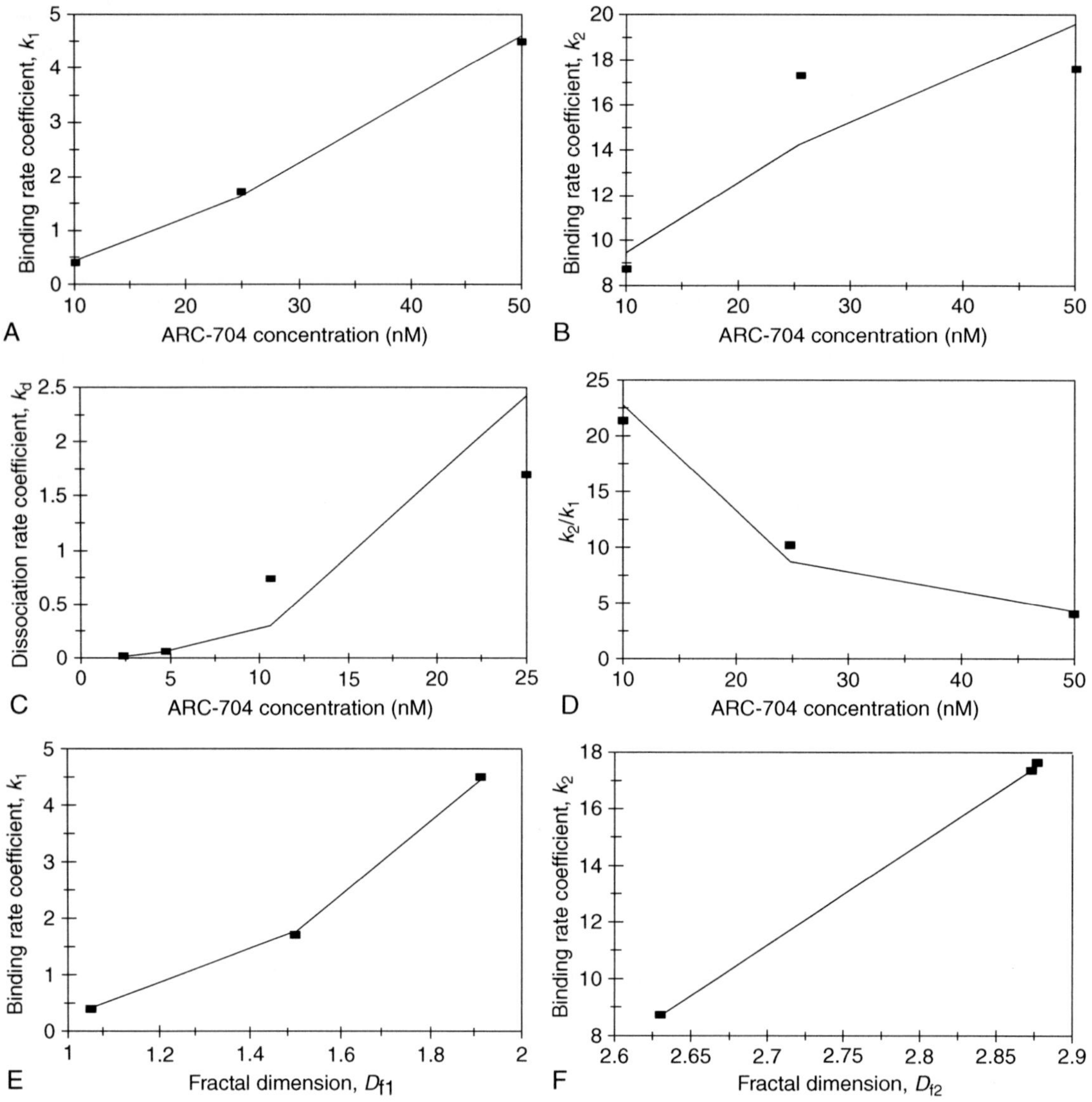

Figure 4.4

(a) Increase in the binding rate coefficient, k_1 with an increase in the ARC-704 concentration (in nM) in solution. (b) Increase in the binding rate coefficient, k_2, with an increase in the ARC-704 concentration (in nM) in solution. (c) Increase in the dissociation rate coefficient, k_d, with an increase in theARC-704 concentration (in nM) in solution. (d) Decrease in the binding rate coefficient ratio, k_2/k_1 with an increase in the ARC-704 concentration (in nM) in solution. (e) Increase in the binding rate coefficient, k_1 with an increase in the fractal dimension, D_{f1}. (f) Increase in the binding rate coefficient, k_2 with an increase in the fractal dimension, D_{f2}.

Continued

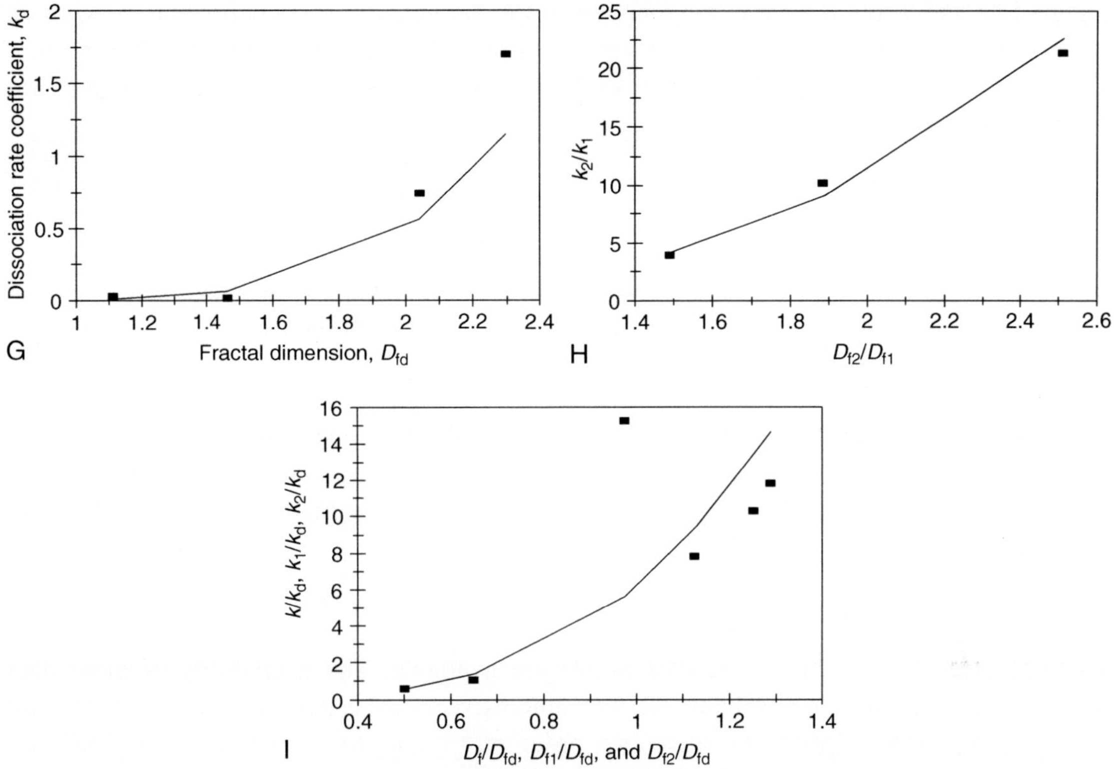

Figure 4.4—Cont'd

(g) Increase in the dissociation rate coefficient ratio, k_d with an increase in the fractal dimension D_{fd}. (h) Increase in the binding rate coefficient ratio, $(=k_2/k_1)$ with an increase in the fractal dimension ratio, D_{f2}/D_{f1}. (i) Increase in the ratio, k/k_{d1}, k_1/k_d, and k_2/k_d with an increase in the ratio, D_f/D_{fd}, D_{f1}/D_{fd}, and D_{f2}/D_{fd}.

The fit is very good. Only three data points are available. The availability of more data points would lead to a more reliable fit. The binding rate coefficient, k_1, exhibits an order of dependence close to one and a half (equal to 1.4928) on the ARC-704 concentration in solution in the 0-50 nM concentration range.

Figure 4.4b and Table 4.3 show the increase in the binding rate coefficient, k_2, with an increase in the ARC-704 concentration (in nM) in the 0-50 nM range in solution for a dual-fractal analysis. For the data shown in Figure 4.4b, the binding rate coefficient, k_2, is given by:

$$k_2 = (3.3348 \pm 0.877)[\text{ARC} - 704, \text{nM}]^{0.4528 \pm 0.2043} \qquad (4.5b)$$

The fit is very good. Only three data points are available. The availability of more data points would lead to a more reliable fit. The binding rate coefficient, k_2, exhibits an order of dependence less than one-half (equal to 0.4528) on the ARC-704 concentration in solution in the 0-50 nM concentration range.

Figure 4.4c and Table 4.3 show the increase in the dissociation rate coefficient, k_d, with an increase in the ARC-704 concentration (in nM) in the 0-50 nM range in solution for a single-fractal analysis. For the data shown in Figure 4.4c, the dissociation rate coefficient, k_d, is given by:

$$k_d = (0.00159 \pm 0.00192)[\text{ARC} - 704, \text{nM}]^{2.279 \pm 0.4669} \tag{4.5c}$$

The fit is very good. Only four data points are available. The availability of more data points would lead to a more reliable fit. The dissociation rate coefficient, k_d, exhibits an order of dependence between two and two and a half (equal to 2.279) on the ARC-704 concentration in solution in the 0-50 nM concentration range.

Figure 4.4d and Table 4.4 show for a dual-fractal analysis the decrease in the binding rate coefficient ratio, k_2/k_1, with an increase in the ARC-704 concentration (in nM) in the 0-50 nM range in solution for a dual-fractal analysis. For the data shown in Figure 4.4d, the binding rate coefficient, k_2/k_1, ratio is given by:

$$k_2/k_1 = (248.68 \pm 48.38)[\text{ARC} - 704, \text{nM}]^{-1.0398 \pm 0.1556} \tag{4.5d}$$

The fit is reasonable. Only three data points are available. The availability of more data points would lead to a more reliable fit. The binding rate coefficient ratio, k_2/k_1, exhibits an order of dependence close to negative first order zero (equal to -1.0398) on the ARC-704 concentration in solution in the 0-50 nM range.

Figure 4.4e and Tables 4.3 and 4.4 show the increase in the binding rate coefficient, k_1, with an increase in the fractal dimension, D_{f1} for the ARC-704 concentration (in nM) in the 0-50 nM range in solution for a dual-fractal analysis. For the data shown in Figure 4.4e, the binding rate coefficient, k_1, is given by:

$$k_1 = (0.3357 \pm 0.0109) D_{f1}^{3.975 \pm 0.0741} \tag{4.5e}$$

The fit is good. Only three data points are available. The availability of more data points would lead to a more reliable fit. The binding rate coefficient, k_1, exhibits an order of dependence close to four (equal to 3.975) on the fractal dimension, D_{f1}, for the ARC-704 concentration in solution in the 0-50 nM concentration range.

Figure 4.4f and Tables 4.3 and 4.4 show the increase in the binding rate coefficient, k_2, with an increase in the fractal dimension, D_{f2}, for the ARC-704 concentration (in nM) in the 0-50 nM range in solution for a dual-fractal analysis. For the data shown in Figure 4.4g, the binding rate coefficient, k_2, is given by:

$$k_2 = (0.004431 \pm 0.000022) D_{f2}^{7.838 \pm 0.06768} \tag{4.5f}$$

The fit is good. Only three data points are available. The availability of more data points would lead to a more reliable fit. The binding rate coefficient, k_2, exhibits an order of dependence between seven and a half and eight (equal to 7.838) on the fractal dimension, D_{f2}, for the ARC-704 concentration in solution in the 0-50 nM concentration range.

Figure 4.4g and Tables 4.3 and 4.4 show the increase in the dissociation rate coefficient, k_d, with an increase in the fractal dimension in the dissociation phase, D_{fd}, for the ARC-704 concentration (in nM) in the 0-50 nM range in solution for a single-fractal analysis. For the data shown in Figure 4.2g, the dissociation rate coefficient, k_d, is given by:

$$k_d = (0.006713 \pm 0.0207) D_{fd}^{6.171 \pm 2.468} \tag{4.5g}$$

The fit is good. Only four data points are available. The availability of more data points would lead to a more reliable fit. The dissociation rate coefficient, k_d, exhibits an order of dependence close to six (equal to 6.171) on the fractal dimension, D_{fd}, for the ARC-704 concentration in solution in the 0-50 nM concentration range.

Figure 4.4h and Tables 4.3 and 4.4 show the increase in the ratio of the binding rate coefficients, k_2/k_1, with an increase in the fractal dimension ratio, D_{f2}/D_{f2}, for a dual-fractal analysis. For the data shown in Figure 4.4h, the ratio of the binding rate coefficients, k_2/k_1, is given by:

$$k_2/k_1 = (1.0937 \pm 0.1783)(D_{f2}/D_{f1})^{3.287 \pm 0.4171} \tag{4.5h}$$

The fit is good. Only three data points are available. The availability of more data points would lead to a more reliable fit. The ratio of the binding rate coefficients, k_2/k_1, is sensitive to the ratio of the fractal dimensions, D_{f2}/D_{f1}, as noted by the close to three and three and a half (equal to 3.287) order of dependence exhibited.

Figure 4.4i and Tables 4.3 and 4.4 show the increase in the affinity, K_1 ($=k_1/k_d$), with an increase in the fractal dimension ratio, D_{f1}/D_{fd}, for a dual-fractal analysis. For the data shown in Figure 4.4i, the affinity, K_1, is given by:

$$K_1 = (1.856 \pm 0.458)(D_{f1}/D_{fd})^{3.441 \pm 0.629} \tag{4.5i}$$

The fit is good. Only four data points are available. The availability of more data points would lead to a more reliable fit. The affinity, K_1, is sensitive to the ratio of the fractal dimensions, D_{f1}/D_{fd}, as noted by the close to three and a half (equal to 3.441) order of dependence exhibited.

Figure 4.4j and Tables 4.3 and 4.4 show the increase in the affinity, K_2 ($=k_2/k_d$), with an increase in the fractal dimension ratio, D_{f2}/D_{fd}, for a dual-fractal analysis. For the data shown in Figure 4.4j, the affinity, K_2, is given by:

$$K_2 = (1.404 \pm 0.13)(D_{f2}/D_{fd})^{4.408 \pm 0.8925} \tag{4.5j}$$

Once again, the fit is good. Only four data points are available. The availability of more data points would lead to a more reliable fit. The affinity, K_2, is very sensitive to the ratio of the fractal dimensions, D_{f2}/D_{fd}, as noted by the close to four and a half (equal to 4.408) order of dependence exhibited.

Only four data points were available and plotted in Figures 4.4i and Figure 4.4j for the affinity values, K_1 and K_2, respectively. It is perhaps instructive to plot them together since so few data points were available for K_1 and K_2 separately. Figure 4.2k shows the plot for K_1 and K_2 versus D_{f1}/D_{fd} and D_{f2}/D_{fd}, respectively. For the combined (K_1 and K_2) data shown in Figure 4.4k, the affinity, K_1 or K_2 is given by:

$$K_1 \text{ or } K_2 = (1.849 \pm 0.283)[(D_{f1}/D_{fd}) \text{ or } (D_{f2}/D_{fd})]^{3.444 \pm 0.1818} \quad (4.5k)$$

Considering that two different sets of data for K_1 and K_2 are plotted together the fit is very good. In this case now eight data points are available. The affinity, K_1 or K_2 is sensitive to the ratio of fractal dimensions, D_{f1}/D_{fd} or D_{f2}/D_{fd}, respectively as noted by the close to three and a half (equal to 3.444) order of dependence exhibited. It is of interest to note that the order of dependence exhibited (equal to 3.444) when K_1 and K_2 are plotted together is very close to the order of dependence exhibited (equal to 3.441) when K_1 is plotted alone. The difference is only in the fourth decimal place. This would indicate that in some way the affinity, K_1 dominates the affinity, K_2 at least in this case as far as the heterogeneities on the biosensor chip are concerned.

Okoh et al. (2006) indicate that inorganic phosphate (P_i) is a product of many enzymic reactions. These authors indicate that protein phosphatases are involved in the control of the phosphorylation state of many proteins. Furthermore, the authors state that the measurement of P_i is an important target for the understanding of cellular activities involving such proteins. Webb (2003) has summarized the various kinetic analyses of phosphatases. Okoh et al. (2006) have developed a fluorescence-based biosensor based on PBP. This PBP was obtained from *Escherichia coli* (Brune et al., 1994). They (Okoh et al., 2006) report that the PBP has two domains that are hinged. These close around the phosphate that binds in the cleft between them (Luecke and Quiocho, 1990). They (Okoh et al., 2006) point out that this conformational change provides a mechanism for the development of a biosensor. Furthermore, these authors indicate that the PBP is very specific to inorganic phosphate and binds phosphate esters and anhydrides very weakly.

Figure 4.5a shows the binding of a 0.63 micromole inorganic phosphate ion (P_i) in solution to a rhodamine-PBP phosphate biosensor (Okoh et al., 2006). A dual-fractal analysis is required to adequately describe the binding kinetics. The values of (a) the binding rate coefficient, k, and the fractal dimension, D_f, for a single-fractal analysis, and (b) the binding rate coefficients, k_1 and k_2, and the fractal dimensions, D_{f1} and D_{f2}, for a dual-fractal analysis

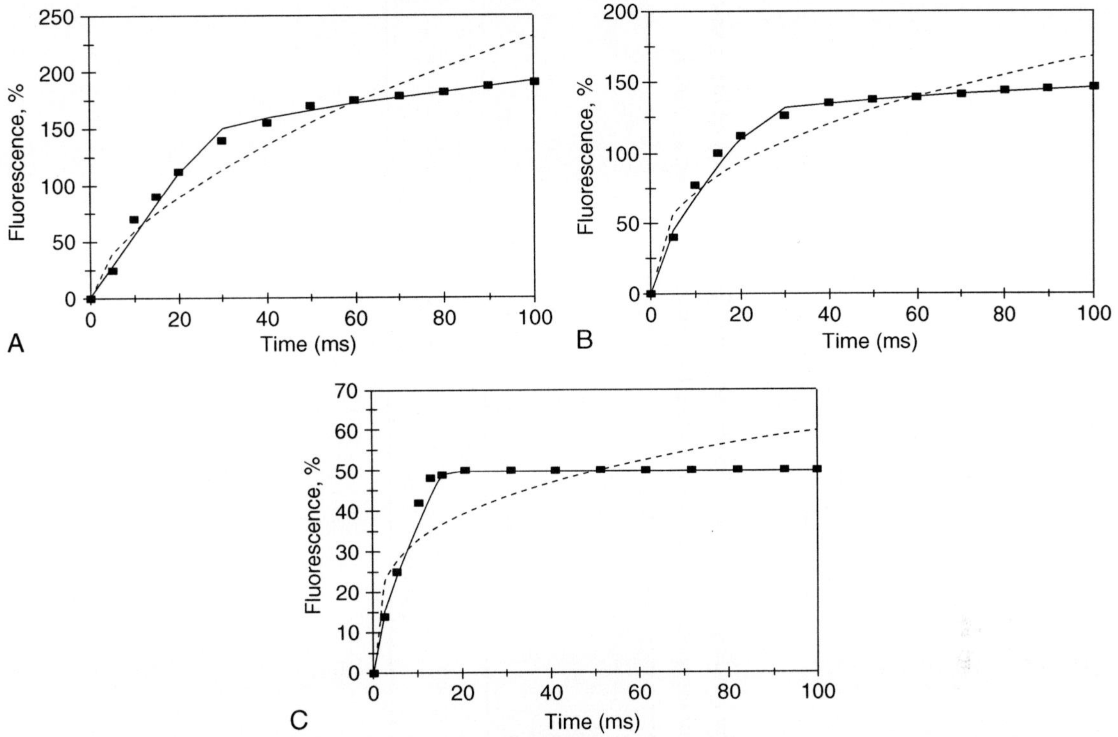

Figure 4.5

Binding of different concentrations (in micromole) of inorganic phosphate, P_i to a biosensor surface (Okoh et al., 2006): (a) 0.63 (b) 1.25 (c) 5. If only a solid line is used (____) then a single-fractal analysis applies. If both a dashed (----) and a solid (____) line are used then the solid line is the best-fit line.

are given in Table 4.5. It is of interest to note that as the fractal dimension increases by a factor of 2.36 from a value of D_{f1} equal to 1.0986 to D_{f2} equal to 2.5928, the binding rate coefficient increases by a factor of 11.77 from a value of k_1 equal to 6.362 to k_2 equal to 74.911 for a dual-fractal analysis. Note that an increase in the degree of heterogeneity or the fractal dimension on the sensor chip surface leads to an increase in the binding rate coefficient.

Figure 4.5b shows the binding of 1.25 micromole inorganic phosphate ion (P_i) in solution to a rhodamine-PBP phosphate biosensor (Okoh et al., 2006). A dual-fractal analysis is required to adequately describe the binding kinetics. The values of (a) the binding rate coefficient, k, and the fractal dimension, D_f, for a single-fractal analysis, and (b) the binding rate coefficients, k_1 and k_2, and the fractal dimensions, D_{f1} and D_{f2}, for a dual-fractal analysis are given in Table 4.5. It is of interest to note, once again, that as the fractal dimension

Table 4.5: Binding rate coefficients and fractal dimensions for the binding of different concentrations (in micromole) of inorganic phosphate (P_i) in solution to rhodamine-PBP (phosphate binding protein) as a phosphate biosensor (Okoh et al.,2006).

Inorganic phosphate (P_i) concentration in solution, micromole	k	k_1	k_2	D_f	D_{f1}	D_{f2}
0.63	15.516 ± 3.738	6.362 ± 1.246	74.911 ± 1.26	1.8252 ± 0.1365	1.0986 ± 0.2610	2.5928 ± 0.0416
1.25	31.305 ± 5.504	15.664 ± 2.06	97.008 ± 0.282	2.2698 ± 0.1024	1.7038 ± 0.1802	2.8228 ± 0.00722
5.0	18.335 ± 4.838	8.425 ± 1.086	50 ± 0.0	2.4838 ± 0.1138	1.7006 ± 0.1398	$3.0\text{-}2\times10^{-15}$

increases by a factor of 1.66 from a value of D_{f1} equal to 1.7038 to D_{f2} equal to 2.8228, the binding rate coefficient increases by a factor of 6.19 from a value of k_1 equal to 15.644 to k_2 equal to 97.008 for a dual-fractal analysis. Note, once again, that an increase in the degree of heterogeneity or the fractal dimension on the sensor chip surface leads to an increase in the binding rate coefficient.

Figure 4.5c shows the binding of 5.0 micromole inorganic phosphate ion (P_i) in solution to a rhodamine-PBP phosphate biosensor (Okoh et al., 2006). A dual-fractal analysis is required to adequately describe the binding kinetics. The values of (a) the binding rate coefficient, k, and the fractal dimension, D_f, for a single-fractal analysis, and (b) the binding rate coefficients, k_1 and k_2, and the fractal dimensions, D_{f1} and D_{f2}, for a dual-fractal analysis are given in Table 4.5. It is of interest to note, once again, that as the fractal dimension increases by a factor of 1.76 from a value of D_{f1} equal to 1.70 to D_{f2} equal to 3.0, the binding rate coefficient increases by a factor of 5.93 from a value of k_1 equal to 8.425 to k_2 equal to 50 for a dual-fractal analysis. Note, once again, that an increase in the degree of heterogeneity or the fractal dimension on the sensor chip surface leads to an increase in the binding rate coefficient.

Figure 4.6a and Table 4.5 show the increase in the fractal dimension, D_{f1}, with an increase in the phosphate ion, P_i in solution in the 0.2-5.0 micromole concentration range for a dual-fractal analysis. For the data shown in Figure 4.6a, the fractal dimension, D_{f1}, is given by:

$$D_{f1} = (1.354 \pm 0.361)[P_i]^{0.1800 \pm 0.1580} \tag{4.6a}$$

The fit is reasonable. Only three data points are available. The availability of more data points would lead to a more reliable fit. The fractal dimension, D_{f1}, exhibits only a very mild dependence on the P_i concentration in solution as noted by the 0.1800 order exhibited. The fractal dimension is based on a log scale, and even very small changes in the fractal dimension value indicate significant changes in the degree of heterogeneity on the biosensor chip surface.

Figure 4.6b and Table 4.5 show the increase in the fractal dimension, D_{f2}, with an increase in the phosphate ion, P_i in solution in the 0.2-5.0 micromole concentration range for a dual-fractal analysis. For the data shown in Figure 4.6b, the fractal dimension, D_{f2}, is given by:

$$D_{f2} = (2.716 \pm 0.081)[P_i]^{0.06656 \pm 0.0197} \tag{4.6b}$$

The fit is good. Only three data points are available. The availability of more data points would lead to a more reliable fit. The fractal dimension, D_{f2}, exhibits only a very mild dependence on the P_i concentration in solution as noted by the 0.06656 order exhibited. Once again, the fractal dimension is based on a log scale, and even very small changes in the fractal dimension value indicate significant changes in the degree of heterogeneity on the biosensor chip surface.

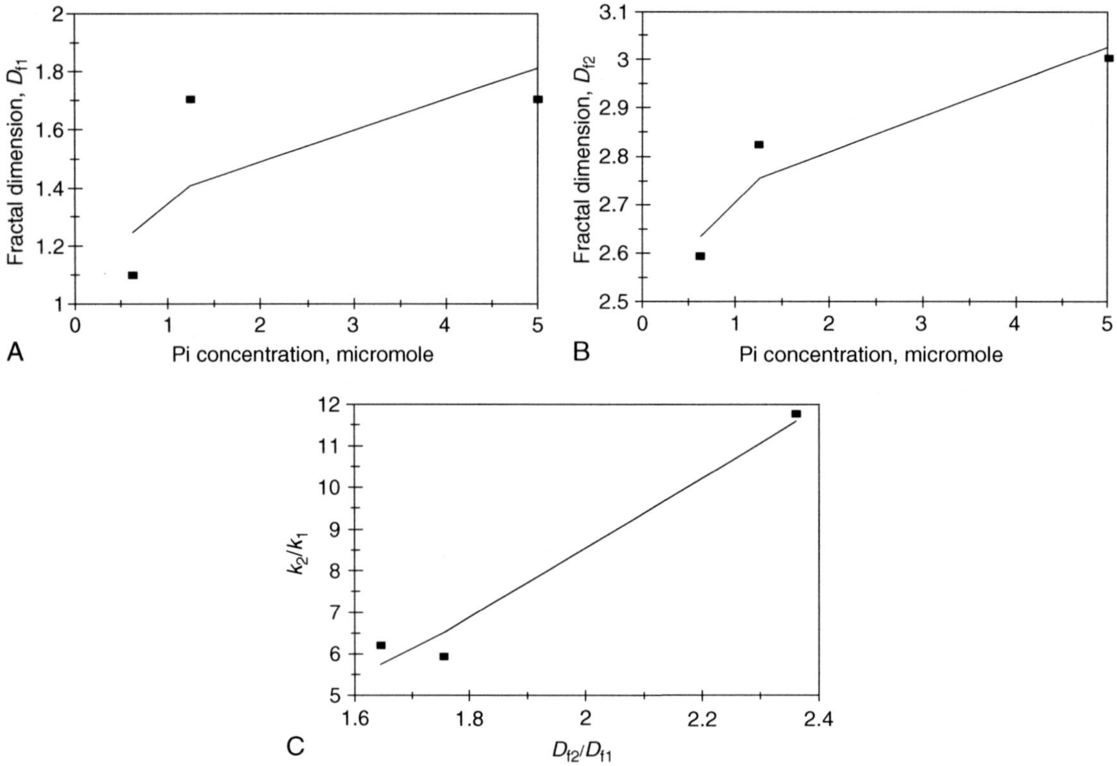

Figure 4.6

(a) Increase in the fractal dimension, D_{f1} with an increase in the organic phosphate, P_i concentration (in micromole) in solution. (b) Increase in the fractal dimension, D_{f2} with an increase in the organic phosphate, P_i concentration (in micromole) in solution. (c) Increase in the binding rate coefficient ratio, k_2/k_1 with an increase in the ratio of the fractal dimensions, D_{f2}/D_{f1}.

Figure 4.6c and Table 4.5 show the increase in the ratio of the binding rate coefficients, k_2/k_1, with an increase in the ratio of the fractal dimensions, D_{f2}/D_{f1}, for a dual-fractal analysis. For the data shown in Figure 4.6c, the ratio of the binding rate coefficients, k_2/k_1, is given by:

$$k_2/k_1 = (2.115 \pm 0.271)(D_{f2}/D_{f1})^{1.979 \pm 0.450} \tag{4.6c}$$

The fit is good. Only three data points are available. The availability of more data points would lead to a more reliable fit. The ratio of the binding rate coefficients, k_2/k_1, exhibits very close to a second (equal to 1.979) order of dependence on the ratio of the fractal dimensions, (D_{f2}/D_{f1}) present on the biosensor chip surface.

Figure 4.7a shows the binding of 0.05 mU of MET-AMC in solution in the cSPA (Forbes et al., 2007). A single-fractal analysis is adequate to describe the binding kinetics. The values

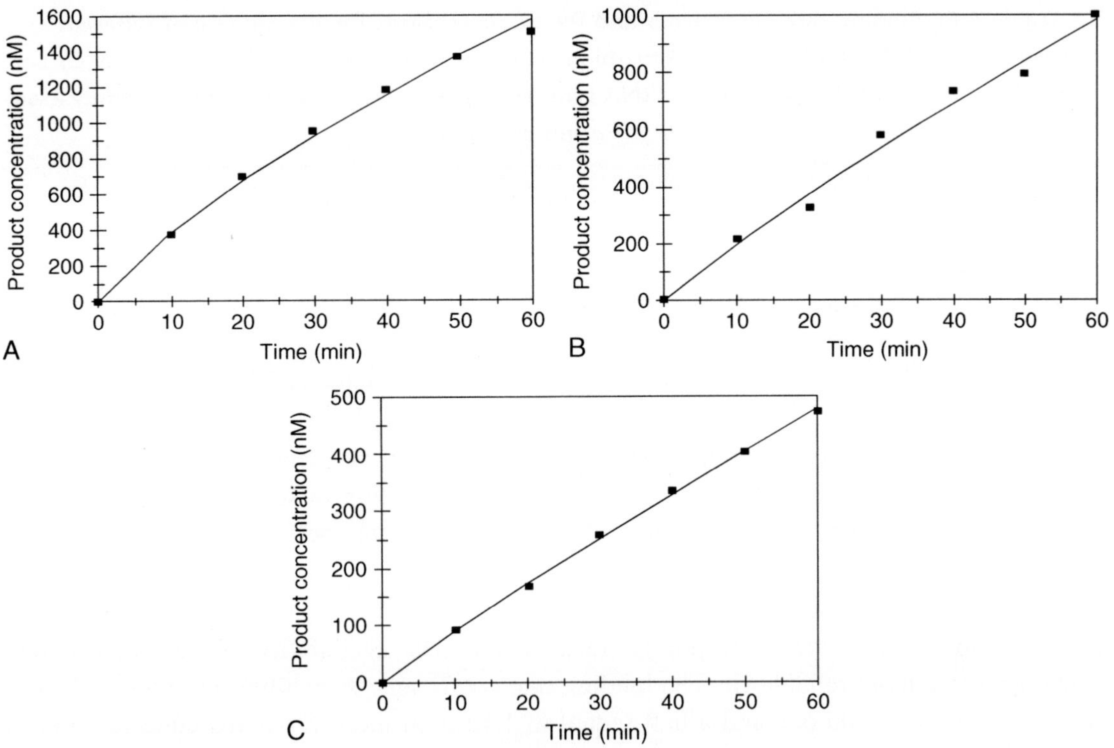

Figure 4.7

Binding of different concentrations (in mU) of methionine-7-amido-4-methylcoumarin (MET-AMC) in solution in the cSPA (competitive scintillation proximity aminoacyl-tRNA synthetase charging assay) (Forbes et al., 2007): (a) 0.05 (b) 0.025 (c) 0.01.

of the binding rate coefficient, k, and the fractal dimension, D_f, for a single-fractal analysis are given in Table 4.6.

Figure 4.7b shows the binding of 0.025 mU of MET-AMC in solution in the cSPA (Forbes et al., 2007). A single-fractal analysis is adequate to describe the binding kinetics. The values of the binding rate coefficient, k, and the fractal dimension, D_f, for a single-fractal analysis are given in Table 4.6. A decrease in the MET-AMC concentration in solution by a factor of 2 from a value of 0.05 to 0.025 mU leads to a decrease in the binding rate coefficient by a factor of 2.55 from a value of 66.248 to 26.008.

Figure 4.7c shows the binding of 0.01 mU of MET-AMC in solution in the cSPA (Forbes et al., 2007). A single-fractal analysis is adequate to describe the binding kinetics. The values of the binding rate coefficient, k, and the fractal dimension, D_f, for a single-fractal analysis are given in Table 4.6. In this case, a decrease in the MET-AMC concentration in solution by factor of 2.5 from a value of 0.025 to 0.01 mU in solution leads to a decrease in the binding rate coefficient, k, by a factor of 2.44 from a value of 26.008 to 10.644.

Table 4.6: Binding rate coefficients and fractal dimensions for the binding of different concentrations (0.01-0.05 mU) of (a) methionine-T-amido-4-methylcoumarin (MET-AMC) and (b) methionine in an aaRS (aminoacyl-tRNA synthetase) competitive scintillation proximity assay (cSPA) (Forbes et al.,2007).

Analyte concentration, mU	Binding rate coefficient, k	Fractal dimension, D_f
(a) MET-AMC, 0.05	66.248 ± 2.767	1.4522 ± 0.05528
MET-AMC, 0.025	26.008 ± 2.751	1.2260 ± 0.1357
MET-AMC, 0.01	10.644 ± 0.255	1.1422 ± 0.3206
(b) Methionine, 0.05	44.50 ± 3.44	1.2268 ± 0.100
Methionine, 0.025	7.393 ± 1.264	1.2181 ± 0.1151
Methionine, 0.01	7.230 ± 1.1812	1.2512 ± 0.1572

Figure 4.8a and Table 4.6 show for a single-fractal analysis the increase in the binding rate coefficient, k, with an increase in the MET-AMC concentration in solution in the 0.01-0.05 mU range. For the data shown in Figure 4.8a, the binding rate coefficient, k, is given by:

$$k = (1836.93 \pm 234.39)[\mathrm{MET}-\mathrm{AMC}]^{1.127\pm0.105} \tag{4.7a}$$

The fit is good. Only three data points are available. The availability of more data points would lead to a more reliable fit. The binding rate coefficient, k, exhibits an order of dependence between first and one and a half (equal to 1.127) on the MET-AMC concentration in solution in the 0.01-0.05 mU concentration range. The nonintegral order of dependence exhibited by the binding rate coefficient, k, on the MET-AMC concentration in solution in the 0.01-0.05 mU range lends support to the fractal nature of the system.

Figure 4.8b and Table 4.6 show the increase in the binding rate coefficient, k, with an increase in the fractal dimension, D_f, for the MET-AMC concentration in solution in the 0.01-0.05 mU range, for a single-fractal analysis. For the data shown in Figure 4.8b, the binding rate coefficient, k, is given by:

$$k = (4.764 \pm 1.548)D_f^{7.232\pm1.613} \tag{4.7b}$$

The fit is very good. Only three data points are available. The availability of more data points would lead to a more reliable fit. The binding rate coefficient, k, is very sensitive to the fractal dimension, D_f, or the degree of heterogeneity that exists on the biosensor chip surface as noted by the greater than seven (equal to 7.232) order of dependence exhibited.

Figure 4.8c and Table 4.6 show the increase in the fractal dimension, D_f, with an increase in the MET-AMC concentration in solution in the 0.01-0.05 mU range for a single-fractal analysis. For the data shown in Figure 4.8c, the fractal dimension, D_f, is given by:

$$D_f = (2.190 \pm 0.120)[\mathrm{MET}-\mathrm{AMC}]^{0.1454\pm0.04698} \tag{4.7c}$$

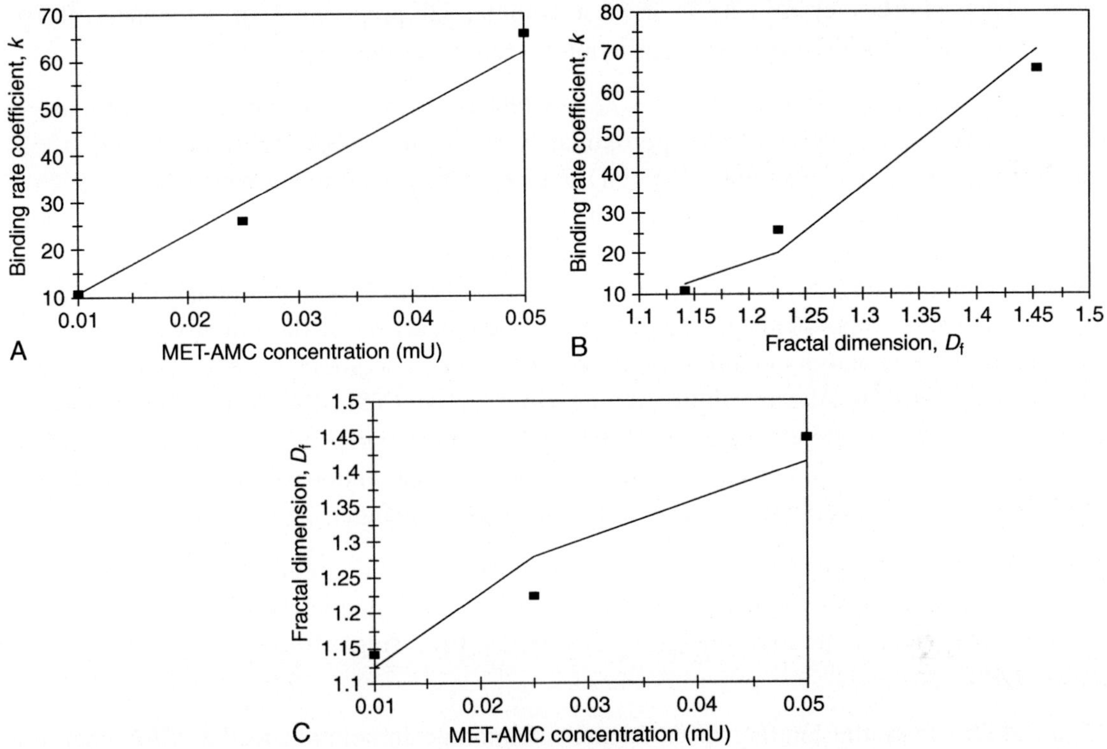

Figure 4.8

(a) Increase in the binding rate coefficient, k with an increase in the MET-AMC concentration (in mU) in solution. (b) Increase in the binding rate coefficient, k with an increase in the fractal dimension, D_f. (c) Increase in the fractal dimension, D_f with an increase in the MET-AMC concentration (in mU) in solution.

The fit is reasonable. Only three data points are available. The availability of more data points would lead to a more reliable fit. The fractal dimension, D_f, exhibits a weak dependence on the MET-AMC concentration in solution as noted by the 0.1454 order of dependence exhibited. Once again, the fractal dimension is based on a log scale, and even very small changes in the fractal dimension on the biosensor surface lead to significant changes in the degree of heterogeneity on the biosensor surface.

Forbes et al. (2007) recently reported that the quantitative detection of amino acids is essential for drug discovery in the pharmaceutical industry. These authors point out that there is a need for an amino-acid detection method that is high throughput, and is capable of detecting a singe amino acid in the presence of other amino acids. They have recently presented a HTS-compatible method for measuring the concentration of most naturally occurring amino acids. They (Forbes et al., 2007) used a commercial preparation of aminoacyl-tRNA-synthetases (aaRSs), tRNAs, and scintillation proximity technology.

Forbes et al. (2007) emphasize that in *E. coli* synthetase enzymes must be able to recognize a few selective cognate tRNAs from approximately 80 distinct tRNA species. Loftfield (1963) and Loftfield and Vansderjagt (1972) point out that synthesized proteins have a low frequency of errors and thus there is a high degree of substrate recognition. Thus, Forbes et al. (2007) emphasize that a method that employs aaRS-catalyzed charging of a cognate tRNA to detect a particular amino acid will be very specific for an amino acid. This is distinguished by its exact side chain. These authors indicate that Macarron and coworkers (2000) initially demonstrated that aaRS charging activity could be assayed under acidic conditions whereby charged tRNA interacts with yttrium silicate scintillation proximity assay (YSi SPA) beads. The method of Forbes et al. (2007) extends the use of aaRSs as a general tool to measure the concentration of a particular amino acid present in a sample consisting of a mixture of many different amino acids. Their method is able to do this in a high-throughput manner

Figure 4.9a shows the binding of 0.05 mU methionine in solution to the cSPA charging assay. A single-fractal analysis is adequate to describe the binding kinetics. The values of the binding rate coefficient, k, and the fractal dimension, D_f, for a single-fractal analysis is given in Table 4.6b.

Figure 4.9b shows the binding of 0.025 mU methionine in solution to the cSPA charging assay. A single-fractal analysis is adequate to describe the binding kinetics. The values of the binding rate coefficient, k, and the fractal dimension, D_f, for a single-fractal analysis is given in Table 4.6b.

Figure 4.9c shows the binding of 0.01 mU methionine in solution to the cSPA charging assay. A single-fractal analysis is adequate to describe the binding kinetics. The values of the binding rate coefficient, k, and the fractal dimension, D_f, for a single-fractal analysis is given in Table 4.6b. Table 4.6b indicates that as the methionine concentration in solution decreases by a factor of 5 from a value of 0.05 to 0.01 mU, the binding rate coefficient, k, decreases by a factor of 6.15 from a value of 44.50 to 7.230. This seems almost like a linear decrease for the binding rate coefficient, k, with the methionine concentration in the 0.01-0.05 mU range.

Figure 4.10 and Table 4.6 show for a single-fractal analysis the increase in the binding rate coefficient, k, with an increase in the methionine concentration in solution in the 0.01-0.05 mU range. For the data shown in Figure 4.10, the binding rate coefficient, k, is given by:

$$k = (752.13 \pm 962.36)[\text{methionine, mU}]^{1.071 \pm 0.722} \tag{4.8}$$

The fit is poor, and this is reflected in the error reported for the binding rate coefficient, k. Only the positive value of the error is presented since the binding rate coefficient, k, cannot

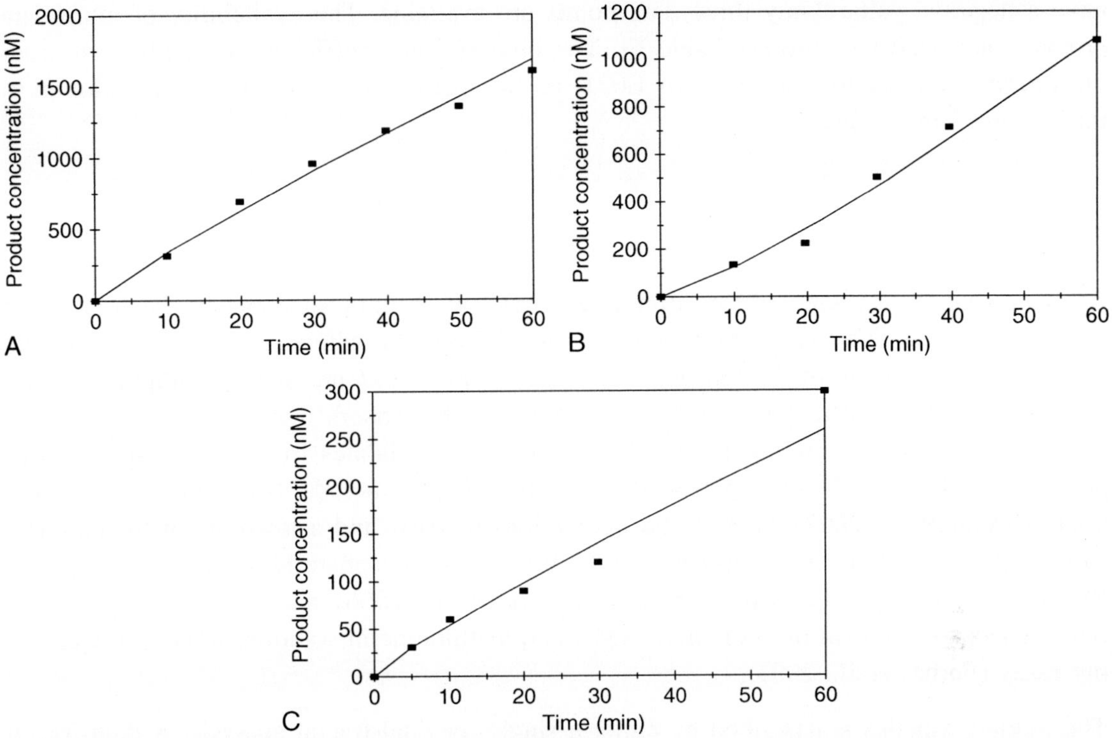

Figure 4.9

Binding of different concentrations (in mU) of methionine in solution in the cSPA (competitive scintillation proximity aminoacyl-tRNA synthetase charging assay) (Forbes et al., 2007): (a) 0.05 (b) 0.025 (c) 0.01.

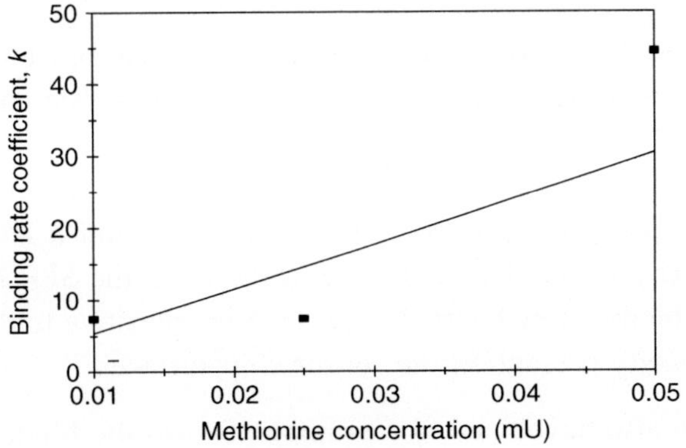

Figure 4.10

Increase in the binding rate coefficient, *k* with an increase in the methionine concentration (in mU) in solution.

have a negative value. Only three data points are available. The availability of more data points would lead to a more reliable fit. The binding rate coefficient, k, exhibits an order of dependence close to one (equal to 1.071) on the methionine concentration in solution in the 0.01-0.05 mU range.

4.4 Conclusions

A fractal analysis is presented for different examples wherein biosensors have been involved in drug discovery. This is an important area of investigation wherein there is a continually increasing application of biosensors, and where biosensors are making important contributions. This is particularly so for cases wherein biosensors may be used as an HTS method to quickly narrow down the possible suitable candidates from a wide spectrum of potential candidates. The examples analyzed in this chapter include: (a) inhibitors of protein kinases (Viht et al., 2007) wherein the interactions of ARC and a isoforms of the catalytic subunit (Cα) of CAPK are examined, (b) the binding of phosphate ion (P_i) to a rhodamine-PBP fluorescence-based phosphate biosensor (Okoh et al., 2006), and (c) the binding of different concentrations (in mU) of MET-AMC and methionine in solution in the cSPA charging assay (Forbes et al., 2007).

The binding kinetics is described by either a single- or dual-fractal analysis. A dual-fractal analysis is only used when a single-fractal analysis does not provide an adequate fit. This is done using Corel Quattro Pro 8.0 (1997). The fractal dimension provides a quantitative measure of the degree of heterogeneity present on the biosensor chip surface. Note that, and as indicated in the earlier chapters in the book, the fractal dimension for the binding and the dissociation phase, D_f and D_{fd}, respectively, is not a typical independent variable, such as analyte concentration, that may be directly manipulated. It is estimated from Equations (4.1-3), and one may consider it as a derived variable.

An increase in the fractal dimension value or the degree of heterogeneity on the surface leads, in general, to an increase in the binding and in the dissociation rate coefficient(s). For example, for the binding of Ca in solution to ARC-704 immobilized on a SPR biosensor chip surface (Viht et al., 2007), and for a dual-fractal analysis, the binding rate coefficient, k_1, exhibits an order of dependence between seven and seven and a half (equal to 7.351) on the fractal dimension, D_{f1}, or the degree of heterogeneity on the SPR biosensor chip surface. This indicates that the binding rate coefficient, k_1, is very sensitive to the fractal dimension or the degree of heterogeneity present on the sensor chip surface.

Predictive relations are also developed, for example, for (a) the binding rate coefficients, k_1 and k_2, as a function of the ARC-704 concentration (in nM) in solution (Viht et al., 2007), (b) the dissociation rate coefficient, k_d, and the ratio of the binding rate coefficients, k_2/k_1, as a function of the ARC-704 concentration (in μg/mL) in solution, (c) the binding rate coefficient, k_2, for a dual-fractal analysis as a function of the fractal dimension, D_{f2}, (d) the dissociation rate coefficient, k_d, as a function of the fractal dimension, D_{fd}, (e) the affinities, K_1 and K_2, as a function of the ratio of fractal dimensions, D_{f1}/D_{fd} and D_{f2}/D_{fd}, respectively, (f) the ratio of the binding rate coefficients, k_2/k_1, as a function of the ratio of fractal dimension, D_{f2}/D_{f1}, for the binding of inorganic phosphate ion (P_i) in solution to a rhodamine-PBP phosphate biosensor (Okoh et al., 2006), (g) the binding rate coefficient, k, for a single-fractal analysis as a function of MET-AMC concentration in solution to the cSPA charging assay (Forbes et al., 2007), and (h) the binding rate coefficient, k, for a single-fractal analysis as a function of the fractal dimension, D_f, or the degree of heterogeneity that exists on the biosensor chip surface.

The three different examples presented in this chapter, emphasize that the degree of heterogeneity that exists on the biosensor surface does significantly affect, in general, the rate coefficient and affinity values, and subsequently the kinetics in general. These are just a few of the representative examples available in the literature. More such studies are required to determine whether the binding and the dissociation rate coefficient(s), and subsequently the affinity values are sensitive to their respective fractal dimensions on the biosensor chip surface with regard to drug discovery.

A better understanding of all possible parameters that influence the kinetics of binding and dissociation of different analyte-receptor systems on biosensor surfaces is critical. This will be of considerable assistance, for example, to help select the correct drug of choice from a list of possible candidates. More often than not, the influence of diffusion and heterogeneity on the biosensor surface is neglected. As indicated in this chapter and elsewhere in the book, the degree of heterogeneity significantly influences, in general, the binding as well as the dissociation kinetics occurring on biosensor surfaces. It would behove the practicing biosensorists to start paying more attention to this aspect of kinetics on biosensor surfaces. One may perhaps argue that the influence of diffusional limitations may be minimized or perhaps even be eliminated if the biosensor is run properly.

In fact, ideally one should really analyze separately the influence of the degree of heterogeneity and the diffusional aspects on the kinetics of analyte-receptor reactions occurring on biosensor surfaces. This is not possible at the present time by the manner in which the fractal analysis is done. If one is able to separate the influence of diffusional limitations and heterogeneities on the biosensor surface, and analyze the influence of each on the analyte-receptor reactions occurring on biosensor surfaces, then one may be able to better manage these analyte-receptor interactions to advantage. This should very significantly impact the different biosensor performance parameters such as sensitivity, selectivity, and stability, and permit one to optimize these parameters in desired directions for the different analyte-receptor reactions occurring on biosensor surfaces.

References

Adams JA, Kinetic and catalytic mechanisms of protein kinases, *Chemical Reviews*, **101**, 2271–2290 (2001).

Boozer C, G Kim, S Cong, H Guan, and T Lundgren, Looking towards label-free biomolecular interaction analysis in a high throughput format: A review of new surface plasmon resonance technologies, *Current Opinion in Biotechnology*, **17**, 400–405 (2006).

Brune M, JL Hunter, JET Corrie, and MR Webb, Direct real-time measurement of rapid inorganic phosphate release using a novel fluorescent probe and its application to actomyosin subfragment 1 ATPase, *Biochemistry*, **33**, 8262–8271 (1994).

Cohen P, Protein kinases: The major drug targets of the twenty-first century?, *National Reviews and Drug Discovery*, **1**, 309–315 (2002).

Fischer PM, The design of drug candidate molecules as selective inhibitors of therapeutic relevant kinases, *Current Medicinal Chemistry*, **11**, 1563–1583 (2004).

Forbes CD, J Myung, and JA Landro, A high-throughput competitive scintillation proximity aminoacyl-tRNA synthetase charging assay to measure amino acid concentration, *Analytical Biochemistry*, **363**, 246–254 (2007).

Havlin S, Molecular diffusion and reaction. In *The Fractal Approach to Heterogeneous Chemistry: Surfaces, Colloids, Polymers*, Avnir D (Ed.), Wiley, New York, 1989, pp. 251–269.

Karlsson R, SPR for molecular interaction analysis: A review of emerging application areas, *Journal of Molecular Recognition*, **17**, 151–161 (2004).

Lee CK and SL Lee, Multi-fractal scaling analysis of reactions over fractal surfaces, *Surface Science*, **325**, 294–310 (1995).

Loftfield RB, The frequency of errors in protein biosynthesis, *Biochem Journal*, **89**, 82–92 (1963).

Loftfield RB and D Van der jagt, The frequency of errors in protein biosynthesis, *Biochem Journal*, **128**, 1353–1356 (1972).

Loog M, A Uri, J Jarv, and P Ek, Bi-substrate analogue ligands for affinity chromatography of protein kinases, *FEBS Letters*, **480**, 244–248 (2000).

Luecke H and FA Quiocho, High specificity of a phosphate transport protein determined by hydrogen bonds, *Nature*, **347**, 402–406 (1990).

Macarron R, L Mensah, C Cid, C Carranza, N Benson, AJ Pope, and E Diez, A homogeneous method to measure aminoacyl-tRNA synthetase aminoacylation activity using scintillation proximity assay technology, *Analytical Biochemistry*, **284**, 183–190 (2000).

Okoh MP, JL Hunter, JET Corrie, and MR Webb, A biosensor for inorganic phosphate using a rhodamine-labeled phosphate binding protein, *Biochemistry*, **45**, 14674–14771 (2006).

Viht K, K Padari, G Raiaru, J Subbi, I Tammiste, M Pooga, and A Uri, Liquid-phase synthesis of a pegylated adenosine-oligoarginine conjugate, cell membrane inhibitor of cAMP-dependent protein kinase, *Bioorganic Medicinal Chemistry Letters*, **13**, 3035–3039 (2003).

Viht K, S Schweinsberg, M Lust, A Vaasa, G Raidaru, D Lavogina, A Uri, and W Herberg, Surface-plasmon-resonance-based biosensor-based biosensor with immobilized bisubstrate analog inhibitor for the determination of affinities of ATP- and protein-competitive ligands of CAMP-dependent protein kinase, *Analytical Biochemistry*, **362**, 268–277 (2007).

Webb MR, A fluorescent sensor to assay inorganic phosphate. In *Kinetic Analysis: A Practical Approach*, Johnson KA (Ed.), Oxford University Press, Oxford, 2003, pp. 131–152.

CHAPTER 5

Nanobiosensors

Chapter Outline

5.1 Introduction

Nanotechnology and its applications in diverse areas have recently generated tremendous interest; utilizing this advantage is the focus of research by different groups throughout the world. In this chapter we present briefly the application of nanotechnology in different biosensor applications, and indicate the advantages of enhancing the biosensor parameters such as sensitivity, selectivity, economics, and response time. Some of the examples presented and analyzed include:

(a) Multifunction cross-linked Au nanoaggregates for optical DNA detection (Li et al., 2009).

Handbook of Biosensors and Biosensor Kinetics. DOI: 10.1016/B978-0-444-53262-6.00005-X

(b) Nano-bio-chips (NBCs) for high performance multiplexed protein detection: determination of cancer markers in serum and saliva using quantum dot (QD) bioconjugate labels (Jokerst et al., 2009).
(c) Nonenzymatic glucose sensor based on renewable electrospun Ni nanoparticle-loaded carbon nanofiber paste (NiCFP) electrode (Liu et al., 2009).
(d) Use of gold nanoparticles in a sensitive immunochromatographic assay for the detection of PSA (prostate specific antigen) in serum (Nagatani et al., 2006)
(e) The development and *in vitro* characterization of an intracellular microsensor responsive to reactive oxygen species (ROS) (Henderson et al., 2009).
(f) Combined fluorescence and SERS molecular beacon assay to detect human viral RNA (Sha et al., 2007).
(g) Nanotube-based biosensor for the detection of disease-specific autoantibodies in human serum (Drouvalakis et al., 2008).
(h) Gold nanoparticle amperometric immunosensor (biosensor) for osteoproteogerin (OPG) (Singh et al., 2008)
(i) Label-free antigen-antibody binding on a gold nanoparticle sensor array (Olkhov and Shaw, 2008).
(j) Electrochemical immunosensing using magnetic beads and gold nanoparticles (Selvaraju et al., 2007)

The examples analyzed have been selected at random. They have appeared recently in the literature. The intention is to provide an idea or perspective of the current trends in the application of nanotechnology in the area of biosensors. Other recent applications of nanotechnology for biosensor applications have been presented at recent conferences. They include:

(a) Enhanced emission of gold nanoparticle due to electron transfer from surface bound molecules and its use in pH sensing (Lee et al., 2009).
(b) Periodic plasmonic nanostructures as efficient SERS (Surface Enhanced Raman Spectra) substrates for biosensing (Lin et al., 2009).
(c) Composition effect of Ag-Cu alloy nanoparticles on luminescence enhancement/quenching of vicinal luminophores (Chowdhury et al., 2009).
(d) Nanoengineered transport metallic nanofibrous membrane and its application for humidity sensing (Jia et al., 2009).
(e) DNA hybridization detection using spectral changes of zinc selenide nanocrystals (Wang et al., 2009).
(f) Ultrafast, highly sensitive label-free pathogen detection via chemically-modified grapheme (CMG) sensors (Nagaraja et al., 2009).
(g) Layer-by-layer assembly of multiwall carbon nanotube ultrathin films for biosensing applications (Manther et al., 2009).

(h) Investigating enzyme kinetics using nanofluidic devices (Goluch et al., 2009).
(i) Single-molecule aptamer-target interactions for sensor applications (Zhang et al., 2009).
(j) Antibody-conjugated gold nanoclusters (nanoroses) for targeted cancer cellular imaging and therapy (Ma et al., 2009).
(k) Combining aptamer technique with nanotechnology for ultrasensitive small molecule and protein detection (Zhou et al., 2009).
(l) Electrochemical biosensor with self-assembled peptide nanotubes encapsulated horse-radish peroxidase (Park et al., 2009).
(m) Bimimetic fabrication of optical nanobiosensor for hydrogen peroxide detection (Tian and Dale, 2009).
(n) Gold nanoparticle assays: towards sensitive detection of unamplified DNA (Verdold et al., 2009).
(o) Carbon nanoparticles as signal labels in protein microarrays (van Amerongen et al., 2009).
(p) Streptavidin conjugated Fe_2O_3 magnetic nanoparticles for *in-situ* biomedical applications (Chomuchka et al., 2009).
(q) Nanostructured silicon-based biosensors for the determination of some mycotoxins (Starodub et al., 2009).
(r) Conductive polyaniline (PANI) nanostructures for sensing applications (Berti et al., 2009).
(s) Gold nanoparticles nanostructured electrochemical biosensors for applications in the food industry (Gonzalez-Cortes et al., 2009).
(t) Glucose biosensors based on novel carbon nanotube fiber microelectrode (Zhu et al., 2009).
(u) Electrochemical immunoassay for cardiac markers as a solid phase and silver nanoparticles as an electroactive label (Szymanski and Porter, 2009).
(v) Carbon nanotube electrode for biosensors (Razumiene et al., 2009).
(w) Antibody immobilization on gold and silver nanoparticles for use in electrochemical immunoassays (Szymanski et al., 2009).

5.2 Optical DNA Detection by Multifunctional Cross-Linked Au Nanoaggregates (Li et al., 2009)

Li et al. (2009) have recently used multifunctional cross-linked Au nanoaggregates to develop an optical DNA detection biosensor. These authors report that oligonucleotide bridged Au (gold) nanoparticles (Au NPs) aggregate provide a red-to-blue color change (Elghanian et al., 1997; Mucic et al., 1998; Storhoff et al., 2000; Jiang et al., 2005). Rosi and Mirkin (2005) initially suggested that this may be used as a DNA detection agent. Since then Li et al. (2009) have reported that Au NP-based or DNA detection assays have been developed (Mirkin et al., 1996; Storhoff et al., 1998; Patolsky et al., 2000; Reynolds et al., 2000;

Liu and Lu, 2003, 2006a,b; Wang et al., 2007; Claridge et al., 2008). Li et al. (2009) confirm that the DNA cross-linked Au nanoaggregates provide a higher surface area when compared with individual Au NPs. Thus, these DNA cross-linked Au nanoaggregates have been used in a SERS (surface enhanced Raman scattering) application (Graham et al., 2008), as a conductive tag (Fang et al., 2008), and as a conductive matrix (Li et al., 2008). Li et al. (2009) indicate that DNA cross-linked Au NP aggregates very significantly enhance the sensitivity of biosensors.

Recently, Li et al. (2008) developed a multicomponent Au NP-based nano probe. This nano probe includes the following functions: DNA recognition, signal amplification (HRP; horse radish peroxidase) and blocking of nonspecific binding (NSB) by using BSA (bovine serum albumin). The authors used a sandwich type assay (Nam et al., 2004; Polsky et al., 2006; Zhang et al., 2006a). They confirm that their capture probe enabled the target DNA along with the detection probe to come close to the magnetic particles. The complexes were further magnetically separated for optical detection. They (Li et al., 2009) have recently improved on their initial method by hybridizing their detection probe Au NP with another cross-linking probe DNA (which they call a cross-linking nano probe, CNP). This permits them to load higher amounts of HRP. The HRP that is on the surface of the Au aggregates catalyzes the enzyme substrate and generates an optical signal. The generated enzymatic signal is used as a read out for the target DNA. The authors used wild-type breast cancer-related BRCA-1 gene as the target (Stenson et al., 2009).

Li et al. (2009) used the following oligonucleotides in their analysis:

Capture probe: 5′(biotin)-TTTTTTTTTTTTCCCACCAACGCTG-3′
Detection probe: 5′ ATCAATTCCACAGTTTTCGCTTTTTTTTTTTTTT-(CH_2)6-SH
Cross-linking probe: 5′ SH-$(CH_2)_6$-TTTTTTTTTTTTGATTATTCATAC3′
target probe: 5′ GAGCATACATAGGGTTTCTCTTGGTTTCTTTGATTATAAT
One-base mismatched DNA: 5′ ACACGCTTGGTAGACTTTTTTTTTTAGCATCGAT AACGTT
blocking DNA: 5′ TTTTTTTTTT 3′

Li et al. (2009) point out that Au NPs have a high surface-to-volume ratio. This permits the attachment of multiple kinds of biomolecules to the single nanoparticle surface (Niemeyer, 2001; Niemeyer and Ceyhan, 2001; Katz and Willner, 2004; Hazarika et al., 2006; Hill and Mirkin, 2006; Becker et al., 2007). They attached HRP (horse radish peroxidase), thiolated oligonucleotide (detection probe or cross-linking probe), and BSA to the surface of the Au NPs in a sequential fashion. They suggest that the treatment with BSA as a nonspecific blocker is essential as it minimizes the background noise. They also show that the cross-linked Au NPs aggregates exhibit an average height thirty times that of the single Au NP (600 nm when compared with 20 nm). This eventually leads to a higher HRP loading for the DNA bridged Au aggregates and to a higher sensitivity for the Au NP aggregates when compared with the single Au NPs. They further add that their Au NPs aggregates are sensitive enough to discriminate single mismatches. They were also able to confirm the presence of cross-linked Au aggregates using light-scattering spectroscopy (Du et al., 2006).

Finally, Li et al. (2009) affirm that their aggregate Au NP optical detection method exhibits high sensitivity, and that their diagnostic device has the potential to be effective, especially for small clinics in developing countries where resources are inherently limited.

5.3 High-Performance Multiplexed Determination of Proteins: Determinations of Cancer Biomarkers in Serum and Saliva Using QD Bioconjugate Labels (Jokerst et al., 2009)

Jokerst et al. (2009) have recently developed a multiplexed biosensor for detecting the three cancer biomarkers, carcinoembryonic antigen (CEA), cancer antigen 125 (CA125), and Her-2/Neu (C-erb-2). They integrated semiconductor nanoparticle QDs into a molecular microfluidic biosensor. They were able to use this biosensor for both serum and whole saliva specimens. Their nano biochips used a fluorescence transduction signal with QDs-labeled detecting antibody in combination with antigen capture by a microporous agarose bead array. They used a sandwich type immunoassay. The authors report that their QD biosensor yields (a) a signal amplification by a factor of 30 when compared with standard molecular fluorophores, and (b) a decrease in the limit of detection (LOD) by approximately two orders of magnitude (0.02 ng/ml; 0.11 pM CEA) relative to ELISA (enzyme-linked immunosorbent assay).

Jokerst et al. (2009) point out that the detection of cancer (protein) biomarkers exhibits promise in the screening, treatment, metastasis evaluation, and determining the response to pharmacologic intervention (Lee et al., 2008). ACS (2008) reports that tests such as PAP smear and PSA are efficient screening tools for cancer detection in asymptomatic individuals. Sofer et al. (2006) mention that breast cancer detection tests based on Her-2/Neu and alpha-fetoprotein tests for testicular cancer have also been developed. Das and Bast (2008) report the development of protein biomarkers to monitor colon (by CEA, carcinoembryonic antigen) and ovarian cancer (by cancer antigen 125 (CA125).

Harris et al. (2007) are of the opinion that identifying at-risk individuals for cancer frequently requires four or more biomarkers. ELISA-based methods are not easily multiplexed. Furthermore, Bhasin et al. (2008) report that the current LOD values for cancer biomarker detection during cancer in its nascency and diagnostic decision values are close to each other; Jokerst et al. (2009) remark that this makes it difficult to help evaluate patients during the early stages of cancer development. The authors also point out that POC (point-of-care) systems further facilitate early, routine and frequent access to diagnostic test results. They further report that POC methods have started using saliva as a diagnostic medium (Wong, 2008).

Mandel (1993) and Malamud (2006) both report that saliva contains biomarkers that offer information about both oral and systemic disease. Jokerst et al. (2009) further affirm that useful genetic and proteomic profile information is also available in the saliva (Hu et al., 2007; Oppenheim et al., 2007). Hu et al. (2007) and Tan et al. (2008) concur that saliva testing minimizes testing antipathy and promotes frequent testing. Streckfus et al. (2006) caution, however, that saliva is more heterogeneous than serum and is not without problems whilst testing for biomarkers. Furthermore, the expression levels for these disease-related biomarkers may be expressed at several orders of magnitude less than those expressed in serum; hence the need for efficient sensing systems which can distinguish between background noise and target-specific signals.

Jokerst et al. (2009) report that QD fluorophores exhibit the potential to help minimize or eliminate these types of problems (Liu et al., 2008a). Sofer et al. (2006) have listed the difficulties in linking QDs to bioligands for POC cancer diagnostics. Furthermore, Resch-Genger et al. (2008) point out that the integration of QDs as quantitative imaging tools, especially with small analyzer platforms, is yet to be demonstrated.

In order that these types of problems may be mitigated new systems are under development for the early detection of neoplastic diseases (Ali et al., 2003; Christodoulides et al., 2007; Weigum et al., 2007). The Jokerst et al. (2009) analysis presents the incorporation of QDs as detection elements in the NBC biosensors for the quantitative measurement of the well-characterized cancer biomarkers, CA125, CEA, and Her-2/Neu (C-erB-2).

Jokerst et al. (2009) claim that a MEMS (micro-electromechanical systems) platform that includes microfluidics elements and bead containers supports their integrated NBC assay. The authors used QDs as a detection moiety in their NBC, and their QD-based platform includes all the components of a sandwich immunoassay, with agarose as a solid phase support. The capture and detection bodies comprise different clones of monoclonal antibody. The authors mention that the beads are layered within their NBCs between precision cut layers of laminate adhesive which facilitate reagent delivery. Their fluorophore conjugated detection antibody yields the signal that is recorded by a CCD camera. They report that the fluorescence signal relates directly to the antigen present in a sample.

Jokerst et al. (2009) explain that for using the QDs in their immunoassay, care was taken to retain their optical properties, hydrophilicity, and the recognition moiety specificity by linking them to a recognition element such as an antibody by different covalent and noncovalent strategies. The QDs were characterized based on background, signal, and nonspecific signal (noise).

Jokerst et al. (2009) compared the performance of their QDs to traditional fluorophores. They used Alexa Fluor 488 (AF 488), one of the brightest fluorophores available for this comparison. The authors concluded from their experiments that the QDs provide 25 times more signal compared to that provided by AF 488 at isomolar concentrations.

Jokerst et al. (2009) further report that the ELISA method, which has been the gold standard to detect proteins since 1971, requires an analysis time of 254.4 min (4.24 h), but their QD-based NBC method requires an analysis time of just 27 min. This represents a significant decrease in the time required, by a factor of 9.4.

Jokerst et al. (2009) emphasize the importance of LOD in detecting cancer biomarkers in saliva as compared to serum. This, as indicated earlier, is due to the much lower concentrations of these biomarkers in saliva. They report that their QD-based NBC permits a high signal at relatively low concentrations of their detection limit, which helps minimize background noise and supports their excellent LOD measurements. On analyzing clinical samples of saliva and serum for these cancer biomarkers, they state that their proof-of-concept pilot studies yield strong correlation results when their NBC method is compared with standard laboratory-based methods. Of special interest is the low LOD values observed, which should positively affect patient outcomes. Finally, Jokerst et al. (2009) report that (a) their QD-based NBC assay was reproducible, (b) it was able to detect CEA-specific signals in clinical samples of saliva, and (c) it exhibited multiplexing capabilities by detecting all three cancer biomarkers, CEA, CA 125, and Her-2/Neu (C-erB-2) after the beads were sensitized to either of the above mentioned cancer biomarkers. For example, they prove that their NBC biosensor is able to detect CEA-specific antigen, and is independent of the presence of 0-400 U/ml CA 125.

They conclude by asserting that their QD-based NBC is an important step towards realizing a harmonized nano-bio analysis system. Their study is the first in which QDs provide quantitative information with regard to cancer biomarkers from saliva samples. Also, their integrated NBC system is disposable, which is an added advantage.

5.4 Carbon Nanofiber Paste Electrode Nonenzymatic Glucose Sensor (Liu et al., 2009)

Liu et al. (2009) have recently developed a nonenzymatic glucose sensor which consists of a renewable NiCFP electrode. The authors used an electrospinning technique combined with thermal treatment to make this electrode. They noted that significant amounts of spherical nanoparticles were either well dispersed or embedded in the carbon nanofibers. These nanoparticles were comprised of Ni and NiO. They also noted that their electrospun electrodes when mixed with mineral oil exhibited a rapid response and were also not poisoned by the chloride ions. The authors point out that as the Ni nanoparticles are firmly embedded and their carbon based electrode is inert, they are highly sensitive and stable biosensors. Also, the method of fabrication of their biosensor is simple and inexpensive.

Liu et al. (2009) also mention that ever since Clark and Lyons (1962) proposed the enzyme electrode for glucose determination, there has been considerable interest for over five decades in helping to detect glucose in analytical solutions (Newman and Turner, 2005; Wilson and Gifford, 2005). Electrochemical sensors have been in the forefront in this area of research (Wang, 2008). Nanomaterials exhibit unique electronic and mechanical properties; therefore, not surprisingly, there has been considerable work with regard to the applications of nanomaterials for glucose detection in solutions (Lu et al., 2006, 2007; Vamvakaki et al., 2006; Wu et al., 2007a,b; Deng et al., 2008; Jeykumari and Narayan, 2008; Liu et al., 2008a,b; Zhou et al. 2008). Katakis and Dominguez (1995), however, point out the instabilities associated with the intrinsic nature of the enzymes and the tedious fabrication procedures involved. There is a need to overcome these shortcomings, perhaps by using nonenzymatic biosensors, which will help eliminate the problems associated with the use of enzymes.

Thus, considerable effort has been spent on developing nonenzymatic electrodes for glucose determination (Vassilyev et al., 1985; Adzic et al., 1989; Sun et al., 2001). Liu et al. (2009) report that the development of nanotechnology has also contributed to this effort by providing various materials for electrodes, including macroporous platinum film (Song et al., 2005), ordered platinum arrays (Yuan et al., 2005), nanoporous platinum-palladium networks (Wang et al., 2008), three-dimensional silver film (Bai et al., 2008), and platinum-ruthenium nanoparticles (Li et al., 2008). However, even here there is a shortcoming, in spite of the high electrocatalytic activity exhibited by the precious metal electrodes, due to the chemisorption of intermediates which subsequently leads to fouling of the electrodes. This results in poor operational stability. Furthermore, as Liu et al. (2009) emphasize, the high cost of these precious metals prohibits their use for commercial applications where low cost, as they correctly point out, is a major requirement.

Liu et al. (2009) report that metal oxide and complex catalysts may be used to modify the electrodes for the nonenzymatic detection of glucose (Chen et al., 1993, 2008; Kang et al., 2007; Ozcan et al., 2008). Ni-based materials are a good low-cost catalyst that may be used for the electrocatalytic oxidation of analytes other than glucose. For example, they have been used in the electrocatalytic oxidation of carbohydrates (Reim and Effen, 1986; You et al., 2003; Casella and Gatta, 2001; Ojani et al., 2008), insulin (Salimi et al., 2007), aspirin (McAuley and Wildgoose, 2008), and ethanol (Casella et al., 1993; Wang et al., 2004a,b). The nickel-based materials apparently catalyze the oxidation processes via the formation of a high valent oxyhydroxide species.

Liu et al. (2009) point out that the electrospinning technique has received attention in both the academic and industrial environments. This is useful, particularly to get nanofibers with different architectures and compositions. For example, Hou and Reneker (2004) combined the electrospinning technique with the carbonization process to obtain a novel hierarchical structure of carbon nanotubes on nanofibers. Electrospinning of three-dimensional nanofibrous tubes with controllable architectures has also been reported (Zhang and Chang, 2008).

Huang et al. (2008) have recently developed a novel carbon nanofiber material, which contains electrospun palladium nanoparticles. In their most recent analysis, Liu et al. (2009) have mixed the electrospun nickel nanoparticle-loaded carbon nanofibers with mineral oil. This is their NiCFP electrode. Their intention was to determine glucose concentrations, *sans* a mediator or an enzyme.

Liu et al. (2009) report that the NiCF composite was made by carbonizing the electrospun PAN (polymer)/NiAA composite fibers. This nanocomposite was then mixed with mineral oil in the ratio of 60/40 w/w. They also report that two steps are required to make these nanocomposite fibers. They include (a) synthesis of the carbon fiber and (b) deposition of the Ni catalysts. It is difficult to discharge the nanoparticles from these fibers as they are deeply embedded in them. Their method, which includes a thermal treatment step prepares these nanofibers directly and effectively. They further indicate that their nanofibers do not consist of pure nickel alone, but are a mix of nickel and nickel oxide. An x-ray diffraction pattern result also indicates that the nanoparticles of nickel and nickel oxide coexist. Their NiCFP electrode combines the properties of the CF (electrochemical performance such as good conductivity) with the electrocatalytic activity of the nickel nanoparticles. They demonstrated clearly that the nickel and nickel oxide particles play an important role in the electrode as the CFP electrode alone did not indicate any current response.

Liu et al. (2009) also report that their simple fabrication method, along with the high electrocatalytic performance exhibited, offers a biosensor platform that is effective for the determination of glucose. Furthermore, their biosensor is able to determine 95% of the glucose concentration within 5 s, and the detection limit is as low as 1 μM. This, according to them, is better than that obtained by using a nickel complex (Ojani et al., 2008), CNT (carbon nanotube)-based complex (Chen et al., 2008; Deng et al., 2008), and alloy-modified electrodes (Li et al., 2008). Their biosensor also demonstrated a wide linear range (2 μM-2.5 mM). The sensitivity exhibited is 3.3 μA per mM.

Another advantage of the Liu et al. (2009) biosensor is the antifouling activity of its NiCFP electrode, due to the resistance offered to surface fouling by chloride ions. Other metal and alloy electrodes are subject to fouling by chloride ions (Sun et al., 2001). An additional advantage of the Liu et al. (2009) biosensor is its good operational stability. The response that the NiCFP electrode demonstrated was consistent, and not self-inhibited by glucose or any of its oxidation products on the electrode's surface. Liu et al. (2009) further report that the high electrocatalytic activity exhibited by the NiCFP electrodes leads to a sensitivity that is 1.5 times that exhibited by a Ni bulk electrode. This is primarily due to the large electroactive surface area prevalent in the NiCFP electrode.

Liu et al. (2009) also report that their biosensor exhibited good reproducibility properties. Also, when their biosensor was stored in a desiccator at room temperature for a month, it lost only 2% of its response to a 1% glucose solution. Their NiCFP electrode also permitted a continuous oxidation of the reactive intermediates formed at the electrode surface. This was clearly demonstrated, according to the authors, by the addition of glucose twenty times in a row. In contrast, there was a 50% reduction in the current signal when the nickel bulk electrode was used under similar circumstances.

The authors conclude that their NiCFP electrode has significant advantages that include good analytical performance as well as a simple preparation process. It also demonstrates important biosensor parameters, namely high sensitivity, a fast amperometric response, a low detection limit, a wide linear range, and good resistance to surface fouling, leading to better operational stability. Therefore they claim that their biosensor exhibits good potential as a platform for mass production at low cost.

5.5 Use of Gold Nanoparticles in a Sensitive Immunochromatographic Assay for the Detection of PSA in Serum (Nagatani et al., 2006)

Nagatani et al. (2006) have developed a sensitive immunochromatographic assay for detecting PSA in serum. PSA was first identified by Hava et al. (1969). Nagatani et al. (2006) report that PSA is a 33 kDa intracellular glycoprotein that is formed only in the prostate gland. PSA in serum exists both in the complex form (to α_1-chymotrypsin; PSA-ACT) and in free soluble form (f-PSA) (Lilija et al., 1991; Zhang et al., 1999; Balk et al., 2003). The authors claim that PSA in serum is a reliable marker for the early detection of prostate cancer. Clinicians suggest that levels greater than 4 ng/ml are an indicator of a potential prostatic abnormality (Caplan and Kratz, 2002; Balk et al., 2003). Rather than total PSA (t-PSA), the ratio of free PSA (f-PSA) to t-PSA is a better indicator of prostate cancer (Prestigiacomo and Stamey, 1995; Aslan et al., 2003; McArdle et al., 2004). They point out that both ELISA (enzyme-linked immunsorbent assay) and CLEIA (chemiluminescent enzyme immunoassay) have been routinely used to detect PSA, but that there is still a need to develop a portable and low-cost device to detect PSA.

Nagatani et al. (2006) explain that an immunosorbent membrane strip is a unique low-cost analytical device to detect analytes of interest, for example, in pregnancy diagnosis (Bhaskar et al., 1996; Chiao et al., 2004; Sato et al., 2004; Xiulan et al., 2005). Tanaka et al. (2006) have recently used gold nanoparticles in the test line of a chromatographic strip. They claim that this is a novel method to enhance the detection technique. They have used their novel immunochromatographic adsorbent strip to detect PSA in serum. They point out that their immunosorbent immunochromatographic strip uses less antibody in their test line than a normal immunochromatographic test strip and CLEIA. The antibody is an expensive component of the detection device, and their method facilitates a lower cost.

They also declare that the procedure to develop their sensitive immunochromatographic test strip for using less antibody is simple compared to the conventional procedure. They report that the primary antibody conjugated gold nanoparticles and the nonconjugated primary antibody solution mixture were immobilized on the test line of the immunochromatographic test strip. Following the immobilization procedure, the antigen and the gold nanoparticle-conjugated secondary antibody were introduced onto the conventional and sensitive immunochromatographic test strips. The authors explain that the nanoparticle-conjugated primary antibodies and the nanoparticle-conjugated secondary antibodies accumulate because of immunogenic reactions. Sonnichsen et al. (2005) point out that the LSPR (localized surface plasmon reaction) absorbance of the gold nanoparticles depends on the distance between the gold nanoparticles. Since the antigen-antibody reaction brings a high amount of gold nanoparticles together, Nagatani et al. (2006) report that a distinct red color may be seen with the naked eye.

Nagatani et al. (2006) report that by using their sensitive immunochromatographic test strip for the detection of PSA

(a) they were able to decrease the amount of antibody used by 50% for the same sensitivity. This facilitates a lower cost,
(b) their method provided a higher sensitivity in the test line when compared with the conventional method used now,
(c) their method could use the naked eye to screen PSA concentrations with a color sample sheet for first screening.

They conclude that their sensitive method exhibits the potential for entering the commercial market as a diagnostic device to detect PSA in the near future.

5.6 In Vitro Characterization of an Intracellular Nanosensor for ROS (Henderson et al., 2009)

Henderson et al. (2009) have recently developed an intracellular nanosensor for real-time monitoring of ROS. Their nanosensor is based on PEBBLE (probes encapsulated by biologically localized embedding) technology. The authors used a sensitive ROS probe dihydrorhodamine 123 (DHR 123) as the sensing element of the PEBBLE device. This was done by entrapment within a porous, bio-inert polyacrylamide nanostructure. According to the authors, this permitted a passive monitoring of the free radical species. They believe that though previous analyses have used fluorescent probes to analyze ROS, they have been found to be cytotoxic and can influence cellular metabolism and adversely affect *in vitro* data. The authors explain that they were able to stimulate the PEBBLE loaded NR8383 cells with PMA (phorbol-12-myristate-13-acetate), which permitted them to carry out real-time monitoring of ROS without affecting cellular viability. They report that their PEBBLE technology provides distinct advantages over existing technologies for monitoring the intracellular environment.

Henderson et al. (2009) report that ROS play a significant role in cellular systems, especially with regard to physiological and pathological processes, and have been the focus of much research interest (Babior et al., 1973; Babior, 1984; Aitken et al., 2007; Brechard and Tschirhart, 2008; Drose and Brandt 2008; Goud et al., 2008). ROS are important for the immune system (De Ravin et al., 2008; Feld et al., 2008; Kang and Malech, 2009). Henderson et al. (2009) show that during periods of oxidative stress, the levels of ROS increase, and the increased ROS levels lead to negative effects. The tissues are unable to deal effectively with this, and this leads subsequently to migraine, ischemic injury, and cardiovascular disease (Cross et al. 1987; Bindokas et al. 1996; Zulueta et al., 1997; Khalil and Khodr, 2001; Khodr and Khalil, 2001). Henderson et al. (2009) report that analytical systems have been developed to directly monitor ROS in real time (McNeil et al., 1989, 1992; Manning et al., 1998; Scheller et al., 1999; McNeil and Manning, 2002; Chang et al., 2005a,b).

They also report that Clark et al. (1998) had developed a nanosensor technology called PEBBLE that was capable of detecting changes in the intracellular environment using a fluorescent probe. The PEBBLE technology has been used to detect quite a few different analytes (Clark et al., 1999a,b; Summer et al., 2002; Xu et al., 2002; Hammond et al., 2008). They have recently used the commercially available fluorescent probe 123 DHR to monitor ROS as indicated above. This probe was previously used to detect a number of analytes including superoxide, hydrogen peroxide, and peroxynitrite (Roesler et al., 1991; Henderson and Chappell, 1993; Crow, 1997; Chang et al., 2005a,b; Goud et al., 2008; Qin et al., 2008).

Henderson et al. (2009) report that permeable dyes and particle-conjugated probes have been used in the literature (Henderson and Chappell, 1993; Jiang et al., 1997; Chang et al., 2005a,b; Palazolo-Ballane et al., 2007. They also report that fluorescent dyes encapsulated in PEBBLE technology offer significant advantages over the conventional dyes used in intracellular environments (Clark et al., 1998; Webster et al., 2005). A big advantage of PEBBLE technology, according to them, is that its inert matrix protects the intracellular environment, and interactions between nonspecific interferents are minimized. This limits false positives (Summer et al., 2002). Webster et al. (2007) have reviewed the different methods by which PEBLE technology may be delivered into the intracellular environment. Henderson et al. (2009) in their recent analysis have delivered the ROS sensitive PEBBLE technology into macrophage cells through phagocytosis.

Henderson et al. (2009) report that polyacrylamide PEBBLE nanosensors were synthesized using a modified version of the method available in the literature (Clark et al., 1999; Webster et al., 2005; Coupland et al., 2008). The PEBBLE nanosensor was calibrated using the generation of the ROS species got by the reaction of the substrate xanthine with the enzyme xanthine oxidase (XOD). With regard to PEBBLE delivery into the cellular environment, they report that confocal microscopy was used to confirm that the NR8383 cell line used in the study was able to internalize the ROS responsive PEBBLE technology through phagocytosis.

This was in spite of the nanoscale dimensions of the nanosensors. Measurements of the intracellular ROS were carried out by incubating the PEBBLE loaded cells with different concentrations of PMA (phorbol-12-myristate-13-acetate).

Henderson et al. (2009) confirm that the intracellular environment was not affected by the PEBBLE method used. This was demonstrated by the MTT assays. Furthermore, the PEBBLE nanosensors have the capacity to monitor normal cellular function in a passive sense without affecting the normal intracellular function. They report that the generation of nitic oxide (NO) as well as hydrogen peroxide (H_2O_2) was not affected by the introduction of the PEBBLE nanosensor into the intracellular environment. The authors propose to use their PEBBLE nanosensor in the future to other cell types besides those cells capable of phagocytosis. Perhaps alternate delivery methods may be used. The authors feel that the continuous ability to monitor ROS in intracellular environments would permit a better understanding of the involvement of ROS in quite a few different biological processes, especially those that are of a physiological and pathological nature. Finally, Henderson et al. (2009) conclude by emphasizing two distinct advantages of their PEBBLE nanosensor:

(a) The biocompatability of their nanosensor permits the long-term monitoring of the ROS species in intracellular environments, and (b) Even though other ROS measurement techniques are available, the PEBBLE nanaosensor directly measures the changes in ROS, whereas other techniques have to depend on reaction products or adducts.

5.7 Combined Fluorescence and SERS Molecular Beacon Assay to Detect Human Viral RNA (Sha et al., 2007)

Sha et al. (2007) have recently developed a dual-mode molecular beacon to measure unlabeled human viral DNA. Their detection system comprises a combined SERS (surface-enhanced Raman scattering) and fluorescent molecular beacon assay on nanobarcode^{TM} particles. These authors indicate that SERS has been used to detect biological analytes ever since Cotton's (1980) initial work on the detection of cytochrome c and myoglobin. This led to the detection of anthrax (Zhang et al., 2005), cancer (Culha et al., 2003), glucose (Shafer-Peltier et al., 2003), DNA and RNA (Cao et al., 2003; Faulds et al., 2004, 2005; Wabuyele and Vo-Dinh, 2005), and protein immunoassays (Cao et al., 2003).

The authors report that due to the uniqueness of the molecular Raman spectra, the detection of several species simultaneously (multiplexed detection) is possible. These authors indicate that due to the SERS enhancement and real-time response, a high sensitivity in biomolecular detection is possible (Kneipp et al., 1997; Nie and Emory, 1997; Fe Ru et al., 2006).

Tyagi and Kramer (1996) initially developed the fluorescent molecular beacon. Sha et al. (2007) explain that the molecular beacon is a single-stranded "loop-and-stem" DNA oligonucleotide that carries both a fluorochrome and a non fluorescent quencher chromophore at opposite ends of a strand. The two ends are near each other in the absence of a complementary strand. As a complementary nucleic acid binds to the loop, the molecular conformation is changed, thereby removing the quencher from the fluorochrome's vicinity. An unquenched fluorescence results. They further explain that DABCYL (a nonfluorescent chromophore; (4-((1-4-dimethylamino)-phenyl)-azo-benzoic acid) is used typically to quench the fluorescence.

Sha et al. (2007) report that metal nanoparticles may also be used to quench fluorescence rather than any organic moiety (Dubertret and Kramer, 2001; Maxwell et al., 2002; Du et al., 2003; Sha et al., 2005; Stoermer et al., 2006). Lakowicz (2001) has indicated that fluorescence from dyes located within a few nanometers from a surface may be quenched. Sha et al. (2007) have used Nanobarcodes™ (NBC) as the metal surface. They have developed a SERS beacon assay for the detection of viral RNA by combining the sensitivity of SERS technology and the multiplexing potential of NBC. The dimensions of NBC are about 300 nm × 6-9 μm. NBC has been synthesized by the sequential electrochemical deposition of metal ions into templates with uniformly sized pores (Nicewarner-Pena et al., 2001; Reiss et al., 2002; Walton et al., 2002).

The authors report that their dual-mode design uses a molecular beacon probe with a hairpin structure. This holds the dye at the 3′end close to the nanoparticle surface when the probe is assembled to it by a 5′-thiol group. As the dye is close to the metal nanoparticle surface its fluorescence is quenched (Sha et al., 2005; Stoermer et al., 2006). SERS spectra are observed when the dye is held near the metal surface. They point out that as the probe hybridizes to the target sequence

(a) The Raman label is forced to separate from the metal surface, and
(b) The SERS signal intensity drops.

This is because, as Wolkow and Moskovits (1987), and Campion and Kambhampati (1998) explain, the Raman enhancement is strongly dependent on the distance between the Raman label and the metal nanoparticle. An inverse 12th order of dependence is exhibited here. Sha et al. (2007) report that their technique permits them to detect a wide variety of oligonucleotide analytes that include viral DNA, DNA biothreat agents, food borne pathogens, SNP (single nuclear polymorphism) or mutant detection, gene expression products, and even biomarker proteins. Thus, their technique is quite flexible for detection purposes.

Sha et al. (2007) show that the HCV (hepatitis C virus) probe sequence is $(CH_2)6$ gcggag CAT AGT GGT CTG CGG AAD CGG TGActcgc $(CH_2)_7$ Cy5-3′ and the oligo sequence is TCA CCG GTT CCG CAG ACC ACT ATG.

The nanobarcodes were made according to the procedure described previously (Nicewarner-Pena et al., 2001; Reiss et al., 2002; Walton et al., 2002). Alternating layers of gold and silver were electroplated into the pores of an alumina template. The template was then subjected to treatment by a strong base. This released the nanoparticles with dimensions of 6 nm × 250 nm.

Sha et al. (2007) report that their Cy5-terminated HCV oligo probes emitted a strong SERS spectrum when assembled on the nanobarcodes. The SERS spectrum emitted from a pure Cy5 is similar to that emitted from a Cy5-HCV probe. The authors report that upon binding to the target sequence a significant decrease in the SERS signal is measured. When a target is present, the SERS signal is relatively unchanged. These authors were careful to note that the SERS signal was not due to other causes. For example, the authors confirmed that the oligo remained attached to the NBC, and the dye-labeled oligo was not stripped from the surface of the particle.

Sha et al. (2007) point out that the dual-modality of their technique is that when the HCV target is present (a) the SERS signal decreases, and (b) the fluorescence signal increases. This dual-modality, the authors report, could serve as a built-in control for rapid diagnostic tests. Also, their NBC substrates could be used in multiplex testing, wherein different analytes (targets) may be detected simultaneously.

The authors have demonstrated the strong inverse correlations between their SERS and fluorescence detection mechanisms by adding different concentrations of the target to the NBC-SERS beacon. The authors conclude that their technique clearly shows the possibility of developing a multiplexed SERS beacon using a Raman microscope with reflectance and fluorescence capabilities. Their assay should be able to detect pathogens, and be useful in monitoring (a) the environment, (b) health-care, and (c) bio- and chemical terrorism. Additional advantages are that their method possesses multiplexing capabilities and is label-free.

5.8 Nanotube-Based Biosensor for the Detection of Disease-Specific Autoantibodies in Human Serum (Drouvalakis et al., 2008)

Drouvalakis et al. (2008) have recently developed a peptide-coated nanotube biosensor for the detection of rheumatoid arthritis (RA)-specific autoantibodies. The authors immobilized an RA-specific citrulline-containing peptide to functionalized SWCNT (single-walled carbon nanotubes) on a QCM (quartz crystal microbalance) sensing crystal. Using QCM sensing, the authors detected the antibody binding from RA patients' serum using their nanotube-based biosensor. They report that their nanosensor is sensitive enough to detect in the femtomol range. Besides, their nanosensor is more sensitive than standard ELISA and current microarray assay methods. For example, they were able to detect 30% more (34.4 and 37.5) of anti-citrullinated peptide from RA-patients using ELISA and microarrays, respectively. The authors report that their nanosensor may be used for both diagnostic and research purposes.

Drouvalakis et al. (2008) report that SWCNTs (single-walled carbon nanotubes) exhibit electronic, mechanical, and optical properties (Kroto et al., 1985; Lijima, 1991; Ebbesen and

Ajayan, 1992; Harmada et al., 1992). These SWNTs have been used as nanoscale probes and sensors in electronic (Niu et al., 1997; Baughman et al., 1999; An et al., 2001), and in biological devices (Mattson et al., 2000, Williams et al., 2002). Drouvalakis et al. (2008) point out other applications of SWNTs, including membrane channels (Hummer et al., 2001; Park et al., 2003; Zhu and Schulten, 2003), molecular tweezers (Kim and Lieber, 1999), probes for imaging biomolecules (Wong et al., 1998; Woolley et al., 2000), and for biosensors (Ng et al., 2001; Chen et al., 2003; Sotiropoulos et al., 2003).

The authors report that immobilization of proteins on nanotubes is simple. No specific chemical linkages which may alter the peptide conformation and subsequent functionality are required, as in nanowires (Zheng et al., 2005) and self-assembled monolayers (Chou et al., 2002; Shen et al., 2005). Thus, protein bioactivity may be retained (Fu et al., 2002; Chen et al., 2003). They further report that their biosensors using antigen-coated nanotubes may be used to detect specific autoantibodies in serum. This, they point out, is a critical advantage, especially the detection of specific auto-antibodies in the serum of patients afflicted with the autoimmune disease, RA. Arnett et al. (1998) report that the diagnosis is based on the detection of biomarkers (Rheumatoid factor, RF) and other clinical features. Studies (Schellekens et al., 1998; Girbal-Niehauser et al., 1999; Nakamura, 2000), show the relevance of citrullinated peptides (citrulline-containing peptides) in RA.

The measurement of these citrullinated peptides, Drouvalakis et al. (2008) point out, is performed by Western Blotting and ELISA, using various synthetic cyclic citrullinated peptides (CCPs) or modified proteins as antigens (Jaroszewski et al., 1996; Schellekens et al., 1998; Nogueira et al., 2001). Drouvalakis et al. (2008) also report that their QCM-based nanosensor, using the peptide-coated nanotube technique, assays human serum, and exhibits better performance than that exhibited by the gold-standard technique, ELISA, or currently available microarray technology (Hueber et al., 2005).

Drouvalakis et al. (2008) selected a cohort of RA patients for testing serum (32 samples) which had been recently diagnosed with RA (less than a year), in accordance with the ACR criteria. They also used two control groups for comparison:

(a) A normal control group consisting 13 serum samples from a normal group
(b) 11 serum samples from patients who had osteoarthritis (nonautoimmune arthritis)

The authors formed a carbon nanotube film on a QCM sensing device by depositing a total of 100 μl of 50 μg/ml SWNT suspension in chloroform drop-wise. This was followed by baking for an hour at 50 °C. The next step involved the immobilization of the antigen on top of the film, using a layer of carboxy-terminated Tween 20 (Tween-COOH) which contained polyethylene glycol (PEG) units and carboxylic acid groups. The PEG minimized NSB and the carboxylic acid groups provided the subsequent attachment. PEG also served as a spacer; it separated the antigen from the denaturing effect of the hydrophobic nanotube surface.

Drouvalakis et al. (2008) report that the detection of the serum antibodies binding to the antigen-functionalized carbon nanotubes was undertaken at the third harmonic resonance of the sensing crystal (QCM). The binding of the antibodies to the immobilized antigen resulted in a mass uptake, which subsequently decreased the natural vibration of the nanotubes. This was then measured as a change in the frequency of the quartz crystal.

They also report that the antibody reactive epitope of the citrullinated peptide is conserved upon immobilization to the nanotubes. This preserves the conformation of the antibody, minimizing the loss of relevant binding sites. This was noted by the higher reactivity of the RA serum to the citrullinated peptide than to the noncitrullinated peptide.

They presented further evidence of the preservation of antibody recognition sites on the nanotube surface. They noted a higher number of RA patients with antibodies to the citrullinated peptides (71.8%) compared to healthy patients (7.7%) and osteoarthritis patients. They report that autoantibodies may also be found in 15% of healthy individuals. However, less than 1% of healthy individuals possess autoantibodies to citrullinated antigens.

Drouvalakis et al. (2008) report that their nanotube-based biosensor is more sensitive than either ELISA or the microarray systems, as a greater number of RA patients were found to be positive for anti-citrullinated peptide antibodies by their nanotube method, compared with the two detection methods mentioned above. For example, their test for sensitivity, using the nanotube method, was 71.9% for the detection of anti-citrullinated peptide antibodies, compared to ELISA (37.5%) and microarray systems (34.4%). They conclude by pointing out that due to the high sensitivity and specificity demonstrated by their nanotube-based biosensor, this is an effective method for the detection of RA-specific autoantibodies. Their method is also more time efficient and is a cost-effective assay as it excludes (a) the need for a secondary molecule, and (b) the labeling of sera or a protein component to detect the antigen-antibody binding as required by either ELISA or the microarray system. Also, their nanotube biosensor detects proteins in the femtomolar range, which exceeds that detected by ELISA.

5.9 Gold Nanoparticle Amperometric Immunosensor (Biosensor) for OPG (Singh et al., 2008)

Singh et al. (2008) have recently fabricated an amperometric immunosensor for the detection of osteoproreogerin (OPG) by depositing gold nanoparticles (Au NPs) on a functionalized conducting polymer (CP) (5,2′5′,2″-terthiophene-3′-carboxylic acid). Monoclonal antibody (anti-OPG) was covalently attached to the Au NPs. Cyclic voltametry was used to electrochemically deposit the Au NPs on the CP. The authors used X-ray photoelectron spectroscopy (XPS) to confirm the immobilization of the anti-OPG. Their biosensor is based on a competitive immunoassay between free-OPG and labeled-OPG for the active sites of anti-OPG. They report that their biosensor exhibited a linear range between 2.5 and 25 pg (pictogram)/ml. The detection limit was 2 pg/ml. The authors also used their biosensor to detect OPG in human samples.

The authors report that OPG is a glycoprotein and a member of the TNF (tumor necrosis factor) receptor super family (Hofbauer and Schoppet, 2001). It is related to bone growth. OPG reduces or prevents increased bone resorption (Blair et al., 2006). Increase in bone resorption increases bone mineral density and bone volume (Hofbauer and Schoppet, 2004). Boyle et al. 2003 have reported that OPG is a biomarker for lytic bone metases. Singh et al. (2008) also indicate that OPG is linked to osteoporosis (OP), as a result of which the bones become porous and fragile (Atkinson, 1964; Trueta, 1966; Singh et al., 2006). The very low levels of OPG present in serum make it very difficult for the diagnosis and prognosis of OP (Boyle et al., 2003). Though ELISA is used to detect OPG (Chen et al., 2001), it is time-consuming and tedious, and requires professional technicians. Singh et al. (2008) also point out that a continuous monitoring of OPG in serum is required. Turner (1997) reported as early as 1997 that immunosensors may be used as an alternate method to detect OPG in serum. Skladal et al. (2005) also recommended the use of RANKL (receptor activator of NF-KB ligand)-based biosensors and real-time piezoelectric immunosensors for OPG detection *in lieu* of the ELISA method.

Singh et al. (2008) also report that an electrochemical immunosensor may be a good alternate to ELISA for sensing OPG because of its low cost, simplicity, rapid measurement capability, and portability (Heineman and Halsall, 1985). Darain et al. (2003) have pointed out that an amperometric biosensor with enzyme label is a good method to detect biomolecules because of its quick response and high sensitivity.

Singh et al. (2008) have described the application of CPs to optical devices, energy conversion devices, and biosensors (Cosnier, 1991; Park, 1997; Doblhofer and Rajeshwar, 1998). CPs are attractive for the fabrication of biosensors because they have functional groups such as –COOH or $-NH_2$. Biomolecules may be easily attached to these groups (Cosnier, 1999; Lee and Shim, 2001; Ban et al. 2004; Rahman et al., 2005; Kwon et al., 2006a). Singh et al. (2008) point out that to fabricate a sensitive biosensor, stable immobilization procedures for the active biomolecules are necessary.

They have developed an amperometric OPG biosensor (immunosensor) by covalently immobilizing an antibody onto silver nanoparticles (Au NPs) deposited poly TTCA (5,2′5′,2″-terthiophene-3′-carboxylic acid) modified electrodes. Theyreport that Au NPs deposited CP layers exhibit high electrocatalytic activity, increasing conductivity, and sensitivity (Ivnitski and Rishpon, 1996; Katz and Willner, 2004; Wang et al., 2006; Shiddiky et al., 2007). They have optimized the performance of their OPG biosensor by analyzing the following biosensor parameters such as anti-OPG amount, incubation time, pH, and applied potential in chronoamperometric measurements.

Singh et al. (2008) coated the antibody-immobilized Au NPs/poly TTCA on a GCE (glassy carbon electrode). Ag/AgCl (in saturated KCl) and a platinum (Pt) wire were used as reference and counter electrodes, respectively. The measurements were carried out in a batch-wise mode. The authors carried out the chronoamperometric measurements at a potential of −0.4 V applied to the HRP (horse radish peroxidase)-OPG/anti-OPG/Au NPs/poly TCCA. This permitted the reduction of hydrogen peroxide added to the cell. The authors state that the basic principle of their immunosensor is that "the added hydrogen peroxide reduced by the labeled HRP generates a cathodic current. This current decreases with the addition of free-OPG. This is because of the competition between the free- and labeled-OPG for the active site of the antibody" (Rahman et al., 2004).

Singh et al. (2008) have confirmed the labeling of the horse radish peroxide to the OPG by recording cyclic voltagram (CV) for a HR-OPG/anti-OPG/poly TCCA modified GCE in PBS solution. They attempted to optimize the incubation time in the 5-35 min range, and noted that at 30-min a maximum signal is obtained. Apparently, the biosensor surface was saturated with labeled antigens. Therefore, a 30-min incubation time was used for further studies.

Similarly, a pH range of 5.5-8.5 was analyzed on their biosensor response. The authors obtained the best response at pH 7.4. They report that after a pH of 7.4, the response is poor (less than the maximum). The reasons are:

(a) Loss of anti-OPOG or HRP activity.
(b) The electrocatalytic reduction of hydrogen peroxide by the horseradish peroxidase may be decreased due to poor enzyme activity. Also, the maximum response was obtained at an applied potential of −0.4 V.

Singh et al. (2008) report that they have successfully fabricated an OPG biosensor (immunosensor) by covalently immobilizing anti-OPG on the Au NPs deposited poly TTCA (5,2′,5′,2″-terthiophene-5′-carboxylic acid) layer .The authors were able to apply their fabricated biosensor to detect OPG from serum samples of OP patients. This is a seriously debilitating and gradually-progressing disease. Any efforts spent in helping detect this disease in its early stages will make for a better prognosis. Of course, one also has to keep in mind the simplicity and the cost-effectiveness of this technique, especially if continuous monitoring is required.

5.10 Label-Free Antigen-Antibody Binding on a Gold Nanoparticle Sensor Array (Olkhov and Shaw, 2008)

Olkhov and Shaw (2008) have recently developed a gold nanoparticle sensor array. They used light-scattering properties to demonstrate the changes in the local refractive index of the surface. The authors bound fibrinogen or BSA to each array spot. Scattered radiation was used to monitor the change in the refractive index as the antibodies bound in a label-free fashion to the antigens immobilized on the gold nanoparticle sensor surface. They noted that the maximum sensitivity for detecting antibody concentration is 100 nM. A kinetic analysis was obtained for the antigen-antibody binding.

The authors report that the gold surface plasmon resonance biosensor is popular. McFarland et al. (2003) and Stuart et al. (2004) have suggested that nanoparticle-based biosensors may serve as a useful alternative to detect a wide variety of analytes. Olkhov and Shaw (2008) point out that when the photon frequency is in resonance with a localized surface plasmon resonance mode in the conduction band of nanoparticles, these noble metal nanoparticles exhibit a strong optical extinction in the visible region of the electromagnetic spectrum. Both scatter and absorption contribute to this optical extinction. Noguez (2007) has indicated that these nanoparticles may be tuned and thereby optimized for a given application by changing the fabrication methods for nanoparticles. Note that Noguez (2007) reports that the optical properties of the nanoparticles may be changed by changing their composition, shape, and size. Researchers (Heaton et al., 2001; Haes et al., 2004) have indicated that biological binding events may be monitored by biosensor platforms that use either nanoparticle arrays or single particles.

Olkhov and Shaw (2008) affirm that high-throughput screening is an area where nanoparticle arrays may be used effectively. They further confirm that label-free surface plasmon detection techniques provide the flexibility of not only noting the presence or absence of a particular biomolecule, but also help determine its concentration. Furthermore, the advantage of an array is that it helps determine the concentrations of different molecules simultaneously. The authors indicate that a pattern of molecular expression may thus be generated (for example, biomarkers). This, they point out, could facilitate personalized medicine for patients. Haes and van Duyne (2004) indicate that there is a large body of information for single-target molecule analysis. For multiple analyte detection, Olkhov and Shaw (2008) emphasize that there is much less information available. Lee et al. (2006) and Phillips et al. (2006) have used the (SPR) surface plasmon resonance imaging technique to determine protein concentrations in the 1 nM range. Olkhov and Shaw (2008) have fabricated multiple-target sensor arrays using gold nanoparticles. They were able to detect antibodies in whole anti-sera by using a light-scattering sensor array reader. They point out that pure light scattering techniques have not been used previously in microarray imaging applications.

Olkhov and Shaw (2008) fabricated arrays of gold nanoparticles on glass slides functionalized with target molecules. They report that the array was imaged in a near-field configuration and the scattered light collected by a camera. A real-time kinetic analysis was possible by noting the changes in the scattered radiation intensity, compared with a control spot intensity. This was done as the target analytes flow over the entire array surface. The authors synthesized the biophotinic surface on the glass surface of a microscopic slide using the two-step mediated growth method. This is similar to the wet-chemical synthesis of gold-rod shaped nanoparticles proposed earlier (Jana et al., 2001; Murphy et al., 2005). They point out that the three most important variables during surface synthesis are:

(a) The electrochemical coupling between gold, silver, and ascorbic acid,
(b) time,
(c) and temperature.

Noguez (2007) had reported earlier that these variables affect the size and shape of the nanoparticles, and subsequently their optical properties. Murphy et al. (2005) have pointed out that the presence of silver ions in solution during the colloidal growth phase helps improve the yield of the gold nanorods.

Olkhov and Shaw (2008) analyzed the antigen-antibody binding between BSA and anti-BSA, and between human fibrinogen (HFG) and anti-HFG. They functionalized the biophotonic surface using the Hermanson et al. (1992) two-step DTSP (dithiobis-succinimidyl propionate) coupling chemistry procedure. They also constructed a simple reader to monitor the intensity of the scattered light from each array. They reported that the sensitivities in their array-based system is a trade-off between resolution and throughput (Rich and Myszka, 2007).

They also point out that there are intensity variations in each of the array spots due to the following reasons:

(a) bulk solution and transparency,
(b) laser power fluctuations, and
(c) camera collection efficiency, for example, shutter-opening time.

NSB may also lead to biological noise. The authors report that there is no decrease in signal when the sensor array exposed to the antibody solution is exposed to buffer. This indicates that antibody-antigen dissociation is negligible. Using a simple kinetic model, Olkhov and Shaw (2008) were able to show that the binding (adsorption, k_a) constants were $(2.5 \pm 0.6) \times 10^3\ M^{-1}s^{-1}$ and $(6.6 \pm 0.6) \times 10^3\ M^{-1}s^{-1}$, respectively. The heterogeneity or the fractal dimension on the sensor surface was not taken into account.

Olkhov and Shaw (2008) report in conclusion that their array reader shows that nanoparticle light scattering may be used to interrogate a biomarker printed on an array spot on a bio-photonic surface. They emphasize that the sensitivity to NSB is reduced due to the short-range nature of the nanoparticle plasmon field which typically penetrates one radius into the medium above the particle. In comparison, the continuous surface plasmon resonance propagates typically 300 nm into the medium above the metal. They also point out that the localization of the particle plasmon offers distinct advantages over the continuous surface biosensor.

5.11 Electrochemical Immunosensing Using Magnetic Beads and Gold Nanocatalysts (Selvaraju et al., 2007)

Selvaraju et al. (2007) have recently developed a nanosensor using a nanocatalyst-based electrochemical immunoassay, with magnetic beads and gold nanocatalysts. The magnetic beads were conjugated with IgG and permitted the easy separation of the target proteins, and allowed for a rapid immunosensing reaction. Note that the gold nanoparticles conjugated with the IgG permit the amplification of the electrocatalytic species via the catalytic reaction of the gold nanoparticles. The immunosensing complex that is formed is bound to an indium tin oxide (ITO) electrode which is modified by ferrocenyl-tethered dendrimers (Fc-Ds). This attraction was due to an external magnet. The authors report that their nanosensor exhibits an extremely low detection limit of 1 fg/ml in cyclic voltametric experiments.

The authors also report that there has recently been improvement in sensitivity and a reduction in detection time in the immunosensors (Bange et al., 2005; Marquette and Blum, 2006). The use of magnetic beads as well as gold nanoparticles is the primary reason for this improvement (Nam et al., 2003; Rosi and Mirkin, 2005; Zhang et al., 2006a,b; Willner et al., 2007). According to them, the advantages of using magnetic beads include: (a) easy separation and localization of target proteins by an external magnet, (b) fast antigen-antibody reactions, and (c) low NSB due to surface modification (Richardson et al., 2001; Matsunga and Okamura, 2002; Hsing et al., 2007). They point out that the gold nanoparticles permit a sensitive detection due to the unique optical and catalytic properties exhibited. There is also an easy conjugation of the biomolecules (Castaneda et al., 2007; Zacco et al., 2006).

Selvaraju et al. (2007) explain that the bio-barcode is a highly sensitive tool for the detection of proteins and DNA. It uses both a magnetic bead and a gold nanoparticle. In other words, the magnetic bead is conjugated with a capture molecule and a gold nanoparticle. This also has a high ratio of signaling DNA to probe (Nam et al., 2003; Jaffrezic-Renault et al., 2007). The two particles help sandwich the target molecule.

Porstmann and Kiessig (1992) report that enzymes may be used as labels to generate colorimetric, fluorometric, and chemiluminometric signals. Bangs (1996) reports that magnetic beads-based assays have become standard formats in high-throughput assays, and are able to detect analytes in the pictogram/ml range. Electrochemical immunosensors using magnetic beads have been able to achieve low detection limits. The advantages of using electrochemical methods are ease of operation and miniaturization. For example, Wang et al. (2004a,b) were able to measure electroactive labels such as purine bases of DNA at 2 pg/ml Metal ions originating from metal nanoparticles were detected at 0.1 ng/ml (Mao et al., 2007). Furthermore, detection of electroactive species amplified by enzyme labels was achievable at 26 ng/ml detection limit (Thomas et al., 2003). Using carbon nanotube labels containing multiple enzymes, a detection limit of 10-500 fg/ml was achieved.

Selvaraju et al. (2007) point out that rapid and highly selective reactions can be applied to signal amplification in biosensors using semi-heterogeneous nanocatalysts (Daniel and Astruc, 2004; Astruc et al., 2005). Das et al. (2006) and Lin et al. (2006) have used catalytic reactions mediated by a nanocatalyst to detect proteins and nucleic acids. Selvaraju et al. (2007) report that an ultralow detection limit is achievable using their magnetic beads assay. Their method exploits the ease of separation and immunoreactions of the magnetic beds, followed by signal amplification by the gold nanoparticles.

The authors used 8.1 $\pm$ 0.8 nm size nanoparticles. They also covalently conjugated IgG to the magnetic bead-Tosyl. According to the manufacturers, 40 μg/ml per mg of magnetic beads is optimal for the conjugation of IgG to the magnetic beads. They further indicate that an anti-mouse IgG-gold nanoparticle conjugate was prepared by direct adsorption of the IgG on the 8.1 nm gold nanoparticles (Roth, 1982; Hermanson et al., 1996; Katz and Willner, 2004).

They used 8.1 $\pm$ 0.8 nm gold nanoparticles. They report that a concentration of 40 μg IgG per I mg of magnetic beads was generally used to conjugate IgG to the magnetic beads. They further report that the IgG-magnetic beads conjugate is stable for at least 3 months if stored at 4 °C. IgG was also directly adsorbed on the 8.1 nm gold nanoparticles to prepare the anti-mouse IgG-AuN conjugate (Roth, 1982; Hermanson, 1996; Katz and Willner, 2004). The authors noticed that there was no significant change in the activity of the IgG-AuN (gold nanoparticles) for about thirty days when the conjugate was used in the immunoassay.

To prepare the Fc-D (ferrocenyl-tethered dendrimer) modified electrode, the Fc-D was immobilized on an ITO, using covalent bonding between the dendrimer amines and the carboxylic acids of a phosphonate dendrimer assembled monolayer (Kim et al., 2003; Kwon et al., 2006a,b). Selvaraju et al. (2007) point out that bare ITO electrodes are not good for the *p*-aminophenol (AP) as electrooxidation occurs at high potentials, where the background current for electrolytes is high. Fc-D may be used to shift the oxidation potential to a less positive value. These authors also indicate that the target protein is captured by both an IgG-magnetic bead conjugate and an IgG-AuN conjugate. This forms the immunosensing complex. This immunosensing complex is attracted to the Fc-D modified ITO electrode by an external magnet. Praharaj et al. (2004) and Das et al. (2006) have reported that the gold nanoparticles of the immunosensing complex generates *p*-AP by the catalytic reduction of *p*-nitrophenol. This is a very fast reaction. This, along with the redox cycling of AP, note Selvaraju et al. (2008), promotes signal amplification.

They recommend that to obtain low detection limits, the NSB between the Ig-G-gold nanoparticles (AuN) and the Ig-G-magnetic beads (MB) along with the background current should be minimized. They also report that the higher density of the immunosensing complexes near the Fc-D modified electrode leads to a higher concentration of the *p*-AP generated by the gold nanoparticles (Au-N). Subsequently, that increases the anodic current of the AP.

The authors report that their gold nanoparticle nanosensor exhibits a detection limit for mouse IgG of 100 ag/ml (0.7 attamoles). This according to them is comparable to that exhibited by the bio-barcode assay (Nam et al., 2003), and is of an order of magnitude lower than that exhibited by cyclic voltametry.

Selvaraju et al. (2007) claim in concluding that they have developed a nanosensor that uses magnetic beads for a nanocatalyst-based immunoassay. Their sensor exhibits an efficient immunoreaction and separation along with ultrasensitive detection. Furthermore, their biosensor not only provides a low level of NSB of the IgG-AuN conjugate but also facilitates the capture of the IgG. Also, the IgG-AuN permit a high signal amplification via the catalytic reduction of *p*-nitrophenol. Their nanosensor may be used to detect a wide range of concentrations in a single assay format.

5.12 Conclusions

Examples of nanobiosensors that have recently appeared in the open literature and those that have been recently presented in a couple of conferences have been presented here, as well as some that have recently appeared in the recent literature. As indicated earlier, the examples were selected at random, and those that were briefly analyzed may be classified into three board categories depending on the analyte detected, as shown below:

(a) Medical Applications	
Cancer	Jokerst et al. (2009)
Glucose	Liu et al. (2009)
PSA	Nagatani et al. (2006)
Reactive oxygen species	Henderson et al. (2009)
Autoantibodies	Drouvalakis et al. (2008)
(b) DNA/RNA Applications	
DNA/RNA	Li et al. (2009)
Viral RNA	Sha et al. (2007)
(c) Other Applications	
Antigen-antibody binding	Olkhov and Shaw (2008)
Electrochemical immunosensing	Selvaraju et al. (2007)

These examples provide a perspective of the different nanotechnology techniques and processes that have been recently used to develop effective nanobiosensors. It should be emphasized that the examples were selected at random, and the brief classification presented above may not be a "complete picture" of the different categories where nanobiosensors are presently being used. Many more examples, and perhaps other categories, may be included, that have recently or in the -too-distant past have appeared in the open literature. Nevertheless, with the classification presented here, the predominance of medical applications is obvious. With the current emphasis on the detection of biomarkers for different diseases such as MI (myocardial infarction), and autoimmune diseases such as RA (rheumatoid arthritis) and SLE (systemic lupus erythematosus), medical applications of biosensors will continue to dominate the market, and the application of nanobiosensors in this area is not to be entirely unexpected. Of course, the quantitative detection of glucose in blood to manage DM (diabetes mellitus) will, as expected, continue to dominate the biosensor market, thanks to the increasing obesity and minimum, if any, time spent by individuals on exercise.

Similarly, the papers that have been presented recently (in late 2009) at conferences may be classified into similar categories for the detection of different analytes:

(a) Medical Applications	
Biomedical applications	Chomuchka et al. (2009)
Cancer	Ma et al. (2009)
Glucose	Zhu et al. (2009)
Cardiac markers	Szymanski and Porter (2009)
(b) DNA	
Unamplified DNA	Verdold et al. (2009)
(c) Other Applications	
pH sensing	Lee et al. (2009)
SERS substrates	Lin et al. (2009)
Luminescence enhancement	Chowdhury et al. (2009)
Humidity sensing	Jia et al. (2009)
DNA hybridization	Wang et al. (2009)
Pathogen detection	Nagaraja et al. (2009)
Biosensing applications	Manther et al. (2009)
Enzyme kinetics	Goluch et al. (2009)
Aptamer-target interactions	Zhang et al. (2009)
Small molecule and protein detection	Zhou et al. (2009)
Hydrogen peroxide	Tian and Dale (2009)
Protein microarrays	van Amerongen et al. (2009)
Mycotoxins	Starodub et al. (2009)

Continued

Sensor applications	Berti et al. (2009)
Food industry	Gonzalez-Cortes et al. (2009)
Biosensor applications	Razumiene et al. (2009)
Electrochemical immubnoassays	Szymanski et al. (2009)

Clearly, in the papers presented at two very recent conferences, the biosensor applications in the "other" category are predominant. Many more examples, selected preferably at random, need to be picked up and placed in appropriate categories. Nevertheless, if one were to be cautious, one may state that it seems that applications of biosensors are expanding, especially in the nonmedical category. This is not to be entirely unexpected, considering the ease of application of biosensors to various areas and fields.

References

ACS, *Cancer Facts and Figures 2008*, American Cancer Society, Atlanta, GA, 2008.

Adzic RR, MW Hsiao, and EB Yaeger, *Journal of Electroanalytical Chemistry*, **266**, 475–485 (1989).

Aitken GR, JR Henderson, SC Chang, CJ McNeill, and MA Birch-Machin, *Clinical Experimental Dermatology*, **32**(6), 722–727 (2007).

Ali MF, R Kirbi, AP Goodney, MD Rodriguez, AD Elllington, DP Nelkirk, and JT Modevitti, DNA hybridization and discrimination of single nucleotide mismatches using chip-based microbead arrays, *Analytical Chemistry*, **75**, 4732–4739 (2003).

An KH, WS Kim, YS Park, JM Moon, DJ Bae, SC Lim, YS Lee, and YH Lee, *Advances in Functional Materials*, **11**, 387–392 (2001).

Arnett FC, SM Edworthy, DA Block, DJ McShane, JF Fries, NS Cooper, LA Healey, SR Kaplan, MH Laing, and HS Luthra, *Arthritis and Rheumatology*, **31**, 9–14 (1998).

Aslan G, B Irer, A Kefi, I Celebi, K Yorukoyla, and A Esen, *International Urology and Nephrology*, **37**, 383–391 (2003).

Astruc D, F Lu, and JR Aranzaes, *Agnew Chemistry Intenational Edition*, **44**, 7852–7872 (2005).

Atkinson PJ, Quantitative analysis of osteoporosis in cortical bone, *Nature*, **201**, 373–375 (1964).

Babior BM, Oxidants from phagocytes: agents of defense and destruction, *Blood*, **64**(5), 959–966 (1984).

Babior BM, RS Kipnes, and JT Curnutte, *Journal of Clinical Investigation*, **52**(3), 741–744 (1973).

Bai Y, Y Sun, and C Sun, *Biosensors and Bioelectronics*, **24**, 579–585 (2008).

Balk SP, YJ Ko, and GJ Bubley, *Journal of Clinical Oncology*, **21**, 383–391 (2003).

Ban C, S Chung, DS Park, and YB Shim, *Nucleic Acids Research*, **32**(13), e110/1–e110/8 (2004).

Bange A, HB Halsall, and WR Heineman, *Biosensors & Bioelectronics*, **20**, 2488–2503 (2005).

Baughman RH, C Cui, AA Zakhidov, Z Iqbal, JN Barisci, GM Spinks, GG Wallace, A Mazzoldi, D de Rossi, AG Rinzier, O Jaschinski, S Roth, and M Kertesz, *Science*, **284**, 1340–1344 (1999).

Becker CFW, Y Marsac, P Hazarika, J Moer, RS Goody, and CM Nemeyer, *Chembiochem*, **8**(1), 32–36 (2007).

Berti F, S Todoros, G Marrazza, G Faglia, D Lakshmi, I Chanella, and APF Turner, Conductive polyaniline (PANI) nanostructures for sensing applications. P2.3.14, *FirstBio-Sensing Conference*, Bristol, United Kingdom, November 10-12, 2009.

Bhasin S, A Zhang, A Coviello, R Jasuja, J Ulloor, R Singh, H Vester, and RS Vasan, *Steroids*, **73**, 1311–1317 (2008).

Bhaskar S, S Singh, and M Sharma, *Journal of Immunological Methods*, **196**, 193–198 (1996).

Bindokas VP, J Jordan, CC Lee, and BJ Miller, *Journal of Neuroscience*, **16**(4), 1324–1336 (1996).

Blair JM, H Zhou, MJ Seibel, and CR Dunstan, *Nature Clinical Practice and Oncology*, **3**, 41–49 (2006).

Boyle WJ, WS Simonet, and DL Lacey, Review article Osteoclast differentiation and activation, *Nature*, **423**, 337–342 (2003).

Brechard S and EJ Tschirhart, *Journal of Leukemia Biology*, **84**(5), 1223–1237 (2008).

Campion A and P Kambhampati, *Chemical Society Reviews*, **27**, 241–250 (1998).

Cao YC, R Jin, J Nam, S Thaxton, and CR Mirkin, *Journal of the American Chemical Society*, **125**, 14676–14677 (2003).

Caplan A and A Kratz, *American Journal of Chemical Pathology*, **117**, 104–108 (2002).

Casella JG, R Tommaso, I Cataldi, AM Salvi, and E Desimoni, Electrocatalytic oxidation and liquid chromatographic detection of aliphatic alcohols at nickel-based glassy carbon modified electrode, *Analytical Chemistry*, **65**, 3143–3150 (1993).

Casella IG and M Gatta, *Electroanalysis*, **13**, 549–554 (2001).

Castaneda MT, S Alegret, and A Merkoci, *Electroanalysis*, **19**, 743–753 (2007).

Chang SC, N Pereora-Rodrigues, JR Henderson, A Cole, F Bedioui, and CJ McNeill, *Biosensors & Bioelectronics*, **21**(6), 917–922 (2005a).

Chang SC, N Pereora-Rodrigues, N Zurgil, JR Henderson, F Bedioui, CJ McNeill, and M Deutsch, *Biochemical and Biophysical Research Communications*, **327**, 979–984 (2005b).

Chen D, NA Sarikaya, H Gunn, SW Martin, and JD Young, *Clinical Chemistry*, **47**, 747–749 (2001).

Chen RJ, S Bangsaruntip, KA Drouvalakis, NW Kam, M Shim, Y Li, W Kim, PJ Utz, and H Dai, *Proceedings of the National Academy of Sciences*, **100**, 4984–4989 (2003).

Chen J, W Zhang, and J Ye, *Electrochemical Communications*, **10**, 1268–1271 (2008).

Chiao DJ, RH Shyu, CS Hu, HY Chiang, and SS Tang, *Journal of Chromatography*, **809**, 37–41 (2004).

Chomuchka J, J Drbohlavova, D Huska, V Adam, R Kizek, and O Schneeweiss, Streptavidin conjugated Fe_2O_3 magnetic nanoparticles for *in situ* biomedical applications. P2.3.10. In *First Bio-sensing Conference*, Bristol, United Kingdom, November 10-12, 2009.

Chowdhury S, V Bhattanobotla, and R Sen, Composition effect of Ag-Cu alloy nanoparticles on luminescence enhancement quenching of vicinal luminophores, paper 419c. In *Annual American Chemical Engineers Meeting*, Nashville, Tennessee, November 8-13, 2009.

Christodoulides N, PN Foriano, CS Miller, JL Ebersole, S Mohanty, P Dharshan, M Griffin, A Lennart, KLM Ballard, CP King Jr., MC Langub, RJ Kryshio, MV Thomas, and JT Modevitti, *Annals New York Academy of Sciences*, **1098**, 411–428 (2007).

Claridge SA, AJ Mastroianni, YB Au, HW Laing, CM Micheel, JMJ Frechet, and AP Alivisatos, Enzymatic ligation creates discrete multinanoparticle building blocks for self-assembly, *Journal of the American Chemical Society*, **130**(29), 9598–9605 (2008).

Clark Jr., LC and C Lyons, Electrode systems for continuous monitoring in cardiovascular surgery, *Annals* of the *New York Academy of Sciences*, **102**, 29–45 (1962).

Clark HA, SLR Barker, M Brasuel, MT Miller, E Monson, S Parus, ZY Shi, A Song, B Thorsrud, R Kopelman, A Ade, W Meixner, B Athey, M Hoyer, D Hill, R Lightie, and MA Philbert, *Sensors & Actuators B*, **51**, 12–16 (1998).

Clark HA, M Hoyer, MA Philbert, and R Kopelman, *Analytical Chemistry*, **71**, 4831–4836 (1999a).

Clark HA, R Kopelman, R Tjalkens, and MA Philbert, *Analytical Chemistry*, **71**, 4837–4843 (1999b).

Cosnier S, Biomolecule immobilization on electrode surfaces by entrapment or attachment to electrochemically polymerized films. A review, *Biosensors & Bioelectronics*, **14**, 443–456 (1999).

Cotton TM, SG Schulz, and RP Van Duyne, *Journal of the American Chemical Society*, **102**, 7960–7962 (1980).

Coupland PG, KA Fisher, DR Jones, and JW Aylott, *Journal of Controlled Release*, **130**(2), 115–120 (2008).

Cross CE, B Halliwell, ET Borish, WA Pryor, BN Ames, RL Saul, JM McCord, and D Harman, *Annals of Internal Medicine*, **107**(4), 526–545 (1987).

Crow JP, *Nitric Oxide*, **1**(2), 145–157 (1997).

Culha M, D Stokes, D Allain, and T Vo-Dinh, *Analytical Chemistry*, **75**, 6196–6201 (2003).

Daniel MC and D Astruc, *Chemical Reviews*, **104**, 293–346 (2004).

Darain F, SU Park, and YB Shim, *Biosensors & Bioelectronics*, **18**, 773–780 (2003).
Das PM RC Bast Jr., *Biomarkers in Medicine*, **2**, 291–303 (2008).
Das J, MA Aziz, and H Yang, *Journal of the American Chemical Society*, **128**, 16022–16023 (2006).
Deng C, J Chen, X Chen, C Xiao, L Nie, and S Yao, Direct electrochemistry of glucose oxidase and biosensing for glucose based on boron-doped carbon nanotubes modified electrode, *Biosensors & Bioelectronics*, **23**, 1272–1277 (2008).
De Ravin SS, N Naumann, EW Cowen, J Friend, D Hilligoss, M Marquesen, JE barlow, KS Barron, ML Turner, JI Gallin, and HL Malech, *Journal of Allergy and Clinical Immunology*, **122**(6), 1097–1103 (2008).
Doblhofer K and K Rajeshwar, *Handbook of Conducting Polymers*, Marcel Dekker, Chapter 20, 1998.
Drouvalakis KA, S Sangsaruntip, W Hueber, LG Kozar, PJ Utz, and H Dai, Peptide coated nanotube-based biosensor for the detection of disease-specific autoantibodies in human serum, *Biosensors & Bioelectronics*, 15 May, pp. 1413–1421 (2008).
Drose S and U Brandt, *Journal of Biological Chemistry*, **283**(31), 21649–21654 (2008).
Du H, M Disney, B Miller, and T Krauss, *Journal of the American Chemical Society*, **125**, 4012–4013 (2003).
Du BA, ZP Li, and CH Liu, *Angew Chemistry International Edition*, **45**(47), 8022–8025 (2006).
Dubertret S and FR Kramer, *Nature Biotechnology*, **19**, 365–370 (2001).
Ebbesen TW and PM Ajayan, *Nature*, **358**, 220–222 (1992).
Elghanian R, JJ Storhoff, RC Mucic, RL Letsinger, and CA Mirkin, Selective calorimetric detection of polynucleotides based on distance-dependent optical properties of gold nanoparticles, *Science*, **277**(5329), 1078–1081 (1997).
Faulds K, WE Smith, and WE Graham, *Analytical Chemistry*, **76**, 412–417 (2004).
Faulds K, L Fruk, DC Robson, TA Enright, E Smith, and D Graham, *Faraday Transactions*, 1–8(2005).
Feld JJ, N Hussain, EC Wright, DE Kleiner, JH Hoofnagle, A Ahlawat, V Anderson, D Hilligoss, JI Gallin, TJ Liang, HL Malech, SM Holland, and T Heller, *Gasteroenterology*, **134**(7), 1917–1926 (2008).
Fe Ru RC, M Meyer, and PC Etchegoin, *Journal of Physical Chemistry*, **110**, 1944–1948 (2006).
Fu K, W Huang, Y Lin, D Zhang, TW Hanks, AM Rao, and YP Sun, *Journal of Neuroscience and Nanotechnology*, **2**, 457–461 (2002).
Girbal-Niehauser E, JJ Duriex, M Arnaud, P Dalbon, M Sebbag, C Vincent, M Simon, T Senshu, C Masion-Bessiere, C Jolivet-Reynaud, M Jolivet, and G Serre, *Journal of Immunology*, **162**, 585 (1999).
Goluch ED, MAG Zeverbersen, B Wolfrum, PS Singh, AWJW Tepper, HA Heering, GW Centers, and SG Lemay, Investigating enzyme kinetics using nanofluidic devices. paper 47d. In *Annual American Institute of Chemical Engineers Meeting*, Nashville, Tennessee, November 8-13, 2009.
Gonzalez-Cortes A, P Yanez-Sederio, and JMP Pigmarron, Gold nanoparticles nanostructured electrochemical biosensors for application in the food industry. P1.2.04. In *First Bio-sensing Conference*, Bristol, United Kingdom, November 10-12, 2009.
Goud AP, PT Goud, MP Diamond, B Gonik, and HM Abu-Soud, *Free Radicals in Biology and Medicine*, **44**(7), 1295–1304 (2008).
Graham D, DG Thompson, WE Smith, and K Faulds, Control of enhanced Raman scattering using a DNA-based assembly process of dye-coded nanoparticles, *Nature Nanotechnology*, **3**(9), 548–551 (2008).
Haes AJ and RP van Duyne, *Analytical and Bioanalytical Chemistry*, **379**(7-8), 920–930 (2004).
Haes AJ, DA Stuart, S Nie, and RP van Duyne, *Journal of Fluorescence*, **14**(4), 355–367 (2004).
Harmada N, S Sawada, and A Oshiyama, *Physical Reviews Letters*, **68**, 1579–1581 (1992).
Harris L, H Fritsche, R Mennel, L Norton, P Raodin, S Tante, MR Somerfeld, DF Hayes, RC Bast Jr., American society of clinical oncology 2007 update of recommendations for the use of tumor markers in breast cancer, *Journal of Clinical Oncology*, **25**, 5287–5312 (2007).
Hava M, T Inoue, Y Koyanagi, T Fukuyama, and H Iki, *Nippon Hoigaku Zasski*, **23**, 222 (1969) (abstract).
Hazarika P, F Kukolka, and C Niemeyer, *Angew Chemistry International Edition*, **45**(41), 6827–6830 (2006).
Heaton RJ, AW Peterson, and RM Georgiadis, *Proceedings of the National Academy of Sciences* USA, **98**(7), 3701–3704 (2001).
Heineman WR and HB Halsall, *Analytical Chemistry*, **57**, 1321A–1331A (1985).
Henderson LM and JB Chappell, *European Journal of Biochemistry*, **217**(3), 973–980 (1993).

Henderson JR, DA Fulton, CJ McNeill, and P Manning, The development and *in vitro* characterization of an intracellular nanosensor responsive to reactive oxygen species, *Biosensors & Bioelectronics*, **24**(12), 15 August, 3608–3614 (2009).

Hermanson GT, AK Mallia, and PK Smith, Immobilized Affinity Ligand Techniques, Academic Press (1992).

Hermanson GT, *Bioconjugate Techniques*, First edition, Academic Press, San Diego, 1996 pp. 593–601.

Hill HD and CA Mirkin, *Nature Protocols*, **1**(1), 324–336 (2006).

Hofbauer LC and M Schoppet, *Journal of the American Medical Association*, **292**, 490–495 (2004).

Hsing IM, Y Xu, and W Zhao, *Electroanalysis*, **19**, 755–768 (2007).

Hu S, JA Loo, and DR Wong, Salivary proteome and its genetic polymorphisms, *Annals of the New York Academy of Sciences*, **1048**, 323–325 (2007).

Hueber W, BA Kidd, BH Toomka, BJ Lee, B Bruce, JF Fries, G Sonderstrup, P Monach, JW Drifhout, WJ van Venrrij, PJ Utz, MC Genovese, and WH Robinson, *Arthritis and Rheumatology*, **52**, 2645 (2005).

Hummer G, JC Rasaiah, and JP Noworyta, *Nature*, **414**, 188–190 (2001).

Ivnitski D and J Rishpon, *Biosensors & Bioelectronics*, **11**, 409–417 (1996).

Jaffrezic-Renault X, C Martelet, Y Chevolot, and JP Cloarec, *Sensors*, **7**, 589–614 (2007).

Jana NR, L Gearhart, and CJ Murphy, *Journal of Physical Chemistry B*, **105**, 4065–4067 (2001).

Jeykumari DRS and SS Narayan, A novel nanobiocomposite based glucose biosensor using neutral red functionalized carbon nanotubes, *Biosensors & Bioelectronics*, **23**, 1404–1411 (2008).

Jia W, Y Wang, J Basu, T Strout, C Barry Carter, and W Lei, Nanoengineered transparent metallic nanofibrous membrane and its application to humidity sensing. paper 419d. In *Annual American Institute of Chemical Engineers Meeting*, Nashville, Tennessee, November 8-13, 2009.

Jiang Q, DA Griffin, DF Barofsky, and K Hurst, *Chemical Research and Toxicology*, **10**(10), 1080–1089 (1997).

Jiang L, HX Zhang, JQ Zhuang, BQ Yang, WS Yang, TJ Li, and CC Sun, *Advances in Materials*, **17**(17), 2066–2070 (2005).

Jokerst JV, A Ramanathan, N Christodoulides, PN Floriano, AA Pollard, GW Simmons, J Wong, C Gage, WB Furmaga, SW Redding, and JT McDevitt, Nano-bio-chips for high performance multiplexed protein detection: determinations of cancer biomarkers in serum and saliva using quantum dot bioconjugate labels, *Biosensors & Bioelectronics*, 15 August, 3622–3629 (2009).

Kang EM and HL Malech, *Immunological Research*, **43**(1-3), 77–84 (2009).

Kang X, Z Mai, X Zou, P Cai, and J Mo, *Analytical Biochemistry*, **363**, 143–150 (2007).

Katakis I and E Dominguez, *Trends in Analytical Chemistry*, **14**, 310–319 (1995).

Katz E and I Willner, *Angew Chemistry International Edition*, **43**(45), 6042–6108 (2004).

Khalil Z and B Khodr, *Free Radicals in Biology and Medicine*, **31**(4), 430–439 (2001).

Khodr B and Z Khalil, *Free Radicals in Biology and Medicine*, **30**(1), 1–8 (2001).

Kim P and CM Lieber, *Science*, **286**, 2148–2150 (1999).

Kim E, K Kim, H Yang, YT Kim, and J Kwak, *Analytical Chemistry*, **75**, 5665–5672 (2003).

Kneipp K, Y Wang, H Kneipp, IT Perelama, I Ltzkan, and RR Dasari, *Physical Reviews Letters*, **78**, 1667–1670 (1997).

Kroto HW, JR Heath, SC O'Brien, RF Curl, and RE Smalley, *Nature*, **318**, 162–163 (1985).

Kwon SJ, E Kim, H Yang, and J Kwak, *Analyst*, **131**, 402–406 (2006a).

Kwon NH, MA Rahman, MS Won, and YB Shim, *Analytical Chemistry*, **78**, 52–60 (2006b).

Lakowicz JR, *Analytical Biochemistry*, **298**, 1–24 (2001).

Lee TY and YB Shim, *Analytical Chemistry*, **73**, 5629–5632 (2001).

Lee HJ, D Nedelkov, and RM Corn, *Analytical Chemistry*, **78**, 6504–6510 (2006).

Lee HJ, AW Wark, and RM Corn, *Analyst*, **133**, 975–983 (2008).

Lee DW, C Takugi, and A Ostafin, Enhanced emission of gold nanoparticles due to electron transfer from surface bound molecules and use in pH sensing, paper 419a. In *Annual American Institute of Chemical Engineers Meeting*, Nashville, Tennessee, November 8-13, 2009.

Li J, S Song, D Li, Y Su, Q Huang, Y Zhao, and C Fan, Multi-functional cross-linked Au nanoaggregates for the amplified optical DNA detection, *Biosensors & Bioelectronics*, **24**(1), 15 July, 3311–3315 (2009).

Lilija H, A Christensson, and U Dahlen, *Clinical Chemistry*, **37**, 1618–1625 (1991).

Lin CH, N Linn, K Liu, J Jun, B Jung, and P Jiang, Periodic plasmonic nanostructures as efficient SERS substrates for biosensing, paper 419b. In *Annual American Institute of Chemical Engineers Meeting*, November 8-13, 2009.

Liu JW and Y Lu, A colorimetric lead biosensor using DNAzyme-directed assembly of gold nanoparticles, *Journal of the American Chemical Society*, **125**(22), 6642–6643 (2003).

Liu J and Y Lu, *Natural Products*, **1**(1), 246–252 (2006a).

Liu J and Y Lu, *Angew Chemistry International Edition*, **125**(22), 6642–6643 (2006b).

Liu W, M Horvath, AB Greytak, Y Zhang, DG Nocera, AY Ting, and MG Bawendi, Compact biocompatible quantum dots functionalized for cellular imaging, *Journal of the American Chemical Society*, **130**, 1274–1284 (2008a).

Liu X, L Shi, W Niu, H Li, and G Xi, Amperometric glucose biosensor based on single-walled carbon nanotubes, *Biosensors & Bioelectronics*, **23**, 187–190 (2008b).

Liu Y, H Teng, H Hou, and T You, Nonenzymatic glucose sensor based on renewable electrospun Ni nanoparticles-loaded carbon nanofiber paste electrode, *Biosensors & Bioelectronics*, 15 July, 3329–3334 (2009).

Lu Y, Y Mei, M Drechsler, and M Ballauf, *Agnew Chemistry International Edition*, **45**, 813–816 (2006).

Lu J, I Do, LT Drzal, RM Worden, and J Lee, *Chemistry of Materials*, **19**, 6240–62346 (2007).

Ma LL, JO Tam, J Qiu, T Wang, JT Jenkins, GD Clarke, K Yokolov, M Feldman, TE Millner, and KP Johnston, Antibody-coated gold nanoclusters (nanoroses) for targeted cancer cellular imaging and therapy, paper 633c. In *Annual American Institute of Chemical Engineers Meeting*, Nashville, Tennessee, November 8-13, 2009.

Malamud D, *Journal of the American Dental Association*, **137**, 284–286 (2006).

Mandel ID, *Journal of the American Dental Association*, **124**, 85–87 (1993).

Manning P, CJ McNeill, JM Cooper, and EW Hillhouse, *Free Radicals in Medicine and Biology*, **24**, 1304–1309 (1998).

Manther S, VA Pedrossa, VA Davis, and DAL Simonian, Layer-by-layer assembly of multiwall nanotube ultrathin films for biosensng applications, paper 467c. In *Annual American Institute of Chemical Engineers Meeting*, Nashville, Tennessee, November 8-13, 2009.

Marquette CA and LJ Blum, *Biosensors & Bioelectronics*, **21**, 1424–1433 (2006).

Matsunga T and Y Okamura, *International Journal of Neuroscience*, **1**, 383–389 (2002).

Mattson MP, RC Haddon, and AM Rao, *Journal of Molecular Neuroscience*, **14**, 175–182 (2000).

Maxwell DJ, JR Taylor, and S Nie, *Journal of the American Chemical Society*, **124**, 906–9612 (2002).

McAuley CB and GG Wildgoose, *Electrochemistry Communications*, **10**, 1129–1131 (2008).

McFarland AD and RP van Duyne, *Nano Letters*, **3**(8), 1057–1062 (2003).

McNeil CJ and P Manning, *Reviews of Molecular Biotechnology*, **82**, 443–455 (2002).

McNeil CJ, KR Greenough, PA Weeks, CH Self, and JM Cooper, *Free Radicals Research Communications*, **17**, 399–406 (1992).

Mirkin CA, RL Letsinger, RC Mucic, and JJ Storhoff, A DNA-based method for rationally assembling nanoparticles into macroscopic materials, *Nature*, **382**(6592), 607–609 (1996).

Mucic RC, JJ Storhoff, CA Mirkin, and RL Letsinger, DNA-directed synthesis of binary nanoparticle network materials, *Journal of the American Chemical Society*, **120**(48), 12674–12675 (1998).

Murphy CJ, TK Sau, AM Gole, CJ Orendorff, J Gao, L Gou, SE Hunyadi, and T Li, *Journal of Physical Chemistry B*, **109**, 13857–13870 (2005).

Nagaraja A, NM Mohanty, and V Berry, Ultrafast highly sensitive label-free pathogen detection via chemically-modified grapheme (GMC) sensors, paper 467b. In *Annual American Institute of Chemical Engineers Meeting*, Nashville, Tennessee, November 8-13, 2009.

Nakamura RM, *Journal of Clinical Laboratory Analysis*, **14**, 305–313 (2000).

Nam JM, CS Thaxton, and CA Mirkin, *Science*, **301**, 1884–1886 (2003).

Nam JM, SI Stoeva, and CA Mirkin, Bio-bar-code based DNA detection with PCR-like sensitivity, *Journal of the American Chemical Society*, **126**(19), 5932–5933 (2004).

Newman JD and APF Turner, *Biosensors & Bioelectronics*, **20**, 2435–2453 (2005).

Ng HT, A Fang, J Li, and SF Li, *Nanoscience and Nanotechnology*, **1**, 375–379 (2001).

Nicewarner-Pena SR, RG Freeman, BD Reiss, L He, DJ Pena, and JD Walton, *Science*, **294**, 137–141 (2001).

Nie S and SR Emory, *Science*, **275**, 1102–1106 (1997).

Niemeyer CM, *Angew Chemistry International Edition*, **40**(19), 3685–3688 (2001).

Niemeyer CM and B Ceyhan, *Angew Chemistry International Edition*, **40**(19), 3685–3688 (2001).

Niu C, EK Sichel, R Hoch, D Moy, and H Tennent, *Applied Physics Letters*, **70**, 1480–1482 (1997).

Nogueira L, M Sebbagg, C Vincent, M Arnaud, B Fournie, A Cntagrel, M Jolivet, and C Serre, *Annals of Rheumatoid Disorders*, **60**, 882–887 (2001).

Noguez C, *Journal of Physical Chemistry C*, **111**(10), 3806–3819 (2007).

Ojani R, J Raoof, and S Fathi, *Electroanalysis*, **20**, 1825–1830 (2008).

Olkhov RV and AM Shaw, Label-free antibody-antigen binding detection by optical sensor array based on surface-synthesized gold nanoparticles, *Biosensor & Bioelectronics*, 14 March, pp. 1298–1302 (2008).

Oppenheim F, G Salih, E SiqueiraWalter, L Zhang, and LW Heimerhorst Eva, *Journal of the New York Academy of Sciences*, **1098**, 22–50 (2007).

Ozcan L, Y Shahin, and H Turk, Non-enzymatic glucose biosensor based on over oxidized polypyrrole nanofiber electrode modified with cobalt (II) phthalocyanine tetrasulfonate, *Biosensors & Bioelectronics*, **24**, 512–517 (2008).

Palazolo-Ballane AM, C Suquet, and JK Hurst, *Biochemistry*, **48**(25), 7536–7548 (2007).

Park SM, In *Handbook of Organic Conductive Molecules and Polymers*, **Vol. 3**, HS Nalwa (Ed.), Wiley, Chichester, United Kingdom, 1997.

Park KH, M Chhowalla, Z Iqbal, and F Sesti, *Journal of Biological Chemistry*, **278**, 50212–50216 (2003).

Park BW, DY Yoon, and DS Kim, Electrochemical biosensor with self-assembled peptide nanotubes encapsulated horseradish peroxidase, paper 647e. In *Annual American Institute of Chemical Engineers Meeting*, Nashville, Tennessee, November 8-13, 2009.

Patolsky F, KT Ranjit, A Lichtenstein, and I Willner, *Chemical Communications*, **12**, 1025–1026 (2000).

Phillips KS, T Wilkop, JJ Wu, RO Al-Kaysi, and Q Cheng, *Journal of the American Chemical Society*, **128**, 9590–9591 (2006).

Polsky R, R Gill, L Kagarnovsky, and I Willner, Nucleic-acid functionalized Pt nanoparticles: catalytic labels for the amplified electrochemical detection of biomolecules, *Analytical Chemistry*, **78**(7), 2268–2271 (2006).

Porstmann T and ST Kiessig, *Journal of Immunological Methods*, **150**, 5–21 (1992).

Praharaj S, S Nath, SK Ghosh, S Kundu, and T Pal, *Langmuir*, **20**, 9889–9892 (2004).

Prestigiacomo AF and TA Stamey, *Scandinavian Journal of Clinical Laboratory Investigation (Supplement)*, **221**, 32–34 (1995).

Qin Y, M Lu, and X Gong, *Cellular Biology International*, **32**(2), 224–228 (2008).

Rahman MA, DS Park, and YB Shim, *Biosensors & Bioelectronics*, **19**, 1565–1571 (2004).

Rahman MA, NH Kwon, MS Won, ES Choe, and YB Shim, *Analytical Chemistry*, **77**, 4854–4860 (2005).

Razumiene J, V Gureviciene, E Voitechovic, J Barkauskas, V Bukauskas, and A Setkus, Carbon nanotube electrode for biosensors, P1.2.39. In *First Bio-sensing Conference*, Bristol, United Kingdom, November 10-12, 2009.

Reim RE and RMV Effen, Determination of carbohydrates by liquid chromatography with oxidation at a nickel (II) oxide electrode, *Analytical Chemistry*, **58**, 3203–3207 (1986).

Reiss BD, RG Freeman, ID Walton, SM Norton, PC Smith, and WG Stonas, *Journal of Electrochemistry Communications*, **522**, 95–103 (2002).

Resch-Genger U, M Grabelle, S Cavaliere-Jaricot, R Nitschke, and T Nann, *Nature Methods*, **8**, 763–775 (2008).

Reynolds RA, CA Mirkin, and RL Letsinger, Homogeneous, nanoparticle-based quantitative colorimetric detection of oligonucleotides, *Journal of the American Chemical Society*, **122**(15), 3795–3796 (2000).

Rich RL and DG Myszka, *Analytical Biochemistry*, **361**, 106 (2007).

Richardson J, P Hawkins, and R Luxton, *Biosensors & Bioelectronics*, **16**, 989–993 (2001).
Roesler J, M Hecht, J Freihorst, ML Lohmann-Matthes, and A Emmendorfer, *European Journal of Pediatrics*, **150**(3), 161–165 (1991).
Rosi NL and CA Mirkin, *Chemical Reviews*, **105**(4), 1547–1562 (2005).
Roth J, *Journal of Histochemistry and Cytochemistry*, **30**, 691–696 (1982).
Salimi A, M Roushani, S Soltanian, and R Hallaj, *Analytical Chemistry*, **79**, 7431–7438 (2007).
Sato I, K Kojima, and T Yamasaki, *Journal of Immunological Methods*, **287**, 137–145 (2004).
Schellekens GA, BAW de Jong, FHJ van den Hoogen, LBA van de Putte, and WJ vanVenrrooij, *Journal of Clinical Investigation*, **101**, 273–281 (1998).
Scheller W, W Jin, E Ehrentreich-Forst, B Ge, F Lisdat, R Buttemeier, U Wollenberger, and FW Scheller, *Electroanalysis*, **11**(10-11), 703–706 (1999).
Sha MY, M Yamanaka, JD Walton, SM Norton, R Stoermer, and CD Keating, *Nanobiotechnology*, **1**, 327–335 (2005).
Shafer-Peltier KE, CL Laynes, M Glucksberg, and RP Van Duyne, *Journal of the American Chemical Society*, **125**, 588–593 (2003).
Shen Z, GA Stryker, RL Mernaugh, L Yu, H Yan, and X Zeng, *Analytical Chemistry*, **77**, 797–805 (2005).
Shiddiky MJA, MA Rahman, and YB Shim, *Analytical Chemistry*, **79**, 6886–6890 (2007).
Singh K, SH Lee, and KC Kim, *Journal of Mechanical Engineering Science and Technology*, **20**, 2265–2283 (2006).
Singh K, A Rahman, JI Sun, KC Kim, and YB Shim, An amperometric immunosensor for osteoproteogerin based on gold nanoparticles deposited conducting polymer, *Biosensors & Bioelectronics,* 15 June, pp. 1595–1601 (2008).
Skladal P, Z Jilkova, I Svoboda, and V Kolar, *Biosensors & Bioelectronics*, **20**, 2027–2034 (2005).
Sofer SA, K Brown, A Ellington, B Frazier, G Garcia-Manero, V Gau, SI Gutman, DF Hayes, B Korte, JL Landers, DK Larson, F Ligler, A Majumdar, N Mascini, D Nolte, Z Rosenzweig, J Wang, and D Wilson, *Biosensors & Bioelectronics*, **21**, 1932–1942 (2006).
Song Y, D Zhang, W Gao, and X Xia, *Chemistry European Journal*, **11**, 2177–2182 (2005).
Sonnichsen C, BM Reinhard, J Liphardt, and AP Alivisatos, *Nature Biotechnology*, **23**, 741–745 (2005).
Sotiropoulos S, V Gavalas, V Vamvakaki, and N Chaniotakis, *Biosensors & Bioelectronics*, **18**, 211–215 (2003).
Starodub NF, N Mel'nichenko, AN Shmireva, IV Pytipenko, and LV Pytipenko, Nanostructured silicon-based biosensor for the determination of some mycotoxins, P2.3.22. In *First Bio-sensing Conference*, Bristol, United Kingdom, November 10-12, 2009.
Stenson PD, M Mort, E Ball, K Howells, A Phillips, and NST Thomas, *Genome Medicine*, **1**(1), Art. 13, 2009.
Stoermer R, K Cederqiust, S McFarland, M Sha, S Penn, and C Keating, *Journal of the American Chemical Society*, **128**, 16892–16903 (2006).
Storhoff JJ, R Elghanian, RC Mucic, CA Mirkin, and RL Letsinger, *Journal of the American Chemical Society*, **120**(9), 1959–1964 (1998).
Storhoff JJ, R Elghanian, RC Mucic, CA Mirkin, RL Letsinger, and GC Schatz, What controls the optical properties of linked nanoparticle assemblies, *Journal of the American Chemical Society*, **122**(19), 4640–4650 (2000).
Streckfus CF, CR Bigler, and M Zwick, *Journal of Oral Pathology and Medicine*, **35**, 292–300 (2006).
Stuart DA, A Haes, AD Mcfarland, S Nie, and RP van Duyne, *Proceedings SPIE*, **5327**, 60–73 (2004).
Summer JP, JW Aylott, E Monson, and R Kopelman, *Analyst*, **127**(1), 11–16 (2002).
Sun Y, H Buck, and TS Mallouk, Combinatorial discovery of alloy electrocatalysts for amperometric glucose sensors, *Analytical Chemistry*, **73**, 1599–1604 (2001).
Szymanski MS and RA Porter, Elecrochemical immunoassay for cardiac markers with magnetic particles as a solid phase and silver nanoparticles as an electroactive label, P1.2.05. In *First Bio-sensing Conference*, Bristol, United Kingdom, November 10-12, 2009.

Szymanski MZ, SL Atrle, and RA Porter, Antibody-immobilization on gold and silver nanoparticles for the use in electrochemical immunosensors. In *First Bio-sensing Conference*, Bristol, United Kingdom, November 10-12, 2009.

Tan W, L Sabet, Y Li, T Yu, PR Klokkevoid, DT Wong, and CM Ho, Optical protein sensor for detecting cancer markers in saliva, *Biosensors & Bioelectronics*, **24**, 266–271 (2008).

Tanaka R, T Yuhi, and N Nagatani, *Analytical and Bioanalytical Chemistry*, **38**, 1414–1420 (2006).

Tian F and N Dale, Bimimetic fabrication of optical nanobiosensor for hydrogen peroxide detection, P2.1.01. In *First Bio-sensing Conference*, Bristol, United Kingdom, November 10-12, 2009.

Trueta J, *Nature*, **209**, 118–119 (1966).

Turner APF, *Nature Biotechnology*, **15**, 421 (1997).

Tyagi S and FR Kramer, *Nature Biotechnology*, **14**, 303–308 (1996).

Vamvakaki V, K Tsagarchi, and N Chaniotakis, Carbon nanofiber-based glucose biosensor, *Analytical Chemistry*, **78**, 5538–5542 (2006).

van Amerongen A, P Noguera, and T Posthuma, Carbon nanoparticles as signal labels in protein microassays, P2.3.04. In *First Bio-sensing Conference*, Bristol, United Kingdom, November 10-12, 2009.

Vassilyev YB, OA Khazova, and NN Nikolaeva, *Journal of Electroanalytical Chemistry*, **196**, 105–125 (1985).

Verdold R, F Ungureanu, D Wasserberg, and RPH Kooyman, Gold nanoparticle assays: towards sensitive detection of unamplified DNA, P2.2.02. In *First Bio-sensing Conference*, Bristol, United Kingdom, November 10-12, 2009.

Wabuyele M and T Vo-Dinh, *Analytical Chemistry*, **77**, 7810–7815 (2005).

Walton ID, SM Norton, A Balasingham, L He, DF Oviso, JD Gupta, and PA Raju, *Analytical Chemistry*, **74**, 2240–2247 (2002).

Wang J, *Chemical Reviews*, **108**, 814–825 (2008).

Wang J, G Liu, and MR Jan, *Journal of the American Chemical Society*, **126**, 3010–3011 (2004a).

Wang J, G Liu, B Munge, L Lin, and Q Zhu, *Angew Chemistry International Edition*, **43**, 2158–2161 (2004b).

Wang HL, W Li, Q Jia, and E Akhadov, *Chemical Materials*, **19**, 52–525 (2006).

Wang ZP, JQ Hu, Y Jin, X Yao, and JH Li, *Clinical Chemistry*, **52**(10), 1958–1961 (2007).

Wang J, DF Turner, and A Chen, Nonenzynmatic electrochemical glucose sensor based on nanoporous PtPb methods, *Analytical Chemistry*, **80**, 997–1004 (2008).

Wang J, T Heckler, P Lei, S Andreadis, and TJ Montzaris, DNA hybridization detection using spectral changes of zinc selenide nanocrystals, paper 419e. In *Annual American Institute of Chemical Engineers Meeting*, Nashville, Tennessee, November 8-13, 2009.

Webster A, SJ Compton, and JW Aylott, *Analyst*, **130**(2), 163–170 (2005).

Webster A, P Coupland, FD Houghton, HJ Leese, and JW Aylott, *Biochemical Society Transactions*, **35**(pt 3), 538–543 (2007).

Weigum SE, PN Floriano, N Christodoulides, and JT Modevitt, *Lab on a Chip*, **7**, 995–1003 (2007).

Williams KA, PTM Veenhulzen, BG de la Tore, R Eritja, and C Decker, *Nature*, **420**, 761 (2002).

Willner I, R Baron, and B Willner, *Biosensors & Bioelectronics*, **9**, 1573–1577 (2007).

Wilson GS and R Gifford, *Biosensors & Bioelectronics*, **20**, 2388–2403 (2005).

Wolkow RA and MJ Moskovits, *Journal of Chemical Physics*, **87**, 5858–5869 (1987).

Wong DT, (Ed.), *Salivary Diagnostics*, Wiley-Blackwell, Hoboken, NJ, 2008.

Wong SS, E Joseivich, HT Woolley, CL Cheung, and CM Lieber, *Nature*, **420**, 52–55 (1998).

Woolley AT, CL Cheung, JH Hafner, and CM Lieber, *Chemical Biology*, **7**, 193–204 (2000).

Wu H, G Liu, J Wang, and Y Lin, *Electrochemistry Communications*, **9**, 1573–1577 (2007a).

Wu L, X Zhang, and H Ju, Amperometric glucose sensor based on catalytic reduction of dissolved oxygen at soluble carbon nanofiber, *Biosensors & Bioelectronics*, **23**, 479–484 (2007b).

Xiulan S, Z Xiaolian, T Jian, J Zhou, and FS Chu, *International Journal of Food Microbiology*, **99**, 185–194 (2005).

Xu H, JW Aylott, and R Kopelman, Fluorescent nano-PEBBLE sensors for the Real-time measurement of glucose inside living cells, *Analyst*, **127**(11), 1471–1477 (2002).

Yuan J, K Wang, and X Xia, *Advances in Functional Materials*, **15**, 803–809 (2005).
Zacco E, M VIdori, S Alegret, R Gaive, and MP Marco, *Analytical Chemistry*, **78**, 1780–1788 (2006).
Zhang WM, P Finne, J Lennonen, S Veaslainen, S Nordling, and UH Stenman, *Clinical Chemistry*, **45**, 814–821 (1999).
Zhang X, MA Young, O Lyandres, and RP Van Duyne, *Journal of the American Chemical Society*, **127**, 4484–4489 (2005).
Zhang H, X Cheng, M Richter, and MI Greene, *Nature Medicine*, **12**, 473–477 (2006a).
Zhang J, SP Song, LY Zhang, LH Wang, HP Wu, D Pan, and C Fan, Sequence-specific detection of femtomolar DNA via chromocoulometic DNA sensor (CDS): effects of nanoparticle mediated amplification and nanoscale control of DNA assembly at electrodes, *Journal of the American Chemical Society*, **128**(26), 8575–8580 (2006b).
Zhang D and J Chang, *Nano Letters*, **8**, 3283–3287 (2008).
Zheng G, F Patolsky, Y Cui, WU Wang, and CM Lieber, *Nature Biotechnology*, **23**, 1294–1301 (2005).
Zhou M, L Shang, B Li, L Huang, and S Dong, Highly ordered mesoporous carbons as electrode material for the construction of electrochemical dehydrogenase and oxidase-based biosensors, *Biosensors & Bioelectronics*, **24**, 442–447 (2008).
Zhu F and K Schulten, *Biophysical Journal*, **85**, 236–244 (2003).
Zhu Z, K Bunigapali, W Song, Y Li, and DF Moussy, Glucose biosensors based on novel carbon nanotube fiber microelectrode, P1.2.06. In *First Bio-sensing Conference*, Bristol, United Kingdom, November 10-12, 2009.
Zulueta JJ, R Sawhney, FS Yu, CC Cote, and PM Hassoun, *American Journal of Physiology*, **272**(5 Pt 1), 897–902 (1997).

CHAPTER 6

Binding of the Same Analyte to Different Biosensor Surfaces

Chapter Outline

6.1 Introduction

A fractal analysis is presented on the binding and dissociation (if applicable) kinetics of the same analyte to different biosensor systems. The intent is to note if (a) further physical insight into these types of systems could be gained, and (b) if there were some reasons why particular ranges of fractal dimensions (degree of heterogeneity) existed on the biosensor surface and in the binding rate coefficients. These physical insights could be of value for analyte-receptor systems, in general, but particularly so in those analyte-receptor systems that exhibit biomedical/medical applications such as glucose, thrombin, α-fetoprotein (AFP) and carcinogenic analytes or those analytes that may be used as cancer biomarkers.

The examples selected for analysis and comparison were done so at random from those that were available in the literature. Simple criteria were used in the selection procedure. There should, of course, be at least two examples available for the same analyte for comparison purposes. More than two examples of the same analyte would be even better for comparison purposes. Concentration (or any other measurable entity made quantitative by the biosensor for that particular analyte-receptor biosensor system) versus time should be available, and

Handbook of Biosensors and Biosensor Kinetics. DOI: 10.1016/B978-0-444-53262-6.00006-1

also be amenable for the fractal-type analysis which is presented in the chapters throughout this book. The fractal analysis of the examples provides the binding and dissociation (if applicable) rate coefficients and fractal dimension values. The comparison of the binding rate coefficients and fractal dimension values for the same analyte interacting on different types of biosensor surfaces as the case may be yields fresh physical insights which is the goal of this particular chapter. Needless to say, because the examples are selected from the literature, one is constrained by what is available in the literature. Often, one hopes that more examples of the binding of the same analyte to other biosensor systems were available in order that one may elucidate or prove a point. It is hoped that this trend catches on, and further analysis of other types of examples would either reinforce the ideas presented in this chapter, or help develop newer ideas which in turn will provide further physical insights into these interactions occurring on the biosensor surfaces. The examples selected at random from the literature for comparison purposes include:

(a) The amperometric detection of *Escherichia coli* using electropolymerization and coating of glass carbon electrodes (GCEs) with pyrrole amine (Pyy-NH_2) (Abu-Rabesh et al., 2009), and the real-time PCR (polymerase chain reaction) amplification of aptamers for the detection of *E. coli* (Lee et al., 2009)
(b) An electrochemical aptamer-based assay coupled to magnetic beads or the detection of thrombin (Centi et al., 2008), and an electrochemical impedance spectroscopy (EIS) biosensor for analyzing aptamer-thrombin interfacial interactions (Li et al., 2008)
(c) An ultrasensitive enhanced CL (chemiluminescence) enzyme immunoassay for detecting AFP which was amplified by double-codified gold nanoparticle (GNP) labels (Yang et al., 2009), and a localized surface plasmon resonance coupled fluorescence (LSPCF) fiber-optic biosensor to detect AFP in human serum (Chang et al., 2009)
(d) A novel biosensor using a modified GCE for the detection of glucose (Sheng et al., 2008), the binding of glucose in solution to the electroless plated Au/Ni/copper low electrical resistance electrode (Lee et al., 2008), the long-term stability of a glucose biosensor based on inserted barrel plating gold electrodes (Hsu et al., 2009a,b,c), and a percutaneous fiber-optic sensor for chronic glucose monitoring *in vivo* (Liao et al., 2008)
(e) An implantable diagnostic device for cancer monitoring (Daniel et al., 2008), and the binding of cancer antigen (CA) 123 in solution to anti-CA antibody immobilized on an SPR biosensor chip surface

Many more of these types of examples are available in the literature. It is hoped that these five sets of examples to be presented together will help set the stage for the analysis of other examples. The intent, as indicated previously, is to obtain better physical insights into these types of examples. Of course, any further insight that may be obtained for biomedically-medically oriented analytes will prove invaluable.

6.2 Theory

6.2.1 Single-Fractal Analysis

Binding Rate Coefficient

Havlin (1989) reports that the diffusion of a particle (analyte [Ag]) from a homogeneous solution to a solid surface (e.g., receptor [Ab]-coated surface) on which it reacts to form a product (analyte-receptor complex; (Ab·Ag)) is given by:

$$(\text{Ab}\cdot\text{Ag}) \approx \begin{cases} t^{(3-D_{\text{f,bind}})/2} = t^p, & t < t_c \\ t^{1/2}, & t > t_c \end{cases} \tag{6.1}$$

Here $D_{\text{f,bind}}$ or D_f is the fractal dimension of the surface during the binding step. t_c is the cross-over value. Havlin (1989) points out that the cross-over value may be determined by $r_c^2 \sim t_c$. Above the characteristic length, r_c, the self-similarity of the surface is lost and the surface may be considered homogeneous. Above time, t_c, the surface may be considered homogeneous, as the self-similarity property disappears, and "regular" diffusion is now present. For a homogeneous surface where D_f is equal to 2, and when only diffusional limitations are present, $p = ½$ as it should be. Another way of looking at the $p = ½$ case (where $D_{\text{f,bind}}$ is equal to 2) is that the analyte in solution views the fractal object, in our case, the receptor-coated biosensor surface, from a "large distance." In essence, in the association process, the diffusion of the analyte from the solution to the receptor surface creates a depletion layer of width $(Ðt)^{½}$ where $Ð$ is the diffusion constant. This gives rise to the fractal power law, (Analyte·Receptor) $\sim t^{(3-D_{\text{f,bind}})/2}$. In the present analysis, t_c is arbitrarily chosen and we assume that the value of t_c is not reached. One may consider the approach as an intermediate "heuristic" approach that may be used in the future to develop an autonomous (and not time-dependent) model for diffusion-controlled kinetics.

Dissociation Rate Coefficient

The diffusion of the dissociated particle (receptor [Ab] or analyte [Ag]) from the solid surface (e.g., analyte [Ag]-receptor [Ab] complex coated surface) into solution may be given, as a first approximation by:

$$(\text{Ab}\cdot\text{Ag}) \approx -t^{3-D_{\text{f,diss}})/2} = -t^p, \quad t > t_{\text{diss}} \tag{6.2}$$

Here $D_{\text{f,diss}}$ is the fractal dimension of the surface for the dissociation step. This corresponds to the highest concentration of the analyte-receptor complex on the surface.

Henceforth, its concentration only decreases. The dissociation kinetics may be analyzed in a manner "similar" to the binding kinetics.

6.2.2 Dual-Fractal Analysis

Binding Rate Coefficient

Sometimes, the binding curve exhibits complexities and two parameters (k, D_f) are not sufficient to adequately describe the binding kinetics. This is further corroborated by low values of r^2 factor (goodness-of-fit). In that case, one resorts to a dual-fractal analysis (four parameters; k_1, k_2, D_{f1}, and D_{f2}) to adequately describe the binding kinetics. The single-fractal analysis presented above is thus extended to include two fractal dimensions. At present, the time ($t = t_1$) at which the "first" fractal dimension "changes" to the "second" fractal dimension is arbitrary and empirical. For the most part, it is dictated by the data analyzed and experience gained by handling a single-fractal analysis. A smoother curve is obtained in the "transition" region, if care is taken to select the correct number of points for the two regions. In this case, the product (antibody-antigen or analyte-receptor complex, Ab·Ag or analyte·receptor) is given by:

$$(\text{Ab}\cdot\text{Ag}) \approx \begin{cases} t^{(3-D_{f1,\,\text{bind}})/2} = t^{p1}, & t < t_1 \\ t^{(3-D_{f2,\,\text{bind}})/2} = t^{p2}, & t_1 < t < t_2 = t_c \\ t^{1/2}, & t > t_c \end{cases} \tag{6.3}$$

In some cases, as mentioned above, a triple-fractal analysis with six parameters (k_1, k_2, k_3, D_{f1}, D_{f2}, and D_{f3}) may be required to adequately model the binding kinetics. This is when the binding curve exhibits convolutions and complexities in its shape due perhaps to the very dilute nature of the analyte (in some of the cases to be presented) or for some other reasons. Also, in some cases, a dual-fractal analysis may be required to describe the dissociation kinetics.

6.3 Results

A fractal analysis is applied to the binding and dissociation (if applicable) kinetics of different analyte-receptor reactions occurring on different biosensor surfaces. Understandably, alternative expressions for fitting the data that include saturation, first-order reaction, and no diffusion limitations are available, but these expressions are apparently deficient in describing the heterogeneity that inherently exists on the surface. Another advantage of this technique is that the analyte-receptor binding (as well as the dissociation reaction) is a complex reaction, and the fractal analysis via the fractal dimension and the rate coefficient provides a useful lumped parameter(s) analysis of the diffusion-limited reaction occurring on a heterogeneous surface.

In the classical situation to demonstrate fractality, one should make a log-log plot, and one should definitely have a large amount of data. It may be useful to compare the fit to some other forms, such as an exponential form or one involving saturation. At present, we do not present any independent proof or physical evidence of fractals in the examples presented. It is a convenient means (since it provides a lumped parameter) to make the degree of heterogeneity that exists on the surface more quantitative. Thus, there is some arbitrariness in the fractal model to be presented. One might justifiably argue that appropriate modeling may be achieved by using a Langmuirian or other approach. The Langmuirian approach has a major drawback because it does not allow for or accommodate the heterogeneity that exists on the surface.

E. coli has recently been detected by different biosensor configurations. Some of the more recent ones that have appeared in the literature include:

(a) Single-cell level detection of *E. coli* in microfluidic device (Han et al., 2008)
(b) Gold screen-printed based impedimetric immunosensors for direct and sensitive *E. coli* quantization (Escamilla-Gomez et al., 2008)
(c) Cy5-labeled antimicrobial peptides for enhanced detection of *E. coli* 0157:H7 (Arcidiacono et al., 2008)
(d) Magnetoresistive immunosensor for the detection of *E. coli* 0157:H7 including a microfluidic network (Mujika et al., 2009)
(e) Nano-silver-modified PQC/DNA biosensor for detecting *E. coli* in environmental water (Sun et al., 2009)
(f) Disposable amperometric immunosensing strips fabricated by Au nanoparticles-modified screen-printed carbon electrodes for the detection of food borne pathogen *E. coli* 0157:H7 (Lin et al., 2008a,b,c)

Abu-Rabesh et al. (2009) have recently developed a highly sensitive disposable amperometric immunosensor for the detection of *E. coli*. The authors used a double layered configuration at the transducer surface that included a polypyrole NH_2-anti *E. coli* antibody (PAE) inner layer and an alginate-polypyrole (Alg-Ppy) outer layer. The authors indicate that in the presence of the substrate PAPG (*p*-aminophenyl β-D-galactopyranoside), the bacterial enzyme, *β*-D-galactosidase produces the PAP (*p*-aminophenol) product. The amperometric signal is generated due to the electrooxidation of the PAP product.

Abu-Rabesh et al. (2009) emphasize the need to detect microorganisms for food and water safety. The presence of *E. coli* may be used as a potential indicator of pathogen presence that originates from humans and animals (Tryland and Fiksdal, 1998). The estimation of coliforms in water samples is essential for the prevention of enteric disease (Buchanan, 1997; Tokarsky and Marshall, 2008) from polluted water supplies that cause public health concerns (Tokarsky and Marshall, 2008).

Abu-Rabesh et al. (2009) indicate that the conventional method to detect *E. coli* based on cell cultures grown in differential agar media followed by counting of the target organism takes 1-3 days (George et al., 2000; Lin et al., 2008a,b,c). This length of time taken to detect the pathogens motivates one to develop a rapid and accurate method for their detection.

Abu-Rabesh et al. (2009) point out that biosensors have been used for the detection of pathogens including *E. coli* (Ivnitskii et al., 1999; Leonard et al., 2003; Tokarsky and Marshall, 2008). Electrochemical biosensors have also been used previously to detect *E. coli* (Abdel-Hamid et al., 1999a,b; Mittelmann et al., 2002; Palenzuela et al., 2004; Lin et al., 2008a,b,c). Abu-Rabesh et al. (2009) indicate that the number of electrochemical platforms for *E. coli* detection by biosensors has grown rapidly (Bakker, 2004; Mehavir and Abidi, 2004).

Abu-Rabesh et al. (2009) have developed an alternative immunosensing approach for the amperometric detection of *E. coli*. Their method is based on electropolymerization and coating of GCEs with polypyrole-amine (Pyy-NH_2). Figure 6.1a shows the binding of 10^7 CFU *E. coli* with lysozyme to the amperometric immunosensor with PAPG. A single-fractal analysis is adequate to describe the binding kinetics. The values of the binding rate coefficient, k, and the fractal dimension, D_f, for a single-fractal analysis are given in Table 6.1.

Figure 6.1b shows the binding of 10^7 CFU *E. coli* without lysozyme (control experiment) to the amperometric immunosensor with PAPG. Once again, a single-fractal analysis is adequate to describe the binding kinetics. It is interesting to note that as the fractal dimension decreases by a factor of 3.2 from a value of D_f equal to 1.6998 to 0.5270, the binding rate coefficient for *E. coli* decreases by a factor of 1.74 from a value of k equal to 0.8083 to 0.4647. Note that changes in the degree of heterogeneity or the fractal dimension and in the binding rate coefficient are in the same direction.

Lee et al. (2009) have developed a method to detect *E. coli* based on immunomagnetic separation and real-time PCR amplification of aptamers. These authors indicate that aptamers are single-stranded DNA or RNA molecules, which can bind to their targets with high specificity and affinities (Ellington and Szostak, 1990; Hermann and Patel, 2000). Lee et al. (2009) point

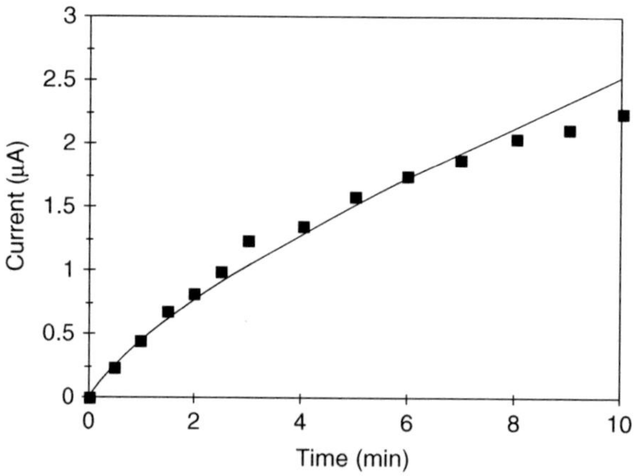

Figure 6.1
Binding of 10^7 CFU/ml *E. coli* to the amperometric biosensor with PAPG as a substrate (Abu-Rabesh et al., 2009).

Table 6.1: Binding and dissociation rate coefficients and fractal dimensions for the binding and the dissociation phase for *Escherichia coli* in solution by two different biosensors.

Analyte in Solution/ Receptor on Surface	k	k_d	D_f	D_{fd}
10^7 CFU/ml *E. coli* with lysozyme/ immunosensor with PAPG	0.4647 ± 0.0520	na	0.5270 ± 0.06676	na
10^7 CFU/ml *E. coli* without lysozyme/immunosensor with PAPG	0.8083 ± 0.0709	na	1.6998 ± 0.0839	na
10^8 CFU/ml *E. coli*/ conjugated magnetic beads	20.063 ± 2.197	3.0656 ± 0.1817	2.0082 ± 0.0760	1.9878 ± 0.0618

out that aptamers can play the role of capture molecules like antibodies in different types of biosensor schemes (Liu and Lu, 2006; Bang et al., 2008). Lee et al. (2009) report an additional advantage of using aptamers vis-à-vis antibodies in the sense that the PCR may be used to increase the detection sensitivity (Zhang et al., 2006).

Lee et al. (2009) point out that though microorganism-specific aptamers to enhance the detection of microorganisms is limited, several aptamers for the detection of *E. coli* are available (So et al., 2008). They selected an RNA aptamer for the detection of *E. coli*. They captured the target *E. coli* on antibody-conjugated magnetic beads, and the aptamers were bound to the surface of the captured *E. coli* by a sandwich method.

Figure 6.2 shows the binding of 10^8 CFU/ml *E. coli* in solution to the antibody-conjugated magnetic beads. The aptamers were amplified by PCR (Lee et al., 2009). A single-fractal analysis is adequate to describe the binding and dissociation kinetics. The binding rate coefficient, k, and the fractal dimension D_f, for a single-fractal analysis, and the dissociation rate coefficient, k_d, and the fractal dimension for dissociation, D_{fd}, are given in Table 6.1.

The aim of this chapter is to compare the binding and the dissociation (if applicable) rate coefficients and the corresponding fractal dimensions for the same analyte on different biosensor surfaces. It is hoped that further physical insights may be gained by such an analysis. It may be noted that the binding of *E. coli* to both the immunosensor with PAPG (Abu-Rabesh et al., 2009), and to the antibody-conjugated magnetic beads is described by a single-fractal analysis. This would indicate that the binding (and dissociation) mechanism is not complex. There is an order of magnitude change in the *E. coli* concentration in solution as one goes from the immunosensor with PAPG (10^7 CFU/ml) to the antibody-conjugated magnetic beads (10^8 CFU/ml) (Lee et al., 2009). Also, no dissociation is exhibited during the binding of 10^7 CFU/ml *E. coli* in solution to the immunosensor. This would indicate that either the binding in this case is strong enough to prevent the dissociation or the *E. coli*

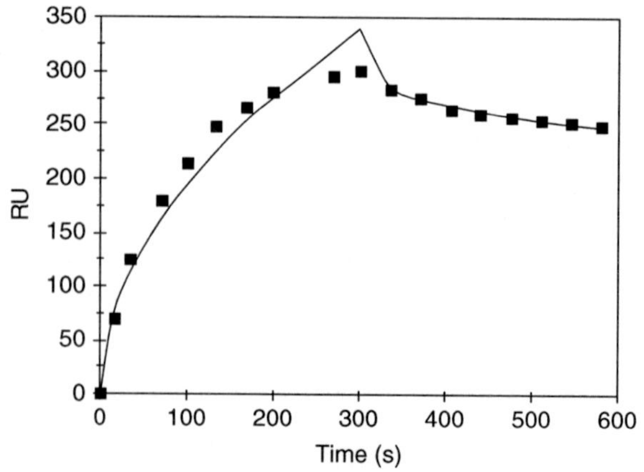

Figure 6.2
Binding and dissociation of the *E. coli* to the aptamer antibody-conjugated magnetic bead biosensor, and amplified by real-time PCR (Lee et al., 2009).

concentration is low. Note also that the fractal dimension exhibited during the binding of 10^8 CFU/ml *E. coli* in solution to the antibody-conjugated magnetic beads is higher by 18% when compared with the fractal dimension exhibited during the binding of 10^7 CFU/ml in solution to the immunosensor. Note that the fractal dimension is based on a log scale and even small changes in the value of the fractal dimension represent significant changes in the degree of heterogeneity on the biosensor surface.

The heterogeneity on the biosensor surface could be the result of various factors. They could include (among others) the heterogeneity of the sensor surface itself, heterogeneities due to the receptors on the surface, or the heterogeneities that arise during the binding of the analyte or in the analyte itself. No attempt is made here to delineate these different causes, or identify them, except to point out that all of these possible heterogeneities on the sensing surface are combined together and described by a single value, the fractal dimension, D_f.

Centi et al. (2008) have recently developed an electrochemical aptamer-based assay coupled to magnetic beads for the detection of thrombin. These authors developed a direct and an indirect competitive assay by immobilizing both an aptamer and a protein. Electrochemical transduction coupled with innovative use of magnetic beads was used. These authors were able to achieve a detection limit of 430 nM of thrombin. Also, these authors were able to achieve a lower limit of 175 nM by detecting the product catalyzed enzymatically by thrombin.

Centi et al. (2008) indicate that aptamers are nucleic acids that may be generated against amino acids, drugs, proteins, and molecules. SELEX (systematic evolution of ligands by experimental enrichment), an iterative procedure that uses binding, separation, and amplification may be used to isolate these aptamers. Centi et al. (2008) point out that though aptamers have appeared in recent literature (Ellington an Szostak, 1990; Tuerk and Gold, 1990;

Tombelli et al., 2004, 2007) and their use for biosensor application as recognition elements have been suggested, very few, if any, practical applications have appeared in the literature. Centi et al. (2008) selected the aptamer (15-mer, GGTTGGTGTGGTTGG-3′) for the binding of thrombin. They also report that the binding interaction between thrombin and the aptamer has been considered a model system ever since the thrombin-binding aptamer G quartet was established (Macaya et al., 1993; Smirnov and Shafer, 2000) and the binding site was identified (Paborsky et al., 1993). This aptamer has been coupled to different transduction platforms to demonstrate its wide applicability for biosensor applications as a bioreceptor (Baldrich et al., 2004; Bang et al., 2005; Gronewold et al. 2005; Hianik et al., 2005; Ikebukoro et al., 2005; Radi et al., 2005; Cai et al., 2006; Mir et al., 2006; Radi et al., 2006; Yoshida et al., 2006; Zhang et al., 2006; Le Floch et al., 2008).

Centi et al. (2008) examined different formats for aptamer-based electrochemical assay, and demonstrated the different optimal assay conditions that may be used with different assay formats for the binding of thrombin to the aptamer coupled to magnetic beads. They emphasize that their technique may also be automated to decrease the assay time (Rashkovetsky et al., 1997).

Figure 6.3a shows the binding of 40 ppm antithrombin in solution to the biotinylated thrombin (aptamer) immobilized on a CMB chip surface (Centi et al., 2008). A single-fractal analysis is adequate to describe the binding and the dissociation kinetics. The values of the binding rate coefficient, k, and the fractal dimension, D_f, for a single-fractal analysis are given in Table 6.2.

Figure 6.3b shows the binding of 20 nM antithrombin in solution to the biotinylated thrombin aptamer immobilized on the surface of a CMB chip. Once again, a single-fractal analysis is adequate to describe the binding kinetics. The values of the binding rate coefficient, k, and the fractal dimension, D_f, for a single-fractal analysis are given in Table 6.2. Note that the values of the binding rate coefficient, k, and the fractal dimension, D_f, for the binding of antithrombin in solution to the thrombin aptamer immobilized on the CMB chip surface are much lower than the corresponding values of binding rate coefficient, k, and the fractal dimension, D_f, for the binding of thrombin in solution to the thrombin aptamer immobilized on the CMB chip surface.

Li et al. (2008) have recently developed an EIS biosensor for analyzing aptamer-thrombin interfacial interactions. These authors have used a thrombin-binding aptamer as the recognition element. The aptamer was immobilized on a GCE platform. They achieved signal enrichment by using GNPs. These GNPs were electrodeposited on the GCE. The interaction of the thrombin was observed by a change in the interfacial electron transfer resistance of their biosensor using a redox couple of $[Fe(CN)_6]^{3-/4-}$ as the probe. The authors noted that the increase in electron transfer resistance of the biosensor is linear in the thrombin concentration in the range of 0.12-30 nM.

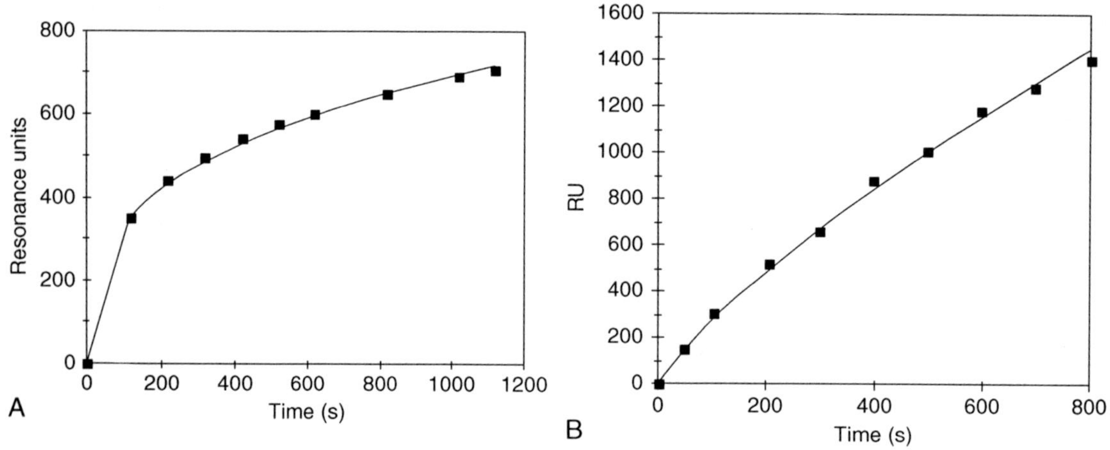

Figure 6.3

(a) Binding of 40 ppm antithrombin in solution to the immobilized biotinylated thrombin on a CMB chip. (b) Binding of 20 nM thrombin in solution to the immobilized biotinylated aptamer on a CMB chip (Centi et al., 2008).

Table 6.2: Binding of (a) 40 ppm antithrombin antibody to the biotinylated thrombin immobilized on a CMB chip (Centi et al., 2008), (b) 20 nM thrombin in solution to an immobilized biotinylated aptamer on a CMB chip (Centi et al., 2008), and (c) binding and dissociation of 60 nM thrombin in solution to an electrochemical impedance spectroscopy (EIS) biosensor (Li et al., 2008).

Analyte in Solution/ Receptor on Surface	k	k_d	D_f	D_{fd}	References
40 ppm antithrombin/ biotinylated thrombin on CMB chip	83.044 ± 1.593	na	2.3862 ± 0.0182	na	Centi et al. (2008)
20 nM thrombin/ biotinylated aptamer immobilized on a CMB chip	7.9747 ± 0.424	na	1.4424 ± 0.03892	na	Centi et al. (2008)
60 nM thrombin/EIS biosensor	122.88 ± 15.23	95.145 ± 22.52	1.810 ± 0.1696	1.716 ± 0.264	Li et al. (2008)

Li et al. (2008) report that antibodies have been widely used as biological recognition elements in biosensors. However, they do have limitations such as their production *in vivo*, limited target analytes, limited shelf life, as well as being subject to thermal denaturation (O'Sullivan, 2002). O'Sullivan et al. (1997) point out that aptamers may be used as biological recognition elements for drugs, enzymes, peptides, and proteins. Ho and Leclerc (2004) and Pavlov et al. (2004) point out that aptamers have been used in optical biosensors.

Li et al. (2008) report that aptamer-based electrochemical biosensors have been used to detect proteins, and are based on aptamers that have been labeled using redox compounds such as methylene blue (Cao et al., 2005). Catalysts such as HRP (horse radish peroxidase) (Monica et al., 2006), and platinum nanoparticles (Polsky et al., 2006) have also been used as signal-producing labels.

Li et al. (2005) and Liu et al. (2005) report that direct electrochemical deposition of GNPs onto an electrode surface is an efficient method of creating a nanomaterial platform for DNA biosensor applications. Li et al. (2008) assert that thrombin plays a significant role in a number of cardiovascular diseases and regulates quite a few processes in inflammation and tissue repair at the vessel wall. The detection of thrombin is important with regard to blood coagulation levels. Zhang et al. (2009a,b) have used an amplified electrochemical aptasensor for thrombin based on the bio-barcode. Li et al. (2008) have developed a GNP deposited GCE EIS biosensor to investigate aptamer-thrombin interactions.

Figure 6.4 shows the binding of 60 nM thrombin in solution to the EIS biosensor (Li et al., 2008). A single-fractal analysis is adequate to describe the binding and the dissociation kinetics. The values of the binding and the dissociation rate coefficients, and the corresponding fractal dimension values are given in Table 6.2.

On comparing the binding and dissociation of 60 nM thrombin in solution to an EIS biosensor (Figure 6.4; Li et al., 2008) with the binding of 20 nM thrombin in solution to the biotinylated aptamer immobilized on a CMB chip (Centi et al., 2008), it is seen that in one case there is binding and dissociation, and in the other just binding. On comparing the binding phase kinetics of both of these biosensor systems that are adequately described by a single-fractal analysis, it is observed that as one goes from the CMB chip (Centi et al., 2008) to the EIS biosensor (Li et al., 2008), the fractal dimension in the binding phase decreases by a

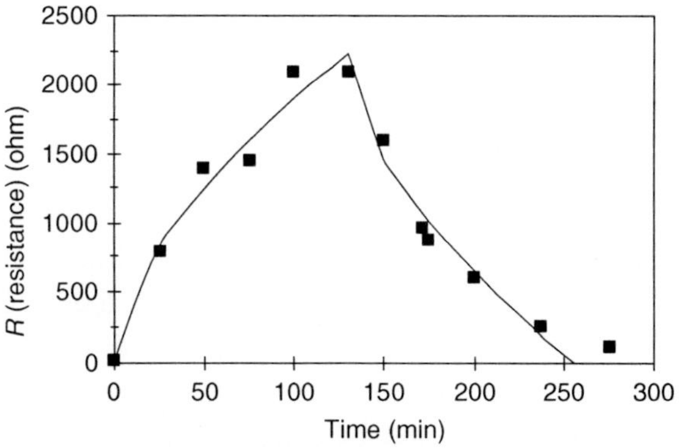

Figure 6.4
Binding of 60 nM thrombin in solution to an aptamer immobilized on a electrochemical impedance spectroscopy (EIS) biosensor (Li et al., 2008).

factor of 1.32 from a value of D_f equal to 2.38632 to D_f equal to 1.810, and the corresponding binding rate coefficient increases by a factor of 1.48 from a value of k equal to 83.044 to k equal to 122.88. In this case, a lower fractal dimension leads to a higher binding rate coefficient. This is contrary to the general trend presented in the different chapters throughout the book. However, one should bear in mind that we are comparing and analyzing the performance of two different biosensors which are detecting the same analyte (albeit with different concentrations; 20 and 60 nm).

Yang et al. (2009) have recently developed an ultrasensitive enhanced CL enzyme immunoassay for detecting AFP which was amplified by double-codified GNP labels. Their method used 4-(4′-iodo) phenylphenol (IPPI) as a signal amplifier and double-codified GNP (DC-AuNPs) labels modified by HRP-conjugated anti-AFP which were used for signal amplification.

Yang et al. (2009) report that gold nanoparticles (AuNPs) have been used in bioassays (Rossi and Mirkin, 2005; Ao et al., 2006; Chen and Zu, 2007; Gomez-Henz et al., 2008; Selvaraju et al., 2008). Yang et al. (2009) also point out that owing to the advantages of the CL technique that includes rapid detection, simple instrumentation, and a wide dynamic range it has been used as a detection technique in biotechnology, pharmacology, molecular biology, and in the environmental area (Kuruoda et al., 2000; Konty et al., 2005; Elzbag et al., 2008; Zhou et al., 2008).

Yang et al. (2009) have developed an ultrasensitive CL assay using IPP as a signal enhancer and double-codified gold nanopaticle labels for further signal amplification to detect AFP which is a tumor marker for the management of heptocellular carcinoma.

Figure 6.5a shows the binding of AFP in the presence of IPP in solution to their ultrasensitive enhanced CL enzyme biosensor using double-codified GNP labels as amplification agents. A single-fractal analysis is adequate to describe the binding kinetics. A dual-fractal analysis is required to adequately describe the dissociation kinetics. The values of (a) the binding rate coefficient, k, and the fractal dimension, D_f, for a single-fractal analysis, and (c) the dissociation rate coefficients, k_{d1} and k_{d2}, and the fractal dimensions, D_{fd1} and D_{fd2}, for a dual-fractal analysis are given in Tables 6.3 and 6.4.

Figure 6.5b shows the binding of AFP in the presence of PIP (*p*-iodophenol) in solution to their ultrasensitive enhanced CL enzyme biosensor using double-codified GNP labels as amplification agents. A dual-fractal analysis is required to adequately describe the binding kinetics. A single-fractal analysis is adequate to describe the dissociation kinetics. The values of (a) the binding rate coefficient, k, and the fractal dimension, D_f, for a single-fractal analysis, (b) the binding rate coefficients, k_1 and k_2, and the fractal dimensions, D_{f1} and D_{f2}, for a dual-fractal analysis, and (c) the dissociation rate coefficient, k_d, and the fractal dimension, D_{fd}, for a single-fractal analysis are given in Tables 6.3 and 6.4.

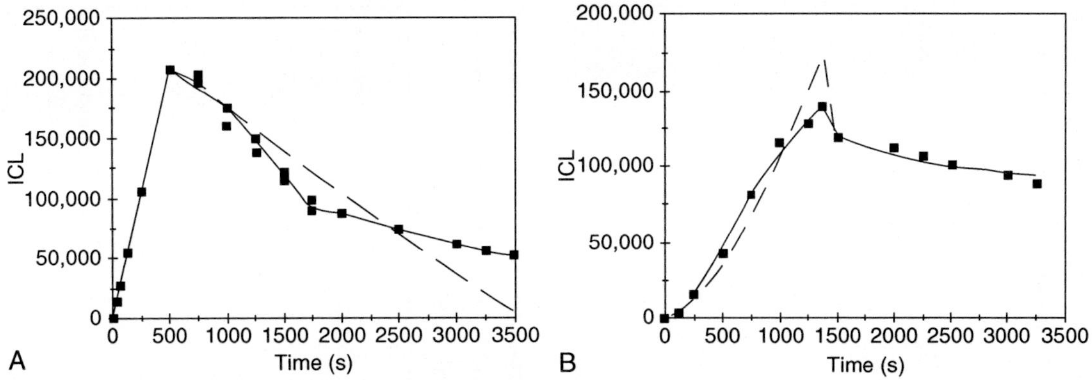

Figure 6.5

(a) Binding of α-fetoprotein in solution with 4-(4′-iodophenylphenol (IPP) to an ultrasensitive enhanced chemiluminescence enzyme immunoassay biosensor amplified by double-codified gold nanoparticles. (b) Binding of α-fetoprotein in solution with PIP to an ultrasensitive immunoassay biosensor amplified with double-codified gold nanoparticles (Yang et al., 2008).

Chang et al. (2009) have recently developed a LSPCF fiber-optic biosensor to detect AFP in human serum. These authors report that evanescent wave-excited fluorescence (EWF) has been widely used in immunoassays. These types of optical biosensors are disposable, inexpensive, and have a simple geometry (Wolfbeis, 2006). A distinct disadvantage of these types of biosensors when compared to ELISA (enzyme-linked immunosorbent assay) and RIA (radioimmunoassay) is the lower detection sensitivity (Ao et al., 2006). However, Chang et al. (2009) point out that gold nanoparticles (AuNPs) have been used in biosensors that effectively enhance the sensitivity detection limits (Matsui et al., 2005; Chau et al., 2006; Mao et al., 2006; Hsieh et al., 2007). For example, Martina et al. (2007) have been able to detect less than 1 picomole of DNA by using AuNPs in biosensors, and Georganopoulos et al. (2005) were able to detect *in vitro* amyloid-β-derived diffusive ligands in cerebrospinal fluid at concentrations lower than 1 picomole exhibited during the early stages of Alzheimer disease.

Chang et al. (2005) have developed a LSPCF fiber-optic biosensor for the clinical diagnosis of AFP in human serum. These authors report that AFP is a 70 kDa oncofetal glycoprotein which is a tumor marker for hepatocellular carcinoma and germ cell tumor (Tsai and Lin, 2005; Fu et al., 2006; Xu et al., 2006).

Figure 6.6a shows the binding and dissociation of 0.1 mg/ml AFP in solution to anti-AFP immobilized on a surface plasmon resonance (SPR) biosensor surface (Chang et al., 2005). A dual-fractal analysis is required to adequately describe the dissociation kinetics. The values of (a) the binding rate coefficient, k, and the fractal dimension, D_f, for a single-fractal analysis, (b) the binding rate coefficients, k_1 and k_2, and the fractal dimensions, D_{f1} and D_{f2}, for a dual-fractal analysis, and (c) the dissociation rate coefficient, k_d, and the fractal dimension, D_{fd}, for a single-fractal analysis are given in Tables 6.3 and 6.4.

Table 6.3: Binding and dissociation rate coefficients for α-fetoprotein to (a) ultrasensitive enhanced chemiluminescence enzyme immunoassay amplified by double-codified gold nanoparticle labels (Yang et al., 2008), and (b) localized surface plasmon resonance coupled fluorescence fiber-optic biosensor (Chang et al., 2008).

Analyte in Solution/Receptor on Surface	k	k_1	k_2	k_d	k_{d1}	k_{d2}	References
AFP/IPP + double-codified gold nanoparticles (DC-AuNPs) labels modified by HRP-conjugated anti-AFP	510.546 ± 12.51	na	na	89.697 ± 34.682	7.803 ± 1.558	7339.97 ± 52.09	Yang et al. (2008)
AFP/PIP + double-codified gold nanoparticles (DC-AuNPs) labels modified by HRP-conjugated anti-AFP	2.009 ± 0.5497	0.3367 ± 0.0899	277.56 ± 16.74	4157.05 ± 531.89	na	na	Yang et al. (2008)
0.1 ng/mL AFP/anti-AFP on a fiber-optic SPR biosensor surface	0.0157 ± 0.0110	0.01856 ± 0.0014	0.0009 ± 0.0005	0.01784 ± 0.00543	na	na	Chang et al. (2008)

Table 6.4: Fractal dimensions for the binding and dissociation of α-fetoprotein to (a) ultrasensitive enhanced chemiluminescence enzyme immunoassay amplified by double-conjugated gold nanoparticle labels (Yang et al., 2008), and (b) localized surface plasmon resonance coupled fluorescence fiber-optic biosensor (Chang et al., 2008).

Analyte in Solution/Receptor on surface	D_f	D_{f1}	D_{f2}	D_{fd}	D_{fd1}	D_{fd2}	References
AFP/IPP + double-codified gold nanoparticles (DC-AuNPs) labels modified by HRP-conjugated anti-AFP	1.8068 ± 0.0218	na	na	1.0652 ± 0.2702	0.2946 + 0.3826	2.2308 ± 0.0468	Yang et al. (2008)
AFP/PIP + double-codified gold nanoparticles (DC-AuNPs) labels modified by HRP-conjugated anti-AFP	0 ± 0.2182	0 + 0.4828	1.2752 ± 0.2450	2.3654 ± 0.1093	na	na	Yang et al. (2008)
0.1 ng/mL AFP/anti-AFP on fiber-optic SPR biosensor surface	1.5302 ± 0.4318	1.2632 ± 0.1518	0 + 1.816	1.8546 ± 0.2258	na	na	Chang et al. (2005)

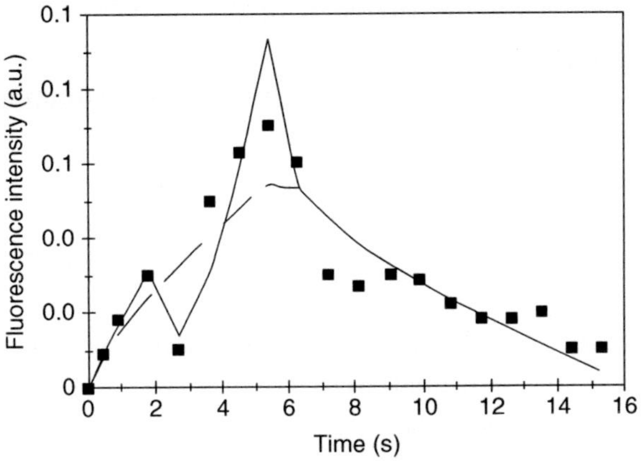

Figure 6.6
Binding and dissociation of 0.1 ng/ml α-fetoprotein (AFP) in solution to capture antibody on a fiber-optic biosensor surface (Chang et al., 2008).

It is of interest to compare the binding and dissociation rate coefficients and the corresponding fractal dimension values for the binding and dissociation of AFP in solution + PIP to the double-codified gold nanoparticles labels modified by HRP-conjugated anti-AFP (Yang et al., 2009) with the binding and dissociation of 0.1 ng/mL AFP to the anti-AFP immobilized on a SPR biosensor surface (Chang et al., 2005). Note that in both of these cases a dual-fractal analysis is required to adequately describe the binding kinetics, whereas the dissociation kinetics may be described by a single-fractal analysis.

It is noteworthy that as one goes from the binding of AFP in solution to anti-AFP immobilized on a SPR biosensor surface (Chang et al., 2005), to the binding of AFP in solution with PIP to the double-codified HRP-conjugated anti-AFP (Yang et al., 2009) the fractal dimension, D_{f1}, decreases by a factor of 2.62 from a value of D_{f1} equal to 1.2632 to D_{f2} equal to 0.4828 for a dual-fractal analysis. The binding rate coefficient, k_1, decreases by a factor of 18.14 from a value of k_1 equal to 0.3367 to 0.01856. However, in this case as the fractal dimension, D_{f2}, increases from a value of 0.0 to 1.2752 as one goes from the binding of AFP in solution to the anti-AFP immobilized on a SPR biosensor surface (Chang et al., 2005) to the binding of AFP in solution with PIP to the double-codified gold nanoparicle (DC-AuNP) labeled modified HRP-conjugated anti-AFP, the binding rate coefficient, k_2, increases by a factor of 313273 from a value of k_2, equal to 0.000886 to k_2 equal to 277.56. This is a substantial increase of more than five orders of magnitude. This is because the fractal dimension on the SPR biosensor surface was 0.0, which represents a Cantor-like dust.

As one might very reasonably expect, quite a lot of effort has been spent on the detection of glucose in solution. Some of the more recent biosensors that have been used to detect glucose in solution include:

(a) Fixture-reduced method for the synthesis of Cu_2OMWCNTs nanocomposites and its application as an enzyme-free glucose sensor (Zhang et al., 2009a,b).
(b) A novel nonenzymatic ECL sensor for glucose using palladium nanoparticles supported on functional carbon nanotubes (Chen et al., 2009).
(c) A glucose biosensor based on the room-temperature phosphorescence of TiO_2/SiO_2 nanocomposite (Li et al., 2009).
(d) A disposable glucose biosensor using electroless-plated Au/Ni/copper low electrical resistance electrodes (Lee et al., 2009).
(e) A glucose biosensor based on the direct electrochemistry of glucose-oxidase immobilized on nitrogen-doped carbon nanotubes (Deng et al., 2009).
(f) A glucose biosensor with superior long-term stability based on inserted barrel plating gold electrodes (Hsu et al., 2009a,b,c).
(g) A glucose biosensor based on Prussian blue/chitosan hybrid film (Wang et al., 2009a,b).
(h) A glucose biosensor using a reusable enzyme-modified ion track membrane reactor (Fink et al., 2009).
(i) An elecrospun poly(vinylidene fluoride)/poly(aminophenylboronic acid) composite nanofibrous membrane as a novel glucose sensor (Manesh et al., 2007).
(j) A glucose sensor for flow injection analysis of serum glucose based on immobilization of glucose oxidase in titania sol-gel membrane (Yu et al., 2003).
(k) A disposable glucose sensor employing engineered glucose dehydrogenases (2000).
(l) A disposable amperometric glucose sensor electrode with enzyme-immobilized Nitrocellulose strip (Cui, 2001).
(m) A noninvasive continuous monitoring of physiological glucose using a monosaccharide-sensing contact lens (Badugu et al., 2004).
(n) A glucose-sensing electrode preparation based on a glucose-attached polyion complex membrane containing microperoxidase and ferocene (Yabuki et al., 2000).
(o) A nonenzymatic glucose detector using mesoporosus membrane (Park et al., 2003).
(p) A functionalized hydrogel-optical fiber biosensor for determining glucose levels: toward continuous monitoring of blood glucose *in vivo* (Tierney et al., 2009a,b).
(q) A sensitive nonenzymatic glucose sensor in alkaline media with a copper Nanocluster/multiwall carbon nanotube-modified GCE (Kang et al., 2007).
(r) A poly(vinylpyrrolidone)-doped nitric oxide-releasing xerogels as glucose biosensor membranes (Schoenfisch et al., 2006).
(s) Design of molecular wires based on supramolecular structures for application in glucose biosensors (Alves et al., 2006).

Sheng et al. (2008) have recently developed a novel biosensor using a modified GCE for the detection of glucose. They have used an enzymatically induced formation of neodymium hexacyanoferate nanoparticles on a glass oxidase/chitosan-modified GCE. The authors report that chitosan not only acts as an immobilizer for the enzyme but also provides the active sites for the nanoparticles to grow.

Sheng et al. (2008) point out that biocatalysis has provided an avenue for the synthesis and enlargement of nanoparticles. This biocatalytic concept has been used for the design of simple and sensitive electrochemical and optical biosensors (Katz et al., 2004; Medintz et al., 2005; Moller et al., 2005). These biocatalytically induced particle growth processes have been shown to be used for enzyme assays (Xiao et al., 2004; Baron et al., 2005a,b; Shlyahovsky et al., 2005; Xiao et al., 2005a,b; Zayata et al., 2005). Sheng et al. (2008) point out that Hwang et al. (2005) initially reported the metal deposition by the enzymatically produced reducing agent. They did this for the electrochemical detection of DNA. A self-assembled monolayer modified electrode with a biologically catalytic reaction for the biosensing of cholesterol has been reported (Zhou et al., 2006a,b). Sheng et al. (2008) report that this reduction and deposition of metal onto the nanoparticles by an enzymatically generated reducing agent has been used for the monitoring of quite a few enzymatic reactions. Sheng et al. (2008) report that the biocatalyzed synthesis of metallic nanoparticles and the biocatalyzed synthesis of semiconductor or magnetic nanoparticles is an attractive area of research (Willner et al., 2006).

Sheng et al. (2008) have developed a novel method to help improve biosensor detection by using metal hexacyanoferrate nanoparticles from the biocatalyzed synthesis of real earth hexacyanoferrate (neodymium hexacyanide). They have used this as a model example to show the application of their protocol.

Figure 6.7 shows the binding of glucose in solution to a neodymium hexacyanoferate nanoparticle on the glucose oxidase/chitosan-modified GCE (Sheng et al., 2008). A dual-fractal analysis is required to adequately describe the binding kinetics. The values of (a) the binding rate coefficient, k, and the fractal the fractal dimension, D_f, for a single-fractal analysis, and (b) the binding rate coefficients, k_1 and k_2, and the fractal dimensions, D_{f1} and D_{f2}, for a dual-fractal analysis are given in Table 6.5. As expected, and as noted throughout the different chapters for different analyte-receptor systems binding (and dissociation) on biosensor surfaces, an increase in the degree of heterogeneity or the fractal dimension on the biosensor surface leads to an increase in the binding rate coefficient.

Lee et al. (2008) have recently developed a disposable glucose biosensor. These authors used an electroless-plated Au/Ni/copper low electrical resistance electrode. They point out that glucose control is required not only for quality control in the food industry, but in particular for managing diabetes (American Diabetic Association, 1994, 1999). Also, the monitoring of blood glucose level is essential in the case of diabetic patients (Owen, 1985; Nakamura and Karube, 2003; Neumann and Turner, 2005). Standard conditions for monitoring of glucose

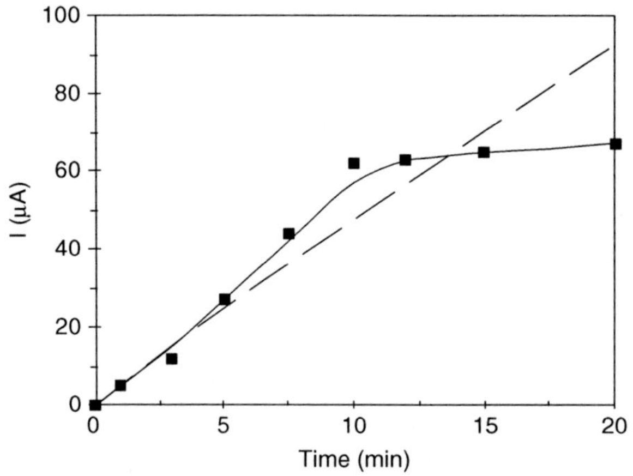

Figure 6.7

Binding of glucose in solution to a neodymium hexacyanoferrate (NDHCF) nanoparticle (NP) on a glucose/chitosan-modified glass carbon electrode (GCE) biosensor (Sheng et al., 2008).

Table 6.5a: Binding rate coefficients for glucose in solution to two different biosensor surfaces

Analyte in solution/ Receptor on surface	k	k_1	k_2	References
SAME as in Table 6.5b	5.212 ± 1.285	4.459 ± 0.761	46.857 ± 1.666	Sheng et al. (2008)
	29.942 ± 1.193	na	na	Lee et al. (2008)
	27.987 ± 0.042	na	na	Lee et al. (2008)
	25.107 ± 0.362	na	na	Lee et al. (2008)

Table 6.5b: Fractal dimensions for glucose in solution to two different biosensor surfaces.

Analyte in Solution/Receptor on Surface	D_f	D_{f1}	D_{f2}	References
Glucose/neodymium-Hexacyanoferrate Nanoparticle on the glucose oxidase/chitosan-modified glass carbon electrode	1.078 ± 0.1702	0.772 ± 0.1786	2.604 ± 0.0196	Sheng et al. (2008)
53 mM glucose/electroless-plated Au/Ni/copper low electrical resistance electrode	2.880 ± 0.01630	na	na	Lee et al. (2008)
27 mM glucose/electroless-plated Au/Ni/copper low electrical resistance electrode	2.9788 ± 0.00376	na	na	Lee et al. (2008)
21 mM glucose/electroless-plated Au/Ni/copper low electrical resistance electrode	3.0 - 0.0364	na	na	Lee et al. (2008)

levels have been set and are available (ISO 15197, 2003). Lee et al. (2008) report that electrochemical glucose sensors based on modified electrodes with immobilized glucose oxidase are available (Sulak et al., 2006). Disposable biosensors have also been developed (Lee et al., 2008).

Different electrode fabrication methods, including screen printing (Crouch et al., 2005) and other deposition techniques (Hashimoto et al., 2007), have been used to develop these biosensors. Lee et al. (2008) point out that carbon-paste electrodes (CPE) have been widely used for voltametric measurements (Yabuki et al., 1992; Ge et al., 1998; Bouquita et al., 2000; Hart et al., 2002; Hong et al., 2002; Fahnrich et al., 2003; Gao et al., 2003, 2005; Honeychurch et al., 2003; Darain et al., 2005). Lee et al. (2008) report that the application of carbon paste to make the electrode is a simple procedure and hence economical. However, up until now the work on low electrical resistance has not been reported. Lee et al. (2008) report that they have developed a biosensor with low electrical resistance besides making the biosensor disposable. The disposable biosensor, according to them, offers a very low uniform electrical resistance of about 0.01 ohms. They have done this by attaching a Au/Ni/copper structure to a plastic film substrate using a laminating procedure.

Figure 6.8a shows the binding of 53 nM glucose in solution to the electroless-plated Au/Ni/copper low electrical resistance electrode (Lee et al., 2008). A single-fractal analysis is adequate to describe the binding kinetics. The values of the binding rate coefficient, k, and the fractal dimension, D_f, for a single-fractal analysis are given in Table 6.5 a and b.

Figure 6.8b and c show the binding of 27 and 21 mM glucose, respectively, to the electroless-plated Au/Ni/copper low resistance electrode (Lee et al., 2008). Once again, a single-fractal analysis is adequate to describe the binding kinetics. The values of the binding rate coefficient, k, and the fractal dimension, D_f, for a single-fractal analysis are given in Table 6.5 a and b.

It is of interest to compare the binding of glucose in solution to the neodymium hexacyanoferrate nanoparticle on the glucose oxidase/chitosan-modified GCE (Sheng et al., 2008) with that of the binding of 1-53 mM glucose in solution to the electroless-plated Au/Ni/copper low electrical resistance electrode (Lee et al., 2008). In the first case a dual-fractal analysis is required to adequately describe the binding case presented. In the second case, a single-fractal analysis is adequate to describe the binding kinetics for all of the three concentrations of glucose (21, 27, and 53 mM) analyzed. In the first case, a complex binding mechanism is involved since a dual-fractal analysis is required to adequately describe the binding kinetics, whereas in the second case a simple binding mechanism is involved since a single-fractal analysis is adequate to describe the binding kinetics. Also, in the second case, even though the fractal dimension or the degree of heterogeneity on the biosensor surface, D_f equal to 2.8850, 2.9788, and 3.0 are higher than those in the second case for the second phase (D_{f2} equal to 2.604), the binding rate coefficient, k, for the second case is lower than the binding rate coefficient, k_2. This is due, of course, to the different biosensor systems involved.

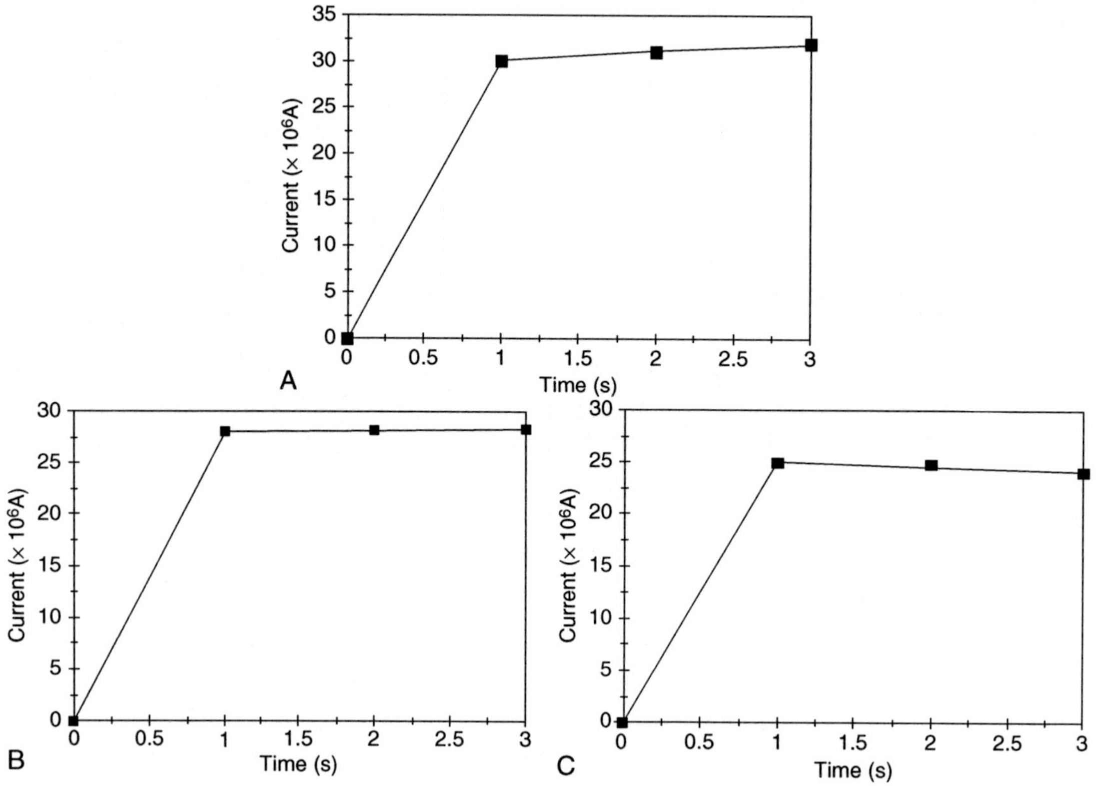

Figure 6.8
Binding of different concentrations of glucose in solution to a disposable biosensor using electroless-plated Au/Ni/copper low electrical resistance electrodes (Lee et al., 2008): (a) 53 mM (b) 27 mM (c) 21 mM.

Hsu et al. (2009a,b,c) have recently analyzed the long-term stability of a glucose biosensor based on inserted barrel plating gold electrodes. These authors point out that glucose one time strips are regularly used by diabetics to monitor their sugar levels. They report that their performance varies. They have critically evaluated the performance of the long-term stability of these strips. They further assert, as indicated elsewhere in this book, that exact glycemic control is essential for the control and management of DM (diabetes mellitus). One shot strips that can be used for determining the glucose levels in blood are cost effective and stable, and may be mass produced. They also point out that an essential component is fast electron transfer and good reproducibility. They state further that glucose oxidase is frequently used in portable glucose sensors. However, in glucose oxidase, the active site is well-embedded in the enzyme, and direct electron transfer from the active site to the sensing electrodes is hindered (Wilson and Turner, 1992; Hecht et al., 1993). Various strategies have been attempted according to Hsu et al. (2009a,b,c) to help enhance this electron relay (Albareda-Savant et al., 2000; Miscoria et al., 2006; Jia et al., 2007; Renedo et al., 2007; Kuwahara et al., 2008). The transducing elements that are in current use include carbonaceous materials (Lim et al., 2005). Hsu et al. (2009a,b,c) have recently reported a simple process for fabricating an amperometric glucose biosensor by inserting two barrel plating gold electrodes onto an injection-molding plastic base. Herein, the authors' objective was to examine different biosensing platforms to determine the one that would provide the most stability.

Figure 6.9 shows the long-term stability of the glucose SPCE biosensor based on the insertion of barrel plating gold electrodes (Hsu et al., 2009a,b,c). A single-fractal analysis is adequate to describe the binding kinetics. The binding rate coefficient (k) value is 6.868 $\pm$ 1.982 and the fractal dimension value is 3.0 – 0.5312. Note that the fractal dimension cannot have a value greater than 3.0, thus only the negative error is given. If one were to take the liberty of comparing the binding rate coefficient (k) values and the fractal dimension value obtained from the long stability studies mentioned above with the previous studies shown for the binding of different concentrations of glucose in solution to the disposable biosensor (Figure 6.8a–c; Lee et al., 2008), it is seen that for long-term stability studies the fractal dimension value is close to the maximum (max 3.0, values—2.88, 2.97, and 3.0) as compared to the previous case and its binding rate coefficient (k) value is about four times less than that of the disposable biosensor.

Liao et al. (2008) have very recently developed a percutaneous fiber-optic sensor for chronic glucose monitoring *in vivo*. Their biosensor is disposable, minimally invasive, and is capable of monitoring the blood glucose level for several weeks. Their blood glucose sensor contains a percutaneous fiber-optic that permits spectroscopic measurements of the chemical

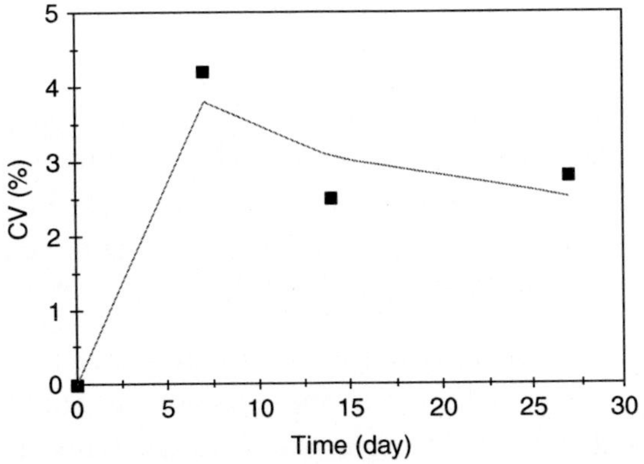

Figure 6.9
Long-term stability of the glucose SPCEbiosensor (Hsu et al., 2009a,b,c).

reactions. This is done by attaching a nanoengineered polymeric matrix attached to the implanted end of the fiber. These authors indicate that the successful management of DM requires (as is well known) the frequent measurement of glucose measurement (Diabetes Control and Complications Trials, 1993; UK Prospective Diabetes Study Group, 1998).

Liao et al. (2008) emphasize that even though insulin delivery technology is well established, there are still problems that exist with glucose sensors (Stell et al., 2004). Some of the properties of an effective biosensor according to them include: low cost to operate (which includes cost of the sensor itself, of installation, and the frequency at which it would have to be changed), it should be minimally obstructive, and should permit frequent measurements. Liao et al. (2008) report that the present method of monitoring blood glucose levels (MBG) is by self monitoring (e.g., by pricking one's finger followed by subsequent measurement by a 95-99% accurate meter, the Clark meter (Clark, 2005)).

The assay method uses glucose oxidase to consume glucose. However, in the body this enzyme, glucose oxidase, deteriorates with time, and is also sensitive to pH and temperature (Usmani and Akmiai, 1994; Tamada et al., 2002; Wenholt et al., 2006). According to Liao et al. (2008) there is a further drawback due to limited or uncertain stability. Liao et al. (2008) are developing a family of disposable, minimally invasive *in vivo* sensors that could measure different analytes in diabetic patients over several weeks. Their biosensor measures glucose concentrations using the fluorescence resonance energy transfer (FRET) assay based on the selective binding of saccharides by the jack bean lecithin Concanavalin A (con A) (Meadows and Schultz, 1993). Their present biosensor design (Liao et al., 2008) is an improvement on their initial design (Liao et al., 2005), because it permits the sensor to measure glucose concentrations in the physiological range of 0-500 mg/dL. Liao et al. (2008) assert that their biosensor is the first step towards the development of a successful sensor. The next step would be the validation of their results against a gold standard such as intravenous blood glucose, followed by other questions such as degradation of materials (e.g., the PEG matrix), and speed and accuracy.

Figure 6.10a and b shows the binding of blood glucose to the percutaneous fiber-optic biosensor used for chronic glucose monitoring (Liao et al., 2008). In both these cases a dual-fractal analysis is required to adequately describe the binding kinetics. The values of (a) the binding rate coefficient, k, and the fractal dimension, D_f, for a single-fractal analysis, and (b) the binding rate coefficients, k_1 and k_2, and the fractal dimensions, D_{f1} and D_{f2}, for a dual-fractal analysis are given in Tables 6.6 and 6.7.

It is of interest to compare the binding of glucose to the percutaneous fiber-optic biosensor (Liao et al., 2008) with the binding of the glucose in solution to the neodymium hexacyanoferrate nanoparticles on the glucose oxidase/chitosan-modified GCE (Sheng et al., 2008). In both these cases, a dual-fractal analysis is required to adequately describe the binding kinetics. It is seen that as one goes from the percutaneous fiber-optic biosensor

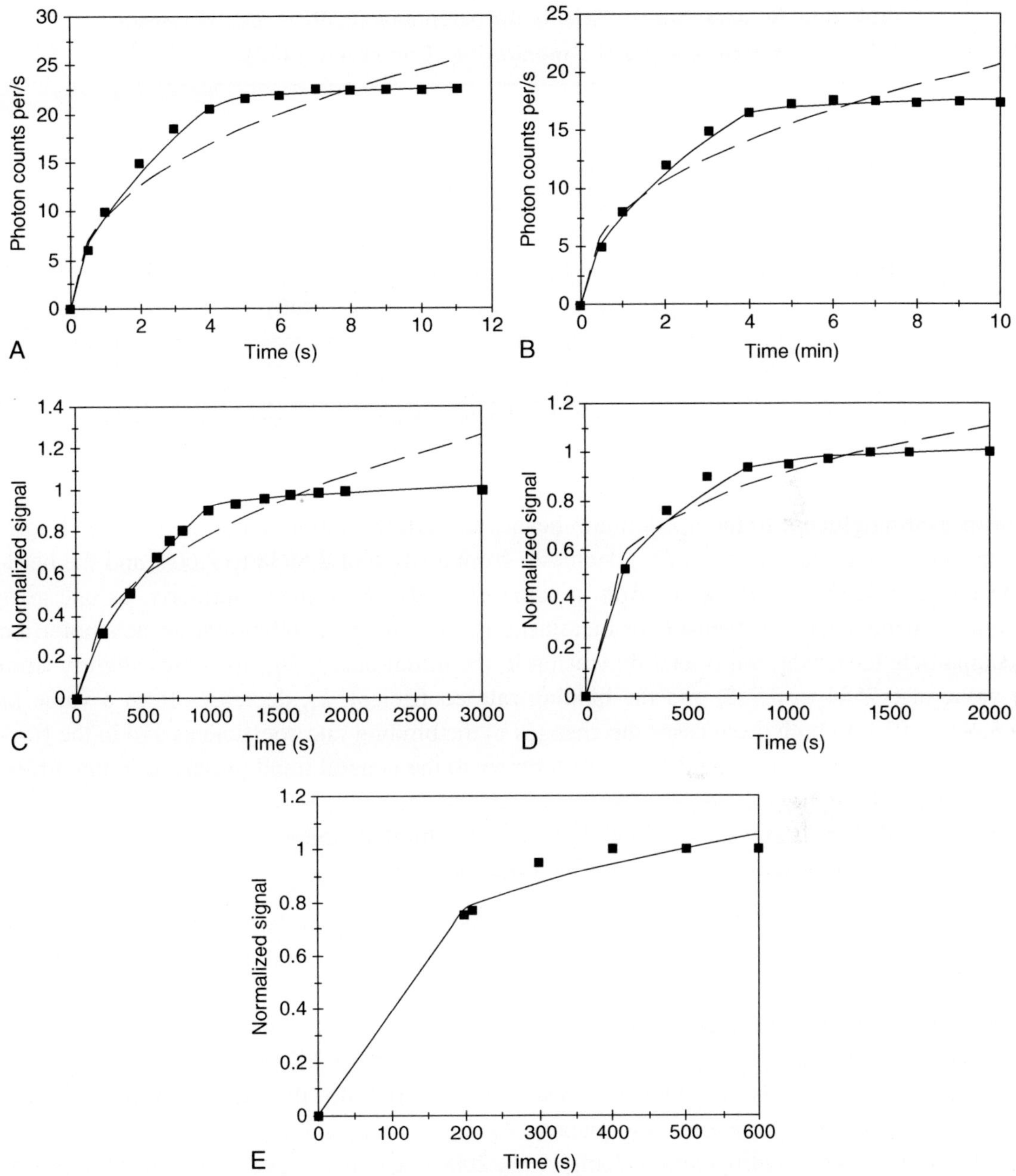

Figure 6.10

(a) and (b) Binding of blood glucose to the percutaneous fiber-optic biosensor (Liao et al., 2008). Binding of 2 mM increase in glucose in phosphate-buffered saline at different temperatures (in degrees centigrade): (c) 25 (d) 31 (e) 37 (Tierney et al., 2009a,b).

Table 6.6: Binding rate coefficients for percutaneous fiber-optic biosensor for chronic glucose monitoring (Liao et al., 2008).

Analyte Concentration	k	k_1	k_2
Percutaneous	9.9604 ± 1.4599	9.509 ± 0.737	19.975 ± 0.199
blood glucose	8.0479 ± 0.1744	7.7473 ± 0.5424	15.722 ± 0.2383

Table 6.7: Fractal dimensions for percutaneous fiber-optic biosensor for chronic glucose monitoring (Liao et al., 2008).

Analyte Concentration	D_f	D_{f1}	D_{f2}
Percutaneous	2.1816 ± 0.8508	0.7584 ± 0.0756	2.8938 ± 0.02868
blood glucose	2.1826 ± 0.08936	1.9072 ± 1.9028	2.9005 ± 0.0375

for measuring glucose to the neodymium hexacyanoferrate nanoparticle biosensor the fractal dimension in the second phase, D_{f2}, decreases from a value of 2.8938 to 2.604, and the binding rate coefficient, k_2, increases from a value of 19.975 to 46.857. Similarly, as one goes from the percutaneous biosensor for measuring glucose to the neodymium hexacyanoferrate nanoparticle biosensor, the fractal dimension in the initial phase, D_{f1}, increases slightly from a value of 0.7584 to 0.772, and the binding rate coefficient, k_1, decreases from a value of 9.509 to 4.959. In both these cases the changes in the binding rate coefficients and in the fractal dimensions are in opposite directions (contrary to the general trend presented in the different chapters throughout the book); but this is because the analysis is of two different systems here, to see if there is any sort of trend that may be exhibited, and what may be learnt from it. Furthermore, one system is *in vivo* and the other is *in vitro*.

Tierney et al. (2009a,b) point out the need for glucose testing in diabetics. They report that most instruments presently available are for intermittent testing. Kondepatu and Heise (2007) report that there are a few instruments available for the continuous monitoring of glucose, which, Tierney et al. (2009a,b) report, would be required for the self-regulating insulin system. Tierney et al. (2007) have developed an optical fiber biosensor for the continuous monitoring of glucose *in vivo* which, is necessary for critically ill patients (Williams et al., 2002). They further point out that glucose oxidase (Malitesta et al., 1990; Hale et al., 1991) and glucose dehydrogenase (Zhang et al., 2004) have been used as also phenylboronic acid (PBA) (James and Shinkai, 2002).

Hydrogels with receptors have been used as detection devices (Asher et al., 2003; Kabilan et al., 2005). Asher and his group have pioneered these hydrogel-based detection devices (Alexeev et al., 2004; Ben-Moshe et al., 2006). In their most recent publication Tierney et al. (2009a,b) have used a responsive hydrogel-based sensor, and have used dimethylaminopropylacrylamide, a tertiary amine, in their sol-gel matrix to enhance the glucose sensitivity and selectivity, and the authors point out that this is a highly selective technique (Tierney et al., 2008).

Figure 6.10c shows the binding of a 2 mM increase in glucose in phosphate-buffered saline to a 8 mol% 3-PBA and 10 mol% DMAPAA (*N*-(3-dimethylaminopropyl)-acrylamide) hydrogel for continuous monitoring *in vivo* at 25 °C (Tierney et al., 2009a,b). The response presented is normalized. A dual-fractal analysis is required to adequately describe the binding kinetics. The values of (a) the binding rate coefficient, k, and the fractal dimension, D_f, for a single-fractal analysis, and (b) the binding rate coefficients, k_1 and k_2, and the fractal dimensions, D_{f1} and D_{f2}, are given in Tables 6.8 and 6.9.

Figure 6.10d shows the binding of a 2 mM increase in glucose in phosphate-buffered saline to a 8 mol% 3-PBA and 10 mol% DMAPAA hydrogel for continuous monitoring *in vivo* at 31 °C (Tierney et al., 2009a,b). The response presented is normalized. Once again, a dual-fractal analysis is required to adequately describe the binding kinetics. The values of (a) the binding rate coefficient, k, and the fractal dimension, D_f, for a single-fractal analysis, and (b) the binding rate coefficients, k_1 and k_2, and the fractal dimensions, D_{f1} and D_{f2}, for a dual-fractal analysis are given in Tables 6.8 and 6.9.

Figure 6.10e shows the binding of a 2 mM increase in glucose in phosphate-buffered saline to a 8 mol% 3-PBA and 10 mol% DMAPAA hydrogel for continuous monitoring *in vivo* at 37 °C (Tierney et al., 2009a,b). The response presented is normalized. In this case, a single-fractal analysis is adequate to describe the binding kinetics. The values of the binding rate coefficient, k, and the fractal dimension, D_f, for a single-fractal analysis are given in Tables 6.8 and 6.9.

Table 6.8: Binding rate coefficients for glucose in solution to a functionalized hydrogel-optical fiber sensor (Tierney et al., 2009a,b).

Temperature (°C)	k	k_1	k_2
25	0.0407 ± 0.0056	0.009881 ± 0.00026	0.5560 ± 0.0095
31	0.1435 ± 0.0134	0.0709 ± 0.0055	0.6799 ± 0.0086
37	0.1819 ± 0.0118	na	na

Influence of temperature in the 25-37 °C temperature range.

Table 6.9: Fractal dimensions for the binding of glucose in solution to a functionalized hydrogel-optical fiber sensor (Tierney et al., 2009a,b).

Temperature (°C)	D_f	D_{f1}	D_{f2}
25	2.1412 ± 0.1032	1.993 ± 0.0400	2.9407 ± 0.0470
31	2.4622 ± 0.0859	2.2288 ± 0.1168	2.8968 ± 0.0676
37	2.4452 ± 0.1238	na	na

Influence of temperature in the 25-37 °C temperature range.

It is of interest to compare the chronic monitoring of glucose *in vivo* using a functionalized hydrogel-optical fiber biosensor (Tierney et al., 2009a,b) with the chronic monitoring of glucose *in vivo* using a percutaneous fiber-optic sensor (Liao et al., 2008). Both cases involve *in vivo* monitoring, but two different biosensor techniques are involved. It is important to mention at the outset, that some of the differences in the binding rate coefficients and fractals that may arise because of the details in the biosensor monitoring techniques that are involved, are not presented or accounted for here. However, both cases to be considered and compared involve the use of dual-fractal analysis, necessary to adequately model the binding kinetics.

It is of interest to note that when a dual-fractal analysis is used, and for the initial phase of binding as one goes from the hydrogel fiber-optic biosensor (Tierney et al., 2009a,b) to the percutaneous fiber-optic sensor for the continuous glucose monitoring *in vivo* (Liao et al., 2008), the fractal dimension decreases from a value of 1.6830 to 0.7854, and the binding rate coefficient, k_1, however, increases from a rather low value of 0.009881 (Tierney et al., 2009a, b) to 9.509 (Liao et al., 2008). This trend is differently presented, in general, throughout the different chapters in the book, when the same analyte-receptor biosensor system is being analyzed. In this case, however, clearly two different analyte-receptor biosensor systems are being analyzed and compared (which is the purpose of this chapter) and hence that general trend is not followed.

Similarly, for the second phase of binding as one goes from the hydrogel-fiber-optic biosensor at 25 °C (Tierney e al., 2009a,b) to the percutaneous fiber-optic sensor for continuous glucose monitoring *in vivo* (Liao et al., 2008), the fractal dimension, D_{f2}, increases very slightly from a value of 2.8498 to 2.8938 (a change in the third decimal place), and the binding rate coefficient, k_2, exhibits a very substantial change in value from a value of 0.5569 to 19.975. A very small change in the fractal dimension value leads to about a 34 times change in the binding rate coefficient. This is primarily attributed to the different analyte-receptor biosensor systems being compared and analyzed.

It is of interest to compare the binding of glucose in a hydrogel fiber-optic biosensor at 37 °C (Tierney et al., 2009a,b) to a disposable fiber-optic biosensor (Lee et al., 2008). The binding curve in both these cases is adequately described by a single-fractal analysis. It is seen that as one goes from the hydrogel fiber-optic biosensor to the disposable glucose biosensor the fractal dimension increases slightly from a value of 2.4497 to 2.8849, the binding rate coefficient increases by a factor of 100.59 from a value of 0.1819 to 29.942. A small change in the degree of heterogeneity or the fractal dimension on the biosensor surface leads to an increase of more than two orders of magnitude in the binding rate coefficient. Once again, a change in the analyte-receptor biosensor system leads to a very significant change in the binding rate coefficient even though the change in the degree of heterogeneity on the biosensor surface is small. In this case, the disposable glucose biosensor exhibits the higher binding rate coefficient, which should be helpful in the monitoring of glucose levels in diabetic patients.

Biosensors have been used recently to analyze and detect the different aspects of cancer, PSA (prostate specific antigen, tumors, cancer biomarkers, cancer cell proliferation), etc. Some of these studies include those on:

(a) A nanoparticle label/immunochromatographic electrochemical biosensor for rapid and sensitive detection of PSA (Lin et al., 2008a,b,c),
(b) Multilayers enzyme-coated carbon nanotubes as a biolabel for ultrasensitive CL immunoassay of cancer biomarker (Bi et al., 2009),
(c) Simultaneous detection of free and total PSA on a screen-printed electrochemical dual sensor (Escamilla-Gomez et al., 2009),
(d) Cell-based immobilization strategy for sensitive piezoelectric immunoassay of total PSA (Ding et al., 2008),
(e) Optical protein sensor for detecting cancer markers in saliva (Tan et al., 2008),
(f) Enhancement of sensitivity and specificity by surface modification of carbon nanotubes in diagnosis of prostate cancer based on carbon nanotube field effect transistors (Kim et al., 2009),
(g) Amperometric microimmunosensor for the detection of tumor biomarker (Prabhulkar et al., 2009),
(h) Nano-bio-chips for high performance multiplexed protein detection: determination of cancer biomarkers in serum and saliva using quantum dot bioconjugate labels (Jokerst et al., 2009),
(i) Tracking cancer cell proliferation on a CMOS capacitance sensor chip (Prakash and Abshire (2008)),
(j) A new immunosensor for breast cancer cell detection using antibody-coated long alkylsilane self-assembled monolayers in a parallel plate flow chamber (Ehrhart et al., 2008),
(k) Colorimetric multiplexed immunoassay for sequential detection of tumor markers (Wang et al., 2009a,b), and
(l) Impedance studies of biobehavior and chemosensitivity of cancer cells by microelectrode arrays (Liu et al., 2009).

Daniel et al. (2008) have recently developed an implantable diagnostic device for cancer monitoring. They have used this to monitor cancer in different mice. They assert that biopsies are essential to provide information for cancer diagnosis. But, because of their invasiveness, alternate forms of cancer monitoring are required, especially for the local environment of the cancer. These authors report that their device could be left behind in the body during the biopsy. Their device uses a semipermeable membrane that contains the nanoparticle magnetic relaxation switches. They further point out the transverse relaxation time of their implantable device in tumor-bearing mice was $20 \pm 10\%$ lower than those observed in control mice after 1 day by mRI (magnetic resonance imaging). These authors report that the short-term applications for their implantable device are far-reaching.

Daniel et al. (2008) in their recent paper have reported the importance of probing the local environment after tumor removal. Sampling of the local environment would indicate if the neoplastic tissue has been completely removed. They point out that PTH (intraoperative parathyroid hormone) is used in this manner. PTH levels decrease quickly, within 5-10 min, once the hypersecreting fluid has been removed. This is a clear marker of further removal being necessary (Lo et al., 2002; Sokol et al., 2004). Daniel et al. (2008) report that local biomarker concentrations are good indicators of the successful removal of different types of tumors (Sodlaczek et al., 2002; Baron et al., 2005a,b). The implantable device that Daniel et al. (2008) have developed could be left behind in the body after tumor resection. They further point out that new tumor growth is difficult to identify from foreign lesions (Gomez-Rio et al., 2004, 2008). Future devices that could be implanted for long periods of time according to Daniel et al. (2008) will permit the evaluation of targeted therapies and facilitate personalized cancer treatment (Chen, 2007; Carney, 2007; Agarwal et al., 2008; Takeguchi et al., 2008).

Figure 6.11a shows the binding (monitoring) of cancer in Mouse 1 to the implantable device (Daniel et al., 2009). A dual-fractal analysis is required to adequately describe the binding kinetics. The values of (a) the binding rate coefficient, k, and the fractal dimension, D_f, for a single-fractal analysis, and (b) the binding rate coefficients, k_1 and k_2, and the fractal dimensions, D_{f1} and D_{f2}, for a dual-fractal analysis are given in Tables 6.10 and 6.11.

Figure 6.11b and c show the binding (monitoring) of cancer in Mouse 2 and 3, respectively to the implantable device (Daniel et al., 2009). A single-fractal analysis is adequate, in these cases, to model the binding kinetics. The values of the binding rate coefficient, k, and the fractal dimension, D_f, for a single-fractal analysis are given in Tables 6.10 and 6.11.

Suwansara-ard et al. (2009) have recently compared the performance of a SPR biosensor with that of a capacitive immunosensor for the detection of CA 125 in human serum samples. These authors point out that CA 125 is a tumor marker for ovarian cancer. They report that CA 125 circulates in the serum of patients (Endo et al., 1988). For healthy humans, the concentration of CA 125 is less than 35 units/ml (Wilder et al., 2003). Suwansara-ard et al. (2009) assert that the determination of CA 125 concentrations in human serum provides information on the disease stage and aid in the monitoring of ovarian cancer.

Techniques have been used to detect and measure CA 125 levels that include radiometric (Marquette et al., 1987, 1997), enzyme immunoassay (Yan et al., 1999; Biomerieux, 2004; Wu et al., 2006). Suwansara-ard et al. (2009) point out that these techniques are time-consuming, require several separation steps and specially equipped laboratories and skilled personnel (He et al., 2003; Lin and Ju, 2005). Thus, the need for developing a simple detection technique, for example, an immunosensor for detecting CA 125 as suggested by Suwansara-ard et al. (2009) becomes imperative.

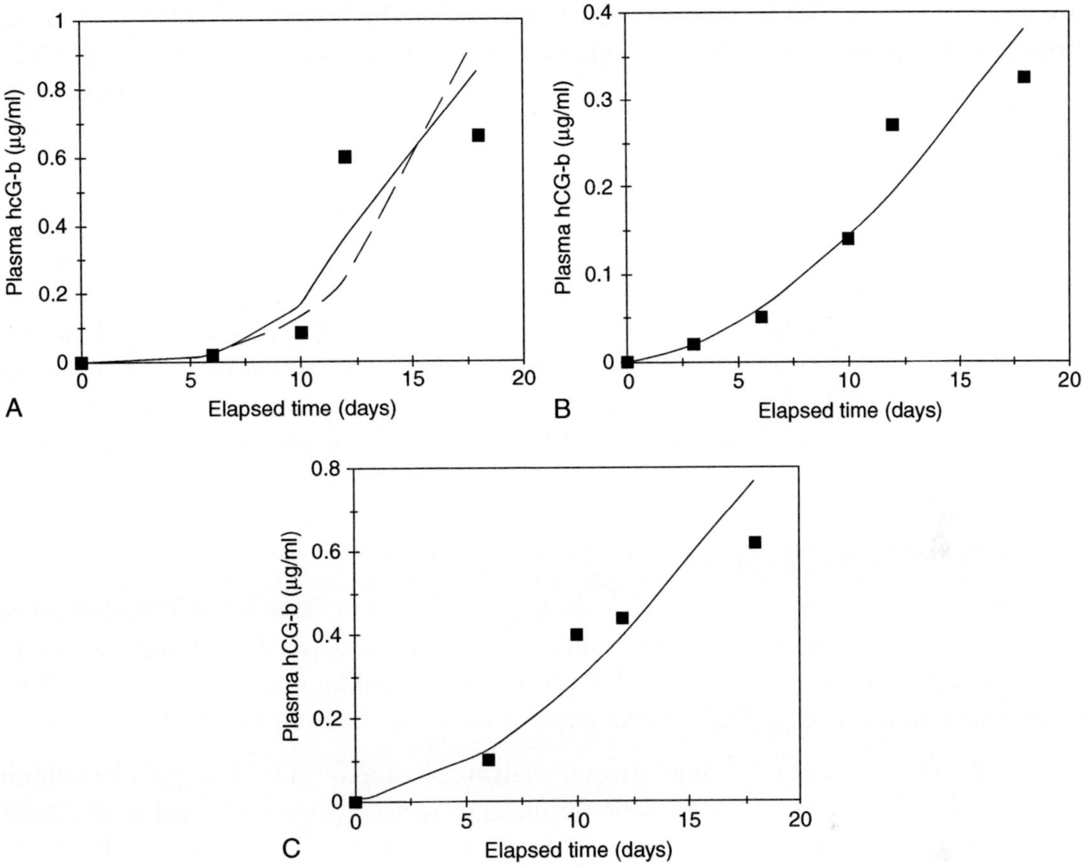

Figure 6.11

Binding of cancer cells to the implantable device: (a) Mouse 1 (b) Mouse 2 (c) Mouse 3.

Table 6.10: Binding rate coefficients for cancer monitoring in different mice (Daniel et al., 2009).

Mouse Number	k	k_1	k_2
#1	0.0000418 + 0.000056	0.000057 + 0.000073	0.000172 + 0.00032
#2	0.003 + 0.00084	na	na
#3	0.006243 ± 0.002340	na	na

Table 6.11: Fractal dimensions for the binding phase for cancer monitoring in mice (Daniel et al., 2009).

Mouse Number	D_f	D_{f1}	D_{f2}
#1	0 + 1.9242	0	0
#2	0 + 0.3512	na	na
#3	0 + 0.8212	na	na

Suwansara-ard et al. (2009) report that the immunosensors method of detecting CA 125 involves labeling with an enzyme and different steps (Dai et al., 2003; Bange et al., 2005; Wu et al., 2006; Fu et al., 2008). Suwansara-ard et al. (2009) point out the importance of label-free detection of CA 125 because it directly permits the detection of changes in the physical properties involved in the antigen-antibody interactions on the transduction surface. Tang et al. (2006) have obtained a detection limit of CA 125 of 1.8 units/ml.

Suwansara-ard et al. (2009) report that the SPR biosensors and capacitive immunosensors have recently attracted substantial attention as direct detection devices (Berggren and Johansson, 1997; Limbut et al., 2006; Yin et al., 2006; Dudak and Boyad, 2007; Takamura and Iwata, 2007; Mazumdar et al., 2008). Suwansara-ard et al. (2009) have compared the performance of the SPR biosensor with that of a capacitive immunosensor for the detection of CA 125. Here only the analysis for the detection of CA 125 by the SPR biosensor is presented. Anti-CA 125 was immobilized on the gold surface of the SPR biosensor chip using a self-assembled monolayer. The performance of the different parameters were optimized. The human serum samples were also analyzed under optimum conditions (Table 6.12).

Figure 6.12 shows the binding of the CA antigen in solution to the anti-CA immobilized on the SPR sensor chip surface (Suwansara-ard et al., 2009). A single-fractal analysis is adequate to describe the binding kinetics. The values of the binding rate coefficient, k, and the fractal dimension, D_f, for a single-fractal analysis are given in Table 6.2.

It is of interest to compare the single-fractal analysis binding of the CA antigen in solution to the anti-CA immobilized on the SPR sensor chip surface (Suwansara-ard et al., 2009) with the binding (cancer monitoring) in mouse 2 and 3 (Daniel et al., 2009). At the outset it should be noted that the comparison is being made between cancer monitoring in mice (*in vivo*) and the binding of CA in solution to anti-CA immobilized on a SPR biosensor chip surface. Furthermore, the fractal dimension values obtained for mouse 2 and 3 are equal to zero. This indicates that the surface acts like a "Cantor-like" dust. The fractal dimension for the *in vitro* binding of the CA is 2.1056. As one goes from the *in vivo* monitoring in mice to the *in vitro* binding of CA, both the fractal dimension and the binding rate

Table 6.12: Binding rate coefficients and fractal dimension for CA 125 binding to anti-CA 125 immobilized on a surface prasmon resonance biosensor (Suwansara-ard et al., 2009).

Analyte in Solution/ Receptor on Surface	k	D_f
CA 125/anti-CA 125	2.7070 ± 0.2822	2.1056 ± 0.0892

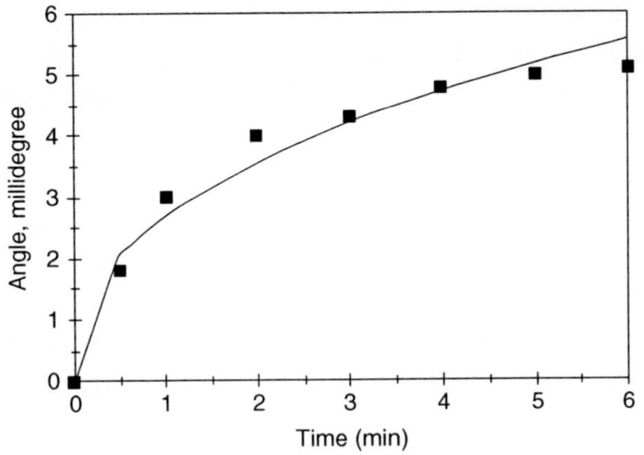

Figure 6.12
Binding of CA antigen in solution to the anti-CA immobilized on the SPR sensor chip.

coefficient increase. This is in accord with the results presented in the different chapters throughout the book wherein, generally, increases in the fractal dimension lead to increases in the binding rate coefficient. This is for the same analyte-receptor system. Here an attempt is made to compare the binding of the same analyte (albeit not quite the same, but cancer in general) to different types of biosensor systems. The intent is to provide fresh physical insights into the binding of cancer and other analytes to different biosensor systems so that a better understanding of the interactions occurring on the different surfaces may be obtained, leading to a better control and management of these interactions in desired directions.

6.5 Conclusions

A fractal analysis is presented for the binding and dissociation (if applicable) of the same analyte to different biosensor surfaces. The examples are analyzed in such a manner as to provide fresh physical insights into these types of analyte-receptor interactions occurring on biosensor surfaces. The intent of the analysis was to provide a fresh and new avenue by which a better understanding of these analyte-receptor interactions selected as model reactions would help pave the way to promote a clearer and better understanding of these reactions occurring on surfaces, in general.

The five sets of reactions selected at random from the literature include:

(1) The amperometric detection of *E. coli* using electropolymerization and coating of GCEs with pyrrole amine (Pyy-NH_2) (Abu-Rabesh et al., 2009), and the real-time PCR amplification of aptamers for the detection of *E. coli* (Lee et al., 2009).

The binding of *E*. coli to both the immunosensor with PAPG (Abu-Rabesh et al., 2009), and to the antibody-conjugated magnetic beads (Lee et al., 2009) is described by a single-fractal analysis. This indicates that the binding (and dissociation) mechanism is not complex. There is an order of magnitude change in the *E. coli* concentration in solution as one goes from the immunosensor with PAPG (10^7 CFU/ml) (Abu-Rabesh et al., 2009) to the antibody-conjugated magnetic beads (10^8 CFU/ml) (Lee et al., 2009). Also, no dissociation is exhibited during the binding of 10^7 CFU/ml *E. coli* in solution to the immunosensor. This indicates that either the binding in this case is strong enough to prevent the dissociation or the *E. coli* concentration is low. It is also seen that the fractal dimension exhibited during the binding of 10^8 CFU/ml *E. coli* in solution to the antibody-conjugated magnetic beads is higher by 18.14% when compared with the fractal dimension exhibited during the binding of 10^7 CFU/ml in solution to the immunosensor. The fractal dimension is based on a log scale and even small changes in the value of the fractal dimension represent significant changes in the degree of heterogeneity on the biosensor surface.

(2) An electrochemical aptamer-based assay coupled to magnetic beads or the detection of thrombin (Centi et al., 2008), and an EIS biosensor for analyzing aptamer-thrombin interfacial interactions (Li et al., 2008)

A comparison of the binding and dissociation of 60 nM thrombin in solution to an EIS biosensor (Figure 6.4; Li et al., 2008) with the binding of 20 nM thrombin in solution to the biotinylated aptamer immobilized on a CMB chip (Centi et al., 2008) indicates that in one case there is binding and dissociation, and in the other just binding. On comparing the binding phase kinetics for both these biosensor systems that are adequately described by a single-fractal analysis, it is observed that as one goes from the CMB chip (Centi et al., 2008) to the EIS biosensor (Li et al., 2008), the fractal dimension in the binding phase decreases by 24.1% from a value of D_f equal to 2.38632 to D_f equal to 1.810, and the corresponding binding rate coefficient increases by a factor of 47.9% from a value of k equal to 83.044 to k equal to 122.88. In this case, a lower fractal dimension leads to a higher binding rate coefficient. This is contrary to the general trend presented in the different chapters throughout the book. However, it should be borne in mind that we are comparing and analyzing the performance of two different biosensors which are detecting the same analyte (albeit of different concentrations; 20 and 60 nm).

(3) An ultrasensitive enhanced CL enzyme immunoassay for detecting AFP which was amplified by double-codified gold nanoparticle labels (Yang et al., 2009), and a LSPCF fiber-optic biosensor to detect AFP in human serum (Chang et al., 2009).

A comparison of the binding and dissociation rate coefficients and the corresponding fractal dimension values for the binding and dissociation of AFP in solution + PIP to the double-codified gold nanoparticle labels modified by HRP-conjugated anti-AFP (Yang et al.,

2009) with the binding and dissociation of 0.1 ng/mL AFP to the anti-AFP immobilized on a SPR biosensor surface (Chang et al., 2005) indicates that in both these cases a dual-fractal analysis is required to adequately describe the binding kinetics, and the dissociation kinetics may be described by a single-fractal analysis.

As one goes from the binding of AFP in solution to anti-AFP immobilized on a SPR biosensor surface (Chang et al., 2005), to the binding of AFP in solution with PIP to the double-codified HRP-conjugated anti-AFP (Yang et al., 2009) the fractal dimension, D_{f1}, decreases by a factor of 2.62 from a value of D_{f1} equal to 1.2632 to D_{f1} equal to 0.4828 for a dual-fractal analysis. The binding rate coefficient, k_1, decreases by a factor of 18.14 from a value of k_1 equal to 0.3367 to 0.01856. It must be noted, however, that in this case as the fractal dimension, D_{f2}, increases from a value of 0.0 to 1.2752 as one goes from the binding of AFP in solution to the anti-AFP immobilized on a SPR biosensor surface (Chang et al., 2005) to the binding of AFP in solution with PIP to the double-codified gold nanoparicle (DC-AuNPs) labeled modified HRP-conjugated anti-AFP, the binding rate coefficient, k_2, increases by a factor of 31327 from a value of k_2 equal to 0.000886 to k_2 equal to 277.56. This is a substantial increase of more than five orders of magnitude. This is because the fractal dimension on the SPR biosensor surface was equal to 0.0, which represents a Cantor-like dust.

(4) A novel biosensor using a modified GCE for the detection of glucose (Sheng et al., 2008), the binding of glucose in solution to the electroless plated Au/Ni/copper low electrical resistance electrode (Lee et al., 2008), the long-term stability of a glucose biosensor based on the insertion of barrel plating gold electrodes (Hsu et al., 2009a,b,c), and a percutaneous fiber-optic sensor for chronic glucose monitoring *in vivo* (Liao et al., 2008).

A comparison of the binding of glucose in solution to the neodymium hexacyanoferrate nanoparticle on the glucose oxidase/chitosan-modified GCE (Sheng et al., 2008) with that of the binding of 1-53 mM glucose in solution to the electroless-plated Au/Ni/copper low electrical resistance electrode (Lee et al., 2008) indicates that in the first case a dual-fractal analysis is required to adequately describe the binding case presented. In the second case, a single-fractal analysis is adequate to describe the binding kinetics for all the three concentrations of glucose (21, 27, and 53 mM) analyzed. In the first case, a complex binding mechanism is involved since a dual-fractal analysis is required to adequately describe the binding kinetics, whereas in the second case a simple binding mechanism is involved since a single-fractal analysis is adequate to describe the binding kinetics. Also, in the second case, even though the fractal dimension or the degree of heterogeneity on the biosensor surface, D_f, equal to 2.8850, 2.9788, and 3.0 are higher than those in the second case for the second phase (D_{f2} equal to 2.604), the binding rate coefficient, k, for the second case is lower than the binding rate coefficient, k_2. This is due, of course, to the different biosensor systems involved.

(5) An implantable diagnostic device for cancer monitoring (Daniel et al., 2008), and the binding of CA 123 in solution to anti-CA antibody immobilized on an SPR biosensor chip surface (Suwansara-ard et al., 2009).

A comparison is made of the single-fractal analysis binding of the CA antigen in solution to the anti-CA immobilized on the SPR sensor chip surface (Suwansara-ard et al., 2009) with the binding (cancer monitoring) in mouse 2 and 3 (Daniel et al., 2009). At the outset it should be noted that the comparison is being made between cancer monitoring in mice (*in vivo*) and the binding of CA in solution to anti-CA immobilized on a SPR biosensor chip surface. Furthermore, the fractal dimension value obtained from mouse 2 and 3 is equal to zero. This indicates that the surface acts like a "Cantor-like" dust. The fractal dimension for the *in vitro* binding of the CA is 2.1056. As one goes from the *in vivo* monitoring in mice to the *in vitro* binding of CA, both the fractal dimension and the binding rate coefficient increase. This is in accord with the results presented in the different chapters throughout the book wherein, generally, increases in the fractal dimension lead to increases in the binding rate coefficient. This is for the same analyte-receptor system. Here an attempt is made to compare the binding of the same analyte (albeit not quite the same, but cancer in general) to different types of biosensor systems. The intent is to provide fresh physical insights into the binding of cancer and other analytes to different biosensor systems so that a better understanding of the interactions occurring on the different surfaces may be obtained, which should lead to a better control and management of these interactions in desired directions.

Many more of these types of examples are available in the literature. It is hoped that these five sets of examples to be presented together will help set the stage for the analysis of other examples. The intent, as indicated previously, is to obtain better physical insights into these types of examples. Of course, any further insight that may be obtained for biomedically-medically oriented analytes will prove invaluable.

References

Abdel-Hamid I, D Ivnitskii, P Alanasov, and E Wilkins, *Biosensors & Bioelectronics*, **14**(3), 309–316 (1999a).

Abdel-Hamid I, D Ivnitskii, P Alanasov, and E Wilkins, *Analytica Chimica Acta*, **199**(1-2), 99–108 (1999b).

Abu-Rabesh K, A Ashkenazi, D Atlas, L Amir, and RS Marks, Highly sensitive amperometric immunosensor for the detection of *Escherichia coli*, *Biosensors & Bioelectronics*, **24**(12), 3461–3468 (2009).

Agarwal PK, PC Black, and AM Kamat, *World Journal of Urology*, **26**, 39–44 (2008).

Albareda-Savant M, A Merkocci, and S Alegret, *Sensors & Actuators B*, **69**, 153–163 (2000).

Alexeev VL, S Das, DN Finegold, and SA Asher, *Clinical Chemistry*, **50**, 3353–3360 (2004).

Alves WA, PA Florito, SI Cordosa de Torresi, and RM Torresi, Design of molecular wires based on supramolecular structures for application in glucose biosensors, *Biosensors & Bioelectronics*, February 14, 2006.

American Diabetes Association ,American diabetes association, *Diabetic Care*, **17**, 81–86 (1994).

American Diabetes Association ,American diabetes association, *Diabetic Care*, **22**, 877–878 (1999).

Ao LM, F Gao, BF Pan, R Ho, and PX Cui, *Analytical Chemistry*, **76**, 1104–1106 (2006).

Asher SA, VL Alexeev, AV Gopenenko, AC Sharma, IK Lednev, CS Wilcox, and DN Finegold, *Journal of the American Chemical Society*, **125**, 3322–3329 (2003).

Badugu R, JR Lakowicz, and CD Geddes, Noninvasive continuous monitoring of physiological glucose using a monosaccharide-sensing contact lens, *Analytical Chemistry*, **76**, 610–618 (2004).

Bakker E, *Analytical Chemistry*, **76**(12), 3285–3298 (2004).

Baldrich E, A Restrepo, and K Sullivan, *Analytical Chemistry*, **76**, 7053–7063 (2004).

Bang GS, S Cho, and BG Kim, *Biosensors & Bioelectronics*, **21**, 863–870 (2005).

Bange A, HB Halsall, and WR Heinemann, *Biosensors & Bioelectronics*, **20**, 2488–2503 (2005).

Baron AT, CH Beardman, JM Lafky, A Rademaker, DC Liu, DA Fishman, KC Pockratz, and NJ Mahle, *Cancer Epidemiology Biomarkers and Prevention*, **14**, 1583 (2005a).

Baron R, M Zayata, and I Willner, *Analytical Chemistry*, **77**, 1566–1571 (2005b).

Ben-Moshe M, VL Alexeev, and SA Asher, *Analytical Chemistry*, **78**, 3149–3157 (2006).

Berggren C and G Johansson, *Analytical Chemistry*, **69**, 3651–3657 (1997).

Bi S, H Zhou, and H Zhang, Multilayers enzyme-coated carbon nanotubes as biolabel for ultrasensitive chemiluminescence immunoassay of cancer biomarker, *Biosensors & Bioelectronics*, **24**(10), 2961–2966 (2009).

Biomerieux ,2004, VIDAS, CA 125, France.

Bouquita M, JP Hart, and R Pittson, *Biosensors & Bioelectronics*, **15**(5-6), 57–263 (2000).

Buchanan RL, *Emerging Infectious Diseases*, **3**(4), 517–521 (1997).

Cai H, T Ming-Huang, and IM Huang, *Sensors & Actuators B Chemical*, **114**, 433–4437 (2006).

Carney WP, *Expert Review of Molecular Diagnostics*, **7**, 613–622 (2007).

Centi S, G Messina, S Tombelli, I Palchetti, and M Mascini, Different approaches for the detection of thrombin by an electrochemical aptamer-based assay coupled to magnetic beads, *Biosensors & Bioelectronics*, **23**(11), 1602–1609 (2008).

Chang YF, RC Chen, YJ Lee, SC Chao, YC Li, and C Chou, Localized surface plasmon coupled fluorescence fiber-optic biosensor for alpha-fetoprotein detection in human serum, *Biosensors & Bioelectronics*, **24**(6), 1810–1814 (2008).

Chang YF, RC Chen, YJ Lee, SC Chao, LC Su, YC Li, and C Chou, Localized surface plasmon coupled fluorescence fiber-optic biosensor for alpha-fetoprotein detection in human serum, *Biosensors and Bioelectronics*, **24**(6), 1610–1614 (2009).

Chau LK, YF Lin, GF Cheng, and TJ Lin, *Sensors & Actuators, Chemical*, **113**, 100–105 (2006).

Chen XM, Z Cai, ZJ Lin, TT Jia, HZ Liu, Y Jia, and X Chen, A novel non-enzymatic ECL sensor for glucose using palladium nanoparticle supported on functional carbon nanotubes, *Biosensors & Bioelectronics*, **24**(12), 3475–3480 (2009).

Chen ZF and YB Zu, *Langmuir*, **23**, 11387–11390 (2007).

Clark WL, *Diabetes Technology and Therapy*, **7**, 776–779 (2005).

Crouch E, DC Cowell, S Hawkins, RW Pittson, and JP Hart, *Biosensors & Bioelectronics*, **21**(5), 712–716 (2005).

Cui G, JH Yoo, BW Woo, SS Kim, GS Cha, and H Nam, Disposable amperometric glucose sensor electrode with enzyme-immobilized nitrocellulose strip, *Talanta*, **54**, 1105–1111 (2001).

Dai Z, F Yan, J Chen, and H Ju, *Analytical Chemistry*, **75**, 5429–5434 (2003).

Daniel KD, GY Kim, CC Cassallou, M Galindo, AR Gulmarses, R Weissleder, AJ Charest, R Langer, and MJ Ciena, Implantable diagnostic device for cancer monitoring, *Biosensors & Bioelectronics*, **24**(11), 3252–3257 (2009).

Darain F, DS Park, JS Park, SC Chang, and YB Shim, *Biosensors & Bioelectronics*, **20**(9), 1780–1787 (2005).

Deng S, G Jian, J Lei, Z Hu, and H Ju, A glucose biosensor based on direct electrochemistry of glucose oxidase immobilized on nitrogen-doped carbon nanotubes, *Biosensors & Bioelectronics*, **25**(2), 373–377 (2009).

Ding Y, H Liu, G Shi, J Liu, G Shen, and S Yu, Cell-based immobilization strategy for sensitive piezoelectric immunoassay of total specific antigen, *Biosensors & Bioelectronics*, **24**(2), 228–232 (2008).

Dudak FC and IH Boyad, *Food Research International*, **40**, 803–807 (2007).

Ehrhart JC, B Bennetau, L Renauld, JP Madrnge, L Thomas, J Morisot, A Brosseau, S Allano, P Tauc, and PL Tan, A new immunosensor for breast cancer detection using antibody-coated long alkylsolane self-assembled monolayers in a parallel plate flow chamber, *Biosensors & Bioelectronics*, **24**(3), 467–474 (2008).

Ellington AD and JW Szostak, *Nature*, **346**, 818–822 (1990).

Elzbag J, B Szlyahovsky, and I Willner, *Chemical Communications*, **13**, 1569–1571 (2008).

Endo K, Y Mtsuoka, T Nakashima, S Fuji, M Kumimatsu, T Saga, Y Watanabe, Y Kawamura, M Koizumi, K Konishi, K Toizuka, and RC Bast, *Journal of Tumor Marker Oncology*, **3**, 65–71 (1988).

Escamilla-Gomez V, D Hernandez-Sntos, MB Gonzalez-Garfcia, JM Pigarron-Carrazon, and A Costa-Garcia, Simultaneous detection of free and total prostate specific antigen on screen-printed electrochemical dual sensor, *Biosensors & Bioelectronics*, **24**(8), 2678–2683 (2009).

Fahnrich KA, M Pravada, and GG Guibault, *Biosensors & Bioelectronics*, **19**(1), 73–82 (2003).

Fink D, I Klinkovich, D Bukelman, RS Marks, A Kiv, D Fuks, WR Fahmer, and L Alfonta, Glucose determination using a re-usable enzyme-modified ion track membrane reactor, *Biosensors & Bioelectronics*, **24**(8), 2702–2706 (2009).

Fu ZF, C Hao, XG Fei, and HX Tu, *Journal of Immunological Methods*, **312**, 61–67 (2006).

Fu Z, F Yan, H Liu, J Lin, and H Ju, *Biosensors & Bioelectronics*, **23**, 1422–1428 (2008).

Gao C, K Cui, F Yang, Y Ma, and K Yang, *Biosensors & Bioelectronics*, **19**(3), 277–282 (2003).

Ge F, KE Zhang, ZP Zhang, and Z Xiao-Mei, *Biosensors & Bioelectronics*, **13**(3-4), 333–339 (1998).

Georganopoulos DG, L Chang, JM Nam, GS Thaxton, CJ Mudson, WL Klein, and CA Mirkin, *Proceedings of the National Academy of Sciences USA*, **102**, 2273–2276 (2005).

George I, M Petit, and P Servais, *Journal of Applied Microbiology*, **88**(3), 404–413 (2000).

Gomez-Henz A, JM Fernandez-Romero, and MP Aguiliar-Carballos, *Trends in Analytical Chemistry*, **27**, 394–406 (2008).

Gomez-Rio M, DMD Torres, A Rodriguez-Fernandez, JM Llamas-Elvira, SO Lozanao, CR Font, EL Ramirez, and M Katall, *Journal of Nuclear Medicine and Molecular Imaging*, **31**, 1237–1243 (2004).

Gronewold TMA, S Glass, E Quiandt, and MM Famulok, *Biosensors & Bioelectronics*, **20**, 2044–2052 (2005).

Hale PD, LI Bogushavsky, T Inagaki, HI Karan, HS Lee, TA Skothelm, and TA Okamato, *Analytical Chemistry*, **63**, 677–682 (1991).

Hart JP, AK Abasse, and D Cowell, *Biosensors & Bioelectronics*, **17**(5), 389–394 (2002).

Hashimoto M, N Sakamoto, S Upadhyay, J Fukuda, and H Suzuki, *Biosensors & Bioelectronics*, **22**(12), 3154–3160 (2007).

Havlin S, Molecular diffusion and reaction. In *The Fractal Approach to Heterogeneous Chemistry: Surfaces, Colloids, Polymers*, Avnir D (Ed.), Wiley, New York, 1989, pp. 251–269.

He Z, N Gao, and W Jin, *Analytica Chimica Acta*, **497**, 75–81 (2003).

Hecht MJ, D Schoenberg, H Kallasz, and RD Schmid, *Biosensors & Bioelectronics*, **8**, 197–203 (1993).

Hermann T. and DJ Patel, *Science*, 287 (2000).

Hianik V, V Ostana, Z Zajcova, and G Evtugyin, *Bioorganic Medical Chemistry Letters*, **16**, 291–295 (2005).

Ho HA and M Leclerc, *Journal of the American Chemical Society*, **126**, 1384–1387 (2004).

Honeychurch KC, JP Hart, PRJ Pritchard, SJ Hawkins, and NM Ratcliffe, *Biosensors & Bioelectronics*, **19**(4), 305–312 (2003).

Hong MY, JY Chang, HC Yoon, and HS Kim, *Biosensors & Bioelectronics*, **17**(1-2), 13–18 (2002).

Hsieh BY, YF Cheng, MY Ng, WC Liu, CH Lin, HT Wu, and C Chau, *Analytical Chemistry*, **79**, 3487–3493 (2007).

Hsu CT, HH Chung, DM Tsai, MY Fang, HC Hsiao, and JM Zen, *Analytical Chemistry*, **81**, 515–518 (2009a).

Hsu CT, HC Hsiao, MY Fang, and JM Zen, Superior long-term stability of a glucose biosensor based on inserted barrel-plating gold electrodes, *Biosensors & Bioelectronics*, **25**(2), 383–387 (2009b).

Hsu CT, HC Hsiao, MS Lee, SF Chang, TC Lee, YS Tsai, and JM Zen, *Clinica Chimica Acta*, **402**, 119–123 (2009c).

Hwang S, E Kim, and J Kovak, *Analytical Chemistry*, **77**, 579–584 (2005).

Ikebukoro K, C Kiyohara, and K Sode, *Biosensors & Bioelectronics*, **20**, 2168–2172 (2005).

ISO , 2003, ISO 15197, 2003.

Ivnitskii D, T Abdel-Hamid, P Alanasov, and E Wilkins, *Biosensors & Bioelectronics*, **14**(7), 599–624 (1999).

James TD and S Shinkai, In *Host-Guest Chemistry*, S Penades (Ed.), Springer Verlag, Berlin, Germany, 2002, Vol 218, pp 159–200.

Jia WZ, K Wang, ZJ Zhu, HT Song, and XH Xia, *Langmuir*, **23**, 11896–11900 (2007).

Jokerst JV, A Ramananthan, N Christoulides, PN Floriano, AA Pollard, CW Simmons, J Wong, C Gage, WB Furmaga, SW Redding, and JT McDevitt, *Biosensors & Bioelectronics*, **24**(12), 3622–3629 (2009).

Kabilan S, AJ Marshall, FK Sartain, MC Lee, A Hussain, XP Yang, J Blyth, N Krangu, K James, J Zeng, D Smith, A Domischke, and CR Lowe, *Biosensors & Bioelectronics*, **20**, 1602–1610 (2005).

Kang X, Z Mai, X Zou, P Cai, and J Mo, A sensitive nonenzymatic glucose sensor in alkaline media with a copper nanocluster/multiwall carbo nanotube-modified glassy carbon electrode, *Analytical Biochemistry*, **363**, 143–150 (2007).

Katz E, I Willner, and J Wang, *Electroanalysis*, **16**, 19–44 (2004).

Kim JP, BY Lee, J Lee, S Hong, and SJ Sim, Enhancement of sensitivity and specificity by surface modification of carbon nanotubes in diagnosis of prostate cancer based on carbon nanotube field effect transistors, *Biosensors & Bioelectronics*, **24**(11), 3372–3378 (2009).

Kondepatu VR and HM Heise, *Analytical and Bioanalytical Chemistry*, **388**, 545–563 (2007).

Konty A, Y Novoa, N Shemer-Avni, N Hnuka, S Cosnier, A Lepeller, and RS Marks, *Analytical Chemistry*, **77**, 1771–1779 (2005).

Kuruoda N, P Shimoda, M Wada, and S Nakashima, *Analytica Chimica Acta*, **403**, 131–136 (2000).

Kuwahara T, H Ohta, M Kondo, and M Shimomura, *Bioelectrochemistry*, **74**, 68–72 (2008).

Lee SR, YT Lee, K Rawada, H Tako, and M Ishida, Development of a disposable glucose biosensor using electroless-plated Au/Ni/copper low electrical resistance electrodes, *Biosensors & Bioelectronics*, **24**(3), 410–414 (2008).

Lee HJ, DC Kim, KW Kim, YK Kim, J Kim, and MK Koh, A sensitive method to detect *Escherichia coli* based on immunomagnetic separation and real-time PCR amplification of aptamers, *Biosensors & Bioelectronics*, **24**(12), 3550–3555 (2009).

Le Floch F, HA Ho, and M Leclerc, *Analytical Chemistry*, **78**, 4727–4731 (2008).

Leonard P, S Hearty, J Brennan, L Dunne, J Quinn, T Chakraborty, and R O'Kennedy, *Enzyme and Microbial Technology*, **329**(1), 3–13 (2003).

Li CZ, Y Liu, and JHT Luong, *Analytical Chemistry*, **77**, 478–485 (2005).

Li X, L Shen, D Zhang, H Qi, Q Gao, F Ma, and Q Zheng, Electrochemical impedance spectroscopy for study of aptamer-thrombin interfacial interactions, *Biosensors & Bioelectronics*, **23**(11), 1624–1630 (2008).

Li Y, X Liu, H Yuan, and D Xiao, Glucose biosensor based on the room-temperature phosphorescence of TiO_2/SiO_2 nanocomposite, *Biosensors & Bioelectronics*, **24**(12), 3706–3710 (2009).

Liao KC, T Hogen-Esch, FJ Richmond, L Marcu, and GE Loeb, *Optical Fiber and Sensors for Medical Applications, V. Proceedings of SPIE*, **5691**, 129–145 (2005).

Liao KC, T Hogen-Esch, FJ Richmond, L Marcu, W Clifton, and GE Loeb, Percutaneous fiber-optic sensor for chronic glucose monitoring *in vivo*, *Biosensors & Bioelectronics*, **23**(10), 1458–1465 (2008).

Lim SH, J Wei, J Lin, Q Li, and K Jin, *Biosensors & Bioelectronics*, **20**, 2341–2346 (2005).

Limbut W, P Kanatharana, B Mattiasson, P Asawatratankul, and P Thavanungkul, *Biosensors & Bioelectronics*, **22**(2), 233–240 (2006).

Lin J and H Ju, *Biosensors & Bioelectronics*, **20**, 1461–1470 (2005).

Lin YH, H Chen, and YC Chuang, *Biosensors & Bioelectronics*, **23**(12), 1832–1837 (2008a).

Lin YH, SH Chen, YC Chuang, YC Lu, TY Shen, CA Chang, and CH Lin, Disposable amperometric immunosensing strips fabricated by Au nanoparticles-modified screen-printed carbon electrodes for the detection of foodborne pathogen *Escherichia coli* O157:H7, *Biosensors & Bioelectronics*, **23**(12), 1832–1837 (2008b).

Lin YY, J Wang, G Liu, H Wu, CM Wai, and Y Lin, A nanoparticle label/immunochromatographic electrochemical biosensor for rapid and sensitive detection of prostate-specific antigen, *Biosensors & Bioelectronics*, **23**(11), 659–665 (2008c).

Liu JW and Y Lu, *Advances in Materials*, **18**, 1667–1671 (2006).

Liu S, Y Li, J Li, and L Jiang, *Biosensors & Bioelectronics*, **21**, 789–795 (2005).

Liu Q, J Yu, L Xiao, J Cheuk, O Tang, Y Zhang, P Wang, and M Yang, Impedance studies of bio-behavior and chemosensitivity of cancer cells by micro-electrode arrays, *Biosensors & Bioelectronics*, **24**(5), 1305–1310 (2009).

Lo CY, JM Luk, and C Tam, *Annals of Surgery*, **236**, 564–569 (2002).

Macaya RF, P Schultze, F Smith, JA Rose, and J Feigon, *Proceedings of the National Academy of Sciences USA*, **90**, 3745–3749 (1993).

Malitesta C, F Palmisano, I Torai, and PG Zambonin, *Analytical Chemistry*, **62**, 2735–2740 (1990).

Manesh KM, P Santhosh, A Gopalan, and KP Lee, Electrospun poly(vinylidene fluoride)/Poly (aminophenylboronic acid) composite nanofibrous membrane as a novel glucose sensor, *Analytical Biochemistry*, **360**, 189–195 (2007).

Mao KL, LJ Yang, XL Su, and YB Li, *Biosensors & Bioelectronics*, **21**, 1178–1185 (2006).

Marquette M, M Pinto, H larry, MD Bernstein, A Dennis, MP Brojan, and C Elaine, *Cancer*, **59**, 218–222 (1987).

Marquette SA, RP Barron, A Miessien, R Madgasakan, W Korz, TR Sykes, CJ Sykes, G Hor, AJB Maewan, and AA Noujalin, *Nuclear Medicine Communications*, **18**, 878–886 (1997).

Martina R, P Baptista, L Rainero, G Doria, L Silva, R Franco, and E Fortunato, *Applied Physics Letters*, **90**, 023903 (2007).

Matsui J, K Akamatsu, N Hara, D Myoshi, H Nawashune, K Tamaki, and N Sugimoto, *Analytical Chemistry*, **77**, 4282–4285 (2005).

Mazumdar SD, B Barlen, T Kramer, and M Keusgen, *Journal of Microbiological Methods*, **75**, 545–550 (2008).

Meadows D and J Schultz, *Analytica Chimica Acta*, **280**, 21–30 (1993).

Medintz IL, HT Hyeda, ER Goldman, and N Malthousi, *Nature Materials*, **4**, 435–447 (2005).

Mehavir M and M Abidi, *Analytical Science*, **20**(8), 1113–1116 (2004).

Mir M, M Vreeke, and I Katakis, *Electrochemical Communications*, **8**, 505–511 (2006).

Miscoria SA, J Desbrieres, GD Barrera, P Labbe, and GA Rivas, *Analytica Chimica Acta*, **678**, 137–144 (2006).

Mittelmann AS, EZ Ron, and J Rishpon, *Analytical Chemistry*, **74**(4), 903–907 (2002).

Moller R, RD Powell, JF Heinfeld, and W Fritzsche, *Nano Letters*, **5**, 1475–1482 (2005).

Monica M, V Mark, and I Katakis, *Electrochemical Communication*, **8**, 505–511 (2006).

Mujika M, S Arana, E Castano, M Tijero, R Vilares, JM Ruano-Lopez, A Cruz, L Sainz, and J Berganza, Magnetoresistive immunosensor for the detection of *Escherichia coli* O157:H7 including a microfluidic network, *Biosensors & Bioelectronics*, **24**(5), 1253–1258 (2009).

Nakamura N and I Karube, *Analytical and Bioanalytical Chemistry*, **377**(3), 465–468 (2003).

Neumann JD and APF Turner, *Biosensors & Bioelectronics*, **20**(12), 2435–2453 (2005).

O'Sullivan CK, *Analytical and Bioanalytical Communication*, **372**, 44–48 (2002).

Owen VM, *Biosensors & Bioelectronics*, **10**(3-4), iv (1985).

Paborsky LR, SN McCurdy, LC Griffin, JJ Toole, and LL Leung, *Journal of Biological Chemistry*, **268**, 20808–20811 (1993).

Palenzuela B, BM Simonet, RM Garda, A Rios, and M Valcarcel, *Analytica Chimica Acta*, **524**(1-2), 167–174 (2004).

Park S, TD Chung, and HC Kim, Nonenzymatic glucose detection using mesoporous platinum, *Analytical Chemistry*, **75**, 3046–3049 (2003).

Pavlov V, Y Xiao, B Shlyahovsky, and I Willner, *Journal of the American Chemical Society*, **126**, 11768–11769 (2004).

Polsky R, R Gill, L Kaganovsky, and I Willner, *Analytical Chemistry*, **78**, 2268–2271 (2006).

Prabhulkar S, S Alwarappan, G Liu, and CZ Li, Amperometric micro-immunosensor for the detection of tumor marker, *Biosensors & Bioelectronics*, **24**(12), 3524–3530 (2009).

Prakash SB and P Abshire, Tracking cancer cell proliferation on a CMOS capacitance sensor chip, *Biosensors & Bioelectronics*, **23**(10), 1449–1457 (2008).

Radi AE, JL Acero Sanchez, E Baldrich, and CK O'Sullivan, *Analytical Chemistry*, **77**, 6320–6323 (2005).

Radi AE, JL Acero-Sanchez, E Baldrich, and CK O'Sullivan, *Journal of the American Chemical Society*, **128**, 117–124 (2006).

Rashkovetsky LG, YV Lyubarskaya, F Foret, DE Hughes, and BL Karger, *Journal of Chromatography A*, **781**, 197–204 (1997).

Renedo OD, MA Alfonso-Lomillo, and MJA Martinez, *Talanta*, **73**, 202–219 (2007).

Rossi ML and CA Mirkin, *Chemical Reviews*, **105**, 1547–1562 (2005).

Schoenfisch MH, AR Rothrock, Jh Shin, MA Polizzi, MF Brinkley, and KP Dobmeier, Poly(vinylpyrrolidone)-doped nitric oxide-releasing xerogels as glucose biosensor membranes, *Biosensors & Boelectronics*, **22**(2), 306–312 (2006).

Selvaraju T, J Dasa, SW Harib, and H Yang, *Biosensors & Bioelectronics*, **23**, 932–938 (2008).

Sheng G, K Luo, J Zheng, and H Zhang, Enzymatically induced formation of neodymium hexacyanoferrate nanoparticles on the glucose oxidase/chitosan modified glass carbon electrode for the detection of glucose, *Biosensors & Bioelectronics*, **24**(2), 429–434 (2008).

Shlyahovsky S, E Katz, Y Xiao, V Pakov, and I Willner, *Small*, **1**, 213–216 (2005).

Smirnov I and RH Shafer, *Biochemistry*, **39**, 1462–1468 (2000).

So HM, DW Park, EK Jeon, YH Kim, BS Kim, CK Lee, SY Choi, SC Kim, H Chang, and JO Lee, *Small*, **4**, 197–201 (2008).

Sodlaczek P, I Frydecka, M Gabrys, A van Dalen, R Einarsson, and A Harlozhinska, *Cancer*, **85**, 1885–1893 (2002).

Sokol LJ, FH Williams, and AT Ramaley, *Clinical Chemistry*, **50**, 1126–1135 (2004).

Stell GM, AE Partelson, and K Rebrin, *Advances in Drug Delivery Reviews*, **56**, 125–144 (2004).

Sulak MT, O Gokdagan, A Golice, and H Golice, *Biosensors & Bioelectronics*, **21**(9), 1719–1726 (2006).

Sun H, TS Choy, DR Zhu, WC Yam, and YS Fung, Nano-silver-modified PQC/DNA biosensor for detecting *E. coli* in environmental water, *Biosensors & Bioelectronics*, **24**(5), 1405–1410 (2009).

Suwansara-ard S, P Kanatharana, and P Asawateratanakul, Comparison of surface plasmon resonance and capacitive immunosensors for cancer antigen 125 detection in human srum samples, *Biosensors & Bioelectronics*, **24**(12), 3436–3441 (2009).

Takamura Y and H Iwata, *Analytical Biochemistry*, **365**, 201–207 (2007).

Takeguchi H, M Kaajima, and Y Kitagawa, *Cancer Science*, **99**, 441–450 (2008).

Tan W, L Sabet, Y Li, T Yu, PR Klokkevold, DT Wong, and CM Ho, Optical sensor for detecting cancer markers in saliva, *Biosensors & Bioelectronics*, **24**(2), 268–271 (2008).

Tang D, R Yuan, and Y Chai, *Analytica Chimica Acta*, **564**, 156–165 (2006).

Tierney S, DR Hjelme, and BT Stokke, *Analytical Chemistry*, **80**, 5086–5093 (2008).

Tierney S, BMH Falch, DR Hjelme, and BT Stokke, Determination of glucose levels using a functionalized hydrogel-optical fiber biosensor: Toward continuous monitoring of blood glucose *in vivo*, *Analytical Chemistry*, **81**, 3630–3636 (2009a).

Tierney S, S Volden, and BT Stokke, *Biosensors & Bioelectronics*, **24**, 3034–3039 (2009b).

Tokarsky O and DL Marshall, *Food Microbiology*, **25**(1), 1–12 (2008).

Tombelli S, M Minunni, and M Mascini, *Biosensors & Bioelectronics*, **20**, 224 (2004).

Tombelli S, M Minunni, and M Mascini, *Biomolecular Engineering*, **24**, 191–200 (2007).

Tryland I and L Fiksdal, *Applied Environmental Microbiology*, **64**(3), 1018–1023 (1998).

Tsai WC and IC Lin, *Sensors & Actuators B, Chemical*, **106**, 455–460 (2005).

Tuerk C and L Gold, *Science*, **249**, 505 (1990).

UK Prospective Diabetes Study Group ,UKPDS, *Lancet*, **352**, 837–853 (1998).

Usmani A and N Akmiai, *Diagnostic Sensor Polymers*, Washington, 1994, American Chemical Society.

Wang J, Y Cao, Y Xu, and G Li, Colorimetric multiplexed immunoassay for sequential detection of tumor markers, *Biosensors & Bioelectronics*, **25**(2), 532–536 (2009a).

Wang X, H Gu, F Yin, and Y Tu, A glucose biosensor based on Prussian-blue/chitosan hybrid film, *Biosensors & Bioelectronics*, **24**(5), 1527–1530 (2009b).

Wenholt RI, J Hoekstra, and J DeVries, *Diabetes Care*, **29**, 1805–1811 (2006).

Wilder JL, E Pavik, JM Straughn, T Kirby, RV Higgins, PD DePriest, FR Ueland, RJ Kryschio, RJ Whitley, and JV Nagell, *Gynecology and Oncology*, **69**, 233–235 (2003).

Williams L, S Rotich, J Qi, R Fineberg, N Espay, A Bruno, A Fineberg, and SE Tierney, *Neurology*, **59**, 67–71 (2002).

Willner I, R Baron, and B Willner, *Advances in Materials*, **18**, 1109–1120 (2006).

Wilson R and APF Turner, *Biosensors & Bioelectronics*, **7**, 165–185 (1992).

Wolfbeis O, *Analytical Chemistry*, **76**, 3859–3874 (2006).

Wu L, J Chen, D Du, and H Ju, *Electrochimica Acta*, **51**, 1208–1214 (2006).

Xiao Y, V Pavlov, S Levine, T Naizov, G Markovitch, and I Willner, *Angewandte Chemistry International Edition*, **43**, 4519–4522 (2004).

Xiao Y, V Pavlov, B Shlyovsky, and I Willner, *Chemistry European Journal*, **11**, 2695–2704 (2005a).

Xiao Y, BD Plorek, and KW Plaxco, *Journal of the American Chemical Society*, **127**, 17990–17991 (2005b).

Xu YY, C Blau, SF Chan, and SH Xia, *Analytica Chimica Acta*, **561**, 48–54 (2006).

Yabuki S, F Mizutani, and Y Kimura, *Biosensors & Bioelectronics*, **7**(10), 695–700 (1992).

Yan G, H Ju, Z Lang, and T Zhang, *Journal of Immunological Methods*, **225**, 1–8 (1999).

Yang KY, YS Guo, S Bi, and SS Zhang, Ultrasensitive enhanced chemilum in escence enzyme immunoassay for the determination of α-fetoprotein amplified by double-codified gold nanoparticle labels, *Biosensors and Bioelectronics*, **24**(8), 2707–2711 (2008).

Yang XY, YS Guo, S Bi, and SS Zhang, Ultrasensitive enhanced chemilum in escence enzyme immunoassay for the determination of α-fetoprotein amplified by double-codified gold nanoparticle labels, *Biosensors and Bioelectronics*, **24**(8), 2707–2711 (2009).

Yin T, W Wei, L Yang, K Gao, and Y Gao, *Sensors & Actuators B*, **117**, 286–294 (2006).

Yoshida W, K Sode, and K Ikebukuro, *Biochemical & Biophysical Research Communications*, **348**, 245–252 (2006).

Yu J, S Liu, and H Ju, Glucose sensor for flow injection analysis of serum glucose based on immobilization of glucose oxidase in titania sol-gel membrane, *Biosensors & Bioelectronics*, **19**, 401–409 (2003).

Zayata M, R Brown, I Popov, and I Willner, *Nano Letters*, **5**, 21–25 (2005).

Zhang MG, A Smith, and W Gorski, *Analytical Chemistry*, **76**, 5045–5050 (2004).

Zhang H, Z Wang, XF Li, and XC Le, *Angewandte Chemistry International Edition*, **45**, 1576–1580 (2006).

Zhang J, W Feng, L Lin, F Zhang, G Cheng, P He, and Y Fang, *Biosensors & Bioelectronics*, **23**, 314–337 (2007).

Zhang X, B Qi, Y Li, and S Zhang, Amplified electrochemical aptasensor for thrombin based on bio-barcode method, *Biosensors & Bioelectronics*, **25**(1), 259–262 (2009a).

Zhang X, G Wang, W Zhang, Y Wei, and B Fang, Fixture-reduce method for the synthesis of Cu_2O/MWCNTs nanocomposites and its application as enzyme-free glucose sensor, *Biosensors & Bioelectronics*, **24**(11), 395–3398 (2009b).

Zhou ND, J Wang, T Chen, and X Li, *Analytical Chemistry*, **78**, 5227–5230 (2006a).

Zhou Y, Y Zhang, C Lau, and J Lu, *Analytical Chemistry*, **78**, 5920–5924 (2006b).

CHAPTER 7

Binding of the Same Analyte (Glucose) to Different Biosensor Surfaces: A Fractal Analysis of the Kinetics

Chapter Outline

7.1 Introduction

Different analytes have been detected in solution and in air by different types of biosensors. It would be of interest to compare and contrast, if possible, the detection of the same analyte by different types of biosensors that are available. It would also provide one with a perspective of the current trends in technology, for example, nanotechnology and nanobiotechnology, and their applications to biosensor development. In essence, are current technology trends being used to arrive at better and better biosensors as far as stability, sensitivity, selectivity, response time, and robustness are concerned?

The initial task is to select the analyte for the purpose of detection. This is not a difficult task as apparently the detection of glucose in solution gave rise to the development of the first biosensor. Thus, in this chapter the detection of glucose in solution by different types of biosensors that have recently appeared in the literature will be analyzed. All examples selected for the analysis of the kinetics of binding and dissociation using fractals were selected at random. There was no particular emphasis on the selection and analysis of one type of biosensor for the detection of glucose in solution. It is the aim of this chapter to

Handbook of Biosensors and Biosensor Kinetics. DOI: 10.1016/B978-0-444-53262-6.00007-3

present the kinetic analysis of the detection of glucose from a wide range of biosensors ranging from the traditional electrochemical methods to the more recent and modern methods involving nanotechnology and nanobiotechnology.

Of course, there are other analytes besides glucose whose detection has been undertaken by different types of biosensors. However, this chapter will concentrate only on the detection of glucose by different biosensors.

Initially, some of the examples for the detection of glucose in solution that have appeared in the more recent literature will be presented. It is important to note that this is not an all-inclusive list, by any means. To help remove any sort of bias, these examples are presented in random order. These examples are: (a) fabrication of microband glucose biosensors using a screen-printed water-based (WB) carbon ink for serum analysis (Pemberton et al., 2009), (b) Poly(vinylpyrrolidine)-doped nitric oxide-releasing xerogels as glucose biosensor membranes (Schoenfisch et al., 2006), (c) a glucose biosensor based on immobilization of glucose oxidase onto multiwall carbon nanotube (CNT)-polyelectrolyte loaded electrospun nanofibrous membrane (Manesh et al., 2008), (d) urine glucose meter based on a microplaner amperometric biosensor and its clinical application for the self-monitoring of urine glucose (Miyashita et al., 2009), (e) carbon post-microarrays for glucose sensors (Xu et al., 2008), (f) glucose sensing electrodes based on a poly(3,4-ethylenedioxythiophene)/Prussian blue bilayer and multiwalled CNTs (Chin et al., 2008), (g) fabrication of a glucose sensor based on a novel nanocomposite electrode (Safavi et al., 2009), (h) the evolution of commercialized glucose sensors in China (Hu, 2009), (i) a sensitive nonenzymatic glucose sensor in alkaline media with a copper nanocluster/multiwall CNT-modified glassy carbon electrode (GCE) (Kang et al., 2007), (j) Pt-Pb nanowire array electrode (NAE) for enzyme-free glucose detection (Bai et al., 2008), (k) preparation of functionalized copper nanoparticles and fabrication of a glucose sensor (Xu et al., 2006), (l) the role of H_2O_2 outer diffusion on the performance of implantable glucose sensors (Vaddiraju et al., 2008), (m) a glucose biosensor based on Prussian blue/chitosan hybrid film (Wang et al., 2009), (n) a novel system based on impedance spectroscopy for noninvasive glucose monitoring in patients with diabetes (Caduff et al., 2006), and (o) home blood glucose biosensors: a commercial perspective (Newman and Turner, 2005).

Out of the examples presented above a few will be selected to determine the binding and dissociation (if applicable) kinetics of glucose in solution to their respective biosensors using fractal analysis. Once again, there is no particular bias in selecting one example over the other for the kinetic analysis.

7.2 Theory

Havlin (1989) has reviewed and analyzed the diffusion of reactants toward fractal surfaces. The details of the theory and the equations involved for the binding and the dissociation phases for analyte-receptor binding are available (Sadana, 2001) in the literature. The details are not repeated here except that the equations are given to permit easier reading. These equations have been applied to other biosensor systems also (Ramakrishnan and Sadana, 2001; Sadana, 2001; Sadana, 2005). For most applications, a single- or a dual-fractal analysis is often adequate to describe the binding and the dissociation kinetics. Peculiarities in the values of the binding and dissociation rate coefficients, as well as in the values of the fractal dimensions with regard to the dilute analyte systems being analyzed will be carefully noted, if applicable.

7.2.1 Single-Fractal Analysis

Binding Rate Coefficient

Havlin (1989) reports that the diffusion of a particle (analyte [Ag]) from a homogeneous solution to a solid surface (e.g. receptor [Ab]-coated surface) on which it reacts to form a product (analyte-receptor complex; (Ab·Ag)) is given by:

$$(\mathrm{Ab{\cdot}Ag}) \approx \begin{cases} t^{(3-D_{\mathrm{f,bind}})/2} = t^p, & t<t_c \\ t^{1/2}, & t>t_c \end{cases} \tag{7.1}$$

Here $D_{\mathrm{f,bind}}$ or D_f (as it is used later on in the Chapter) is the fractal dimension of the surface during the binding step. t_c is the cross-over value. Havlin (1989) points out that the cross-over value may be determined by $r_c^2 \sim t_c$. Above the characteristic length, r_c, the self-similarity of the surface is lost and the surface may be considered homogeneous. Above time, t_c, the surface may be considered homogeneous, since the self-similarity property disappears, and "regular" diffusion is now present. For a homogeneous surface where D_f is equal to 2, and when only diffusional limitations are present, $p = ½$ as it should be. Another way of looking at the $p = ½$ case (where $D_{\mathrm{f,bind}}$ is equal to 2) is that the analyte in solution views the fractal object, in this case, the receptor-coated biosensor surface, from a "large distance." In essence, in the association process, the diffusion of the analyte from the solution to the receptor surface creates a depletion layer of width $(Đt)^{½}$ where $Đ$ is the diffusion constant. This gives rise to the fractal power law, (Analyte·Receptor) $\sim t^{(3-D_{\mathrm{f,bind}})/2}$. For the present analysis, t_c is arbitrarily chosen and we assume that the value of t_c is not reached. One may consider the approach as an intermediate "heuristic" approach that may be used in the future to develop an autonomous (and not time-dependent) model for diffusion-controlled kinetics.

Dissociation Rate Coefficient

The diffusion of the dissociated particle (receptor [Ab] or analyte [Ag]) from the solid surface (e.g., analyte [Ag]-receptor [Ab] complex coated surface) into solution may be given, as a first approximation by:

$$(\text{Ab}\cdot\text{Ag}) \approx -t^{(3-D_{\text{f,diss}})/2} = -t^{p}, \quad t > t_{\text{diss}} \tag{7.2}$$

Here $D_{\text{f,diss}}$ is the fractal dimension of the surface for the dissociation step. This corresponds to the highest concentration of the analyte-receptor complex on the surface. Henceforth, its concentration only decreases. The dissociation kinetics may be analyzed in a manner "similar" to the binding kinetics.

7.2.2 Dual-Fractal Analysis

Binding Rate Coefficient

Sometimes, the binding curve exhibits complexities and two parameters (k, D_{f}) are not sufficient to adequately describe the binding kinetics. This is further corroborated by low values of r^2 factor (goodness-of-fit). In that case, one resorts to a dual-fractal analysis (four parameters; k_1, k_2, D_{f1}, and D_{f2}) to adequately describe the binding kinetics. The single-fractal analysis presented above is thus extended to include two fractal dimensions. The time ($t = t_1$) at which the "first" fractal dimension "changes" to the "second" fractal dimension is arbitrary and empirical. For the most part, it is dictated by the data analyzed and experience gained by handling a single-fractal analysis. A smoother curve is obtained in the "transition" region, if care is taken to select the correct number of points for the two regions. In this case, the product (antibody-antigen; or analyte-receptor complex, Ab·Ag or analyte·receptor) is given by:

$$(\text{Ab}\cdot\text{Ag}) \approx \begin{cases} t^{(3-D_{\text{f1,bind}})/2} = t^{p1}, & t < t_1 \\ t^{(3-D_{\text{f2,bind}})/2} = t^{p2}, & t_1 < t < t_2 = t_{\text{c}} \\ t^{1/2}, & t > t_{\text{c}} \end{cases} \tag{7.3}$$

In some cases, as mentioned above, a triple-fractal analysis with six parameters (k_1, k_2, k_3, D_{f1}, D_{f2}, and D_{f3}) may be required to adequately model the binding kinetics. This is when the binding curve exhibits convolutions and complexities in its shape due perhaps to the very dilute nature of the analyte (in some cases to be presented) or for some other reasons. Also, in some cases, a dual-fractal analysis may be required to describe the dissociation kinetics.

7.3 Results

In this chapter we use fractal analysis to analyze the binding and dissociation (if applicable) kinetics of (a) binding of 1 mM glucose in solution (using CV, cyclic voltametry) to Nf (nafion)-CNTs-Cu (Cu-CNT-GCE) (Kang et al., 2007), (b) Influence of repeat runs on the binding and dissociation of 0.1 mM glucose to DMG (dimethylglycoxime)-CuNP (copper nanoparticles) CME (copper-based chemically modified electrodes) (Xu et al., 2006), (c) binding of different concentrations (in mM) of glucose in 0.1 M PBS solution to Pt-Pb

NAE (Bai et al., 2008), (d) binding of 400 mg/dl glucose in solution (hematocrit adjusted) to 40% FAD-GDH (flavin adenine dinucleotide-dependent glucose dehydrogenase) that was screen-printed on a biosensor strip, (e) binding of blood glucose and urine glucose in postmeal testing to a urine glucose meter (Miyashita et al., 2009), and (f) binding of 6 mM glucose in solution to screen-printed WB CoPC (cobalt phthalocyanine) (electrocatalyst) microband biosensor (Pemberton et al., 2008).

Figure 7.1a-c shows the binding and dissociation of 1 mM glucose in solution (using CV) to Nf (Nafion)-CNTs-Cu(CU-CNT-GCE) (Kang et al., 2007). The influence of repeat runs in chronological order, a,b, and c was analyzed by these authors. They have recently developed a sensitive nonenzymatic glucose sensor in alkaline media with a copper nanocluster /multiwall nanotube-modified GCE.

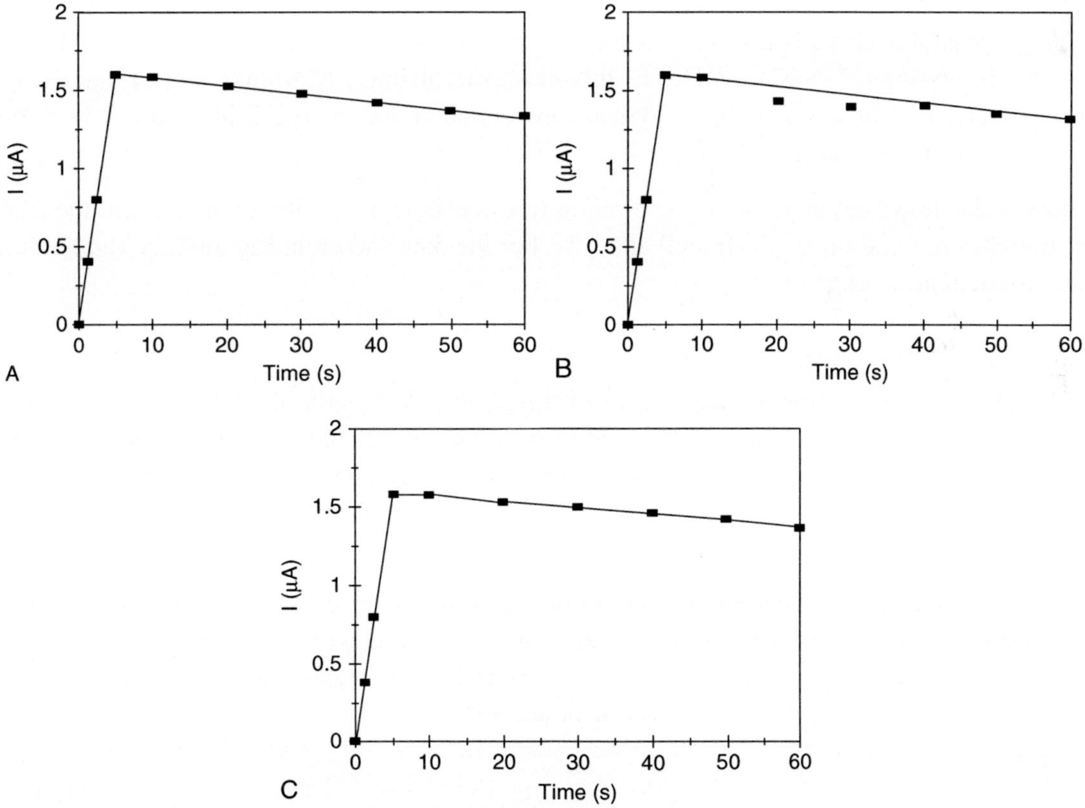

Figure 7.1
Binding of 1 mM glucose in solution (using CV, cyclic voltametry) to Nf (Nafion)-CNTs (carbon nanotube)-Cu (Cu-CNT-GCE) (glassy carbon electrode) (Kang et al., 2007). Influence of repeat runs in chronological order, a, b, and c.

Kang et al. (2007) report that the activity of the enzyme glucose oxides (GO_x) can be easily affected by temperature, humidity, and toxic chemicals (Wilson and Turner, 1992). Thus, the enzymeless glucose sensor is an attractive alternative. Kang et al. (2007) point out that recently CNTs and transition metallic nanoparticles (NPs) have been used in biosensors. Owing to their high surface/volume ratio CNTs are attractive materials for electroanalysis (Merkoci et al., 2005; Zhao and Ju, 2006). Kang et al. (2007) have developed an effective nonenzymatic glucose sensor by using catalytic oxidation to electrode-post the copper nanoclusters onto the electrode modified with Nafion (Nf)-solubilized CNTs. They used this sensor to analyze glucose in real blood serum samples.

A single-fractal analysis is adequate to describe the binding and the dissociation kinetics in Figure 7.1a. The values of (a) the binding rate coefficient, k, and the fractal dimension, D_f, for the binding phase for a single-fractal analysis, and (b) the dissociation rate coefficient, k_d, and the fractal dimension, D_{fd}, for a single-fractal analysis are given in Table 7.1. The affinity, K, $(=k/k_d)$ value is 86.48. This is the first run in the sequence.

A single-fractal analysis is also adequate to describe the binding and the dissociation kinetics for the chronological runs b and c. In this case, the affinity, K, values are 64 and 74.73, respectively. For these three chronological runs depicted and analyzed in Figure 7.1a-c, the average affinity, K, value is 75.07.

Figure 7.2a shows the increase in the binding rate coefficient, k, with an increase in the fractal dimension, D_f, for a single-fractal analysis. For the data shown in Figure 7.2a, the binding rate coefficient, k, is given by:

$$k = (0.32 \pm 0)D_f^{0.8872 \pm 5\times10^{-15}} \quad (7.4a)$$

Only three data points are available. Two of the points are exactly identical. The fit is good. The availability of more data points would lead to a more reliable fit. The binding rate coefficient, k, exhibits less than first (equal to 0.8872) order of dependence on the fractal dimension, D_f, or the degree of heterogeneity on the biosensor surface.

Table 7.1: Binding and dissociation rate coefficients and affinity values, and fractal dimensions for the binding and the dissociation phase using cyclic voltametry (CV) for 1 mM glucose in solution to Nf(nafion)-CNTs (carbon nanotube)-GCE (glass carbon electrode) (Kang et al., 2007).

Run Number	k	k_d	k/k_d	D_f	D_{fd}	D_f/D_{fd}
1	0.32 ± 0	0.0037 ± 0.0002	86.48	1.0 ± 10^{-16}	0.6650 ± 0.0518	1.503
2	0.32 ± 0	0.05 ± 0.0	64	1.0 ± 10^{-16}	1.0	1
3	0.2989 ± 0.305	0.004 ± 0.0	74.73	0.926 ± 0.0428	$1.0 \pm 3.6\times10^{-15}$	0.926

Repeat runs under the same conditions in chronological sequence 1, 2 and 3.

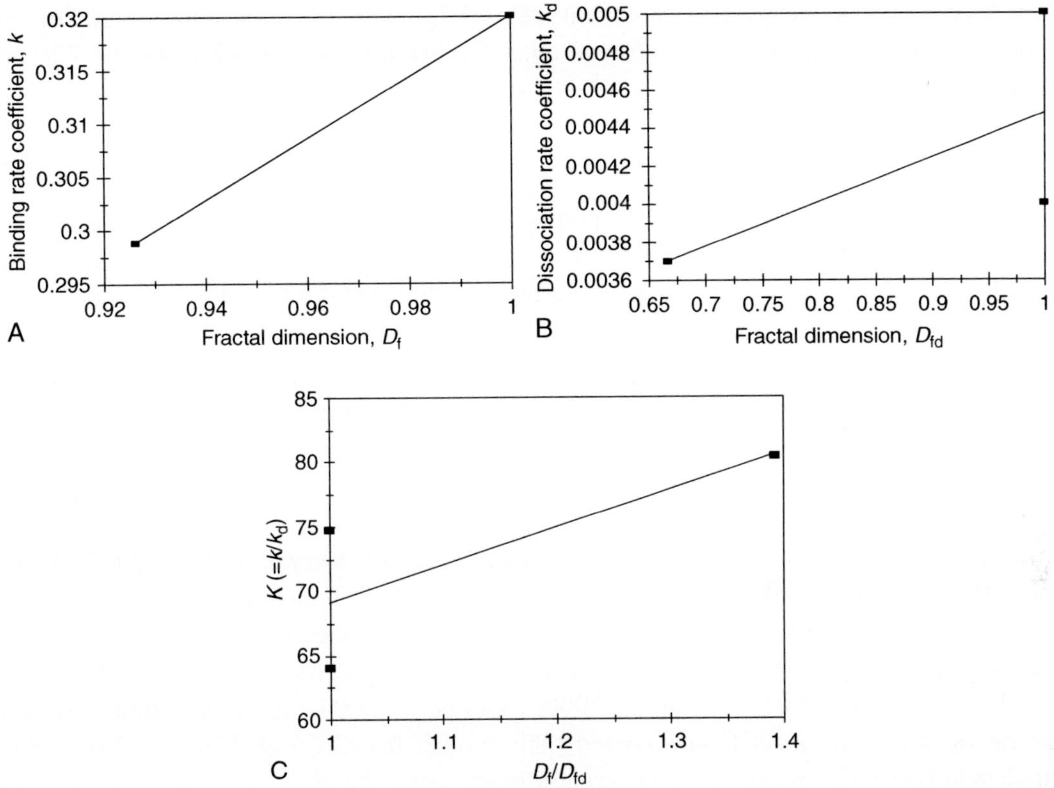

Figure 7.2

(a) Increase in the binding rate coefficient, k with an increase in the fractal dimension, D_f.
(b) Increase in the dissociation rate coefficient, k_d with an increase in the fractal dimension, D_{fd}.
(c) Increase in the affinity, $K(=k/k_d)$ with an increase in the ratio of fractal dimensions in the binding and in the dissociation phases (D_f/D_{fd}).

Figure 7.2b shows the increase in the dissociation rate coefficient, k_d, with an increase in the fractal dimension, D_{fd}, for a single-fractal analysis. For the data shown in Figure 7.2b, the dissociation rate coefficient, k_d, is given by:

$$k_d = (0.004472 \pm 0.000764) D_{fd}^{0.465 \pm 0.474} \tag{7.4b}$$

There is scatter in the data. The fit is not good. Only three data points are available. The availability of more data points would lead to a more reliable fit. The dissociation rate coefficient, k_d, for a single-fractal analysis exhibits less than one-half (equal to 0.465) order of dependence on the fractal dimension, D_{fd}, in the dissociation phase.

Figure 7.2c shows the increase in the affinity, K $(=k/k_d)$, with an increase in the ratio of fractal dimensions, D_f/D_{fd}, for a single-fractal analysis. For the data shown in Figure 7.2c, the affinity, K, is given by:

$$K(=k/k_d) = (69.155 \pm 8.006)(D_f/D_{fd})^{0.457 \pm 0.406} \tag{7.4c}$$

The fit is reasonable. Only three data points are available. The availability of more data points would lead to a more reliable fit. The affinity, K $(=k/k_d)$, exhibits less than one-half (equal to 0.457) order of dependence on the ratio of the fractal dimensions, D_f/D_{fd}.

Xu et al. (2006) have recently used DMG functionalized nanoparticles (DMG-CuNPs) to construct a new glucose sensor. Their results indicated that DMG could be used to encapsulate the copper nanoparticles to control their growth. The sensor that they developed exhibited good selectivity and sensitivity, besides a wide linearity range and low detection limits. Xu et al. (2006) further point out that due to the simple operability and instrumentation electrochemical detection is popular (Cui et al., 2000; Karyakin et al., 2002; Yabuki et al., 2003). Xu et al. (2006) also report that Cu-based CMEs have the advantages of ease of operation and commercial availability. Furthermore, copper nanoparticles which have unusual properties have been frequently studied for their potential applications (Lu et al., 2000; Eastman et al., 2001; You et al., 2002; Mala et al., 2004). Xu et al. (2006) modified their nanoparticles on a GCE and coated their modified GCE with Nafion. This Nafion-coated modified GCE prevented interference from acetaminophen, ascorbic acid, and uric acid.

Xu et al. (2006) analyzed the influence of binding and dissociation of 0.1 mM glucose to DMG-CuNPs CME in sequential runs. Figure 7.3a-c shows the binding and dissociation in three consecutive runs. Figure 7.3a shows that the binding may be adequately described by a single-fractal analysis. A dual-fractal analysis is required to adequately describe the dissociation kinetics. The values of (a) the binding rate coefficients and the fractal dimensions for the binding phase, D_f, for a single-fractal analysis (b) the dissociation rate coefficient, k_d, and the fractal dimension for the dissociation phase, D_{fd}, for a single-fractal analysis, and (c) the dissociation rate coefficients, k_{d1} and k_{d2}, and the fractal dimensions in the dissociation phase, D_{fd1} and D_{fd2}, for a dual-fractal analysis are given in Tables 7.2 and 7.3.

Figure 7.3b shows that the binding may be adequately described by a single-fractal analysis. Once again, a dual-fractal analysis is required to adequately describe the dissociation kinetics. The values of (a) the binding rate coefficients and the fractal dimensions for the binding phase, D_f, for a single-fractal analysis, (b) the dissociation rate coefficient, k_d, and the fractal dimension for the dissociation phase, D_{fd}, for a single-fractal analysis, and (c) the

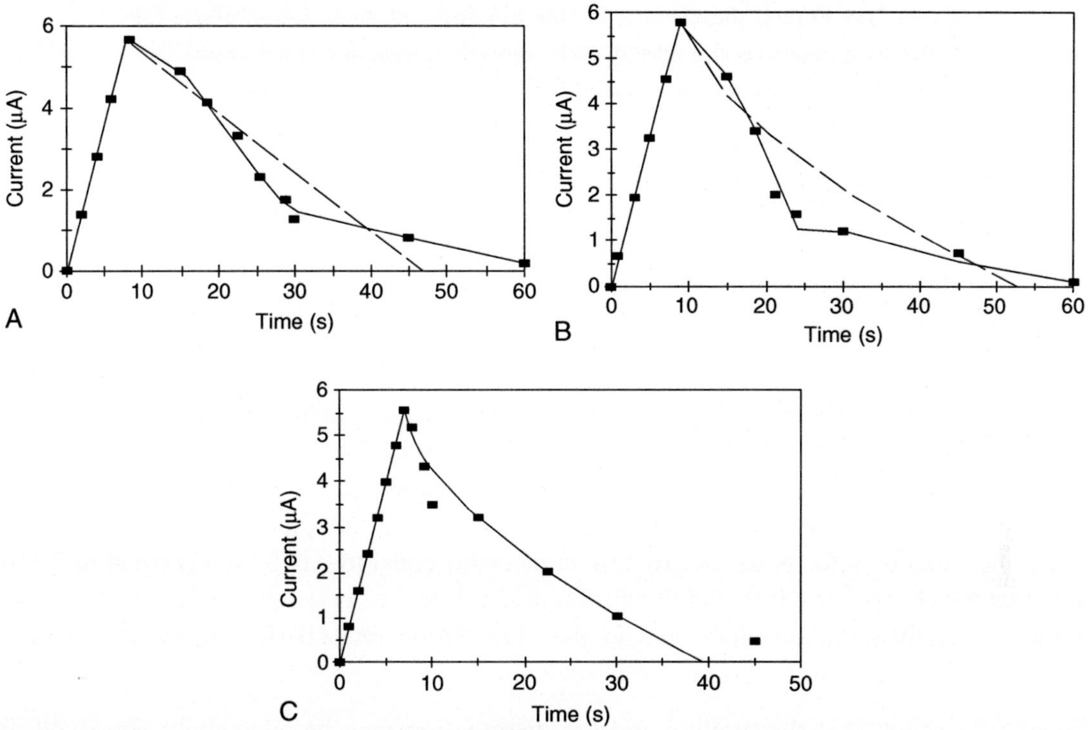

Figure 7.3
Binding and dissociation phases of 0.1 mM glucose to DMG (dimethylglycoxime)-CuNP (copper nanoparticles) CME (copper-based chemically modified electrodes) (Xu et al., 2006). Influence of repeat runs in chronological order, a, b, and c.

Table 7.2: Binding and dissociation rate coefficients for the binding and dissociation of 0.1 M glucose to the DMGCuNP (dimethylglycoxime functionalized copper nanoparticles) (Xu et al., 2006).

Run	k	k_d	k_{d1}	k_{d2}
A	0.693 ± 0.005	$0.180 + 2.905$	0.0384 ± 0.0013	1.559 ± 0.095
B	0.650 ± 0.003	0.541 ± 0.146	0.0957 ± 0.0125	2.116 ± 0.056
C	0.8 ± 0	0.673 ± 0.273	na	na

Influence of repeat runs.

Table 7.3: Fractal dimensions for the binding and dissociation phase for 0.1 M glucose to the DMGCuNP (dimethylglycoxime functionalized copper nanoparticles) (Xu et al., 2006).

Run	D_f	D_{fd}	D_{fd1}	D_{fd2}
A	0.982 ± 0.0157	1.110 ± 0.307	0 + 0.914	2.3604 ± 0.1554
B	1.006 ± 0.0042	1.740 ± 0.260	0.142 + 0.362	2.502 ± 0.0836
C	1.0 ± 0	1.780 ± 0.206	na	na

Influence of repeat runs.

dissociation rate coefficients, k_{d1} and k_{d2}, and the fractal dimensions in the dissociation phase, D_{fd1} and D_{fd2}, for a dual-fractal analysis are given in Tables 7.2 and 7.3. It is of interest to note that as the fractal dimension in the dissociation phase increases by a factor of 17.62 from a value of D_{fd1} equal to 0.142 to D_{fd2} equal to 2.3604, the dissociation rate coefficient increases by a factor of 22.1 from a value of k_{d1} equal to 0.0957 to k_{d2} equal to 2.116. Note that the changes in the fractal dimension or the degree of heterogeneity on the biosensor surface in the dissociation phase and in the dissociation rate coefficient are in the same direction.

Figure 7.3c shows that the binding and the dissociation may be adequately described, in this case, by a single-fractal analysis. In this case, the affinity, K $(=k/k_d)$, value is equal to 1.19. Figure 7.4 shows the increase in the dissociation rate coefficient, k_d or k_{d2}, with an increase in the fractal dimension, D_{fd} or D_{fd2}. The two different pieces of data

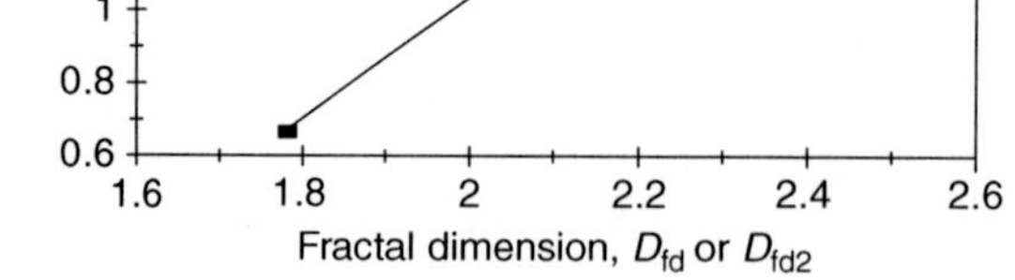

Figure 7.4

Increase in the dissociation rate coefficient, k_d or k_{d2} with an increase in the fractal dimension, D_{fd} or D_{fd2} (Xu et al., 2006).

are plotted together owing to the lack of sufficient data (a minimum number of three data points are required). For the data shown in Figure 7.4, the dissociation rate coefficient is given by:

$$k_{\rm d} \text{ or } k_{\rm d2} = (0.102 \pm 0.009)(D_{\rm fd} \text{ or } D_{\rm fd2})^{3.24 \pm 0.323} \tag{7.5a}$$

The fit is good in spite of the fact that the two different pieces of data are plotted together. Only three data points are available for the combined data. The availability of more data points would lead to a more reliable fit. The dissociation rate coefficient, $k_{\rm d}$ or $k_{\rm d2}$, is sensitive to the degree of heterogeneity that exists on the biosensor surface in the dissociation phase, since it exhibits an order of dependence between three (equal to 3.24) and three and a half on the fractal dimension in the dissociation phase.

Figure 7.5 shows the increase in the dissociation rate coefficient, k or $k_{\rm d1}$, with an increase in the fractal dimension, $D_{\rm fd}$ or $D_{\rm fd1}$. The two different pieces of data are, once again, plotted together owing to the lack of sufficient data (a minimum number of three data points are required). For the data shown in Figure 7.5, the dissociation rate coefficient is given by:

$$k_{\rm d} \text{ or } k_{\rm d1} = (0.361 + 0.433)(D_{\rm fd} \text{ or } D_{\rm fd1})^{0.355 \pm 0.146} \tag{7.5b}$$

The fit is good in spite of the fact that the two different pieces of data are plotted together. Only three data points are available for the combined data. The availability of more data points would lead to a more reliable fit. The dissociation rate coefficient, $k_{\rm d}$ or $k_{\rm d1}$, exhibits less than one-half (equal to 0.355) order of dependence on the degree of heterogeneity or the fractal dimension that exists on the biosensor surface.

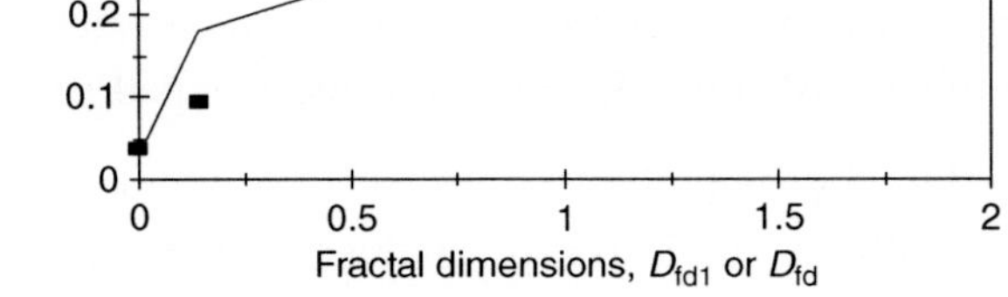

Figure 7.5
Increase in the dissociation rate coefficient, $k_{\rm d}$ or $k_{\rm d1}$with an increase in the fractal dimension, $D_{\rm fd}$ or $D_{\rm fd1}$.

Bai et al. (2008) have recently developed a Pt-Pb NAE for enzyme-free glucose detection. These authors point out that there is an inherent instability of enzymatic sensors for glucose detection as well as the interfering effects of some other electro-oxidizable species (Wilson and Turner, 1992; Shoji and Freund, 2001; Park et al., 2006). Pt-Pb alloy electrodes exhibited stable, reproducible, and larger responses compared to Pt electrodes alone (Sun et al., 2001). Furthermore, mesoporous platinum (Park et al., 2005) with high surface roughness (higher fractal dimension) provided not only a better amperometric response but also an effectively lower interference effect from other electrostatic species. This indicated to Bai et al. (2008) that both the component as well as the surface structure significantly affects the catalytic oxidation of glucose.

Figure 7.6a shows the binding and dissociation of 1 mM glucose in 0.1 M PBS in solution to the pt-Pb NAE (Bai et al., 2008). A single-fractal analysis is adequate to describe the binding kinetics. A dual-fractal analysis is required to adequately describe the dissociation kinetics. The values of (a) the binding rate coefficient, k, and the fractal dimension, D_f, for a single-fractal analysis, (b) the dissociation rate coefficient, k_d, and the fractal dimension, D_{fd}, for a single-fractal analysis, and (c) the dissociation rate coefficients, k_{d1} and k_{d2}, and the fractal dimensions, D_{fd1} and D_{fd2}, for a dual-fractal analysis are given in Tables 7.4 and 7.5.

It is of interest to note that as the fractal dimension for dissociation increases by a factor of 1.88 from a value of D_{fd1} equal to 1.5932 to D_{fd2} equal to 3.0 (the maximum value), the dissociation rate coefficient increases by a factor of 6.0 from a value of k_{d1} equal to 0.001 to k_{d2} equal to 0.006. It is seen that an increase in the degree of heterogeneity or the fractal dimension on the biosensor surface in the dissociation phase leads to an increase in the dissociation rate coefficient. This is consistent with the statement of Bai et al. (2008) that the surface structure significantly affects the catalytic oxidation of glucose. Besides, this result indicates that the dissociation rate coefficient is sensitive to the degree of heterogeneity on the biosensor surface. The affinity, K_1 ($=k/k_{d1}$) and K_2 ($=k/k_{d2}$), values are equal to 5.2 and 0.867, respectively.

Figure 7.6b shows the binding and dissociation of 1 mM glucose in 0.1 M PBS to the Pt-PbNAE (Bai et al., 2008). A single-fractal analysis is adequate to describe the binding and the dissociation kinetics. The values of (a) the binding rate coefficient, k, and the fractal dimension, D_f, for a single-fractal analysis, and (b) the dissociation rate coefficient, k_d, and the fractal dimension for dissociation, D_{fd}, for a single-fractal analysis are given in Tables 7.4 and 7.5. In this the affinity, K ($=k/k_d$), value is equal to 29.3.

Figure 7.6c shows the binding and dissociation of 1 mM glucose in 0.1 M PBS to the Pt-PbNAE (Bai et al., 2008). A single-fractal analysis is once again adequate to describe the binding and the dissociation kinetics. The values of (a) the binding rate coefficient, k, and the fractal dimension, D_f, for a single-fractal analysis, and (b) the dissociation rate

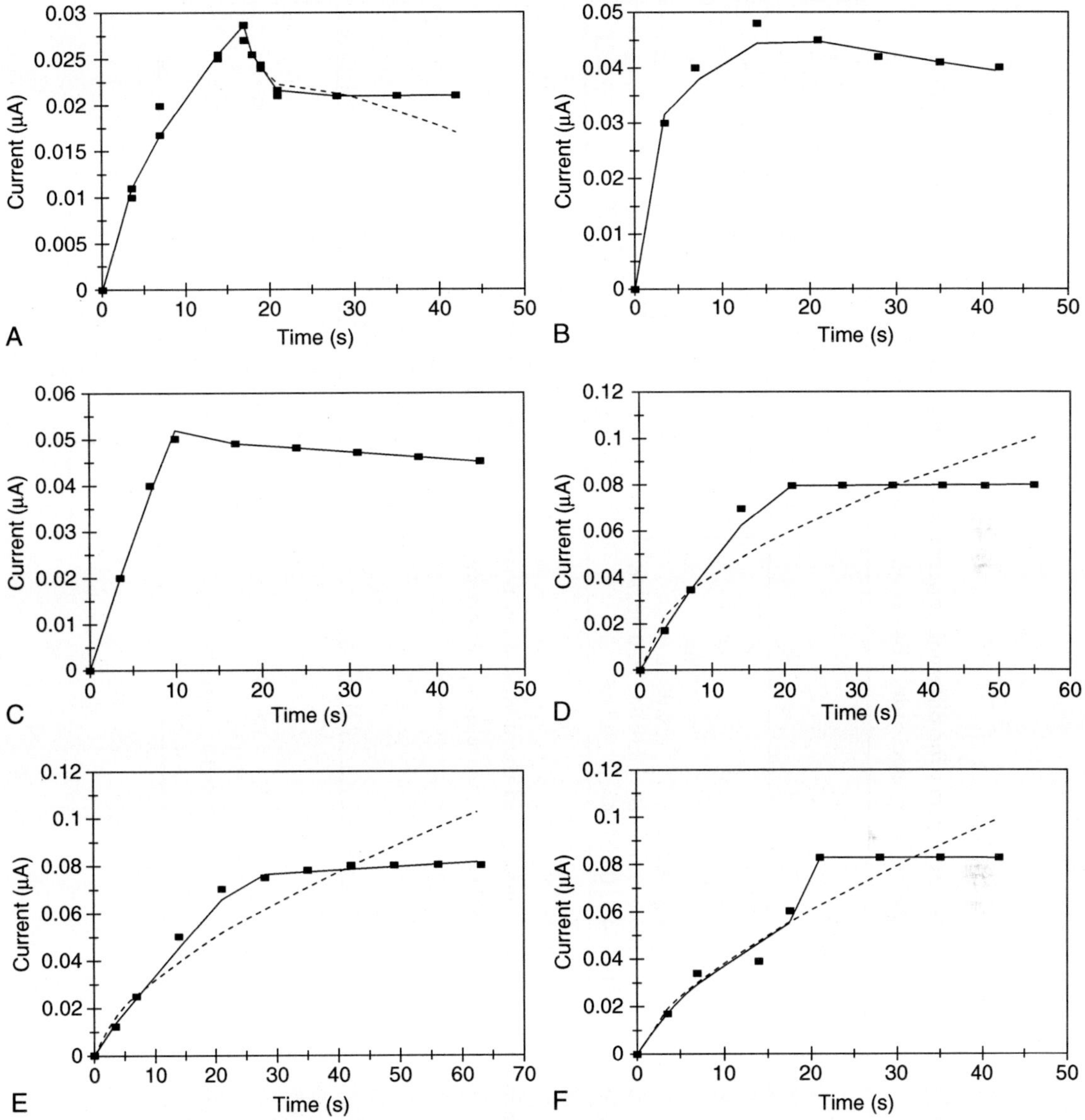

Figure 7.6
Binding of different concentrations (in mM) of glucose in 0.1 M PBS solution to Pt-Pb NAE (nanowire array electrode) (Bai et al., 2008): (a) 1 (b) 1 (c) 1 (d) 2 (e) 3 (f) 4.

Table 7.4: Binding and dissociation rate coefficients for different concentrations of glucose (in mM) to Pt-PbNAE (nanowire array electrode) (Bai et al., 2008).

Glucose Concentration (mM)	k	k_1	k_2	k_d	k_{d1}	k_{d2}
1	0.0052 ± 0.0007	na	na	0.0012 ± 0.0003	0.001 ± 0.0001	0.006 ± 0
1	0.02344 ± 0.00249	na	na	0.0008 ± 0.0001	na	na
1	0.006685 ± 0.00049	na	na	0.000143 ± 0	na	na
2	0.0123 ± 0.0033	0.00609 ± 0.00078	0.08 ± 0	na	na	na
3	0.007453 ± 0.001811	0.004288 ± 0.000460	0.05917 ± 0.00086	na	na	na
4	0.00849 ± 0.00143	0.007864 ± 0.001649	0.085 ± 0	na	na	na

Table 7.5: Fractal dimensions for the binding and the dissociation phase for different concentrations of glucose (in mM) to Pt-PbNAE (nanowire array electrode) (Bai et al., 2008).

Glucose Concentration (mM)	D_f	D_{f1}	D_{f2}	D_{fd}	D_{fd1}	D_{fd2}
1	1.790 ± 0.238	na	na	1.854 ± 0.224	1.5932 ± 0.1878	3.0 ± 0
1	2.5158 ± 0.1482	na	na	1.5836 ± 0.2248	na	na
1	1.2198 ± 0.1875	na	na	1.0 ± 0	na	na
2	1.9506 ± 0.1796	1.2402 ± 0.1749	3.0 ± 0	na	na	na
3	1.7316 ± 0.1522	1.20848 ± 0.1206	1.8484 ± 0.04266	na	na	na
4	1.6690 ± 0.1636	1.6244 ± 0.3020	3.0 ± 0.0	na	na	na

coefficient, k_d, and the fractal dimension for dissociation, D_{fd}, are given in Tables 7.4 and 7.5. In this case the affinity, K $(=k/k_d)$, value is equal to 46.75.

Figure 7.6d shows the binding of 2 mM glucose in 0.1 M PBS in solution to the Pt-PBHNAE (Bai et al., 2008). A dual-fractal analysis is required to adequately describe the binding kinetics. The values of (a) the binding rate coefficient, k, and the fractal dimension, D_f, for a single-fractal analysis, and (b) the binding rate coefficients, k_1 and k_2, and the fractal dimensions, D_{f1} and D_{f2}, for a dual-fractal analysis are given in Table 7.4 and 7.5. It is of interest to note that as the fractal dimension increases by a factor of 2.42 from a value of D_{f1} equal to 1.2402 to D_{f2} equal to 3.0, the binding rate coefficient increases by a factor of 13.14 from a value of k_1 equal to 0.00609 to k_2 equal to 0.08.

Figure 7.6e shows the binding of 3 mM glucose in 0.1 M PBS in solution to the Pt-PBHNAE (Bai et al., 2008). A dual-fractal analysis is required to adequately describe the binding kinetics. The values of (a) the binding rate coefficient, k, and the fractal dimension, D_f, for a single-fractal analysis, and (b) the binding rate coefficients, k_1 and k_2, and the fractal dimensions, D_{f1} and D_{f2}, for a dual-fractal analysis are given in Table 7.4 and 7.5. It is of interest to note that as the fractal dimension increases by a factor of 1.53 from a value of D_{f1} equal to 1.2084 to D_{f2} equal to 1.8484, the binding rate coefficient increases by a factor of 13.80 from a value of k_1 equal to 0.004288 to k_2 equal to 0.05917.

Figure 7.6f shows the binding of 4 mM glucose in 0.1 M PBS in solution to the Pt-PBHNAE (Bai et al., 2008). A dual-fractal analysis is required to adequately describe the binding kinetics. The values of (a) the binding rate coefficient, k, and the fractal dimension, D_f, for a single-fractal analysis, and (b) the binding rate coefficients, k_1 and k_2, and the fractal dimensions, D_{f1} and D_{f2}, for a dual-fractal analysis are given in Tables 7.4 and 7.5. It is of interest to note that as the fractal dimension increases by a factor of 1.85 from a value of D_{f1} equal to 1.6244 to D_{f2} equal to 3.0, the binding rate coefficient increases by a factor of 10.81 from a value of k_1 equal to 0.007864 to k_2 equal to 0.085.

Figure 7.7a and Tables 7.4 and 7.5 show the increase in the binding rate coefficient, k, with an increase in the fractal dimension, D_f, for a single-fractal analysis. For the data shown in Figure 7.7a, the binding rate coefficient, k, is given by:

$$k = (0.00359 + 0.00399)D_f^{1.683} \tag{7.6a}$$

There is scatter in the data, and this is reflected in the error in the binding rate coefficient. Only the positive error is presented since the binding rate coefficient cannot be a negative value. Only three data points are available. The availability of more data points would lead to a more reliable fit. The binding rate coefficient, k, is sensitive to the degree of heterogeneity on the biosensor surface or the fractal dimension since it exhibits an order of dependence between one and a half and two (equal to 1.683) on the fractal dimension on the biosensor surface.

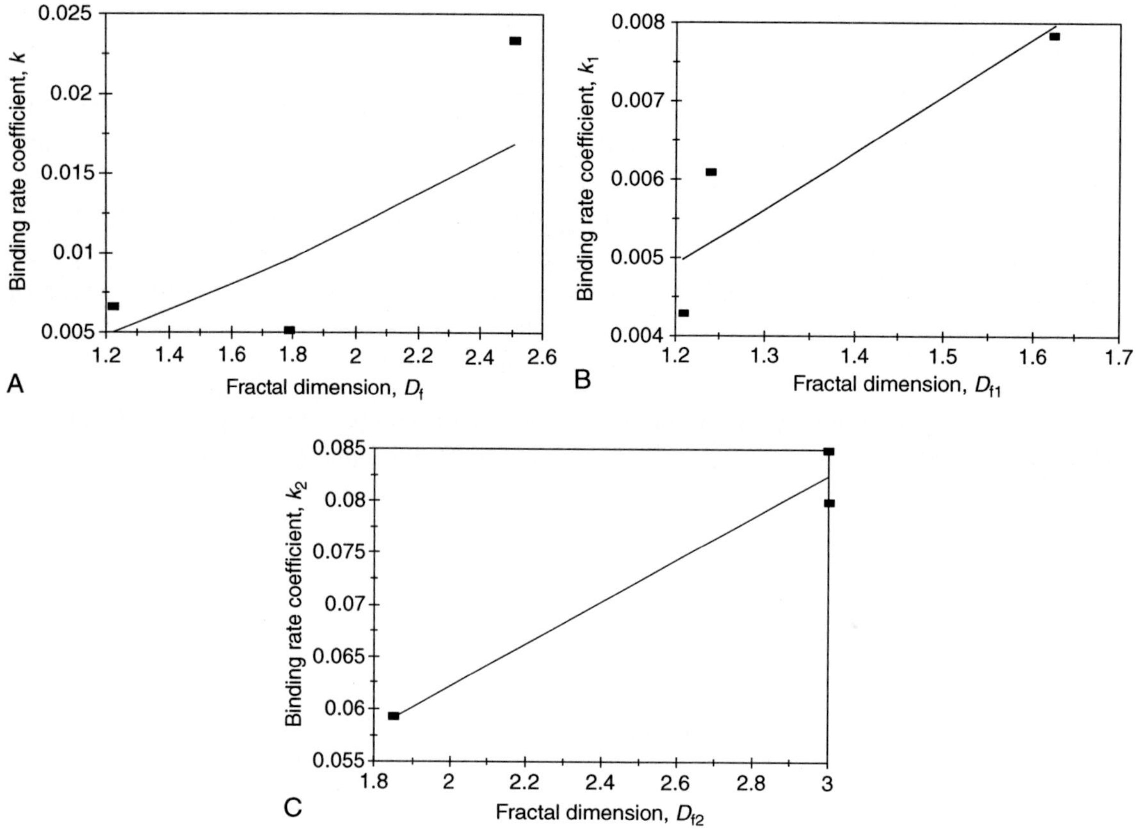

Figure 7.7

(a) Increase in the binding rate coefficient, k with an increase in the fractal dimension, D_f.
(b) Increase in the binding rate coefficient, k_1 with an increase in the fractal dimension, D_{f1}.
(c) Increase in the binding rate coefficient, k_2 with an increase in the fractal dimension, D_{f2}.

Figure 7.7b and Tables 7.4 and 7.5 show the increase in the binding rate coefficient, k_1, with an increase in the fractal dimension, D_{f1}, for a dual-fractal analysis. For the data shown in Figure 7.7b, the binding rate coefficient, k_1, is given by:

$$k_1 = (0.00367 + 0.00090)D_{f1}^{1.598 \pm 0.946} \tag{7.6b}$$

The fit is reasonable. Only three data points are available. The availability of more data points would lead to a more reliable fit. The binding rate coefficient, k_1, is sensitive to the degree of heterogeneity on the biosensor surface or the fractal dimension since it exhibits an order of dependence between one and one-half and two (equal to 1.598) on the fractal dimension, D_{f1} on the biosensor surface.

Figure 7.7c and Tables 7.4 and 7.5 show the increase in the binding rate coefficient, k_2, with an increase in the fractal dimension, D_{f2}, for a dual-fractal analysis. For the data shown in Figure 7.7c, the binding rate coefficient, k_2, is given by:

$$k_2 = (0.03884 + 0.00170)D_{f2}^{0.685\pm0.108} \tag{7.6c}$$

The fit is reasonable. Only three data points are available. The availability of more data points would lead to a more reliable fit. The binding rate coefficient, k_2, is only mildly sensitive to the degree of heterogeneity on the biosensor surface or the fractal dimension since it exhibits an order of dependence between one-half and one (equal to 0.685) on the fractal dimension on the biosensor surface.

Hu (2009) has recently reviewed commercialized glucose sensors in China. According to a report issued by the Chinese CDC (Center of Disease Control, 2006), Ministry of Health, there were 40 million diabetics in China. Williams et al. (1970) first demonstrated the amperometric blood glucose determination using a redox couple-mediated, glucose oxidase (GO_x)-catalyzed reaction. However, Hu (2009) points out that this research did not lead to the self-monitoring blood glucose levels at home. He points out that many studies on blood glucose determination have been conducted since the 1980s. These include the studies by Cass et al. (1984), Neuman et al. (1970), Wang et al. (1995), Cui et al. (2000), Gao et al. (2005), Neuman and Turner (2005), and Heller and Feldman (2008).

Figure 7.8 shows the binding and dissociation of whole blood glucose with a glucose concentration of 400 mg/dl to a FAD-GDH that was screen-printed on the working side of the biosensor strip. The operating potential is 300 mV. The hematocrit was adjusted to 40%

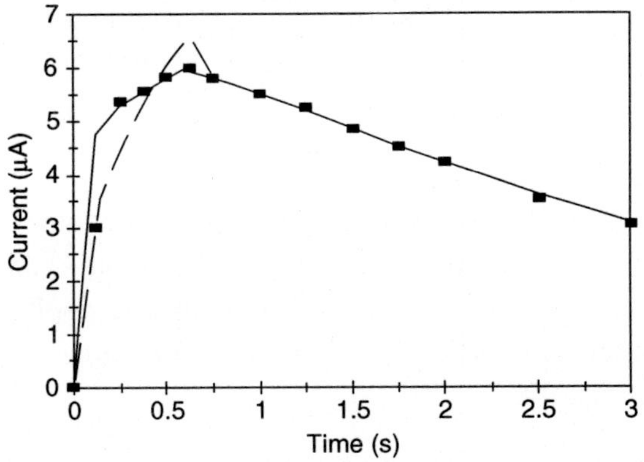

Figure 7.8
Binding of 400 mg/dl glucose in solution (hematocrit adjusted to 40%) to FADGDH (flavin adenine dinucleotide-dependent glucose dehydrogenase)/ferricyanide that was screen-printed on a biosensor strip (Hu, 2009).

(Hu, 2009). A single-fractal analysis is adequate to describe the binding and the dissociation kinetics. The values of (a) the binding rate coefficient, k, and the fractal dimension, D_f, for a single-fractal analysis and (b) the dissociation rate coefficient, k_d, and the fractal dimension for dissociation, D_{fd}, for a single-fractal analysis are given in Tables 7.6 and 7.7.

Hu (2009) further reports that to improve the properties of the disposable glucose sensor new materials and improved screen-printing techniques have been developed in recent years. He refers to the development of a fabrication procedure for a disposable amperometric glucose sensor using WB enzyme ink containing binder, stabilizer, mediator, and surfactant.

Recently, Miyashita et al. (2009) pointed out that according to the 2007 autumn report issued by the International Diabetic Foundation (IDF) the estimated number of diabetics was 246 million people. This number according to the IDF report is expected to grow to 380 million by the year 2025. This is more than a 50% increase in about 18 years. These authors report that the first personal urine glucose meter was introduced in the Japanese market in 1996 (Nakijama et al., 1996). This glucose sensor monitored urine glucose levels semi-quantitatively in the range 0-400 mg/dl. Miyashita et al. (2009) report on the development of a highly sensitive microplaner urine glucose meter. This glucose meter contains a biosensor that amperometrically detects glucose by an immobilized glucose oxidase (GO_x) on a planer peroxide membrane. Photolithography is used to fabricate the electrode.

Miyashita et al. (2009) draw attention to the influence of postmeal blood glucose monitoring by postmeal urine glucose level monitoring. There is an apparent increase in the postmeal blood glucose level though it is transient. This level is generally restored after 2 hours. Miyashita et al. (2009) assert that estimation of postmeal plasma glucose level is important in achieving lower hemoglobin A1c (HbA1c) levels, which reflects diabetes development (Monnier et al., 2003, 2007; Woerle et al., 2007). They point out that SMBG (self-monitoring of blood glucose) with a single measurement may not provide a precise measurement of the highest level of blood glucose after a meal. Frequent SMBG is not recommended owing to the inconveniences associated with this.

Figure 7.9a shows the binding of blood glucose in postmeal testing to the urine glucose meter. The binding and the dissociation kinetics may be adequately described by a single-fractal analysis. The values of (a) the binding rate coefficient, k, and the fractal dimension, D_f, for a single-fractal analysis, and (b) the dissociation rate coefficient, k_d, and the fractal dimension, D_{fd}, for a single-fractal analysis are given in Tables 7.6 and 7.7. In this case, the affinity, K ($=k/k_d$) is equal to 12.29.

Figure 7.10 shows the binding of urine glucose to the urine glucose meter for postmeal testing using an elemental diet. A single-fractal analysis is adequate to describe the binding kinetics. The values of (a) the binding rate coefficient, k, and the fractal dimension, D_f, for a single-fractal analysis are given in Tables 7.6 and 7.7.

Table 7.6: Binding and dissociation (if applicable) rate coefficients for (a) a commercial glucose sensor in China (Hu, 2009), (b) urine glucose meter based on a microplaner amperometric biosensor (Miyashita et al., 2009), (c) and an implantable glucose sensor (Vaddiraju et al., 2009).

Analyte/Receptor	k	k_1	k_2	k_d	k_{d1}	k_{d2}	References
(a) 400 mg/dl glucose/screen-printed test strip	7.9664 ± 1.202	na	na	1.307 ± 0.091	na	na	Hu (2009)
(b) Blood glucose level in postmeal testing/glucose meter	51.210 ± 1.410	na	na	4.168 ± 0.74	na	na	Miyashita et al. (2009)
Urine glucose level in postmeal testing/ glucose meter	31.852 ± 7.988	na	na	na	na	na	Miyashita et al. (2009)
(c) Glucose/ implantable glucose meter	1.5745 ± 0.4710	0.6473 ± 0.1342	18.5 ± 0	0.0561 ± 0.01785	0.0207 ± 0.0005	1.1979 ± 0.0787	Vaddiraju et al. (2009)

Table 7.7: Binding and dissociation (if applicable) phases for (a) a commercial glucose sensor in China (Hu, 2009), (b) urine glucose meter based on microplaner amperometric biosensor (Miyashita et al., 2009), (c) and an implantable glucose sensor (Vaddiraju et al., 2009).

Analyte/Receptor	D_f	D_{f1}	D_{f2}	D_{fd}	D_{fd1}	D_{fd2}	References
(a) 400 mg/dl glucose/ screen-printed test strip	2.174 ± 0.222	na	na	1.1206 ± 0.0534	na	na	Hu (2009)
(b) Blood glucose level in postmeal testing/ glucose meter	2.2514 ± 0.0692	na	na	1.5874 ± 0.1786	na	na	Miyashita et al. (2009)
Urine glucose level in postmeal testing/ glucose meter	2.6040 ± 0.4566	na	na	na	na	na	Miyashita et al. (2009)
(c) Glucose/ implantable glucose meter 2	1.9688 ± 0.2386	1.4508 ± 0.2400	3.0 ± 0	1.3090 ± 0.1970	0.898 ± 0.0287	2.2282 ± 0.2778	Vaddiraju et al. (2009)

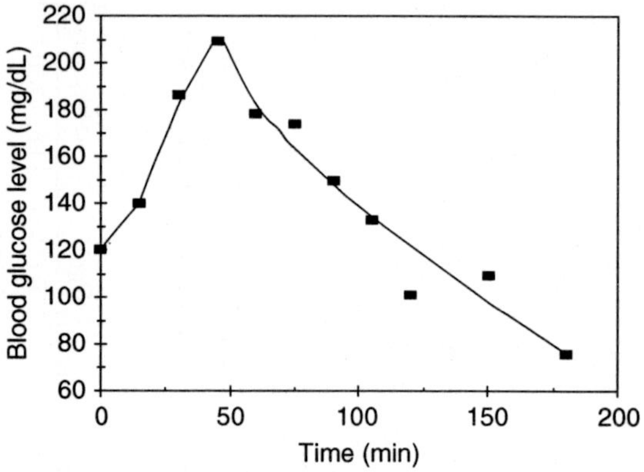

Figure 7.9
Binding of blood glucose in postmeal testing to a urine glucose meter (Miyashita et al., 2009).

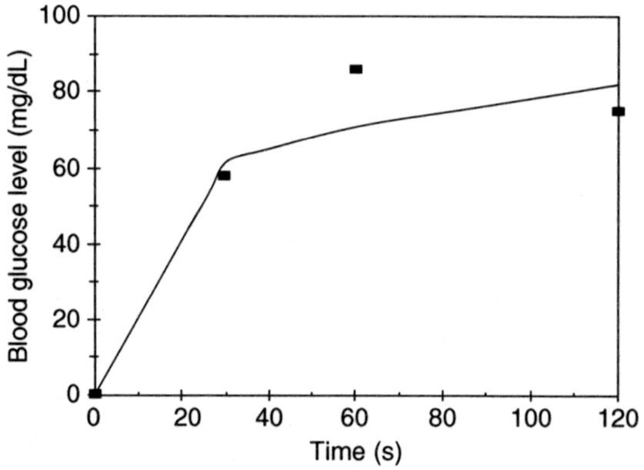

Figure 7.10
Binding of urine glucose in postmeal testing (elemental diet) (Miyashita et al., 2009).

It is of interest to compare the binding rate coefficient, k, and the fractal dimension, D_f, observed for the postmeal testing of blood glucose and urine glucose by this urine glucose meter. On going from the blood glucose to the urine glucose measurement, the single measurement indicates a 15.56% increase in the fractal dimension value from a D_f value equal to 2.2514 to 2.6040, which leads to a decrease in the binding rate coefficient, k value by a factor of 0.622, from a value of k equal to 51.210 to 31.852. In this case, an increase in the fractal dimension on the urine glucose sensor surface leads to a decrease in the binding rate coefficient. It is observed, however, that two different systems (glucose in blood and glucose in urine) are being analyzed by the same urine glucose meter.

Vaddiraju et al. (2009) have very recently analyzed the role of H_2O_2 outer diffusion on implantable glucose sensor performance. These authors point out that the outer membrane in implantable glucose sensors significantly influences the performance of the sensors by governing the diffusion of various participating species. These authors correlated the role of the outward H_2O_2 diffusion through the outer membrane of glucose sensors to their sensitivity. In electrochemical biosensors, particularly first-generation Clark-type glucose sensors, the flavoenzyme glucose oxidase (GO_x) is immobilized on a working electrode. The FAD redox cofactor of GO_x catalyzes the oxidation of glucose to glucolactone (Vaddiraju et al., 2009):

$$\text{Glucose} + GO_x(\text{FAD}) \rightarrow \text{gluoclactone} + GO_x(\text{FADH}_2) \tag{7.7a}$$

$$GO_x(\text{FADH}_2) \rightarrow O_2 + GO_x(\text{FAD}) + H_2O_2 \tag{7.7b}$$

The H_2O_2 generated is detected amperometrically on a working electrode, which correlates the current generated to the glucose concentration.

In *in vivo* applications the concentration of oxygen is rather low which leads to signal saturation for higher glucose concentration determination.

Vaddiraju et al. (2009) report that the use of outer membranes leads to:

(a) a decrease in biofouling (Wilson and Gifford, 2005) and
(b) a minimization of temperature-induced variations in sensor responses as a result of enzyme reaction kinetics (Jablecki and Gough, 2000).

The use of outer membranes, however, does lead to an increase in the response time and a decrease in sensitivity (Wilson and Gifford, 2005). Furthermore, Mercado and Moussey (1998) draw attention to calcification-induced permeability changes and degradation of the outer membrane.

Figure 7.11 shows the binding and dissociation of 6 mM glucose to the implantable glucose sensor (Vaddiraju et al., 2009). A dual-fractal analysis is required to adequately describe the binding and the dissociation kinetics. The values of (a) the binding rate coefficient, k, and the fractal dimension, D_f, for a single-fractal analysis, (b) the binding rate coefficients, k_1 and k_2, and the fractal dimensions, D_{f1} and D_{f2}, for a dual-fractal analysis, (c) the dissociation rate coefficient, k_d, and the fractal dimension, D_{fd}, for a single-fractal analysis, and (d) the dissociation rate coefficients, k_{d1} and k_{d2}, and the fractal dimensions, D_{fd}, and D_{fd2} for a dual-fractal analysis are given in Tables 7.6 and 7.7. It is seen that the affinity, K_1 ($=k/k_d$) and K_2 ($=k_2/k_d$) values are 31.2 and 15.4, respectively.

It is of interest to note that for the binding phase, as the fractal dimension increases by a factor of 2.07 from a value of D_{f1} equal to 1.45 to D_{f2} equal to 3.0, the binding rate

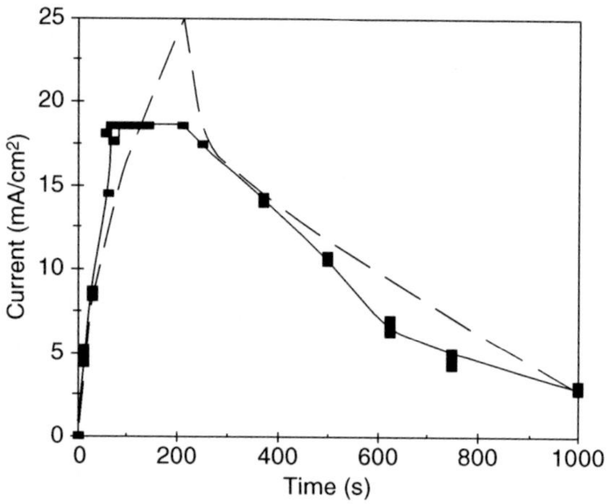

Figure 7.11
Binding of 6 mM glucose to an implantable glucose sensor (Vaddiraju et al., 2008).

coefficient increases by a factor of 28.6 from a value of k_1 equal to 0.6473 to k_2 equal to 18.5. Similarly, for the dissociation phase, as the fractal dimension increases by a factor of 2.48 from a value of D_{fd1} equal to 0.898 to D_{fd2} equal to 2.2282, the dissociation rate coefficient increases by a factor of 57.87 from a value of k_{d1} equal to 0.0207 to k_{d2} equal to 1.1878. In both the cases mentioned above, an increase in the degree of heterogeneity on the biosensor surface (increase in the fractal dimension) leads to a corresponding increase in the rate coefficient.

Recently, Pemberton et al. (2009) have fabricated a microband glucose biosensor using a screen-printed WB carbon ink and used it for serum analysis. They used this sensor for glucose analysis at 25 °C. They optimized the operational pH at 8.0. Steady-state conditions were determined indicating the presence of radial diffusion.

Pemberton et al. (2009) report that the screen-printing approach has been used for biosensor fabrication on an industrial mass production scale (Newman and Setford, 2006). It is particularly suitable for the production of low cost disposable devices. They further point out that they have been successfully employed to make the test strips used by diabetics to monitor their blood glucose levels (Matthews et al., 1987).

Pemberton et al. (2009) point out that to demonstrate microelectrode behavior using screen-printing technology with carbon inks, a working electrode of micron size in at least one dimension is required. Their sensor, fabricated using a solvent-based CoPC-containing carbon ink, demonstrated microelectrode-type behavior. Their sensor was capable of exhibiting steady-state current responses in the presence of H_2O_2 (Rawson et al., 2007).

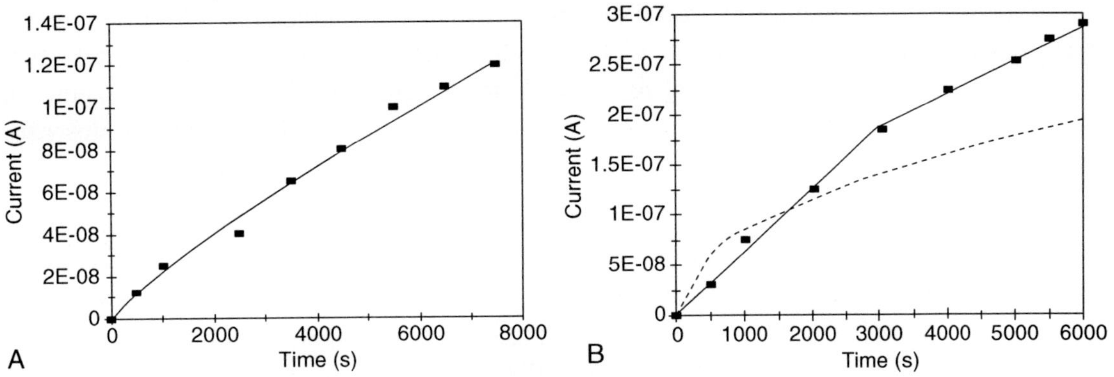

Figure 7.12

Binding of 500 mM glucose in solution to screen-printed water-based (WB) CoPC (cobalt phthlocyanine) (electrocatalyst) microband biosensor (Pemberton et al., 2008). Influence of quiescent and stirred conditions: (a) Quiescent (b) stirred.

Figure 7.12a shows the binding of 500 mM glucose to a WB (water based)-CoPC (cobalt phthalocyanine)-GOD (glucose oxidase) microband electrode at pH 8.0 (phosphate buffer) under quiescent conditions (Pemberton et al., 2009). A single-fractal analysis is adequate to describe the binding kinetics. The values of the binding rate coefficient, k, and the fractal dimension, D_f, for a single-fractal analysis are given in Table 7.8.

Figure 7.12b shows the binding of 500 mM glucose in solution to a WB-CoPC-GOD microband electrode at pH 8.0 under stirred conditions (Pemberton et al., 2009). A dual-fractal analysis is required to adequately describe the binding kinetics. The values of (a) the binding rate coefficient, k, and the fractal dimension, D_f, for a single-fractal analysis, and (b) the binding rate coefficients, k_1 and k_2, and the fractal dimensions, D_{f1} and D_{f2}, for a dual-fractal analysis are given in Table 7.8.

Table 7.8: Binding of 500 mM glucose in solution to WB (water-based)-CoPC (cobalt phthalocyanate)-GOD (glucoseoxidase) microband biosensor at pH 8.0 (phosphate buffer) (Pemberton et al., 2009).

Experimental Condition	k	k_1	k_2	D_f	D_{f1}	D_{f2}
Quiescent	$7.1\times10^{-11} \pm 0.7\times10^{-11}$	na	na	1.3316 ± 0.07396	na	na
Stirred	$3.6\times10^{-9} \pm 2.2\times10^{-9}$	$7.0\times10^{-11} \pm 0.9\times10^{-11}$	$1.2\times10^{-9} \pm 0.0$	2.0848 ± 0.2608	1.0232 ± 0.1814	1.7414 ± 0.0606

Influence of stirring.

It is of interest to note that as the fractal dimension increases by a factor of 1.70 from a value of D_{f1} equal to 1.0232 to D_{f2} equal to 1.7414, the binding rate coefficient increases by a factor of 17.1 from a value of k_1 equal to 7×10^{-11} to k_2 equal to 1.2×10^{-9}. Once again, an increase in the degree of heterogeneity or the fractal dimension on the microband biosensor surface leads to an increase in the binding rate coefficient.

Pemberton et al. (2009) analyzed the binding of 50 mM glucose solution using the microelectrodes under stirred conditions. They analyzed the binding using hydrodynamic voltagrams under different applied potentials. Figure 7.13a shows the hydrodynamic voltagram at an applied potential of 0.3 V. A single-fractal analysis is adequate to describe the binding kinetics. The values of (a) the binding rate coefficient, k, and the fractal dimension, D_f, for a single-fractal analysis is given in Table 7.9.

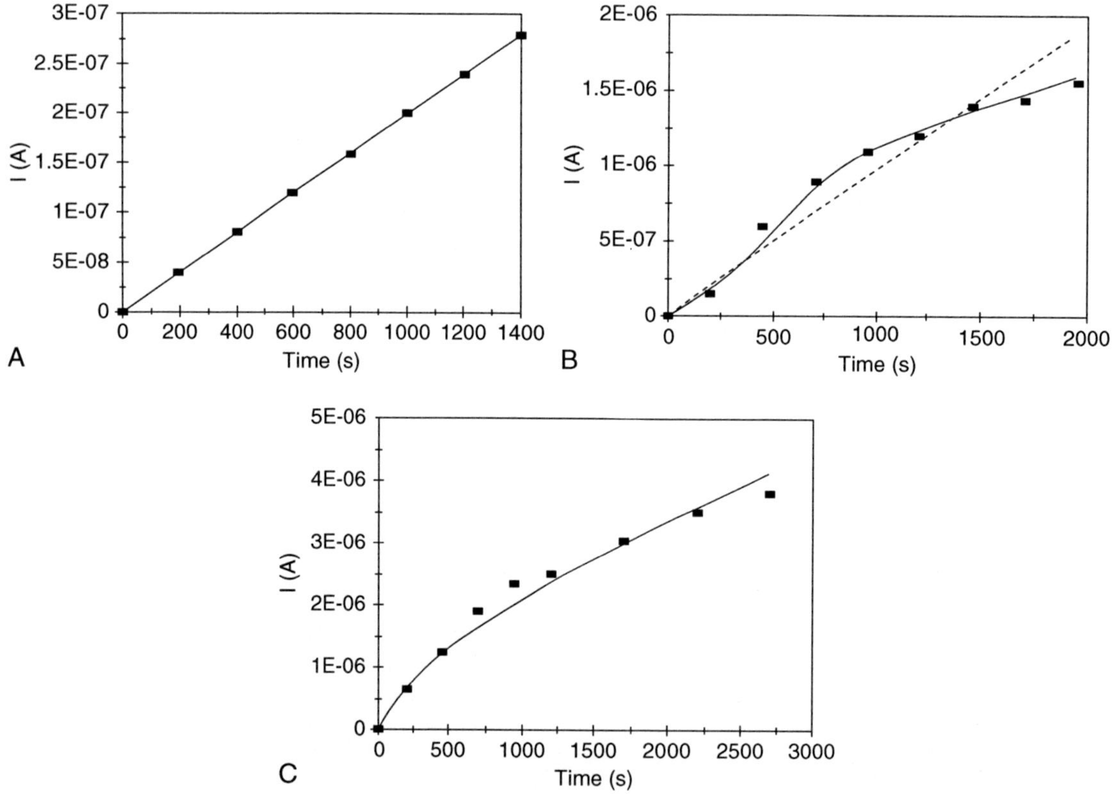

Figure 7.13
Binding of sequential (every 10 s) of 70 μl additions of 50 mM glucose solution to the screen-printed water-based (WB) CoPC microband biosensor using microelectrodes under stirred conditions at different potentials (Pemberton et al., 2009): (a) 0.3V (b) 0.4V (c) 0.5V.

Table 7.9: Binding of 500 mM glucose in solution (sequential addition every 100 s) to the microelectrodes under stirred conditions.

Applied Potential (V)	k	k_1	k_2	D_f	D_{f1}	D_{f2}
0.3	$(2.0\pm0.0)10^{-10}$	na	na	$(1.0\pm2.0)10^{-14}$	na	na
0.4	$(1.2\pm0.3)10^{-9}$	$(1.7\pm0.4)10^{-10}$	$(3.5\pm0.1)10^{-8}$	1.0574 ± 0.2298	0.394 ± 0.351	1.9930 ± 0.1754
0.5	$(2.0\pm0.2)10^{-8}$	na	na	1.6514 ± 0.0828	na	na

Influence of varied applied potentials (Pemberton et al., 2009).

Figure 7.13b shows the hydrodynamic voltagram (binding kinetics) at an applied potential of 0.4 V (Pemberton et al., 2009). A dual-fractal analysis is required to adequately describe the binding kinetics. The values of (a) the binding rate coefficient, k, and the fractal dimension, D_f, for a single-fractal analysis, and (b) the binding rate coefficients, k_1 and k_2 and the fractal dimensions, D_{f1} and D_{f2} for a dual-fractal analysis are given in Table 7.9.

It is of interest to note that as the fractal dimension increases by a factor of 5.06 from a D_{f1} value equal to 0.394 to D_{f2} equal to 1.9930, the binding rate coefficient increases by a factor of 206 from a value of k_1 equal to 1.7×10^{-10} to k_2 equal to 3.5×10^{-8}. Once again, an increase in the degree of heterogeneity or the fractal dimension on the microband sensor surface leads to an increase in the binding rate coefficient.

Figure 7.13c shows the hydrodynamic voltagram at an applied potential of 0.5 V. A single-fractal analysis is adequate to describe the binding kinetics. The values of (a) the binding rate coefficient, k, and the fractal dimension, D_f, for a single-fractal analysis is given in Tables 7.9.

7.4 Conclusions

This chapter has concentrated on and presented different examples for the detection of glucose in solution by different biosensors available in the literature. The kinetics of binding and dissociation (wherever applicable) were analyzed using fractal analysis. As indicated in the beginning of the chapter the detection of glucose in solution was, in a way, the first application that led to the development of biosensors. Over the years more and more sophisticated biosensors have been developed for the detection of glucose in keeping with the current trends in research, such as nanotechnology and nanobiotechnology. The ease of use of biosensors has led to the application of biosensors in different areas such as the detection of pollutants in the environment and pathogens in the early onset of diseases, in biological warfare detection, detection of explosives for aircraft and passenger protection, food preservation and spoilage, etc.

As noted throughout the different chapters in the book, the biosensor development is in an unlimited growth phase, which does not seem likely to slow down in the near future. Chapters or publications such as this can easily be written for the detection of other analyte(s) using different biosensors. Perhaps, some start-up or well-established companies are already considering this aspect before they develop a biosensor and bring it to the market. Of course, the economics of development will and does play a critical part in the selection of the biosensor a particular company wants to bring to the market. It is possible that a particular company with already developed expertise on a particular (biosensor) platform may use it with modifications to detect a particular analyte. The company may also like to explore a novel or a new biosensor platform for the detection of an analyte it has already identified. This chapter will hopefully bring this type of thinking and reasoning to the forefront, before it is too late.

The economics of biosensor development is an extremely important area of investigation, and forms the basis of the last chapter in this book. As might be expected, industrial sources are very well aware of this. This economic aspect is generally paid scant attention to, or at worst completely neglected by my university colleagues. Hopefully, emphasis on economics and alternative and appropriate biosensor platforms for the detection of glucose and other analytes of interest will gradually begin to change the frame of mind of researchers in the university as well as perhaps consolidate this type of thinking in industrial researchers too. Nanotechnology and nanobiotechnology do seem to exhibit a lot of potential in the development of biosensors as well as other areas. It is hoped that this will translate into better biosensors that exhibit better performance characteristics, and ultimately result in the development of more economical biosensors.

It would have been excellent if economic data had been available and presented for the different examples of biosensors used for the detection of glucose. However, at present, such data is unavailable (perhaps it may never be) especially in the universities where most, if not all, of this research is selected from. Perhaps, this type of data on the development of different biosensor platforms for the detection of glucose or other analytes is available in industrial sources. But, as one might very reasonably expect, this type of economic information, if it exists, is well-guarded and will not be available from industrial sources. Perhaps, more university sources need to analyze this area which has scant information available in the open literature.

References

Bai Y, Y Sun, and C Sun, Pt-Pb nanowire array electrode for enzyme-free glucose detection, *Biosensors and Bioelectronics*, **24**(4), 579–585 (2008).

Caduff AF, F Dewarrat, M Talary, G Stadler, L Heinemann, and Yu Feldman, Non-invasive glucose monitoring in patients with diabetes: A novel system based on impedance spectroscopy, *Biosensors & Bioelectronics*, **24**(9), 2778–2784 (2006).

Cass AEG, G Davis, GD Francis, HAO Hill, WJ Aston, IJ Higgins, EV Plotkin, LDL Scott, and APF Turner, Ferrocene-mediated enzyme electrode for amperometric determination of glucose, *Analytical Chemistry*, **56**, 667–671 (1984).

Chin J-Y, C-M Yu, M-J Yen, and L-C Chen, Glucose sensing electrodes based on a poly(3,4-ethylenedioxythiophine)/Prussian blue bilayer and multi-walled carbon nanotubes, *Biosensors and Bioelectronics*, **24**(7), 2015–2020 (2008).

Cui G, SJ Kim, SH Choi, H Nam, GS Cha, and KJ Paeng, A disposable amperometric sensor screen printed on a nitrocellulose strip: A glucose biosensor employing lead oxide as an interference-removing agent, *Analytical Chemistry*, **72**, 1925–1929 (2000).

Eastman JA, S SUS Choi, W Yu Li, and LJ Thompson, Anomalously increased effective thermal conductivities of ethylene glycol-based nanofilms containing copper nanoparticles, *Applied Physics Letters*, **78**, 718–720 (2001).

Gao ZC, F Xie, M Shariff, M Arshad, and JY Yin, A diffusional glucose biosensor based on diffusional mediator dispersed in nanoparticulate membrane on screen-printed carbon electrode, *Sensors & Actuators B*, **111**, 339–345 (2005).

Havlin S, Molecular diffusion and reactions. In *The Fractal Approach to Heterogeneous Chemistry: Surfaces, Colloids, Polymers*, Avnir D (Ed.), Wiley, New York, 1989, pp. 251–269.

Heller A and B Feldman, Electrochemical glucose sensors and their applications in diabetes management, *Chemical Reviews*, **108**, 2482–2505 (2008).

Hu J, The evolution of commercialized glucose sensors in China, *Biosensors and Bioelectronics*, **24**(5), 1083–1089 (2009).

Jablecki M and DA Gough, Simulations of the frequency response of implantable glucose sensors, *Analytical Chemistry*, **72**, 1853–1859 (2000).

Karyakin AA, EA Kotel'nikova, LV Lukachova, EE Karykina, and J Wang, Optimal environment for glucose oxidase in perfluorosulfonated ionomer membranes: Improvement of first-generation biosensors, *Analytical Chemistry*, **74**, 1597–1603 (2002).

Lu L, ML Sui, and K Lu, Superplastic extensibility of nanocrystalline copper at room temperature, *Science*, **287**, 1463–1466 (2000).

Mala KB, S Hrapovic, Y Liu, D Wang, and JHT Luong, Electrochemical detection of carbohydrates using copper nanoparticles and carbon nanotubes, *Analytica Chimica Acta*, **516**, 35–41 (2004).

Manesh KM, HT Kim, P santosh, AJ Gopalana, and K-P Lee, A novel glucose biosensor based on immobilization of glucose oxidase into multiwall carbon nanotubes-polyelectrolyte loaded electrospun nanofibrous membrane, *Biosensors & Bioelectronics*, **23**(6), 771–779 (2008).

Matthews DR, E Brown, A Watson, RR Holman, J Steemson, S Hugghes, and D Scott, *Lancet*, **8536**, 778–779 (1987).

Mercado RC and F Moussey, *In vitro* and *in vivo* mineralization of Nafion membrane for implantable glucose sensors, *Biosensors & Bioelectronics*, **13**(2), 133–145 (1998).

Merkoci A, M Pumera, X Llopis, B Perez, M del Valle, and S Alegret, New materials for electrochemical sensing. VI. Carbon nanotubes, *Trends in Analytical Chemistry*, **24**, 826–838 (2005).

Miyashita M, N Ito, S Ikeda, T Murayama, K Oguma, and J Kimura, Development of urine glucose meter based on micro-planer amperometric biosensor and its clinical application for self-monitoring of urine glucose, *Biosensors and Bioelectronics*, **24**(5), 1336–1340 (2009).

Monnier L, H Lapinski, and C Collette, Contributions of fasting and post-prandial plasma glucose increments to the overall diurnal hyperglycemia of type 2 patients, *Diabetic Care*, **26**, 881–885 (2003).

Monnier L, C Collette, GJ Dunseath, and DR Owens, The loss of postprandial control preceeds stepwise deterioration of fasting with worsening diabetes, *Diabetic Care*, **30**, 263–269 (2007).

Nakijama S, H Endoh, and M Inagami, *OMRON TECHICVS (Japanese)*, **36**, 235–239 (1996).

Neuman JD and APF Turner, *Biosensors & Bioelectronics*, **20**, 2435–2453 (2005).

Neuman JD, APF Turner, and G Marrazza, *Analytica Chimica Acta*, **42**, 118–123 (1970).

Newman JD and SJ Setford, *Molecular Biotechnology*, **32**(3), 249–268 (2006).

Pemberton RM, R Pittson, N Biddle, and JP Hart, Fabrication of micro and glucose biosensors using a screen-printing water-based carbon ink and their application in serum analysis, *Biosensors and Bioelectronics*, **24**(5), 1246–1252 (2009).

Ramakrishnan A and A Sadana, A single fractal analysis of cellular analyte-receptor binding kinetics casing biosensors, *Biosystems*, **59**, 35–51 (2001).

Rawson FJ, WM Purcell, J Xu, DC Cowell, PR Fielden, N Biddle, and JP Hart, Fabrication and characterization of novel screen-printed tubular microband electrodes, and their application to the measurement of hydrogen peroxide, Electrochimica *Acta*, **52**, 7248–7253 (2007).

Sadana A, A fractal analysis for the evaluation of hybridization kinetics in biosensors, *Jorunal of Colloid and Interface Science*, **151**(1), 166–177 (2001).

Sadana A, *Fractal Binding and Dissociation Kinetics for Different Biosensor Applications*, Amsterdam, 2005, Elsevier.

Safavi A, N malaki, and E Farjami, Fabrication of a glucose sensor based on a novel nanocomposite electrode, *Biosensors & Bioelectronics*, **24**(6), 1655–1660 (2009).

Schoenfisch MH, AR Rorhrock, JH Shin, MA Polizzi, MF Brinkley, and KP Dobmeier, Poly(vinyulpyrtolidine)-doped nitric oxide-releasing xerogels as glucose biosensor membranes, *Biosensors & Bioelectronics*, **22**(6), 306–312 (2006).

Shoji F and MS Freund, Potentiometric sensors based on the inductive effect on the pK_a of poly(aniline): A nonenzymatic glucose sensor, *Journal of the American Chemical Society*, **123**, 3383–3384 (2001).

Sun YP, H Buck, and TE Mallouk, Combinatorial discovery of alloy electrocatalysts for amperometric glucose sensors, *Analytical Chemistry*, **73**, 1599–1604 (2001).

Vaddiraju S, DJ Burgess, FC Jain, and F Papadimatrakopoulos, The role of H_2O_2 outer diffusion on the performance of implantable glucose sensors, *Biosensors and Bioelectronics*, **24**(6), 1557–1562 (2009).

Wang X, H Gu, F Yin, and Y Tu, A glucose biosensor based on Prussian blue/chitosan hybrid film, *Biosensors & Bioelectronics*, **24**(5), 1527–1530 (2009).

Williams DL, AR Doig Jr, and A Korosi, Electrochemical-enzymatic analysis of blood glucose and lactate, *Analytical Chemistry*, **42**, 118–123 (1970).

Wilson GS and R Gifford, Biosensors for real-time *in vivo* measurements, *Biosensors & Bioelectronics*, **20**(12), 2388–2403 (2005).

Wilson R and APF Turner, Glucose oxidase: An ideal enzyme, *Biosensors & Bioelectronics*, **7**, 165–185 (1992).

Woerle HJ, C Neumann, S Zachau, S Tenner, A Insigier, J Schirra, JE Gerich, and B Goke, Impact of fasting and postprandial glycemia on overall glycemic control in type 2 diabetes: Importance of postprandial glycemica to achieve HbA1c levels, *Diabetes Research Clinical Practice*, **77**(2), 2808–2850 (2007).

Xu Q, Y Zhao, JZ Xu, and JJ Zhu, Preparation of functionalized copper nanoparticles and fabrication of a glucose sensor, *Sensors & Actuators B*, **114**, 379–3886 (2006).

Xu H, K Malladi, C Wang, L Kulinsky, M Song, and M Madou, Carbon post-microarrays for glucose sensors, *Biosensors & Bioelectronics*, **23**(11), 1637–1644 (2008).

Yabuki S, F Mizutani, Y Satao, and Y Hirata, Immobilization of polyglutamate-glucose oxidase onto a cysteamine-modified gold electrode, *Sensors & Actuators B: Chemical*, **91**, 187–190 (2003).

You TY, O Niwa, M Tomita, H Ando, M Suzuki, and S Hirano, Characterization and electrochemical properties of highly dispersed copper oxide/hydroxide nanoparticles in graphite-like carbon films prepared by RF sputtering method, *Electrochemical Communication*, **4**, 468–471 (2002).

Zhao HT and HX Ju, Multilayer membranes for glucose biosensing via layer-by-layer assembly of multiwall carbon nanotubes and glucose oxidase, *Analytical Biochemistry*, **350**, 138–144 (2006).

CHAPTER 8

Medical Applications of Biosensors

Chapter Outline

8.1 Introduction

Medical applications are the leading applications of biosensors. The diabetes management market (making quantitative glucose levels in the blood) is the biggest market application. The ease of use of biosensors for the detection of analytes that are monitored especially for the onset of the different types of diseases as well as their management has led to the development of different types of biosensors. Some of these include:

(a) Sensitive immunoassay of a biomarker tumor necrosis factor-α (TNF-α) based in poly (guanine)-functionalized silica nanoparticle (NP) label (Wang et al., 2006)
(b) Quantitative measurement of cardiac markers in undiluted serum (Masson et al., 2007)
(c) Novel microfluidic impedance assay for monitoring endothelin-induced cardiac hypertrophy (Yang et al., 2007)
(d) Development of quartz-crystal microbalance-based immunosensor array for clinical immunophenotyping of acute leukemias (Zeng et al., 2006).
(e) Effect of oxazaborolidines on immobilized fructosyltransferase (FTF) analyzed by surface plasmon-resonance (Jabbour et al., 2007). These are involved in dental diseases.
(f) Histone deacylase (HDAC) inhibitor assay based on resonance energy transfer (Riester et al., 2007). These authors report that histone HDACs are important enzymes for the transcriptional regulation of gene expression in eukaryotic cells. Recently, they have become key targets for chemotherapeutic intervention in malignant diseases.

Handbook of Biosensors and Biosensor Kinetics. DOI: 10.1016/B978-0-444-53262-6.00008-5

(g) Development of a screen-printed cholesterol biosensor (Shen and Liu, 2007)
(h) Glucose sensor using nonwoven single-wall carbon nanotube films (Lei and Jia, 2007)
(i) Chemistry for a single walled carbon nanotube fluorescence-based glucose sensor (Barone and Strano, 2007)
(j) Biochip for a rapid and sensitive detection of multiple cancer markers simultaneously (Goluch et al., 2007)
(k) Development of label-free nanopattern-enhanced biosensors for food safety and monitoring and early cancer diagnostics (Jiang, 2007)
(l) Hexagonal saw interleukin-6 biosensor (Cular et al., 2007)
(m) Diamond microneedle electrodes for neurochemical detection (Martin, 2007).

In this chapter we use fractal analysis to analyze the binding (and dissociation) kinetics of (a) the binding of TNF-α in solution to poly(guanine)-functionalized silica NPs (Wang et al., 2006), (b) the binding of different antigens in solution to the anti CD antigen immobilized on a quartz-crystal microbalance (QCM) surface (Zeng et al., 2006), (c) the binding of 50 ng/mL myoglobin in serum to antimyoglobin antibody immobilized on a surface plasmon resonance (SPR) biosensor surface (Masson et al., 2007), (d) the binding and dissociation of cardioimyocytes plus endothelin-1 (ET-1) with and without a DEP (dielectrophoresis) device (Yang et al., 2007), and (e) the binding and dissociation of different concentrations of oxazaborolidine derivatives, BNO1, BNO2, BNO3, and BNO4 + 2 mM sucrose in solution to the enzyme FTF immobilized on a SPR biosensor chip surface (Jabbour et al., 2007). This is just one possible way to analyze the kinetics. The fractal analysis method provides one with the values of the binding and the dissociation rate coefficient values as well as the fractal dimension (the degree of heterogeneity) values on the biosensor chip surface. As indicated earlier, throughout the book, other ways of obtaining the binding and dissociation kinetics are also available; though these other methods do not account for either the heterogeneity on the biosensor surface or the presence of external diffusional limitations.

8.2 Theory

Havlin (1989) has reviewed and analyzed the diffusion of reactants towards fractal surfaces. The details of the theory and the equations involved for the binding and the dissociation phases for analyte-receptor binding are available (Sadana, 2001) in the literature. The details are not repeated here except that the equations are given to permit an easier reading. These equations have been applied to other biosensor systems (Ramakrishnan and Sadana, 2001; Sadana, 2001). For most applications, a single- or a dual-fractal analysis is often adequate to describe the binding and the dissociation kinetics. Peculiarities in the values of the binding and the dissociation rate coefficients, as well as in the values of the fractal dimensions with regard to the dilute analyte systems being analyzed will be carefully noted, if applicable.

8.2.1 Single-Fractal Analysis

Binding Rate Coefficient

Havlin (1989) points out that the diffusion of a particle (analyte [Ag]) from a homogeneous solution to a solid surface (e.g. receptor [Ab]-coated surface) on which it reacts to form a product (analyte-receptor complex; (Ab·Ag)) is given by:

$$(\text{Analyte}\cdot\text{Receptor}) \sim \begin{cases} t^{(3-D_{\mathrm{f,bind}})/2} = t^{p}, & t < t_{\mathrm{c}} \\ t^{1/2}, & t > t_{\mathrm{c}} \end{cases} \tag{8.1a}$$

Here $D_{\mathrm{f,bind}}$ or D_{f} (used later on in the chapter) is the fractal dimension of the surface during the binding step. t_{c} is the cross-over value. Havlin (1989) points out that the cross-over value may be determined by $r_{\mathrm{c}}^{2} \sim t_{\mathrm{c}}$. Above the characteristic length, r_{c}, the self-similarity of the surface is lost and the surface may be considered homogeneous. Above time, t_{c}, the surface may be considered homogeneous, since the self-similarity property disappears, and "regular" diffusion is now present. For a homogeneous surface where D_{f} is equal to 2, and when only diffusional limitations are present, $p = ½$ as it should be. Another way of looking at the $p = ½$ case (where $D_{\mathrm{f,bind}}$ is equal to two) is that the analyte in solution views the fractal object, in our case, the receptor-coated biosensor surface, from a "large distance." In essence, in the association process, the diffusion of the analyte from the solution to the receptor surface creates a depletion layer of width $(Đt)^{½}$ where $Đ$ is the diffusion constant. This gives rise to the fractal power law, $(\text{Analyte}\cdot\text{Receptor}) \sim t^{(3-D_{\mathrm{f,bind}})/2}$. For the present analysis, t_{c} is arbitrarily chosen and we assume that the value of t_{c} is not reached. One may consider the approach as an intermediate "heuristic" approach that may be used in the future to develop an autonomous (and not time-dependent) model for diffusion-controlled kinetics.

Dissociation Rate Coefficient

The diffusion of the dissociated particle (receptor [Ab] or analyte [Ag]) from the solid surface (e.g., analyte [Ag]-receptor [Ab]) complex coated surface into solution may be given, as a first approximation by:

$$\begin{aligned}(\text{Analyte}\cdot\text{Receptor}) &\sim -t^{(3-D_{\mathrm{f,diss}})/2}, \quad t > t_{\mathrm{diss}} \\ &= -k_{\mathrm{diss}} t^{(3-D_{\mathrm{f,diss}})/2}\end{aligned} \tag{8.1b}$$

Here $D_{\mathrm{f,diss}}$ is the fractal dimension of the surface for the dissociation step. This corresponds to the highest concentration of the analyte-receptor complex on the surface. Henceforth, its concentration only decreases. The dissociation kinetics may be analyzed in a manner "similar" to the binding kinetics.

8.2.2 Dual-Fractal Analysis

Binding Rate Coefficient

Sometimes, the binding curve exhibits complexities and two parameters (k, D_f) are not sufficient to adequately describe the binding kinetics. This is further corroborated by low values of the r^2 factor (goodness-of-fit). In that case, one resorts to a dual-fractal analysis (four parameters; k_1, k_2, D_{f1}, and D_{f2}) to adequately describe the binding kinetics. The single-fractal analysis presented above is thus extended to include two fractal dimensions. At present, the time ($t = t_1$) at which the "first" fractal dimension "changes" to the "second" fractal dimension is arbitrary and empirical. For the most part, it is dictated by the data analyzed and experience gained by handling a single-fractal analysis. A smoother curve is obtained in the "transition" region, if care is taken to select the correct number of points for the two regions. In this case, the product (antibody-antigen; or analyte-receptor complex, Ab·Ag or analyte·receptor) is given by:

$$(\text{Ab}\cdot\text{Ag}) \approx \begin{cases} t^{(3-D_{f1,\,\text{bind}})/2} = t^{p1}, & t < t_1 \\ t^{(3-D_{f2,\,\text{bind}})/2} = t^{p2}, & t_1 < t < t_2 = t_c \\ t^{1/2}, \quad t > t_c \end{cases} \tag{8.1c}$$

In some cases, as mentioned above, a triple-fractal analysis with six parameters (k_1, k_2, k_3, D_{f1}, D_{f2}, and D_{f3}) may be required to adequately model the binding kinetics. This is when the binding curve exhibits convolutions and complexities in its shape due perhaps to the very dilute nature of the analyte or for some other reasons. Also, in some cases, a dual-fractal analysis may be required to describe the dissociation kinetics.

8.3 Results

The fractal analysis will be applied to different analyte-receptor reactions occurring on biosensor chip surfaces with the specific medical applications. Attempts will be made to relate particularly changes in the fractal dimension on the biosensor chip surface with the changes in the binding and the dissociation rate coefficients.

At the outset it should be pointed out that alternative expressions for fitting the binding and dissociation data are available that include saturation, first-order reaction, and no diffusional limitations, but these expressions are deficient in describing the heterogeneity that inherently exists on the surface. It is this heterogeneity on the biosensor surface that one is attempting to relate to the different biosensor performance parameters. More specifically the question we wish to answer is how may one change the heterogeneity or the fractal dimension, D_f, on the biosensor chip surface in order that one may be able to enhance the different biosensor performance parameters.

Other modeling attempts also need to be mentioned. One might justifiably argue that appropriate modeling may be achieved by using a Langmuirian or other approach. The Langmuirian approach may be used to model the data presented if one assumes the presence of discrete classes of sites, for example double exponential analysis as compared with the single-fractal analysis. Lee and Lee (1995) report that the fractal approach has been applied to surface science, for example, adsorption and reaction processes. These authors point out that the fractal approach provides a convenient means to represent the different structures and the morphology at the reaction surface. They also draw attention to using the fractal approach to develop optimal structures and as a predictive approach. Another advantage of the fractal technique is that the analyte-receptor association is a complex reaction, and the fractal analysis via the fractal dimension and the rate coefficient provide a useful lumped parameter analysis of the diffusion-limited reaction occurring on a heterogeneous surface.

In a classical situation, to demonstrate fractality, one should make a log-log plot, and one should definitely have a large amount of data. It may be useful to compare the fit to some other forms, such as the exponential form, or one involving saturation, etc. At present, no independent proof or physical evidence of fractals in the examples is presented. Nevertheless, we still use fractals and the degree of heterogeneity on the biosensor surface to gain insights into enhancing the different biosensor performance parameters. The fractal approach is a convenient means (since it is a lumped parameter) to make the degree of heterogeneity that exists on the surface more quantitative. Thus, there is some arbitrariness in the fractal approach to be presented. The fractal approach provides additional information about interactions that may not be obtained by a conventional analysis of biosensor data. In this chapter as mentioned above, an attempt is made to relate the fractal dimension, D_f, or the degree of heterogeneity on the biosensor surface with different biosensor performance parameters. More specifically, the interest is in finding out how changes in the fractal dimension or the degree of heterogeneity on the biosensor chip surface affect the different biosensor parameters of interest.

Unless specifically mentioned there is no nonselective adsorption of the analyte. In other words, nonspecific binding is ignored. Nonselective adsorption would skew the results obtained very significantly. In these types of systems, it is imperative to minimize this nonselective adsorption. It is also recognized that, in some cases, this nonselective adsorption may not be a significant component of the adsorbed material and that the rate of association, which is of a temporal nature would depend on surface availability.

Wang et al. (2006) have recently developed a sensitive immunoassay for the biomarker, TNF-α based on a poly(guanine)-functionalized silica NP label. These authors report that TNF-α is an extremely potent peptide cytokine which serves as an endogeneous mediator of inflammatory, immunodefense, and host defense function (Old, 1985, 1987; Jones et al., 1989). Wang et al. (2006) point out that TNF-α is involved in a wide variety of pathological and physiological processes. High TNF concentrations in serum have been observed in pathological states such as endotoxic shock, graft rejection, HIV infection, and rheumatoid arthritis (De Kossodo et al., 1995). Wang et al. (2006) assert that the electrochemical technique is an attractive method for the immunoassay of biomarkers, especially since TNF-α, in general, is observed to be at very low levels in biological samples such as serum.

Wang et al. (2006) report that the electrochemical immunoassay technique is highly sensitive and inherently simple. Furthermore, it can be miniaturized and is a low cost technique. These authors further state that the inclusion of NPs in electrochemical sensors exhibits a significant potential for the detection of trace biomolecules (Daniel and Astruc, 2004; Wang et al., 2004). Wang et al. (2006) have developed and used poly(guanine)(poly[G]-)-functionalized silica NPs along with mediator-induced catalytic oxidation of guanine for the amplified electrochemical immunoassay of TNF-α.

Figure 8.1 shows the binding of 1.0 ng/mL of TNF-α in solution to the poly(guanine)-functionalized silica NPs (Wang et al., 2006). A single-fractal analysis is adequate to describe the binding kinetics. These authors report that at 45 min incubation time the interaction of TNF-α in solution and the TNF-α antibody immobilized on the electrochemical sensor reaches saturation. The value of the binding rate coefficient, k, and the fractal dimension, D_f, for a single-fractal analysis are given in Table 8.1.

Zeng et al. (2006) report that recently there has been an increasing demand to improve the sensitivity and reaction rate parameters for both automated and miniaturized clinical analysis. These authors have very recently developed a quartz-crystal microbalance-based

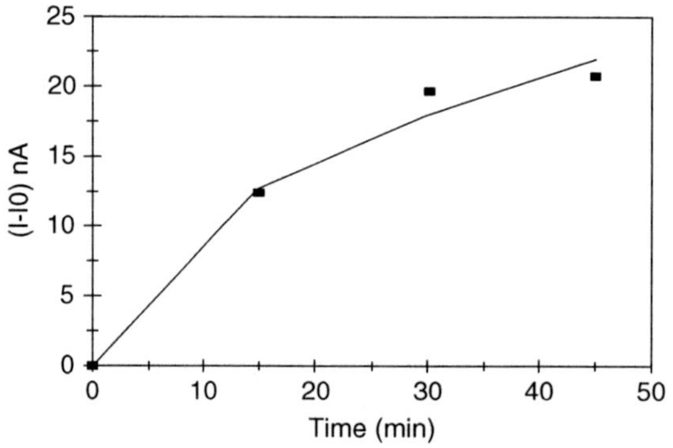

Figure 8.1
Binding of 1.0 ng/mL of TNF-α (tumor necrosis factor) in solution to the poly(guanine)-functionalized silica nanoparticles (NPs) (Wang et al., 2006).

Table 8.1: Binding rate coefficients and fractal dimensions for (a) the binding of 1.0 ng/mL TNF-α in solution to the poly (guanine)-functionalized silica nanoparticles (NPs) (Wang et al., 2006), and (b) the binding of different antigens CD3, CD5, and CD7 to the antibodies anti-CD3, anti-CD5, and anti-CD7, respectively, immobilized on a plasma-polymerized film (PPF) of n-butylamine, nanogold particles, and protein A (PA) on a QCM (quartz-crystal microbalance) sensor surface (Zeng et al., 2006).

Analyte in Solution/ Receptor on Surface	D_f	D_{f1}	D_{f2}	k	k_1	k_2
(a)						
1.0 ng/mL TNF-α/ functionalized silica nanoparticles (NPs)	2.0160 ± 0.2806	na	na	3.381 ± 0.394	na	na
(b)						
CD7/anti-CD7 on a QCM surface	1.5688 ± 0.101	1.1436 ± 0.0936	2.8362 ± 0.0363	25.253 ± 4.898	19.275 ± 1.964	154.1 ± 1.54
CD5/anti-CD5 on a QCM surface	1.3196 ± 0.155	0.8498 ± 0.272	2.3474 ± 0.1249	14.868 ± 4.210	11.074 ± 2.651	62.33 ± 4.93
CD3/anti-CD3 on a QCM surface	1.5558 ± 0.119	1.0768 ± 0.151	2.4426 ± 0.0943	13.927 ± 3.063	10.081 ± 1.408	46.636 ± 2.852

immunosensor array for the clinical immunophenotyping of acute leukemias. These authors point out that immunophenotyping usually utilizes antibodies to recognize various differentiated antigens of leukocytes. This, they state is a vitally important means for defining certain phenotype linkages or subsets of acute leukemias (Traweek, 1993; Ba, 1998; Hoffman et al., 2000).

Zeng et al. (2006) report that the QCM immunosensor is an active area of investigation for bioassays. The QCM immunosensor exhibits high sensitivity and specificity, is simple to use, and is cost-effective. These authors fabricated QCM crystal probes first with plasma-polymerized film (PPF) and NPs (Zeng et al., 2006). Then, protein A (PA) was utilized to orient the different immobilized leukemia-linkage-associated CD antibodies. This permitted the formation of a QCM-based immunosensor array for probing the degrees of expression of differentiated antigens on corresponding leukocytes. These authors used their QCM array to immunophenotype 120 human bone marrow (BM) samples. They emphasize that their proposed analytical procedure is direct and simple. Also, there are no multiple labeling and separation steps.

Figure 8.2a shows the binding of CD7 antigen in solution to the anti CD7 antibody immobilized on the QCM immunosensor surface (Zeng et al., 2006). A dual-fractal analysis is required to adequately describe the binding kinetics. The values of (a) the binding rate coefficient, k, and the fractal dimension, D_f, for a single-fractal analysis, and (b) the binding rate coefficients, k_1 and k_2, and the fractal dimensions, D_{f1} and D_{f2}, for a dual-fractal analysis are given in Table 8.1. It is of interest to note that as the fractal dimension increases by a factor of 2.48 from a value of D_{f1} equal to 1.1436 to D_{f2} equal to 2.8362, the binding rate coefficient increases by a factor of 7.99 from a value of k_1 equal to 19.275 to k_2 equal to 154.1. It is seen that changes in the degree of heterogeneity or the fractal dimension on the QCM surface and in the binding rate coefficient are in the same direction.

Figure 8.2b shows the binding of CD5 antigen in solution to the anti CD5 antibody immobilized on the QCM immunosensor surface (Zeng et al., 2006). A dual-fractal analysis is required to adequately describe the binding kinetics. The values of (a) the binding rate coefficient, k, and the fractal dimension, D_f, for a single-fractal analysis, and (b) the binding rate coefficients, k_1 and k_2, and the fractal dimensions, D_{f1} and D_{f2}, for a dual-fractal analysis are given in Table 8.1. It is of interest to note that as the fractal dimension increases by a factor of 2.76 from a value of D_{f1} equal to 0.8498 to D_{f2} equal to 2.3474, the binding rate coefficient increases by a factor of 5.63 from a value of k_1 equal to 11.074 to k_2 equal to 62.33. It is seen that, once again, changes in the degree of heterogeneity or the fractal dimension on the QCM surface and in the binding rate coefficient are in the same direction.

Figure 8.2c shows the binding of CD3 antigen in solution to the anti CD3 antibody immobilized on the QCM immunosensor surface (Zeng et al., 2006). A dual-fractal analysis is required to adequately describe the binding kinetics. The values of (a) the binding rate

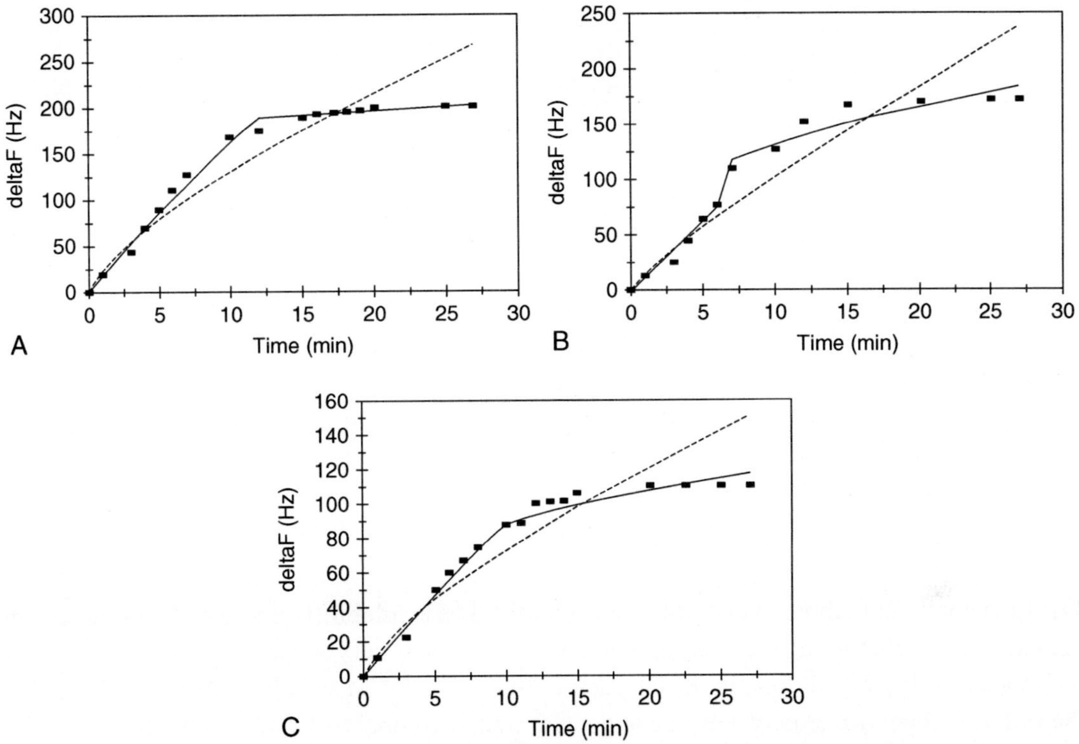

Figure 8.2

Binding of different antigens in solution to the anti CD7, anti CD5, and anti CD3, antibody, respectively immobilized on the Quartz Crystal Microbalance (QCM) surface (Zeng et al., 2006): (a) CD7 (b) CD5 (c) CD3. When only a solid line (–) is used then a single-fractal analysis applies. When both a dashed (- - -) and a solid (–) line are used then the dashed line represents a single-fractal analysis and the solid line represents a dual-fractal analysis.

coefficient, k, and the fractal dimension, D_f, for a single-fractal analysis, and (b) the binding rate coefficients, k_1 and k_2, and the fractal dimensions, D_{f1} and D_{f2}, for a dual-fractal analysis are given in Table 8.1. It is of interest to note that as the fractal dimension increases by a factor of 2.27 from a value of D_{f1} equal to 1.0768 to D_{f2} equal to 2.4426 the binding rate coefficient increases by a factor of 4.63 from a value of k_1 equal to 10.081 to k_2 equal to 46.636. It is seen that, once again, changes in the degree of heterogeneity or the fractal dimension on the QCM surface and in the binding rate coefficient are in the same direction.

Figure 8.3 and Table 8.1 show for the binding of the different CD antigens, CD1, CD3, and CD5 in solution to their corresponding antibodies, anti-CD3, anti-CD5, and anti-CD7, respectively, and immobilized on a QCM biosensor surface, the increase in the binding rate coefficient, k_2, with an increase in the fractal dimension, D_{f2}. The binding rate coefficient, k_2, is given by:

$$k_2 = (0.394 \pm 0.178) D_{f2}^{5.667 \pm 2.637} \tag{8.2}$$

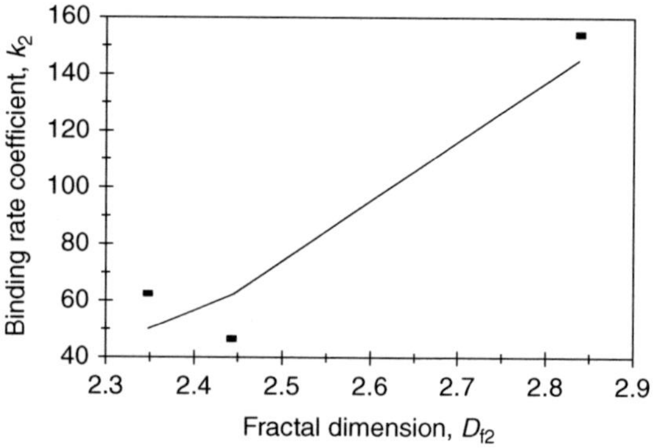

Figure 8.3
Increase in the binding rate coefficient, k_2, with an increase in the fractal dimension, D_{f2}.

The fit is good. Only three data points are available. The availability of more data points would lead to a more reliable fit. The binding rate coefficient, k_2, is very sensitive to the degree of heterogeneity or the fractal dimension, D_{f2}, that exists on the QCM sensor surface as indicated by the order of dependence between five and a half and six (equal to 5.667) exhibited.

Masson et al. (2007) point out the need to quantitatively measure the expression of biological markers to better understand their roles in disease progression. The measurement of biological markers in serum is difficult primarily because of the nonspecific adsorption of serum proteins. Battaglia et al. (2005) have indicated that the stability of SPR biosensors was improved in a cell culture media during the direct measurement of biomarkers for wound healing by using a coating of *N*-hydroxysuccinimide-activated 16-mercaptohexadecanoic acid (NHS-MHA). Masson et al. (2007) have extended this concept to measure the biomarkers of myocardial infraction (MI) in undiluted serum by using NHS-MHA as a nonfouling coating on a SPR biosensor surface. Masson et al. (2004, 2005) have shown that nonspecific binding in SPR biosensors may be reduced as a result of the recent developments in surface coatings used for SPR biosensing surfaces. These coatings permit the reduction of nonspecific binding while simultaneously improving the specific signals. This is achieved by increasing the number of antibodies on the gold surface of the SPR biosensor.

Myoglobin (MG) and cardiac troponin I (cTnI) are two MI markers. Masson et al. (2007) have detected MG in undiluted serum without sample pretreatment using the SPR biosensor. Figure 8.4 shows the binding of 50 ng/mL myoglobin in solution to antimyoglobin antibody immobilized on a SPR biosensor surface. A dual-fractal analysis is required to adequately describe the binding kinetics. The values of (a) the binding rate coefficient, k, and the fractal dimension, D_f, for a single-fractal analysis, and (b) the binding rate coefficients, k_1 and k_2, and the fractal dimensions, D_{f1} and D_{f2}, for a dual-fractal analysis are given in

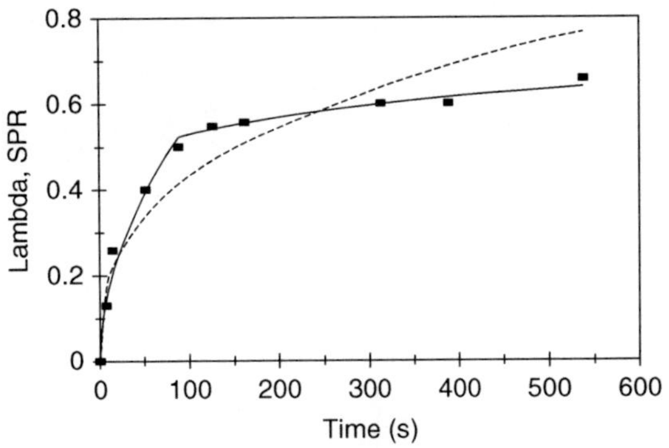

Figure 8.4

Binding of 50 ng/mL myoglobin in serum to antimyoglobin antibody immobilized on a SPR biosensor surface (Masson et al., 2007). When only a solid line (—) is used then a single-fractal analysis applies. When both a dashed (- - -) and a solid (—) line are used then the dashed line represents a single-fractal analysis and the solid line represents a dual-fractal analysis.

Tables 8.2 and 8.3. It is of interest to note that as the fractal dimension increases by a factor of 1.39 from a value of D_{f1} equal to 1.9884 to D_{f2} equal to 2.7716, the binding rate coefficient increases by a factor of 5.78 from a value of k_1 equal to 0.0541 to k_2 equal to 0.3129. It is seen that changes in the fractal dimension or the degree of heterogeneity on the biosensor surface and in the binding rate coefficient are in the same direction.

Yang et al. (2007) report that cardiac hypertrophy is an independent risk factor for the development of heart failure and sudden cardiac death (Levy et al., 1990; Ruskoaho, 1992). In cardiac hypertrophy there is an increase in heart size and/or myofibrillar volume without a change in myocyte number (Yang et al., 2007). Lovell and Carabello (2000) and Diane et al. (2000) point out that the cell morphology (cell size increases and there is myofibrillar re-organization) is altered. Yang et al. (2007) report that there is a need to monitor cardiac myocyte hypertrophy noninvasively and in real time. These authors have developed a novel microfluidic impedance assay for monitoring endothelin-induced cardiomyocyte hypertrophy.

Yang et al. (2007) developed and fabricated a DEP microfluidic device. This device is capable of concentrating cells from a dilute sample to form a confluent cell monolayer on a microelectrode surface. This device, the authors claim can increase the sensitivity of their impedance system besides significantly reducing the detection time required. They also treated their cardiomyocytes with ET-1, which is a known hypertrophic agent.

Figure 8.5a shows the binding and dissociation of the cardiac myocytes (heart muscle cells) in the presence of 4.6 $\times$ $10^3 mL^{-1}$ of ET-1 and on using the DEP device to the microelectrodes (comprising the impedance assay). A single-fractal analysis is adequate to describe the binding and the dissociation kinetics. The values of (a) the binding rate coefficient, k, and the fractal dimension, D_f, for a single-fractal analysis, and the dissociation rate coefficient, k_d, and the fractal dimension for dissociation, D_{fd}, for a single-fractal analysis are given in Tables 8.2 and 8.3. In this case the affinity, K $(=k/k_d)$, value is equal to 2.2×10^{-3}.

Table 8.2: Binding and dissociation rate coefficients for (a) the binding of 50 ng/mL of myoglobin in serum to antimyoglobin antibody immobilized on a SPR biosensor surface (Masson et al., 2007), and (b) binding and dissociation of cardiomyocytes with endothelin-1 (ET-1), and with and without a DEP (dielectrophoresis) microfluidic device that is capable of concentrating cells from a dilute sample to form a confluent cell monolayer on the microelectrode surface (Yang et al., 2007).

Analyte in Solution/Receptor on Surface	k	k_1	k_2	k_d
(a)				
50 ng/mL myoglobin/antimyoglobin antibody on SPR biosensor surface	0.0897 ± 0.0187	0.0541 ± 0.0105	0.3129 ± 0.0079	na
(b)				
4.6×10^3 cardiomyocytescells/mL + ET1 + DEP/electrode	$4.1 \times 10^{-05} \pm 1.1 \times 10^{-05}$	na	na	0.01856 ± 0.00356
5.3×10^3 cardiomyocytescells/mL + ET1/electrode	0.00103 ± 0.00002	na	na	na

Table 8.3: Fractal dimensions for the binding and the dissociation phase for (a) the binding of 50 ng/mL of myoglobin in serum to antimyoglobin antibody immobilized on a SPR biosensor surface (Masson et al., 2007), and (b) binding and dissociation of cardiomyocytes with endothelin-1 (ET-1), and with and without a DEP (dielectrophoresis) microfluidic device that is capable of concentrating cells from a dilute sample to form a confluent cell monolayer on the microelectrode surface (Yang et al., 2007).

Analyte in Solution/Receptor on Surface	D_f	D_{f1}	D_{f2}	D_{fd}
(a)				
50 ng/mL myoglobin/antimyoglobin antibody on SPR biosensor surface	2.3156 ± 0.0919	1.9884 ± 0.1804	2.7716 ± 0.0412	na
(b)				
4.6×10^3 cardiomyocytescells/mL + ET1 + DEP/electrode	0 + 0.2048	na	na	2.6970 ± 0.1335
5.3×10^3 cardiomyocytescells/mL + ET1/electrode	1.7272 ± 0.1190	na	na	na

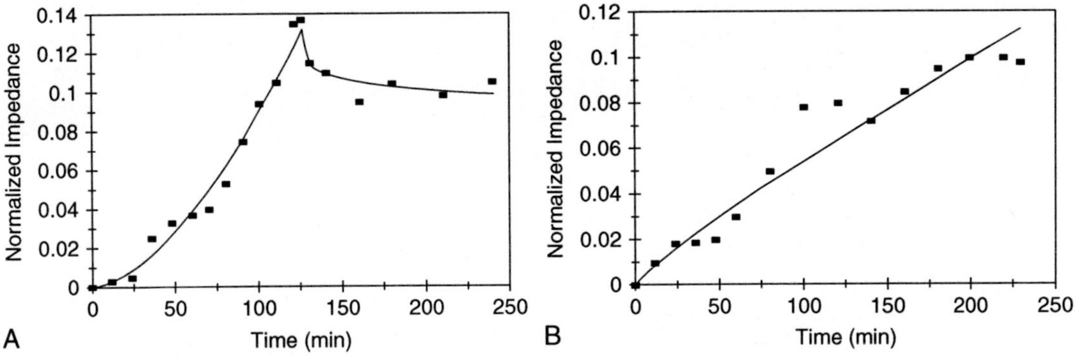

Figure 8.5

Binding and dissociation of cardiomyocytes plus endothelin-1 (a) with and (b) without a DEP (dielectrophoresis) device (Yang et al., 2007).

Figure 8.5b shows the binding of the cardiac myocytes (heart muscle cells), in the presence of 5.3×10^3/mL of ET-1 and without using the DEP device, to the microelectrodes (comprising the impedance assay). A single-fractal analysis is adequate to describe the binding kinetics. The values of the binding rate coefficient, k, and the fractal dimension, D_f, for a single-fractal analysis are given in Tables 8.2 and 8.3.

Jabbour et al. (2007) have recently used the SPR biosensor to analyze the effect of oxazaborolidines on immobilized frutosyltransferase (FTF). FTF are extracellular enzymes of several oral bacteria. Jabbour et al. (2007) report that FTF are associated with the formation of fructans which play a role in the biofilm formation and oral bacteria physiology. Fructans are a matrix of extracellular polysaccharides. The oxazaborolidines apparently inhibit biofilm formation. Steinberg (2000) point out that dental plaque formation is the major pathogenic factor associated with peridontal diseases. Allen and Bowen (1990) report that FTF is one of the extracellular enzymes secreted by several different species of oral bacteria. Various authors have pointed out the virulence of FTF in the oral cavity because of its enzymatic activity, effect of bacterial adhesion, and biofilm formation on dental surfaces (Burne, 1991; Rozen et al., 2001; Steinberg, 2000).

Jabbour et al. (2007) point out that the biological role of boron has been the subject of many analyses (Shelp and Gupta, 1993; Blevins and Lukaszewki, 1998). Jabbour et al. (2004) point out that oxazaborolidines possess antibacterial activity. Also, these compounds affect bacterial adhesion to hydroxapatite, and may influence biofilm formation (Jabbour et al., 2005). Thus, Jabbour et al. (2007) attempted to analyze the influence of oxazaborolidines on the real time synthesis of fructans immobilized by FTF using a SPR biosensor.

Figure 8.6a shows the binding and dissociation of 0 μM oxazaborolidine derivative, BNO1 + 2 mM sucrose in solution to FTF immobilized on a SPR sensor chip surface (Jabbour et al.,

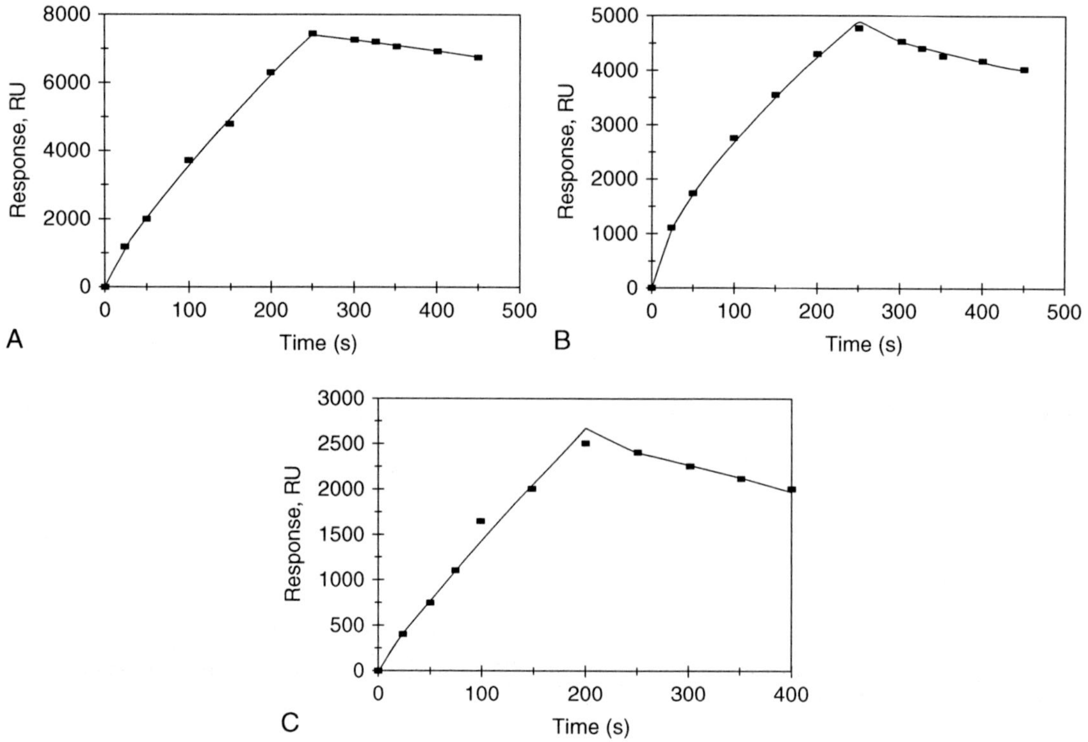

Figure 8.6

Binding and dissociation of different concentrations (in μM) of oxazaborolidine derivative (BNO1) + 2 μM sucrose in solution to fructosyltransferase (FTF) immobilized on a SPR sensor chip surface (Jabbour et al., 2007): (a) 0 (b) 60 (c) 600.

2007). A single-fractal analysis is adequate to describe the binding and the dissociation kinetics. The values of (a) the binding rate coefficient, k, and the fractal dimension, D_f, for a single-fractal analysis, and (b) the dissociation rate coefficient, k_d, and the fractal dimension for the dissociation phase, D_{fd}, for a single-fractal analysis are given in Tables 8.4 and 8.5. In this case, the affinity, K $(=k/k_d)$, value is 19.1.

Figure 8.6b shows the binding and dissociation of 60 μM oxazaborolidine derivative, BNO1 + 2 mM sucrose in solution to FTF immobilized on a SPR sensor chip surface (Jabbour et al., 2007). A single-fractal analysis is adequate to describe the binding and the dissociation kinetics. The values of (a) the binding rate coefficient, k, and the fractal dimension, D_f, for a single-fractal analysis, and (b) the dissociation rate coefficient, k_d, and the fractal dimension for the dissociation phase, D_{fd}, for a single-fractal analysis are given in Tables 8.4 and 8.5. In this case, the affinity, K $(=k/k_d)$, value is 11.82.

Figure 8.6c shows the binding and dissociation of 600 μM oxazaborolidine derivative, BNO1 + 2 mM sucrose in solution to FTF immobilized on a SPR sensor chip surface

Table 8.4: Binding and dissociation rate coefficients for the binding and dissociation phase for (a) the binding and dissociation of 2 mM sucrose and different concentrations of (a) BNO1, (b) BNO2, (c) BNO3 and (d) BNO4 in solution to fructosyltransferase (FTF) immobilized on a surface plasmon resonance (SPR) biosensor surface (Jabbour et al., 2007).

Analyte in Solution	k	k_1	k_2	k_d
(a)				
2 mM sucrose + 0 μM BNO1	90.83 ± 2.60	na	na	4.753 ± 0.431
2 mM sucrose + 60 μM BNO1	140.9 ± 2.76	na	na	11.92 ± 0.89
2 mM sucrose + 600 μM BNO1	22.84 ± 1.95	na	na	1.022 ± 0.082
(b)				
2 mM sucrose + 15 nM BNO2	2.88 ± 0.295	na	na	0.138 ± 0.0003
2 mM sucrose + 30 nM BNO2	3.895 ± 0.066	na	na	4.9586 ± 0.496
(c)				
2 mM sucrose + 0.006 nM BNO3	6.478 ± 0.813	na	na	0.5 ± 0.0
2 mM sucrose + 30 nM BNO3	1.825 ± 0.313	1.515 ± 0.241	180.0 ± 0.0	na
(d)				
2 mM sucrose + 60 μM BNO4	5.884 ± 0.247	na	na	1.125 ± 0.149
2 mM sucrose + 150 μM BNO4	10.713 ± 1.57	16.640 ± 1.051	0.8287 ± 0.0008	4.372 ± 0.646

(Jabbour et al., 2007). A single-fractal analysis is adequate to describe the binding and the dissociation kinetics. The values of (a) the binding rate coefficient, k, and the fractal dimension, D_f, for a single-fractal analysis, and (b) the dissociation rate coefficient, k_d, and the fractal dimension for the dissociation phase, D_{fd}, for a single-fractal analysis are given in Tables 8.4 and 8.5. In this case, the affinity, $K(=k/k_d)$, value is 22.35.

Figure 8.7a shows the increase in the binding rate coefficient, k, for a single-fractal analysis with an increase in the fractal dimension, D_f, for the binding of 0-600 μM oxazaborolidine, BNO1 in solution to the FTF immobilized on a SPR biosensor chip surface. The binding rate coefficient, k, is given by:

$$k = (11.269 \pm 6.926)D_f^{4.996 \pm 1.908} \tag{8.3a}$$

The fit is good. Only three data points are available. The availability of more data points would lead to a more reliable fit. The binding rate coefficient, k, exhibits close to a fifth (equal to 4.996) order of dependence on the fractal dimension, D_f, or the degree of heterogeneity that exists on the SPR biosensor chip surface.

Figure 8.7b shows the increase in the affinity, $K(=k/k_d)$, for a single-fractal analysis with an increase in the ratio of fractal dimensions, (D_f/D_{fd}), for the binding of 0-600 μM

Table 8.5: Fractal dimensions for the binding and dissociation phase for (a) the binding and dissociation of 2 mM sucrose and different concentrations of (a) BNO1, (b) BNO2, (c) BNO3 and (d) BNO4 in solution to fructosyltransferase (FTF) immobilized on a surface plasmon resonance (SPR) biosensor surface (Jabbour et al., 2007).

Analyte in Solution	D_f	D_{f1}	D_{f2}	D_{fd}
(a)				
2 mM sucrose + 0 μM BNO1	1.404 ± 0.029	na	na	1.109 ± 0.0158
2 mM sucrose + 60 μM BNO1	1.715 ± 0.0196	na	na	1.425 ± 0.131
2 mM sucrose + 600 μM BNO1	1.204 ± 0.0967	na	na	0.637 ± 0.148
(b)				
2 mM sucrose + 15 nM BNO2	0.858 ± 0.133	na	na	0.9884 ± 0.0019
2 mM sucrose + 30 nM BNO2	1.9034 ± 0.02314	na	na	1.9894 ± 0.496
(c)				
2 mM sucrose + 0.006 nM BNO3	1.0644 ± 0.0978	na	na	1.0 ± 0.0
2 mM sucrose + 30 nM BNO3	1.094 ± 0.101	0.9748 ± 0.126	3.0-4.2 × 10^{-15}	na
(d)				
2 mM sucrose + 60 μM BNO4	1.500 ± 0.0358	na	na	1.698 ± 0.1050
2 mM sucrose + 150 μM BNO4	2.118 ± 0.0894	2.3816 ± 0.0776	1.1550 ± 0.0712	2.152 ± 0.166

oxazaborolidine, BNO1 in solution to the FTF immobilized on a SPR biosensor chip surface. The affinity, $K(=k/k_d)$, is given by:

$$K(=k/k_d) = 11.965 \pm 4.255)(D_f/D_{fd})^{1.021\pm0.869} \quad (8.3b)$$

The fit is reasonable. Only three data points are available. The availability of more data points would lead to a more reliable fit. The affinity, K $(=k/k_d)$, exhibits close to a first (equal to 1.021) order of dependence on the ratio of fractal dimensions, (D_f/D_{fd}), that exists on the SPR biosensor chip surface.

Figure 8.7c shows the increase in the dissociation rate coefficient, k_d, for a single-fractal analysis with an increase in the fractal dimension, D_{fd}, for the dissociation of 0-600 μM

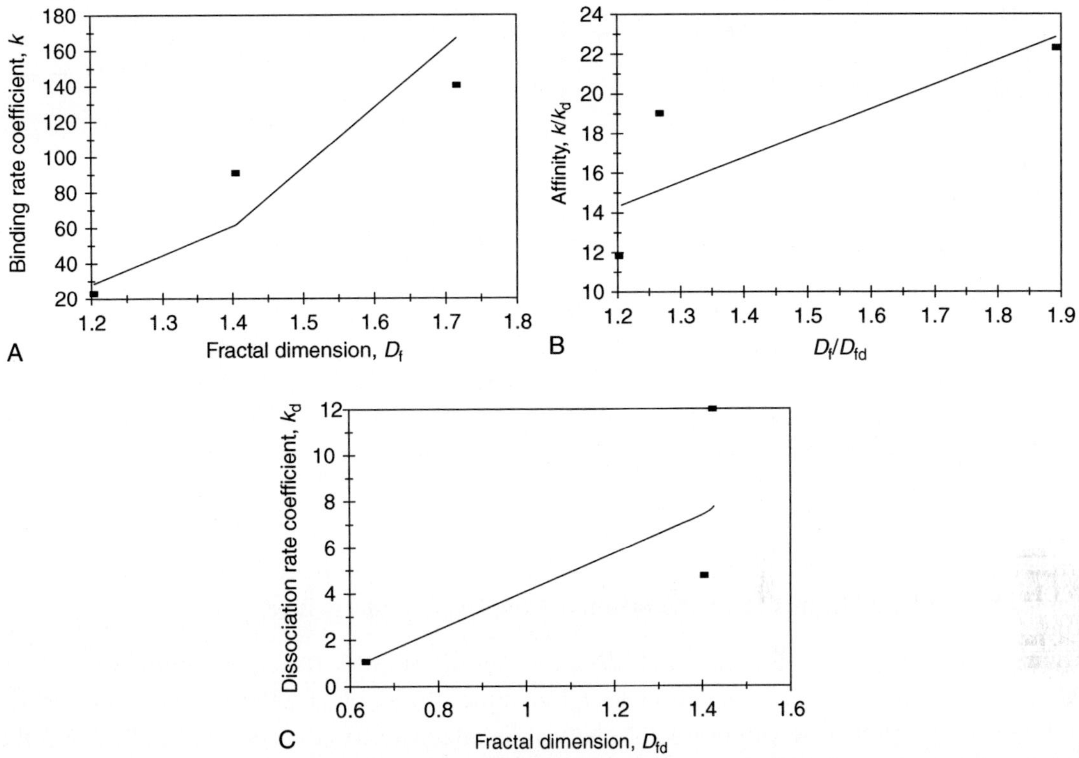

Figure 8.7
(a) Increase in the binding rate coefficient, k, with an increase in the fractal dimension, D_f.
(b) Increase in the affinity, $K(=k/k_d)$, with an increase in the fractal dimension dimension ratio, D_f/D_{fd}.
(c) Increase in the dissociation rate coefficient, k_d, with an increase in the fractal dimension, D_{fd}.

oxazaborolidine, BNO1 in solution to the FTF immobilized on a SPR biosensor chip surface. The dissociation rate coefficient, k_d, is given by:

$$k_d = (3.155 \pm 2.732)(D_{fd})^{2.516 \pm 0.957} \tag{8.3c}$$

The fit is reasonable. Only three data points are available. The availability of more data points would lead to a more reliable fit. The dissociation rate coefficient, k_d, exhibits very close to a two and a half (equal to 2.516) order of dependence on the fractal dimension, D_{fd}, in the dissociation phase or the degree of heterogeneity that exists on the SPR biosensor chip surface.

Figure 8.8a shows the binding and dissociation of 15 nM oxazaborolidine derivative, BNO2 + 2 mM sucrose in solution to FTF immobilized on a SPR sensor chip surface (Jabbour et al., 2007). A single-fractal analysis is adequate to describe the binding and the dissociation kinetics. The values of (a) the binding rate coefficient, k, and the fractal dimension, D_f, for a single-fractal analysis, and (b) the dissociation rate coefficient, k_d, and the

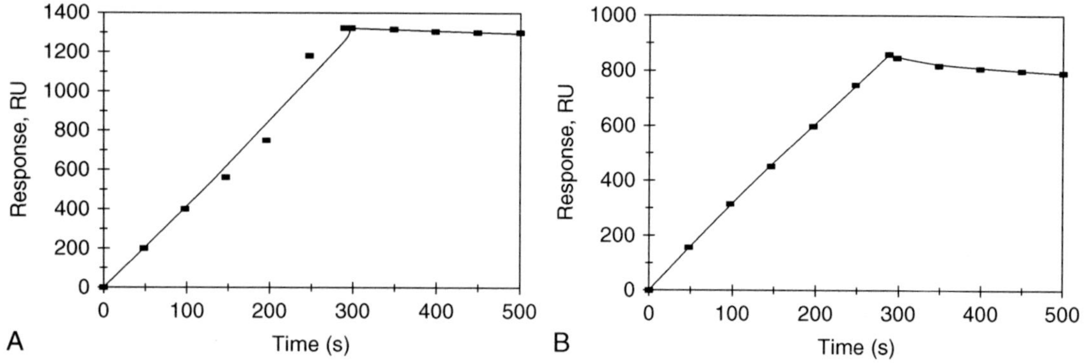

Figure 8.8
Binding and dissociation of different concentrations (in nM) of oxazaborolidine derivative (BNO2) + 2 mM sucrose in solution to fructosyltransferase (FTF) immobilized on a SPR sensor chip surface (Jabbour et al., 2007): (a) 15 (b) 30.

fractal dimension for the dissociation phase, D_{fd}, for a single-fractal analysis are given in Tables 8.4 and 8.5. In this case, the affinity, K ($=k/k_d$), value is 20.87.

Figure 8.8b shows the binding and dissociation of 30 nM oxazaborolidine derivative, BNO2 + 2 mM sucrose in solution to FTF immobilized on a SPR sensor chip surface (Jabbour et al., 2007). A single-fractal analysis is adequate to describe the binding and the dissociation kinetics. The values of (a) the binding rate coefficient, k, and the fractal dimension, D_f, for a single-fractal analysis, and (b) the dissociation rate coefficient, k_d, and the fractal dimension for the dissociation phase, D_{fd}, for a single-fractal analysis are given in Tables 8.4 and 8.5. In this case, the affinity, K ($=k/k_d$), value is 0.786.

Figure 8.9a shows the increase in the binding rate coefficient, k, for a single-fractal analysis with an increase in the fractal dimension, D_f, for the binding of 0-30 nM oxazaborolidine, BNO2 in solution to the FTF immobilized on a SPR biosensor chip surface. The binding rate coefficient, k, is given by:

$$k = (5.247 \pm 11.43) D_f^{7.01 \pm 3.32} \tag{8.4a}$$

The fit is reasonable. Only three data points are available. The availability of more data points would lead to a more reliable fit. There is scatter in the data, and this is reflected in the error in the binding rate coefficient. Only the positive error is given since the binding rate coefficient cannot have a negative value. The binding rate coefficient, k, exhibits close to a seventh (equal to 7.01) order of dependence on the fractal dimension, D_f, or the degree of heterogeneity that exists on the SPR biosensor chip surface.

Figure 8.9b shows the increase in the dissociation rate coefficient, k_d, for a single-fractal analysis with an increase in the fractal dimension, D_{fd}, for the dissociation of 0-30 nM

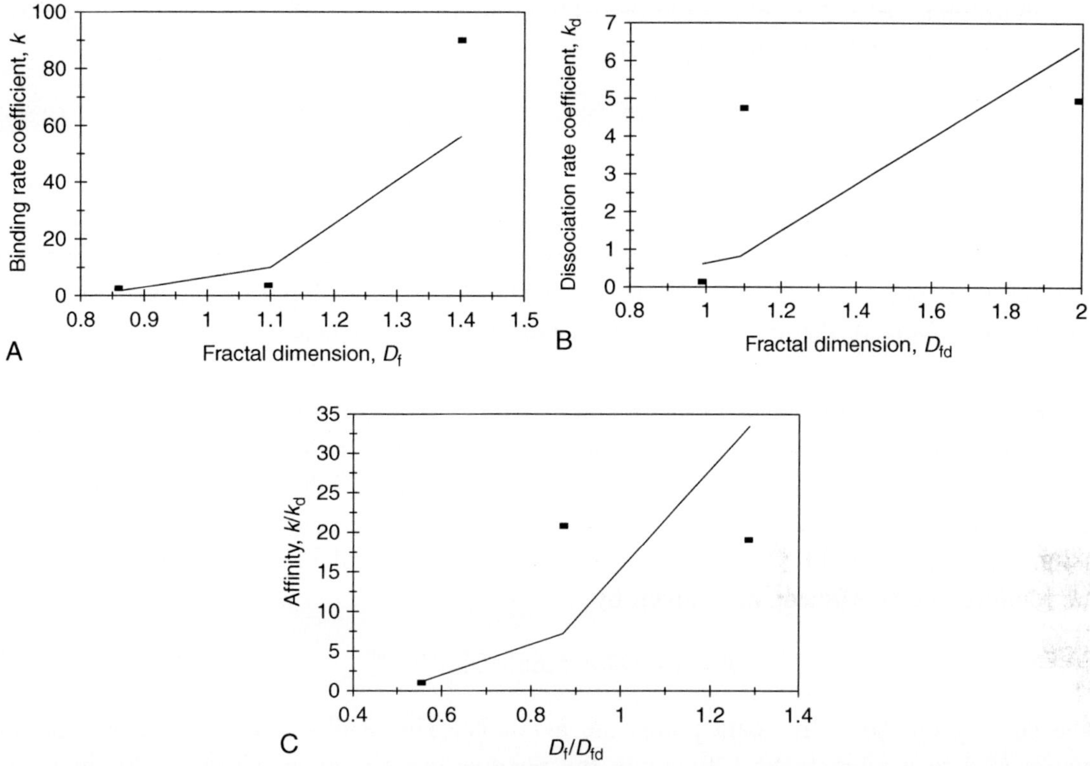

Figure 8.9

(a) Increase in the binding rate coefficient, k, with an increase in the fractal dimension, D_f. (b) Increase in the dissociation rate coefficient, k_d, with an increase in the fractal dimension, D_{fd}. (c) Increase in the affinity, K $(=k/k_d)$, with an increase in the fractal dimension dimension ratio, D_f/D_{fd}.

oxazaborolidine, BNO2 in solution to the FTF immobilized on a SPR biosensor chip surface. The dissociation rate coefficient, k_d, is given by:

$$k_d = (3.155 \pm 2.732)(D_{fd})^{2.516 \pm 0.957} \tag{8.4b}$$

The fit is reasonable. Only three data points are available. The availability of more data points would lead to a more reliable fit. There is scatter in the data. The dissociation rate coefficient, k_d, exhibits an order of dependence slightly higher than two and one-half (equal to 2.516) on the fractal dimension, D_{fd}, in the dissociation phase or the degree of heterogeneity that exists on the SPR biosensor chip surface.

Figure 8.9c shows the increase in the affinity, K $(=k/k_d)$, for a single-fractal analysis with an increase in the ratio of fractal dimensions, (D_f/D_{fd}), for the binding of 0-30 nM oxazaborolidine, BNO2 in solution to the FTF immobilized on a SPR biosensor chip surface. The affinity, K ($=k/k_d$), is given by:

$$K(=k/k_d) = (12.71 + 32.96)(D_f/D_{fd})^{3.87\pm2.14} \tag{8.4c}$$

The fit is reasonable. There is scatter in the data. Only three data points are available. The availability of more data points would lead to a more reliable fit. Only the positive error is given since the affinity, K, cannot have a negative value. The affinity, K ($=k/k_d$), exhibits between three and a half and four (equal to 3.87) order of dependence on the ratio of fractal dimensions, (D_f/D_{fd}), that exists on the SPR biosensor chip surface.

Figure 8.10 shows the increase in the binding rate coefficient, k, with an increase in the fractal dimension, D_f, on the SPR biosensor chip surface for both the results presented in Figure 8.7a (oxazaborolidine derivative, BNO1) and in Figure 8.7b (oxazaborolidine derivative, BNO2). In this case, there are five data points available compared to only three data points when BNO1 and BNO2 were plotted separately. For the data presented in Figure 8.10, the binding rate coefficient, k, is given by:

$$k = (5.577 + 5.688)D_f^{6.386\pm1.349} \tag{8.5}$$

The fit is good. Only five data points are available. The availability of more data points would lead to a more reliable fit. Only the positive error is given since the binding rate coefficient, k, cannot have a negative value. The fact that two data sets are plotted together (for BNO1 and BNO2), and that the fit is good is encouraging. The binding rate coefficient, k, exhibits an order of dependence between six and six and a half (equal to 6.386) on the fractal dimension, D_f, or the degree of heterogeneity that exists on the SPR biosensor surface for the two cases presented together. The higher than sixth order dependence exhibited indicates that the binding rate coefficient, k, is very sensitive to the degree of heterogeneity that exists on the SPR biosensor chip surface.

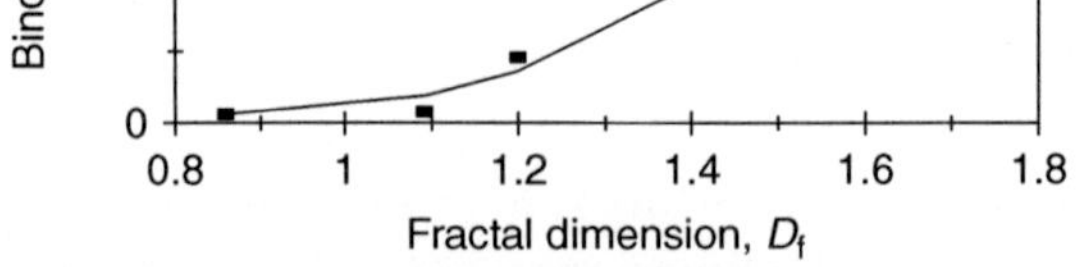

Figure 8.10
Increase in the binding rate coefficient, k, with an increase in the fractal dimension, D_f.

It is seen that, as expected, the order of dependence exhibited in Equation (8.5) above (equal to 6.386) is in between that exhibited in Equation (8.3a) (for BNO1) equal to 4.996 and Equation (8.4a) (for BNO2) equal to 7.01. It would be of interest to plot how the binding rate coefficient, k, changes with the fractal dimension, D_f, separately and together for the other oxazaborolidine derivatives. If all or most of the data points fit nicely together on a single plot, then this indicates that the surface of the SPR biosensor chip is one of the major determining factors in influencing the quantitative value of the binding rate coefficient, k.

Figure 8.11a shows the binding and dissociation of 30 nM oxazaborolidine derivative, BNO3 + 2 mM sucrose in solution to FTF immobilized on a SPR sensor chip surface (Jabbour et al., 2007). A dual-fractal analysis is required to describe the binding kinetics. A single-fractal analysis is adequate to describe the dissociation kinetics. The values of (a) the binding rate coefficient, k, and the fractal dimension, D_f, for a single-fractal analysis, and (b) the binding rate coefficients, k_1 and k_2, and the fractal dimensions, D_{f1} and D_{f2}, for a dual-fractal analysis, and (c) the dissociation rate coefficient, k_d, and the fractal dimension, D_{fd}, for a single-fractal analysis are given in Tables 8.4 and 8.5. Note that an increase in the fractal dimension by a factor of 3.08 from a value of D_{f1} equal to 0.9748 to D_{f2} equal to 3.0 (the maximum value) leads to an increase in the binding rate coefficient by a factor of 118.81 from a value of k_1 equal to 1.515 to k_2 equal to 180.0. It is seen that an increase in the fractal dimension or the degree of heterogeneity on the SPR biosensor chip surface leads to an increase in the binding rate coefficient in the same direction.

Figure 8.11b shows the binding and dissociation of 0.006 nM oxazaborolidine derivative, BNO3 + 2 mM sucrose in solution to FTF immobilized on a SPR sensor chip surface (Jabbour et al., 2007). A single-fractal analysis is adequate to describe the binding and the

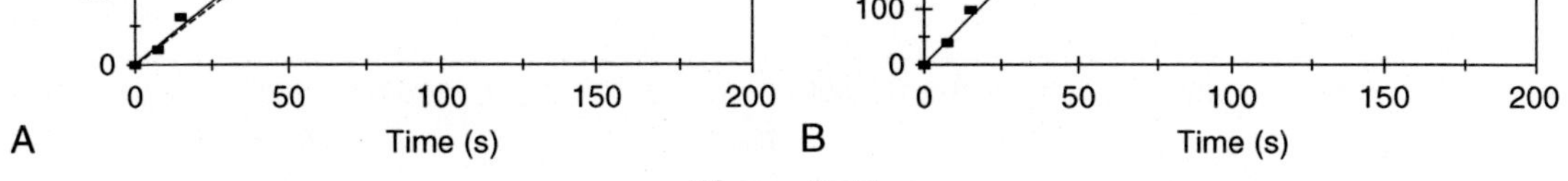

Figure 8.11

Binding and dissociation of different concentrations (in nM) oxazaborolidine derivative (BNO3) + 2 mM sucrose in solution to fructosyltransferase (FTF) immobilized on a SPR sensor chip surface (Jabbour et al., 2007): (a) 30 (b) 0.006. When only a solid line (—) is used then a single-fractal analysis applies. When both a dashed (- - -) and a solid (—) line are used then the dashed line represents a single-fractal analysis and the solid line represents a dual-fractal analysis.

dissociation kinetics. The values of (a) the binding rate coefficient, k, and the fractal dimension, D_f, for a single-fractal analysis, and (b) the dissociation rate coefficient, k_d, and the fractal dimension for the dissociation phase, D_{fd}, for a single-fractal analysis are given in Tables 8.4 and 8.5. In this case, the affinity, K ($=k/k_d$), value is 12.96.

Figure 8.12a shows the binding and dissociation of 60 mM oxazaborolidine derivative, BNO4 + 2 mM sucrose in solution to FTF immobilized on a SPR sensor chip surface (Jabbour et al., 2007). A single-fractal analysis is adequate to describe the binding and the dissociation kinetics. The values of (a) the binding rate coefficient, k, and the fractal dimension, D_f, for a single-fractal analysis, and (b) the dissociation rate coefficient, k_d, and the fractal dimension for the dissociation phase, D_{fd}, for a single-fractal analysis are given in Tables 8.4 and 8.5. In this case, the affinity, K ($=k/k_d$), value is 5.23.

Figure 8.12b shows the binding and dissociation of 150 μM oxazaborolidine derivative, BNO3 + 2 mM sucrose in solution to FTF immobilized on a SPR sensor chip surface (Jabbour et al., 2007). A dual-fractal analysis is required to describe the binding kinetics. A single-fractal analysis is adequate to describe the dissociation kinetics. The values of (a) the binding rate coefficient, k, and the fractal dimension, D_f, for a single-fractal analysis, (b) the binding rate coefficients, k_1 and k_2, and the fractal dimensions, D_{f1} and D_{f2}, for a dual-fractal analysis, and (c) the dissociation rate coefficient, k_d, and the fractal dimension, D_{fd}, for a single-fractal analysis are given in Tables 8.4 and 8.5. It is seen that a decrease in the fractal dimension by a 51.5% from a value of D_{f1} equal to 2.3816 to D_{f2} equal to 1.1550 leads to a decrease in the binding rate coefficient by a factor of 20.08 from a value of k_1 equal to 16.640 to k_2 equal to 0.8287. It is seen that a decrease in the fractal dimension

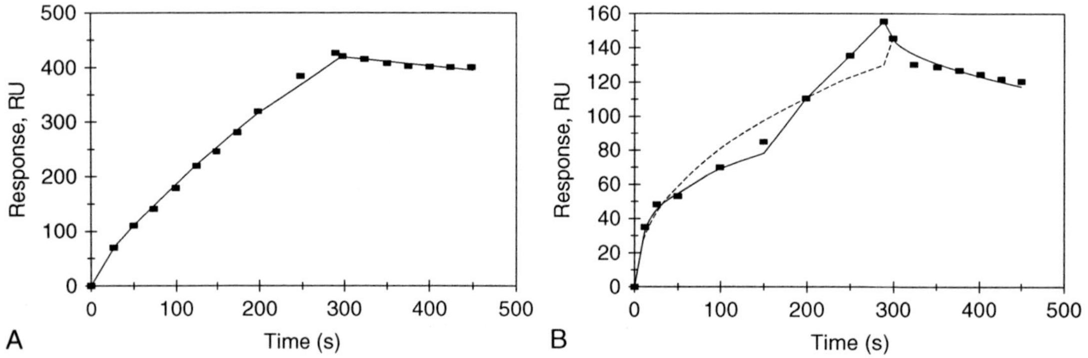

Figure 8.12
Binding and dissociation of different concentrations (in μM) oxazaborolidine derivative (BNO4) + 2 mM sucrose in solution to fructosyltransferase (FTF) immobilized on a SPR sensor chip surface (Jabbour et al., 2007): (a) 60 (b) 150. When only a solid line (—) is used then a single-fractal analysis applies. When both a dashed (- - -) and a solid (—) line are used then the dashed line represents a single-fractal analysis and the solid line represents a dual-fractal analysis.

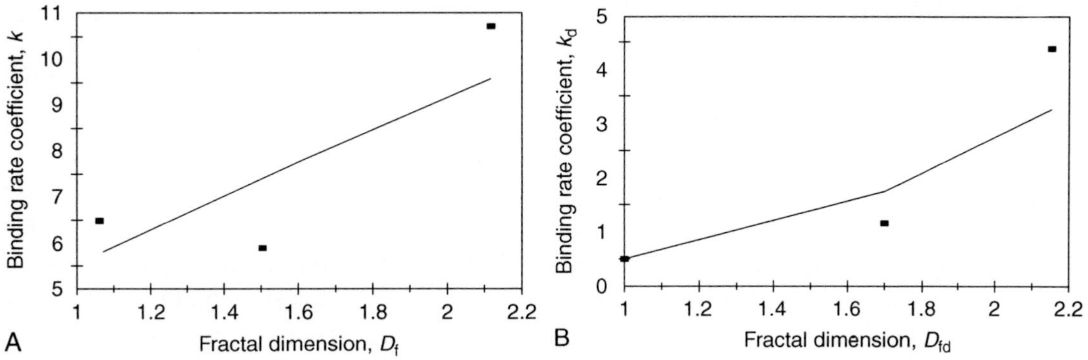

Figure 8.13

(a) Increase in the binding rate coefficient, k, with an increase in the fractal dimension, D_f
(b) Increase in the dissociation rate coefficient, k_d, with an increase in the fractal dimension, D_{fd}.

or the degree of heterogeneity on the SPR biosensor chip surface leads to a decrease in the binding rate coefficient in the same direction.

Figure 8.13a shows the increase in the binding rate coefficient, k, with an increase in the fractal dimension, D_f, on the SPR biosensor chip surface for the binding of 0-150 μM oxazaborolidine derivative, BNO4 in solution to the FTF immobilized on the SPR sensor chip surface. For the data presented in Figure 8.13a, the binding rate coefficient, k is given by:

$$k = (5.511 \pm 1.804) D_f^{0.7432 \pm 0.582} \tag{8.6a}$$

The fit is reasonable. Only three data points are available. The availability of more data points would lead to a more reliable fit. The binding rate coefficient, k, exhibits less than a first (equal to 0.732) order of dependence on the fractal dimension, D_f, or the degree of heterogeneity that exists on the SPR biosensor surface. This indicates that the binding rate coefficient is sensitive to the degree of heterogeneity that exists on the SPR biosensor chip surface.

Figure 8.13b shows the increase in the dissociation rate coefficient, k_d, for a single-fractal analysis with an increase in the fractal dimension, D_{fd}, for the dissociation of 0-30 nM oxazaborolidine, BNO4 in solution to the FTF immobilized on a SPR biosensor chip surface. The dissociation rate coefficient, k_d, is given by:

$$k_d = (0.4368 \pm 0.319)(D_{fd})^{2.611 \pm 0.987} \tag{8.6b}$$

The fit is reasonable. Only three data points are available. The availability of more data points would lead to a more reliable fit. There is scatter in the data. The dissociation rate coefficient, k_d, exhibits between two and a half and three (equal to 2.611) order of dependence on the fractal dimensions, D_{fd}, in the dissociation phase or the degree of heterogeneity that exists on the SPR biosensor chip surface.

8.4 Conclusions

A fractal analysis is presented for (a) the binding of TNF-α in solution to poly(guanine)-functionalized silica NPs (Wang et al., 2006), (b) the binding of different antigens in solution to the anti CD antigen immobilized on a QCM surface (Zeng et al., 2006), (c) the binding of 50 ng/mL myoglobin in serum to antimyoglobin antibody immobilized on a SPR biosensor surface (Masson et al., 2007), (d) the binding and dissociation of cardioimyocytes plus ET-1 with and without a DEP device (Yanfg et al., 2007), and (e) the binding and dissociation of different concentrations of oxazaborolidine derivatives, BNO1, BNO2, BNO3, and BNO4 + 2 mM sucrose in solution to the enzyme FTF immobilized on a SPR biosensor chip surface (Jabbour et al., 2007). Both, a single- and a dual-fractal analysis are used to model the binding and the dissociation (if applicable) kinetics. The dual-fractal analysis is used only if the single-fractal analysis does not provide an adequate fit.

The fractal dimension is not a classical independent variable such as analyte (antigen, antibody, or other biological molecule) concentration in solution. Nevertheless, the expressions obtained for the binding (and the dissociation) rate coefficients for a single- and a dual-fractal analysis as a function of the fractal dimension indicate a high sensitivity of these rate coefficients on their respective fractal dimensions on the SPR sensor chip surface. It can be seen that the data analysis in itself does not provide any evidence for surface roughness or heterogeneity, and the existence of surface roughness or heterogeneity assumed may not be correct. However, considering the complexity involved on the SPR chip surface, this is not an unreasonable assumption. Furthermore, there is deviation in the data that may be minimized by providing a correction for the depletion of the analyte.

Predictive relations are presented for the binding rate coefficient, k_2, as a function of the fractal dimension D_{f2}, for the binding of different CD antigens, CD1, CD2, and CD5 in solution to their corresponding antibodies, anti-CD1, anti-CD3, and anti-CD5 immobilized on a QCM biosensor surface (Zeng et al., 2006). In this case the binding rate coefficient, k_2 is very sensitive to the fractal dimension, D_{f2}, or the degree of heterogeneity that exists on the QCM biosensor surface as noted by the order of dependence between five and a half and six (equal to 5.667) exhibited.

Predictive relations are also presented for (a) the binding rate coefficient, k, for a single-fractal analysis as a function of the fractal dimension, D_f, for the binding of 0-600 μM oxazaborolidine, BNO1 in solution to the enzyme, FTF immobilized on a SPR biosensor chip surface and (b) for the affinity, K ($=k/k_d$), for a single-fractal analysis as a function of the ratio of fractal dimensions, D_f/D_{fd} (Zeng et al., 2006), and (c) the dissociation rate coefficient, k_d, for a single-fractal analysis as a function of the fractal dimension, D_{fd} (Zeng et al., 2006). For case (a) close to a fifth (equal to 4.996) order of dependence is exhibited, for case (b) close to a first (equal to 1.021) order is exhibited, and for case (c) close to a two and a half (equal to 2.516) order is exhibited.

References

Allen PZ and WH Bowen, *Archives of Oral Biology*, **35**, 55–62 (1990).

Ba DN, *Modern Immunological Techniques and Applications*, Beijing Medical University Press, Beijing, 1998, pp 877–883 (in Chinese).

Barone PW and MS Strano, Chemistry for a single-walled carbon nanotube fluorescence based glucose sensor, paper 187c, In *Annual American Institute of Chemical Engineers Meeting*, Salt Lake City, Utah, November 4-9, 2007.

Battaglia TM, JF Masson, M Sierks, S Beaudoin, J Rogers, KN Foster, GA Holloway, and K Booksh, Quantification of cytokines involved in wound healing using surface plasmon resonance, *Analytical Chemistry*, **77**, 7016–7023 (2005).

Blevins DG and KM Lukaszewki, *Annual Reviews of Plant Physiology*, **49**, 481–500 (1998).

Burne RA, In *Oral Ecology Disasters: The Role of Short-Term Extracellular Storage Polysaccharides*, WH Bowen WH and LA Tabak LA (Eds.), University of Rochester Press, Rochester, NY, 1991, pp 351–364.

Cular S, VR Bhethanabotla, DW Branch, and JA Strom, Hexagonal saw interleukin-6 biosensor. In *Annual American Institute of Chemical Engineers Meeting*, Salt Lake City, Utah, November 4-9 2007.

Daniel MC and D Astruc, Gold nanoparticles: Assembly, supramolecular chemistry, quantum-size-related properties, and applications toward biology, catalysis, and nanotechnology, *Chemical Reviews*, **104**, 293–346 (2004).

De Kossodo S, V Houba, and GE Grau, WCS Group: *Journal of Immunological Methods*, **182**, 107–114 (1995).

Diane ME, BS James, G Geetha, L Jueren, LB Kenneth, and MS Allen, *American Journal of Physiology: Heart Circulation and Physiology*, **278**, H1695–H1707 (2000).

Goluch ED, S Stoeva, KA Shaikh, SS Szegedi, J-S Lee, TN Chiesl, AE Barron, CA Mirkin, and C Liu, A biochip for rapid and sensitive detection of multiple cancer markers simultaneously. In *Annual American Institute of Chemical Engineers Meeting*, paper 523a, Salt Lake City, Utah, November 4-9 2007.

Havlin S, Molecular diffusion and reactions. In *The Fractal Approach to Heterogeneous Chemistry: Surfaces, Colloids, Polymers*, Avnir D (Ed.), Wiley, New York, 1989, pp. 251–269.

Hoffman R, J Edward, and J Banz, et al., In *Hematology: Basic Principles and Practice*, R Hoffman (Ed.), Third edition, Churchill Livingstone Press, New York, NY, 2000, pp. 1007–1008, 1075.

Jabbour A, D Steinberg, VM Demibitsky, A Moussaieff, B Zaks, and M Srebnik, *Journal of Medicinal Chemistry*, **47**, 2409–2410 (2004).

Jabbour A, M Srebnik, B Zaks, VM Dembitsky, and DM Steinberg, *International Journal of Antimicrobial Agents*, **26**, 491–496 (2005).

Jabbour A, M Shemesh, M Srebnik, B Zaks, and D Steinberg, Effect of oxazaborolidines on immobilized fructosyltransferase analyzed by surface plasmon resonance, *Biosensors & Bioelectronics*, **22**, 1658–11633 (2007).

Jiang S, Development of label-free nanopattern-enhanced biosensors or food safety monitoring and early cancer diagnostics, paper 523c, In *Annual American Institute of Chemical Engineers Meeting*, Salt Lake City, Utah, November 4-9, 2007.

Jones EY, DI Stuart, and NPC Walker, *Nature*, **338**, 225–228 (1989).

Lee CK and SL Lee, Multi-fractal scaling and analysis of reactions over fractal surfaces, *Surface Science*, **325**, 294–310 (1995).

Lei Y and W Jia, Glucose sensor using non-woven single-wall carbon nanotube films, paper 187b, In *Annual American Institute of Chemical Engineers Meeting*, Salt Lake City, Utah, November 4-9, 2007.

Levy D, RJ Garrison, DD Savaga, WB Kannel, and WP Castelli, Prognostic implications of echocardiographically determined left ventricular mass in the Framingham heart study, *New England Journal of Medicine*, **322**, 1561–1566 (1990).

Lovell BH and BA Carabello, Left ventricular hypertrophy: Pathogenisis, detection, and prognosis, *Circulation*, **102**, 470–479 (2000).

Martin HB, Diamond microneedle electrodes for neurochemical detection, paper 576c, In *Annual American Institute of Chemical Engineers Meeting*, Salt Lake City, Utah, November 4-9, 2007.

Masson JF, TM Battaglia, YC Kim, A Prakash, S Beaudoin, and KS Booksh, Preparation of anlayte-sensitive polymeric supports for biochemical sensors, *Talanta*, **64**, 716–725 (2004).

Masson JF, TM Battaglia, MJ Davidson, YC Kim, AMC Prakash, S Beaudoin, and KS Booksh, Biocompatible polymers for antibody support on gold particles, *Talanta*, **67**, 918–925 (2005).

Masson JF, TM Battaglia, P Khairallah, S Beaudoin, and KS Booksh, Quantitative measurement of cardiac markers in undiluted serum, *Analytical Chemistry*, **79**, 612–619 (2007).

Old LJ, Tumor necrosis factor (TNF), *Science*, **230**, 630–632 (1985).

Old LJ, *Nature*, **326**, 330–331 (1987).

Ramakrishnan A and A Sadana, A single fractal analysis of cellular analyte-receptor binding kinetics using biosensors, *Biosystems*, **59**, 35–51 (2001).

Riester D, C Hildmann, A Schweinhorst, and FJ Meyer-Almes, Histone deacylase inhibitor assay based on fluorescence resonance energy transfer, *Analytical Biochemistry*, **362**, 136–141 (2007).

Rozen R, G Bachrach, M Broshteyn, I Gedalia, and D Steinberg, *FEMS Microbiology Letters*, **195**, 205–210 (2001).

Ruskoaho H, *Pharmacologial Reviews*, **44**, 479–602 (1992).

Sadana A, A fractal analysis for the evaluation of hybridization kinetics in biosensors, *Journal of Colloid and Interface Science*, **151**(1), 166–177 (2001).

Sadana A, *Fractal Binding and Dissociation Kinetics for Different Biosensor Applications*, Amsterdam, Elsevier, 2005.

Shelp BJ and UC Gupta, CRC Press: Boca Raton, pp. 53-85, (1993).

Shen J and C-C Liu, Development of a screen-printed cholesterol biosensor: Comparing the performance of gold and platinum as the working electrode material and fabrication using a self-assembly approach, *Sensors & Actuators* B, **120**, 417–425 (2007).

Steinberg D. In *Handbook of Bacterial Adhesion: Principles, Methods, and Applications*, YH An YH and RJ Friedman RJ (Eds.), Humana Press, Totowa, NJ, 2000, pp. 353–370.

Steinberg D, R Rozen, M Bronshteyn, B Zaks, I Gedalia, and G Bachrach, *Carbohydrate Research*, **337**, 701–710 (2000).

Traweek ST, Immunophenotypic analysis of acute leukemias, *American Journal of Clinical Pathology*, **99**, 504–512 (1993).

Wang H, ZM Zeng, and HY Liu, Immunophenotyping of acute leukemia using an integrated piezoelectric immunosensor array, *Analytical Chemistry*, **76**, 2203–2209 (2004).

Wang J, G Liu, MH Engelhard, and Y Lin, Sensitive immunoassay of a biomarker tumor necrosis factor-α based on poly(guanine)-functionalized silica nanoparticle label, *Analytical Chemistry*, **78**, 6974–6979 (2006).

Yang M, CC Lim, R Liao, and X Zheng, A novel microfluidic impedance assay for monitoring endothelin-induced cardiomyocyte hypertrophy, *Biosensors and Bioelectronics*, **22**, 1688–1693 (2007).

Zeng H, H Wang, F Chen, H Xin, G Wang, L Xiao, K Song, D Wu, Q He, and G Shen, Development of quartz crystal-microbalance-based immunosensor array for clinical immunophenotyping of acute leukemias, *Analytical Biochemistry*, (2006).

CHAPTER 9

Physiological Cellular Reactions Detection on Biosensor Surfaces: A Fractal Analysis

Chapter Outline

9.1 Introduction

The real-time monitoring of cells should help reveal the mysteries of modern cell biology (Ziblat et al., 2006). These authors point out that though the surface plasmon resonance (SPR) is a useful technique, it is not well suited for analyzing the reactions *in vivo* or *in situ*. They, therefore emphasize the need to develop new experimental techniques that would monitor quantitatively in real time the dynamics between molecules and their cognate receptors in cells. They have developed a novel SPR method based on FTIR.

Fang et al. (2006) also reinforce this view by pointing out that the ability to analyze living cells in their natural and physiological state is essential to understand the biological functions of cellular targets. This should also assist in drug discovery and in its development. These authors have summarized the principles that are involved in current cell-base arrays that include measuring of a specific cellular event that includes second-messenger generation to the translocation of a particular target tagged with a fluorescent label. One may also include the expression of a reporter gene, and the alteration of a particular phenotype (Blake, 2001; Taylor et al., 2001; Abraham et al., 2004). Fang et al. (2006) report that optical biosensors

Handbook of Biosensors and Biosensor Kinetics. DOI: 10.1016/B978-0-444-53262-6.00009-7

that use evanescent waves have been used effectively in research. The resonant waveguide grating (RWG) biosensor has been used to help determine affinities and the kinetics of target analytes in a simple binding to the biological receptors immobilized on a sensor surface. These authors report that there is increasing interest in the activities of living cells including cell adhesion and spreading, toxicity, and proliferation (Ramsden et al., 1994; Voros et al., 2000; Quinn et al., 2000; Hide et al., 2002).

In this chapter we use fractal analysis to analyze the binding and dissociation (if applicable) kinetics for (a) the binding and dissociation of different concentrations of bradykinin to a bradykinin B_2 receptor on a RWG biosensor, (Fang et al., 2006), (b) the binding and dissociation of mβCD cholesterol to HeLa cells cultivated on a gold-plated prism (Ziblat et al., 2006), and (c) binding and dissociation (if applicable) of calcium + FRET-based calcium biosensor employing troponin (Mank et al., 2006) for the binding and dissociation of TXNL in solution to the sensor-chip surface.

9.2 Theory

9.2.1 Single-Fractal Analysis

Binding Rate Coefficient

Havlin (1989) points out that the diffusion of a particle (analyte [Ag]) from a homogeneous solution to a solid surface (e.g., receptor [Ab]-coated surface) on which it reacts to form a product (analyte-receptor complex; (Ab · Ag)) is given by:

$$(\mathrm{Ab} \cdot \mathrm{Ag}) \approx \begin{cases} t^{(3-D_{\mathrm{f,bind}})/2} = t^p, & t < t_{\mathrm{c}} \\ t^{1/2}, & t > t_{\mathrm{c}} \end{cases}. \tag{9.1}$$

Here $D_{\mathrm{f,bind}}$ or D_{f} is the fractal dimension of the surface during the binding step. t_{c} is the cross-over value. Havlin (1989) points out that the cross-over value may be determined by $r_{\mathrm{c}}^2 \sim t_{\mathrm{c}}$. Above the characteristic length, r_{c}, the self-similarity of the surface is lost and the surface may be considered homogeneous. Above time, t_{c} the surface may be considered homogeneous, since the self-similarity property disappears, and "regular" diffusion is now present. For a homogeneous surface where D_{f} is equal to 2, and when only diffusional limitations are present, $p = ½$ as it should be. Another way of looking at the $p = ½$ case (where $D_{\mathrm{f,bind}}$ is equal to 2) is that the analyte in solution views the fractal object, in our case, the receptor-coated biosensor surface, from a "large distance." In essence, in the association process, the diffusion of the analyte from the solution to the receptor surface creates a depletion layer of width $(Ɖt)^{½}$ where $Ɖ$ is the diffusion constant. This gives rise to the fractal power law, (Analyte · Receptor) $\sim t^{(3-D_{\mathrm{f,bind}})/2}$. For the present analysis, t_{c} is arbitrarily chosen and we assume that the value of t_{c} is not reached. One may consider the approach as an intermediate "heuristic" approach that may be used in the future to develop an autonomous (and not time-dependent) model for diffusion-controlled kinetics.

Dissociation Rate Coefficient

The diffusion of the dissociated particle (receptor [Ab] or analyte [Ag]) from the solid surface (e.g., analyte [Ag]-receptor [Ab] complex coated surface) into solution may be given, as a first approximation by:

$$(\mathrm{Ab}\cdot\mathrm{Ag}) \approx -t^{(3-D_{\mathrm{f,diss}})/2} = -t^{p}, \quad t > t_{\mathrm{diss}} \tag{9.2}$$

Here $D_{\mathrm{f,diss}}$ is the fractal dimension of the surface for the dissociation step. This corresponds to the highest concentration of the analyte-receptor complex on the surface. Henceforth, its concentration only decreases. The dissociation kinetics may be analyzed in a manner "similar" to the binding kinetics.

9.2.2 Dual-Fractal Analysis

Binding Rate Coefficient

Sometimes, the binding curve exhibits complexities and two parameters (k, D_{f}) are not sufficient to adequately describe the binding kinetics. This is further corroborated by low values of the r^2 factor (goodness-of-fit). In that case, one resorts to a dual-fractal analysis (four parameters; k_1, k_2, D_{f1}, and D_{f2}) to adequately describe the binding kinetics. The single-fractal analysis presented above is thus extended to include two fractal dimensions. At present, the time ($t = t_1$) at which the "first" fractal dimension "changes" to the "second" fractal dimension is arbitrary and empirical. For the most part, it is dictated by the data analyzed and experience gained by handling a single-fractal analysis. A smoother curve is obtained in the "transition" region, if care is taken to select the correct number of points for the two regions. In this case, the product (antibody-antigen; or analyte-receptor complex, Ab · Ag or analyte · receptor) is given by:

$$(\mathrm{Ab}\cdot\mathrm{Ag}) \approx \begin{cases} t^{(3-D_{\mathrm{f1,bind}})/2} = t^{p1}, & t < t_1 \\ t^{(3-D_{\mathrm{f2,bind}})/2} = t^{p2}, & t_1 < t < t_2 = t_{\mathrm{c}} \\ t^{1/2}, & t > t_{\mathrm{c}} \end{cases} \tag{9.3}$$

In some cases, as mentioned above, a triple-fractal analysis with six parameters (k_1, k_2, k_3, D_{f1}, D_{f2}, and D_{f3}) may be required to adequately model the binding kinetics. This is when the binding curve exhibits convolutions and complexities in its shape due perhaps to the very dilute nature of the analyte (in some of the cases to be presented) or for some other reasons. Also, in some cases, a dual-fractal analysis may be required to describe the dissociation kinetics.

9.3 Results

The fractal analysis will be applied to the binding and dissociation of bradykinin concentrations (in nM) in solution to bradykinin B_2 receptors immobilized on a RWG biosensor (Fang et al., 2006), binding and dissociation of 20 nM mβCD-cholesterol cells exposed to two cycles of cholesterol enrichment (Ziblat et al., 2006), and to the binding and dissociation on a calcium + FRET-based calcium biosensor employing troponin C (Mank et al., 2006).

At the outset it should be pointed out that alternative expressions for fitting the binding and dissociation data are available that include saturation, first-order reaction, and no diffusional limitations, but these expressions are deficient in describing the heterogeneity that inherently exists on the surface. It is this heterogeneity on the biosensor surface that one is attempting to relate to the different biosensor performance parameters. More specifically the question we wish to answer is how may one change the heterogeneity or the fractal dimension, D_f on the biosensor chip surface in order that one may be able to enhance the different biosensor performance parameters.

Other modeling attempts also need to be mentioned. One might justifiably argue that appropriate modeling may be achieved by using a Langmuirian or other approach. The Langmuirian approach may be used to model the data presented if one assumes the presence of discrete classes of sites, for example double exponential analysis as compared with the single-fractal analysis. Lee and Lee (1995) report that the fractal approach has been applied to surface science, for example, adsorption and reaction processes. These authors point out that the fractal approach provides a convenient means to represent the different structures and the morphology at the reaction surface. They also draw attention to using the fractal approach to develop optimal structures and as a predictive approach. Another advantage of the fractal technique is that the analyte-receptor association is a complex reaction, and the fractal analysis via the fractal dimension and the rate coefficient provide a useful lumped parameter analysis of the diffusion-limited reaction occurring on a heterogeneous surface.

In a classical situation, to demonstrate fractality, one should make a log-log plot, and one should definitely have a large amount of data. It may be useful to compare the fit to some other forms, such as exponential form, or one involving saturation, etc. At present, no independent proof or physical evidence of fractals in the examples is presented. Nevertheless, fractals and the degree of heterogeneity on the biosensor surface are still used to gain insights into enhancing the different biosensor performance parameters. The fractal approach is a convenient means (since it is a lumped parameter) to make the degree of heterogeneity that exists on the surface more quantitative. Thus, there is some arbitrariness in the fractal approach to be presented. The fractal approach provides additional information about interactions that may not be obtained by a conventional analysis of biosensor data. In this chapter as mentioned above, an attempt is made to relate the fractal dimension, D_f, or the degree of heterogeneity on the biosensor surface with different biosensor performance parameters. More specifically, the interest is in finding out how changes in the fractal dimension or the degree of heterogeneity on the biosensor chip surface affect the different biosensor parameters of interest.

Unless specifically mentioned there is no nonselective adsorption of the analyte. In other words, nonspecific binding is ignored. Nonselective adsorption would skew the results obtained very significantly. In these types of systems, it is imperative to minimize this nonselective adsorption. It is also recognized that, in some cases, this nonselective adsorption may not be a significant component of the adsorbed material and that the rate of association, which is of a temporal nature would depend on surface availability.

Fang et al. (2006) recently reported that cell-based assays that help monitor the activities and health of living cells are important in drug discovery and development. These cells-based assays measure specific molecular cellular events (Blake, 2001; Taylor et al., 2001; Abraham et al., 2004). Fang et al. (2006) emphasize the need for a cell-based assay to provide a noninvasive and continuous record of cellular activity. Besides, a high sensitivity would be helpful. These authors have recently applied RWG biosensors to analyze cytoskeleton modulation (Fang et al., 2005a,b), cell signaling mediated through epidermal growth factor (EGF) receptor (Fang et al., 2005b), or a G-protein-coupled receptor (GPCR) bradykinin B_2 receptor (Fang et al., 2005c). They point out that these studies have led to the development of MRCAT (mass redistribution cell assay technology). In their latest publication Fang et al. (2006) have introduced multiple optical readouts for cell sensing using RWG biosensors. Theoretical analysis as well as experimental data is presented with emphasis on the sensitivities of these optical readouts as the nature of the dynamic mass distribution values.

Fang et al. (2006) analyzed the binding and dissociation of different concentrations of bradykinin to a bradykinin B_2 receptor on a RWG biosensor to help characterize stimulation-mediated cell responses including signaling. Figure 9.1a shows the binding and dissociation of 128 nM bradykinin concentration in solution to the bradykinin B_2 on a RWG biosensor surface. A dual-fractal analysis is required to adequately describe the binding and the dissociation kinetics. The values of (a) the binding rate coefficient, k, and the fractal dimension, D_f, for a single-fractal analysis, (b) the binding rate coefficients, k_1 and k_2, and the fractal dimensions, D_{f1} and D_{f2}, for a dual-fractal analysis, and (c) the dissociation rate coefficient, k_d, and the fractal dimension, D_{fd}, for a single-fractal analysis, and (d) the dissociation rate coefficients, k_{d1} and k_{d2}, and the fractal dimensions, D_{fd1} and D_{fd2}, for a dual-fractal analysis are given in Tables 9.1 and 9.2.

It is of interest to note that for a dual-fractal analysis as the fractal dimension increases by a factor of 1.965 from a value of D_{f1} equal to 1.402 to D_{f2} equal to 2.7544, the binding rate coefficient increases by a factor of 12.98 from a value of k_1 equal to 0.0199 to k_2 equal to 0.3579.

Figure 9.1b shows the binding and dissociation of 64 nM bradykinin concentration in solution to the bradykinin B_2 on a RWG biosensor surface (Fang et al., 2006). Once again, a

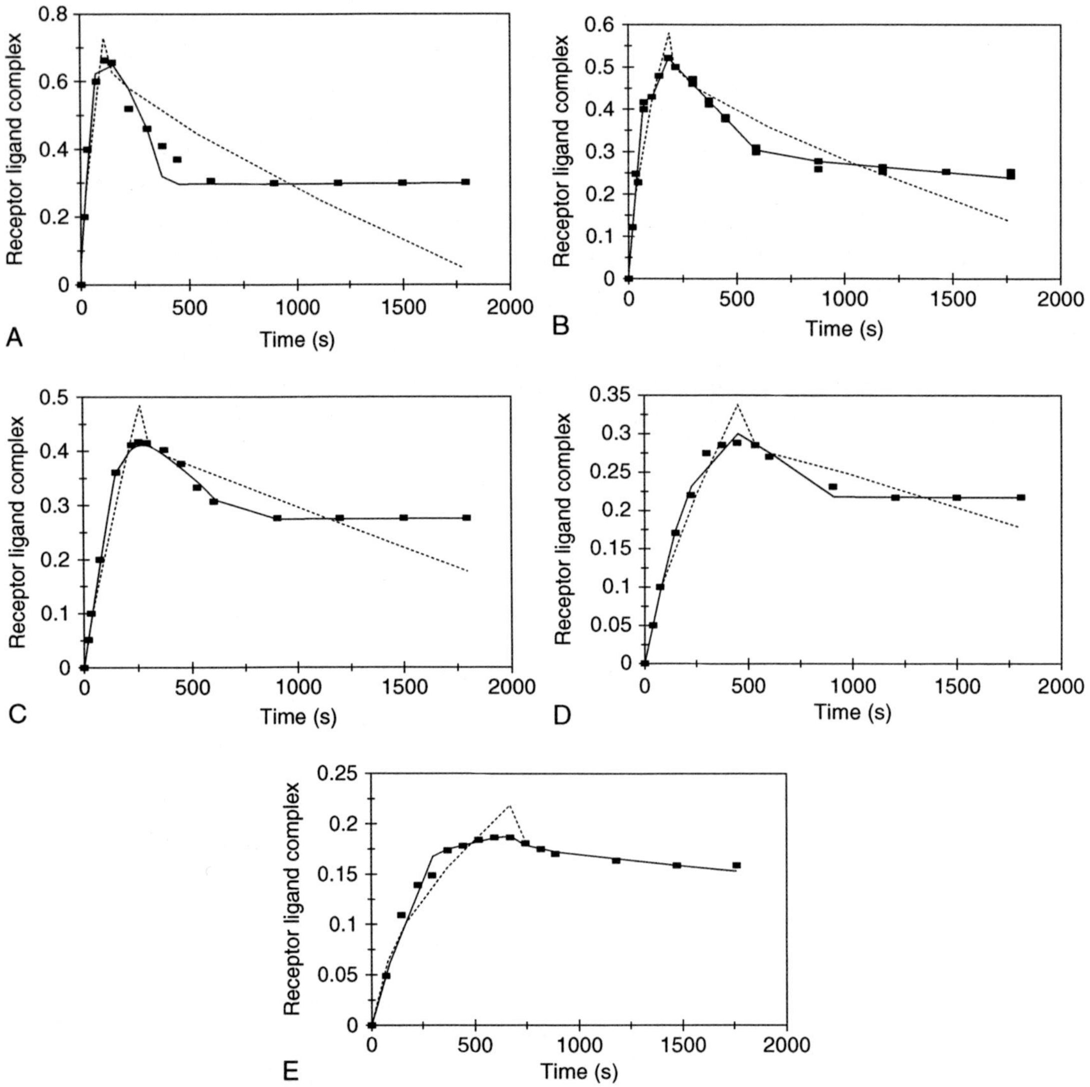

Figure 9.1

Binding and dissociation of different bradykinin concentrations (in nM) to bradykinin B_2 receptors immobilized on a resonant waveguide grating (RWG) biosensor (Fang et al., 2006): (a) 128 (b) 64 (c) 32 (d) 16 (e) 8. When only a solid line (—) is used then a single-fractal analysis applies. When both a dashed (- - -) and a solid (—) line are used then the dashed line represents a single-fractal analysis and the solid line represents a dual-fractal analysis.

dual-fractal analysis is required to adequately describe the binding and the dissociation kinetics. The values of (a) the binding rate coefficient, k, and the fractal dimension, D_f, for a single-fractal analysis, (b) the binding rate coefficients, k_1 and k_2, and the fractal dimensions, D_{f1} and D_{f2}, for a dual-fractal analysis, and (c) the dissociation rate coefficient, k_d, and the fractal dimension, D_{fd}, for a single-fractal analysis, and (d) the dissociation rate coefficients, k_{d1} and k_{d2}, and the fractal dimensions, D_{fd1} and D_{fd2}, for a dual-fractal analysis are given in Tables 9.1 and 9.2.

Table 9.1: Binding and dissociation rate coefficients for different bradykinin concentrations in solution to a resonant waveguide grating (RWG) biosensor (Fang et al., 2006).

Analyte in Solution/Receptor on Surface	k	k_1	k_2	k_d	k_{d1}	k_{d2}
128 nM Bradykinin/RWG biosensor	0.0300 ± 0.0048	0.0199 ± 0.0027	0.3579 ± 0.0146	0.0020 ± 0.0019	$2.5\times10^{-5} \pm 1.6\times10^{-5}$	0.36 ± 0
64 nM Bradykinin/RWG biosensor	0.02331 ± 0.00402	0.00952 ± 0.0011	0.07470 ± 0.00011	0.002138 ± 0.00074	0.000471 ± 0.00006	0.07464 ± 0.00451
32 nM Bradykinin/RWG biosensor	0.000491 + 0.00076	0.002564 ± 0.000014	0.0921 ± 0.0018	0.000248 ± 0.00015	$1.8\times10^{-5} \pm 0.4\times10^{-5}$	0.14 ± 0
16 nM Bradykinin/RWG biosensor	0.00420 ± 0.00053	0.002239 ± 0.000144	0.02938 ± 0.00204	$6.2\times10^{-5} \pm 4.4\times10^{-5}$	$4.3\times10^{-6} \pm 3.2\times10^{-6}$	0.72 ± 0
8 nM Bradykinin/RWG biosensor	0.00556 ± 0.00091	0.00161 ± 0.00025	0.08324 ± 0.00059	0.000695 ± 0.00017	na	na

Table 9.2: Fractal dimensions for the binding and the dissociation phase for different bradykinin concentrations in solution to a resonant waveguide grating (RWG) biosensor (Fang et al., 2006).

Analyte in Solution/Receptor on Surface	D_f	D_{f1}	D_{f2}	D_{fd}	D_{fd1}	D_{fd2}
128 nM Bradykinin/RWG biosensor	1.646 ± 0.2188	1.402 ± 0.2508	2.7544 ± 0.1618	1.4594 ± 0.3538	0 + 0.65025	$3.0\text{-}2.4 \times 10^{-15}$
64 nM Bradykinin/RWG biosensor	1.7698 ± 0.1624	1.2476 ± 0.2316	2.2580 ± 0.00758	1.5936 ± 0.1687	0.9608 ± 0.1304	2.6938 ± 0.1090
32 nM Bradykinin/RWG biosensor	1.3746 + 0.1237	0.9808 ± 0.01125	2.24544 ± 0.09738	1.1312 ± 0.2792	0.030 + 0.2746	$3.0\text{-}3.4 \times 10^{-14}$
16 nM Bradykinin/RWG biosensor	1.5642 ± 0.1074	1.2346 ± 0.1254	2.2400 ± 0.2586	0.926 ± 0.411	0 + 0.8956	$3.0\text{-}3.4 \times 10^{-14}$
8 nM Bradykinin/RWG biosensor	1.8708 ± 0.1498	1.3706 ± 0.2794	2.7478 ± 0.0305	1.8908 ± 0.1843	na	na

It is of interest to note that for a dual-fractal analysis as the fractal dimension increases by a factor of 1.809 from a value of D_{f1} equal to 1.2476 to D_{f2} equal to 2.2580, the binding rate coefficient increases by a factor of 7.85 from a value of k_1 equal to 0.00952 to k_2 equal to 0.0747. Also, as the fractal dimension in the dissociation phase increases by a factor of 2.80 from a value of D_{fd1} equal to 0.9608 to D_{fd2} equal to 2.6938 the dissociation rate coefficient increases by a factor of 158.5 from a value k_{d1} equal to 0.000471 to k_{d2} equal to 0.07464.

Figure 9.1c shows the binding and dissociation of 32 nM bradykinin concentration in solution to the bradykinin B_2 on a RWG biosensor surface. A dual-fractal analysis is once again, required to adequately describe the binding and the dissociation kinetics. The values of (a) the binding rate coefficient, k and the fractal dimension, D_f for a single-fractal analysis, (b) the binding rate coefficients, k_1 and k_2, and the fractal dimensions, D_{f1} and D_{f2}, for a dual-fractal analysis, (c) the dissociation rate coefficient, k_d, and the fractal dimension, D_{fd}, for a single-fractal analysis, and (d) the dissociation rate coefficients, k_{d1} and k_{d2}, and the fractal dimensions, D_{fd1} and D_{fd2}, for a dual-fractal analysis are given in Tables 9.1 and 9.2.

It is of interest to note that for a dual-fractal analysis as the fractal dimension increases by a factor of 2.29 from a value of D_{f1} equal to 0.9808 to D_{f2} equal to 2.2454, the binding rate coefficient increases by a factor of 35.92 from a value of k_1 equal to 0.002564 to k_2 equal to 0.0921. Also, as the fractal dimension in the dissociation phase increases by a factor of 100 from a value of D_{fd1} equal to 0.030 to D_{fd2} equal to 3.0, the dissociation rate coefficient increases by a factor of 16744 from a value k_{d1} equal to 4.3×10^{-6} to k_{d2} equal to 0.72.

Figure 9.1d shows the binding and dissociation of 16 nM bradykinin concentration in solution to the bradykinin B_2 on a RWG biosensor surface. A dual-fractal analysis is once again, required to adequately describe the binding and the dissociation kinetics. The values of (a) the binding rate coefficient, k, and the fractal dimension, D_f, for a single-fractal analysis, (b) the binding rate coefficients, k_1 and k_2, and the fractal dimensions, D_{f1} and D_{f2}, for a dual-fractal analysis, (c) the dissociation rate coefficient, k_d, and the fractal dimension, D_{fd}, for a single-fractal analysis, and (d) the dissociation rate coefficients, k_{d1} and k_{d2}, and the fractal dimensions, D_{fd1} and D_{fd2}, for a dual-fractal analysis are given in Tables 9.1 and 9.2.

It is of interest to note that for a dual-fractal analysis as the fractal dimension increases by a factor of 1.814 from a value of D_{f1} equal to 1.2346 to D_{f2} equal to 2.2400, the binding rate coefficient increases by a factor of 13.12 from a value of k_1 equal to 0.002239 to k_2 equal to 0.02938.

Figure 9.1e shows the binding and dissociation of 8 nM bradykinin concentration in solution to the bradykinin B_2 on a RWG biosensor surface. A dual-fractal analysis is once again, required to adequately describe the binding and the dissociation kinetics. The values of (a) the binding rate coefficient, k, and the fractal dimension, D_f, for a single-fractal analysis, (b) the binding rate coefficients, k_1 and k_2, and the fractal dimensions, D_{f1} and D_{f2}, for a dual-fractal analysis, and (c) the dissociation rate coefficient, k_d, and the fractal dimension, D_{fd}, for a single-fractal analysis are given in Tables 9.1 and 9.2.

It is of interest to note that for a dual-fractal analysis as the fractal dimension increases by a factor of 2 from a value of D_{f1} equal to 1.3706 to D_{f2} equal to 2.7478, the binding rate coefficient increases by a factor of 51.70 from a value of k_1 equal to 0.00161 to k_2 equal to 0.08324.

Figure 9.2a shows for a dual-fractal analysis the increase in the binding rate coefficient, k_1 with an increase in the bradykinin concentration in solution in the 8-128 nM range. For the data shown in Figure 9.2a, the binding rate coefficient, k_1 is given by:

$$k_1 = (0.000175 \pm 0.000081)[\text{Bradykinin, nM}]^{0.934 \pm 0.174} \tag{9.4a}$$

The fit is very good. Only five data points are available. The availability of more data points would lead to a more reliable fit. The binding rate coefficient, k_1 exhibits close to a first (equal to 0.934) order of dependence on the bradykinin concentration in solution. The noninteger order of dependence exhibited lends support to the fractal nature of the system.

Figure 9.2b and Tables 9.1 and 9.2 show the increase in the binding rate coefficient, k_2, with an increase in the fractal dimension, D_{f2}, in the binding phase for a dual-fractal analysis. For the data shown in Figure 9.2b the binding rate coefficient, k_2 is given by:

$$k_2 = (0.000231 \pm 0.000230) D_{f2}^{6.571 \pm 3.429} \tag{9.4b}$$

There is scatter in the data. Only five data points are available. The availability of more data points would lead to a more reliable fit. The binding rate coefficient, k_2, is extremely sensitive to the fractal dimension, D_{f2}, or the degree of heterogeneity that exists on the biosensor chip surface as noted by the close to six and a half order of dependence exhibited.

Figure 9.2c shows the increase in the affinity, K_2 ($= k_2/k_d$), with an increase in the fractal dimension ratio, D_{f2}/D_{fd}. For the data shown in Figure 9.2c, the affinity, K_2 is given by:

$$K_2(= k_2/k_d) = (0.001667 + 0.002999)(D_{f2}/D_{fd})^{1.44 \pm 0.664} \tag{9.4c}$$

There is scatter in the data. Only four data points are available. The availability of more data points would lead to a more reliable fit. Only the positive value of the error is given since the affinity cannot have a negative value. The affinity, K_2 exhibits close to a one and a half (equal to 1.44) order of dependence on the ratio of the fractal dimensions in the binding and in the dissociation phases, ($= D_{f2}/D_{fd}$), respectively.

Fang et al. (2006) also analyzed the binding and the dissociation of different Bradykinin concentrations (16-128 nM) in solution to internalized receptors through endocytes or diffusion into the cytoplasm of cells. These internalized receptors were immobilized on a RWG biosensor surface. Figure 9.3a shows the binding and the dissociation of 128 nM Bradykinin in solution to the internalized receptor immobilized on a RWG biosensor surface. A dual-fractal analysis is required to adequately describe the binding and the dissociation kinetics. The values of (a) the

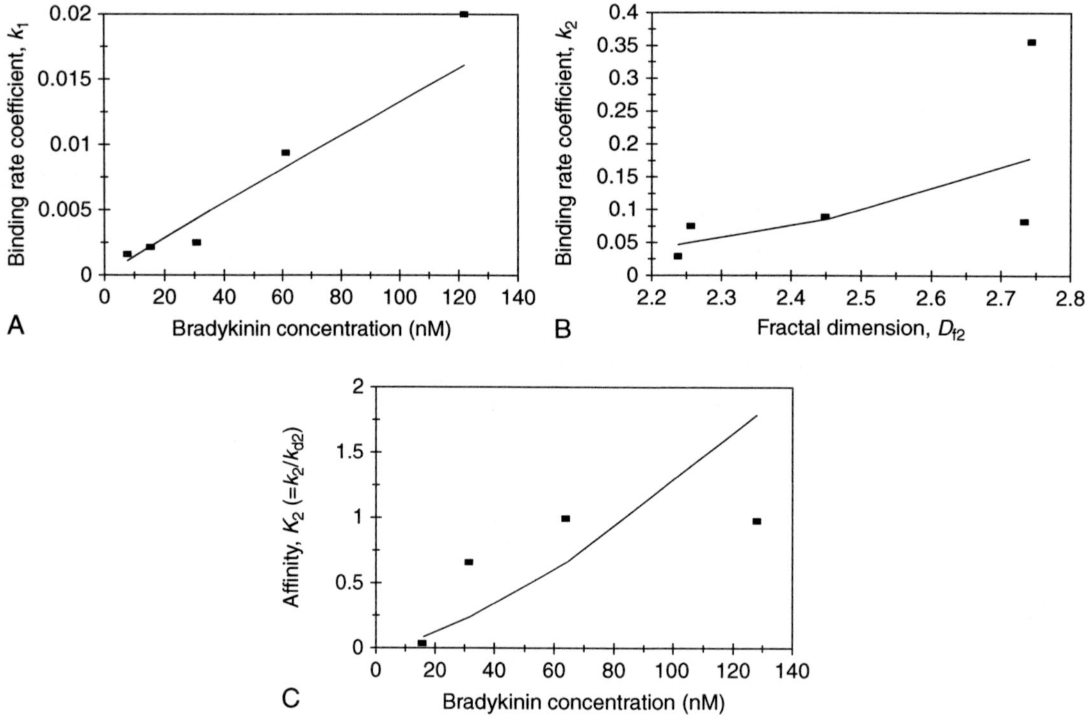

Figure 9.2

(a) Increase in the binding rate coefficient, k_1 with an increase in the bradykinin concentration (in nM) in solution. (b) Increase in the binding rate coefficient, k_2 with an increase in the fractal dimension, D_{f2}. (c) Increase in the affinity, K_2 (k_2/k_d) with an increase in the bradykinin concentration (in nM) in solution.

binding rate coefficient, k, and the fractal dimension, D_f, for a single-fractal analysis, (b) the binding rate coefficients, k_1 and k_2, and the fractal dimensions, D_{f1} and D_{f2}, for a dual-fractal analysis, (c) the dissociation rate coefficient, k_d, and the fractal dimension for the dissociation phase, D_{fd}, for a single-fractal analysis and (d) the dissociation rate coefficients, k_{d1} and k_{d2}, and the fractal dimensions, D_{fd1} and D_{fd2}, for a dual-fractal analysis are given in Tables 9.3 and 9.4.

It is of interest to note that for a dual-fractal analysis as the fractal dimension increases from a value of D_{f1} equal to zero to D_{f2} equal to 2.4906, the binding rate coefficient increases by a factor of 552.6 from a value of k_1 equal to 0.001816 to k_2 equal to 1.0036. An increase in the degree of heterogeneity or the fractal dimension on the RWG biosensor surface leads to an increase in the binding rate coefficient.

Similarly, an increase in the degree of heterogeneity or the fractal dimension on the RWG biosensor surface in the dissociation phase from a value of D_{d1} equal to zero to D_{fd2} equal to 2.1566 leads to an increase in the dissociation rate coefficient by a factor of 44982 from a value of k_{d1} equal to 5.6×10^{-6} to k_{d2} equal to 0.2519.

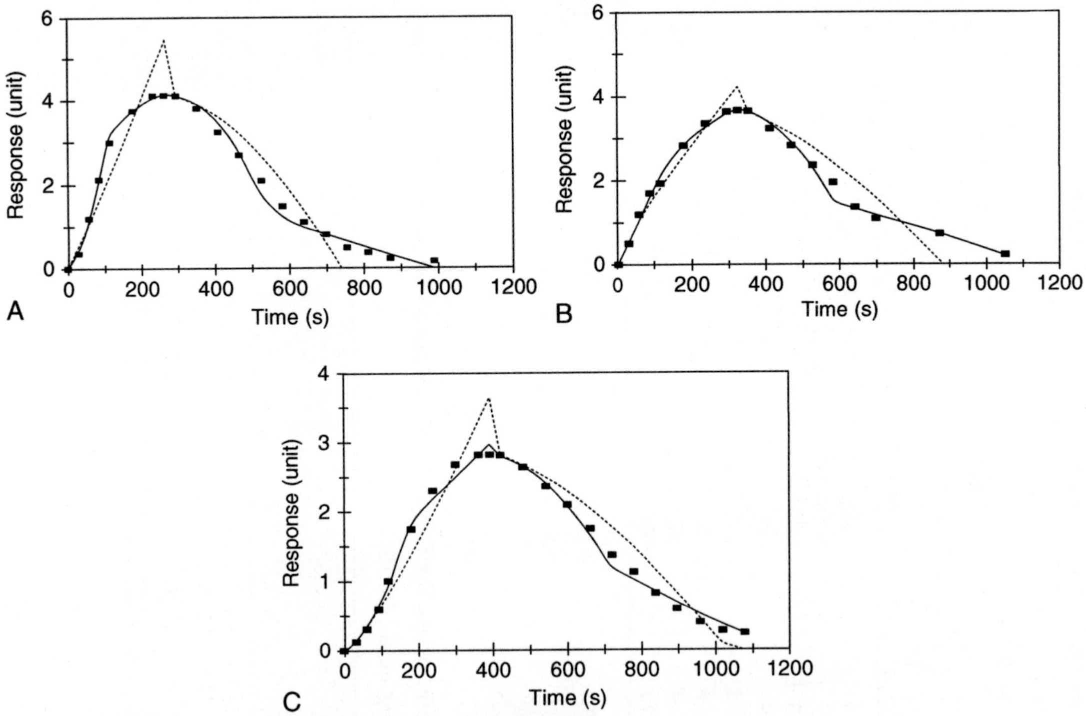

Figure 9.3

Binding and dissociation of DMR signals mediated by different bradykinin concentrations (in nM) in solution (Fang et al., 2006): (a) 128 (b) 64 (c) 32 (d) 16. When only a solid line (—) is used then a single-fractal analysis applies. When both a dashed (- - -) and a solid (—) line are used then the dashed line represents a single-fractal analysis and the solid line represents a dual-fractal analysis.

Figure 9.3b shows the binding and the dissociation of 64 nM Bradykinin in solution to the internalized receptor immobilized on a RWG biosensor surface. A dual-fractal analysis is once again required to adequately describe the binding and the dissociation kinetics. The values of (a) the binding rate coefficient, k, and the fractal dimension, D_f, for a single-fractal analysis, (b) the binding rate coefficients, k_1 and k_2, and the fractal dimensions, D_{f1} and D_{f2}, for a dual-fractal analysis, (c) the dissociation rate coefficient, k_d, and the fractal dimension for the dissociation phase, D_{fd}, for a single-fractal analysis and (d) the dissociation rate coefficients, k_{d1} and k_{d2}, and the fractal dimensions, D_{fd1} and D_{fd2}, for a dual-fractal analysis are given in Tables 9.3 and 9.4.

It is of interest to note that for a dual-fractal analysis as the fractal dimension increases by a factor of 2.15 from a value of D_{f1} equal to 0.992 to D_{f2} equal to 2.1368, the binding rate coefficient increases by a factor of 1.75 from a value of k_1 equal to 0.1757 to k_2 equal to 0.3078. An increase in the degree of heterogeneity or the fractal dimension on the RWG biosensor surface, once again, leads to an increase in the binding rate coefficient.

Table 9.3: Binding and dissociation rate coefficients for DMR signals mediated by different bradykinin concentrations (Fang et al., 2006).

Analyte in Solution/ Receptor on Surface (nM)	k	k_1	k_2	k_d	k_{d1}	k_{d2}
128	0.01181 ± 0.0038	0.001816 ± 0.000181	1.0036 ± 0.0207	0.00007 ± 0.000051	5.6×10^{-6} ± 2.8×10^{-6}	0.2519 ± 0.0093
64	0.03889 ± 0.00559	0.1757 ± 0.00227	0.3078 ± 0.0089	0.000541 ± 0.000322	6.9×10^{-5} ± 2.6×10^{-5}	0.2285 ± 0.0054
32	0.001796 ± 0.000379	0.000643 ± 0.000057	0.06998 ± 0.00418	0.000216 ± 0.000071	5.2×10^{-5} ± 0.8×10^{-5}	0.04512 ± 0.00233
16	0.00209 ± 0.000295	0.00148 ± 0.00021	0.03584 ± 0.00219	0.000233 ± 0.00048	9.8×10^{-5} ± 0.7×10^{-5}	0.00744 ± 0.000542

Table 9.4: Fractal dimensions for the binding and the dissociation phase for DMR signals mediated by different bradykinin concentrations (Fang et al., 2006).

Analyte in Solution/Receptor on Surface (nM)	D_f	D_{f1}	D_{f2}	D_{fd}	D_{fd1}	D_{fd2}
128	0.8092 ± 0.2878	0 + 0.1824	2.4906 ± 0.1378	0 + 0.3544	0 + 0.4236	2.1566 ± 0.1364
64	1.3818 ± 0.1208	0.992 ± 0.2338	2.1368 ± 0.1216	0.2066 + 0.3336	0 + 0.3544	2.1960 ± 0.0618
32	0.4442 ± 0.1542	0 + 0.1614	1.7420 ± 0.1841	0.0432 + 0.136	0 + 0.14486	1.7582 ± 0.1631
16	0.0696 ± 0.0973	0.5376 ± 0.1982	1.6696 ± 0.2318	0.2410 ± 0.1051	0 + 0.07376	1.3452 ± 0.2144

Similarly, an increase in the degree of heterogeneity or the fractal dimension on the RWG biosensor surface in the dissociation phase from a value of D_{d1} equal to zero to D_{fd2} equal to 2.1960 leads to an increase in the dissociation rate coefficient by a factor of 3312 from a value of k_{d1} equal to 6.9×10^{-5} to k_{d2} equal to 0.2285.

Figure 9.3c shows the binding and the dissociation of 32 nM Bradykinin in solution to the internalized receptor immobilized on a RWG biosensor surface. A dual-fractal analysis is once again required to adequately describe the binding and the dissociation kinetics. The values of (a) the binding rate coefficient, k and the fractal dimension, D_f, for a single-fractal analysis, (b) the binding rate coefficients, k_1 and k_2, and the fractal dimensions, D_{f1} and D_{f2}, for a dual-fractal analysis, (c) the dissociation rate coefficient, k_d, and the fractal dimension for the dissociation phase, D_{fd}, for a single-fractal analysis and (d) the dissociation rate coefficients, k_{d1} and k_{d2}, and the fractal dimensions, D_{fd1} and D_{fd2}, for a dual-fractal analysis are given in Tables 9.3 and 9.4.

It is of interest to note that for a dual-fractal analysis as the fractal dimension increases from a value of D_{f1} equal to 0 to D_{f2} equal to 1.7420, the binding rate coefficient increases by a factor of 108.8 from a value of k_1 equal to 0.000643 to k_2 equal to 0.06998. An increase in the degree of heterogeneity or the fractal dimension on the RWG biosensor surface, once again, leads to an increase in the binding rate coefficient.

Similarly, an increase in the degree of heterogeneity or the fractal dimension on the RWG biosensor surface in the dissociation phase from a value of D_{d1} equal to zero to D_{fd2} equal to 1.7582 leads to an increase in the dissociation rate coefficient by a factor of 867.7 from a value of k_{d1} equal to 5.2×10^{-5} to k_{d2} equal to 0.04512.

Figure 9.3d shows the binding and the dissociation of 16 nM Bradykinin in solution to the internalized receptor immobilized on a RWG biosensor surface. A dual-fractal analysis is once again required to adequately describe the binding and the dissociation kinetics. The values of (a) the binding rate coefficient, k, and the fractal dimension, D_f, for a single-fractal analysis, (b) the binding rate coefficients, k_1 and k_2, and the fractal dimensions, D_{f1} and D_{f2}, for a dual-fractal analysis, (c) the dissociation rate coefficient, k_d, and the fractal dimension for the dissociation phase, D_{fd}, for a single-fractal analysis and (d) the dissociation rate coefficients, k_{d1} and k_{d2}, and the fractal dimensions, D_{fd1} and D_{fd2}, for a dual-fractal analysis are given in Tables 9.3 and 9.4.

It is of interest to note that for a dual-fractal analysis as the fractal dimension increases by a factor of 3.10 from a value of D_{f1} equal to 0.5376 to D_{f2} equal to 1.6696, the binding rate coefficient increases by a factor of 24.2 from a value of k_1 equal to 0.00148 to k_2 equal to 0.03584. An increase in the degree of heterogeneity or the fractal dimension on the RWG biosensor surface, once again, leads to an increase in the binding rate coefficient.

Similarly, an increase in the degree of heterogeneity or the fractal dimension on the RWG biosensor surface in the dissociation phase from a value of D_{d1} equal to zero to D_{fd2} equal to 1.3452 leads to an increase in the dissociation rate coefficient by a factor of 75.9 from a value of k_{d1} equal to 5.2×10^{-5} to k_{d2} equal to 0.04512.

Figure 9.4a and Table 9.3 show the increase in the binding rate coefficient, k_2, with an increase in the Bradykinin concentration (in nM) in solution for a dual-fractal analysis. For the data shown in Figure 9.4a, the binding rate coefficient, k_2, is given by:

$$k_2 = (0.000302 \pm 0.000087)[\text{Bradykinin, nM}]^{1.656\pm0.163} \tag{9.3a}$$

The fit is very good. Only four data points are available. The availability of more data points would lead to a stronger fit. For the dual-fractal analysis the binding rate coefficient, k_2, exhibits an order of dependence slightly greater than one and a half (equal to 1.656) order of dependence on the Bradykinin concentration in the 16-128 nM concentration range in solution. The nonintegral order of dependence exhibited lends support to the fractal nature of the system.

Figure 9.4b and Table 9.3 show the increase in the dissociation rate coefficient, k_{d2}, with an increase in the Bradykinin concentration (in nM) in solution for a dual-fractal analysis. For the data shown in Figure 9.4b, the dissociation rate coefficient, k_{d2}, is given by:

$$k_{d2} = (8.1 \pm 1.9)10^{-5}[\text{Bradykinin, nM}]^{1.758\pm0.412} \tag{9.3b}$$

The fit is reasonable. Only four data points are available. The availability of more data points would lead to a better fit. For the dual-fractal analysis the binding rate coefficient, k_2, exhibits an order of dependence between one and a half and two (equal to 1.758) on the Bradykinin concentration in the 16-128 nM concentration range in solution. The nonintegral order of dependence exhibited, once again, lends support to the fractal nature of the system.

Figure 9.4c and Table 9.3 show the increase in the binding rate coefficient, k_2, with an increase in the fractal dimension, D_{f2}, in the binding phase for a dual-fractal analysis. For the data shown in Figure 9.4c, the binding rate coefficient, k_2, is given by:

$$k_2 = (0.000694 \pm 0.000125)D_{f2}^{8.00\pm0.516} \tag{9.3c}$$

The fit is very good. Only four data points are available. The availability of more data points would lead to a stronger fit. For the dual-fractal analysis the binding rate coefficient, k_2, is very sensitive to the fractal dimension, D_{f2}, or the degree of heterogeneity that exists on the sensor surface as noted by the eighth order of dependence exhibited.

Figure 9.4d and Tables 9.3 and 9.4 show the increase in the dissociation rate coefficient, k_{d2}, with an increase in the fractal dimension in the dissociation phase, D_{fd2}, for a dual-fractal analysis. For the data shown in Figure 9.4d, the dissociation rate coefficient, k_{d2}, is given by:

$$k_{d2} = (0.000836 \pm 0.000125)D_{fd2}^{7.24\pm0.352} \tag{9.3d}$$

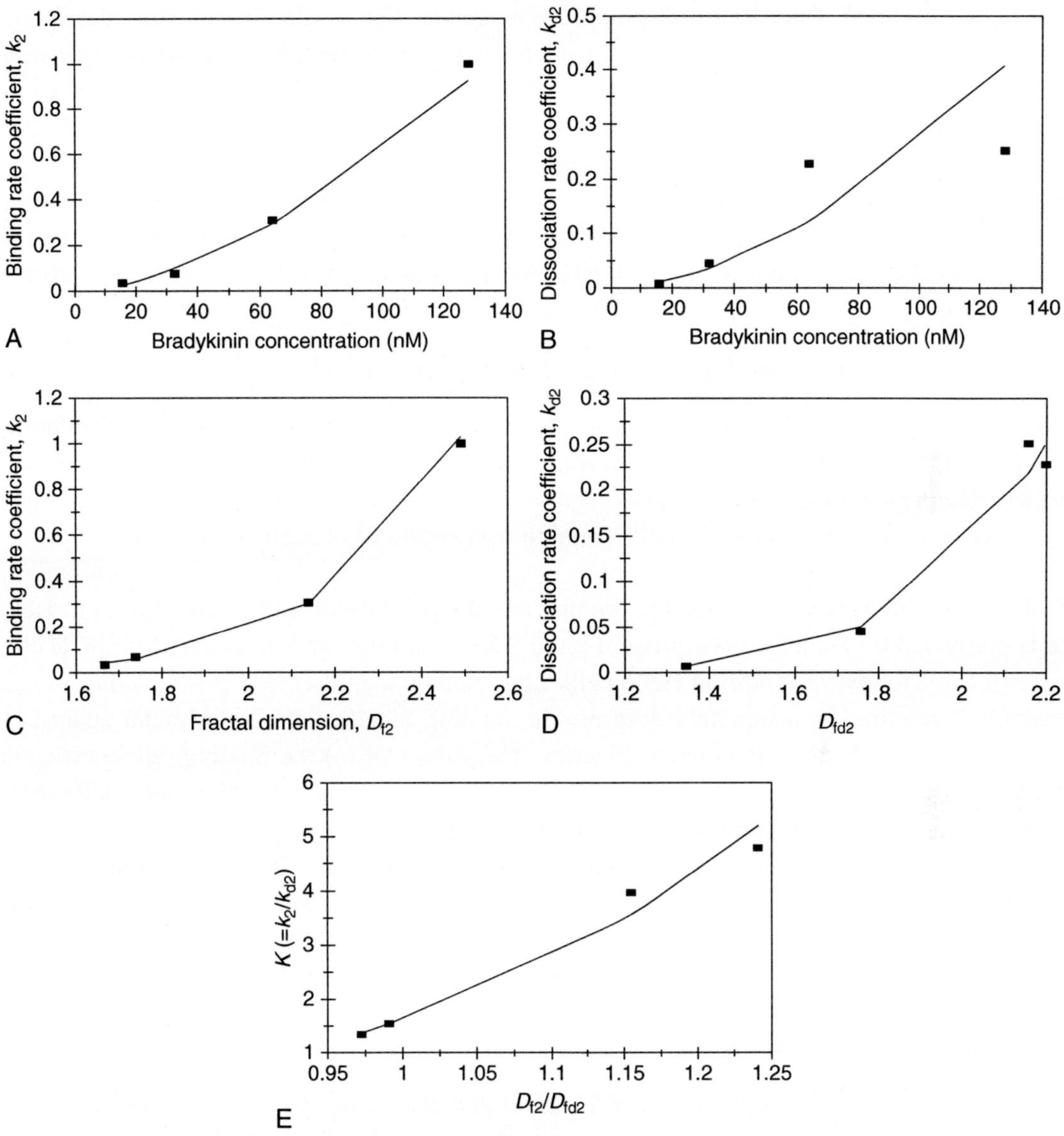

Figure 9.4

(a) Increase in the binding rate coefficient, k_2 with an increase in the bradykinin concentration (in nM) in solution. (b) Increase in the dissociation rate coefficient, k_{d2} with an increase in the bradykinin concentration (in nM) in solution. (c) Increase in the binding rate coefficient, k_2 with an increase in the fractal dimension, D_{f2}. (d) Increase in the dissociation rate coefficient, k_{d2} with an increase in the fractal dimension, D_{fd2}. (e) Increase in the affinity, K_2 $(= k_2/k_{d2})$ with an increase in the ratio of the fractal dimensions, D_{f2}/D_{fd2}.

The fit is very good. Only four data points are available. The availability of more data points would lead to a more reliable fit. The dissociation rate coefficient, k_{d2}, is extremely sensitive to the degree of heterogeneity or the fractal dimension that exists in the dissociation phase, D_{fd2}, as noted by the order of dependence between seven and seven and a half exhibited.

Figure 9.4e and Tables 9.3 and 9.4 show the increase in the affinity, K_2 (= k_2/k_{d2}), with an increase in the ratio of fractal dimensions present in the binding and in the dissociation phases, D_{f2}/D_{fd2}, for a dual-fractal analysis. For the data shown in Figure 9.4e, the affinity, K_2 is given by:

$$K_2(= k_2/k_{d2}) = (1.6236 \pm 0.1802)(D_{f2}/D_{fd2})^{5.41 \pm 0.513} \tag{9.3e}$$

The fit is very good. Only four data points are available. The availability of more data points would lead to a more reliable fit. The affinity, K_2, is extremely sensitive to the ratio of the fractal dimensions present in the binding and in the dissociation phases, respectively, D_{f2}/D_{fd2}, as noted by the close to five and five and a half (equal to 5.41) order of dependence exhibited.

Ziblat et al. (2006) have analyzed the binding and dissociation of mβCD cholesterol to HeLa cells cultivated on a gold-coated prism. Figure 9.5a shows the binding of 20 nM mβCD cholesterol to cyclodextrin-modified HeLa cells cultivated on a gold-coated prism. A dual-fractal analysis is required to adequately describe the binding kinetics. A single-fractal analysis is adequate to describe the dissociation kinetics. The values of (a) the binding rate coefficient, k, and the fractal dimension, D_f, for a single-fractal analysis, (b) the binding rate coefficients, k_1 and k_2, and the fractal dimensions, D_{f1} and D_{f2}, for a dual-fractal analysis, and (c) the dissociation rate coefficient, k_d, and the fractal dimension D_{fd}, for a single-fractal analysis are given in Tables 9.3 and 9.4. It is of interest to note that as the fractal dimension increases by a factor of 1.50 from a value of D_{f1} equal to 1.892 to D_{f2} equal to 2.8457, the binding rate coefficient increases by a factor of 1.83 from a value of k_1 equal to 2.913 to k_2 equal to 5.3312. An increase in the degree of heterogeneity or the fractal dimension on the sensor chip surface leads to an increase in the binding rate coefficient.

Figure 9.5b shows the binding of 20 nM mβCD cholesterol to cyclodextrin modified HeLa cells cultivated on a gold-coated prism (second cycle of binding). A dual-fractal analysis is, once again, required to adequately describe the binding kinetics. The values of (a) the binding rate coefficient, k, and the fractal dimension, D_f, for a single-fractal analysis, and (b) the binding rate coefficients, k_1 and k_2, and the fractal dimensions, D_{f1} and D_{f2}, for a dual-fractal analysis, are given in Tables 9.5 and 9.6. It is of interest to note that as the fractal dimension increases from a value of D_{f1} equal to zero to D_{f2} equal to 2.8392, the binding rate coefficient increases by a factor of 6.253 from a value of k_1 equal to 1.1117 to k_2 equal to 6.952. An increase in the degree of heterogeneity or the fractal dimension on the sensor chip surface, once again, leads to an increase in the binding rate coefficient.

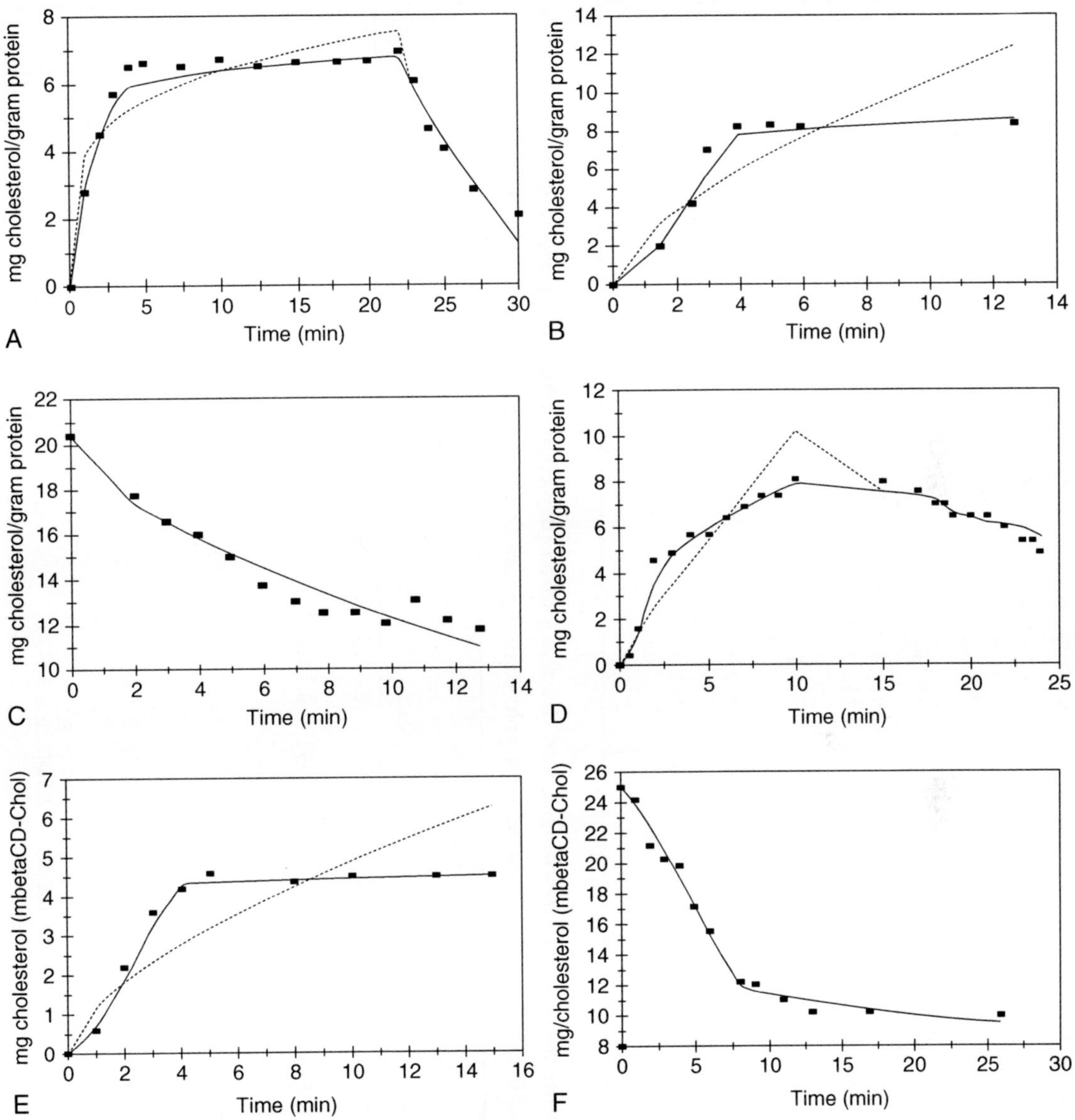

Figure 9.5

(a) Binding and dissociation of 20 nM mβCD-chol cells exposed to two cycles of cholesterol enrichment (Ziblat et al., 2006). (b) Binding and dissociation of 20 nM mβCD-chol cells exposed to two cycles of cholesterol enrichment (second cycle) (Ziblat et al., 2006). (c) Dissociation of 10 nM mβCD-chol cells (Ziblat et al., 2006). (d) Dissociation of 10 nM mβCD-chol cells followed by inactive analog of mβCD (3 nM). (e) Binding of 10 mM mβCD-chol mβCD, αCD (3 nM). (f) Dissociation of 3 nM inactive analog of mβCD, αCD. When only a solid line (—) is used then a single-fractal analysis applies. When both a dashed (- - -) and a solid (—) line are used then the dashed line represents a single-fractal analysis and the solid line represents a dual-fractal analysis.

Table 9.5: Binding (enrichment) and dissociation (depletion) rate coefficients of cholesterol to cyclodextrin modified HeLa cells cultivated on a gold-coated prism (Ziblat et al., 2006).

Analyte in Solution/ Cyclodextrin Modified HeLa Cells	k	k_1	k_2	k_d	k_{d1}	k_{d2}
20 nM mβCD cholesterol	3.883 ± 0.668	2.913 ± 0.211	5.3312 ± 0.0884	1.0877 ± 0.2302	na	na
20 nM mβCD cholesterol (second cycle)	2.5156 ± 1.100	1.1117 ± 0.1808	6.952 ± 0.472	na	na	na
10 nM mβCD dissociation (depletion) only	na	na	na	1.954 ± 0.211	na	na
10 nM mβCD dissociation (depletion) only	1.3620 ± 0.5581	1.3086 ± 0.5019	3.1336 ± 0.1086	0.5472 ± 0.1052	na	na
10 mM mβCD-chol	1.208 ± 0.61	0.6715 ± 0.1529	4.1405 ± 0.0344	na	na	na
3mM inactive analog of mβCD, αCD	na	na	na	1.478 ± 0.651	0.9638 ± 0.1938	9.668 ± 0.373

Table 9.6: Fractal dimensions for the binding (enrichment) and the dissociation (depletion) phase for cholesterol to cyclodextrin modified HeLa cells cultivated on a gold-coated prism (Ziblat et al., 2006).

Analyte in Solution/ Cyclodextrin Modified HeLa Cells	D_f	D_{f1}	D_{f2}	D_{fd}	D_{fd1}	D_{fd2}
20 nM mβCD cholesterol	2.5720 ± 0.09632	1.892 ± 0.1093	2.8457 ± 0.07252	1.4094 ± 0.2308	na	na
20 nM mβCD cholesterol (second cycle)	1.7630 ± 0.4284	0 + 0.4218	2.8392 ± 0.1187	na	na	na
10 nM mβCD dissociation (depletion) only	na	na	na	1.7696 ± 0.1055	na	na
10 nM mβCD dissociation (depletion) only	1.2468 ± 0.2258	0.0002 + 0.4752	2.1936 ± 0.0848	1.6064 ± 0.1521	na	na
10 mM mβCD-chol	1.78 ± 0.3192	0.1118 + 0.3940	2.9347 ± 0.03426	na	na	na
3mM inactive analog of mβCD, αCD	na	na	na	1.2174 ± 0.2366	0.4128 ± 0.3058	2.7130 ± 0.07898

Figure 9.5c shows the depletion (only) binding of 10 nM mβCD cholesterol to cyclodextrin modified HeLa cells cultivated on a gold-coated prism. A single-fractal analysis is adequate to describe the dissociation (depletion) kinetics. The values of the dissociation (depletion) rate coefficient, k_d, and the fractal dimension D_{fd}, for a single-fractal analysis are given in Tables 9.5 and 9.6.

Figure 9.5d shows the binding and dissociation (depletion) of 10 nM mβCD cholesterol followed by inactive analog of 3 nM of mβCD to cyclodextrin modified HeLa cells cultivated on a gold-coated prism. A dual-fractal analysis is, once again, required to adequately describe the binding kinetics. A single-fractal analysis is adequate to describe the dissociation kinetics. The values of (a) the binding rate coefficient, k, and the fractal dimension, D_f, for a single-fractal analysis, (b) the binding rate coefficients, k_1 and k_2, and the fractal dimensions, D_{f1} and D_{f2}, for a dual-fractal analysis, and (c) the dissociation rate coefficient, k_d, and the fractal dimension D_{fd}, for a single-fractal analysis are given in Tables 9.5 and 9.6. It is of interest to note that as the fractal dimension increases from a value of D_{f1} equal to 0.0002 to D_{f2} equal to 2.1936, the binding rate coefficient increases by a factor of 2.39 from a value of k_1 equal to 1.3086 to k_2 equal to 3.1336. Once again, an increase in the degree of heterogeneity or the fractal dimension on the sensor chip surface, leads to an increase in the binding rate coefficient.

Figure 9.5e shows the binding of 10 nM mβCD cholesterol to cyclodextrin modified HeLa cells cultivated on a gold-coated prism (second cycle of binding). A dual-fractal analysis is, once again, required to adequately describe the binding kinetics. The values of (a) the binding rate coefficient, k, and the fractal dimension, D_f, for a single-fractal analysis, and (b) the binding rate coefficients, k_1 and k_2, and the fractal dimensions, D_{f1} and D_{f2}, for a dual-fractal analysis, are given in Tables 9.5 and 9.6. It is of interest to note that as the fractal dimension increases by a factor of 26.24 from a value of D_{f1} equal to 0.1118 to D_{f2} equal to 2.9347, the binding rate coefficient increases by a factor of 6.17 from a value of k_1 equal to 0.6715 to k_2 equal to 4.1405. An increase in the degree of heterogeneity or the fractal dimension on the sensor chip surface, once again, leads to an increase in the binding rate coefficient.

Figure 9.5f shows the dissociation (depletion only) of 3nM inactive analog of mβCD, αCD from the cyclodextrin HeLa cells cultivated on a gold-coated prism to the solution. A dual-fractal analysis is required to adequately describe the dissociation kinetics. The values of (a) the dissociation rate coefficient, k_d, and the fractal dimension in the dissociation phase, D_{fd}, for a single-fractal analysis, and (b) the dissociation rate coefficients, k_{d1} and k_{d2}, and the fractal dimensions, D_{fd1} and D_{fd2}, for a dual-fractal analysis are given in Tables 9.5 and 9.6. It is of interest to note that for a dual-fractal analysis, as the fractal dimension in the dissociation phase increases by a factor of 6.572 from a value of D_{fd1} equal to 0.4128 to D_{fd2} equal to 2.7130, the dissociation rate coefficient increases by a factor of 10.03 from a value of k_{d1} equal to 0.9638 to k_{d2} equal to 9.668. Once again, an increase in the degree of

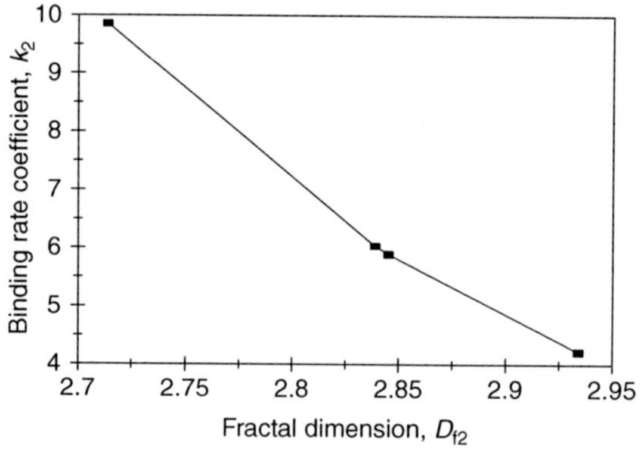

Figure 9.6
Decrease in the binding rate coefficient, k_2, with an increase in the fractal dimension, D_{f2}.

heterogeneity on the sensor chip surface (in this case in the dissociation phase) leads to an increase in the dissociation rate coefficient.

Figure 9.6 and Tables 9.5 and 9.6 show the increase in the binding rate coefficient, k_2, with an increase in the fractal dimension, D_{f2}, for a dual-fractal analysis. For the data shown in Figure 9.6, the binding rate coefficient, k_2, is given by:

$$k_2 = (466298 \pm 63304)D_{f2}^{-10.783 \pm 2.271} \tag{9.4}$$

The fit is very good. Only four data points are available. The availability of more data points would lead to a more reliable fit. The binding rate coefficient, k_2, is extremely sensitive to the fractal dimension, D_{f2}, or the degree of heterogeneity that exists on the sensor surface as noted by the greater than negative order of dependence between 10 and a half and 11 (equal to -10.783) exhibited.

Mank et al. (2006) recently pointed out that biosensors based on the green fluorescent protein (GFP) are important tools in cell biology and neuroscience (Zhang et al., 2002; Miyawaki, 2003; Griesbeck, 2004). Mank et al. (2006) have developed a new generation of calcium biosensors that use variants of troponin C (TnC) (a specialized calcium sensor of skeletal and cardiac muscle) as calcium binding moieties. These authors emphasize that their FRET (fluorescence resonance energy transfer)-based calcium biosensor which includes increased ion selectivity, moderate calcium sensitivity, and strongly increased maximum fluorescence is a useful tool for *in vivo* imaging experiments. They emphasize that their biosensor is fast, and is stable in imaging experiments. Also, their biosensor exhibits enhanced fluorescence change.

Mank et al. (2006) report that the most critical test for their biosensor is its *in vivo* performance in cells and subcellular compartments of interest. These authors engineered transgenic

Drosophila flies that permitted the expression of TN-XL (for X-large) under the control of an upstream activation sequence (UAS) (Brand and Perrimon, 1993). They did this to analyze the exhibited changes of TN-XL in neurons of living animals. These authors noted that electric stimulation of the nerve harboring the innervating axon evoked reliable fluorescent changes.

Figure 9.7a shows the binding and the dissociation phases for the 1.5 mM calcium + FRET-based calcium biosensor employing troponin C (Mank et al., 2006). The external stimulation frequency was 160 Hz for 2.2 s. A dual-fractal analysis is required to adequately describe the kinetics in the binding phase. A single-fractal analysis is adequate to describe the kinetics in the dissociation phase. The values of (a) the binding rate coefficient, k, and the fractal dimension, D_f, for a single-fractal analysis, (b) the binding rate coefficients, k_1 and k_2, and the fractal dimensions, D_{f1} and D_{f2}, for a dual-fractal analysis, and (c) the dissociation rate coefficient, k_d, and the fractal dimension, D_{fd}, in the dissociation phase are given in Tables 9.7 and 9.8.

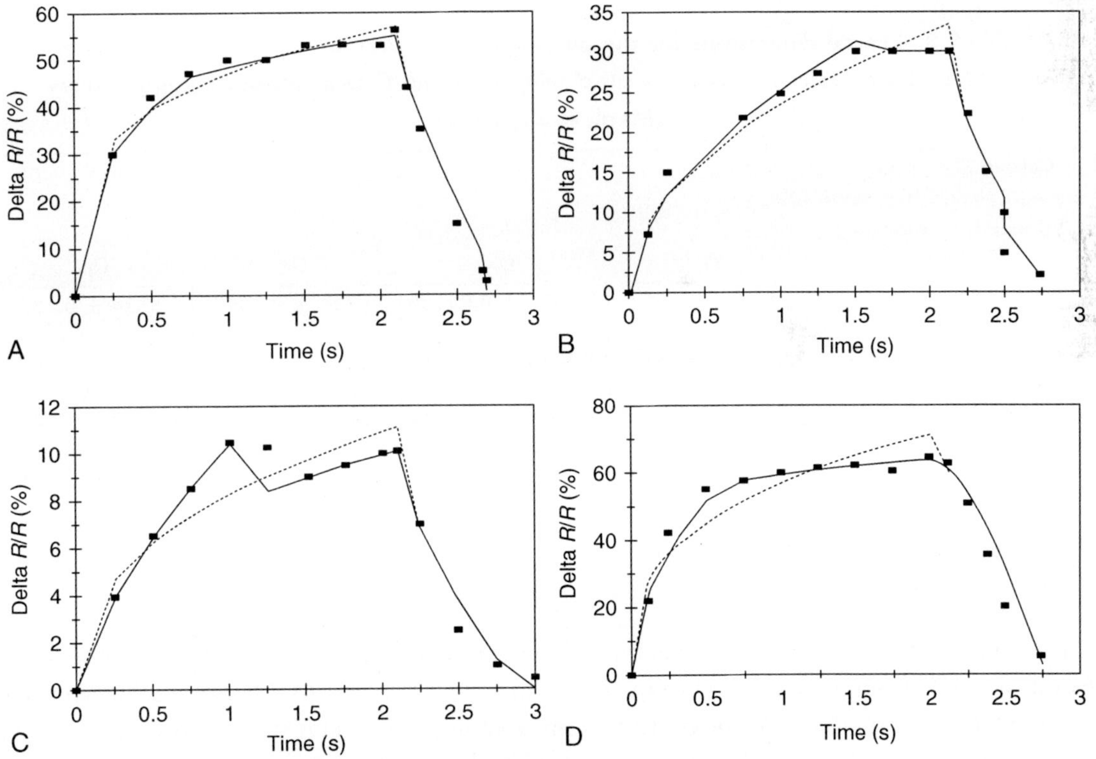

Figure 9.7

Binding and dissociation of calcium+ FRET-based calcium biosensor employing troponin C (Mank et al., 2006). TN-XL fluorescence in *Drosophila in vivo*. 1.5 mM external calcium (a-c). Effect of different external stimulation frequencies for 2.2 s. (a) 160 Hz (b) 80 Hz (c) 40 Hz (d) 160 Hz + 10 mM external calcium. When only a solid line (—) is used then a single-fractal analysis applies. When both a dashed (- - -) and a solid (—) line are used then the dashed line represents a single-fractal analysis and the solid line represents a dual-fractal analysis.

Table 9.7: Binding and dissociation rate coefficients for calcium to a FRET-based calcium biosensor employing troponin C as a calcium-binding moiety (Mank et al., 2006).

Calcium/Troponin C Based Calcium Biosensor Stimulation Frequency (Hz)	k	k_1	k_2	k_d
1.5 mM external calcium;160 Hz for 2.2 s	47.167 ± 2.994	51.736 ± 2.435	48.642 ± 1.147	71.562 ± 10.229
1.5 mM external calcium; 80 Hz for 2.2 s	23.762 ± 2.876	25.515 ± 3.623	30.2 ± 0.0	34.191 ± 5.136
1.5 mM external calcium; 40 Hz for 2.2 s	8.293 ± 1.326	10.458 ± 0.075	7.808 ± 0.029	11.633 ± 2.063
10 mM external calcium; 160 Hz for 2.2 s	56.591 ± 9.545	74.569 ± 12.857	59.148 ± 1.224	97.878 ± 54.11

TN-XL (biosensor) fluorescence changes in *Drosophila in vivo*.

Table 9.8: Fractal dimensions for the binding and the dissociation phase for calcium to a FRET-based calcium biosensor employing troponin C as a calcium-binding moiety (Mank et al., 2006).

Calcium/Troponin C Based Calcium Biosensor Stimulation Frequency (Hz)	D_f	D_{f1}	D_{f2}	D_{fd}
1.5 mM external calcium; 160 Hz for 2.2 s	2.4844 ± 0.0612	2.2534 ± 0.09186	2.6754 ± 0.1093	1.5324 ± 0.1374
1.5 mM external calcium; 80 Hz for 2.2 s	2.0796 ± 0.0812	1.9644 ± 0.1328	3.0 ± 0.0	1.7234 ± 0.1627
1.5 mM external calcium; 40 Hz for 2.2 s	2.2062 ± 0.07384	1.6144 ± 0.01372	2.2976 ± 0.0278	1.6948 ± 0.2412
10 mM external calcium; 160 Hz for 2.2 s	2.3482 ± 0.1159	1.9486 ± 0.2280	2.8129 ± 0.03394	0 + 0.5116

TN-XL (biosensor) fluorescence changes in *Drosophila in vivo*.

It is of interest to note that for a dual-fractal analysis as the fractal dimension increases by a factor of 1.187 from a value of D_{f1} equal to 2.2534 to D_{f2} equal to 2.6754, the binding rate coefficient decreases by a factor of 1.06 from a value of k_1 equal to 51.736 to k_2 equal to 48.642. In this case, and this is rare, an increase in the degree of heterogeneity or the fractal dimension on the biosensor surface leads to a decrease in the binding rate coefficient.

Figure 9.7b shows the binding and the dissociation phases for the 1.5 mM calcium + FRET-based calcium biosensor employing troponin C (Mank et al., 2006). The external stimulation frequency was 80 Hz for 2.2 s. Once again, a dual-fractal analysis is required to adequately describe the kinetics in the binding phase. A single-fractal analysis is adequate to describe the kinetics in the dissociation phase. The values of (a) the binding rate coefficient, k, and the fractal dimension, D_f, for a single-fractal analysis, (b) the binding rate coefficients, k_1 and k_2, and the fractal dimensions, D_{f1} and D_{f2}, for a dual-fractal analysis, and (c) the dissociation rate coefficient, k_d, and the fractal dimension, D_{fd}, in the dissociation phase are given in Tables 9.7 and 9.8.

In this case, it is of interest to note that for a dual-fractal analysis as the fractal dimension increases by a factor of 1.527 from a value of D_{f1} equal to 1.9644 to D_{f2} equal to 3.0, the binding rate coefficient increases by a factor of 1.18 from a value of k_1 equal to 25.515 to k_2 equal to 30.2. In this case, an increase in the degree of heterogeneity or the fractal dimension on the biosensor surface leads to an increase in the binding rate coefficient.

Figure 9.7c shows the binding and the dissociation phases for the 1.5 mM calcium + FRET-based calcium biosensor employing troponin C (Mank et al., 2006). The external stimulation frequency was 40 Hz for 2.2 s. Once again, a dual-fractal analysis is required to adequately describe the kinetics in the binding phase. A single-fractal analysis is adequate to describe the kinetics in the dissociation phase. The values of (a) the binding rate coefficient, k, and the fractal dimension, D_f, for a single-fractal analysis, (b) the binding rate coefficients, k_1 and k_2, and the fractal dimensions, D_{f1} and D_{f2}, for a dual-fractal analysis, and (c) the dissociation rate coefficient, k_d, and the fractal dimension, D_{fd}, in the dissociation phase are given in Tables 9.7 and 9.8.

In this case, it is of interest to note that for a dual-fractal analysis as the fractal dimension increases by a factor of 1.423 from a value of D_{f1} equal to 1.6144 to D_{f2} equal to 2.2976, the binding rate coefficient decreases by a factor of 0.747 from a value of k_1 equal to 10.458 to k_2 equal to 7.808. In this case, an increase in the degree of heterogeneity or the fractal dimension on the biosensor surface leads to a decrease in the binding rate coefficient.

Figure 9.7d shows the binding and the dissociation phases for the 10 mM calcium + FRET-based calcium biosensor employing troponin C (Mank et al., 2006). The external stimulation frequency was 160 Hz for 2.2 s. Once again, a dual-fractal analysis is required to adequately describe the kinetics in the binding phase. A single-fractal analysis is adequate to describe the kinetics in the dissociation phase. The values of (a) the binding rate coefficient, k, and the fractal dimension, D_f, for a single-fractal analysis, (b) the binding rate coefficients, k_1 and k_2, and the fractal dimensions, D_{f1} and D_{f2}, for a dual-fractal analysis, and (c) the dissociation rate coefficient, k_d, and the fractal dimension, D_{fd}, in the dissociation phase are given in Tables 9.7 and 9.8.

In this case, it is of interest to note that for a dual-fractal analysis as the fractal dimension increases by a factor of 1.44 from a value of D_{f1} equal to 1.9486 to D_{f2} equal to 2.8129, the binding rate coefficient decreases by a factor of 0.793 from a value of k_1 equal to 74.568 to k_2 equal to 59.148. In this case, an increase in the degree of heterogeneity or the fractal dimension on the biosensor surface leads to a decrease in the binding rate coefficient.

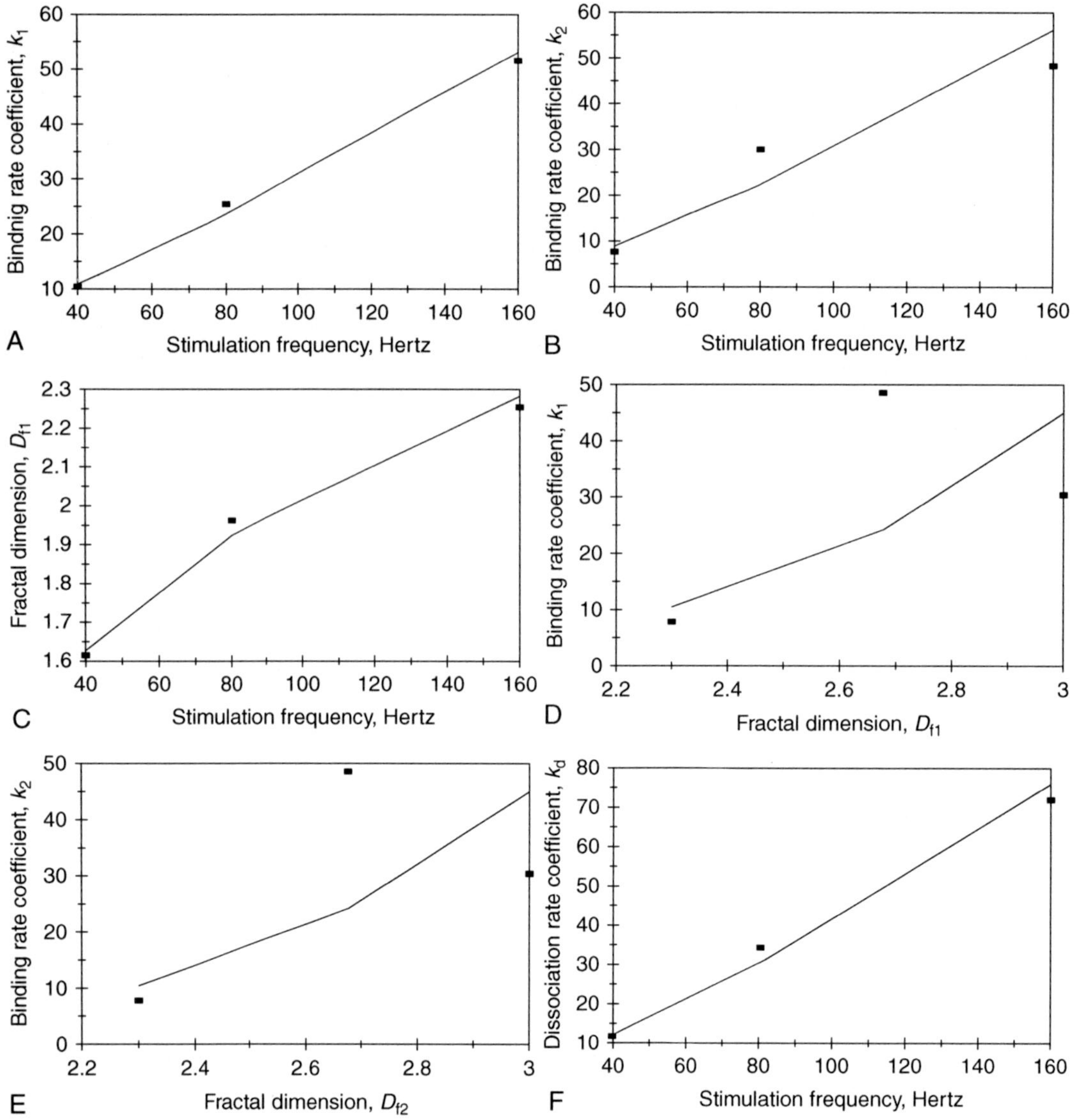

Figure 9.8

(a) Increase in the binding rate coefficient, k_1 with an increase in the stimulation frequency (in Hz). (b) Increase in the binding rate coefficient, k_2 with an increase in the stimulation frequency (in Hz). (c) Increase in the fractal dimension, D_{f1} with an increase in the stimulation frequency in (Hz). (d) Increase in the binding rate coefficient, k_1 with an increase in the fractal dimension, D_{f1}. (e) Increase in the binding rate coefficient, k_2 with an increase in the fractal dimension, D_{f2}. (f) Increase in the dissociation rate coefficient, k_d with an increase in the stimulation frequency (in Hz).

continued

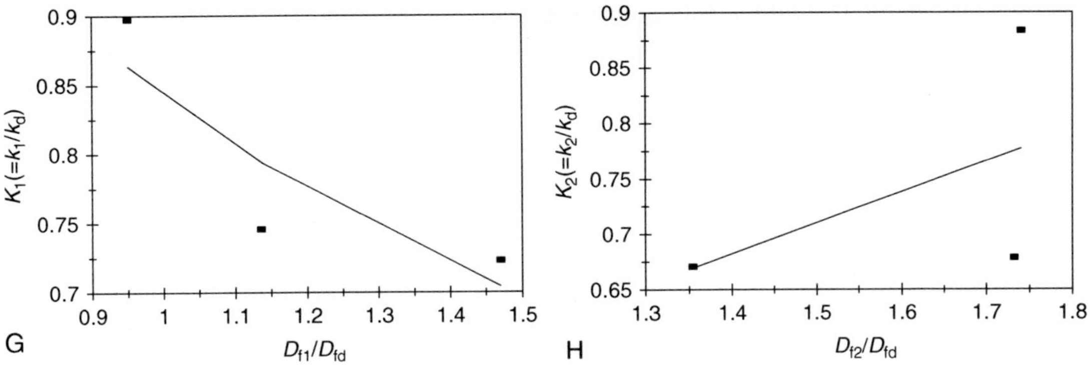

Figure 9.8—cont'd

(g) Decrease in the affinity, k_1/k_d with an increase in the fractal dimension ratio, D_{f1}/D_{fd}.
(h) Increase in the affinity, k_2/k_d with an increase in the fractal dimension ratio, D_{f2}/D_{fd}.

Figure 9.8a and Table 9.7 show the increase in the binding rate coefficient, k_1, with an increase in the stimulation frequency in Hz for a dual-fractal analysis. For the data shown in Figure 9.8a and in the 40-160 Hz range, the binding rate coefficient, k_1, is given by:

$$k_1 = (0.153 \pm 0.012)[\text{Hz}]^{1.153 \pm 0.0771} \tag{9.5a}$$

The fit is very good. Only three data points are available. The availability of more data points would lead to a more reliable fit. The binding rate coefficient, k_1 exhibits slightly higher than first (equal to 1.157) order of dependence on the stimulation frequency in Hz in the 40-160 Hz range.

Figure 9.8b and Table 9.7 show for a dual-fractal analysis the increase in the binding rate coefficient, k_1 with an increase in the stimulation frequency in Hz. For the data shown in Figure 9.8b and in the 40-160 Hz range, the binding rate coefficient, k_2 is given by:

$$k_2 = (0.0695 \pm 0.0298)[\text{Hz}]^{1.319 \pm 0.365} \tag{9.5b}$$

The fit is very good. Only three data points are available. The availability of more data points would lead to a more reliable fit. The binding rate coefficient, k_2, exhibits slightly higher than first (equal to 1.153) order of dependence on the stimulation frequency in Hz in the 40-160 Hz range. It is of interest to note that the binding rate coefficient, k_2, exhibits a slightly higher order of dependence (equal to 1.319) than the binding rate coefficient, k_1, on the stimulation frequency in Hz in the 40-160 Hz range.

Figure 9.8c and Table 9.7 show the increase in the fractal dimension, D_{f1}, with an increase in the stimulation frequency in Hz for a dual-fractal analysis. For the data shown in Figure 9.8c and in the 40-160 Hz range, the fractal dimension, D_{f1}, is given by:

$$D_{f1} = (0.6709 \pm 0.0155)[\text{Hz}]^{0.2405 \pm 0.02329} \tag{9.5c}$$

The fit is very good. Only three data points are available. The availability of more data points would lead to a more reliable fit. The fractal dimension, D_{f1}, exhibits only a very slight (equal to 0.240) order of dependence on the stimulation frequency in Hz in the 40-160 Hz range.

Figure 9.8d and Table 9.7 show the increase in the binding rate coefficient, k_1, with an increase in the fractal dimension, D_{f1}, for a dual-fractal analysis. For the data shown in Figure 9.8d and in the 40-160 Hz range, the binding rate coefficient, k_1, is given by:

$$k_1 = (1.047 \pm 0.036)[D_{f1}]^{4.78 \pm 0.142} \tag{9.5d}$$

The fit is reasonable. Only three data points are available. The availability of more data points would lead to a more reliable fit. The binding rate coefficient, k_1, exhibits a very strong order of dependence on the fractal dimension, D_f in the 40-160 Hz range.

Figure 9.8e and Table 9.7 show the increase in the binding rate coefficient, k_2, with an increase in the fractal dimension, D_{f2}, for a dual-fractal analysis. For the data shown in Figure 9.8d and in the 40-160 Hz range, the binding rate coefficient, k_2, is given by:

$$k_2 = (0.1140 \pm 0.1555)[D_{f2}]^{5.442 \pm 4.548} \tag{9.5e}$$

There is scatter in the data. This is reflected in the error in the values of the rate coefficients exhibited. Only three data points are available. The availability of more data points would lead to a more reliable fit. The binding rate coefficient, k_2, is very sensitive to the degree of heterogeneity that exists on the FRET-based calcium sensor surface as noted by the order of dependence between five and five and a half exhibited.

Figure 9.8f and Table 9.9 show the increase in the dissociation rate coefficient, k_d, with an increase in the stimulation frequency, in Hz for a single-fractal analysis. For the data shown in Figure 9.8f, the dissociation rate coefficient, k_d, is given by:

$$k_d = (0.0979 \pm 0.0146)[\mathrm{Hz}]^{1.310 \pm 0.141} \tag{9.5f}$$

The fit is very good. Only three data points are available. The availability of more data points would lead to a more reliable fit. The dissociation rate coefficient, k_d, exhibits an order of dependence between first and one and a half (equal to 1.310) on the stimulation frequency, in Hz.

Figure 9.8g and Tables 9.9 and 9.10 show the decrease in the affinity, K_1 $(= k_1/k_d)$, with an increase in the fractal dimension ratio, D_{f1}/D_{fd}, for a dual-fractal analysis. For the data shown in Figure 9.8g, the affinity, K_1, is given by:

$$K_1 (= k_1/k_d) = (0.845 \pm 0.069)(D_{f1}/D_{fd})^{-0.473 \pm 0.253} \tag{9.5g}$$

Table 9.9 Binding and dissociation rate coefficients for TN-XL and two other GECI (YC2.0 and GCaMP1.6) response kinetics using identical stimulus conditions (AP-frequency 40 Hz, 2.2 s) (Mank et al., 2006).

Analyte	k	k_1	k_2	k_d	k_{d1}	k_{d2}
TN-XL	1.1497 ± 0.0044	na	na	0.2413 ± 0.5659	0.03368 ± 0.5639	0.3943 ± 0.0360
YC 2.0	0.6703 ± 0.0829	na	na	0.2965 ± 0.1259	0.3682 ± 0.1259	0.5697 ± 0.0770
GCAMP13	0.7729 ± 0.1234	na	na	0.3790 ± 0.2237	0.4984 ± 0.0537	0.9046 ± 0.0092

Table 9.10 Fractal dimensions for the binding and the dissociation phase for TN-XL and two other GECI (YC2.0 and GCaMP1.6) response kinetics using identical stimulus conditions (AP-frequency 40 Hz, 2.2 s) (Mank et al., 2006).

Analyte	D_f	D_{f1}	D_{f2}	D_{fd}	D_{fd1}	D_{fd2}
TN-XL	1.6630 ± 0.0044	na	na	0 + 1.4778	0 + 0.4146	0. + 0.3994
YC 2.0	1.1896 ± 0.3912	na	na	1.1322 ± 0.2666	0.254 + 0.2578	2.3436 ± 0.04462
GCAMP13	0.9670 ± 0.1999	na	na	0 + 0.731	0 + 0.3030	2.7930 ± 0.05614

The fit is reasonable. Only three data points are available. The availability of more data points would lead to a more reliable fit. The affinity, $K_1(= k_1/k_d)$, exhibits close to a negative one-half (equal to −0.473) order of dependence on the stimulation frequency (in Hz).

Figure 9.8h and Tables 9.9 and 9.10 show the increase in the affinity, K_2 $(= k_2/k_d)$, with an increase in the fractal dimension ratio, D_{f2}/D_{fd}, for a dual-fractal analysis. For the data shown in Figure 9.8h, the affinity, K_2, is given by:

$$K_2(= k_2/k_d) = (0.5581 \pm 0.1124)(D_{f2}/D_{fd})^{0.595+0.909} \tag{9.5h}$$

The fit is poor. Only three data points are available. The availability of more data points would lead to a more reliable fit. The lack of a good fit is clearly reflected in the figure and in the order of dependence equal to 0.595 + 0.909 exhibited. Clearly, as the Figure 9.8h shows the affinity, K_2, increases with the fractal dimension ratio, D_{f2}/D_{fd}. Note that only the positive error (+0.909) is given in this case to correspond to Figure 9.8h.

Figure 9.9a shows the binding and the dissociation of TN-XL in solution to the sensor chip surface. The stimulus conditions were AP-frequency 40 Hz for 2.2 s. A single-fractal analysis is adequate to describe the binding kinetics. A dual-fractal analysis is required to adequately describe the dissociation kinetics. The values of (a) the binding rate coefficient, k, and the fractal dimension, D_f, for a single-fractal analysis, and (b) the binding rate coefficients, k_1 and k_2, and the fractal dimensions, D_{f1} and D_{f2}, for a dual-fractal analysis are given in Tables 9.9 and 9.10.

Figure 9.9b shows the binding and the dissociation of YC 2.0 in solution to the sensor chip surface. The stimulus conditions were AP-frequency 40 Hz for 2.2 s. Once again, a single-fractal analysis is adequate to describe the binding kinetics. A dual-fractal analysis is required to adequately describe the dissociation kinetics. The values of (a) the binding rate coefficient, k, and the fractal dimension, D_f, for a single-fractal analysis, (b) the dissociation rate coefficient, k_d, and the fractal dimension, D_{fd}, for a single-fractal analysis, and (c) the dissociation rate coefficients, k_{d1} and k_{d2}, and the fractal dimensions, D_{fd1} and D_{fd2}, for dual-fractal analysis are given in Tables 9.9 and 9.10. It is of interest to note that for the dissociation phase as the fractal dimension increases by a factor of 9.23 from a value of D_{fd1} equal to 0.254 to D_{fd2} equal to 2.3436 the dissociation rate coefficient increases by a factor of 1.55 from a value of k_{d1} equal to 0.3682 to k_{d2} equal to 0.5697. An increase in the degree of heterogeneity in the dissociation phase leads to an increase in the dissociation rate coefficient.

Figure 9.9c shows the binding and the dissociation of GCAMP13.0 in solution to the sensor chip surface. The stimulus conditions were AP-frequency 40 Hz for 2.2 s. Once again, a single-fractal analysis is adequate to describe the binding kinetics. A dual-fractal analysis is required to adequately describe the dissociation kinetics. The values of (a) the binding rate coefficient, k, and the fractal dimension, D_f, for a single-fractal analysis, (b) the

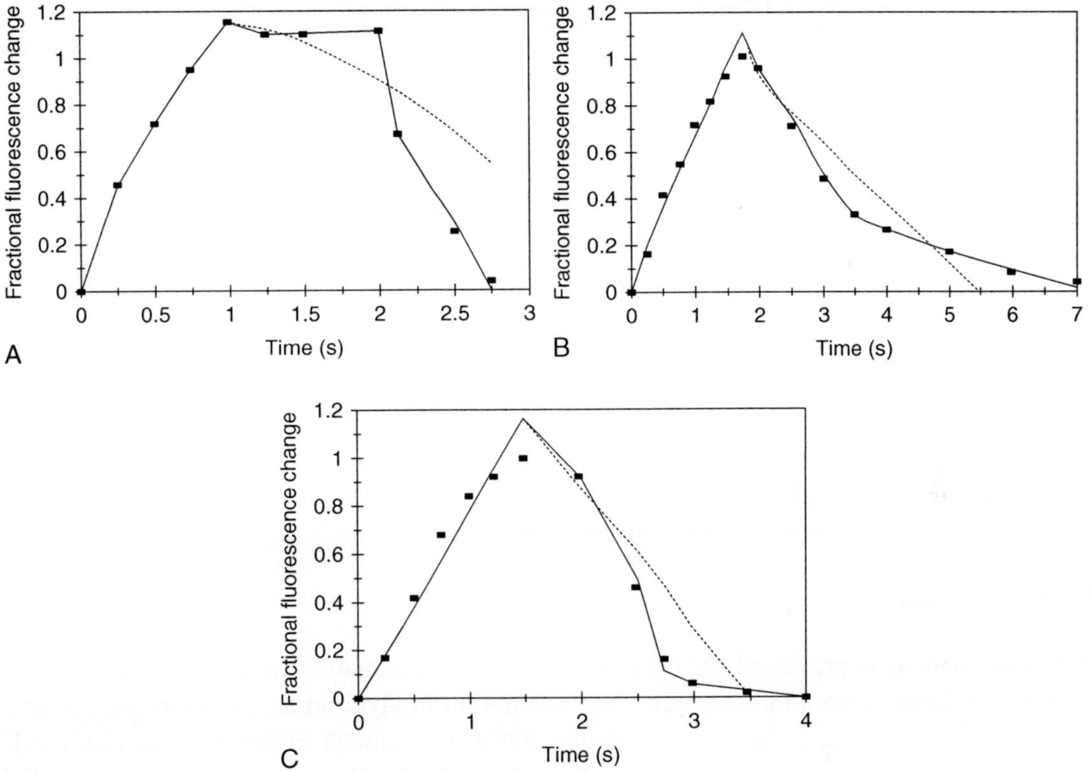

Figure 9.9

Binding and dissociation of TNXL and two other GECI (YC 2.0 and CGAMP1.6). Response kinetics using identical stimulus conditions (AP frequency, 40 Hz, and 2.2 s) (Mank et al., 2006): (a) TNXL (b) YC 2.0 (c) GCAMP1.6. When only a solid line (—) is used then a single-fractal analysis applies. When both a dashed (- - -) and a solid (—) line are used then the dashed line represents a single-fractal analysis and the solid line represents a dual-fractal analysis.

dissociation rate coefficient, k_d, and the fractal dimension, D_{fd}, for a single-fractal analysis, and (c) the dissociation rate coefficients, k_{d1} and k_{d2}, and the fractal dimensions, D_{fd1} and D_{fd2}, for a dual-fractal analysis are given in Tables 9.9 and 9.10. Once again, an increase in the degree of heterogeneity in the dissociation phase leads to an increase in the dissociation rate coefficient.

Figure 9.10 and Tables 9.9 and 9.10 show the increase in the binding rate coefficient, k, with an increase in the fractal dimension, D_f, for a single-fractal analysis. For the data shown in Figure 9.10, the binding rate coefficient, k, is given by:

$$k = (0.7052 \pm 0.1896) D_f^{0.816 \pm 0.615} \tag{9.6}$$

The fit is reasonable. Only three data points are available. The availability of more data points would lead to a more reliable fit. The binding rate coefficient, k, exhibits less than a

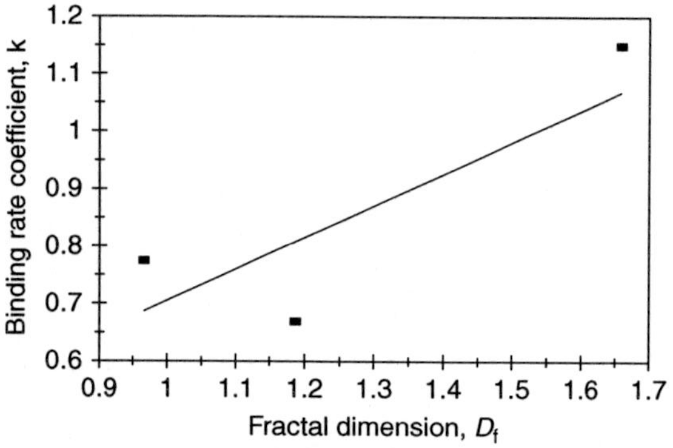

Figure 9.10
Increase in the binding rate coefficient, k with an increase in the fractal dimension, D_f.

first- (equal to 0.816) order of dependence on the fractal dimension, D_f, or the degree of heterogeneity that exists on the sensor chip surface.

9.4 Conclusions

A fractal analysis is presented for the binding and the dissociation (if applicable) of (a) different bradykinin concentrations (in nM) in solution to bradykinin B_2 receptors immobilized on a RWG biosensor surface (Fang et al., 2006), (b) mβCD cholesterol to HeLa cells cultivated on a gold-coated prism surface (Ziblat et al., 2006), and (c) a calcium + FRET-based calcium biosensor employing troponin C. TN-XL fluorescence observed was observed *in vivo* in this case (Mank et al., 2006). Both single-and dual-fractal analysis were used. The dual-fractal analysis was used only when the single-fractal analyses did not provide an adequate fit. This was done using Corel Quattro Pro 8.0 (1999).

The fractal dimension provides a quantitative measure of the degree of heterogeneity present on the biosensor chip surface. The fractal dimension for the binding and the dissociation phase, D_f and D_{fd}, respectively, is not a typical independent variable, such as analyte concentration, that can be directly manipulated. It is estimated from Equations (9.1)–(9.3), and one may consider it as a derived variable.

An increase in the fractal dimension value or the degree of heterogeneity on the biosensor surface leads, in general, to an increase in the binding rate coefficient. For example, for the binding of different bradykinin concentrations in solution in the 8-128 nM range (Fang et al., 2006), and for a dual-fractal analysis the binding rate coefficient, k_2, exhibits a 6.571 order of dependence on the fractal dimension, D_{f2}, or the degree of heterogeneity that exists on the biosensor surface. This indicates that, in this case at least, the binding rate coefficient, k_2, is very sensitive to the fractal dimension or the degree of heterogeneity present on the biosensor chip surface.

Predictive relations are also developed for (a) the binding rate coefficient, k_1 and k_2, for a dual-fractal analysis as a function of the bradykinin concentration (in nM) in solution (Fang et al., 2006), (b) the affinity, K_2 ($= k_2/k_d$) as a function of the ratio of the fractal dimensions, D_{f2}/D_{fd}, present in the binding and the dissociation phases for the different bradykinin concentrations in solution in the 8-128 nM range (Fang et al., 2006), and (c) the binding rate coefficient, k_2, as a function of the fractal dimension, D_{f2}, or the degree of heterogeneity present on the sensor chip surface (Fang et al., 2006). In this case an eighth order of dependence is exhibited, which indicates that the binding rate coefficient, k_2, is extremely sensitive to the degree of heterogeneity that exists ob the biosensor chip surface. (d) The binding rate coefficient, k_2, exhibits a negative 10.783 order of dependence on the fractal dimension, D_{f2}, or the degree of heterogeneity that exists on the biosensor chip surface for the binding of mβCD cholesterol in solution to HeLa cells cultivated on a gold-coated prism (Ziblat et al., 2006), (e) the binding rate coefficient, k_1 and k_2, as a function of the stimulation frequency, in Hz, for the binding and dissociation phases for the calcium-FRET-based calcium biosensor employing troponin C (Mank et al., 2006), and (f) the binding rate coefficient, k_2, as a function of the fractal dimension, D_{f2}, or the degree of heterogeneity that exists on the sensor chip surface.

The three different examples presented in this chapter emphasize that the degree of heterogeneity that exists on the biosensor surface does significantly affect, in general, the rate coefficient and affinity values, and subsequently the kinetics, in general. These are just a few of the representative examples that are available in the literature. More such studies are required to determine whether the binding and the dissociation rate coefficient(s), and subsequently the affinity values are sensitive to their fractal dimensions present on the biosensor surface with regard to these types of reactions.

References

Abraham V, DL Taylor, and JR Haskins, High content screening applied to large-scale cell biology, *Trends in Biotechnology*, **22**, 15–22 (2004).

Blake RA, Cellular screening assays using fluorescence microscopy, *Current Opinions in Pharmacology*, **1**, 533–538 (2001).

Brand AH and N Perrimon, Targeted gene expression as a means of altering cell fates and generating phenotypes, *Development*, **118**, 401–415 (1993).

Corel Quattro Pro 8.0, Corel Corporation, Ottawa, Canada (1997).

Fang Y, AM Ferrie, and G Li, Probing cytoskeleton modulation by optical biosensors, *FEBS Letters*, **579**, 4175–4180 (2005a).

Fang Y, AM Ferrie, NH Fontaine, and PK Yuen, Characteristics of dynamic mass distribution of EGF receptor signaling in living cells measured with label free optical biosensors, *Analytical Chemistry*, **77**, 5720–5725 (2005b).

Fang Y, G Li, and J Peng, Optical biosensor provides insights for bradykinin B_2 receptor signaling in A431 cells, *FEBS Letters*, **579**, 6365–6374 (2005c).

Fang Y, AM Ferrie, NH Fonatine, J Mauro, and J Balakrishnan, Resonant waveguide grating biosensor for living cell biosensing, *Biophysical Journal*, **91**, 1925–1940 (2006).

Griesbeck O, Fluorescent proteins as sensors for cellular functions, *Current Opinions in Neurobiology*, **14**, 636–641 (2004).

Havlin S, Molecular diffusion and reactions. In *The Fractal Approach to Heterogeneous Chemistry: Surfaces, Colloids, Polymers*, Avnir D (Ed.), Wiley, New York, 1989, pp. 251–269.

Hide M, T Tsutsui, H Sato, T Nishimura, K Morimoto, S Yamamoto, and K Yoshizato, Real-time analysis of ligand-induced cell surface and intracellular reactions using mast *cells* using a surface plasmon resonance-based biosensor, *Analytical Biochemistry*, **302**, 28–37 (2002).

Lee CK and SL Lee, Multi-fractal scaling analysis of reactions over fractal surfaces, *Surface Science*, **325**, 294–310 (1995).

Mank M, DF Reiff, N Heim, MW Firedrich, A Borst, and AF Griesbeck, A FRET-based calcium biosensor with fast signal kinetics and high fluorescence change, *Biophysical Journal*, **90**, 1790–1796 (2006).

Miyawaki A, Visualization of the spatial and temporal dynamics of intracellular signaling, *Development Cell*, **4**, 295–305 (2003).

Quinn JG, SO' Neill, A Doyle, C McAtamney, D Diamond, BD MacCraith, and RO' Kennedy, Development and application of surface plasmon resonance-based biosensor for the detection of cell-ligand interactions, *Analytical Biochemistry*, **281**, 135–143 (2000).

Ramsden JJ, SY Li, JE Prenosil, and E Heinzle, Kinetics of adhesion and spreading of animal cells, *Biotechnology & Bioengineering*, **43**, 939–1845 (1994).

Taylor DL, ES Woo, and KA Giuliano, Real-time molecular and cellular analysis: The frontier of drug discovery, *Current Opinion in Pharmacology*, **12**, 75–81 (2001).

Voros J, R Graf, GL Kenausis, A Bruinink, J Mayer, M Textor, E Wintermantel, and ND Spencer, Feasibility study of an online toxicological sensor based on the optical waveguide technique, *Biosensors & Bioelectronics*, **15**, 423–429 (2000).

Zhang J, RE Campbell, AY Ting, and RY Tsien, Creating new fluorescent probes for cell biology, *Nature Reviews in Molecular and Cell Biology*, **3**, 906–918 (2002).

Ziblat R, V Lirstman, D Davidor, and B Aroeti, Infrared surface plasmon resonance: a novel tool for real time sensing of variations in living cells, *Biophysical Journal*, **90**, 2592–2599 (2006).

CHAPTER 10

Detection of Gases on Biosensor Surfaces

Chapter Outline

10.1 Introduction

The detection of different (at least, the harmful ones) gases in the atmosphere is important for environmental awareness and pollution control. For example Cao and Duan (2005) report that the accurate detection of near real-time monitoring of gaseous ammonia has wide applications in environmental science and occupational inspection. These authors draw attention to the presence of ammonia in the atmosphere as a potential hazard for human beings and ecosystems. Tsai et al. (2007) have recently pointed out that there is increasing concern about global climate change and air pollution. Thus, a lot of attention has been paid to using hydrogen as a clean energy source. Hydrogen fuel cells exhibit significant potential for many applications. However, Tsai et al. (2007) warn that hydrogen mixed with oxidants may cause explosions. Thus, there is a critical need to develop a hydrogen sensor for the continuous monitoring of hydrogen gas concentrations in air. Tsai et al. (2007) report that the rapid advances in fabrication and growth technologies have permitted many metal-oxide-semiconductor (MOS) structures to be used as hydrogen sensors including the catalytic metal and different semiconductor materials (Diwedi et al., 2000; Chen et al., 2002; Lu et al., 2003; Medlin et al., 2003).

The above were just a few examples wherein biosensors have been used to detect gases in the atmosphere. In this chapter we use fractal analysis to analyze the binding and dissociation (if applicable) kinetics of (a) the binding of liquid petroleum gas (LPG) to zinc oxide films

Handbook of Biosensors and Biosensor Kinetics. DOI: 10.1016/B978-0-444-53262-6.00010-3

prepared by the spray pyrolysis method onto a glass substrate (Shinde et al., 2007), (b) the binding and dissociation of different NH_3 concentrations in air to a sol-gel derived thin film biosensor (Roy et al., 2005), (c) binding of NH_3 in air to an optical fiber-based evanescent sensor (Cao and Duan, 2005), (d) binding to a nc-Fe_3O_4/Si-NPA (nanocrystal magnetite/silicon nanoporous pillar array) humidity sensor (Wang and Li, 2005), and (e) the binding and dissociation of different methanol concentrations in ppm) to a polyimide thin layer biosensor (Manera et al., 2006).

As discussed elsewhere in the different chapters in the book, fractal analysis is just one possible method to analyze the binding and dissociation kinetics of the different analytes (gases in this case) to the different biosensor surfaces. The distinct advantage of the method is that it provides binding and dissociation (if applicable) rate coefficient values, and the fractal dimension, D_f or the degree of heterogeneity present on the biosensor surface. Furthermore, the analysis attempts to relate these binding and dissociation rate coefficients to the degree of heterogeneity present on the biosensor surface.

10.2 Theory

Havlin (1987) has reviewed and analyzed the diffusion of reactants towards fractal surfaces. The details of the theory and the equations involved for the binding and the dissociation phases for analyte-receptor binding are available (Sadana, 2001) in the literature. The details are not repeated here except that the equations are given to permit an easier reading. These equations have been applied to other biosensor systems (Ramakrishnan and Sadana, 2001; Sadana, 2001, 2005). For most applications, a single- or a dual-fractal analysis is often adequate to describe the binding and the dissociation kinetics. Peculiarities in the values of the binding and the dissociation rate coefficients, as well as in the values of the fractal dimensions with regard to the dilute analyte systems being analyzed will be carefully noted, if applicable.

10.2.1 Single-Fractal Analysis

Binding Rate Coefficient

Havlin (1989) points out that the diffusion of a particle (analyte [Ag]) from a homogeneous solution to a solid surface (e.g., receptor [Ab]-coated surface) on which it reacts to form a product (analyte-receptor complex; Ab·Ag) is given by:

$$(\mathrm{Ab{\cdot}Ag}) \approx \begin{cases} t^{(3-D_{f,\mathrm{bind}})/2} = t^p, & t < t_c \\ t^{1/2}, & t > t_c \end{cases}. \tag{10.1}$$

Here $D_{f,\mathrm{bind}}$ or D_f (used later on in the chapter) is the fractal dimension of the surface during the binding step. t_c is the cross-over value. Havlin (1989) points out that the cross-over value may be determined by $r_c^2 \sim t_c$. Above the characteristic length, r_c, the self-similarity of the surface is lost and the surface may be considered homogeneous. Above time, t_c the surface may be considered homogeneous, since the self-similarity property disappears, and "regular" diffusion is now present. For a homogeneous surface where D_f is equal to 2, and when only diffusional limitations are present, $p = ½$ as it should be. Another way of looking at the $p = ½$ case (where $D_{f,bind}$ is equal to two) is that the analyte in solution views the fractal object, in our case, the receptor-coated biosensor surface, from a "large distance." In essence, in the association process, the diffusion of the analyte from the solution to the receptor surface creates a depletion layer of width $(Đt)^{½}$ where $Đ$ is the diffusion constant. This gives rise to the fractal power law, $(\text{Analyte} \cdot \text{Receptor}) \sim t^{(3-D_{f,\text{bind}})/2}$. For the present analysis, t_c is arbitrarily chosen and we assume that the value of t_c is not reached. One may consider the approach as an intermediate "heuristic" approach that may be used in the future to develop an autonomous (and not time-dependent) model for diffusion-controlled kinetics.

Dissociation Rate Coefficient

The diffusion of the dissociated particle (receptor [Ab] or analyte [Ag]) from the solid surface (e.g., analyte [Ag]-receptor [Ab]) complex coated surface) into solution may be given, as a first approximation by:

$$(\text{Ab}\cdot\text{Ag}) \approx -t^{(3-D_{f,\text{diss}})/2} = -t^p, \quad t > t_{\text{diss}} \tag{10.2}$$

Here $D_{f,diss}$ is the fractal dimension of the surface for the dissociation step. This corresponds to the highest concentration of the analyte-receptor complex on the surface.

Henceforth, its concentration only decreases. The dissociation kinetics may be analyzed in a manner "similar" to the binding kinetics.

10.2.2 Dual-Fractal Analysis

Binding Rate Coefficient

Sometimes, the binding curve exhibits complexities and two parameters (k, D_f) are not sufficient to adequately describe the binding kinetics. This is further corroborated by low values of the r^2 factor (goodness-of-fit). In that case, one resorts to a dual-fractal analysis (four parameters; k_1, k_2, D_{f1}, and D_{f2}) to adequately describe the binding kinetics. The single-fractal analysis presented above is thus extended to include two fractal dimensions. At present, the time ($t = t_1$) at which the "first" fractal dimension "changes" to the "second" fractal dimension is arbitrary and empirical. For the most part, it is dictated by the data analyzed and experience gained by handling a single-fractal analysis. A smoother curve is obtained in the "transition" region, if care is taken to select the correct number of points for the two regions. In this case, the product (antibody-antigen; or analyte-receptor complex, Ab·Ag or analyte·receptor) is given by:

$$(\text{Ab·Ag}) \approx \begin{cases} t^{(3-D_{\text{f1,bind}})/2} = t^{p1}, & t < t_1 \\ t^{(3-D_{\text{f2,bind}})/2} = t^{p2}, & t_1 < t < t_2 = t_c \\ t^{1/2}, \quad t > t_c \end{cases} \tag{10.3}$$

In some cases, as mentioned above, a triple-fractal analysis with six parameters (k_1, k_2, k_3, D_{f1}, D_{f2}, and D_{f3}) may be required to adequately model the binding kinetics. This is when the binding curve exhibits convolutions and complexities in its shape due perhaps to the very dilute nature of the analyte (in some of the cases to be presented) or for some other reasons. Also, in some cases, a dual-fractal analysis may be required to describe the dissociation kinetics.

10.3 Results

A fractal analysis is presented for (a) the binding of LPG to zinc oxide films prepared by the spray pyrolysis method onto a glass substrate (Shinde et al., 2007) (b) the binding and dissociation of different NH_3 concentrations in air to a sol-gel derived thin film (Roy et al., 2005), (c) binding of NH_3 to an optical fiber-based evanescent sensor (Cao and Duan, 2005), (d) binding to a nc-Fe_3O_4/Si-NPA humidity sensor (Wang and Li, 2005), and (e) the binding and dissociation of different methanol concentrations (in ppm) to a polyimide thin layer biosensor (Manera et al., 2006).

Alternative expressions for fitting the data are available that include saturation, first-order reaction, and no diffusion limitations, but these expressions are apparently deficient in describing the heterogeneity that inherently exists on the surface. One might justifiably argue that the appropriate modeling may be achieved by using a Langmuirian or other approach. The Langmuirian approach may be used to model the data presented if one assumes the presence of discrete classes of sites (e.g., double exponential analysis as compared with a single-fractal analysis). Lee and Lee (1995) report that the fractal approach has been applied to surface science, for example, adsorption and reaction processes. These authors point out that the fractal approach provides a convenient means to represent the different structures and the morphology at the reaction surface. They also draw attention to using the fractal approach to develop optimal structures and as a predictive approach. Another advantage of the fractal technique is that the analyte-receptor association (as well as the dissociation reaction) is a complex reaction, and the fractal analysis via the fractal dimension and the rate coefficient provide a useful lumped parameter(s) analysis of the diffusion-limited reaction occurring on a heterogeneous surface.

In a classical situation, to demonstrate fractality, one should make a log-log plot, and one should definitely have a large amount of data. It may be useful to compare the fit to some other forms, such as exponential, or one involving saturation, etc. At present, no independent proof or physical evidence of fractals in the examples is presented. It is a convenient means (since it is a lumped parameter) to make the degree of heterogeneity that exists on the surface more quantitative. Thus, there is some arbitrariness in the fractal model to be presented. The fractal approach provides additional information about interactions that may not be obtained by conventional analysis of biosensor data.

There is no nonselective adsorption of the analyte. The present system (environmental pollutants in the aqueous or the gas phase) being analyzed may be typically very dilute. Non-selective adsorption would skew the results obtained very significantly. In these types of systems, it is imperative to minimize this nonselective adsorption. It is also recognized that, in some cases, this nonselective adsorption may not be a significant component of the adsorbed material and that this rate of association, which is of a temporal nature, would depend on surface availability. If the nonselective adsorption were to be accommodated into the model, there would be an increase in the heterogeneity on the surface, as, by its very nature, non-specific adsorption is more homogeneous than specific adsorption. This would lead to higher fractal dimension values since the fractal dimension is a direct measure of the degree of heterogeneity that exists on the surface.

Shinde et al. (2007) recently investigated the use of ZnO thin films prepared by the spray pyrolysis method for the sensing of LPG. These authors deposited nanocrystalline ZnO films onto glass substrates by the spray pyrolysis of zinc nitrate solution, and used this as a LPG gas sensor. These authors further explain that ZnO has replaced the more toxic and expensive materials such as CdS, TIO_2, GaN, and SiO_2 for applications in gas sensors (Rao and Rao, 1999). Ismail et al. (2001) have shown that owing to its resistivity control in the range 10^{-3} to 10^5 ohm cm ZnO is particularly suitable for gas sensors. Besides, it exhibits high electrochemical stability, absence of toxicity, and is readily available in nature. Shinde et al. (2007) point out that thin films are particularly suited for gas sensors, as the gas sensing properties of metal oxides (a) may be related to the material surface, and (b) gases are readily adsorbed and react with the thin film biosensor surface (Liu et al., 1997). Furthermore, Zhu et al. (1993) and Chai et al. (1995) report that thin film gas sensing materials have good gas sensitivity and selectivity. Patil (1999) has demonstrated the versatility of using the spray pyrolysis method for the deposition of metal oxides.

Shinde et al. (2007) report that the gas sensing properties of oxide materials may be related to surface morphology and are grain size dependent. This surface morphology is what we will attempt to characterize and make quantitative using the fractal analysis method. As indicated in the different chapters throughout the book, the fractal dimension (a) provides a quantitative measure of the degree of heterogeneity on the sensing surface, and (b) an increase in the fractal dimension on the sensing surface indicates an increase in the degree of heterogeneity on the sensing surface.

Shinde et al. (2007) used scanning electron micrographs to analyze the surface morphology studies of the ZnO films sprayed on glass substrates. These authors noted that the grain size increased with an increase in the sprayed precursor solution [S_1 (0.1 M), to S_2 (0.15 M), to S_3 (0.25 M) zinc nitrate solution]. This is consistent with the results obtained by Korotcenko et al. (2001).

Figure 10.1a shows the binding and dissociation of 0.2 volume percent LPG in the gas phase to sample S_1 (0.1M zinc nitrate solution used in the spray pyrolysis method) (Shinde et al., 2007). A dual-fractal analysis is required to adequately describe the binding and the dissociation kinetics. Tables 10.1 and 10.2 show (a) the binding rate coefficient, *k*, and the fractal

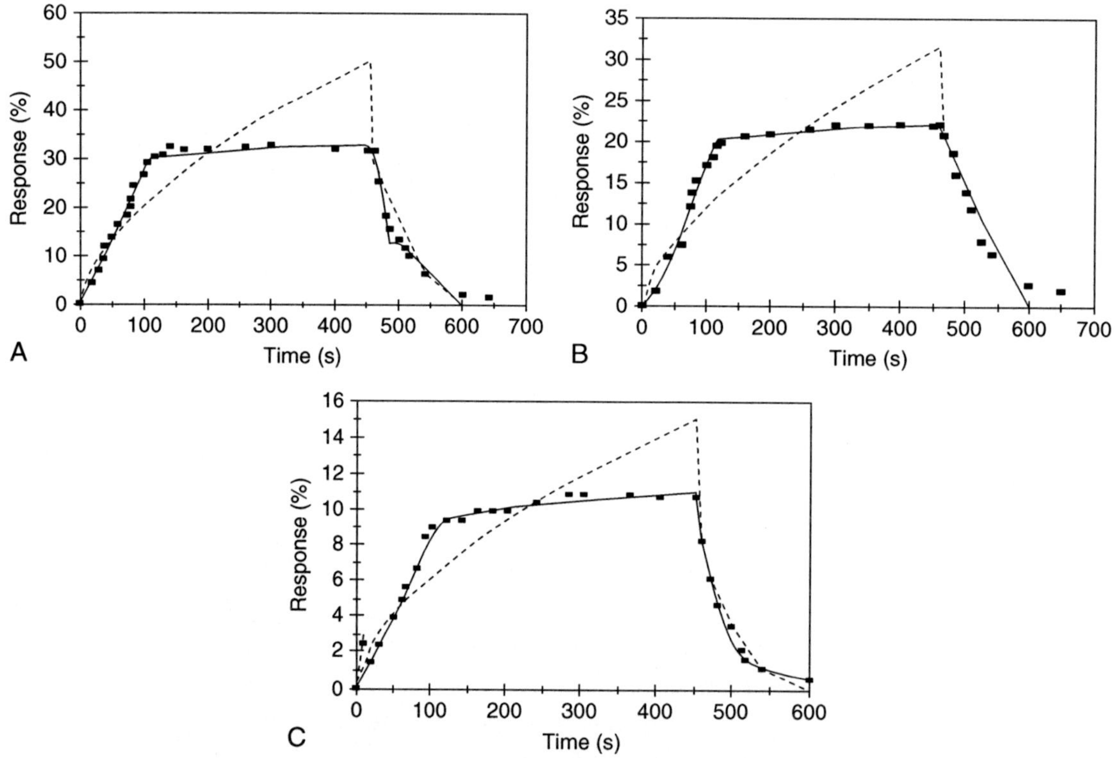

Figure 10.1

Binding of samples S_1, S_2, and S_3 (LPG; liquid petroleum gas) to zinc oxide (ZnO) films prepared by the spray pyrolysis method onto a glass substrate (Shinde et al., 2007). Influence of zinc oxide concentration (0.1-0.3 M) solution: (a) 0.1M, sample S_1, (b) 0.15M, sample S_2, (c) 0.3M, sample S_3. When only a solid line is used (—) then a single-fractal applies. When both a dotted (- - -) and a solid line (—) is used, then the dotted line is for a single-fractal analysis, and the solid line is for a dual-fractal analysis. In this case, the dual-fractal analysis provides the better fit.

Table 10.1: Binding and dissociation rate coefficients for liquid petroleum gas (LPG) on ZnO films prepared by the spray pyrolysis method onto a glass substrate. Influence of zinc nitrate concentration (0.1-0.2 M) solution (Shinde et al., 2007).

Zinc nitrate concentration (M)	k	k_1	k_2	k_d	k_{d1}	k_{d2}
0.1, Sample S1	1.284 ± 0.399	0.1825 ± 0.0166	19.334 ± 0.801	0.2685 ± 0.1731	0.00633 ± 0.00124	3.739 ± 0.163
0.15, Sample S2	0.6835 ± 0.292	0.0351 ± 0.0041	14.066 ± 0.146	0.3514 ± 0.0615	na	na
0.2, Sample S3	0.4449 ± 0.1245	0.06332 ± 0.00393	5.1146 ± 0.1115	0.8868 ± 0.1231	0.6243 ± 0.0590	4.859 ± 0.063

Table 10.2: Fractal dimensions for the binding and the dissociation phase for liquid petroleum gas (LPG) on ZnO films prepared by the spray pyrolysis method onto a glass substrate. Influence of zinc nitrate concentration (0.1-0.2 M) solution (Shinde et al., 2007).

Zinc nitrate concentration (M)	D_f	D_{f1}	D_{f2}	D_{fd}	D_{fd1}	D_{fd2}
0.1, Sample S1	1.7954 ± 0.1444	0.8196 ± 0.1274	2.8210 ± 0.0480	0.9864 ± 0.3674	0 + 0.3764	2.1732 ± 0.0709
0.15, Sample S2	1.7478 ± 0.1969	0.308 ± 0.141	2.849 ± 0.0154	1.3234 ± 0.1073	na	na
0.2, Sample S3	1.8404 ± 0.1312	0.8628 ± 0.0731	2.7406 ± 0.0358	1.9492 ± 0.1124	1.7272 ± 1.2726	2.7066 ± 0.0468

dimension, D_f, for a single-fractal analysis, (b) the dissociation rate coefficient, k_d, and the fractal dimension for the dissociation phase for a single-fractal analysis, (c) the binding rate coefficients, k_1 and k_2, and the fractal dimensions, D_{f1} and D_{f2}, for a dual-fractal analysis, and (d) the dissociation rate coefficients, k_{d1} and k_{d2}, and the fractal dimensions, D_{fd1} and D_{fd2}, for the dissociation phase for a dual-fractal analysis.

The values of the binding and the dissociation rate coefficients and the fractal dimensions for the binding phase presented in Tables 10.1 and 10.2 were obtained from a regression analysis using Corel Quattro Pro 8.0 (1997) to model the data using Equations. (10.1)–(10.3). wherein $(\mathrm{Ab{\cdot}Ag}) = kt^{(3-D_f)}$ for a single-fractal analysis, and $(\mathrm{Ab{\cdot}Ag}) = k_1t^{(3-D_{f1})}$ for time, $t < t_1$, and $(\mathrm{Ab{\cdot}Ag}) = k_2t^{(3-D_{f2})}$ for time, $t = t_1 < t < t_2 = t_c$ for a dual-fractal analysis. The binding and the dissociation rate coefficients presented in Table 10.1 are within 95% limits. For example, for the binding (and dissociation) of LPG to sample S_1 the binding rate coefficient, k_1, is equal to 0.1825 ± 0.0166. The 95% confidence limit indicates that the k_1 value lies between 0.1659 and 0.1991. This indicates that the values are precise and significant.

An increase in the fractal dimension for a dual-fractal analysis by a factor of 3.44 from a value of D_{f1} equal to 0.8196 to D_{f2} equal to 2.8210 leads to an increase in the binding rate coefficient by a factor of 105.93 from a value of k_1 equal to 0.1825 to k_2 equal to 19.334. Note that changes in the fractal dimension or the degree of heterogeneity on the ZnO-glass substrate and in the binding rate coefficient are in the same direction. In other words, and as indicated elsewhere in the different chapters in this book, an increase in the degree of heterogeneity on the sensing surface leads to an increase in the binding rate coefficient.

Figure 10.1b shows the binding and dissociation of 0.15 volume % LPG in the gas phase to sample S_2 (0.15 M zinc nitrate solution used in the spray pyrolysis method) (Shinde et al., 2007). A dual-fractal analysis is required to adequately describe the binding kinetics. A single-fractal analysis is adequate to describe the dissociation kinetics. Tables 10.1 and 10.2 show (a) the binding rate coefficient, k, and the fractal dimension, D_f, for a single-fractal analysis, (b) the dissociation rate coefficient, k_d, and the fractal dimension, D_{fd} for a single fractal analysis, and (c) the binding rate coefficients, k_1 and k_2, and the fractal dimensions, D_{f1} and D_{f2}, for a dual-fractal analysis.

Once again, for the binding phase an increase in the fractal dimension by a factor of 3.44 from a value of D_{f1} equal to 0.308 to D_{f2} equal to 2.849 leads to an increase in the binding rate coefficient by a factor of 400.7 from a value of k_1 equal to 0.0351 to k_2 equal to 14.066. An increase in the degree of heterogeneity on the biosensor surface leads, once again, to an increase in the binding rate coefficient.

Figure 10.1c shows the binding and dissociation of 0.2 volume percent LPG in the gas phase to sample S_3 (0.2 M zinc nitrate solution used in the spray pyrolysis method) (Shinde et al., 2007). A dual-fractal analysis is required to adequately describe the binding and the dissociation kinetics. Tables 10.1 and 10.2 show (a) the binding rate coefficient, k, and the fractal dimension, D_f, for a single-fractal analysis, (b) the dissociation rate coefficient, k_d, and the fractal dimension for the dissociation phase, D_{fd} for a single-fractal analysis, (c) the binding rate coefficients, k_1 and k_2, and the fractal dimensions, D_{f1} and D_{f2}, for a dual-fractal analysis, and (d) the dissociation rate coefficients, k_{d1} and k_{d2}, and the fractal dimensions, D_{fd1} and D_{fd2}, for the dissociation phase for a dual-fractal analysis.

An increase in the fractal dimension for a dual-fractal analysis by a factor of 3.18 from a value of D_{f1} equal to 0.8628 to D_{f2} equal to 2.7406 leads to an increase in the binding rate coefficient by a factor of 80.77 from a value of k_1 equal to 0.06332 to k_2 equal to 5.1146. Note that changes in the fractal dimension or the degree of heterogeneity on the ZnO-glass substrate and in the binding rate coefficient are, once again, in the same direction.

In this case for the dissociation phase and for a dual-fractal analysis an increase in the fractal dimension for dissociation by a factor of 1.57 from a value of D_{fd1} equal to 1.7272 to D_{fd2} equal to 2.7066, the dissociation rate coefficient increases by a factor of 7.78 from a value of k_{d1} equal to 0.6243 to k_{d2} equal to 4.859. Once again, an increase in the degree of heterogeneity on the biosensor surface in the dissociation phase leads to an increase in the dissociation rate coefficient.

Figure 10.2a and Tables 10.1 and 10.2 show the increase in the binding rate coefficient, k_2, with an increase in the fractal dimension, D_{f2}, for a dual-fractal analysis. For the data shown in Figure 10.2a, the binding rate coefficient, k_2, is given by:

$$k_2 = (2.8 \times 10^{-13} \pm 1.5 \times 10^{-13})D_{f2}^{30.4} \tag{10.4a}$$

The fit is reasonable. Only three data points are available. The availability of more data points would lead to a more reliable fit. The binding rate coefficient, k_2, is extremely sensitive to the fractal dimension, D_{f2}, that exists on the biosensor surface as noted by the 30.4 order of dependence exhibited.

Figure 10.2b and Tables 10.1 and 10.2 show the increase in the ratio of the binding rate coefficients, k_2/k_1, with an increase in the fractal dimension ratio, D_{f2}/D_{f1}, for a dual-fractal analysis. For the data shown in Figure 10.2b, the ratio of the binding rate coefficients, k_2/k_1, is given by:

$$k_2/k_1 = (16.43 \pm 1.78)(D_{f2}/D_{f1})^{1.438 \pm 0.122} \tag{10.4b}$$

The fit is very good. Once again, only three data points are available. The availability of more data points would lead to a more reliable fit. The ratio of the binding rate coefficients, k_2/k_1, exhibits close to a one and a half order of dependence on the ratio of fractal dimensions, D_{f2}/D_{f1}.

Figure 10.2c and Tables 10.1 and 10.2 show the increase in the ratio, k/k_d, k_1/k_{d1}, and k_2/k_{d2} with an increase in the fractal dimension ratio, D_f/D_{fd}, D_{f1}/D_{fd1}, and D_{f2}/D_{fd2}.

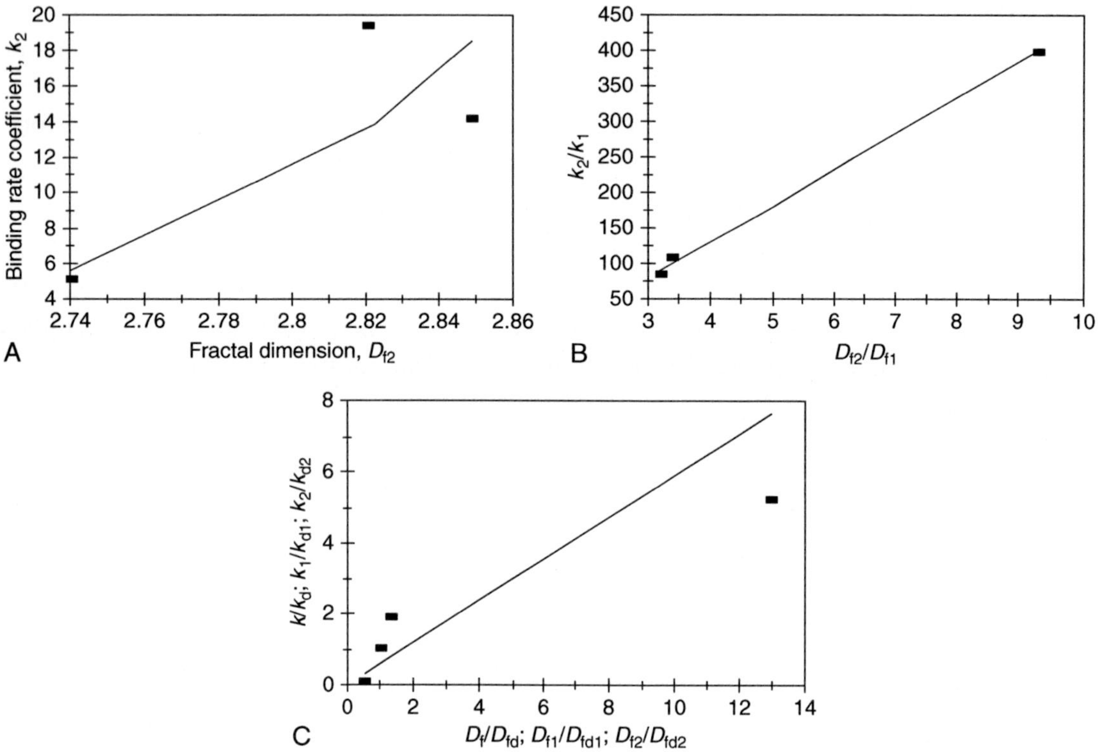

Figure 10.2

(a) Increase in the binding rate coefficient, k_2, with an increase in the fractal dimension, D_{f2}. (b) Increase in the binding rate coefficient ratio, k_2/k_1 with an increase in D_{f2}/D_{f1}. (c) Increase in the ratio, k/k_d, k_1/k_{d1}, and k_2/k_{d2} with an increase in the fractal dimension ratio, D_f/D_{fd}, D_{f1}/D_{fd1}, and D_{f2}/D_{fd2}.

For the data shown in Figure 10.2c, the ratio of the binding rate coefficients, k_2/k_1, is given by:

$$k/k_d, k_1/k_{d1}, \text{and } k_2/k_{d2} = (0.5943 + 1.203)\,(D_f/D_{fd}, D_{f1}/D_{fd1}, \text{and } D_{f2}/D_{fd2}D_{f2}/D_{f1})^{0.997\pm0.453} \quad (10.4c)$$

The fit is reasonable. Only four data points are available. There is scatter in the data, and this is reflected in the error in the ratio of the binding and the dissociation rate coefficient presented. The ratio of three different coefficient values are also presented, and this too contributes to the error. This is done because the number of points are so few. In any case, the three different ratios of the binding and their corresponding dissociation rate coefficients presented exhibit very close to a first (equal to 0.997) order of dependence on the ratio of fractal dimensions exhibited in the binding and in the dissociation phases, respectively.

Roy et al. (2005) recently developed a sol-gel derived thin film biosensor to detect ppm concentrations of NH_3 in air. They analyzed the influence of presintering temperature on gas sensitivity. These authors report that ammonia gas sensors have been used in chemical plants, food technology, fertilizers, in environmental pollution monitoring, and in food technology. Matsuguchi et al. (2002) have used thin films of polyaniline-insulating matrix polymer blend to detect ammonia near room temperature. Roy et al. (2005) point out that polymer-based materials may degrade, and often suffer a limited cycle of operation. They further explain that inorganic materials are better for gas sensing. Sen et al. (2003) have used an elemental thin film of tellurium that was prepared by the thermal evaporation technique for the detection of ammonia at concentrations lower than 100 ppm.

Roy et al. (2005) have recently used the ammonia-sensing behavior of barium strontium titanate (BST) films to detect ammonia. The BST films were deposited by the sol-gel spin coating technique, and these thin films showed an increase in resistance when exposed to ammonia gas. They further state that the sensitivity variation is from 20% to 60%. The lowest detection limit was around 160 ppm.

Figure 10.3a shows the binding and dissociation of NH_3 in air to the sol-gel derived thin film sensor where the presintering was performed at 873 °K (Roy et al., 2005). A single-fractal analysis is adequate to describe the binding and the dissociation kinetics. The values of the rate coefficient and the fractal dimension for the binding and the dissociation phases are given in Table 10.3.

Figure 10.3b shows the binding and dissociation of NH_3 in air to the sol-gel derived thin film sensor where presintering was performed at 773 °K. A single-fractal analysis is adequate to describe the binding and the dissociation kinetics. The values of the rate coefficient and the fractal dimension for the binding and the dissociation phases are given in Table 10.3. In this case, the affinity, K $(=k/k_d)$, value is 4.57.

Figure 10.3c shows the binding and dissociation of NH_3 in air to the sol-gel derived thin film sensor where presintering was performed at 673 °K. A single-fractal analysis is adequate to describe the binding and the dissociation kinetics. The values of the rate coefficient and the fractal dimension for the binding and dissociation phases are given in Table 10.3. In this case, the affinity, K $(=k/k_d)$, value is 0.420. A decrease in temperature leads to a decrease in the affinity, K, value in the range 673-873 °K temperature.

Figure 10.3d and Table 10.3 show the increase in the binding rate coefficient, k, with an increase in the fractal dimension, D_f. For the data shown in Figure 10.3d, the binding rate coefficient, k, is given by:

$$k = (0.1931 \pm 0.0250)D_f^{5.008 \pm 0.3361} \quad (10.5a)$$

The fit is good. Only three data points are available. The availability of more data points would lead to a more reliable fit. The binding rate coefficient, k, for a single-fractal analysis is very sensitive to the fractal dimension, D_f, or the degree of heterogeneity that exists on the biosensor surface as noted by the fifth (equal to 5.008) order of dependence exhibited.

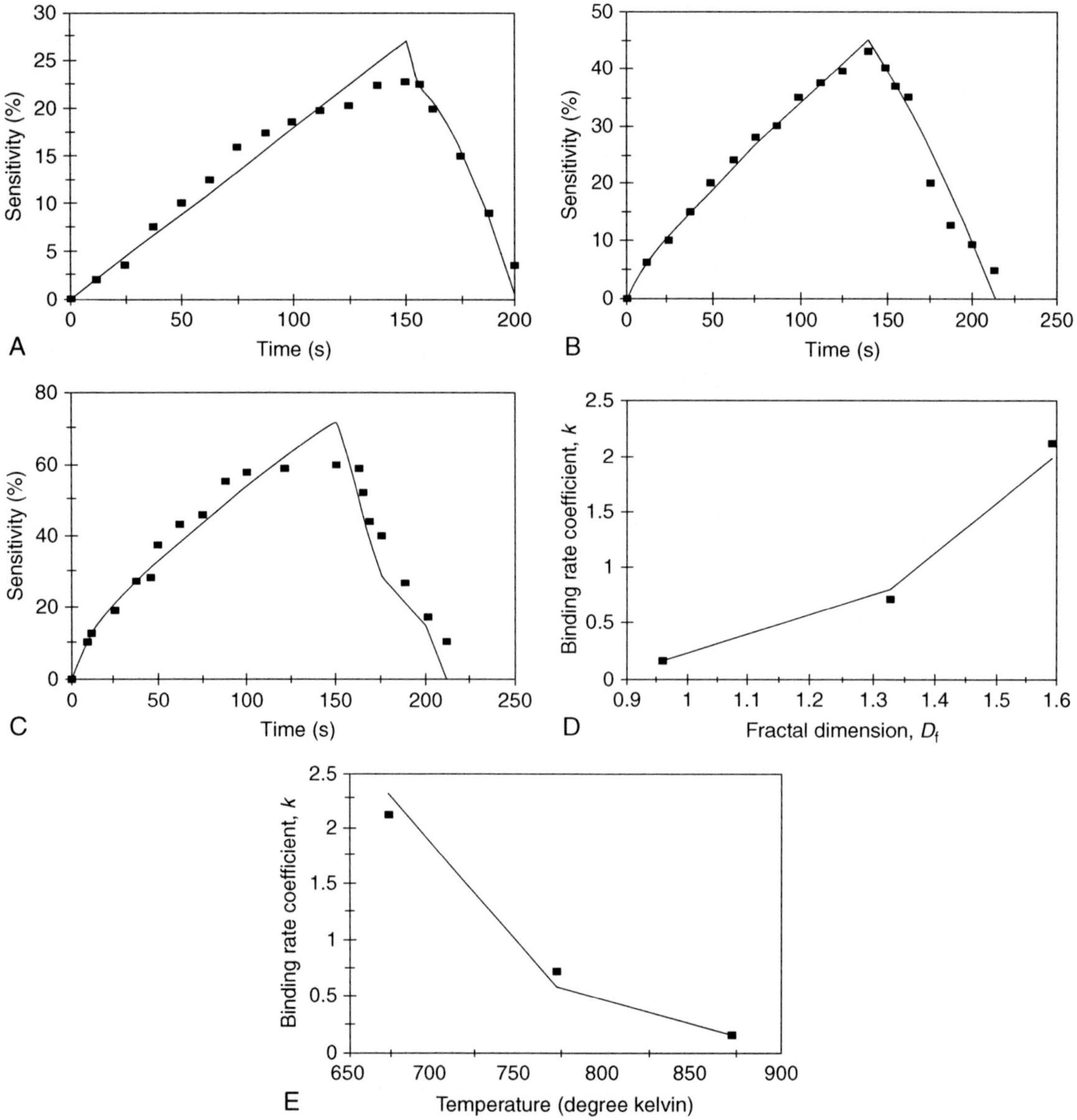

Figure 10.3

Binding and dissociation of NH_3 concentration in air to a sol-gel derived thin film (Roy et al., 2005). Influence of temperature in °K: (a) 873 (b) 773 (c) 673 (d) Increase in the binding rate coefficient, k with an increase in the fractal dimension, D_f (e) Decrease in the binding rate coefficient, k with an increase in the temperature (in °K).

Figure 10.3e and Table 10.3 show the increase in the binding rate coefficient, *k*, with an increase in the temperature range 673-873 °K. For the data shown in Figure 10.3e, the binding rate coefficient, *k*, is given by:

$$k = (1.2 \times 10^{28} \pm 0.3 \times 10^{28})(\text{T, in °K})^{-9.80 \pm 1.26} \qquad (10.5b)$$

Table 10.3: Binding and dissociation rate coefficients and fractal dimensions for the binding and the dissociation phases for NH_3 in air to a sol-gel derived thin film biosensor (Roy et al., 2005). Influence of presintering temperature.

Analyte in air/Presintering	Receptor on surface temperature (°K)	k	k_d	D_f	D_{fd}	$K=k/k_d$
ppm NH_3/sol-gel derived thin film biosensor	873	0.1638 ± 0.0245	0.0303 ± 0.0088	0.9608 ± 0.111	0 + 0.2962	5.40
ppm NH_3/sol-gel derived thin film biosensor	773	0.7220 ± 0.0305	0.1597 ± 0.0353	1.3270 ± 0.035	0.3832 ± 0.2008	4.5
ppm NH_3/sol-gel derived thin film biosensor	673	2.124 ± 0.0621	5.0561 ± 0.649	1.594 ± 0.0654	1.889 ± 0.29	0.42

The fit is good. Only three data points are available. The availability of more data points would lead to a more reliable fit. The binding rate coefficient, *k*, for a single-fractal analysis is very sensitive to the fractal presintering temperature as noted by the negative 9.8 order of dependence exhibited.

At this point, the influence of presintering temperature on the binding rate coefficient, *k*, in the temperature range 673-873 °K is not clear. Perhaps, there is an Arrhenius like dependence of the form, $k = k_0 \exp(-A/RT)$, where A and k_0 are the coefficients as in the Arrhenius expression, R is the universal gas constant, and T is the temperature in degree Kelvin. Thus, the data in Figure 10.3e are plotted again in Figure 10.4a with the reciprocal temperature in $°K^{-1}$ as the independent variable. For the data shown in Figure 10.4a, the binding rate coefficient, *k*, is given by:

$$k = (1.3 \times 10^{28} \pm 0.4 \times 10^{28})(T^{-1}, \text{ in } °K^{-1})^{9.81 \pm 1.26} \quad (10.6a)$$

Note that the coefficients in Equation (10.5b) and in Equation (10.6a) are very close to each other, as expected. Once again, the fit is good.

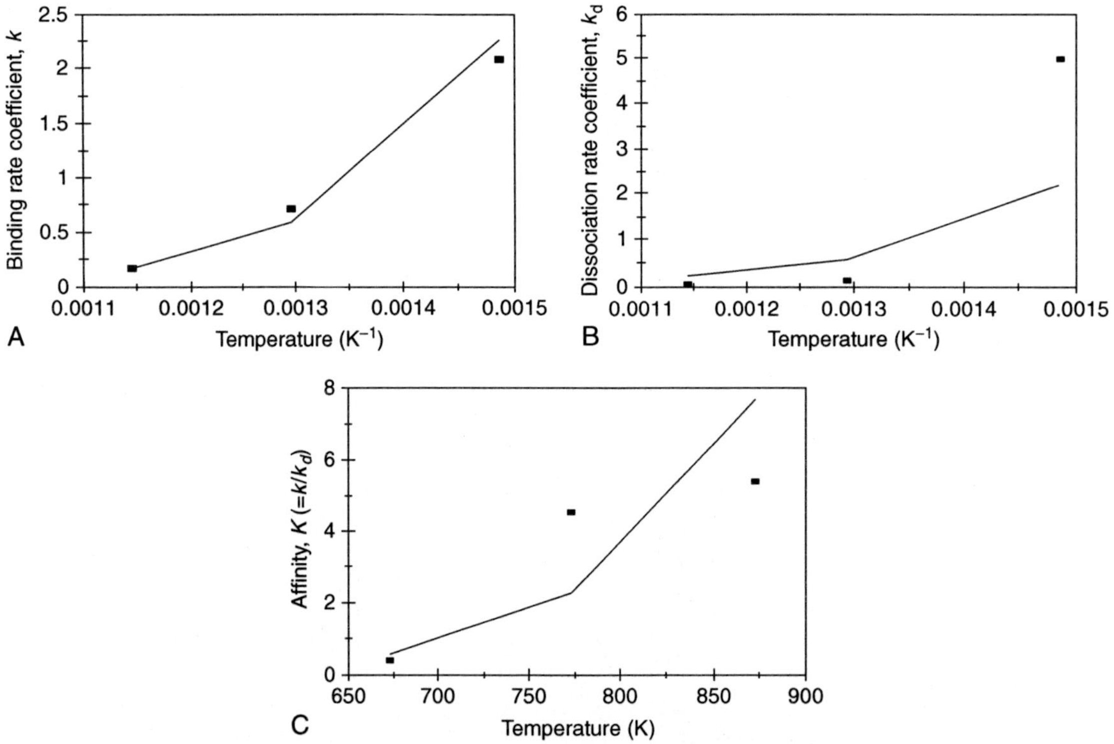

Figure 10.4
(a) Increase in the binding rate coefficient, k, with an increase in the reciprocal temperature, T^{-1}; (b) increase in the dissociation rate coefficient, k_d, with an increase in the reciprocal temperature, T^{-1}; (c) increase in the affinity, K $(=k/k_d)$, with an increase in the temperature, T.

Figure 10.4b and Table 10.3 show the increase in the dissociation rate coefficient, k_d, with an increase in the reciprocal temperature, T in $°K^{-1}$. For the data shown in Figure 10.4b, the dissociation rate coefficient, k_d, is given by:

$$k_d = (3.8 \times 10^{56} \pm 3.1 \times 10^{56})(\mathrm{T}^{-1}, \mathrm{in}\, °\mathrm{K}^{-1})^{19.79 \pm 3.31} \tag{10.6b}$$

There is scatter in the data. This is reflected in the fit and in the error in the binding rate coefficient, k, value. The dissociation rate coefficient, k_d, is very sensitive to the reciprocal temperature, T^{-1}, as reflected in the 19.79 order of dependence exhibited. In other words, both the binding rate coefficient, k, as well as the dissociation rate coefficient, k_d, are both extremely sensitive to the presintering temperature. Note that the dissociation rate coefficient, k_d, exhibits more that twice the order of dependence (equal to 19.79) on temperature, in °K than the binding rate coefficient, k, in the 673-873 °K temperature range.

Figure 10.4c and Table 10.3 show the increase in the affinity, K $(=k/k_d)$, with an increase in the reciprocal temperature, T in $°K^{-1}$. For the data shown in Figure 10.4c, the affinity, K, is given by:

$$K(= k/k_d) = (3.3 \times 10^{-29} \pm 4.3 \times 10^{-29})(\mathrm{T}, \mathrm{in} °\mathrm{K})^{9.99 \pm 4.53} \tag{10.6c}$$

There is scatter in the data. This is reflected in the fit and in the error in the affinity value presented. Only the positive value of the affinity, K, is presented since the affinity cannot have a negative value. The affinity, K, is extremely sensitive to the temperature, T, as reflected in the close to the tenth (equal to 9.99) order of dependence exhibited.

Figure 10.5a and Table 10.3 show the decrease in the fractal dimension, D_f, for a single-fractal analysis with an increase in the temperature, T in °K in the 673-873 °K temperature range. For the data shown in Figure 10.5a, the fractal dimension, D_f, is given by:

$$D_f = (474{,}207.8 \pm 34{,}557.5)(\mathrm{T}\, \mathrm{in}\, °\mathrm{K})^{-1.931 \pm 0.382} \tag{10.6d}$$

The fit is good. Only three data points are available. The availability of more data points would lead to a better fit. The fractal dimension, D_f, or the degree of heterogeneity on the biosensor chip surface is sensitive to presintering temperature as it exhibits close to a second (equal to -1.93) order of dependence. The binding rate coefficient, k, exhibits a decrease with an increase in temperature in the 673-873 °K range. For a 200 °K change in the presintering temperature from 873 to 673 °K, the binding rate coefficient, k, exhibits a 65.9% increase as it goes from a value of 0.9608 to 1.594, respectively. Note that the fractal dimension, D_f, is based on a log scale, and thus this 65.9% change on changing the presintering temperature from 873 to 673 °K represents a very significant change in the degree of heterogeneity on the biosensor surface.

Figure 10.5b and Table 10.3 show the decrease in the fractal dimension in the dissociation phase, D_{fd}, with an increase in the temperature, T in °K in the 673-873 °K temperature range. For the data shown in Figure 10.5b, the fractal dimension, D_{fd}, is given by:

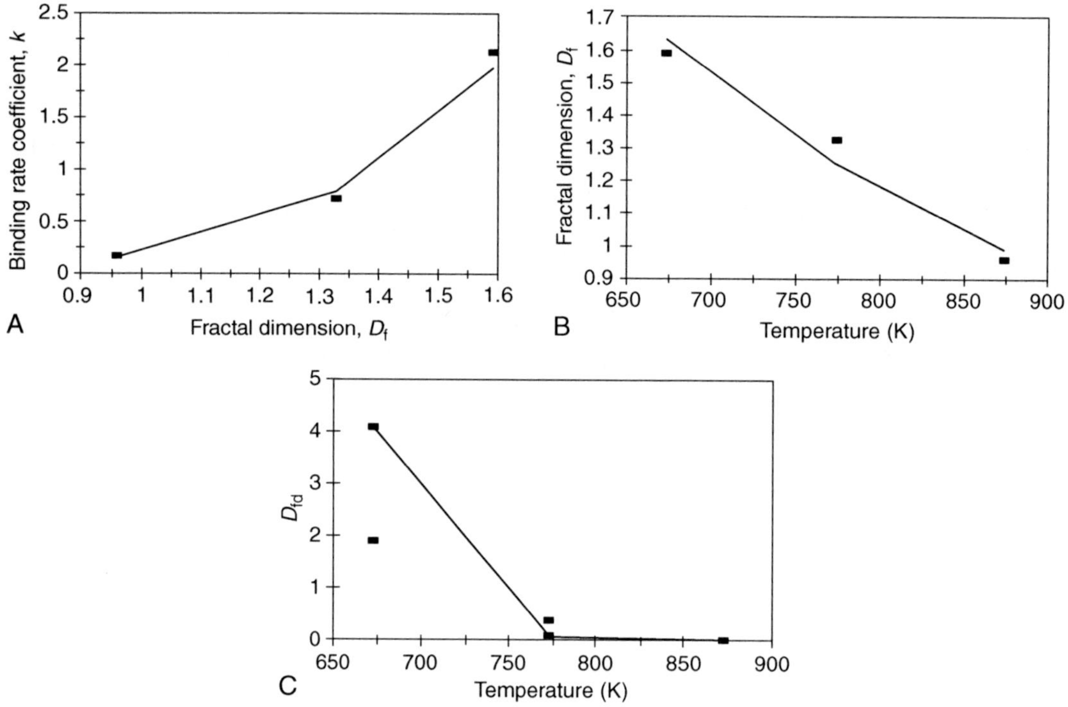

Figure 10.5

(a) Increase in the binding rate coefficient, k, with an increase in the fractal dimension, D_f. (b) Decrease in the fractal dimension, D_f, with an increase in the temperature, T (in °K). (c) Decrease in the fractal dimension for dissociation, D_{fd}, with an increase in the temperature, T (in °K).

$$D_{fd} = (2.9 \times 10^{81} \pm 2.1 \times 10^{81})(\mathrm{T\,in\,^\circ K})^{-28.59} \qquad (10.6e)$$

The fit is reasonable. Only five data points are available. The availability of more data points would lead to a more reliable fit. The fractal dimension in the dissociation phase, D_{fd}, is extremely sensitive to the temperature as it exhibits a negative 28.59 order of dependence on the temperature in the 673-873 °K range. No explanation is offered, at present, to elucidate this extremely high order of dependence exhibited by the fractal dimension in the dissociation phase, D_{fd}, on the temperature, T (in °K) in the 673-873 °K range. It is of interest to note that D_f as well as D_{fd} exhibit decreases as the temperature increases. In other words, the degree of heterogeneity on the biosensor surface, in this case, decreases as the temperature increases in the 673-873 °K range.

Roy et al. (2005) also analyzed the influence of NH_3 concentration in ppm in air on its binding and dissociation kinetics to the sol-gel derived thin film biosensor. Figure 10.6a shows the binding and dissociation of 160 ppm NH_3 in air to the sol-gel derived thin film biosensor. A dual-fractal analysis is required to adequately describe the binding kinetics. A single-fractal analysis is adequate to describe the dissociation kinetics. The values of (a) the binding rate coefficient, k, and the fractal dimension, D_f, for a single-fractal analysis, (b) the dissociation rate coefficient, k_d, and the fractal dimension for the dissociation phase for a single-fractal analysis, D_{fd}, and (c) the binding rate coefficients, k_1 and k_2, and the fractal dimensions, D_{f1} and D_{f2}, for a dual-fractal analysis are given in Tables 10.4 and 10.5.

It is of interest to note that as the fractal dimension increases by a factor of approximately two from a value of D_{f1} equal to 1.0774 to D_{f2} equal to 2.14, the binding rate coefficient increases by a factor of 9.44 from a value of k_1 equal to 0.264 to k_2 equal to 2.4922.

Figure 10.6b shows the binding and dissociation of 320 ppm NH_3 in air to the sol-gel derived thin film biosensor (Roy et al., 2005). A single-fractal analysis is adequate to describe the binding kinetics. A dual-fractal analysis is required to adequately describe the dissociation kinetics. The values of (a) the binding rate coefficient, k, and the fractal dimension, D_f, for a single-fractal analysis, (b) the dissociation rate coefficient, k_d, and the fractal dimension for dissociation, D_{fd}, for a single-fractal analysis, and (c) the dissociation rate coefficients, k_{d1} and k_{d2}, and the fractal dimensions for dissociation, D_{fd1} and D_{fd2}, for a dual-fractal analysis are given in Tables 10.4 and 10.5.

It is of interest to note that for a dual-fractal analysis in the dissociation phase, as the fractal dimension increases by a factor of 4.90 from a value of D_{fd1} equal to 0.1828 to D_{fd2} equal to

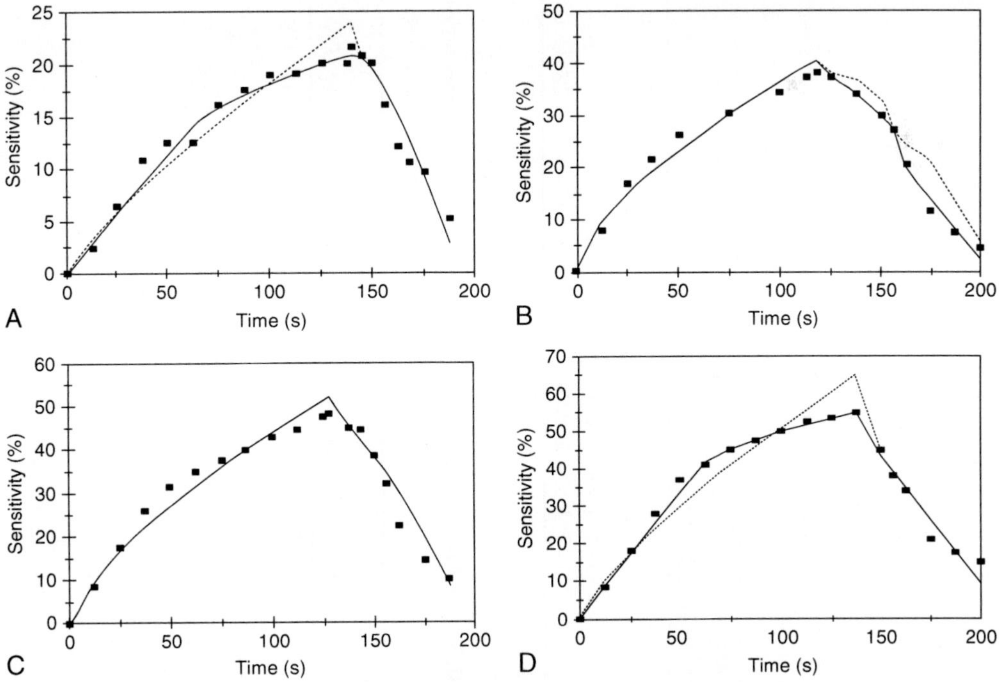

Figure 10.6

Binding and dissociation of different NH_3 concentrations (in ppm) in air to a sol-gel derived thin film (Roy et al., 2005): (a) 160 (b) 320 (c) 640 (d) 1280. When only a solid line (—) is used then a single-fractal analysis applies. When both a dashed (- - -) and a solid (—) line are used then the dashed line represents a single-fractal analysis and the solid line represents a dual-fractal analysis.

Table 10.4: Binding and dissociation rate coefficients for different NH_3 concentrations (in ppm) in air to a sol-gel derived thin film (Roy et al., 2005).

Ammonia concentration (ppm)	k	k_1	k_2	k_d	k_{d1}	k_{d2}
160	0.4511 ± 0.0827	0.2640 ± 0.0483	2.4922 ± 0.0834	0.09162 ± 0.0248	na	na
320	1.873 ± 0.127	na	na	0.0507 ± 0.0071	0.0634 ± 0.0025	0.3433 ± 0.0322
640	1.775 ± 0.212	na	na	0.1176 ± 0.0391	na	na
1280	1.574 ± 0.242	0.8170 ± 0.0744	11.296 ± 0.101	1.2474 ± 0.1617	na	na

Table 10.5: Fractal dimensions for the binding and the dissociation phase for different NH_3 concentrations (in ppm) in air to a sol-gel derived thin film (Roy et al., 2005).

Ammonia concentration (ppm)	D_f	D_{f1}	D_{f2}	D_{fd}	D_{fd1}	D_{fd2}
160	1.3926 ± 0.1354	1.0774 ± 0.2000	2.1400 ± 0.3750	$0.2388 + 0.2546$	na	na
320	1.7092 ± 0.1092	na	na	$0 + 0.1242$	0.1828 ± 0.0584	0.8960 ± 0.3932
640	1.6038 ± 0.09712	na	na	$0.1530 + 0.3126$	na	na
1280	1.4886 ± 0.1212	1.0946 ± 0.1175	2.3548 ± 0.0502	1.2598 ± 0.1776	na	na

0.8960, the dissociation rate coefficient increases by a factor of 5.41 from a value of k_{d1} equal to 0.0634 to k_{d2} equal to 0.3433.

Figure 10.6c shows the binding and dissociation of 640 ppm of NH_3 in air to the sol-gel derived thin film biosensor (Roy et al., 2005). A single-fractal analysis is adequate to describe the binding and the dissociation kinetics. The values of (a) the binding rate coefficient, k, and the fractal dimension, D_f, for a single-fractal analysis, and (b) the dissociation rate coefficient, k_d, and the fractal dimension for the dissociation phase, D_{fd}, for a single-fractal analysis are given in Table 10.4.

Figure 10.6d shows the binding and dissociation of 1280 ppm NH_3 in air to the sol-gel derived thin film biosensor. A dual-fractal analysis is required to adequately describe the binding kinetics. A single-fractal analysis is adequate to describe the dissociation kinetics. The values of (a) the binding rate coefficient, k, and the fractal dimension, D_f, for a single-fractal analysis, (b) the binding rate coefficients, k_1 and k_2 and the fractal dimensions, D_{f1} and D_{f2} for a dual-fractal analysis, and (c) the dissociation rate coefficient, k_d, and the fractal dimension for the dissociation phase for a single-fractal analysis, D_{fd}, are given in Tables 10.4 and 10.5.

It is of interest to note that as the fractal dimension increases by a factor of approximately 2.15 from a value of D_{f1} equal to 1.0946 to D_{f2} equal to 2.3548, the binding rate coefficient increases by a factor of 13.83 from a value of k_1 equal to 0.8170 to k_2 equal to 11.296.

Figure 10.7a and Table 10.4 show the increase in the dissociation rate coefficient, k_d, with an increase in the NH_3 concentration in ppm in air. For the data shown in Figure 10.7a, the dissociation rate coefficient, k_d, is given by:

$$k_d = (0.000248 + 0.000571)[NH_3]^{1.102 \pm 0.798} \tag{10.7a}$$

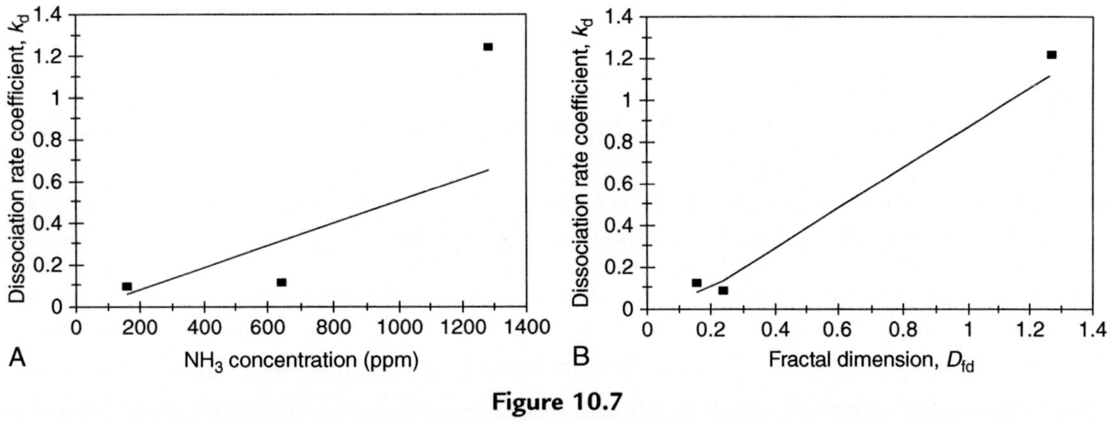

Figure 10.7

(a) Increase in the dissociation rate coefficient, k_d, with an increase in the NH_3 concentration in air (in ppm). (b) Increase in the dissociation rate coefficient, k_d, with an increase in the fractal dimension for dissociation, D_{fd}.

The fit is not good. There is scatter in the data, and this is reflected in the error in the dissociation rate coefficient, k_d. Only the positive error is presented since the dissociation rate coefficient, k_d, cannot exhibit a negative value. The dissociation rate coefficient, k_d, exhibits close to a first (equal to 1.102) order of dependence on the NH_3 concentration in ppm in air. The non-integer order of dependence exhibited by the dissociation rate coefficient, k_d, on the NH_3 concentration in ppm in air lends support to the fractal nature of the system.

Figure 10.7b and Table 10.4 show the increase in the dissociation rate coefficient, k_d, with an increase in the fractal dimension for dissociation, D_{fd}. For the data shown in Figure 10.7b, the dissociation rate coefficient, k_d is given by:

$$k_d = (0.851 \pm 0.669)D_{fd}^{1.243 \pm 0.369} \qquad (10.7b)$$

The fit is very good. Only three data points are available. The availability of more data points would lead to a more reliable fit. The dissociation rate coefficient, k_d, exhibits an order of dependence between one and one and a half (equal to 1.243) on the fractal dimension in the dissociation phase, D_{fd}, or the degree of heterogeneity present on the biosensor surface during the dissociation phase.

Roy et al. (2005) analyzed the influence of film thickness on the binding and the dissociation of NH_3 in ppm in air to the sol-gel derived nM thin film biosensor. Figure 10.8a shows the binding and the dissociation of NH_3 in ppm in air to the sol-gel derived 150 nM thin film biosensor. A dual-fractal analysis is required to adequately describe the binding kinetics. A single-fractal analysis is adequate to describe the dissociation kinetics. The values of (a) the binding rate coefficient, k, and the fractal dimension, D_f, for a single-fractal analysis, (b) the dissociation rate coefficient, k_d, and the fractal dimension D_{fd}, for a single-fractal analysis, and (c) the binding rate coefficients, k_1 and k_2, and the fractal dimensions, D_{f1} and D_{f2}, for a dual-fractal analysis, are given in Tables 10.5 and 10.6.

It is of interest to note that as the fractal dimension increases by a factor of 7.37 from a value of D_{f1} equal to 0.22 to D_{f2} equal to 1.6212, the binding rate coefficient increases by a factor of 16.47 from a value of k_1 equal to 0.05949 to k_2 equal to 0.98. In this case, the affinity, K_1 ($=k_1/k_d$) is equal to 0.523, and K_2 ($=k_2/k_d$) is equal to 0.861.

Figure 10.8b shows the binding and the dissociation of NH_3 in ppm in air to the sol-gel derived 320 nM thin film biosensor. A dual-fractal analysis is required to adequately describe the binding kinetics. A single-fractal analysis is adequate to describe the dissociation kinetics. The values of (a) the binding rate coefficient, k, and the fractal dimension, D_f, for a single-fractal analysis, (b) the dissociation rate coefficient, k_d, and the fractal dimension, D_{fd}, for a single-fractal analysis, and (c) the binding rate coefficients, k_1 and k_2, and

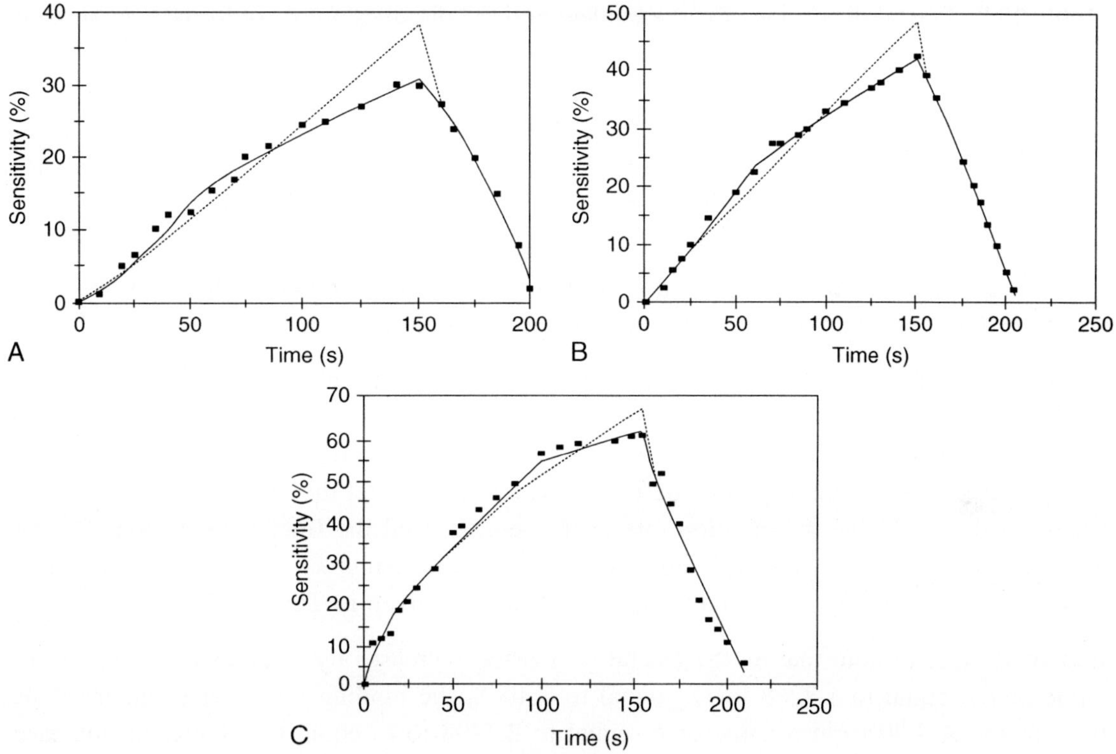

Figure 10.8

Binding and dissociation of NH_3 in air to a sol-gel derived thin film (Roy et al., 2005). Influence of film thickness (in mm): (a) 150 (b) 320 (c) 480. When only a solid line is used (—) then a single-fractal applies. When both a dotted (- - -) and a solid line (—) is used, then the dotted line is for a single-fractal analysis, and the solid line is for a dual-fractal analysis. In this case, the dual-fractal analysis provides the better fit.

Table 10.6: Binding and dissociation rate coefficients for NH_3 in air to a sol-gel derived thin film. Influence of film thickness in mm (Roy et al., 2005).

Film thickness (mm)	k	k_1	k_2	k_d
150	0.1498 ± 0.0428	0.05949 ± 0.01791	0.9800 ± 0.0434	0.1138 ± 0.0149
320	0.4053 ± 0.0631	0.2384 ± 0.0325	1.6449 ± 0.0191	0.5622 ± 0.0213
480	2.9591 ± 0.3446	2.7898 ± 0.3892	13.692 ± 0.540	2.1362 ± 0.5628

the fractal dimensions, D_{f1} and D_{f2}, for a dual-fractal analysis, are given in Tables 10.6 and 10.7.

It is of interest to note that as the fractal dimension increases by a factor of 2.26 from a value of D_{f1} equal to 0.7546 to D_{f2} equal to 1.7064, the binding rate coefficient increases by a

Table 10.7: Fractal dimensions for the binding and the dissociation phase for NH_3 in air to a sol-gel derived thin film. Influence of film thickness in mm (Roy et al., 2005).

Film thickness (mm)	D_f	D_{f1}	D_{f2}	D_{fd}
150	0.7844 ± 0.1715	0.2200 ± 0.3098	1.6212 ± 0.1155	0.2126 ± 0.1728
320	1.0898 ± 0.08686	0.7546 ± 0.1252	1.7064 ± 0.0416	0.8576 ± 0.03100
480	1.7638 ± 0.09774	1.7268 ± 0.09774	2.4048 ± 0.1414	1.3568 ± 0.2092

factor of 6.89 from a value of k_1 equal to 0.2384 to k_2 equal to 1.6449. In this case, the affinity, K_1 $(=k_1/k_d)$ is equal to 0.424, and K_2 $(=k_2/k_d)$ is equal to 2.93.

Figure 10.8c shows the binding and the dissociation of NH_3 in ppm in air to the sol-gel derived 480 nM thin film biosensor. A dual-fractal analysis is required to adequately describe the binding kinetics. A single-fractal analysis is adequate to describe the dissociation kinetics. The values of (a) the binding rate coefficient, k, and the fractal dimension, D_f, for a single-fractal analysis, (b) the dissociation rate coefficient, k_d, and the fractal dimension, D_{fd}, for a single-fractal analysis, and (c) the binding rate coefficients, k_1 and k_2, and the fractal dimensions, D_{f1} and D_{f2}, for a dual-fractal analysis, are given in Tables 10.6 and 10.7.

It is of interest to note that as the fractal dimension increases by a factor of 1.392 from a value of D_{f1} equal to 1.7268 to D_{f2} equal to 2.4048, the binding rate coefficient increases by a factor of 4.90 from a value of k_1 equal to 2.7898 to k_2 equal to 13.692. In this case, the affinity, K_1 $(=k_1/k_d)$ is equal to 1.306, and K_2 $(=k_2/k_d)$ is equal to 6.41.

Figure 10.9a and Tables 10.6 and 10.7 show for a dual-fractal analysis the increase in the binding rate coefficient, k_1, with an increase in the film thickness, in mm. For the data shown in Figure 10.9a, the binding rate coefficient, k_1, for a dual-fractal analysis is given by:

$$k_1 = (7.5 \times 10^{-09} + 10.5 \times 10^{-09})[\text{film thickness, mm}]^{3.12 \pm 1.08} \tag{10.8a}$$

The fit is poor. Only three data points are available. The availability of more data points would lead to a better and more reliable fit. The poor fit is also reflected in the error in the binding rate coefficient, k_1, value presented. Only the positive value is presented since the binding rate coefficient cannot have a negative value. The binding rate coefficient, k_1, is sensitive to the film thickness, in mm since it exhibits a greater than third (equal to 3.12) order of dependence on the film thickness. The non-integer order of dependence exhibited by k_1 on the film thickness lends support to the fractal nature of the system.

Figure 10.9b and Tables 10.6 and 10.7 show the increase in the binding rate coefficient, k_2, with an increase in the film thickness, in mm for a dual-fractal analysis. For the data shown in Figure 10.b, the binding rate coefficient, k_2, for a dual-fractal analysis is given by:

$$k_2 = (2.4 + 3.9) \times 10^{-05}[\text{film thickness, mm}]^{2.065 \pm 1.156} \tag{10.8b}$$

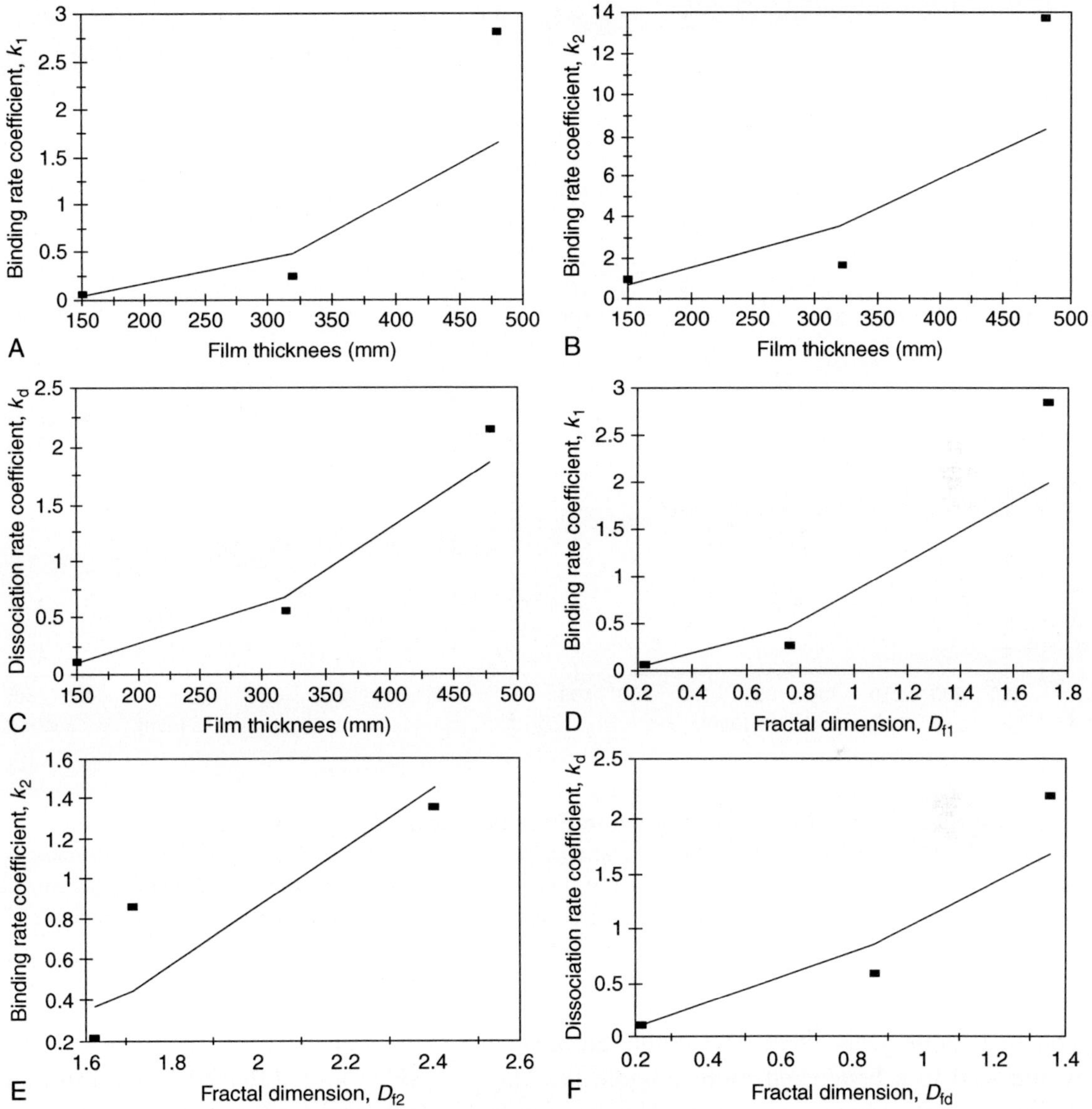

Figure 10.9

(a) Increase in the binding rate coefficient, k_1, with an increase in the film thickness (in mm).
(b) Increase in the binding rate coefficient, k_2, with an increase in the film thickness (in mm).
(c) Increase in the dissociation rate coefficient, k_d, with an increase in the film thickness (in mm).
(d) Increase in the binding rate coefficient, k_1, with an increase in the fractal dimension, D_{f1}.
(e) Increase in the binding rate coefficient, k_2, with an increase in the fractal dimension, D_{f2}.
(f) Increase in the dissociation rate coefficient, k_d, with an increase in the fractal dimension for dissociation, D_{fd}.

continued

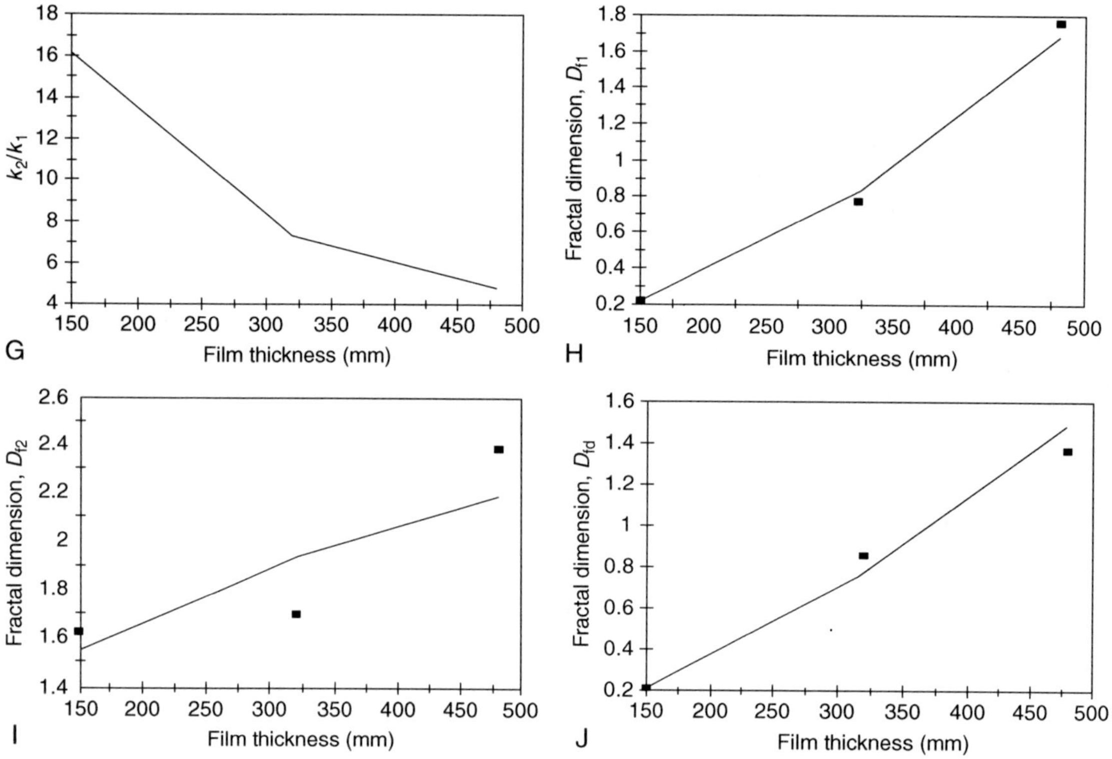

Figure 10.9—cont'd

(g) Decrease in the ratio of the binding rate coefficients, k_2/k_1, with an increase in the film thickness (in mm). (h) Increase in the fractal dimension, D_{f1}, with an increase in the film thickness (in mm). (i) Increase in the fractal dimension, D_{f2}, with an increase in the film thickness (in mm). (j) Increase in the fractal dimension for dissociation, D_{fd}, with an increase in the film thickness (in mm).

The fit is poor. Only three data points are available. The availability of more data points would lead to a better and more reliable fit. The poor fit is also reflected in the error in the value of the binding rate coefficient, k_1, presented. Only the positive value is presented since the binding rate coefficient cannot have a negative value. The binding rate coefficient, k_2, is sensitive to the film thickness, in mm since it exhibits a slightly greater than second (equal to 2.065) order of dependence on the film thickness.

Figure 10.9c and Tables 10.6 and 10.7 show the increase in the dissociation rate coefficient, k_d, with an increase in the film thickness, in mm for a dual-fractal analysis. For the data shown in Figure 10.9c, the dissociation rate coefficient, k_d, is given by:

$$k_d = (4.5 \pm 1.3) \times 10^{-07}[\text{film thickness, in mm}]^{2.47 \pm 0.301} \qquad (10.8c)$$

The fit is good. Only three data points are available. The availability of more data points would lead to a more reliable fit. The dissociation rate coefficient, k_d, for a single-fractal analysis is sensitive to the film thickness, in mm, since it exhibits close to a two and a half order (equal to 2.47) of dependence on the film thickness, in mm. Once again, the non-integer order of dependence exhibited by the dissociation rate coefficient, k_d, on the film thickness, in mm lends support to the fractal nature of the system.

Figure 10.9d and Tables 10.6 and 10.7 show the increase in the binding rate coefficient, k_1, with an increase in the fractal dimension, D_{f1}, for a dual-fractal analysis. For the data shown in Figure 10.9d, the binding rate coefficient, k_1, is given by:

$$k_1 = (0.724 + 0.795)D_{f1}^{1.81\pm0.506} \tag{10.8d}$$

The fit is good. Only three data points are available. The availability of more data points would lead to a more reliable fit. The binding rate coefficient, k_1, exhibits an order of dependence between one and a half and two (equal to 1.81) on the fractal dimension, D_{f1}, or the degree of heterogeneity that exists on the biosensor surface. This indicates that the binding rate coefficient, k_1, is sensitive to the degree of heterogeneity that exists on the biosensor surface.

Figure 10.9e and Tables 10.6 and 10.7 show the increase in the binding rate coefficient, k_2, with an increase in the fractal dimension, D_{f2}, for a dual-fractal analysis. For the data shown in Figure 10.9e, the binding rate coefficient, k_2, is equal to:

$$k_2 = (0.0697 + 0.9961)D_{f2}^{3.48\pm2.86} \tag{10.8e}$$

The fit is not good. There is scatter in the data, and this is reflected in the error in the binding rate coefficient, k_2. Only the positive value of the error is presented since the binding rate coefficient, k_2 cannot have a negative value. Only three data points are available. The availability of more data points would lead to a more reliable fit. The binding rate coefficient, k_2, is sensitive to the fractal dimension, D_{f2}, or the degree of heterogeneity present on the biosensor surface as noted by the close to three and a half (equal to 3.48) order of dependence exhibited.

Figure 10.9f and Tables 10.6 and 10.7 show the increase in the dissociation rate coefficient, k_d, with an increase in the fractal dimension in the dissociation phase, D_{fd}, for a single-fractal analysis. For the data shown in Figure 10.9f, the dissociation rate coefficient, k_d, is given by:

$$k_d = (1.026 \pm 0.528)D_{fd}^{1.48\pm0.349} \tag{10.8f}$$

The fit is reasonable. Only three data points are available. The availability of more data points would lead to a more reliable fit. The dissociation rate coefficient, k_d, exhibits close to a one and a half (equal to 1.480) order of dependence on the fractal dimension in the dissociation phase, D_{fd}, for a single-fractal analysis. This indicates that the dissociation rate coefficient, k_d, is sensitive to the fractal dimension, D_{fd}, or the degree of heterogeneity that exists on the sensor chip surface.

Figure 10.9g shows the decrease in the ratio of the binding rate coefficients, k_2/k_1, with an increase in the film thickness, in mm, for a dual-fractal analysis. For the data shown in Figure 10.9g, the ratio of the binding rate coefficients, k_2/k_1, is given by:

$$k_2/k_1 = (3208.88 \pm 214.51)[\text{film thickness, in mm}]^{-1.056 \pm 0.0775} \qquad (10.8g)$$

The fit is good. Only three data points are available. The availability of more data points would lead to a more reliable and better fit. The ratio of the binding rate coefficients, k_2/k_1, exhibits close to a negative first (equal to -1.056) order of dependence on the film thickness, in mm. This indicates that the ratio of the binding rate coefficients, k_2/k_1, is sensitive to the film thickness in the 150-480 mm range analyzed.

Figure 10.9h and Tables 10.6 and 10.7 show the increase in the fractal dimension, D_{f1}, with an increase in the film thickness, in mm, for a dual-fractal analysis. For the data shown in Figure 10.9h, the fractal dimension, D_{f1}, is given by:

$$D_{f1} = (3.3 \times 10^{-05} + 1.1 \times 10^{-09})[\text{film thickness, in mm}]^{1.753 \pm 0.105} \qquad (10.8h)$$

The fit is very good. Only three data points are available. The availability of more data points would lead to a more reliable fit. For a dual-fractal analysis, the fractal dimension, D_{f1}, exhibits a order of dependence between one and a half and two (equal to 1.753) on the film thickness in the 150-480 mm range.

Figure 10.9i and Tables 10.6 and 10.7 show the increase in the fractal dimension, D_{f2}, with an increase in the film thickness, in mm, for a dual-fractal analysis. For the data shown in Figure 10.9i, the fractal dimension, D_{f2}, is given by:

$$D_{f2} = (0.3368 + 0.0606)[\text{film thickness, in mm}]^{0.3034 \pm 0.198} \qquad (10.8i)$$

The fit is poor. Only three data points are available. The availability of more data points would lead to a more reliable fit. For a dual-fractal analysis, the fractal dimension, D_{f2}, exhibits a 0.3034 order of dependence on the film thickness in the 150-480 mm range. This indicates that the fractal dimension, D_{f2}, is only mildly sensitive to the film thickness, in mm, in the 150-480 mm range.

Cao and Duan (2005) recently developed an optical fiber-based evanescent biosensor for the detection of ammonia. These authors immobilized the sensing dye, bromocresol purple

(BCP) in the substitutional cladding using a sol-gel process. They report that their sensor exhibits good repeatability and reversibility. They used the following gases as carrier gases for ammonia: air, nitrogen, and argon. Out of these three, air gave the best response time and sensitivity.

Malins et al. (1999) point out that ammonia has been widely used for the production of fertilizers, explosives, and as an industrial coolant. It is of importance to detect NH_3 present in air quickly as even a small dose of ammonia vapor can cause acute poisioning in humans on inhalation (Cao and Duan, 2005). Cao and Duan (2005) point out that traditionally ammonia has been detected in laboratories using potentiometric electrodes (Bailescu and Cosofret, 1978; Meyerhoff, 1980; Fraticelli and Meyerhoff, 1981; Morf et al., 1981; West et al., 1992; Morales-Bahnik et al., 1994; Buhlman et al., 1998). Even though these detection methods are accurate, sensitive, and selective, Cao and Duan (2005) explain that these methods are expensive, consume analyte(s), and require the presence of an experienced operator. These authors report that optical fiber chemical sensors (OFCSs) exhibit some distinctive properties. Their small size and sensor design flexibility makes them excellent tools for *in situ* and *in vivo* analysis. These authors used a sol-gel film to encode bromocresol purple on the surface of a bared fiber code. Evanescent absorption was measured through a spectrometer.

Figure 10.10a–c shows the reversibility of the ammonia sensor. Figure 10.10a shows the binding of 145 ppm NH_3 with air as a carrier gas to the optical fiber-based evanescent ammonia sensor (Cao and Duan, 2005). A single-fractal analysis is adequate to describe the binding kinetics. The values of the binding rate coefficient, k, and the fractal dimension, D_f, for a single-fractal analysis are given in Table 10.8.

Figure 10.10b shows the binding of 145 ppm NH_3 with air as a carrier gas (consecutive run #2) to the optical fiber-based evanescent ammonia sensor (Cao and Duan, 2005). Once again, a single-fractal analysis is adequate to describe the binding kinetics. The values of the binding rate coefficient, k, and the fractal dimension, D_f, for a single-fractal analysis are given in Table 10.8.

Figure 10.10c shows the binding of 145 ppm NH_3 with air as a carrier gas (consecutive run #3) to the optical fiber-based evanescent ammonia sensor (Cao and Duan, 2005). Once again, a single-fractal analysis is adequate to describe the binding kinetics. The values of the binding rate coefficient, k and the fractal dimension, D_f for a single-fractal analysis are given in Table 10.8.

The average value of the binding rate coefficient, k, and the fractal dimension, D_f, presented in Table 10.8 for these reversibility runs are 76.895 and 2.279, respectively. This represents a deviation of 23.96% for the binding rate coefficient, k, and 4.87% for the fractal dimension,

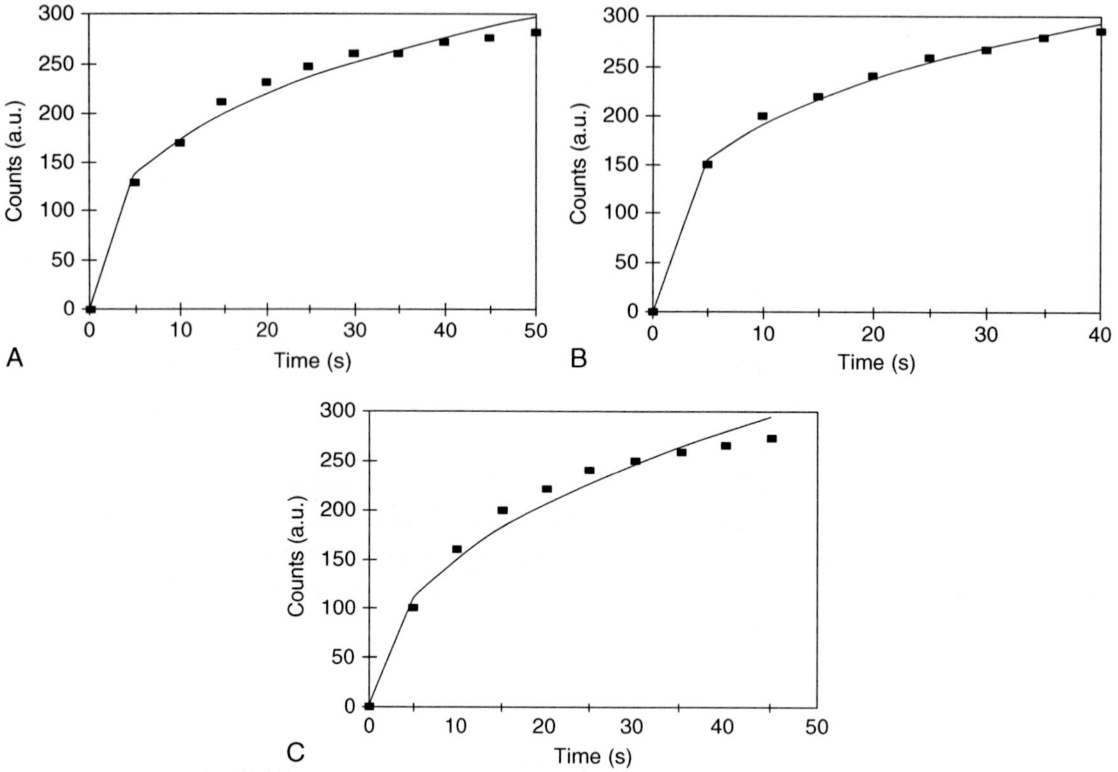

Figure 10.10
Binding of NH_3 in air to an optical fiber-based evanescent sensor. Effect of reversibility (Cao and Duan, 2005): (a) Run # 1 (b) Run # 2 (c) Run # 3.

Table 10.8: Binding rate coefficients for NH_3 in air to an optical fiber-based evanescent ammonia sensor. Effect of reversibility (Cao and Duan, 2005).

Run #	k	D_f
1	80.472 ± 3.843	2.3350 ± 0.0424
2	95.326 ± 2.628	2.3902 ± 0.02922
3	54.887 ± 4.320	2.1132 ± 0.07446

D_f, respectively. The above numbers obtained from the fractal analysis do indicate that the above analyzed runs are reversible. The error exhibited in the fractal dimension is relatively small. Note that the fractal dimension is based on a log scale, and even small changes in the fractal dimension indicate significant changes in the degree of heterogeneity that exists on the biosensor surface.

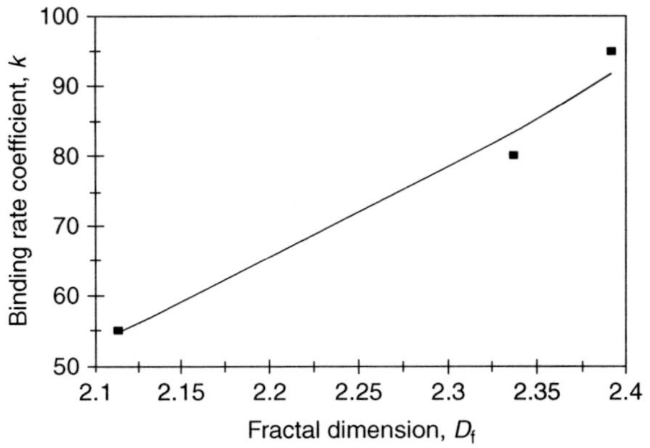

Figure 10.11
Increase in the binding rate coefficient, k, with an increase in the fractal dimension, D_f.

Figure 10.11 and Table 10.8 show the increase in the binding rate coefficient, k, with an increase in the fractal dimension, D_f, for a single-fractal analysis. For the data shown in Figure 10.11, the binding rate coefficient, k, is given by:

$$k = (2.201 \pm 0.112)D_f^{4.289 \pm 0.537} \quad (10.9)$$

The fit is good. Only three data points are available. The availability of more data points would lead to a more reliable fit. The binding rate coefficient, k, (for these reversibility runs) is very sensitive to the fractal dimension, D_f, or the degree of heterogeneity that exists on the biosensor surface as noted by the high order of dependence between four and four and a half (equal to 4.289) exhibited.

Cao and Duan (2005) analyzed the influence of the carrier gases argon, nitrogen and air on the detection of 145 ppm of NH_3 at 22.1°C using their optical fiber-based evanescent ammonia sensor. These authors noted that the response time of the sensor varied clearly for the different gases. They also noted that when air was used as the carrier gas the absorption change was the highest, and when Argon was used as the carrier gas, the absorption change was the lowest.

Figure 10.12a shows the binding of 145 ppm NH_3 in air (carrier gas) to the optical fiber-based evanescent ammonia sensor (Cao and Duan, 2005). A dual-fractal analysis is required to adequately describe the binding kinetics. The values of (a) the binding rate coefficient, k, and the fractal dimension, D_f, for a single-fractal analysis, and (b) the binding rate coefficients, k_1 and k_2, and the fractal dimensions, D_{f1} and D_{f2}, for a dual-fractal analysis are given in Table 10.9. It is of interest to note that as the fractal dimension increases by a factor of 1.53 from a value of D_{f1} equal to 1.9298 to D_{f2} equal to 2.7950, the binding rate coefficient increases by a factor of 2.76 from a value of k_1 equal to 37.39 to k_2 equal to 103.3.

Table 10.9: Binding rate coefficients and fractal dimensions for NH_3 to an optical fiber-based evanescent ammonia sensor (Cao and Duan, 2005). Influence of different carrier gases

Carrier gas	k	k_1	k_2	D_f	D_{f1}	D_{f2}
Air	41.74 ± 5.40	37.387 ± 4.659	103.26 ± 0.39	2.1248 ± 0.0863	1.9298 ± 0.1278	2.7950 ± 0.0237
N_2	32.522 ± 6.873	24.885 ± 0.318	59.642 ± 0.877	1.9152 ± 0.1314	0.9568 ± 0.0286	2.400 ± 0.0408
Ar	24.256 ± 2.438	17.885 ± 1.418	41.489 ± 0.829	2.0862 ± 0.0935	1.6806 ± 0.194	2.4742 ± 0.0509

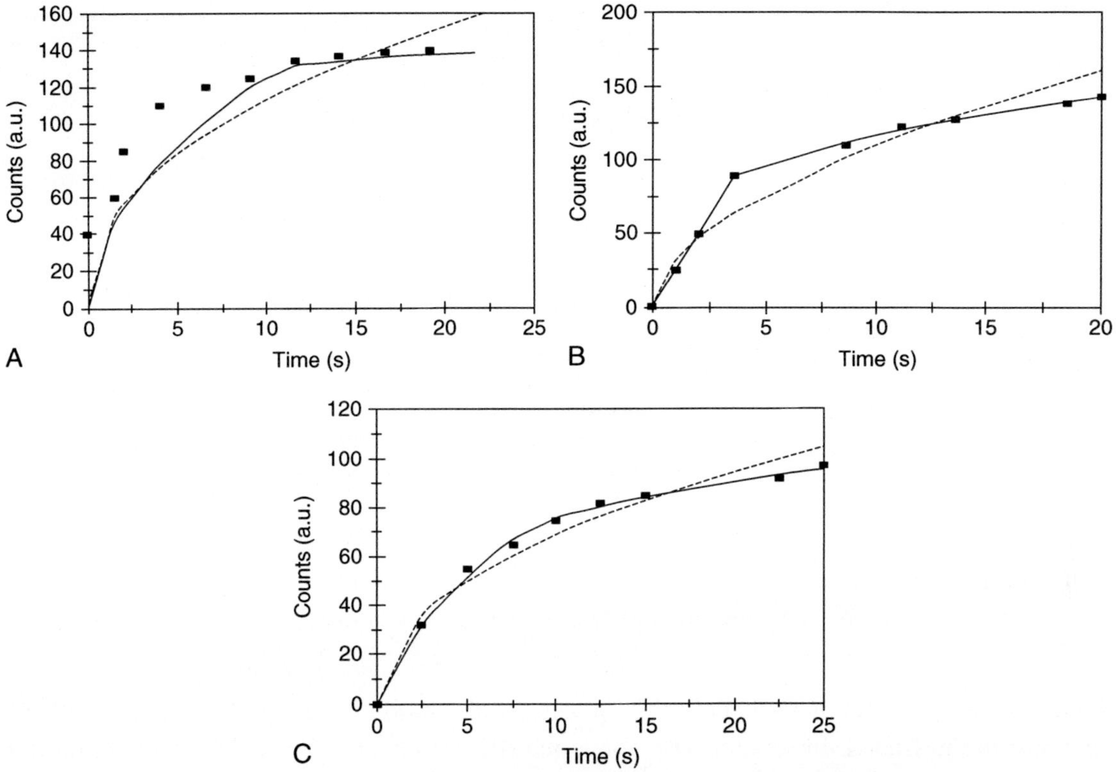

Figure 10.12

Influence of the carrier gas on the binding of NH_3 to an optical fiber-based evanescent sensor (Cao and Duan, 2005): (a) Air (b) N_2 (c) Ar. When only a solid line (—) is used then a single-fractal analysis applies. When both a dashed (- - -) and a solid (—) line are used then the dashed line represents a single-fractal analysis and the solid line represents a dual-fractal analysis.

Figure 10.12b shows the binding of 145 ppm NH_3 in nitrogen (carrier gas) to the optical fiber-based evanescent ammonia sensor (Cao and Duan, 2005). A dual-fractal analysis is required to adequately describe the binding kinetics. The values of (a) the binding rate coefficient, k, and the fractal dimension, D_f, for a single-fractal analysis, and (b) the binding rate coefficients, k_1 and k_2, and the fractal dimensions, D_{f1} and D_{f2}, for a dual-fractal analysis are given in Table 10.9. It is of interest to note that as the fractal dimension increases by a factor of 2.51 from a value of D_{f1} equal to 0.9568 to D_{f2} equal to 2.40, the binding rate coefficient increases by a factor of 2.40 from a value of k_1 equal to 24.885 to k_2 equal to 59.642.

Figure 10.12c shows the binding of 145 ppm NH_3 in Ar (carrier gas) to the optical fiber-based evanescent ammonia sensor (Cao and Duan, 2005). A dual-fractal analysis is required to adequately describe the binding kinetics. The values of (a) the binding rate coefficient, k, and the fractal dimension, D_f, for a single-fractal analysis, and (b) the binding rate coefficients, k_1 and k_2, and the fractal dimensions, D_{f1} and D_{f2}, for a dual-fractal analysis are given in Table 10.9. It is of interest to note that as the fractal dimension increases by a

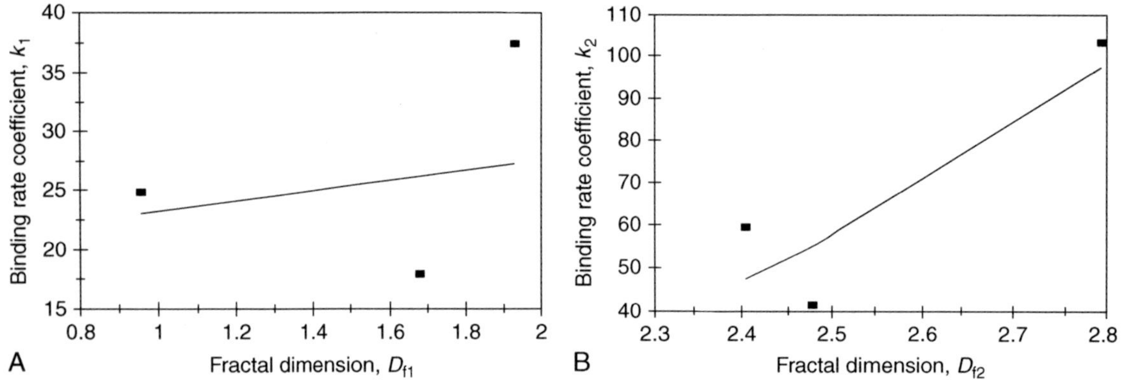

Figure 10.13

(a) Increase in the binding rate coefficient, k_1, with an increase in the fractal dimension, D_{f1}.
(b) Increase in the binding rate coefficient, k_2, with an increase in the fractal dimension, D_{f2}.

factor of 1.47 from a value of D_{f1} equal to 1.6806 to D_{f2} equal to 2.4742, the binding rate coefficient increases by a factor of 2.40 from a value of k_1 equal to 17.885 to k_2 equal to 41.489.

Figure 10.13a and Table 10.9 show the increase in the binding rate coefficient, k_1, with an increase in the fractal dimension, D_{f1}, for a dual-fractal analysis when the different carrier gases, air, nitrogen, and argon are used. For the data shown in Figure 10.13a, the binding rate coefficient, k_1, is given by:

$$k_1 = (23.29 \pm 13.36) D_{f1}^{0.243+0.964} \tag{10.10a}$$

The fit is poor. There is scatter in the data, and this is clearly reflected in the error in the order of the fractal dimension, D_{f1}. Only three data points are available. The availability of more data points would lead to a more reliable and better fit. The binding rate coefficient, k_1, is only mildly sensitive to the fractal dimension, D_{f1}, or the degree of heterogeneity that exists on the biosensor surface as noted by the close to zero (equal to 0.243) order of dependence exhibited.

Figure 10.13b and Table 10.9 show for a dual-fractal analysis the increase in the binding rate coefficient, k_2, with an increase in the fractal dimension, D_{f2}, when the different carrier gases, air, nitrogen, and argon are used. For the data shown in Figure 10.13b, the binding rate coefficient, k_2, is given by:

$$k_2 = (0.770 \pm 0.339) D_{f2}^{4.71 \pm 3.20} \tag{10.10b}$$

The fit is good. Only three data points are available. The availability of more data points would lead to a more reliable and better fit. The binding rate coefficient, k_2, exhibits an order of dependence between four and a half and five (equal to 4.71) on the fractal dimension, D_{f2}, or the degree of heterogeneity that exists on the biosensor surface. This indicates that the binding rate coefficient, k_2, is very sensitive to the degree of heterogeneity that exists on the biosensor surface.

Wang and Li (2005) recently analyzed the structural and capacitive humidity sensing properties of nc-Fe_3O_4/Si-NPA. These authors fabricated a composite thin film by coating nc-Fe_3O_4 on a Si-NPA. These authors report that this showed a regular hierarchical structure that comprised of the pillar array in the micron dimension, and a nanoporous structure in the film of Fe_3O_4. These authors point out that their humidity sensors were made based on the nc-Fe_3O_4/Si-NPA system.

Wang and Li (2005) further explain that several kinds of porous ceramic films have been analyzed to help develop miniaturized and integrated humidity sensors (Traversa, 1995; Traversa et al., 1996; Nahar and Khanna, 1998; Qu et al., 2000). In these cases, Wang and Li (2005) point out that the solution of an appropriate substrate along with suitable morphology is the key to obtaining good sensing properties. For example, Bisi et al. (2000) report that porous silicon fabricated by traditional anodization exhibits an integral sponge structure. This apparently leads to a relatively lengthy response time. Xu et al. (2005) had reported earlier that these response times could be improved by fabricating a micron/nanometer composite structure, that is, the Si-NPA.

Wang and Li (2005) report that magnetite is a traditional humidity sensing material. These authors combined the properties of Fe_3O_4 and the unique structure of Si-NPA. nc-Fe_3O_4 was spin coated on Si-NPA, and this was then used as the humidity sensor.

Figure 10.14 shows the RH (relative humidity) increasing progress (binding), that is the time response of the nc-Fe_3O_4/SI-NPA sensor (Wang and Li, 2005). A single-fractal analysis is adequate to describe the "binding" kinetics. In this case the binding rate coefficient, k, is equal to 0.000996 ± 0.000089, and the fractal dimension, D_f, is equal to 2.2716 ± 0.0938. These authors point out that their nc-Fe_3O_4/SI-NPA sensor exhibits high sensitivity, a high output signal intensity, and short response times.

Manera et al. (2007) recently reported that polymer-based materials are being increasingly used for gas applications. Polymers based on polyaniline, polythiophene, and polypyrrole exhibit good transport and optical sensing properties at room temperature (Rellas et al., 1999; Nicolas-Debarnot and Poncin-Epaillard, 2003; De Melo et al., 2005). Manera et al. (2007) analyzed the sensing properties of polyimide thin films using the surface plasmon resonance (SPR) technique. These authors further indicate that polyimides are a class of organic polymers that find applications in the area of microelectronics industry as films, varnishes, adhesives, and matrix resins. This is because of their thermal and chemical stability as well as their resistance to mechanical deformation at high temperature (Ghosh and Mittal, 1996;

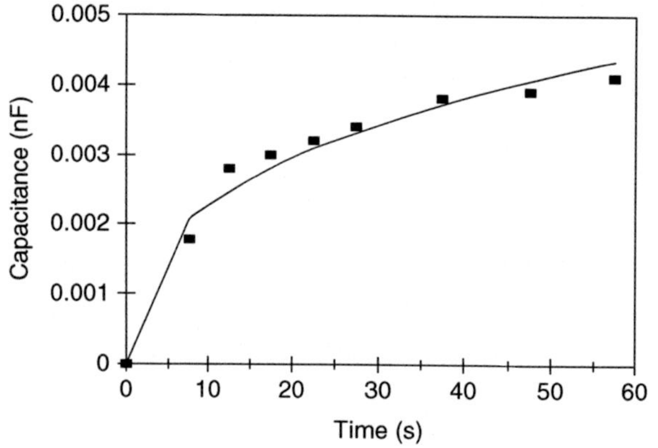

Figure 10.14
Binding to a nc-Fe_3O_4/Si-NPA (nanocrystal magnetite/silicon nanoporous pillar array) (Wang and Li, 2005).

Wessa et al., 1998). Manera et al. (2007) present an analysis of the sensing properties of polyimide films by means of the SPR technique for the detection of volatile organic carbons (VOCs) such as ethanol and methanol. The detection of the vapors of methanol and ethanol represent a key step in the control of food quality (Manera et al., 2007).

Figure 10.15a shows the binding and dissociation of 3320 ppm methanol in air to a polyimide thin layer biosensor (Manera et al., 2007). A single-fractal analysis is adequate to describe the binding and dissociation kinetics. The values of (a) the binding rate coefficient, k, and the fractal dimension, D_f, for a single-fractal analysis, and (b) the dissociation rate coefficient, k_d, and the fractal dimension for the dissociation phase, D_{fd}, are given in Tables 10.10 and 10.11. In this case the affinity, K ($=k/k_d$) is equal to 0.677.

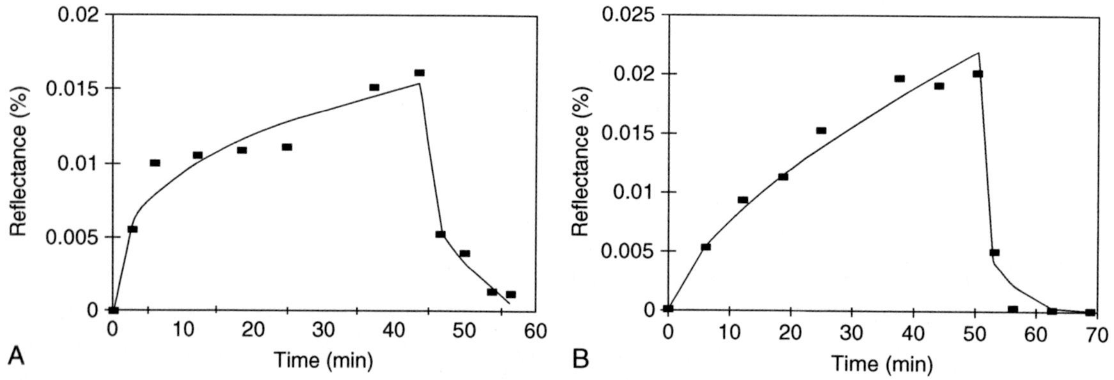

Figure 10.15
Binding and dissociation of different methanol concentrations (in ppm) to a polyimide thin layer biosensor (Manera et al., 2006). Influence of a repeat run: (a) 3320 (b) 3320.

Table 10.10: Binding and dissociation rate coefficients for different methanol concentrations to a polyimide thin layer biosensor (Manera et al., 2006).

Methanol Concentration (ppm)	k	k_1	k_2	k_d
3320	0.004354 ± 0.000648	na	na	0.006427 ± 0.000327
3320	0.001535 ± 0.000131	na	na	0.01363 ± 0.00133
4980	0.02389 ± 0.00117	na	na	0.005807 ± 0.000173
4980	0.007511 ± 0.0014	0.005371 ± 0.00132	0.01690 ± 0.00023	0.007229 ± 0.000492

Table 10.11: Fractal dimensions for the binding and dissociation phase for different methanol concentrations to a polyimide thin layer biosensor (Manera et al., 2006).

Methanol Concentration (ppm)	D_f	D_{f1}	D_{f2}	D_{fd}
3320	2.3352 ± 0.1164	na	na	2.4418 ± 0.09514
3320	1.6418 ± 0.08864	na	na	2.7084 ± 0.1336
4980	2.6894 ± 0.05266	na	na	2.1872 ± 0.05014
4980	2.3242 ± 0.1596	1.7776 ± 0.4490	0.01690 ± 0.00023	2.9855 ± 0.05412

Figure 10.15b shows a repeatability run. In other words, it shows the binding and dissociation of 3320 ppm methanol in air (once again) to a polyimide thin layer biosensor (Manera et al., 2007). Once again, a single-fractal analysis is required to adequately describe the binding and dissociation kinetics. The values of (a) the binding rate coefficient, k, and the fractal dimension, D_f, for a single-fractal analysis, and (b) the dissociation rate coefficient, k_d, and the fractal dimension for the dissociation phase, D_{fd}, are given in Tables 10.10 and 10.11. In this case, the affinity, K ($=k/k_d$) value is 0.113.

The affinity, K values for the two repeatability runs are very different, 0.677 and 0.113, respectively. They differ by a factor of 6.0. As expected, and in accord with the affinity values, the dissociation rate coefficient values (0.004351; 0.00154) for the first and the repeatability runs are also different. They are presented in brackets above.

Figure 10.16a shows the binding and dissociation of 4980 ppm methanol in air to a polyimide thin layer biosensor (Manera et al., 2007). A single-fractal analysis is adequate to describe the binding and dissociation kinetics. The values of (a) the binding rate coefficient, k, and the fractal dimension, D_f, for a single-fractal analysis, and (b) the dissociation rate coefficient, k_d, and the fractal dimension for the dissociation phase, D_{fd}, are given in Tables 10.10 and 10.11. In this case the affinity, K ($=k/k_d$) is equal to 4.11.

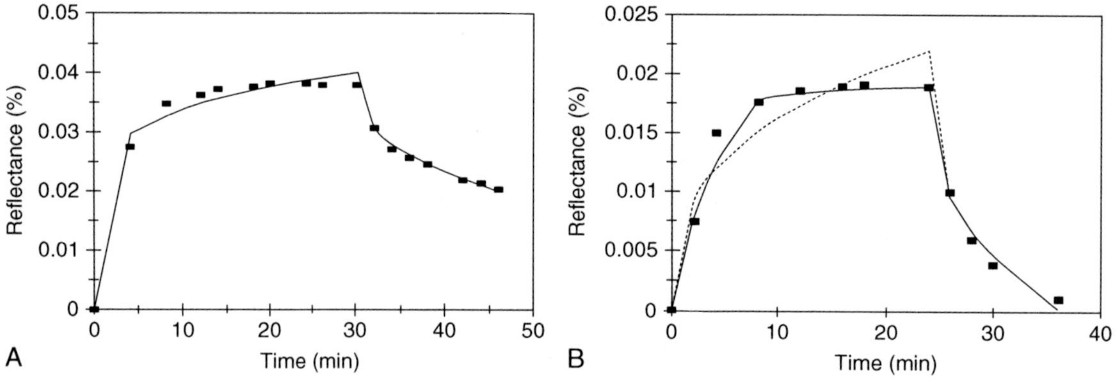

Figure 10.16

Binding and dissociation of different methanol concentrations (in ppm) to a polyimide thin layer biosensor (Manera et al., 2006). Influence of a repeat run: (a) 4980 (b) 4980. When only a solid line is used (—) then a single-fractal applies. When both a dotted (- - -) and a solid line (—) is used, then the dotted line is for a single-fractal analysis, and the solid line is for a dual-fractal analysis. In this case, the dual-fractal analysis provides the better fit.

Figure 10.16b shows the binding and dissociation of 4980 ppm methanol in air to a polyimide thin layer biosensor. In this case, a dual-fractal analysis is required to adequately describe the binding kinetics. A single-fractal analysis is still adequate to describe the dissociation kinetics. Clearly, this cannot be classified as a repeatability run. It is of interest to note that for the dual-fractal analysis that is required to describe the binding kinetics, as the fractal dimension decreases by a factor of 105.18 from a value of D_{f1} equal to 1.776 to D_{f2} equal to 0.1690, the binding rate coefficient, k, increases by a factor of 3.146 from a value of k_1 equal to 0.005381 to k_2 equal to 0.01690. In this case, changes in the degree of heterogeneity on the polyimide thin layer biosensor surface and in the binding rate coefficient are in opposite directions.

It is of interest to note that the dissociation rate coefficients for the two consecutive runs for the 4980 ppm methanol in air are within 23.9% (0.00581; 0.00723) of each other, even though the binding mechanisms for these two cases are different (single-fractal analysis; dual-fractal analysis).

Figure 10.17a and Tables 10.10 and 10.11 show the increase in the binding rate coefficient, k, with an increase in the fractal dimension, D_f, for a single-fractal analysis. For the data shown in Figure 10.17a, the binding rate coefficient, k, is given by:

$$k = (0.000106 + 0.000113)D_f^{5.06\pm2.02} \tag{10.11a}$$

The fit is reasonable. There is scatter in the data, and this is reflected in the error in the binding rate coefficient, k. Only the positive error is presented since the binding rate coefficient, k, cannot have a negative value. Only three data points are available. The availability of more data points would lead to a more reliable fit. For a single-fractal analysis, the binding rate

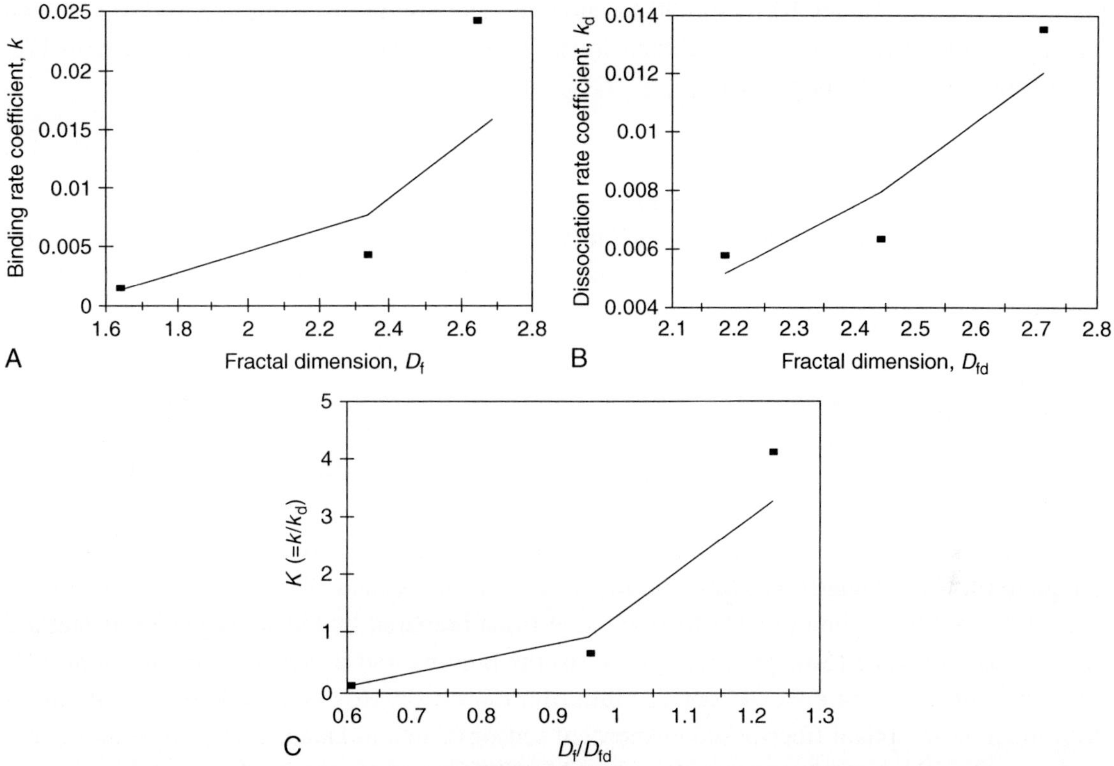

Figure 10.17

(a) Increase in the binding rate coefficient, k, with an increase in the fractal dimension, D_f. (b) Increase in the dissociation rate coefficient, k_d, with an increase in the fractal dimension for the dissociation phase, D_{fd}. (c) Increase in the affinity, K $(=k/k_d)$, with an increase in the fractal dimension ratio, D_f/D_{fd}.

coefficient, k, is very sensitive to the fractal dimension, D_f, or the degree of heterogeneity that exists on the surface as noted by the slightly higher than fifth (equal to 5.06) order of dependence exhibited.

Figure 10.17b and Tables 10.10 and 10.11 show the increase in the dissociation rate coefficient, k_d, with an increase in the fractal dimension, D_{fd}, for a single-fractal analysis. For the data shown in Figure 10.1b, the dissociation rate coefficient, k_d, is given by:

$$k_d = (0.000235 \pm 0.00074)D_{fd}^{3.96 \pm 1.83} \qquad (10.11b)$$

The fit is good. Only three data points are available. The availability of more data points would lead to a more reliable fit. For a single-fractal analysis, the dissociation rate coefficient, k_d, is very sensitive to the fractal dimension, D_{fd}, or the degree of heterogeneity that exists on the surface as noted by the slightly less than fourth (equal to 3.96) order of dependence exhibited.

Figure 10.17c and Tables 10.10 and 10.11 show the increase in the affinity, K ($=k/k_d$) with an increase in the ratio of the fractal dimensions, D_f/D_{fd}. For the data shown in Figure 10.17c, the affinity, K ($=k/k_d$) is given by:

$$K(=k/k_d) = (1.189 \pm 0.629)(D_f/D_{fd})^{4.95\pm0.837} \quad (10.11c)$$

The fit is good. Only three data points are available. The availability of more data points would lead to a more reliable fit. The affinity, K, is very sensitive to the ratio of the fractal dimensions, D_f/D_{fd}, as noted by the close to fifth (equal to 4.95) order of dependence exhibited.

10.4 Conclusions

A fractal analysis is presented for the binding and dissociation (wherever applicable) of different gases on different biosensor surfaces. Both single- and dual-fractal analysis are used. The dual-fractal analysis is used only when the single-fractal analysis does not provide an adequate fit. Corel Quattro Pro 8.0 (1989) was used to fit the data. The fractal analysis is used to analyze (a) the binding of LPG to zinc oxide films prepared by the spray pyrolysis method onto a glass substrate (Shinde et al., 2007), (b) the binding and dissociation of different NH_3 concentrations in air to a sol-gel derived thin film biosensor (Roy et al., 2005), (c) binding of NH_3 in air to an optical fiber-based evanescent sensor (Cao and Duan, 2005), (d) binding to a nc-Fe_3O_4/Si-NPA humidity sensor (Wang and Li, 2005), and (e) the binding and dissociation of different methanol concentrations in ppm) to a polyimide thin layer biosensor (Manera et al., 2006).

The binding rate coefficients are quite sensitive to the degree of heterogeneity or the fractal dimension on, for example, (a) the ZnO films sprayed on glass substrates for the detection of 0.2 volume percent LPG in the gas phase (Shinde et al., 2007), (b) the binding of NH_3 to a sol-gel derived thin film biosensor where the influence of presintering temperature was analyzed (Roy et al., 2005), (c) the binding of NH_3 in air to a sol-gel derived thin film biosensor where the influence of film thickness was analyzed (Roy et al., 2005), (d) during reversibility studies on an ammonia sensor, and during studies that analyzed the influence of different carrier gases (N_2, air, and argon) for the detection of ammonia by an ammonia sensor (Cao and Duan, 2005), (e) during humidity sensing using a composite thin film by coating a nc-Fe_3O_4 on a Si-NPA (Wang and Li, 2005), and (f) the binding of different concentrations of ppm methanol in air to a polyimide thin layer biosensor (Manera et al., 2007). Similar results are presented for the dissociation rate coefficients, wherever applicable.

The detection of different gases (analyte) present in air and in different carrier gases using the different types of biosensors provides one with an idea of what is available in the literature and what gases may be detected. Of course, these constitute only a small section of the different gases that have been detected by the different biosensors and are available in the literature. The examples analyzed and presented here were selected at random. A more detailed study of the detection of different gases available in the literature by the different biosensors should help provide a better perspective of the different gases that may be detected by the different biosensors available at present.

Novel biosensor techniques are continuously being developed for the detection of even "conventional" gases such as ammonia. Hopefully, these novel biosensor techniques should help detect conventional and other gases at lower and lower detection levels, besides becoming more and more sensitive, robust, and reliable detection techniques.

References

Bailescu F and VV Cosofret, *Applications of Ion-Selective Membrane Electrodes in Organic Analysis*, Ellis-Horwood, Chicister (1978).

Bisi O, S Ossicini, and L Pavesci, Porous silicon: a quantum sponge structure for silicon based optoelectronics, *Surface Science Reports*, **38**, 1–126 (2000).

Buhlman P, E Pretsch, and E Bakker, Carrier-based ion-selective electrodes and bulk optodes. 2. Ionophores for potentiometric and optical sensors. *Chemical Reviews*, **98**, 1593–1687 (1998).

Cao W and Y Duan, Optical fiber-based evanescent ammonia sensor, *Sensors & Actuators B*, **110**, 252–259 (2005).

Chai CC, J Peng, and BP Yan, Preparation and gas-sensing properties of Fe_2O_3 thin films, *Journal of Electronic Materials*, **24**, 82–86 (1995).

Chen HI, YI Chou, and CY Chu, A novel high-sensitive Pd/InP hydrogen sensor fabricated by electroless plating, *Sensors & Actuators B*, **85**, 10–18 (2002).

De Melo CP, BB Neto, EG de Lima, LFB de Lira, and JEG de Souza, Use of conducting polypyrrole blends as gas sensors, *Sensors & Actuators B: Chemical*, **109**, 348–354 (2005).

Diwedi D, R Diwedi, and SK Srivastava, Sensing properties of palladium-gate MOS (Pd-MOS) hydrogen sensor-based on plasma grown silicon dioxide, *Sensors & Actuators B*, **71**, 161–168 (2000).

Fraticelli YM and ME Meyerhoff, Automated determination of ammonia with a potentiometric gas sensor and flowing internal electrolyte, *Analytical Chemistry*, **53**, 992–997 (1981).

Ghosh MK and KL Mittal (Eds.), *Polyimides: Fundamentals and Applications*, Marcel Dekker, New York, 1996.

Havlin S, Molecular diffusion and reaction in *The Fractal Approach to Heterogeneous Chemistry: Surfaces, Colloids, Polymers* (ed. D Avnir), Wiley, New york, pp. 251–269 (1989).

Ismail B, MA Abaab, and B Rezig, Structural and electrical properties of ZnO films prepared by screen printing technique, *Thin Solid Films*, **383**, 92–94 (2001).

Korotcenko G, V Brinzari, J Schwank, M DiBattista, and A Visiliev, Pecularities of SnO_2 thin film deposition by spray pyrolysis for gas sensor applications, *Sensors & Actuators B*, **77**, 244–252 (2001).

Lee CK and SL Lee, Multi-fractal scaling analysis of reactions over fractal surfaces, *Surface Science*, **325**, 294–310 (1995).

Liu XQ, SW Tao, and YS Shen, Preparation and characterization of nanocrystalline α-Fe_2O_3 by a sol-gel process, *Sensors & Actuators B*, **40**, 161–165 (1997).

Lu CT, KW Lin, HI Chen, HM Chuang, CY Chen, and WC Liu, A new Pd-oxide-$Al_{0.3}Ga_{0.7}As$ MOS hydrogen sensor, *IEEE Electronic Device Letters*, **24**, 390–392 (2003).

Malins C, A Doyle, BD MacCraith, F Kvasnok, M Landl, P Simon, L Kalvoda, R Lukas, K Pufler, and I Babusik, Personal ammonia sensor for industrial environments, *Journal of Environmental Monitoring*, **1**(5), 417–422 (1999).

Manera MG, ML Lev, R Curri, R Comparelli, R Rella, A Agostiano, and L Vasanelli, Determination of optical parameters of colloidal TiO_2 nanocrystal-based thin films by using surface plasmon resonance measurements for sensing applications, *Sensors & Actuators B: Chemical*, **115**, 365–373 (2006).

Manera MG, CD Fernandez, G Maggoni, G Mattei, S Carturan, A Quaranta, MG Della, R Rella, L Vasanelli, and P Mazzoldi, Binding of 3320 ppm methanol to polyimide thin layer, *Sensors & Actuators B*, **120**, 712–718 (2007).

Matsuguchi M, J Io, G Sugiyama, and Y Sakai, Effect of NH_3 gas on the electrical conductivity of polyaniline blend films, *Synthetic Methods*, **128**, 15–19 (2002).

Medlin JW, AE Lutz, R Bastasz, and AH Mc Daniel, The response of palladium metal-insulator-semiconductor devices to hydrogen-oxygen mixtures: comparisons between kinetic models and experiment, *Sensors & Actuators B*, **96**, 290–297 (2003).

Meyerhoff ME, Polymeric membrane electrode-based potentiometric gas sensor, *Analytical Chemistry*, **52**, 1532–1534 (1980).

Morales-Bahnik A, R Czolk, and HJ Ache, An optochemical ammonia sensor-based on immobilized metalloporphyrins, *Sensors & Actuators B: Chemical*, **19**, 493–496 (1994).

Morf WE, Pungor E, and SW Incsedy (Eds.), *The Principles of Ion-Selective Electrodes and Membrane Transport*, Elsevier, Amsterdam, 1981, pp. 402–406.

Nahar RK and VK Khanna, Ionic doping and inversion of the characteristic of thin film porous Al_2O_3 humidity sensor, *Sensors & Actuators B*, **46**, 35–41 (1998).

Nicolas-Debarnot D and F Poncin-Epaillard, Polyaniline as a new sensitive layer for gas sensors, *Analytica Chimica Acta*, **475**, 1–15 (2003).

Patil PS, Versatility of chemical spray pyrolysis technique, *Material Chemistry and Physics*, **59**, 185–198 (1999).

Qu W, W Wlodarski, and JU Meyer, Comparative study on micromorphology and humidity sensitive properties of thin-film and thick-film humidity sensors based on semiconducting $MnWO_4$, *Sensors & Actuators B*, **64**, 76–82 (2000).

Ramakrishnan A and A Sadana, A single fractal analysis of cellular analyte-receptor binding kinetics using biosensors, *Biosystems*, **59**, 35–51 (2001).

Rao GST and DT Rao, Gas sensitivity of ZnO based thick film sensor to NH_3 at room temperature, *Sensors & Actuators B*, **55**, 166–169 (1999).

Rellas R, P Siciliano, G Toscano, L Vaili, L Schnetti, A Mucci, and D Iarossi, Langmuir-Blodgett films of poly [3-(butylthio)thiophene]: optical properties and electrical measurements in controlled atmosphere, *Sensors & Actuators* B, **57**, 125–129 (1999).

Roy SC, GL Sharma, MC Bhatnagar, and SB Samanta, Novel ammonia sensing phenomena in sol-gel derived $Ba_{0.5}ISr_{0.5}TiO_3$ thin films, *Sensors & Actuators B*, **110**, 299–303 (2005).

Sadana A, A fractal analysis for the evaluation of hybridization kinetics in biosensors, *Journal of Colloid and Interface Science*, **151**(1), 166–177 (2001).

Sadana A, *Fractal Binding and Dissociation Kinetics for Different Biosensor Applications*, Elsevier, Amsterdam (2005).

Sen S, KP Muthe, N Joshi, SC Gadkari, SK Gupta, MR Jagannath, SK Deshpande, and JV Yakhmi, Room temperature operating NH_3 gas sensor based on Tellurium thin film, *Bulletin of the Indian Vacuum Society*, **6**, 3–10 (2003).

Shinde VR, TP Gujar, and LPG CD Lokhande, sensing properties of ZnO films prepared by spray pyrolysis method. Effect of molarity of precursor solution, *Sensors & Actuators B*, **120**, 551–559 (2007).

Traversa E, Ceramic sensors for humidity detection: the state of the art and future developments, *Sensors & Actuators B*, **23**, 135–156 (1995).

Traversa E, G Gnappi, A Montenero, and G Gusmano, Ceramicthin films by sol-gel processing as novel materials for integrated humidity sensors, *Sensors & Actauators B*, **31**, 59–70 (1996).

Tsai YY, CC Cheng, PH Lai, SI Fu, CW Hong, HI Chen, and WC Liu, Comprehensive study of hydrogen sensing characteristics of Pd metal-oxide-semiconductor (MOS) transistors with $Al_{0.24}Ga_{0.76}As$ and $In_{0.49}Ga_{0.51}$ Schottky contact layers, *Sensors & Actuators B*, **120**, 687–693 (2007).

Wang HY and XJ Li, Structural and capacitive sensing properties of nanocrystal magnetite/silicon nanoporous pillar array, *Sensors & Actuators B*, **110**, 260–263 (2005).
Wessa N, M Barie, and HJ Hache, Polyimide, a new shielding layer for sensing applications, *Sensors & Actuators B: Chemical*, **53**, 63–68 (1998).
West SJ, S Ozawa, K Seiler, SSS Tan, and W Simon, Selective ionophore-based optical sensors for ammonia measurement in air, *Analytical Chemistry*, **64**, 533–540 (1992).
Xu YY, XJ Li, JT He, X Hu, and HY Wang, Capacitive humidity sensing properties of hydrothermally-etched silicon nano-porous pillar array, *Sensors & Actuators B*, **105**, 219–222 (2005).

CHAPTER 11

Detection of Analytes on Arrays/ Microarrays/DNA Chips

Chapter Outline

11.1 Introduction

In this chapter we use fractal analysis to analyze (a) the binding and dissociation (hybridization) of different targets (400 nM) in solution to a probe immobilized on a DNA chip surface (Fiche et al., 2007), (b) binding (hybridization) of different concentrations (in nM) of free-DNA in solution to a 22-mer strand (bound DNA) immobilized via a phenylene-diisocyanate linker molecule on a glass substrate (Michel et al., 2007), (c) binding (hybridization) of SA-HRP (streptavidin-horseradish peroxide) in solution to a capture probe on a QCM (quartz crystal microbalance) electrode along with a detection probe (Feng et al., 2007), (d) binding (hybridization) of a complementary and a noncomplementary (three-base mismatch strand) DNA in solution to a 30-mer 3′-thiolated DNA strand immobilized on an electrochemical enzymatic genosensor (Abad-Valle et al., 2007a,b), (e) binding (hybridization) of (i) a perfectly matched oligonucleotide (ODN-P) and (ii) a noncomplementary ODN (ODN-N) to an electrochemical sensor with a EST2-A34 reporter (Wang et al., 2007), (f) binding and dissociation during PNA-DNA hybridization—binding of different concentrations (in μM) of target DNA complementary to CYP2C9*2 (target DNA2) to CYP2C9*2 as a probe PNA immobilized on a ion-sensitive field-effect transistor (IS-FET)-based biosensor (Uno et al., 2007), (g) binding and dissociation during PNA-DNA hybridization—binding of different concentrations (in μM) of target DNA complementary to CYP2C9*2 (target DNA2) to CYP2C9*2 as a probe PNA

Handbook of Biosensors and Biosensor Kinetics. DOI: 10.1016/B978-0-444-53262-6.00011-5

immobilized on an IS-FET-based biosensor (Uno et al., 2007), (h) binding and dissociation of RNA synthesized on a (i) 42 nM template and a (ii) 420 nM template (Blair et al., 2007), and (i) binding (hybridization) of different concentrations of ss DNA in solution preincubated with prehybridized 22-nt FQ duplex to a "broken beacon" immobilized on a sensor surface (Blair et al., 2007). One may consider the fractal analysis as an alternate method of analyzing the kinetics of binding and dissociation during hybridization in these types of analyte-receptor reactions occurring on biosensor surfaces.

11.2 Theory

11.2.1 Single-Fractal Analysis

Binding Rate Coefficient

Havlin (1989) points out that the diffusion of a particle (analyte [Ag]) from a homogeneous solution to a solid surface (e.g., receptor [Ab]-coated surface) on which it reacts to form a product (analyte-receptor complex; Ab·Ag) is given by:

$$(\mathrm{Ab}\cdot\mathrm{Ag}) \approx \begin{cases} t^{(3-D_{\mathrm{f,bind}})/2} = t^{p}, & t < t_{\mathrm{c}} \\ t^{1/2}, & t > t_{\mathrm{c}} \end{cases} \tag{11.1}$$

Here $D_{\mathrm{f,bind}}$ or D_{f} is the fractal dimension of the surface during the binding step. t_{c} is the cross-over value. Havlin (1989) points out that the cross-over value may be determined by $r_{\mathrm{c}}^{2} \sim t_{\mathrm{c}}$. Above the characteristic length, r_{c}, the self-similarity of the surface is lost and the surface may be considered homogeneous. Above time, t_{c} the surface may be considered homogeneous, since the self-similarity property disappears, and "regular" diffusion is now present. For a homogeneous surface where D_{f} is equal to 2, and when only diffusional limitations are present, $p = ½$ as it should be. Another way of looking at the $p = ½$ case (where $D_{\mathrm{f,bind}}$ is equal to two) is that the analyte in solution views the fractal object, in our case, the receptor-coated biosensor surface, from a "large distance." In essence, in the association process, the diffusion of the analyte from the solution to the receptor surface creates a depletion layer of width $(Đt)^{½}$ where Đ is the diffusion constant. This gives rise to the fractal power law, $(\mathrm{Analyte}\cdot\mathrm{Receptor}) \sim t^{(3-D_{\mathrm{f,bind}})/2}$. For the present analysis, t_{c} is arbitrarily chosen and we assume that the value of the t_{c} is not reached. One may consider the approach as an intermediate "heuristic" approach that may be used in the future to develop an autonomous (and not time-dependent) model for diffusion-controlled kinetics.

Dissociation Rate Coefficient

The diffusion of the dissociated particle (receptor [Ab] or analyte [Ag]) from the solid surface (e.g., analyte [Ag]-receptor [Ab]) complex coated surface) into solution may be given, as a first approximation by:

$$(\mathrm{Ab}\cdot\mathrm{Ag}) \approx -t^{(3-D_{\mathrm{f,diss}})/2} = -t^{p}, \quad t > t_{\mathrm{diss}} \tag{11.2}$$

Here $D_{f,diss}$ is the fractal dimension of the surface for the dissociation step. This corresponds to the highest concentration of the analyte-receptor complex on the surface. Henceforth, its concentration only decreases. The dissociation kinetics may be analyzed in a manner "similar" to the binding kinetics.

11.2.2 Dual-Fractal Analysis

Binding Rate Coefficient

Sometimes, the binding curve exhibits complexities and two parameters (k, D_f) are not sufficient to adequately describe the binding kinetics. This is further corroborated by low values of the r^2 factor (goodness-of-fit). In that case, one resorts to a dual-fractal analysis (four parameters; k_1, k_2, D_{f1}, and D_{f2}) to adequately describe the binding kinetics. The single-fractal analysis presented above is thus extended to include two fractal dimensions. At present, the time ($t = t_1$) at which the "first" fractal dimension "changes" to the "second" fractal dimension is arbitrary and empirical. For the most part, it is dictated by the data analyzed and experience gained by handling a single-fractal analysis. A smoother curve is obtained in the "transition" region,if care is taken to select the correct number of points for the two regions. In this case, the product (antibody-antigen; or analyte-receptor complex, Ab·Ag or analyte·receptor) is given by:

$$(\mathrm{Ab}\cdot\mathrm{Ag}) \approx \begin{cases} t^{(3-D_{f1,\mathrm{bind}})/2} = t^{p1}, & t < t_1 \\ t^{(3-D_{f2,\mathrm{bind}})/2} = t^{p2}, & t_1 < t < t_2 = t_c \\ t^{1/2}, & t > t_c \end{cases} \tag{11.3}$$

In some cases, as mentioned above, a triple-fractal analysis with six parameters (k_1, k_2, k_3, D_{f1}, D_{f2}, and D_{f3}) may be required to adequately model the binding kinetics. This is when the binding curve exhibits convolutions and complexities in its shape due perhaps to the very dilute nature of the analyte (in some of the cases to be presented) or for some other reasons. Also, in some cases, a dual-fractal analysis may be required to describe the dissociation kinetics.

11.3 Results

We will use fractal analysis to analyze the binding (hybridization) and dissociation kinetics exhibited by different analyte-receptor reactions occurring on biosensor surfaces. This is just one possible method of analyzing the kinetics of the different analyte-recptor (hybridization reactions) presented in this chapter. Alternative expressions for fitting the data are available that include saturation, first-order reaction, and no diffusion limitations, but these expressions are apparently deficient in describing the heterogeneity that inherently exists on the surface. One might justifiably argue that the appropriate modeling may be achieved by using a Langmuirian or other approach. The Langmuirian approach may be used to model the data presented if one assumes the presence of discrete classes of sites (for example, double exponential analysis as compared with a single-fractal analysis). Lee and Lee (1995) report that the fractal approach has been applied to surface science, for example, adsorption and reaction processes. These authors point out that the fractal approach provides a convenient means to represent the different structures and the morphology at the reaction surface. They also draw attention to the use of the fractal approach to develop optimal structures and as a predictive approach. Another advantage of the fractal technique is that the analyte-receptor association (as well as the dissociation reaction) is a complex reaction, and the fractal analysis via the fractal dimension and the rate coefficient provides a useful lumped parameter(s) analysis of the diffusion-limited reaction occurring on a heterogeneous surface.

In a classical situation, to demonstrate fractality, one should make a log-log plot, and one should definitely have a large amount of data. It may be useful to compare the fit to some other forms, such as exponential, or one involving saturation, etc. At present, no independent proof or physical evidence of fractals in the examples is presented. It is a convenient means (since it is a lumped parameter) to make the degree of heterogeneity that exists on the surface more quantitative. Thus, there is some arbitrariness in the fractal model to be presented. The fractal approach provides additional information about interactions that may not be obtained by conventional analysis of biosensor data.

There is no nonselective adsorption of the analyte. The present system being analyzed may be typically very dilute. Nonselective adsorption would skew the results obtained very significantly. In these types of systems, it is imperative to minimize this nonselective adsorption. It is also recognized that, in some cases, this nonselective adsorption may not be a significant component of the adsorbed material and that this rate of association, which is of a temporal nature, would depend on surface availability. If the nonselective adsorption were to be accommodated into the model, there would be an increase in the heterogeneity on the surface, as, by its very nature, nonspecific adsorption is more homogeneous than specific adsorption. This would lead to higher fractal dimension values since the fractal dimension is a direct measure of the degree of heterogeneity that exists on the surface.

Fiche et al. (2007) recently analyzed hybridization experiments on a DNA chip using surface plasmon resonance imaging (SPRi). These authors point out that to obtain quantitative results it is essential to clarify the heterogeneity of the hybridization of DNA on microarrays. The aim of their experiment was to obtain a detailed account of the equilibrium and kinetics of hybridization on a DNA chip. They point out that experimental results on a DNA chip are commonly analyzed using the Langmuir model (Peterson et al., 2002; Hekstra et al., 2003; Tawa and Knoll, 2004; Yu et al., 2004; Wark et al., 2005; Yao et al., 2005). In most of the above mentioned studies the heterogeneity of the hybridization is generally not taken into account. Furthermore, Fiche et al. (2007) explain that as most of the experiments are done at room temperature effects on the equilibrium and kinetics properties are not taken into account. Thus, these authors analyzed temperature effects on DNA experiments.

Fiche et al. (2007) performed hybridization experiments at different temperatures and target concentrations. All their probes had at their 5′end a 10-thyminespacer and a pyrrole (Py) moiety for the electropolymerization grafting method. Figure 11.1a shows the binding and dissociation of the target T1 to the probe P14 (5′-Py-(T_{10})-GCC.TGG.ACG.ATA.CA-3′) immobilized on the DNA chip. 14 refers to the number of hybridizing bases. A dual-fractal analysis is required to adequately describe the binding kinetics. A single-fractal analysis is adequate to describe the dissociation kinetics. The values of (a) the binding rate coefficient, k, and the fractal dimension, D_f, for a single-fractal analysis, (b) the binding rate coefficients, k_1 and k_2, and the fractal dimensions, D_{f1} and D_{f2}, for a dual-fractal analysis, and the dissociation rate coefficient, k_d and the fractal dimension for dissociation, D_{fd} are given in Table 11.1. The values of the binding and the dissociation rate coefficients, and the fractal dimensions for the binding and the dissociation phase presented in Table 11.1 were obtained from a regression analysis

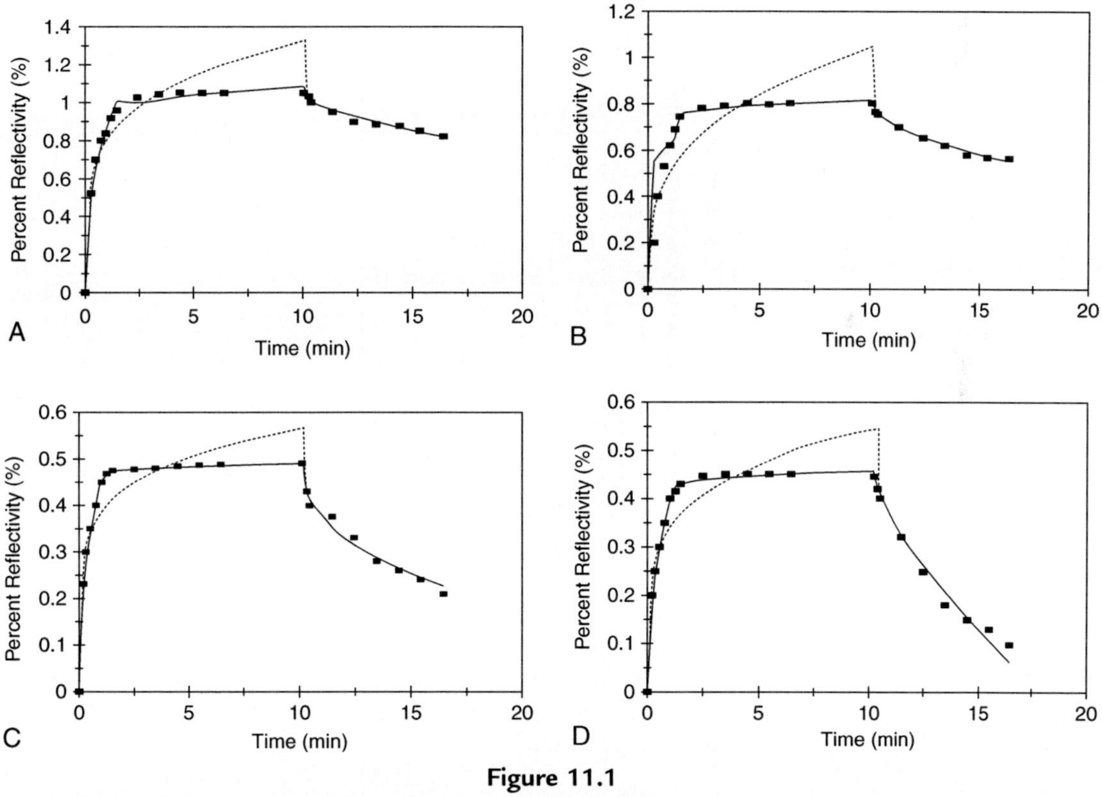

Figure 11.1

Binding and dissociation (hybridization) of different targets (400 nM) in solution to a probe immobilized on a DNA chip surface at 32.5 °C (Fiche et al., 2007): (a) P14. (b) P12. (c) P10. (d) P9. When only a solid line (—) is used then a single-fractal analysis applies. When both a dashed (- - -) and a solid (—) line are used then the dashed line represents a single-fractal analysis and the solid line represents a dual-fractal analysis.

Table 11.1: Binding and dissociation rate coefficients and fractal dimensions for the binding and the dissociation phase during hybridization on a DNA chip (Fiche et al., 2007).

Target/Probe	k	k_1	k_2	k_d	D_f	D_{f1}	D_{f2}	D_{fd}
400 nM T1/P14	0.8028 ± 0.0774	0.8623 ± 0.0445	0.9422 ± 0.0268	2.807 ± 0.051	2.5632 ± 0.06376	2.2344 ± 0.1346	2.8783 ± 0.0292	3.0- 0.0774
400 nM T1/P12	0.5051 ± 0.1408	0.6261 ± 0.0849	0.7441 ± 0.0119	0.0941 ± 0.0052	2.3744 ± 0.1346	2.3744 ± 0.1346	2.9218 ± 0.0204	1.9260 ± 0.0312
400 nM T1/P10	0.1235 ± 0.0152	0.4545 ± 0.021	0.4693 ± 0.0016	0.1235 ± 0.0152	2.6704 ± 0.05926	2.6704 ± 0.0593	2.9594 ± 0.0037	2.1790 ± 0.06688
400 nM T1/P9	0.1057 ± 0.0104	0.3995 ± 0.0096	0.4230 ± 0.0076	0.1057 ± 0.0104	2.6186 ± 0.0606	2.6186 ± 0.0606	2.9301 ± 0.0185	1.5948 ± 0.05398

using Corel Quattro Pro 8.0 (1997) to model the experimental data using Equations (11.1)–(11.3), wherein [analyte·receptor or Ab·Ag] $= kt^p$ for the binding step, and [analyte·receptor or Ab·Ag]$= -kt^p$ for the dissociation step.

The binding and the dissociation rate coefficients presented in Table 11.1 are within 95% confidence limits. For example, for the binding of 400 nM target T1 in solution to the probe P14 immobilized on a DNA sensor chip the binding rate coefficient, k_1, value for a dual-fractal analysis is 0.8623 ± 0.0445. The 95% confidence limit indicates that the k_1 values will lie between 0.8178 and 0.90968. This indicates that the values are precise and significant. To indicate the goodness-of-fit, the r^2 value is provided. In this case the r^2 value is 0.961. This is a typical value obtained.

Figure 11.1b shows the binding and dissociation of the target T1 to the probe P12 (5′-Py-(T_{10})-GCC.TGG.ACG.ATA-3′) immobilized on the DNA chip. 12 refers to the number of hybridizing bases. Once again, a dual-fractal analysis is required to adequately describe the binding kinetics. A single-fractal analysis is adequate to describe the dissociation kinetics. The values of (a) the binding rate coefficient, k, and the fractal dimension, D_f, for a single-fractal analysis, (b) the binding rate coefficients, k_1 and k_2, and the fractal dimensions, D_{f1} and D_{f2}, for a dual-fractal analysis, and the dissociation rate coefficient, k_d and the fractal dimension for dissociation, D_{fd} are given in Table 11.1.

Figure 11.1c shows the binding and dissociation of the target T1 to the probe P10 (5′-Py-(T_{10})-GCC.TGG.ACG.A-3′) immobilized on the DNA chip. 10 refers to the number of hybridizing bases. Once again, a dual-fractal analysis is required to adequately describe the binding kinetics. A single-fractal analysis is adequate to describe the dissociation kinetics. The values of (a) the binding rate coefficient, k, and the fractal dimension, D_f, for a single-fractal analysis, (b) the binding rate coefficients, k_1 and k_2, and the fractal dimensions, D_{f1} and D_{f2}, for a dual-fractal analysis, and the dissociation rate coefficient, k_d and the fractal dimension for dissociation, D_{fd} are given in Table 11.1.

Figure 11.1d shows the binding and dissociation of the target T1 to the probe P9 (5′-Py-(T_{10})-GCC.TGG.ACG-3′) immobilized on the DNA chip. 9 refers to the number of hybridizing bases. Once again, a dual-fractal analysis is required to adequately describe the binding kinetics. A single-fractal analysis is adequate to describe the dissociation kinetics. The values of (a) the binding rate coefficient, k, and the fractal dimension, D_f, for a single-fractal analysis, (b) the binding rate coefficients, k_1 and k_2, and the fractal dimensions, D_{f1} and D_{f2}, for a dual-fractal analysis, and the dissociation rate coefficient, k_d and the fractal dimension for dissociation, D_{fd} are given in Table 11.1.

Figure 11.2a and Table 11.1 show for the binding of different targets (400 nM) in solution to a probe immobilized on a DNA chip surface at 32.5 °C and for a dual-fractal analysis the decrease in the binding rate coefficient, k_2, with an increase in the fractal

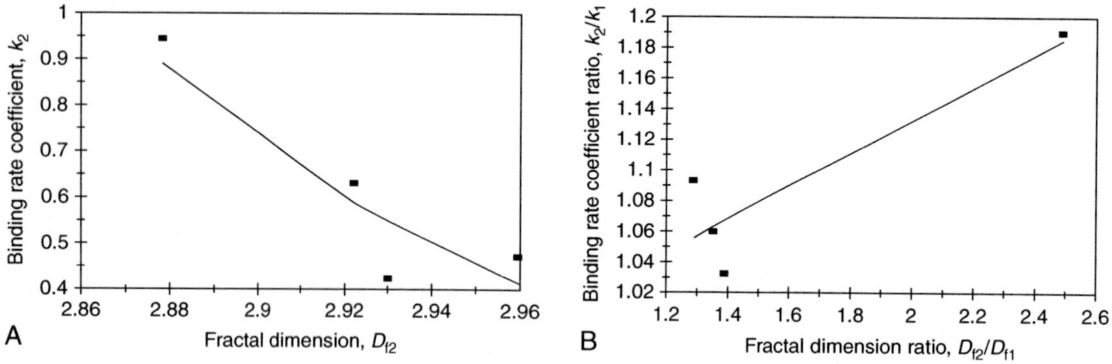

Figure 11.2

(a) Decrease in the binding rate coefficient, k_2, with an increase in the fractal dimension, D_{f2}. Increase in the ratio of the binding rate coefficients, k_2/k_1, with an increase in the ratio of fractal dimensions, D_{f2}/D_{f1}.

dimension, D_{f2}. For the data shown in Figure 11.2a, the binding rate coefficient, k_2, is given by:

$$k_2 = [3.6 \pm 0.8 \times 10^{12}]D_{f2}^{-27.46\pm10.45} \tag{11.4a}$$

The fit is good. Only four data points are available. The availability of more data points would lead to a more reliable fit. The binding rate coefficient, k_2, is extremely sensitive to the degree of heterogeneity on the biosensor surface or the fractal dimension, D_{f2}, as noted by the negative order close to twenty seven and a half (equal to −27.46) exhibited.

Figure 11.2b also shows the increase in the ratio of the binding rate coefficients, k_2/k_1, with an increase in the ratio of the fractal dimensions, D_{f2}/D_{f1}, for the binding of different targets (400 nM) in solution to a probe immobilized on a DNA chip surface at 32.5 °C and for a dual-fractal analysis. For the data shown in Figure 11.2b, the ratio of the binding rate coefficients, k_2/k_1, is given by:

$$k_2/k_1 = (1.0094 \pm 0.0362)(D_{f2}/D_{f1})^{0.174\pm0.0656} \tag{11.4b}$$

The fit is reasonable. Only four data points are available. The availability of more date points would lead to a more reliable fit. The ratio of the binding rate coefficients, k_2/k_1, exhibit only a mild order (equal to 0.174) of dependence on the ratio of the fractal dimensions, D_{f2}/D_{f1}.

Michel et al. (2007) recently reported that microarrays can readily identify DNA sequences simultaneously, and are rapidly becoming major tools for pharmacogenomics and clinical pathology. These authors used an optical method to analyze the DNA surface hybridization. They noted that DNA surface density is a key parameter in microarray hybridization kinetics.

Also, a change in the bulk concentration has a significant impact on hybridization kinetics. They analyzed hybridization kinetics on glass substrates. One 22-mer strand (bound DNA) was immobilized via a phenylene-diisocyanate linker molecule on the glass substrate. The dye-labeled (Cy3) complementary strand was in solution in a reaction chamber. These authors further explain that to work efficiently with microarrays a knowledge of kinetics and thermodynamics is essential.

Figure 11.3a shows the binding of 10 nM free-DNA in solution to a 22-mer strand (bound DNA) immobilized via a phenylene-diisocyanate linker molecule on a glass substrate (Michel et al., 2007). A dual-fractal analysis is required to adequately describe the binding kinetics. The values of (a) the binding rate coefficient, k, and the fractal dimension, D_f, for a single-fractal analysis and (b) the binding rate coefficients, k_1 and k_2, and the fractal dimensions, D_{f1} and D_{f2}, for a dual-fractal analysis are given in Table 11.2.

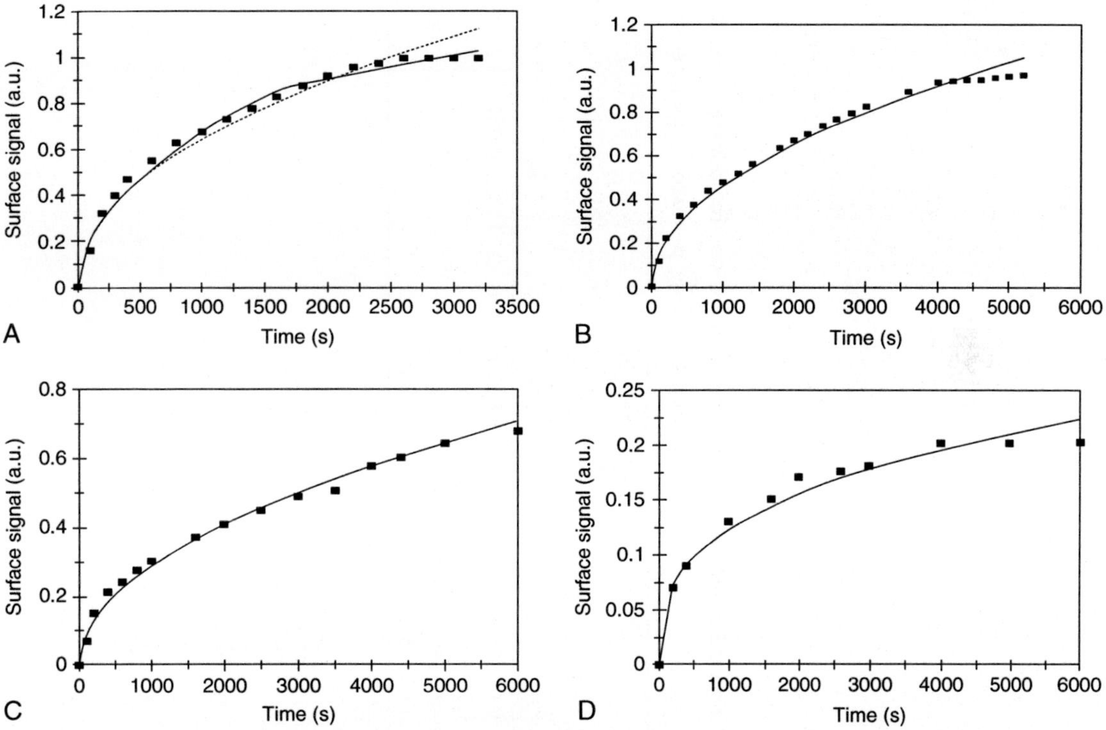

Figure 11.3

Binding (hybridization) of different concentrations (in nM) of free-DNA in solution to a 22-mer strand (bound DNA) immobilized via a phenylene-diisocyanate linker molecule on a glass substrate (Michel et al., 2007): (a) 10. (b) 7.5. (c) 5. (d) 2. When only a solid line (—) is used then a single-fractal analysis applies. When both a dashed (- - -) and a solid (—) line are used then the dashed line represents a single-fractal analysis and the solid line represents a dual-fractal analysis.

Table 11.2: Binding rate coefficients and fractal dimensions for different initial-free DNA concentrations (in nM) on a substrate prepared following protocol A (Michel et al., 2007).

Initial Free-DNA Concentration (nM)	k	k_1	k_2	D_f	D_{f1}	D_{f2}
10	0.02345 ± 0.0019	0.01789 ± 0.00190	0.1193 ± 0.0031	2.0390 ± 0.04794	1.9482 ± 0.0640	2.4652 ± 0.07622
7.5	0.01509 ± 0.001	na	na	2.0062 ± 0.02562	na	na
5	0.009188 ± 0.000959	na	na	2.0004 ± 0.04282	na	na
2	0.01326 ± 0.00083	na	na	2.3518 ± 0.03674	na	na

Note that an increase in the fractal dimension by a factor of 1.265 from a value of D_{f1} equal to 1.9482 to D_{f2} equal to 2.465 leads to an increase in the binding rate coefficient by a factor of 6.67 from a value of k_1 equal to 0.01789 to k_2 equal to 0.1193.

Figure 11.3b shows the binding of 7.5 nM free-DNA in solution to a 22-mer strand (bound DNA) immobilized via a phenylene-diisocyanate linker molecule on a glass substrate (Michel et al., 2007). A dual-fractal analysis is required to adequately describe the binding kinetics. The values of (a) the binding rate coefficient, k, and the fractal dimension, D_f, for a single-fractal analysis and (b) the binding rate coefficients, k_1 and k_2, and the fractal dimensions, D_{f1} and D_{f2}, for a dual-fractal analysis are given in Table 11.2.

Figure 11.3c shows the binding of 5.0 nM free-DNA in solution to a 22-mer strand (bound DNA) immobilized via a phenylene-diisocyanate linker molecule on a glass substrate (Michel et al., 2007). A dual-fractal analysis is required to adequately describe the binding kinetics. The values of (a) the binding rate coefficient, k, and the fractal dimension, D_f, for a single-fractal analysis and (b) the binding rate coefficients, k_1 and k_2, and the fractal dimensions, D_{f1} and D_{f2}, for a dual-fractal analysis are given in Table 11.2.

Figure 11.3d shows the binding of 2.0 nM free-DNA in solution to a 22-mer strand (bound DNA) immobilized via a phenylene-diisocyanate linker molecule on a glass substrate (Michel et al., 2007). A dual-fractal analysis is required to adequately describe the binding kinetics. The values of (a) the binding rate coefficient, k, and the fractal dimension, D_f, for a single-fractal analysis and (b) the binding rate coefficients, k_1 and k_2, and the fractal dimensions, D_{f1} and D_{f2}, for a dual-fractal analysis are given in Table 11.2.

Figure 11.4 and Table 11.2 show for a single-fractal analysis the decrease in the fractal dimension, D_f with an increase in the initial free-DNA concentration in the 2-7.5 nM range in solution. For this 2-7.5 nM concentration range, the fractal dimension, D_f, is given by:

$$D_f = (2.547 \pm 0.107)[\text{initial free} - \text{DNA, in nM}]^{-0.1298 \pm 0.0429} \tag{11.4c}$$

The fit is good. Only three data points are available. The availability of more data points would lead to a more reliable fit. The fractal dimension, D_f, exhibits a very slight negative order (equal to -0.1298) of dependence on the initial free-DNA concentration in solution. The fractal dimension, D_f, is based on a log scale. Thus, even very small changes in the fractal dimension indicate significant changes in the degree of heterogeneity on the biosensor chip surface.

Figure 11.5 shows the binding of 1 nM initial free-DNA concentration in solution at 22 mer strand (bound DNA) immobilized via a phenylene-diisocyanate linker molecule on a glass substrate (Michel et al., 2007). A single-fractal analysis is adequate to describe the binding kinetics. The values of the binding rate coefficient, k, and the fractal dimension, D_f, are given in Table 11.3.

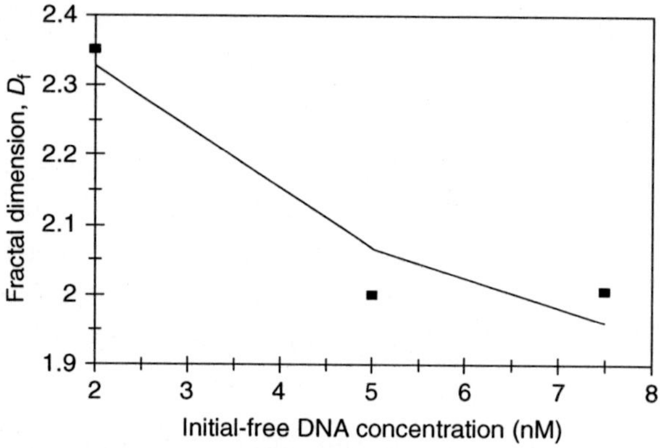

Figure 11.4
Decrease in the fractal dimension, D_f, with an increase in the free-DNA concentration (in nM) in solution.

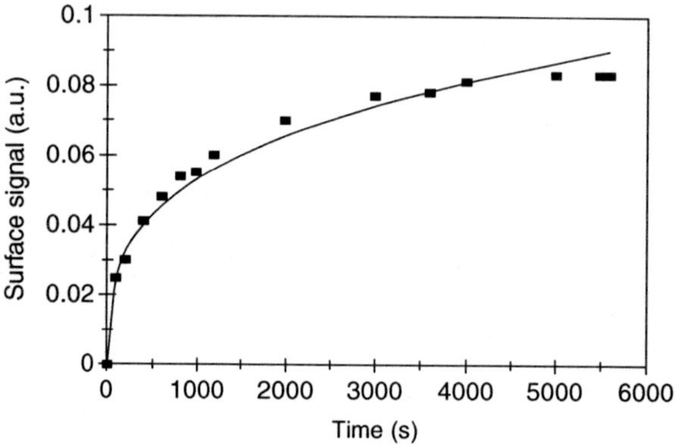

Figure 11.5
Binding (hybridization) of 1 nM free-DNA concentration in solution to a 22 mer strand (bound DNA) immobilized via a phenylene-diisocyanate linker molecule on a glass substrate (Michel et al., 2007).

Figure 11.6a shows the binding of nonmatching, noncomplementary strand m22 5′-Cy3-TGA GCG TTC GTG GTG GGA TAG T-3′ in solution to one strand (bound DNA; i22, 5′-NH_2-C6-TTT TTT TTT TTT TTT TGA TAG GGT GGT GCT GGT GCT TGC GAG T-3′) immobilized on a glass substrate (Michel et al., 2007). A single-fractal analysis is adequate to describe the binding kinetics. The values of the binding rate coefficient, k, and the fractal dimension, D_f, for a single-fractal analysis are given in Table 11.3.

Figure 11.6b shows the binding of a matching sequence, complementary strand p22 5′-Cy3-ACT CGC AAG CAC CAC CCT ATC-A-3′ in solution to one strand (bound DNA; i22, 5′-NH_2-C6 TTT TTT TTT TTT TTT TGA TAG GGT GGT GCT GGT GCT TGC GAG T-3′) immobilized

Table 11.3: Binding rate coefficients and fractal dimensions for (a) bulk concentration of free DNA in solution (1 nM) to sensor surface, (b) nonmatching (non complementary; m22; 5′-Cy3-TGA-GCG-TTC-GTG-GTG-GGA-TAG-T-3′), and matching sequence (complementary; p22; 5′-Cy3-ACT-CGC-AAG-CAC-CAC-CCT-ATC-A-3′) in solution to one strand (bound DNA; i22; 5′-NH_2-C6-TTT-TTT-TTT-TTT-TTT-TGA-TAG-GGT-GGT-GCT-GGT-GCT-TGC-GAG-T-3′) immobilized on a glass substrate (Michel et al., 2007).

Strand in Solution/ Receptor on Surface	k	k_1	k_2	D_f	D_{f1}	D_{f2}
1 nM free DNA/ surface	0.006628 ± 0.000415	na	na	2.3966 ± 0.02612	na	na
Noncomplementary (m22)/i22	$5.7 \times 10^{-05} \pm 3 \times 10^{-05}$	na	na	1.7036 ± 0.1151	na	na
Complementary (p22)/i22	0.06240 ± 0.01266	0.008629 ± 0.000020	0.4462 ± 0.0162	2.2074 ± 0.1243	1.4592 ± 0.005772	2.7794 ± 0.05058

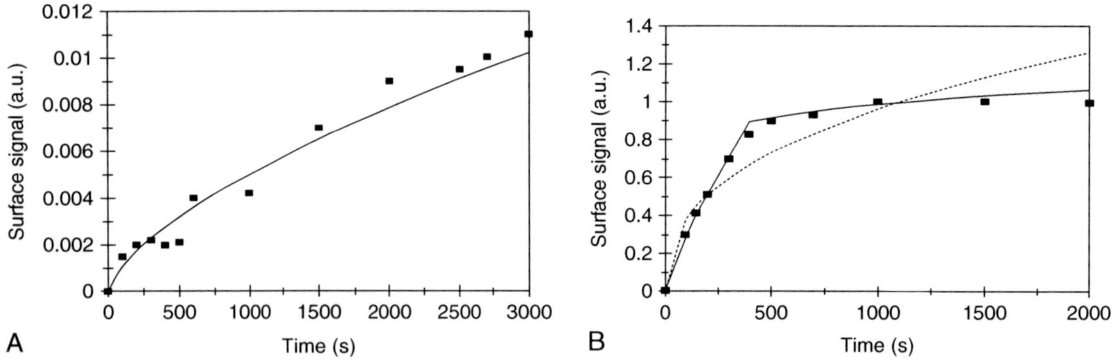

Figure 11.6

Binding of 100 nM free-DNA in solution to a (a) noncomplementary and (b) a complementary 22 mer strand (bound DNA) immobilized via a phenylene-diisocyanate linker molecule on a glass substrate (Michel et al.,2007). When only a solid line (—) is used then a single-fractal analysis applies. When both a dashed (- - -) and a solid (—) line are used then the dashed line represents a single-fractal analysis and the solid line represents a dual-fractal analysis.

on a glass substrate (Michel et al., 2007). A dual-fractal analysis is required to adequately describe the binding kinetics. The values of (a) the binding rate coefficient, k, and the fractal dimension, D_f, for a single-fractal analysis, and the binding rate coefficients, k_1 and k_2, and the fractal dimensions, D_{f1} and D_{f2}, for a dual-fractal analysis are given in Table 11.3. It is of interest to note that for a dual-fractal analysis, as the fractal dimension increases by a factor of 1.905 from a value of D_{f1} equal to 1.4592 to D_{f2} equal to 2.7794, the binding rate coefficient increases by a factor of 51.71 from a value of k_1 equal to 0.00869 to k_2 equal to 0.4462.

Also note that for the binding in the nonmatching (noncomplementary) case a single-fractal analysis is adequate to describe the binding kinetics. However, for the matching (complementary) case a dual-fractal analysis is required to describe the binding kinetics. This would indicate that, at least for this case, the binding of the matching (complementary) case is more complicated than that of the nonmatching (noncomplementary) case. No explanation is offered, at present, to help explain why this is the case.

Feng et al. (2007) recently reported that single nucleotide polymorphisms (SNPs) are important in clinical diagnostics, pathology detection, and genetic diseases. Lin et al. (2005) point out that SNPs are point mutations that include the most-common genetic variation. Wabuyele et al. (2003) have explained that quite a few genetic diseases and cancers are associated with mutation in the sequence of particular genes. Landegren et al. (1988) initially used DNA ligase for the detection of SNPs. Feng et al. (2007) have used the QCM technique coupled with the DNA enzyme-based ligase reaction to sense a point mutation in a DNA target. These authors used a signal amplification method for the quantitative detection of the target gene that included the deposition of an insoluble product of 3,3-diaminobenzidine (DAB) (Karousis et al., 2002) on the electrode supports mediated by SA-HRP conjugate.

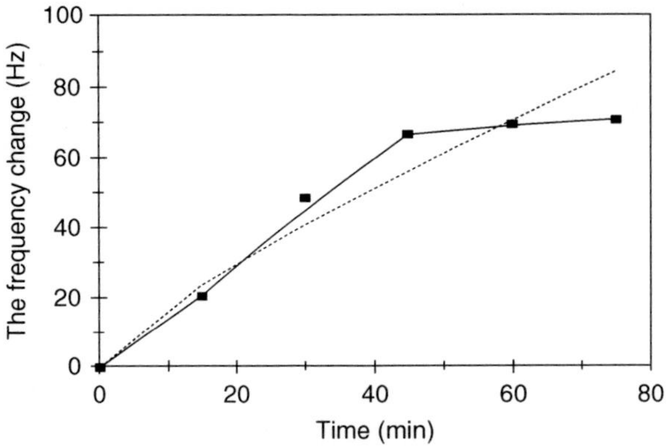

Figure 11.7

Binding (hybridization) of SA-HRP (streptavidin horseradish peroxidase) in solution to a capture probe on a QCM (quartz crystal microbalance) electrode along with a detection probe (Feng et al., 2007). When only a solid line (—) is used then a single-fractal analysis applies. When both a dashed (- - -) and a solid (—) line are used then the dashed line represents a single-fractal analysis and the solid line represents a dual-fractal analysis.

Figure 11.7a shows the binding of SA-HRP and DAB in solution to the capture probe modified QCM electrode along with 1 μM detection probe. A dual-fractal analysis is required to adequately describe the binding kinetics. The values of (a) the binding rate coefficient, k, and the fractal dimension, D_f, for a single-fractal analysis, and (b) the binding rate coefficients, k_1 and k_2 and the fractal dimensions, D_{f1} and D_{f2}, for a dual-fractal analysis are given in Tables 11.4 and 11.5. It is of interest to note that for a dual-fractal analysis as the fractal dimension increases by a factor of 3.51 from a value of D_{f1} equal to 0.7886 to D_{f2} equal to 2.7684, the binding rate coefficient increases by a factor of 41.96 from a value of k_1 equal to 1.0132 to k_2 equal to 42.513.

Abad-Valle et al. (2007a,b) recently used an electrochemical enzymatic genosensor to analyze DNA single-base mismatches. These authors report that electrochemical transducers provide rapid and sensitive measurements. Besides, these devices are simple low cost, and exhibit the potential to be miniaturized. Abad-Valle et al. (2005) further explain that enzyme labels, due to their inherent amplification help permit an increase in assay sensitivity. Caruana and Heller (1999) used a soybean peroxidase label for detecting a single-base mismatch in an 18-base oligonucleotide. Abad-Valle et al. (2005) had previously developed an enzymatic electrochemical genosensor on gold films to analyze the selectivity of DNA hybridization. Abad-Valle et al. (2007a,b) report that they have used a sequence of the SARS (severe acute respiratory syndrome) coronavirus (CoV) as a target. This SARS CoV is the causative agent of an atypical pneumonia. They further point out that it is essential to identify the SARS-CoV quickly and accurately owing to the rate of mortality of patients.

Table 11.4: Binding and dissociation rate coefficients for the hybridization of different analytes in solution to complementary or noncomplementary receptors immobilized on different biosensor surfaces.

Biosensor Type	Analyte in Solution/ Receptor on surface	k	k_1	k_2	k_3	k_d	References
QCM	3,3-diaminobenzidine/ sterptavidin-peroxide horseradish (SA-HRP)	2.7136 ± 0.5831	1.0132 ± 0.1069	42.513 ± 0.141	na	na	Feng et al. (2007)
Electrochemical enzymatic genosensor	Sequence of the SARS (severe acute respiratory syndrome) coronavirus (CoV) SARS-CoV/(c-DNA) 30-mer 3′-thiolated DNA strand	7.0291 ± 1.1079	na	na	na	na	Abad-Valle et al. (2007a,b)
Electrochemical enzymatic genosensor	Sequence of the SARS (severe acute respiratory syndrome) coronavirus (CoV) SARS-CoV/3 base mismatch 30-mer 3′-thiolated DNA strand	23.569 ± 2.627	na	na	na	na	Abad-Valle et al. (2007a,b)
Electrochemical detection	*p*-aminophenylbutyrate/ esterase 2 from *Alicyclobacillus acidocaldarius* plus oligodeoxynucleotide (ODN) in a site-specific manner (perfectly matched; ODN-P)	6.9462 ± 1.9652	2.8403 ± 0.2542	37.626 ± 1.136	na	0.07685 ± 0.00598	Wang et al. (2007)
Electrochemical detection	*p*-aminophenylbutyrate/ esterase 2 from *Alicyclobacillus acidocaldarius* plus oligodeoxynucleotide (ODN) in a site-specific manner (noncomplementary ODN; ODN-N)	0.1261 ± 0.0600	0.0409 ± 0.0135	1.5596 ± 0.1142	4.7240 ± 0.4477	na	Wang et al. (2007)

Table 11.5: Fractal dimensions for the binding and the dissociation phases for the hybridization of different analytes in solution to complementary or noncomplementary receptors immobilized on different biosensor surfaces.

Biosensor Type	Analyte in Solution/ Receptor on Surface	D_f	D_{f1}	D_{f2}	D_{f3}	D_{fd}	References
QCM	3,3-diaminobenzidine/ sterptavidin-peroxide horseradish (SA-HRP)	1.4132 ± 0.3062	0.7886 ± 0.2512	2.7684 ± 0.01818	na	na	Feng et al. (2007
Electrochemical enzymatic genosensor	Sequence of the SARS (severe acute respiratory syndrome) coronavirus (CoV) SARS-CoV/(c-DNA) 30-mer 3′-thiolated DNA strand	1.9290 ± 0.2812	na	na	na	na	Abad-Valle et al. (2007a,b)
Electrochemical enzymatic genosensor	Sequence of the SARS (severe acute respiratory syndrome) coronavirus (CoV) SARS-CoV/3 base mismatch 30-mer 3′-thiolated DNA strand	2.7530 ± 0.2030	na	na	na	na	Abad-Valle et al. (2007a,b)
Electrochemical detection	*p*-aminophenylbutyrate/ esterase 2 from *Alicyclobacillus acidocaldarius* plus oligodeoxynucleotide (ODN) in a site-specific manner (perfectly matched; ODN-P)	1.4252 ± 0.2030	0.4432 ± 0.2182	2.5556 ± 0.1208	na	0.5014 ± 0.1011	Wang et al. (2007)
Electrochemical detection	*p*-aminophenylbutyrate/ esterase 2 from *Alicyclobacillus acidocaldarius* plus oligodeoxynucleotide (ODN) in a site-specific manner (noncomplementary ODN; ODN-N)	0.5602 ± 0.2460	0. + 0.4148	1.78902 ± 0.2730	2.4304 ± 0.1647	na	Wang et al. (2007)

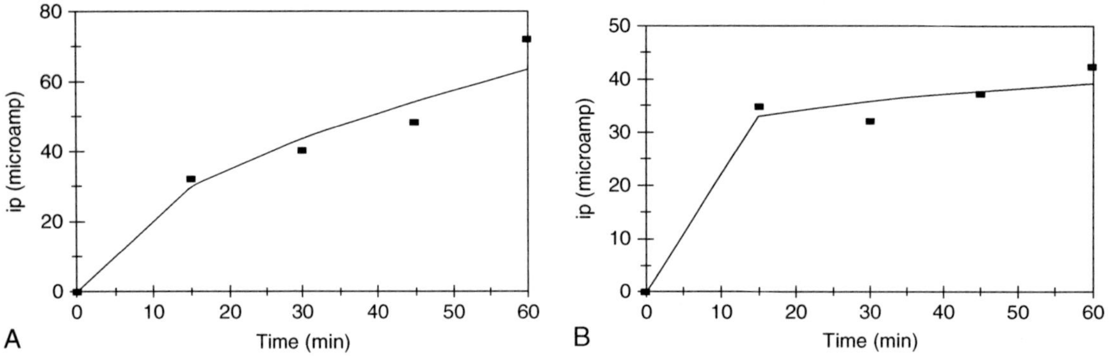

Figure 11.8

Binding (hybridization) of (a) complementary and (b) a noncomplementary (three-base mismatch strand) DNA in solution to a 30-mer 3′-thiolated DNA strand immobilized on an electrochemical enzymatic genosensor (Abad-Valle et al., 2007a,b)

Figure 11.8a shows the binding (hybridization) of a complementary DNA in solution to a 30-mer 3′-thiolated DNA strand immobilized on an electrochemical genosensor (Abad-Valle et al., 2007a,b). A single-fractal analysis is adequate to describe the binding kinetics. The values of the binding rate coefficient, k, and the fractal dimension, D_f, for a single-fractal analysis are given in Tables 11.4 and 11.5.

Figure 11.8b shows the binding (hybridization) of a three-base mismatch DNA strand to a 30-mer 3′-thiolated DNA strand immobilized on an electrochemical genosensor (Abad-Valle et al., 2007a,b). Once again a single-fractal analysis is adequate to describe the binding kinetics. The values of the binding rate coefficient, k, and the fractal dimension, D_f, for a single-fractal analysis are given in Tables 11.4 and 11.5. It is of interest to note that as one goes from the binding of the complementary DNA to the three base-mismatch strand in solution to the 30-mer 3′-thiolated DNA strand immobilized on the electrochemical genosensor, the fractal dimension increases by a factor of 1.427 from a value of D_f equals to 1.9290 to 2.7520, and the binding rate coefficient, k increases by a factor of 3.353 from a value of k equal to 7.0291 to k equal to 23.569. Increases in the degree of heterogeneity or the fractal dimension on the sensor chip surface and in the binding rate coefficient are in the same direction.

Wang et al. (2007) recently analyzed the binding of complementary ODN (ODN-P) (2-diolgonucleotide) and a noncomplementary ODN-N (nonmatching) to an electrochemical sensor with a EST2-A34 reporter. These authors used esterase 2-oligonucleotide conjugate as a sensitive reporter for the electrochemical detection of nucleic acid hybridization. Figure 11.9a shows the binding of p-aminophenylbutyrate/esterase 2 from *Alicyclobacillus acidocaldarius* plus oligonucleotide (ODN) in solution to a site-specific manner ODN-P (perfectly matched; complementary) immobilized on an electrochemical biosensor surface. A dual-fractal analysis is required to adequately describe the binding kinetics. A single-fractal

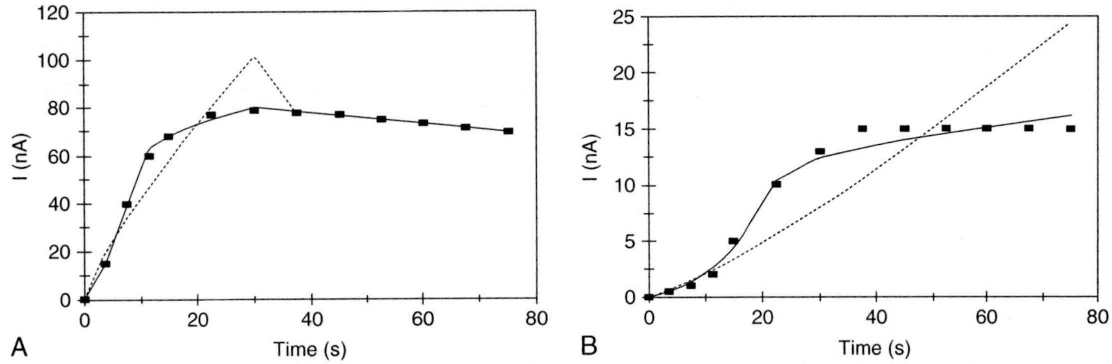

Figure 11.9

Binding (hybridization) of (a) a perfectly matched ODN (ODN-P) and (b) a noncomplementary ODN (ODN-N) to an electrochemical sensor with a EST2-A34 reporter (Wang et al., 2007). When a solid line is only used then a single-fractal analysis applies. When only a solid line (—) is used then a single-fractal analysis applies. When both a dashed (- - -) and a solid (—) line are used then the dashed line represents a single-fractal analysis and the solid line represents a dual-fractal analysis. In Figure (b) a single- and a triple-fractal analysis is shown.

analysis is required to adequately describe the dissociation kinetics. The values of (a) the binding rate coefficient, k, and the fractal dimension, D_f, for a single-fractal analysis, (b) the binding rate coefficients, k_1 and k_2, and the fractal dimensions, D_{f1} and D_{f2}, for a dual-fractal analysis, and (c) the dissociation rate coefficient, k_d, and the fractal dimension, D_{fd}, for a single-fractal analysis are given in Tables 11.4 and 11.5. It is of interest to note that for a dual-fractal analysis, as the fractal dimension increases by a factor of 5.766 from a value of D_{f1} equal 0.4432 to D_{f2} equal to 2.556, the binding rate coefficient increases by a factor of 13.25 from a value of k_1 equal to 2.8403 to k_2 equal to 37.626. Increases in the degree of heterogeneity or the fractal dimension on the electrochemical biosensor surface and in the binding rate coefficient are in the same direction.

Figure 11.9b shows the binding of *p*-aminophenylbutyrate/esterase 2 from *Alicyclobacillus acidocaldarius* plus ODN in solution to a site-specific manner ODN-N (mismatch; noncomplementary) immobilized on an electrochemical biosensor surface. In this case, a triple-fractal analysis is required to adequately describe the binding kinetics. The values of (a) the binding rate coefficient, k, and the fractal dimension, D_f, for a single-fractal analysis, (b) the binding rate coefficients, k_1 and k_2, and the fractal dimensions, D_{f1} and D_{f2}, for a dual-fractal analysis, and (c) the binding rate coefficients, k_1, k_2, and k_3, and the fractal dimensions, D_{f1}, D_{f2} and D_{f3}, for a triple-fractal analysis are given in Tables 11.4 and 11.5. The binding kinetics is a bit more complicated in this case (ODN-N; noncomplementary) when compared to the complementary (ODN-P) case, as for the ODN-N case a triple-fractal analysis is required to adequately describe the binding kinetics whereas for the complementary (ODN-P) case a dual-fractal analysis is adequate to describe the binding kinetics. It is of interest to note that for a triple-fractal analysis, as the fractal dimension increases by a factor of 1.358 from a value of D_{f2} equal 1.78902 to D_{f3} equal to 2.4304, the binding rate coefficient increases by a factor of 3.03 from a value of k_2 equal to 1.5596 to k_3 equal to 4.7240. Increases in the degree of heterogeneity or the fractal dimension on the electrochemical biosensor surface and in the binding rate coefficient are once again in the same direction.

Uno et al. (2007) recently developed a peptide-nucleic acid (PNA)-modified IS-FET-based biosensor that they have used for the direct detection of DNA hybridization. These authors report that their IS-FET based biosensor uses the change in the surface potential on the hybridization of a negatively charged DNA. They explain that the use of PNA in their system permits the highly specific and selective binding at low ionic strength. Uno et al. (2007) point out that IS-FET based biological sensors are attractive in the sense that they are of small size and weight, provide a fast response, are portable, can be mass produced at a low cost, and are highly reliable. They further report that IS-FET-based DNA sensors have exhibited potential in clinical and research applications. Uno et al. (2007) report that the IS-FET can detect surface potential changes due to the surface adsorption of charged molecules in an aqueous environment (Souteyrand et al., 1997; Berney et al., 2000; Frits et al., 2002; Kim et al., 2004; Li et al., 2004). Uno et al. (2004) and Ohtake et al. (2004) have shown that the hybridization of an immobilized PNA with a complementary DNA induces a decrease in the saturation current and a positive shift in the threshold voltage.

Figure 11.10a shows the binding of 5 μM target DNA2 (complementary to CYP2C9*2) in solution to CYP2C9*2 used as a probe and immobilized on a SPR biosensor surface (Uno et al., 2007). This permitted these authors to analyze the molecular recognition at the solution-surface interface. A single-fractal analysis is adequate to describe the binding and the dissociation kinetics. The values of (a) the binding rate coefficient, k, and the fractal dimension, D_f, for a single-fractal analysis, and (b) the dissociation rate coefficient, k_d, and the fractal dimension, D_{fd}, for a single-fractal analysis are given in Table 11.6(a) and (b). In this case, the affinity, K ($=k/k_d$) value is 30,347.2 (an extremely high value).

Figure 11.10b shows the binding of 5 μM target DNA with a single base mismatch 2 (complementary to CYP2C9*2) in solution to CYP2C9*2 used as a probe and immobilized on a SPR biosensor surface (Uno et al., 2007). A single-fractal analysis is once again adequate to describe the binding and the dissociation kinetics. The values of (a) the binding rate coefficient, k, and the fractal dimension, D_f, for a single-fractal analysis, and (b) the dissociation rate coefficient, k_d, and the fractal dimension, D_{fd}, for a single-fractal analysis are given in Table 11.6(a) and (b). In this case, the affinity, K ($=k/k_d$), value is 229.54. It is of interest to note that as the fractal dimension decreases by 2.26% from a value of D_f equal to 2.6306 to 2.5712 the binding rate coefficient, k also decreases by 56.3% from a value of k equal to 131.10 to 57.253. Note that changes in the binding rate coefficient, k, and the fractal dimension, D_f, or the degree of heterogeneity on the sensor surface are in the same direction.

Figure 11.10c shows the binding of 5 μM target complementary DNA (complementary to CYP2C9*2) in solution to CYP2C9*2 used as a probe with a single mismatch (CYP2C9*1)

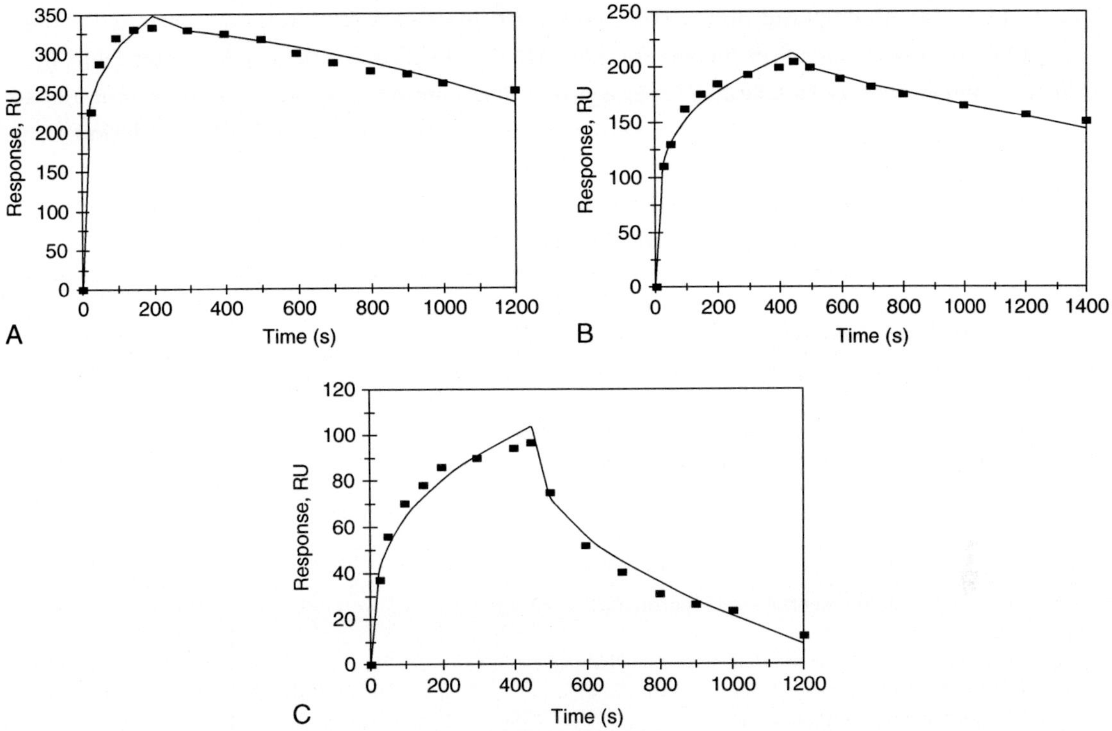

Figure 11.10

Binding and dissociation (hybridization) of 5 μM target in solution (a) complementary to CYP2C9*2, (b) with a single base mismatch to CYP2C9*2 immobilized on an ion-sensitive field-effect transistor-based biosensor, and (c) 5 μM target DNA in solution to a single-mismatch DNA, CYP2C9*1 immobilized on an ion-sensitive field-effect transistor-based biosensor (Uno et al., 2007).

and immobilized on a SPR biosensor surface (Uno et al., 2007). A single-fractal analysis is once again adequate to describe the binding and the dissociation kinetics. The values of (a) the binding rate coefficient, k, and the fractal dimension, D_f, for a single-fractal analysis, and (b) the dissociation rate coefficient, k_d, and the fractal dimension, D_{fd}, for a single-fractal analysis are given in Table 11.6(a) and (b). In this case, the affinity, K ($=k/k_d$), value is 3.954.

Figure 11.11a and Table 11.6(a) and (b) show the increase in the binding rate coefficient, k, with an increase in the fractal dimension, D_f, for a single-fractal analysis. For the data shown in Figure 11.11a and Table 11.6(a) and (b), the binding rate coefficient, k, is given by:

$$k = (4.6 \times 10^{-07} \pm 1.7 \times 10^{-07}) D_f^{10.91 \pm 3.59} \quad (11.5a)$$

The fit is reasonable. Only three data points are available. The availability of more data points would lead to a more reliable fit. The binding rate coefficient, k, for a single-fractal analysis is very sensitive to the fractal dimension, D_f, or the degree of heterogeneity that

Table 11.6: (a) Binding and dissociation rate coefficients and (b) fractal dimensions for the binding and the dissociation phases for PNA-DNA hybridization (a) 5 μM target DNA in solution complementary to CYP2C9*2 as a probe PNA immobilized on a nucleic acid-modified ion-selective field-effect transistor-based biosensor, and (b) target DNA in solution complementary to CYP2C9*2 and with involvement of a single mismatch in either the target DNA or the probe PNA immobilized on the nucleic acid-modified ion-selective field-effect transistor-based biosensor (Uno et al., 2007).

(a)		
Analyte in Solution/Receptor on Surface	**k**	**k_d**
5 μM target DNA complementary to receptor, CYP2C9*2/probe PNA, CYP2C9*2	131.10 ± 8.02	0.00432 ± 0.00099
Target DNA with single base mismatch/probe PNA, CYP2C9*2	57.253 ± 2.398	0.2496 ± 0.0310
Target DNA complementary to receptor/Probe PNSA, CYP2C9*2 with a single mismatch	15.429 ± 1.369	3.9020 ± 0.344
(b)		
Analyte in Solution/Receptor on Surface	**D_f**	**D_{fd}**
5 μM target DNA complementary to receptor, CYP2C9*2/probe PNA, CYP2C9*2	2.6306 ± 0.0730	0.00432 ± 0.00099
Target DNA with single base mismatch/probe PNA, CYP2C9*2	2.5712 ± 0.00304	0.2496 ± 0.0310
Target DNA complementary to receptor/Probe PNSA, CYP2C9*2 with a single mismatch	2.6306 ± 0.0703	0.102 + 0.179

exists on the biosensor surface as noted by the close to eleventh (equal to 10.91) order of dependence exhibited.

Figure 11.11b and Table 11.6(a) and (b) show the increase in the dissociation rate coefficient, k_d with an increase in the fractal dimension, D_{fd}, for a single-fractal analysis. For the data shown in Figure 11.11a and Table 11.6(a) and (b), the dissociation rate coefficient, k_d, is given by:

$$k_d = (0.3678 + 1.10262)D_{fd}^{2.007 \pm 0.6065} \tag{11.5b}$$

The fit is not good. Only three data points are available. The availability of more data points would lead to a more reliable fit. The dissociation rate coefficient, k_d, for a single-fractal analysis exhibits very close to a second (equal to 2.007) order of dependence on the fractal dimension, D_{fd}, or the degree of heterogeneity that exists in the dissociation phase on the biosensor surface.

Figure 11.11c and Table 11.6(a) and (b) show the increase in the affinity, K $(=k/k_d)$, with an increase in the ratio of the fractal dimensions in the binding and in the dissociation phases (D_f/D_{fd}), for a single-fractal analysis. For the data shown in Figure 11.11a and Table 11.6(a) and (b), the affinity, K, is given by:

$$K(= k/k_d) = (13.6 + 124.34)(D_f/D_{fd})^{1.746 \pm 0.978} \tag{11.5c}$$

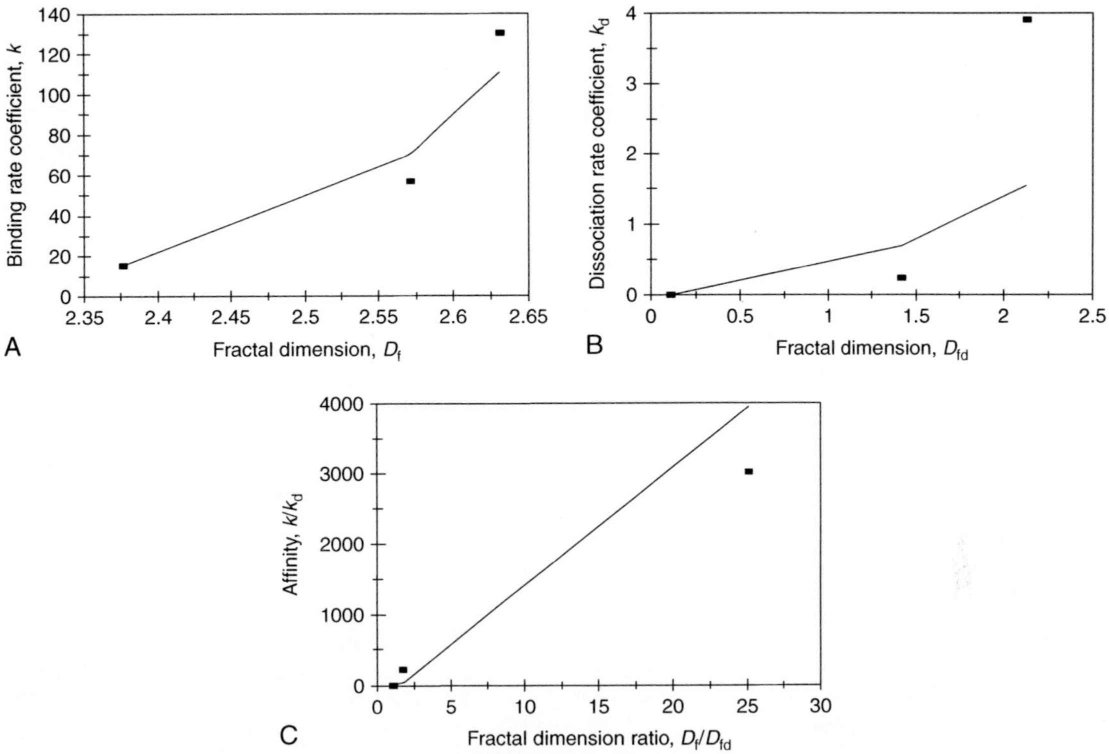

Figure 11.11

(a) Increase in the binding rate coefficient, k, with an increase in the fractal dimension, D_f. (b) Increase in the dissociation rate coefficient, k_d, with an increase in the fractal dimension in the dissociation phase, D_{fd}. (c) Increase in the affinity, K $(=k/k_d)$, with an increase in the fractal dimension ratio, D_f/D_{fd}.

The fit is reasonable. Only three data points are available. The availability of more data points would lead to a more reliable fit. Only the positive error is shown in Equation (11.5c), since the error is large, and the affinity can only have positive values. The affinity, K, exhibits an order of dependence between one and a half and two (equal to 1.746) on the ratio of fractal dimensions, (D_f/D_{fd}), present on the sensor chip surface.

Uno et al. (2007) analyzed the SPR biosensor responses to PNA-DNA hybridization. These authors used CYP2C9*2 as the probe PNA and the target DNA was complementary CYP2C9*2 (target DNA2). They used target DNA concentrations in the 0.1-5.0 μM range. Figure 11.12a shows the binding and the dissociation of 5 μM target DNA concentration in solution to the probe PNA immobilized on the sensor surface. A dual-fractal analysis is required to adequately describe the binding kinetics. A single-fractal analysis is adequate to describe the dissociation kinetics. The values of (a) the binding rate coefficient, k, and the fractal dimension, D_f, for a single-fractal analysis, (b) the binding rate coefficients,

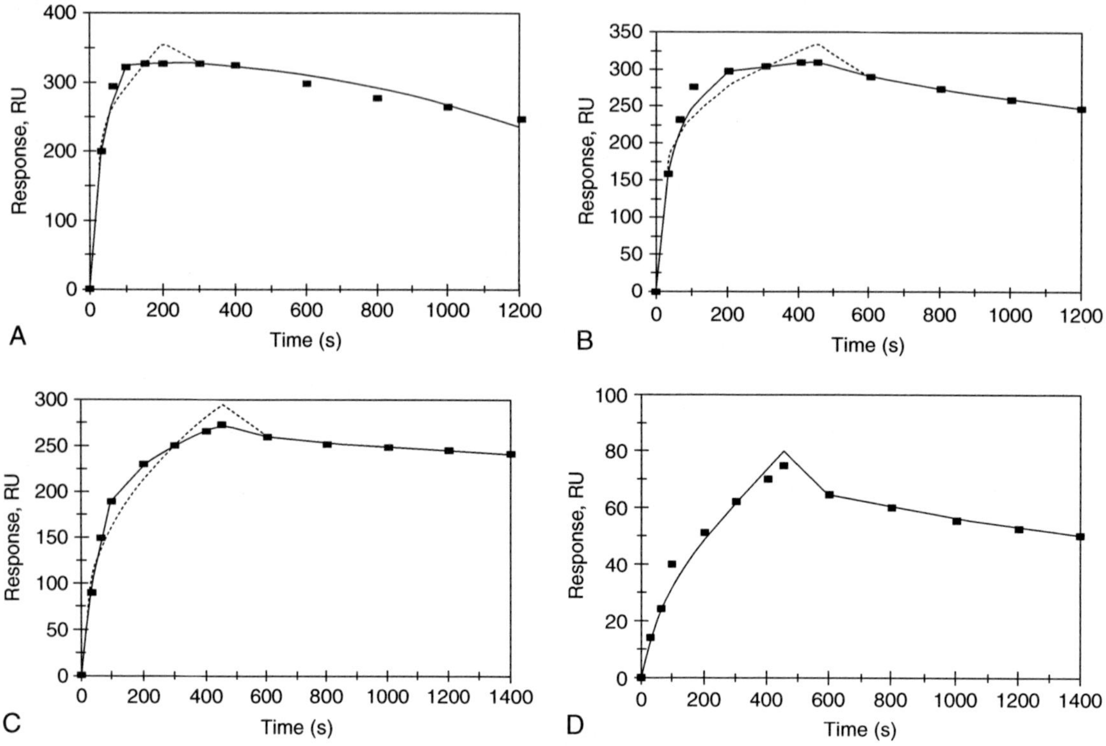

Figure 11.12

Binding and dissociation during PNA-DNA hybridization. Binding of different concentrations (in μM) of target DNA complementary to CYP2C9*2 (target DNA2) to CYP2C9*2 as a probe PNA immobilized on a ion-sensitive field-effect transistor-based biosensor (Uno et al.,2007): (a) 5. (b) 2.5. (c) 1. (d) 0.1. When only a solid line (—) is used then a single-fractal analysis applies. When both a dashed (- - -) and a solid (—) line are used then the dashed line represents a single-fractal analysis and the solid line represents a dual-fractal analysis.

k_1 and k_2, and the fractal dimensions, D_{f1} and D_{f2}, for a dual-fractal analysis, and (c) the dissociation rate coefficient, k_d, and the fractal dimension, D_{fd}, for the dissociation phase for a single-fractal analysis are given in Tables 11.7 and 11.8. Note that for a dual-fractal analysis, as the fractal dimension increases by 38.9% from a value of D_{f1} equal to 2.1140 to D_{f2} equal to 2.9360, the binding rate coefficient increases by a factor of 6.40 from a value of k_1 equal to 43.465 to k_2 equal to 278.35. Note that changes in the fractal dimension or the degree of heterogeneity on the sensor chip surface and in the binding rate coefficient are in the same direction.

Figure 11.12b shows the binding and the dissociation of 2.5 μM target DNA concentration in solution to the probe PNA immobilized on the sensor surface. Once again, a dual-fractal analysis is required to adequately describe the binding kinetics. A single-fractal analysis is adequate to describe the dissociation kinetics. The values of (a) the binding rate coefficient, k, and the fractal dimension, D_f, for a single-fractal analysis, (b) the binding rate coefficients,

Table 11.7: Binding and dissociation rate coefficients for different target complementary DNA concentrations (in μM) in solution to a DNA probe, CYP29*2 immobilized on an ion-sensitive field-effect transistor (IS-FET)-based biosensor (Uno et al., 2007).

Complementary DNA Concentration in Solution (μM)	k	k_1	k_2	k_d
5	85.694 ± 9.653	43.465 ± 3.303	278.35 ± 1.34	0.000256 ± 0.000118
2.5	82.808 ± 9.833	49.086 ± 5.681	225.0 ± 0.6886	0.6096 ± 0.0056
1.0	26.792 ± 3.322	8.427 ± 0.302	74.183 ± 0.174	1.864 ± 0.092
0.1	1.847 ± 0.255	na	na	0.8969 ± 0.126

Table 11.8: Fractal dimensions for the binding and the dissociation phases for the different target complementary DNA concentrations (in μM) in solution to a DNA probe, CYP29*2 immobilized on an ion-sensitive field-effect transistor (ISFET)-based biosensor (Uno et al., 2007).

Complementary DNA Concentration in Solution (μM)	D_f	D_{f1}	D_{f2}	D_{fd}
5	2.4606 ± 0.1505	2.1140 ± 0.1856	2.9360 ± 0.01941	0. + 0.3388
2.5	2.5416 ± 0.09368	2.2890 ± 0.1675	2.8933 ± 0.00977	1.6066 ± 0.0150
1.0	2.2146 ± 0.09662	1.2392 ± 0.08908	2.5728 ± 0.00748	2.1800 ± 0.06648
0.1	1.7662 ± 0.1066	na	na	2.0364 ± 0.01924

k_1 and k_2, and the fractal dimensions, D_{f1} and D_{f2}, for a dual-fractal analysis, and (c) the dissociation rate coefficient, k_d, and the fractal dimension, D_{fd}, for the dissociation phase for a single-fractal analysis are given in Tables 11.7 and 11.8. Note that for a dual-fractal analysis, as the fractal dimension increases by 26.4% from a value of D_{f1} equal to 2.2890 to D_{f2} equal to 2.8933, the binding rate coefficient increases by a factor of 4.58 from a value of k_1 equal to 49.086 to k_2 equal to 225.0. Note that changes in the fractal dimension or the degree of heterogeneity on the sensor chip surface and in the binding rate coefficient are once again in the same direction.

Figure 11.12c shows the binding and the dissociation of 1 μM target DNA concentration in solution to the probe PNA immobilized on the sensor surface. Once again, a dual-fractal analysis is required to adequately describe the binding kinetics. A single-fractal analysis is adequate to describe the dissociation kinetics. The values of (a) the binding rate coefficient, k, and the fractal dimension, D_f, for a single-fractal analysis, (b) the binding rate coefficients, k_1 and k_2, and the fractal dimensions, D_{f1} and D_{f2}, for a dual-fractal analysis, and (c) the dissociation rate coefficient, k_d, and the fractal dimension, D_{fd}, for the dissociation phase for a single-fractal analysis are given in Tables 11.7 and 11.8. Note that for a dual-fractal analysis, as the fractal dimension increases by a factor of 2.076 from a value of D_{f1} equal to 1.2392 to D_{f2} equal to 2.5728, the binding rate coefficient increases by a factor of 8.80 from a value of

k_1 equal to 8.427 to k_2 equal to 74.183. It is seen that changes in the fractal dimension or the degree of heterogeneity on the sensor chip surface and in the binding rate coefficient are once again in the same direction.

Figure 11.12d shows the binding and dissociation of 0.1 μM target DNA concentration in solution to the probe PNA immobilized on the sensor surface. A single-fractal analysis is adequate to describe the binding and the dissociation kinetics. The values of (a) the binding rate coefficient, k, and the fractal dimension, D_f, for a single-fractal analysis, and (b) the dissociation rate coefficient, k_d, and the fractal dimension, D_{fd}, for a single-fractal analysis are given in Tables 11.7 and 11.8. In this case, the affinity, K $(=k/k_d)$, is given by 1.046. No reason is given at present as to why at this lowest DNA target concentration (0.1 μM) in solution a single-fractal analysis is adequate to describe the binding kinetics, whereas at the higher DNA concentration in solution in the 1.0-5.0 μM range, a dual-fractal analysis is required to adequately describe the binding kinetics.

In the entire 0.1-5.0 μM target concentration range a single-fractal analysis is adequate to describe the dissociation kinetics.

Figure 11.13a and Table 11.7 show the increase in the binding rate coefficient, k_1, with an increase in the target DNA concentration in the 1.0-5.0 μM range in solution for a dual-fractal analysis. For the data shown in Figure 11.13a, the binding rate coefficient, k_1, is given by:

$$k_1 = (10.67 \pm 10.27)[\text{target DNA}]^{1.067 \pm 0.590} \tag{11.6a}$$

The fit is not good. There is scatter in the data, and this is reflected in the error in the value of the binding rate coefficient, k_1. The binding rate coefficient, k_1, exhibits close to a first (equal to 1.067) order of dependence on the target DNA concentration in solution. The nonintegral order of dependence exhibited by the binding rate coefficient, k_1, lends support to the fractal nature of the system.

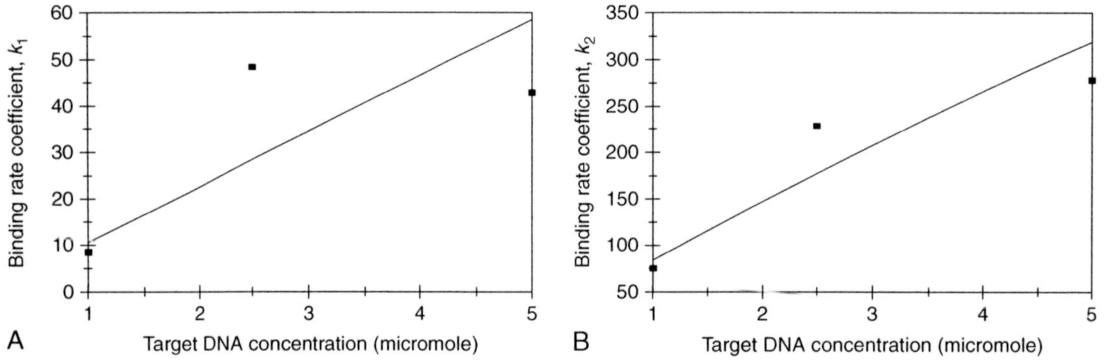

Figure 11.13
Increase in the binding rate coefficient (a) k_1 and (b) k_2 with an increase in the target DNA concentration in (in micromole) in solution.

Figure 11.13b and Table 11.7 show the increase in the binding rate coefficient, k_2, with an increase in the target DNA concentration in the 1.0-5.0 μM range in solution for a dual-fractal analysis. For the data shown in Figure 11.13b the binding rate coefficient, k_2, is given by:

$$k_2 = (82.13 \pm 27.67)[\text{target DNA}]^{0.842 \pm 0.254} \tag{11.6b}$$

The fit is reasonable. There is some scatter in the data. The binding rate coefficient, k_2, exhibits an order of dependence between a half and one (equal to 0.842) on the target DNA concentration in solution. The nonintegral order of dependence exhibited by the binding rate coefficient, k_1, once again lends support to the fractal nature of the system.

It is seen that the binding rate coefficient, k_2, exhibits an order of dependence less than one (equal to 0.842) on the target DNA concentration in solution., and the binding rate coefficient, k_1, exhibits an order of dependence greater than one (equal to 1.067) on the target DNA concentration in solution.

Figure 11.14a and Tables 11.7 and 11.8 show the increase in the binding rate coefficient, k_1, with an increase in the fractal dimension, D_{f1} for a dual-fractal analysis. For the data shown in Figure 11.4a, the binding rate coefficient, k_1, is given by:

$$k_1 = (4.517 \pm 0.377) D_{f1}^{2.944 \pm 0.1698} \tag{11.7a}$$

The fit is good. Only three data points are available. The availability of more data points would lead to a more reliable fit. The binding rate coefficient, k_1, is sensitive to the fractal dimension, D_{f1}, or the degree of heterogeneity that exists on the sensor chip surface as noted by the close to third (equal to 2.944) order of dependence exhibited.

Figure 11.14b and Tables 11.7 and 11.8 show the increase in the binding rate coefficient, k_2, with an increase in the fractal dimension, D_{f2}, for a dual-fractal analysis. For the data shown in Figure 11.4b, the binding rate coefficient, k_2, is given by:

$$k_2 = (0.00754 \pm 0.00038) D_{f2}^{9.734 \pm 0.48728} \tag{11.7b}$$

The fit is good. Only three data points are available. The availability of more data points would lead to a more reliable fit. The binding rate coefficient, k_2, is extremely sensitive to the fractal dimension, D_{f2}, or the degree of heterogeneity that exists on the sensor chip surface as noted by the order of dependence between nine and a half and ten (equal to 9.734) exhibited. No explanation is offered at present to help explain this extremely very high order of dependence exhibited.

Figure 11.14c and Tables 11.7 and 11.8 show the increase in the ratio of the binding rate coefficients, k_2/k_1, with an increase in the fractal dimension ratio, D_{f2}/D_{f1}, for a dual-fractal

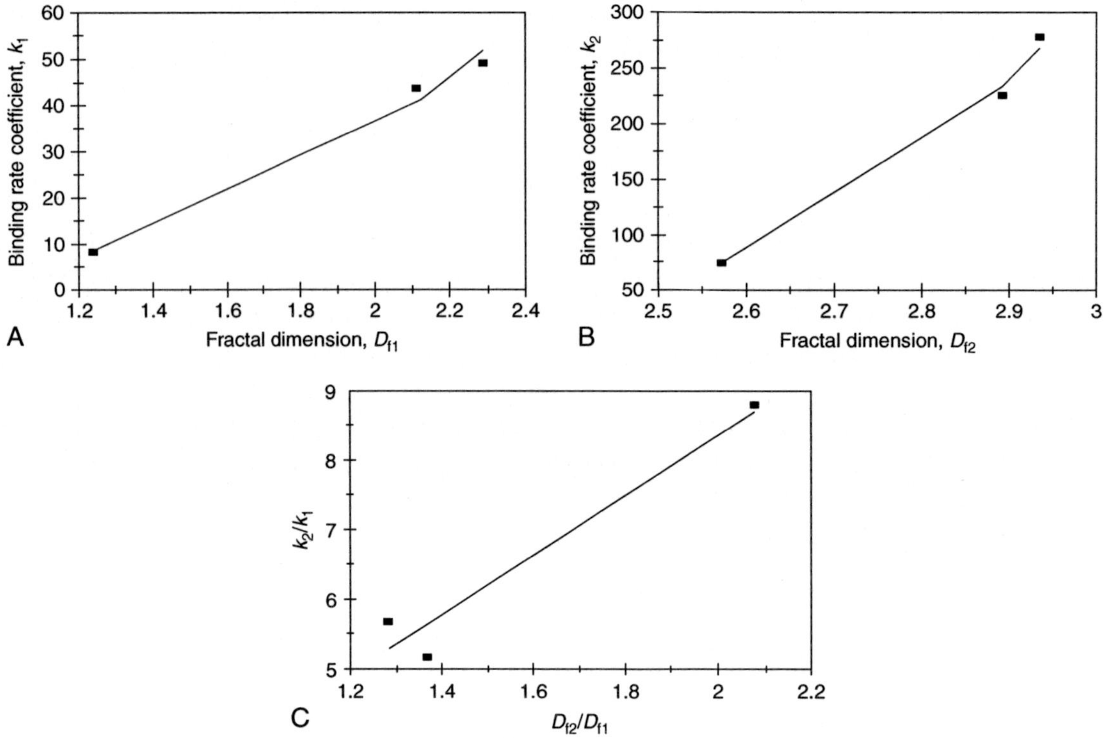

Figure 11.14

(a) Increase in the binding rate coefficient, k_1, with an increase in the fractal dimension, D_{f1}
(b) Increase in the binding rate coefficient, k_2, with an increase in the fractal dimension, D_{f2}
(c) Increase in the ratio of the binding rate coefficients, k_2/k_1, with an increase in the ratio of the fractal dimensions, D_{f2}/D_{f1}.

analysis. For the data shown in Figure 11.14c the ratio of the binding rate coefficients, k_2/k_1, is given by:

$$k_2/k_1 = (4.063 \pm 0.486)(D_{f2}/D_{f1})^{1.043 \pm 0.306} \tag{11.7c}$$

The fit is good. Only three data points are available. The availability of more data points would lead to a more reliable fit. The ratio of binding rate coefficients, k_2/k_1, exhibits close to a first (equal to 1.043) order of dependence on the ratio of fractal dimensions, D_{f2}/D_{f1}, that exists on the sensor chip surface.

Figure 11.15a and Tables 11.7 and 11.8 show the increase in the fractal dimension, D_{f1}, with an increase in the target DNA concentration in solution in the 1-5 μM range. For the data shown in Figure 11.15a the fractal dimension, D_{f1}, is given by:

$$D_{f1} = (1.354 \pm 0.387)[\text{target DNA, in } \mu\text{M}]^{0.349 \pm 0.221} \tag{11.7d}$$

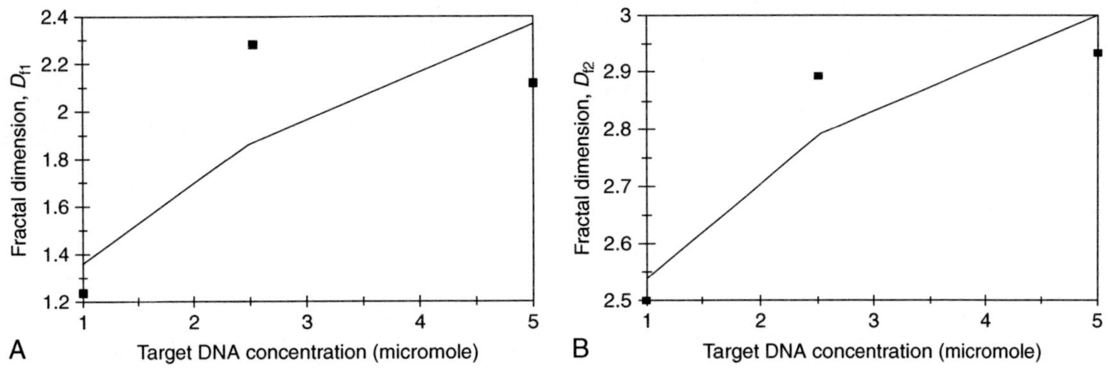

Figure 11.15

Increase in (a) the fractal dimension, D_{f1} and (b) the fractal dimension, D_{f2} with an increase in the target DNA concentration (in μM) in solution.

The fit is reasonable. There is scatter in the data. Only three data points are available. The availability of more data points would lead to a more reliable fit. The fractal dimension, D_{f1}, exhibits only a very mild (equal to 0.349) order of dependence on the target DNA concentration in solution in the 1-5 μM range.

Figure 11.15b and Tables 11.7 and 11.8 show the increase in the fractal dimension, D_{f2}, with an increase in the target DNA concentration in solution in the 1-5 μM range. For the data shown in Figure 11.15b the fractal dimension, D_{f2}, is given by:

$$D_{f1} = (2.539 \pm 0.115)[\text{target DNA, in } \mu\text{M}]^{0.103 \pm 0064} \tag{11.7e}$$

The fit is reasonable. There is scatter in the data. Only three data points are available. The availability of more data points would lead to a more reliable fit. The fractal dimension, D_{f2}, exhibits only a very low (equal to 0.103) order of dependence on the target DNA concentration in solution in the 1-5 μM range. Note that the fractal dimension is based on a log scale, and even small changes in the fractal dimension lead to significant changes in the degree of heterogeneity on the sensor chip surface.

Blair et al. (2007) report that techniques have been used for the *in situ* quantification of DNA during enzymatic synthesis, for example, during a real-time polymerase chain reaction (Watzinger et al., 2006). Blair et al. (2007) point out that hybridization probes involving "molecular beacons" is a DNA quantification method. They have a stem-loop structure with complementary ends that anneal to each other. A fluorophore and a quencher are at opposite ends. Blair et al. (2007) explain that the loop structure is complementary to the target. Thus, as the beacon binds to the target, the fluorophore is separated from the quencher. This leads to an increase in the fluorescence, and is proportional to the amount of the DNA produced

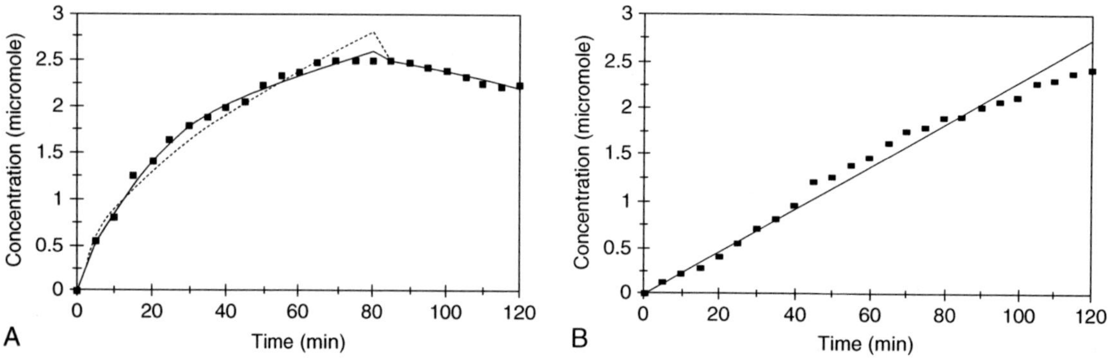

Figure 11.16

Binding and dissociation of RNA synthesized on a (a) 420 nM template and a (b) 42 nM template (Blair et al., 2007). When only a solid line (—) is used then a single-fractal analysis applies. When both a dashed (- - -) and a solid (—) line are used then the dashed line represents a single-fractal analysis and the solid line represents a dual-fractal analysis.

(Tyagi and Kramer, 1996; Leone et al., 1998; Guilietti et al., 2001; Summerer and Marx, 2002; Marras et al., 2006).

Blair et al. (2007) have developed a novel hybridization-based assay for the real-time monitoring of RNA synthesis. In their method complementary nucleotides were used to quantify the amount of RNA production by T7 polymerase. Figure 11.16a shows the binding and dissociation of RNA synthesized on a 420 nM template. The RNA concentration was determined by conversion using a calibration curve. A dual-fractal analysis is required to adequately describe the binding kinetics. A single-fractal analysis is adequate to describe the dissociation kinetics. The values of (a) the binding rate coefficient, k, and the fractal dimension, D_f, for a single-fractal analysis, (b) the binding rate coefficients, k_1 and k_2, and the fractal dimensions, D_{f1} and D_{f2}, for a dual-fractal analysis, and (c) the dissociation rate coefficient, k_d, and the fractal dimension for the dissociation phase, D_{fd}, for a single-fractal analysis are given in Tables 11.9 and 11.10. It is of interest to note that for a dual-fractal analysis as the fractal dimension increases by a factor of 1.35 from a value of D_{f1} equal to 1.6862 to D_{f2}r equal to 2.2812, the binding rate coefficient increases by a factor of 2.79 from a value of k_1 equal to 0.1909 to k_2 equal to 0.5324. The changes in the degree of heterogeneity on the sensor chip surface and in the binding rate coefficient are in the same direction.

Table 11.9: Binding and dissociation rate coefficients for (a) RNA synthesized on a 42 nM and a 420 nM template (Blair et al., 2007).

Template Concentration (nM)	k	k_1	k_2	k_d
420	0.2546 ± 0.0121	0.1909 ± 0.0121	0.5324 ± 0.0120	0.00204 ± 0.00031
42	0.02317 ± 0.00234	na	na	na

Table 11.10: Fractal dimensions in the binding and in the dissociation phase for (a) RNA synthesized on a 42 nM and a 420 nM template (Blair et al., 2007).

Template Concentration (nM)	D_f	D_{f1}	D_{f2}	D_{fd}
420	1.9086 ± 0.04988	1.6862 ± 0.07340	2.2812 ± 0.06678	0.2438 ± 0.1439
42	1.0136 ± 0.04822	na	na	na

Figure 11.16b shows the binding and dissociation of RNA synthesized on a 42 nM template. A single-fractal analysis is adequate to describe the binding kinetics. The values of the binding rate coefficient, k, and the fractal dimension, D_f, for a single-fractal analysis are given in Tables 11.9 and 11.10. Note that for the lower 42 nM template concentration a single-fractal analysis is adequate to describe the binding kinetics, whereas for the higher 420 nM template concentration a dual-fractal analysis is required to adequately describe the binding kinetics. This would seem to indicate that there is a change in the binding mechanism as one goes from the lower (42 nM template) to the higher (420 nM template) concentration.

Blair et al. (2007) analyzed the kinetics of displacement as a function of target concentration. These authors incubated the pre-hybridized 22-nt FQ (fluorophore quencher) complex with varying concentrations of the ss DNA target at 37 °C for 90 min. Figure 11.17a shows the binding of the 500 nM target ss DNAS (T) in solution in a "broken beacon" assay.

It is of interest to note that as the fractal dimension increases by a factor of 1.352 from a value of D_{f1} equal to 1.6862 to D_{f2} equal to 2.2812, the binding rate coefficient increases by a factor of 2.79 from a value of k_1 equal to 0.1909 to k_2 equal to 0.5324. The changes in the binding rate coefficient and in the fractal dimension or the degree of heterogeneity on the sensor chip surface are in the same direction.

Figure 11.17b shows the binding of the 250 nM target ss DNA (T) in solution in a "broken beacon" assay (Blair et al., 2007). A dual-fractal analysis is required to adequately describe the binding kinetics. The values of (a) the binding rate coefficient, k, and the fractal dimension, D_f, for a single-fractal analysis, and (b) the binding rate coefficients, k_1 and k_2, and the fractal dimensions, D_{f1} and D_{f2}, for a dual-fractal analysis are given in Tables 11.11 and 11.12.

Once again, it is of interest to note that as the fractal dimension increases by a factor of 1.129 from a value of D_{f1} equal to 2.5368 to D_{f2} equal to 2.9659, the binding rate coefficient increases by a factor of 1.60 from a value of k_1 equal to 17311.23 to k_2 equal to 27699.83. Once again, it is seen that changes in the binding rate coefficient and in the fractal dimension or the degree of heterogeneity on the sensor chip surface are in the same direction.

Tables 11.11 and 11.12 Binding rate coefficients and fractal dimensions for different concentrations of ss DNA concentration in solution pre-incubated with prehybridized 22-nt FQ duplex to the "broken beacon" (Blair et al., 2007)

ss DNA concentration in solution (in nM)	k	k_1	k_2	D_f	D_{f1}	D_{f2}
500 nM T	27162.52 ± 2112.81	24434.52 ± 2335.01	34530.40 ± 338.91	2.6996 ± 0.03036	2.5368 ± 0.08588	2.8285 ± 0.01352
250 nM T	18844.78 ± 1281.48	17311.23 ± 223.15	27699.83 ± 130.21	2.7582 ± 0.02608	2.6264 ± 0.01132	2.9659 ± 0.00629
100 nM T	10108.8 ± 949.88	9171.31 ± 492.50	14970.66 ± 21.90	2.7772 ± 0.03812	2.6278 ± 0.03964	2.9916 ± 0.00315

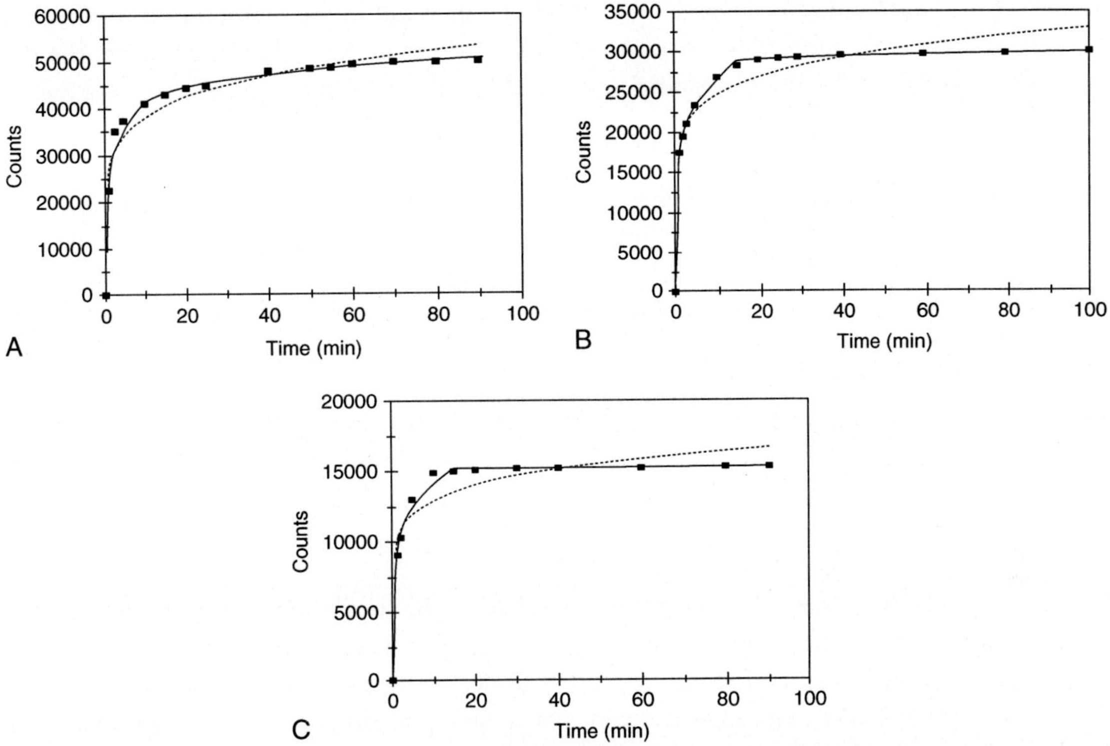

Figure 11.17

Binding (hybridization) of different concentrations of ss DNA in solution preincubated with prehybridized 22-nt FQ duplex to a "broken beacon" immobilized on a sensor surface (Blair et al., 2007): (a) 500 nM T. (b) 250 nM T. (c) 100 nM T. When only a solid line (—) is used then a single-fractal analysis applies. When both a dashed (- - -) and a solid (—) line are used then the dashed line represents a single-fractal analysis and the solid line represents a dual-fractal analysis.

Figure 11.17c shows the binding of the 100 nM target ss DNA (T) in solution in a "broken beacon" assay (Blair et al., 2007). A dual-fractal analysis is required to adequately describe the binding kinetics. The values of (a) the binding rate coefficient, k, and the fractal dimension, D_f, for a single-fractal analysis, and (b) the binding rate coefficients, k_1 and k_2, and the fractal dimensions, D_{f1} and D_{f2}, for a dual-fractal analysis are given in Tables 11.11 and 11.12.

Once again, it is of interest to note that as the fractal dimension increases by a factor of 1.138 from a value of D_{f1} equal to 2.6278 to D_{f2} equal to 2.9916, the binding rate coefficient increases by a factor of 1.632 from a value of k_1 equal to 9171.31 to k_2 equal to 14970.66. Once again, and as above, it is seen that changes in the binding rate coefficient and in the fractal dimension or the degree of heterogeneity on the sensor chip surface are in the same direction.

Figure 11.18a and Tables 11.11 and 11.12 show the increase in the binding rate coefficient, k_1, with an increase in the ss DNA concentration in solution in the 100-500 nM range. For the data shown in Figure 11.18a the binding rate coefficient, k_1, is given by:

$$k_1 = (556.51 \pm 36.18)[\text{ss DNA, in nM}]^{0.6132 \pm 0.05517} \quad (11.8a)$$

The fit is good. Only three data points are available. The availability of more data points would lead to a more reliable fit. The binding rate coefficient, k_1, exhibits an order of dependence between a half and one (equal to 0.6132) on the ss DNA concentration in solution in the 100-500 nM range. The nonintegral order of dependence exhibited lends support to the fractal nature of the system.

Figure 11.18b and Tables 11.11 and 11.12 show the increase in the binding rate coefficient, k_2, with an increase in the ss DNA concentration in solution in the 100-500 nM range. For the data shown in Figure 11.18b the binding rate coefficient, k_2, is given by:

$$k_2 = (1383.75 \pm 8.41)[\text{ss DNA, in nM}]^{0.5443 \pm 0.3197} \quad (11.8b)$$

The fit is good. Only three data points are available. The availability of more data points would lead to a more reliable fit. The binding rate coefficient, k_2, exhibits an order of dependence slightly more than a half (equal to 0.5443) order on the ss DNA concentration in solution in the 100-500 nM range. Once again, the nonintegral order of dependence exhibited lends support to the fractal nature of the system.

Figure 11.18c and Tables 11.11 and 11.12 show the decrease in the binding rate coefficient, k_1, with an increase in the fractal dimension, D_{f1}. For the data shown in Figure 11.18c the binding rate coefficient, k_1, is given by:

$$k_1 = (1.3 \times 10^{-12} \pm 0.8 \times 10^{-12}) D_{f1}^{-19.14 \pm 15.48} \quad (11.8c)$$

The fit is reasonable. Only three data points are available. The availability of more data points would lead to a more reliable fit. The binding rate coefficient, k_1, is extremely sensitive to the fractal dimension, D_{f1}, or the degree of heterogeneity that exists on the sensor surface as noted by the greater than nineteenth (equal to -19.14) order of dependence exhibited.

Figure 11.18d and Tables 11.11 and 11.12 show the decrease in the binding rate coefficient, k_2, with an increase in the fractal dimension, D_{f2}. For the data shown in Figure 11.18d the binding rate coefficient, k_2, is given by:

$$k_2 = (5.4 \times 10^{-12} \pm 2.3 \times 10^{-12}) D_{f2}^{-11.45 \pm 8.644} \quad (11.8d)$$

The fit is reasonable. Only three data points are available. The availability of more data points would lead to a more reliable fit. The binding rate coefficient, k_2, is extremely sensitive to the fractal dimension, D_{f2}, or the degree of heterogeneity that exists on the sensor

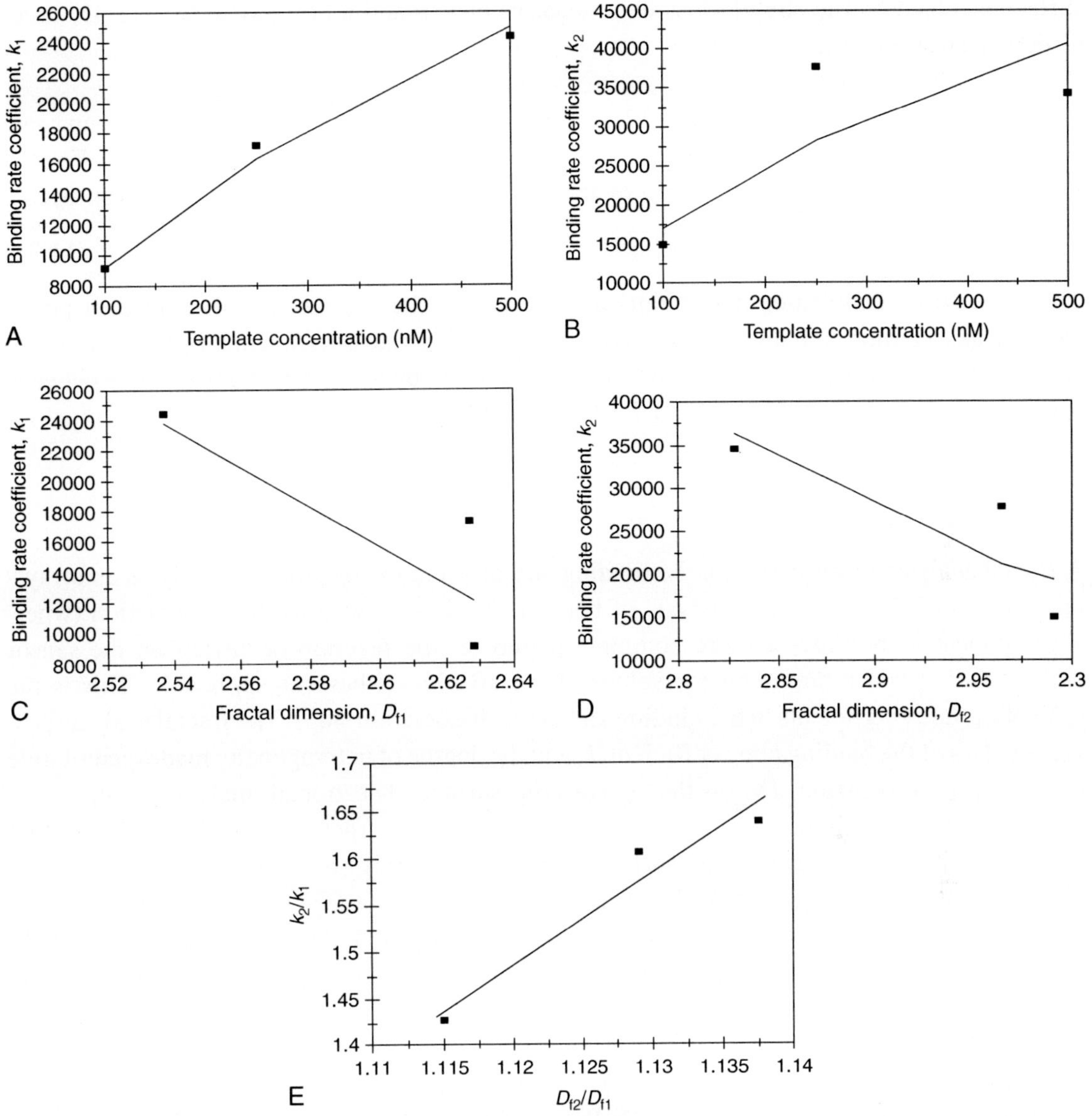

Figure 11.18

(a) Increase in the binding rate coefficient (a) k_1 and (b) k_2 with an increase in the template concentration (in nM) (c) Decrease in the binding rate coefficient, k_1, with an increase in the fractal dimension, D_{f1} (d) Decrease in the binding rate coefficient, k_2, with an increase in the fractal dimension, D_{f2} (e) Increase in the ratio of the binding rate coefficients, k_2/k_1, an increase in the ratio of the fractal dimensions, D_{f2}/D_{f1}.

surface as noted by the slightly less than negative eleven and a half (equal to -11.45) order of dependence exhibited.

Figure 11.18e and Tables 11.11 and 11.12 show the increase in the ratio of the binding rate coefficients, k_2/k_1, with an increase in the ratio of the fractal dimensions, D_{f2}/D_{f1}. For the data shown in Figure 11.18e the ratio of the binding rate coefficients, k_2/k_1, is given by:

$$k_2/k_1 = (0.653 \pm 0.019) D_{f2}/D_{f1}{}^{7.17 \pm 1.93} \quad (11.8e)$$

The fit is good. Only three data points are available. The availability of more data points would lead to a more reliable fit. The ratio of the binding rate coefficients, k_2/k_1, is very sensitive to the ratio of fractal dimensions, D_{f2}/D_{f1}, as noted by the order of dependence between seven and seven and a half (equal to 7.17) exhibited.

11.4 Conclusions

A fractal analysis is presented for the binding and dissociation of different analytes on arrays/microarrays/DNA chips. The analysis of both the binding as well as the dissociation (wherever applicable) provides a more complete picture of the reaction occurring on the sensor chip surface, besides providing for values of the affinities wherever applicable. This is the ratio of the rate coefficients in the binding and in the dissociation steps. The fractal analysis provides values of the binding rate coefficient, k, and the degree of heterogeneity made quantitative by the fractal dimension, D_f, on the sensor chip surface. The fractal analysis is applied to (a) the binding and dissociation (hybridization) of different targets (400 nM) in solution to a probe immobilized on a DNA chip surface (Fiche et al., 2007), (b) binding (hybridization) of different concentrations (in nM) of free-DNA in solution to a 22-mer strand (bound DNA) immobilized via a phenylene-diisocyanate linker molecule on a glass substrate (Michel et al., 2007), (c) binding (hybridization) of SA-HRP in solution to a capture probe on a QCM electrode along with a detecttion probe (Feng et al., 2007), (d) binding (hybridization) of a complementary and a noncomplementary (three-base mismatch strand) DNA in solution to a 30-mer 3′-thiolated DNA strand immobilized on an electrochemical enzymatic genosensor (Abad-Valle et al., 2007a,b), (e) binding (hybridization) of (i) a ODN-P and (ii) a noncomplementary ODN (ODN-N) to an electrochemical sensor with a EST2-A34 reporter (Wang et al., 2007), (f) binding and dissociation during PNA-DNA hybridization—binding of different concentrations (in μM) of target DNA complementary to CYP2C9*2 (target DNA2) to CYP2C9*2 as a probe PNA immobilized on a IS-FET-based biosensor (Uno et al., 2007), (g) binding and dissociation during PNA-DNA hybridization—binding of different concentrations (in μM) of target DNA complementary to CYP2C9*2 (target DNA2) to CYP2C9*2 as a probe PNA immobilized on an IS-FET-based biosensor (Uno et al., 2007), (h) binding and dissociation of RNA synthesized on a (i) 42 nM template and a (ii) 420 nM template (Blair et al., 2007), and (i) binding (hybridization) of different concentrations of ss DNA in solution pre-incubated with pre-hybridized 22-nt FQ duplex to a "broken beacon" immobilized on a sensor surface (Blair et al., 2007).

Both single- and dual-fractal analysis are used to adequately model the binding and dissociation kinetics. The dual-fractal analysis was used only when the single-fractal analysis did not provide an adequate fit (sum of least squares less than 0.97). This was done by the regression analysis provided by Corel Quattro Pro 8.0 (1997). The fractal analysis permits a link between the binding rate coefficient, k, and the degree of heterogeneity, D_f that exists on the biosensor surface. This provides a more complete picture of the reaction kinetics occurring on the sensor chip surface.

It is suggested that the fractal surface (roughness) leads to turbulence, which enhances mixing, decreases diffusional limitations, and leads to an increase in the binding rate coefficient (Martin et al., 1991). For this to occur, the characteristic length of the turbulent boundary layer may have to extend a few monolayers above the sensor chip surface to affect bulk diffusion to and from the surface. However, given the extremely laminar flow regimes in most biosensors this may not actually take place. The sensor chip (arrays/microarrays/DNA chips) surface is characterized by grooves and ridges, and this surface morphology may lead to eddy diffusion. This eddy diffusion can then help to enhance the mixing and extend the characteristic length of the boundary layer to affect the bulk diffusion to and from the surface.

The analysis of the different examples of the detection of analytes on arrays/microarrays/DNA chips should encourage experimentalists to pay more attention to the nature of the surface, and how it may be manipulated in desired directions. Detection of analytes on arrays/microarrays/DNA chips is bound to increase in the future as these "tools" find increasing applications in a wide variety of areas. This is of particular value primarily in the biomedical area, and also in other areas of application. For example, the identification of DNA sequences is of particular value in clinical pathology. A clinical pathologist is a medical doctor responsible for the diagnosis of diseases based on the analysis of body fluids, for example, blood and urine. The earlier one may detect and diagnose the probable onset of diseases the earlier one can begin the medical protocols necessary to help prevent, alleviate, or correct the onset of, especially, debilitating and intractable diseases. It is hoped that fractal analysis should be particularly helpful in providing a better understanding of the onset of diseases, particularly those that are insidious and debilitating. Any insight that is made available by such an analysis that helps in the management of intractable diseases should prove invaluable.

References

Abad-Valle P, MT Fernandez-Abedul, and A Costa-Garcia, DNA single-base mismatch study with an electrochemical enzymatic genosensor, *Biosensors & Bioelectronics*, **20**, 2251–2260 (2007a).

Abad-Valle P, MT Fernandez-Abedul, and A Costa-Garcia, DNA single-base mismatch study with an electrochemical enzymatic genosensor, *Biosensors & Bioelectronics*, **22**, 1642–1650 (2007b).

Blair RH, ES Rosenblum, ED Dawson, RD Kuchta, LR Kuck, and KL Rowlen, Real-time quantification of RNA polymerase activity using a "broken beacon", *Biosensors & Bioelectronics*, **22**, 213–220 (2007).

Caruana DJ and A Heller, Enzyme-amplified amperometric detection of hybridization and of a single-base pair mutation in an 18-base oligonucleotide on a 7 μm-diameter Electrode, *Journal of the American Chemical Society*, **121**, 769–774 (1999).

Feng K, J Li, JH Jiang, GL Shen, and RQ Yu, QCM detection with single-base mutation based on ligase reaction and biocatalyzed deposition amplification, *Biosensors & Bioelectronics*, **22**, 1651–1657 (2007).

Fiche JB, A Buhot, R Calemczuk, and T Livache, Temperature effects on DNA chip experiments from surface plasmon resonance imaging: Isotherms and melting curves, *Biophysical Journal*, **92**, 935–946 (2007).
Guilietti A, L Overbergh, D Valckx, B Decallonne, R Bouillon, and C Mathieu, An overview of real-time quantitative PCR: Applications to quantify cytokine gene expression, *Methods*, **25**, 386–401 (2001).
Havlin S, Molecular diffusion and reactions in *The Fractal Approach to Heterogeneous Chemistry: Surfaces, Colloids, Polymers* (ed. D Avnir), Wiley, New York, 1989, pp. 251–269.
Hekstra D, AR Taussig, M Magnasco, and F Naef, Absolute mRNA concentrations from sequence-specific calibration of oligonucleotide arrays, *Nucleic Acids*, **31**, 1962–1968 (2003).
Karousis NG, S Aouabdi, AS Way, and SM Reddy, Quartz crystal microbalance determination of organophosphorous and carbamate pesticides, *Analytica Chimica Acta*, **469**, 189–196 (2002).
Kim DS, YT Jeong, HK Cyu, HJ Park, HS Kim, JK Shin, P Choi, JH Lee, G Lim, and M Ishida, *Japan Journal of Applied Physics*, **42**, 4111–4115 (2003).
Landegren U, R Kaiser, J Sanders, and L Hood, A ligase-mediated gene detection technique, *Science*, **1241**, 1077–1080 (1988).
Lee CK and SL Lee, Multi-fractal scaling analysis of reactions over fractal surfaces, *Surface Science*, **325**, 294–310 (1995).
Leone G, H van Schijndel, B van Gemen, FR Kramer, and CD Schoen, Molecular beacon probes combined with amplification by NASBA enable homogeneous, real-time detection of RNA, *Nucleic Acid Research*, **26**, 2150–2155 (1998).
Ohtake T, C Hamai, T Uno, H Tabata, and T Kawai, *Japan Journal of Applied Physics*, **43**(9A/B), 11137–11139 (2004).
Marras SAE, S Tyagi, and FR Kramer, Real-time assays with molecular beacons and other fluorescence nucleic acid hybridization probes, *Clinica Chimica Acta*, **363**, 48–60 (2006).
Martin ST, VE Granstuff, and GC Frye, Effect of surface roughness on the response of thickness-shear mode resonators in liquids, *Analytical Chemistry*, **65**, 2910–2922 (1991).
Michel W, T Mai, T Naiser, and A Ott, Optical study of DNA surface hybridization reveals DNA surface density as a key parameter for microarray hybridization kinetics, *Biophysical Journal*, **92**, 994–1004 (2007).
Peterson AW, LK Wolf, and RM Georgiadis, Hybridization of mismatched or partially matched DNA at surfaces, *Journal of the American Chemical Society*, **124**, 14601–14607 (2002).
Summerer D and A Marx, A molecular beacon for quantitative monitoring of the DNA polymerase reaction in real-time, *Angew Chemistry International Edition*, **41**, 3620–3622 (2002).
Tawa K and W Knoll, Mismatching base-pair dependence of the kinetics of DNA-DNA hybridization studied by surface plasmon resonance fluorescence spectroscopy, *Nucleic Acids Research*, **32**, 2372–2377 (2004).
Tyagi S and FR Kramer, Molecular beacons: Probes that fluoresce upon hybridization, *Nature Biotechnology*, **14**, 303–308 (1996).
Uno T, T Ohtake, H Tabata, and T Kawai, *Japan Journal of Applied Physics*, **43**(12B), 11584–11587 (2004).
Uno T, H Tabata, and T Kawai, Peptide-nucleic acid-modified ion-sensitive field-effect based biosensor for direct detection of DNA hybridization, *Analytical Chemistry*, **79**, 52–59 (2007).
Wabuyele MB, H Farquar, W Stryjewski, RP Hammer, SA Sofer, YW Cheng, and F Barany, Approaching real-time molecular diagnostics: Single-pair fluorescence energy transfer (spFRET) detection for the analysis of low abundant point mutations in K-ras oncogenes, *Journal of the American Chemical Society*, **125**, 6937–6945 (2003).
Wang Y, W Gumbrecht, M Humenik, and M Sprinzl, Esterase 2-oligonucleotide conjugates as sensitive reporter for electrochemical detection of nucleic acid hybridization, *Biosensors & Bioelectronics*, **22**, 1798–1806 (2007).
Wark AW, HJ Lee, and RM Corn, Long-range surface plasmon resonance imaging for bioaffinity sensors, *Analytical Chemistry*, **77**, 3904–3907 (2005).
Watzinger F, K Ebner, and T Lion, Detection and monitoring of virus infections by real-time PCR, *Molecular Aspects of Medicine*, **27**, 254–298 (2006).
Yao D, J Kim, F Yu, PE Nielson, EK Sinner, and W Knoll, Surface density dependence of PCR amplicon hybridization on PN1/DNA probe layers, *Biophysical Journal*, **88**, 2745–2751 (2005).
Yu F, D Yao, and W Knoll, Oligonucleotide hybridization studied by a surface plasmon resonance diffraction sensor (SPDS), *Nucleic Acids Research*, **32**, e75 (2004).

CHAPTER 12

Binding and Dissociation Kinetics of Different Analytes on Novel Biosensing Surfaces: A Fractal Analysis

Chapter Outline

12.1 Introduction

The biosensor field has expanded rapidly over the last few years. Well-tested biosensor techniques are being applied for the detection of different analytes. However, novel and effective biosensing techniques are being continuously developed. In this chapter we use the fractal analysis technique to analyze the binding and dissociation (if applicable) kinetics of (a) the binding and dissociation of IgG species to a porous SiO_2 interferometric biosensor coated with protein A (Schwartz et al., 2007), (b) binding (hybridization) using differential surface plasmon resonance (Boecker et al., 2007), (c) binding of glucose to a One Touch II blood glucose meter and a surface-enhanced Raman scattering (SERS) sensor (Stuart et al., 2006), (d) binding of H9 avian virus to cadmium quantum dots (Yun et al., 2007), and (e) the binding of sodium ions of $Na_{0.44-x}MnO_2$ to a selective sodium ion sensor (Sauvage et al., 2007).

Some of the other novel biosensing techniques that have recently appeared in the literature and are not analyzed here by the fractal analysis method include: (a) a novel platform for the oriented buildup of immunoglobulins on a gold surface for a surface plasmon

Handbook of Biosensors and Biosensor Kinetics. DOI: 10.1016/B978-0-444-53262-6.00012-7

resonance imaging microarray (Ha et al., 2007), (b) affinity-based chromatographic assays for thrombin (Zhao et al., 2008), (c) a new platform technology for DNA extraction and fast detection of gram positive bacteria (Aslan et al., 2008), and (d) a highly-selective electrogenerated chemiluminescence (ECL) biosensor for the detection of target single-strand DNA (ss-DNA) using hairpin DNA as the recognition element (Zhang et al., 2008).

12.2 Theory

Havlin (1989) has reviewed and analyzed the diffusion of reactants towards fractal surfaces. The details of the theory and the equations involved for the binding and the dissociation phases for analyte-receptor binding are available in the literature (Sadana, 2001). The details are not repeated here except that the equations are given to permit an easier reading. These equations have been applied to other biosensor systems (Ramakrishnan and Sadana, 2001; Sadana, 2001, 2005). For most applications, a single- or a dual-fractal analysis is often adequate to describe the binding and the dissociation kinetics. Peculiarities in the values of the binding and the dissociation rate coefficients, as well as in the values of the fractal dimensions with regard to the dilute analyte systems being analyzed will be carefully noted, if applicable.

In this chapter we analyze the binding and dissociation kinetics of the binding and dissociation of IgG species to a porous SiO_2 interferometric biosensor coated with protein A (Schwartz et al., 2007), (b) binding (hybridization) using differential surface plasmon resonance (Boecker et al., 2007), (c) binding of glucose to a One Touch II blood glucose meter and a SERS sensor (Stuart et al., 2006), (d) binding of H9 avian virus to cadmium quantum dots (Yun et al., 2007), and (e) the binding of sodium ions of $Na_{0.44-x}MnO_2$ to a selective sodium ion sensor (Sauvage et al., 2007).

12.2.1 Single-Fractal Analysis

Binding Rate Coefficient

Havlin (1989) points out that the diffusion of a particle (analyte [Ag]) from a homogeneous solution to a solid surface (e.g., receptor [Ab]-coated surface) on which it reacts to form a product (analyte-receptor complex; Ab·Ag) is given by:

$$(\text{Ab}\cdot\text{Ag}) \sim \begin{cases} t^{(3-D_{\text{f,bind}})/2} = t^{p}, & t < t_c \\ t^{1/2}, & t > t_c \end{cases} \tag{12.1}$$

Here $D_{\text{f,bind}}$ or D_f (used later on in the manuscript) is the fractal dimension of the surface during the binding step. t_c is the cross-over value. Havlin (1989) points out that the cross-over value may be determined by $r_c^2 \sim t_c$. Above the characteristic length, r_c, the self-similarity of the surface is lost and the surface may be considered homogeneous. Above time, t_c the surface may be considered homogeneous, since the self-similarity property disappears, and "regular" diffusion is now present. For a homogeneous surface where D_f is equal to 2, and when only diffusional limitations are present, $p = ½$ as it should be. Another way of looking at the $p = ½$ case (where $D_{f,bind}$ is equal to two) is that the analyte in solution views the fractal object, in our case, the receptor-coated biosensor surface, from a "large distance." In essence, in the association process, the diffusion of the analyte from the solution to the receptor surface creates a depletion layer of width $(Ðt)^{½}$ where $Ð$ is the diffusion constant. This gives rise to the fractal power law, $(\text{Analyte} \cdot \text{Receptor}) \sim t^{(3-D_{f,bind})/2}$. For the present analysis, t_c is arbitrarily chosen and we assume that the value of the t_c is not reached. One may consider the approach as an intermediate "heuristic" approach that may be used in the future to develop an autonomous (and not time-dependent) model for diffusion-controlled kinetics.

Dissociation Rate Coefficient

The diffusion of the dissociated particle (receptor [Ab] or analyte [Ag]) from the solid surface (e.g., analyte [Ag]-receptor [Ab]) complex coated surface) into solution may be given, as a first approximation by:

$$(\text{Ab}\cdot\text{Ag}) \approx -t^{(3-D_{f,diss})/2} = -t^p, \quad t > t_{diss} \tag{12.2}$$

Here $D_{f,diss}$ is the fractal dimension of the surface for the dissociation step. This corresponds to the highest concentration of the analyte-receptor complex on the surface. Henceforth, its concentration only decreases. The dissociation kinetics may be analyzed in a manner "similar" to the binding kinetics.

12.2.2 Dual-Fractal Analysis

Binding Rate Coefficient

Sometimes, the binding curve exhibits complexities and two parameters (k, D_f) are not sufficient to adequately describe the binding kinetics. This is further corroborated by low values of the r^2 factor (goodness-of-fit). In that case, one resorts to a dual-fractal analysis (four parameters; k_1, k_2, D_{f1}, and D_{f2}) to adequately describe the binding kinetics. The single-fractal analysis presented above is thus extended to include two fractal dimensions. At present, the time ($t = t_1$) at which the "first" fractal dimension "changes" to the "second" fractal dimension is arbitrary and empirical. For the most part, it is dictated by the data analyzed and experience gained by handling a single-fractal analysis. A smoother curve is obtained in the "transition" region, if care is taken to select the correct number of points for the two regions.

In this case, the product (antibody-antigen; or analyte-receptor complex, Ab·Ag or analyte·receptor) is given by:

$$(\text{Ab}\cdot\text{Ag}) \approx \begin{cases} t^{(3-D_{\text{f1,bind}})/2} = t^{p1}, & t < t_1 \\ t^{(3-D_{\text{f2,bind}})/2} = t^{p2}, & t_1 < t < t_2 = t_c \\ t^{1/2}, \quad t > t_c \end{cases} \tag{12.3}$$

In some cases, as mentioned above, a triple-fractal analysis with six parameters (k_1, k_2, k_3, D_{f1}, D_{f2}, and D_{f3}) may be required to adequately model the binding kinetics. This is when the binding curve exhibits convolutions and complexities in its shape due to perhaps to the very dilute nature of the analyte (in some of the cases to be presented) or for some other reasons. Also, in some cases, a dual-fractal analysis may be required to describe the dissociation kinetics.

12.3 Results

The fractal analysis will be applied to the binding and dissociation kinetics. In this chapter we analyze the binding and dissociation kinetics of the binding and dissociation of IgG species to a porous SiO_2 interferometric biosensor coated with protein A (Schwartz et al., 2007), (b) binding (hybridization) using differential surface plasmon resonance (Boecker et al., 2007), (c) binding of glucose to a One Touch II blood glucose meter and a SERS sensor (Stuart et al., 2006), (d) binding of H9 avian virus to cadmium quantum dots (Yun et al., 2007), and (e) the binding of sodium ions of $Na_{0.44-x}MnO_2$ to a selective sodium ion sensor (Sauvage et al., 2007).

Schwartz et al. (2007) recently analyzed the binding kinetics of protein A to immunoglobulins derived from different species using a porous SiO_2 interferometric biosensor. These authors emphasize that it is important to develop biosensing methods that are simple to use, easy to manufacture, are inexpensive, and are portable. Besides, they should readily be incorporated into high-throughput arrays. They emphasize that porous Si interferometers meet these requirements, and different authors have used these instruments for label-free biological sensing (Starodub et al., 1996; Lin et al., 1997; Dancil et al., 1999; Chan et al., 2000; Schwartz et al., 2006). Furthermore, Schwartz et al. (2007) indicate that porous Si is an attractive material for biological sensing due to its compatibility with conventional silicon microfabrication methods.

Figure 12.1a shows the binding of 177 nM human IgG to a Protein A-coated porous SiO_2 surface (Schwartz et al., 2007). A dual-fractal analysis is required to adequately describe the binding kinetics. The values of (a) the binding rate coefficient, k, and the fractal dimension, D_f, for a single-fractal analysis, and (b) the binding rate coefficients, k_1 and k_2, and the fractal dimensions, D_{f1} and D_{f2}, for a dual-fractal analysis are given in Tables 12.1 and 12.2.

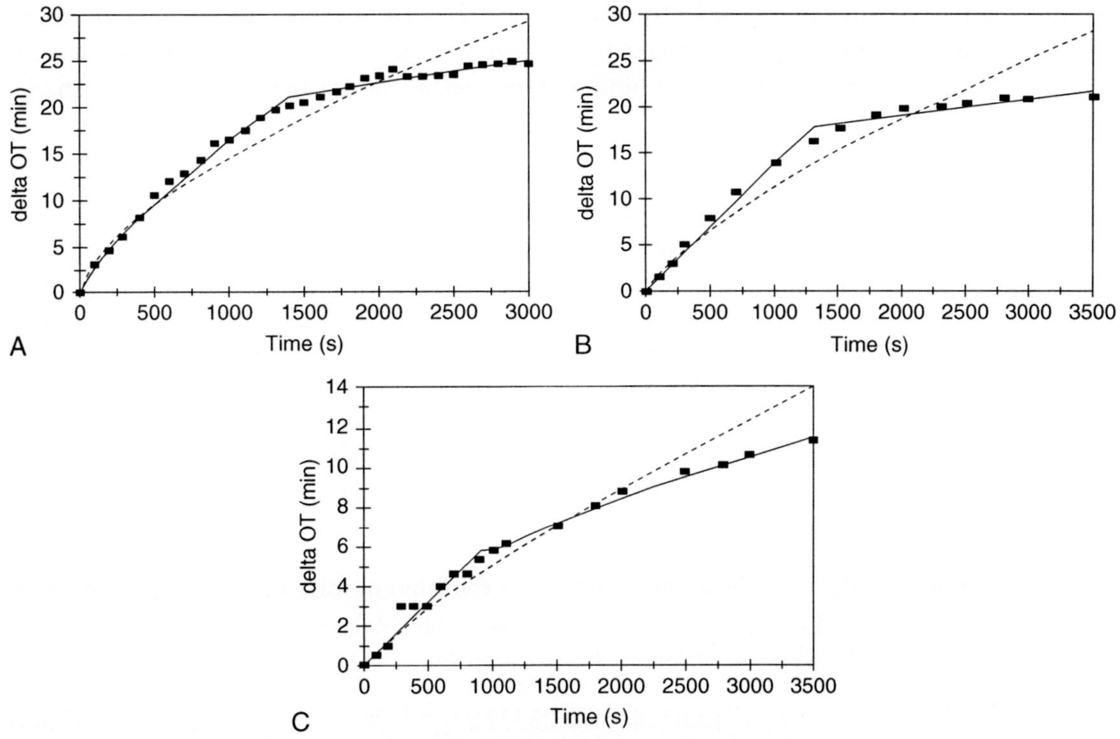

Figure 12.1

Binding of different IgG species to a porous SIO_2 interferometric biosensor coated with protein A (Schwartz et al., 2007): (a) 171 nM protein A, (b) 171 nM rabbit IgG, (c) 171 nM goat IgG. When only a solid line (—) is used then a single-fractal analysis applies. When both a dashed (- - -) and a solid (—) line are used then the dashed line represents a single-fractal analysis and the solid line represents a dual-fractal analysis. In this the solid line provides the better fit.

It is of interest to note that for a dual-fractal analysis, as the fractal dimension increases by a factor of 5.05 from a value of D_{f1} equal to 0.4984 to D_{f2} equal to 2.5188, the binding rate coefficient increases by a factor of 39.8 from a value of k_1 equal to 0.0911 to k_2 equal to 3.629. The changes in the degree of heterogeneity or the fractal dimension on the sensing surface and in the binding rate coefficient are in the same direction.

Figure 12.1b shows the binding of 171 nM rabbit IgG to a Protein A-coated porous SiO_2 surface (Schwartz et al., 2007). A dual-fractal analysis is required to adequately describe the binding kinetics. The values of (a) the binding rate coefficient, k, and the fractal dimension, D_f, for a single-fractal analysis, and (b) the binding rate coefficients, k_1 and k_2, and the fractal dimensions, D_{f1} and D_{f2}, for a dual-fractal analysis are given in Tables 12.1 and 12.2.

Once again, it is of interest to note that for a dual-fractal analysis, as the fractal dimension increases by a factor of 2.32 from a value of D_{f1} equal to 1.1238 to D_{f2} equal to 2.6090, the binding rate coefficient increases by a factor of 202.32 from a value of k_1 equal to 0.0215 to k_2 equal to 4.350. Once again, changes in the degree of heterogeneity or the fractal dimension on the sensing surface and in the binding rate coefficient are in the same direction.

Figure 12.1c shows the binding of 171 nM goat IgG to a Protein A-coated porous SiO_2 surface (Schwartz et al., 2007). A dual-fractal analysis is required to adequately describe the binding kinetics. The values of (a) the binding rate coefficient, k, and the fractal dimension, D_f, for a single-fractal analysis, and (b) the binding rate coefficients, k_1 and k_2, and the fractal dimensions, D_{f1} and D_{f2}, for a dual-fractal analysis are given in Tables 12.1 and 12.2.

Once again, it is of interest to note that for a dual-fractal analysis, as the fractal dimension increases by a factor of 2.06 from a value of D_{f1} equal to 0.9260 to D_{f2} equal to 1.9088, the binding rate coefficient increases by a factor of 26.58 from a value of k_1 equal to 0.0050 to k_2 equal to 0.1329. Note that, one again, changes in the degree of heterogeneity or the fractal dimension on the sensing surface and in the binding rate coefficient are in the same direction.

Figure 12.2a and Table 12.1 show the decrease in the binding rate coefficient, k_1, with an increase in the fractal dimension, D_{f1}, for a dual-fractal analysis. For the data shown in Figure 12.2a, the binding rate coefficient, k_1, is given by:

$$k_1 = (0.01241 + 0.03831)D_{f1}^{-2.486 \pm 2.04} \tag{12.4a}$$

The fit is reasonable. Only three data points are available. There is scatter in the data, and this is reflected in the error in the binding rate coefficient. Only the positive error is presented here since the binding rate coefficient, k_1, cannot be negative. The binding rate coefficient, k_1, is sensitive to the fractal dimension, D_{f1}, since it exhibits close to a negative two and a half (equal to -2.486) order of dependence on the fractal dimension, D_{f1}.

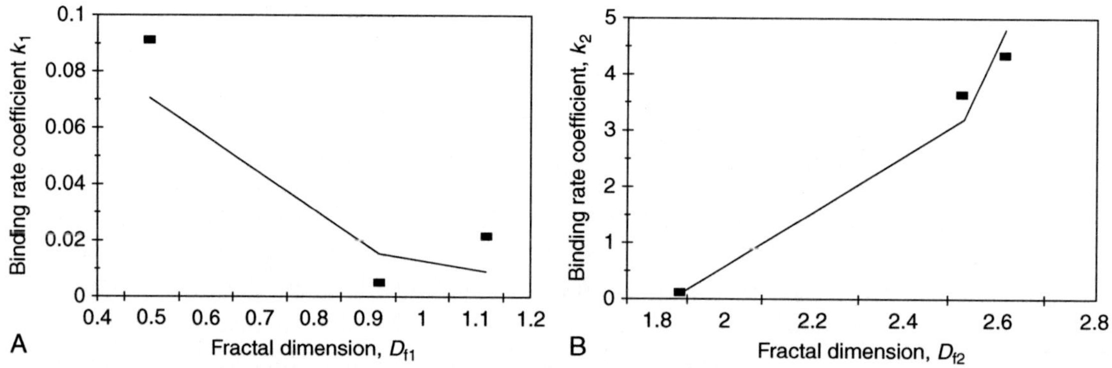

Figure 12.2
(a) Decrease in the binding rate coefficient, k_1, with an increase in the fractal dimension, D_{f1}. (b) Increase in the binding rate coefficient, k_2, with an increase in the fractal dimension, D_{f2}.

Table 12.1: Binding rate coefficients and fractal dimensions for different IgG species to a porous SiO_2 interferometric biosensor coated with protein A (Schwartz et al., 2007).

Analyte in Solution/Protein A on a Porous SiO_2 Interferometric Biosensor	k	k_1	k_2	D_f	D_{f1}	D_{f2}
171 nM human IgG	0.1997 ± 0.0236	0.0911 ± 0.0054	3.6269 ± 0.0779	1.7556 ± 0.0485	0.4984 ± 0.0396	2.5188 ± 0.0503
171 nM rabbit IgG	0.06793 ± 0.0139	0.0215 ± 0.0019	4.350 ± 0.00195	1.5234 ± 0.0923	1.1238 ± 0.0774	2.6090 ± 0.0669
171 nM goat IgG	0.0198 ± 0.00489	0.0050 ± 0.0012	0.1329 ± 0.0028	1.3940 ± 0.1084	0.9260 ± 0.1970	1.9088 ± 0.3298

Table 12.2: Binding and dissociation rate coefficients and fractal dimensions for the binding and the dissociation phase for 51 and 171 nM human IgG in solution to a protein-A coated SiO_2 surface (Schwartz et al., 2007).

Analyte in Solution	k	k_1	k_2	k_d	D_f	D_{f1}	D_{f2}	D_{fd}
51 nM human IgG	0.297 ± 0.055	0.09222 ± 0.00559	5.734 ± 0.075	2.7046 ± 0.2362	2.1226 ± 0.1367	1.7524 ± 0.0862	1.7356 ± 0.0387	2.7046 ± 0.2362
171 nM human IgO	0.0773 ± 0.00066	na	na	0.003078 ± 0.000083	0.9796 ± 0.00778	na	na	1.0644 ± 0.01621

Figure 12.2b and Table 12.1 show the increase in the binding rate coefficient, k_2, with an increase in the fractal dimension, D_{f2}, for a dual-fractal analysis. For the data shown in Figure 12.2b, the binding rate coefficient, k_2 is given by:

$$k_2 = (8.2 \times 10^{-05} \pm 1.4 \times 10^{-05}) D_{f2}^{11.45 \pm 0.6492.04} \quad (12.4b)$$

The fit is very good. Only three data points are available. The availability of more data points would lead to a more reliable fit. The binding rate coefficient, k_2, is extremely sensitive to the fractal dimension, D_{f2}, as noted by the close to eleven and a half (equal to 11.45) order of dependence exhibited on the fractal dimension, D_{f2}, on the sensor surface.

Figure 12.3 shows the binding and dissociation of 171 nM human IgG dosed on a protein-A coated surface (Schwartz et al., 2007). A dual-fractal analysis is required to adequately model the binding kinetics. The dissociation kinetics is adequately described by a single-fractal analysis. The values of (a) the binding rate coefficient, k, and the fractal dimension, D_f, for a single-fractal analysis, (b) the binding rate coefficients, k_1 and k_2, and the fractal dimensions, D_{f1} and D_{f2}, for a dual-fractal analysis and the dissociation rate coefficient, k_d and the fractal dimension, D_{fd} for a single-fractal analysis are given in Table 12.2.

It is of interest to note that as the fractal dimension increases by a factor of 1.53 from value of D_{f1} equal to 1.8666 to D_{f2} equal to 2.8612, the binding rate coefficient increases by a factor of 48.67 from a value of k_1 equal to 0.2832 to k_2 equal to 13.782. The changes in the degree of heterogeneity on the sensing surface and in the binding rate coefficient are in the same direction.

Figure 12.4 shows the binding and dissociation of 51 nM human IgG dosed on a protein-A coated surface (Schwartz et al., 2007). A dual-fractal analysis is required to adequately describe the binding kinetics. A single-fractal analysis is adequate to describe the dissociation kinetics. The values of (a) the binding rate coefficient, k, and the fractal dimension, D_f, for a

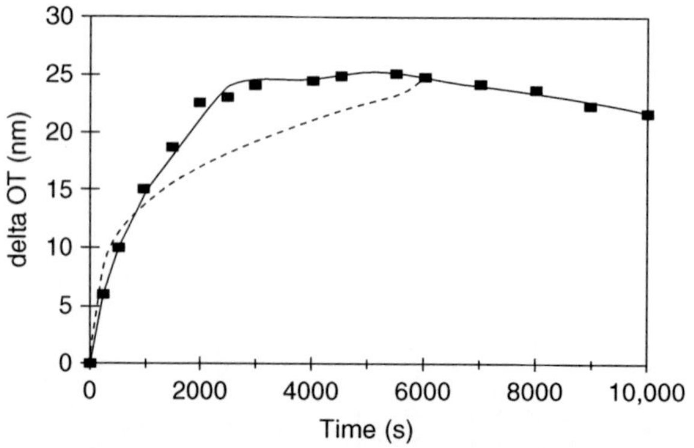

Figure 12.3

Binding and dissociation of human IgG on a protein A-coated biosensor surface (Schwartz et al., 2007). When only a solid line (—) is used then a single-fractal analysis applies. When both a dashed (- - -) and a solid (—) line are used then the dashed line represents a single-fractal analysis and the solid line represents a dual-fractal analysis.

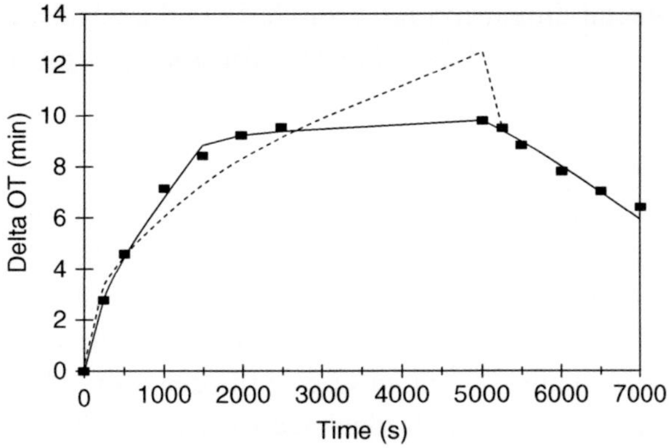

Figure 12.4

Binding and dissociation of 51 nM human IgG dosed on a protein A-coated porous SiO_2 surface (Schwartz et al., 2007). When only a solid line (—) is used then a single-fractal analysis applies. When both a dashed (- - -) and a solid (—) line are used then the dashed line represents a single-fractal analysis and the solid line represents a dual-fractal analysis.

single-fractal analysis, (b) the binding rate coefficients, k_1 and k_2, and the fractal dimensions, D_{f1} and D_{f2}, for a dual-fractal analysis, and (c) the dissociation rate coefficient, k_d, and the fractal dimension in the dissociation phase, D_{fd} for a single-fractal analysis are given in Table 12.2.

It is of interest to note that as the fractal dimension increases by a factor of 1.639 from value of D_{f1} equal to 1.7524 to D_{f2} equal to 2.874, the binding rate coefficient increases by a factor of 20.25 from a value of k_1 equal to 0.09222 to k_2 equal to 5.7354. The changes in the degree of heterogeneity on the sensing surface and in the binding rate coefficient are in the same direction.

Figure 12.5 shows the binding and the dissociation of human 171 nM IgG solution dosed on a protein A-coated porous SiO_2 surface (Schwartz et al., 2007). A single-fractal analysis is adequate to describe the binding and the dissociation kinetics. The values of (a) the binding rate

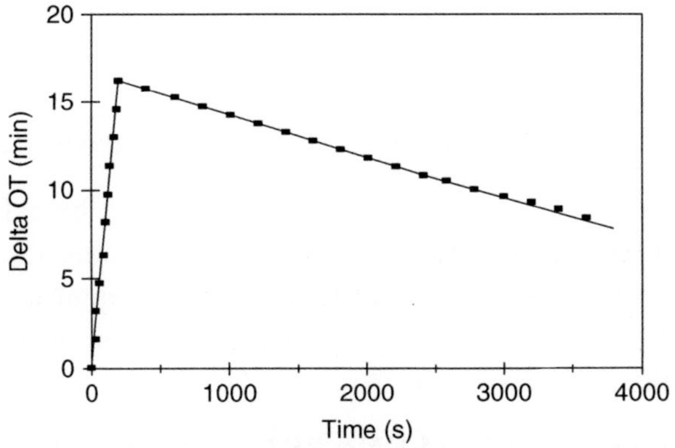

Figure 12.5

Binding and dissociation of 171 nM human IgG solution dosed on a protein A-coated porous SiO_2 surface (Schwartz et al., 2007). When only a solid line (—) is used then a single-fractal analysis applies. When both a dashed (- - -) and a solid (—) line are used then the dashed line represents a single-fractal analysis and the solid line represents a dual-fractal analysis.

coefficient, k, and the fractal dimension for a single-fractal analysis, and (b) the dissociation rate coefficient, k_d, and the fractal dimension D_{fd}, for a single-fractal analysis are given in Table 12.2.

Boecker et al. (2007) recently pointed out that the SPR (surface plasmon resonance) detection of analytes in solution has been an important means for the detection of biomolecules ever since the first application when this instrument was used (Leidberg et al., 1995). However, Boecker et al. (2007) report that for high throughput and for the detection of biomolecules at low detection limits, the SPR instrument poses challenges. These authors describe a differential imaging that reduces factors which interogate the detection of single-wavelength imaging methods. These authors used an additional second laser with a different wavelength for differential image processing. This allowed them to obtain low detection limits in relatively small spots. They used a CCD camera for the simultaneous processing of two images at the different wavelengths provided by the two laser diodes.

Boecker et al. (2007) explain that the SPR imaging technique is based on intensity measurements at a fixed reflection angle. The surface is illuminated by a collimated monochromatic beam which is slightly shifted from the SPR resonance angle. Any shift of the resonance causes a change in the reflected intensity. These intensity variations due to molecular binding may be observed simultaneously by means of a CCD matrix (Shumaker-Parry and Campbell, 2004). Boecker et al. (2007) point out that the technique is affected by possible changes of the resonance curve owing to modification of the surface roughness or to light absorption. Zybin et al. (2005) recently applied two widened, collimated laser beams of different wavelengths combined in one beam irradiating the metal surface of the SPR device.

Boecker et al. (2007) analyzed the hybridization (binding) of different DNA to their complementary DNA by their differential surface plasmon resonance imaging technique. Figure 12.6a shows the binding of 1 μM DNA RS1 in solution to RS1-c immobilized on the sensing surface. A dual-fractal analysis is required to adequately describe the hybridization (binding) kinetics. The values of (a) the binding rate coefficient, k, and the fractal dimension, D_f, for a single-fractal analysis, and (b) the binding rate coefficients, k_1 and k_2, and the fractal dimensions, D_{f1} and D_{f2}, for a dual-fractal analysis are given in Table 12.3.

It is of interest to note that as the fractal dimension increases by a factor of 2.438 from a value of D_{f1} equal to 1.0566 to D_{f2} equal to 2.5766, the binding rate coefficient increases by a factor of 1.62 from a value of k_1 equal to 1.073 to k_2 equal to 1.742. Changes in the degree of heterogeneity on the sensing surface or the fractal dimension and in the binding rate coefficient are in the same direction.

Figure 12.6b shows the binding of 1 μM DNA RS2 in solution to RS2-c immobilized on the sensing surface. A dual-fractal analysis is required to adequately describe the hybridization (binding) kinetics. The values of (a) the binding rate coefficient, k, and the fractal dimension, D_f, for a single-fractal analysis, and (b) the binding rate coefficients, k_1 and k_2, and the fractal dimensions, D_{f1} and D_{f2}, for a dual-fractal analysis are given in Table 12.3.

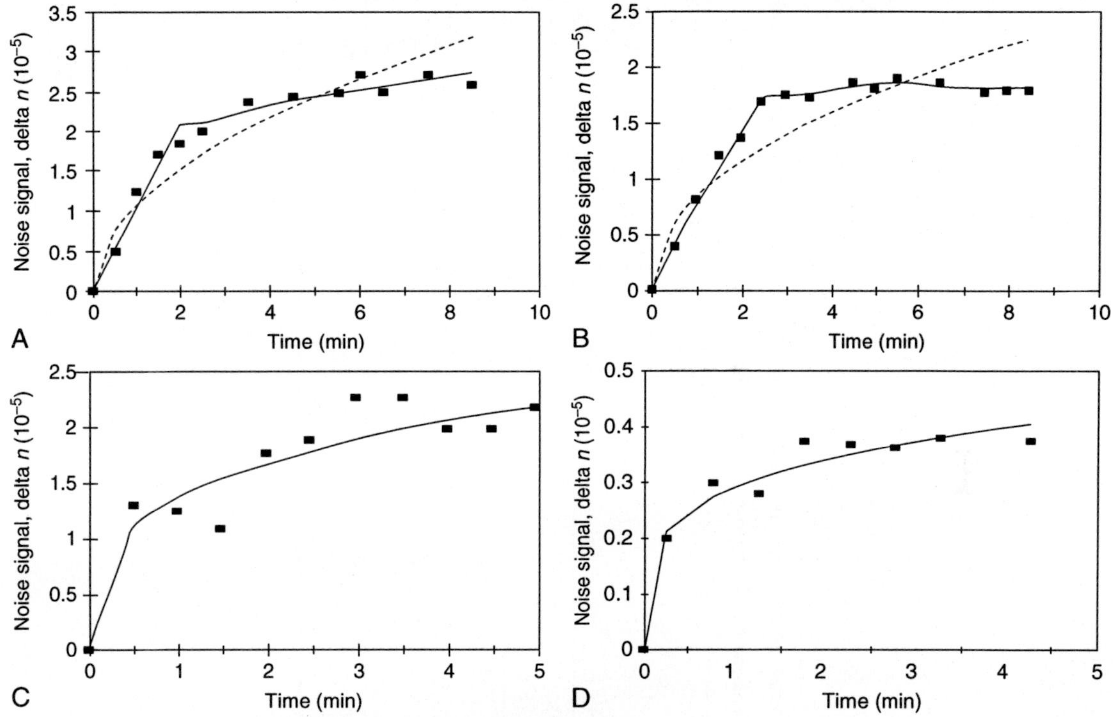

Figure 12.6

Binding (hybridization) in solution using differential surface plasmon resonance: (Boecker et al., 2007): (a) 1 μM DNA RS1 to RS1-c, (b) 1 μM DNA RS2 to RS2-c, (c) 1 μM DNA T2-c to T2, (d)1 μM DNA G-c to G. When only a solid line (—) is used then a single-fractal analysis applies. When both a dashed (- - -) and a solid (—) line are used then the dashed line represents a single-fractal analysis and the solid line represents a dual-fractal analysis.

It is of interest to note, once again, that as the fractal dimension increases by a factor of 2.26 from a value of D_{f1} equal to 1.2218 to D_{f2} equal to 2.7628, the binding rate coefficient increases by a factor of 1.964 from a value of k_1 equal to 0.7768 to k_2 equal to 1.526. Once again, changes in the degree of heterogeneity on the sensing surface or the fractal dimension and in the binding rate coefficient are in the same direction.

Figure 12.6c shows the binding of 1 μM DNA T2-c in solution to T2 immobilized on the sensing surface. A singe-fractal analysis is adequate to describe the hybridization (binding) kinetics. The values of the binding rate coefficient, k, and the fractal dimension, D_f, for a single-fractal analysis are given in Table 12.3.

Figure 12.6d shows the binding of 1 μM DNA G-c in solution to G immobilized on the sensing surface. A singe-fractal analysis is adequate to describe the hybridization (binding) kinetics. The values of the binding rate coefficient, k, and the fractal dimension, D_f, for a single-fractal analysis are given in Table 12.3.

Table 12.3 Binding and dissociation rate coefficients for the hybridization of different analytes in solution to their complementary receptors using differential surface plasmon resonance imaging (Boecker et al., 2007)

Analyte in solution/ Receptor on surface	k	k_1	k_2	k_d	D_f	D_{f1}	D_{f2}	D_{fd}
1 μM DNA RS1/RS1-c	1.0945 ± 0.2452	1.0729 ± 0.188	1.7423 ± 0.0868	na	1.9962 ± 0.137	1.0556 ± 0.3108	2.577 ± 0.09	na
1 μM DNA RS2/RS2-c	0.8431 ± 0.1962	0.7769 ± 0.055	1.5261 ± 0.0224	0.05 ± 0.0	2.0788 ± 0.137	1.2218 ± 0.1079	2.763 ± 0.057	3.0 ± 0
1 μM DNA T2-c/T2	1.3659 ± 0.2356	na	na	na	2.4024 ± 0.144	na	na	na
1 μM DNA G-c/G	0.2924 ± 0.0265	na	na	na	2.5400 ± 0.071	na	na	na

Stuart et al. (2006) recently published the first *in vivo* application of SERS. They used SERS to obtain quantitative *in vivo* glucose measurements from an animal model. They functionalized the silver film covered nanosphere surfaces with a two-component self-assembled monolayer. These authors subcutaneously implanted the nanospheres into a Sprague-Dewey rat, and were able to measure the glucose concentrations of the interstitial fluid by spectroscopically addressing the sensor through an optical window.

The authors were able to develop a technique that addresses the critical problems that limit the use of SERS to glucose measurements, particularly, *in vivo*. They developed (a) strong and stable enhancing SERS-active surfaces, and (b) chemical functionalization of those surfaces with self-assembled monolayers (SAMs). (Yonzon et al., 2004; Lyandres et al., 2005; Stuart et al., 2005). Stuart et al. (2006) further reported that these SAMs improve the glucose signal (Shafer-Peltier et al., 2003; Yonzon et al., 2005; Dieringer et al., 2006).

Stuart et al. (2006) used intermittent intravenous infusion for three hours to vary the glucose in the rat. These authors delivered the glucose infusion via the femoral cannula. Sterile phosphate-buffered saline at a glucose concentration of 1 g/mL was delivered over a period of 5-10 min. A droplet of blood was drawn from the rat and measured with the One Touch II glucometer and the corresponding SERS measurements were made by the SERS sensor.

Figure 12.7a shows the binding of the glucose to the One Touch II glucometer using an *in vivo* analysis. Stuart et al. (2006) pointed out that the standard glucometer effectively tracked the change in the glucose concentration. A single-fractal analysis is adequate to describe the binding kinetics. The values of the binding rate coefficient, k, and the fractal dimension, D_f, for a single-fractal analysis are given in Table 12.4.

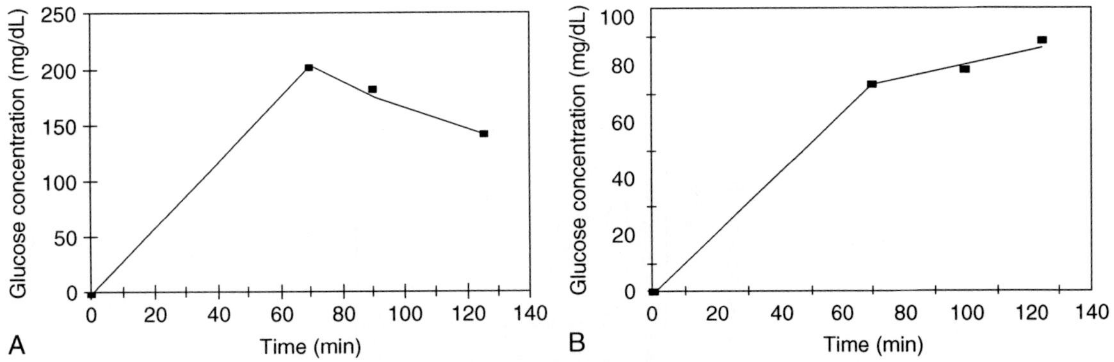

Figure 12.7

(a) Binding indicating the time course of glucose measurement by One Touch II blood glucose meter (Stuart et al., 2006), (b) Binding indicating the time course of glucose measurement by SERS sensor (Stuart et al., 2006).

Table 12.4: Binding and dissociation rate coefficients and fractal dimensions for the binding and the dissociation phase for *in vivo* glucose measurement and for *ex vivo* analysis of response time (Stuart et al., 2006).

Measurement	Instrument	k	k_1	k_2	k_d	D_f	D_{f1}	D_{f2}	D_{fd}
In vivo glucose	One Touch II blood glucose meter	2870.1 ± 117.4	na	na	na	3.0 ± 0.0	na	na	na
In vivo glucose	SERS sensor	19.35 ± 0.778	na	na	na	2.381 ± 0.191	na	na	na
Ex vivo analysis; response to a step change in glucose concentration	SERS sensor	79.07 ± 15.64	12.14 ± 1.17	177.5 ± 7.15	69.13 ± 2.62	2.486 ± 0.152	1.448 ± 0.394	2.825 ± 0.061	2.456 ± 0.029

Figure 12.7b shows the binding of glucose to the SERS-based biosensor using an *in vivo* analysis. Stuart et al. (2006) reported that the SERS-based biosensor also effectively tracked the change in the glucose concentration. A single-fractal analysis is, once again, adequate to describe the binding kinetics. The values of the binding rate coefficient, k, and the fractal dimension, D_f, for a single-fractal analysis are given in Table 12.4.

Figure 12.8 shows the binding of glucose to the SERS biosensor by an *ex vivo analysis*. A dual-fractal analysis is required to adequately describe the binding kinetics. The values of (a) the binding rate coefficient, k, and the fractal dimension, D_f, for a single-fractal analysis, and (b) the binding rate coefficients, k_1 and k_2, and the fractal dimensions, D_{f1} and D_{f2}, for a dual-fractal analysis are given in Table 12.4.

It is of interest to note that as the fractal dimension increases by a factor of 1.95 from a value of D_{f1} equal to 1.484 to D_{f2} equal to 2.8248, the binding rate coefficient increases by a factor of 14.62 from a value of k_1 equal to 12.140 to k_2 equal to 177.54. The changes in the degree of heterogeneity on the sensor surface or in the fractal dimension and in the binding rate coefficient are in the same direction.

Guo et al. (2007) recently reported that noble metal nanoparticles such as Au (silver) or Ag (gold) have been studied for their extraordinary size-dependent optical properties. They point out that these nanoparticles exhibit a strong UV-vis absorption band. They explain that this kind of absorption band is observed when the incident photon frequency is resonant with the collective excitation of the conduction electrons.

Guo et al. (2007) report that they have an improved method for the detection of Concanavlin A (Con A) with label-free optical biosensors. These authors explain that 1-Dodecanethiol (DDT) self-assembled onto gold nanoparticles were deposited on glass slides. Thereafter,

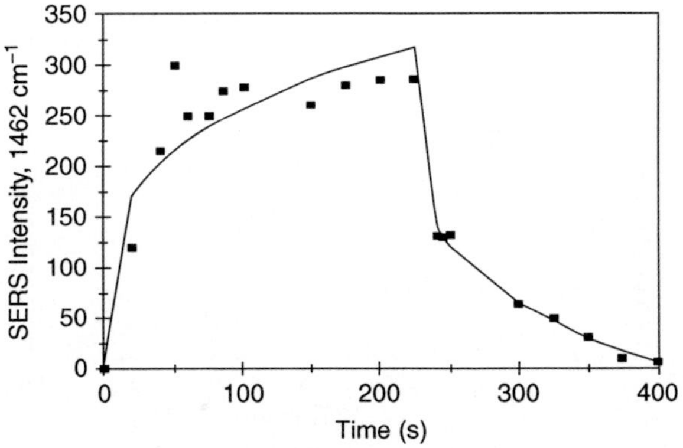

Figure 12.8
Binding and dissociation of *ex vivo* analysis using a real-time SERS sensor following a step change in glucose concentration (Stuart et al., 2006).

glycolipid molecules were inserted into dodecanethiol by physical insertions. Guo et al. (2007) state that the recognition between Con A and carbohydrate was observed by UV-vis spectrophotometry. They state that there is an additional monolayer (i.e., alkylthiol), which is used to tether the reporter (i.e., glycolipid) monolayer and the transducer (i.e., Au colloids). The detection limit for Con A by the biosensor was 0.1 nM.

Figure 12.9a shows the binding and dissociation of 0.1 nM Con A in solution to the gold nanoparticle biosensor (Guo et al., 2007). A dual-fractal analysis is required to adequately describe the binding kinetics. A single-fractal analysis is adequate to describe the dissociation kinetics. The values of (a) the binding rate coefficient, k, and the fractal dimension, D_f, for a single-fractal analysis, and (b) the binding rate coefficients, k_1 and k_2, and the fractal dimensions, D_{f1} and D_{f2}, for a dual-fractal analysis are given in Table 12.5.

It is of interest to note that as the fractal dimension increases by a factor of 2.5438 from a value of D_{f1} equal to 1.0 to D_{f2} equal to 2.5438, the binding rate coefficient increases by a factor of 2.526 (almost a linear increase) from a value of k_1 equal to 1.0 to k_2 equal to

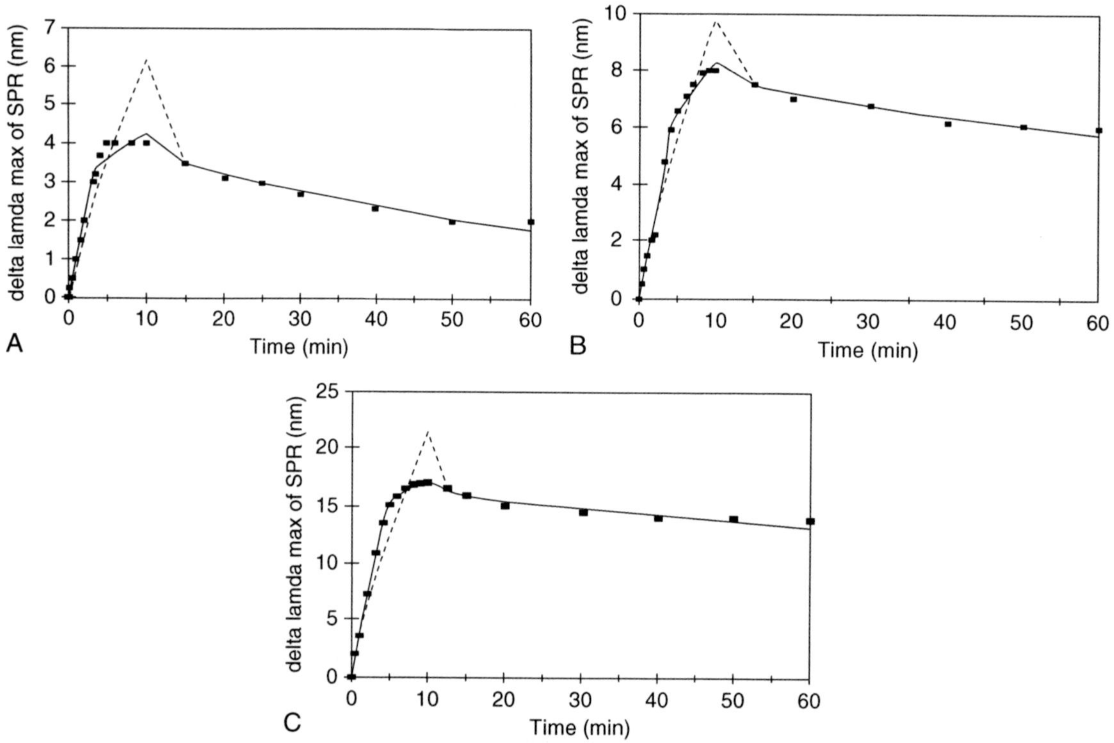

Figure 12.9
Binding of different concentrations (in nM) of Con A to a colloidal gold coated with glycolipid/ dodecanethiol bilayer (Guo et al., 2007): (a) 0.1, (b) 1.2, (c) 4.8.

Table 12.5: Binding and dissociation rate coefficients and fractal dimensions for the binding and the dissociation phase for different concentrations in solution to gold nanoparticles biosensor chips modified with a self-assembled bilayer for the detection of Con A (Guo et al., 2007).

Con A concentration (nM)	k	k_1	k_2	k_d	D_f	D_{f1}	D_{f2}	D_{fd}
0.1	0.9731 ± 0.251	1.0 ± 0	2.526 ± 0.198	0.201 ± 0.016	1.3944 ± 0.123	1.0 ± 0.0	2.5438 ± 0.140	1.7776 ± 0.077
1.2	1.596 ± 0.263	1.5722 ± 0.2945	3.7724 ± 0.104	0.2171 ± 0.0295	1.4224 ± 0.0754	1.3648 ± 0.1665	2.3138 ± 0.067	1.8192 ± 0.129
4.8	3.928 ± 0.701	3.745 ± 0.210	12.439 ± 0.173	0.3865 ± 0.095	1.5232 ± 0.108	1.8150 ± 0.055	2.7190 ± 0.068	1.8304 ± 0.161

2.526. The changes in the fractal dimension or the degree of heterogeneity on the sensing surface and in the binding rate coefficient are in the same direction.

Figure 12.9b shows the binding and dissociation of 1.2 nM Con A in solution to the gold nanoparticle biosensor (Guo et al., 2007). A dual-fractal analysis is required to adequately describe the binding kinetics. A single-fractal analysis is adequate to describe the dissociation kinetics. The values of (a) the binding rate coefficient, k, and the fractal dimension, D_f, for a single-fractal analysis, (b) the binding rate coefficients, k_1 and k_2, and the fractal dimensions, D_{f1} and D_{f2}, for a dual-fractal analysis and the dissociation rate coefficient, k_d and the fractal dimension, D_{fd} for a single-fractal analysis are given in Table 12.5.

It is of interest to note that as the fractal dimension increases by a factor of 1.696 from a value of D_{f1} equal to 1.3648 D_{f2} equal to 2.3138, the binding rate coefficient increases by a factor of 2.40 from a value of k_1 equal to 1.5722 to k_2 equal to 3.7724. The changes in the fractal dimension or the degree of heterogeneity on the sensing surface and in the binding rate coefficient are, once again, in the same direction.

Figure 12.9c shows the binding and dissociation of 4.8 nM Con A in solution to the gold nanoparticle biosensor (Guo et al., 2007). A dual-fractal analysis is required to adequately describe the binding kinetics. A single-fractal analysis is adequate to describe the dissociation kinetics. The values of (a) the binding rate coefficient, k, and the fractal dimension, D_f, for a single-fractal analysis, and (b) the binding rate coefficients, k_1 and k_2, and the fractal dimensions, D_{f1} and D_{f2}, for a dual-fractal analysis and the dissociation rate coefficient, k_d and the fractal dimension, D_{fd} for a single-fractal analysis are given in Table 12.5.

It is of interest to note that as the fractal dimension increase by a factor of 1.498 from a value of D_{f1} equal to 1.8150 to D_{f2} equal to 2.7190, the binding rate coefficient increases by a factor of 3.32 from a value of k_1 equal to 3.745 to k_2 equal to 12.439. The changes in the fractal dimension or the degree of heterogeneity on the sensing surface and in the binding rate coefficient are in the same direction.

Figure 12.10a and Table 12.5 show the increase in the binding rate coefficient, k_1, with an increase in the Con A concentration in the 0.1-5.0 nM range for a dual-fractal analysis. For the data shown in Figure 12.10a, the binding rate coefficient, k_1, is given by:

$$k_1 = (1.916 \pm 0.718)(\text{Con A})^{0.322 \pm 0.115} \tag{12.5a}$$

The fit is good. Only three data points are available. The availability of more data points would lead to a more reliable fit. The binding rate coefficient, k_1, exhibits a low (equal to 0.322) order of dependence on the Con A concentration in solution. The fractional order of dependence exhibited by the binding rate coefficient, k_1, on the Con A concentration in solution lends support to the fractal nature of the system.

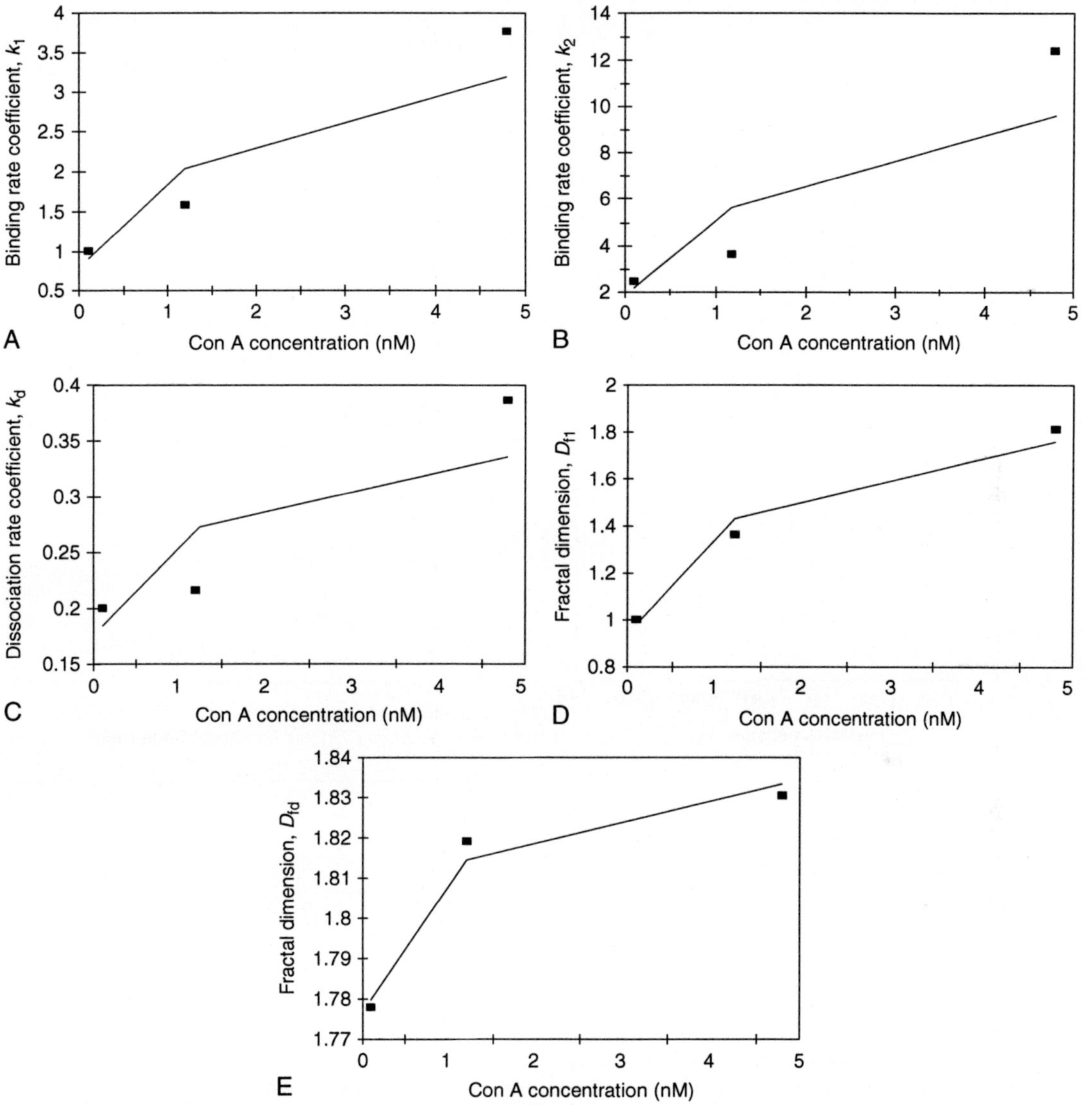

Figure 12.10

(a) Increase in the binding rate coefficient, k_1, with an increase in the Con A concentration in solution. (b) Increase in the binding rate coefficient, k_2, with an increase in the Con A concentration in solution. (c) Increase in the dissociation rate coefficient, k_d, with an increase in the Con A concentration in solution. (d) Increase in the fractal dimension, D_{f1}, with an increase in the Con A concentration in solution. (e) Increase in the fractal dimension for dissociation, D_{fd}, with an increase in the Con A concentration in solution.

continued

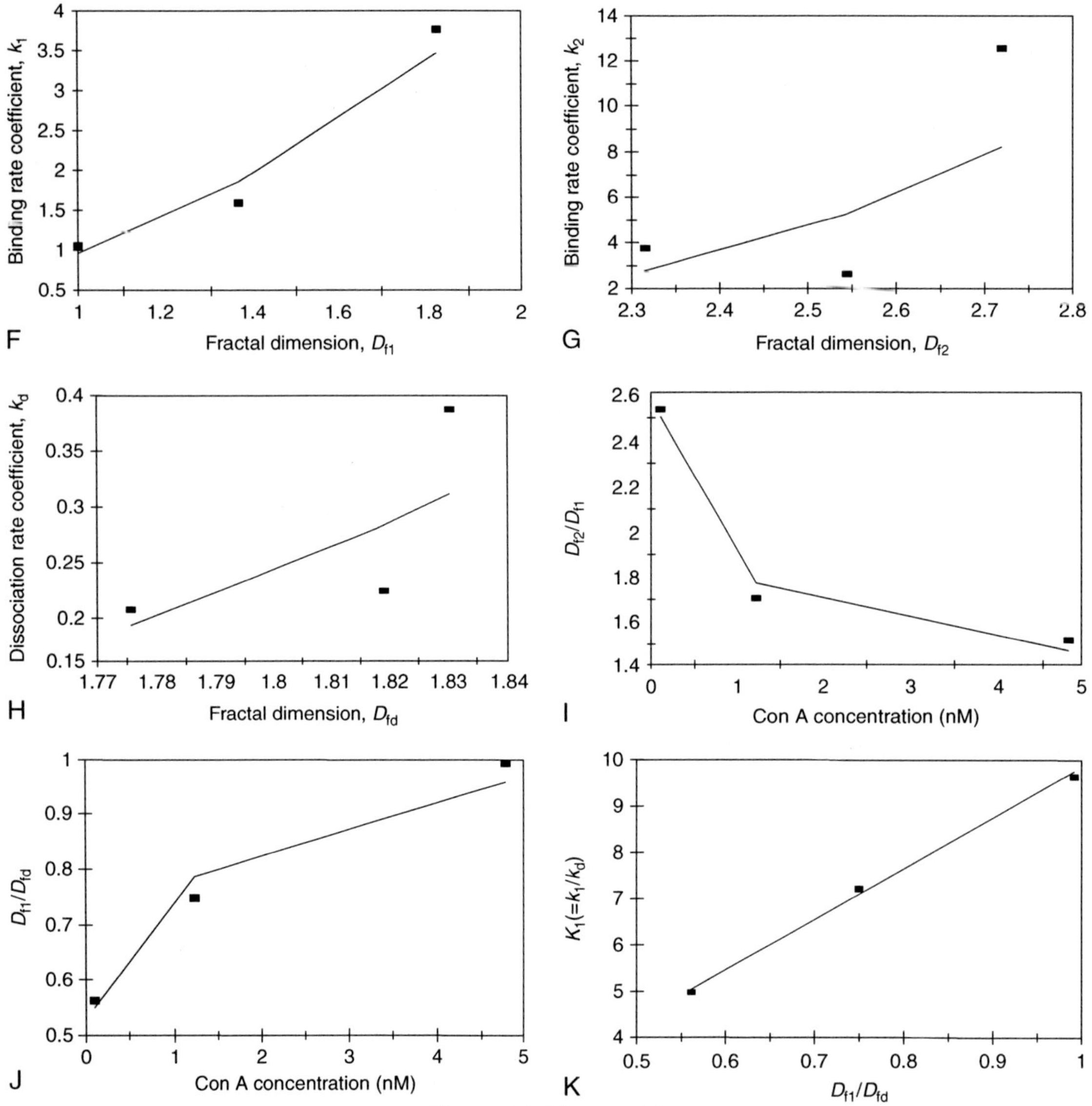

Figure 12.10—cont'd

(f) Increase in the binding rate coefficient, k_1, with an increase in the fractal dimension, D_{f1}. (g) Increase in the binding rate coefficient, k_2, with an increase in the fractal dimension, D_{f2}. (h) Increase in the dissociation rate coefficient, k_d, with an increase in the fractal dimension for dissociation, D_{fd}. (i) Decrease in the ratio of the fractal dimensions, D_{f2}/D_{f1}, with an increase in the Con A concentration in solution. (j) Increase in the ratio of the fractal dimensions, D_{f1}/D_{fd}, with an increase in the Con A concentration in solution. (k) Increase in the affinity, $K_1(=k_1/k_d)$, with an increase in the ratio of fractal dimensions, D_{f1}/D_{fd}.

continued

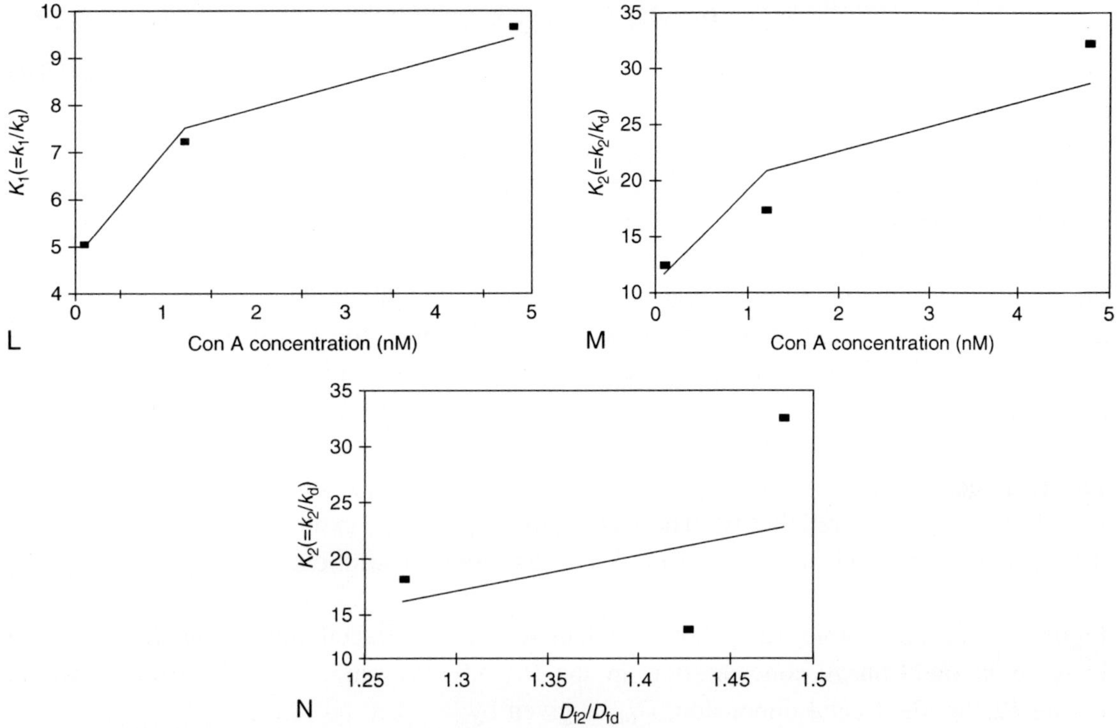

Figure 12.10—cont'd

(l) Increase in the affinity, $K_1(=k_1/k_d)$, with an increase in the Con A concentration in solution. (m) Increase in the affinity, $K_2(=k_2/k_d)$, with an increase in the Con A concentration in solution. (n) Increase in the affinity, $K_2(=k_2/k_d)$, with an increase in the ratio of the fractal dimensions, D_{f2}/D_{fd}.

Figure 12.10b and Table 12.5 show the increase in the binding rate coefficient, k_2, with an increase in the Con A concentration in the 0.1-5.0 nM range for a dual-fractal analysis. For the data shown in Figure 12.10b, the binding rate coefficient, k_2, is given by:

$$k_2 = (5.269 \pm 3.431)(\text{Con A})^{0.382 \pm 0.181} \quad (12.5b)$$

The fit is reasonable. Only three data points are available. The availability of more data points would lead to a more reliable fit. The binding rate coefficient, k_1, exhibits a low (equal to 0.382) order of dependence on the Con A concentration in solution. The fractional order of dependence exhibited by the binding rate coefficient, k_2, on the Con A concentration in solution lends support to the fractal nature of the system.

Figure 12.10c and Table 12.5 show the increase in the dissociation rate coefficient, k_d, with an increase in the Con A concentration in the 0.1-5.0 nM range for a single-fractal analysis. For the data shown in Figure 12.10c, the dissociation rate coefficient, k_d, is given by:

$$k_d = (0.264 \pm 0.083)(\text{Con A})^{0.153 \pm 0.099} \tag{12.5c}$$

The fit is reasonable. Only three data points are available. The availability of more data points would lead to a more reliable fit. The dissociation rate coefficient, k_d, exhibits a very low (equal to 0.153) order of dependence on the Con A concentration in solution. The fractional order of dependence exhibited by the dissociation rate coefficient, k_d, on the Con A concentration in solution lends support to the fractal nature of the system.

Figure 12.10d and Table 12.5 show the increase in the fractal dimension, D_{f1}, with an increase in the Con A concentration in the 0.1-5.0 nM range for a dual-fractal analysis. For the data shown in Figure 12.10d, the fractal dimension, D_{f1}, is given by:

$$D_{f1} = (1.391 \pm 0.083)(\text{Con A})^{0.151 \pm 0.021} \tag{12.5d}$$

The fit is good. Only three data points are available. The availability of more data points would lead to a more reliable fit. The fractal dimension, D_{f1}, exhibits a very low (equal to 0.151) order of dependence on the Con A concentration in solution.

Figure 12.10e and Table 12.5 show the increase in the fractal dimension, D_{fd}, with an increase in the Con A concentration in the 0.1-5.0 nM range. For the data shown in Figure 12.10e, the fractal dimension, D_{fd}, is given by:

$$D_{fd} = (1.811 \pm 0.0063)(\text{Con A})^{0.0077 \pm 0.0012} \tag{12.5e}$$

The fit is good. Only three data points are available. The availability of more data points would lead to a more reliable fit. The fractal dimension for dissociation, D_{fd}, exhibits a very low (equal to 0.0077) order of dependence on the Con A concentration in solution.

Figure 12.10f and Table 12.5 show the increase in the binding rate coefficient, k_1, with an increase in the fractal dimension, D_{f1}, for a dual-fractal analysis. For the data shown in Figure 12.10f the binding rate coefficient, k_1, is given by:

$$k_1 = (0.927 \pm 0.197) D_{f1}^{2.204 \pm 0.458} \tag{12.5f}$$

The fit is reasonable. Only three data points are available. The availability of more data points would lead to a more reliable fit. For a dual-fractal analysis the binding rate coefficient, k_1, exhibits a slightly higher than second (equal to 2.204) order of dependence on the fractal dimension, D_{f1}, or the degree of heterogeneity that exists on the biosensor surface.

Figure 12.10g and Table 12.5 show the increase in the binding rate coefficient, k_2, with an increase in the fractal dimension, D_{f2}, for a dual-fractal analysis. For the data shown in Figure 12.10g the binding rate coefficient, k_2, is given by:

$$k_2 = (0.0109 + 0.0158) D_{f2}^{6.607 + 7.80} \tag{12.5g}$$

The fit is poor. Only three data points are available. The availability of more data points would lead to a more reliable fit. For a dual-fractal analysis the binding rate coefficient, k_2, exhibits a very high (equal to 6.607) order of dependence on the fractal dimension, D_{f2}, or the degree of heterogeneity that exists on the biosensor surface.

Figure 12.10h shows the increase in the dissociation rate coefficient, k_d, with an increase in the fractal dimension for dissociation, D_{fd}. For the data shown in Figure 12.10h, the dissociation rate coefficient, k_d, is given by:

$$k_d = (1.4 \times 10^{-05} \pm 0.6 \times 10^{-05}) D_{fd}^{16.54 \pm 15.19} \tag{12.5h}$$

The fit is poor. Only three data points are available. The availability of more data points would lead to a more reliable fit. The dissociation rate coefficient, k_d, is extremely sensitive to the fractal dimension in the dissociation phase, D_{fd}, noted by the very slightly greater than sixteen and a half (equal to 16.54) order of dependence exhibited on the fractal dimension in the dissociation phase, D_{fd}.

Figure 12.10i shows the decrease in the ratio of the fractal dimensions, D_{f2}/D_{f1}, with an increase in the Con A concentration in solution in the 1-5 nM concentration range for a dual-fractal analysis. For the data shown in Figure 12.10i, the ratio of the fractal dimensions, D_{f2}/D_{f1}, is given by:

$$D_{f2}/D_{f1} = (1.815 \pm 0.099)[\text{Con A}]^{-0.139 \pm 0.0191} \tag{12.5i}$$

The fit is good. Only three data points are available. The availability more data points would lead to a more reliable fit. The ratio of the fractal dimensions, D_{f2}/D_{f1}, exhibits a slight negative order (equal to −0.139) of dependence on the Con A concentration in solution.

Figure 12.10j shows the increase in the ratio of the fractal dimensions, D_{f1}/D_{fd}, with an increase in the Con A concentration in solution in the 1-5 nM concentration range for a dual-fractal analysis. For the data shown in Figure 12.10j, the ratio of the fractal dimensions, D_{f1}/D_{fd}, is given by:

$$D_{f1}/D_{fd} = (0.768 \pm 0.048)[\text{Con A}]^{0.142 \pm 0.0221} \tag{12.5j}$$

The fit is good. Only three data points are available. The availability more data points would lead to a more reliable fit. The ratio of the fractal dimensions, D_{f1}/D_{fd}, exhibits a slight order (equal to 0.142) of dependence on the Con A concentration in solution.

Figure 12.10k shows the increase in the affinity, K_1 ($=k_1/k_d$), with an increase in the ratio of the fractal dimensions, D_{f1}/D_{fd}, for a dual-fractal analysis. For the data shown in Figure 12.10k, the ratio of the affinity, K_1, is given by:

$$K_1 = (9.907 \pm 0.296)[D_{f1}/D_{fd}]^{1.17 \pm 0.074} \tag{12.5k}$$

The fit is very good. Only three data points are available. The availability more data points would lead to a more reliable fit. The affinity, K_1, exhibits slightly higher than first (equal to 1.17) order of dependence on the Con A concentration in solution in the 1-5 nM range. The fractional order of dependence exhibited on the Con A concentration in solution lends support to the fractal nature of the system.

Figure 12.10l shows the increase in the affinity, K_1 $(=k_2/k_d)$, with an increase in the Con A concentration in solution in the 1-5 nM range for a dual-fractal analysis. For the data shown in Figure 12.10l, the ratio of the affinity, K_1, is given by:

$$K_1 = (7.572 \pm 0.311)[\text{Con A}]^{0.1694 \pm 0.01513} \tag{12.5l}$$

The fit is very good. Only three data points are available. The availability more data points would lead to a more reliable fit. The affinity, K_1, exhibits a very low (equal to 0.1694) order of dependence on the Con A concentration in solution in the 1-5 nM range. The fractional order of dependence exhibited on the Con A concentration in solution lends support to the fractal nature of the system.

Figure 12.10m shows the increase in the affinity, K_2 $(=k_2/k_d)$, with an increase in the Con A concentration in solution in the 1-5 nM range for a dual-fractal analysis. For the data shown in Figure 12.10m, the ratio of the affinity, K_2, is given by:

$$K_2 = (19.99 \pm 6.07)[\text{Con A}]^{0.229 \pm 0.0814} \tag{12.5m}$$

The fit is good. Only three data points are available. The availability more data points would lead to a more reliable fit. The affinity, K_2, exhibits only a slight (equal to 0.229) order of dependence on the Con A concentration in solution in the 1-5 nM range. The fractional order of dependence exhibited on the Con A concentration in solution lends support to the fractal nature of the system.

Figure 12.10n shows the increase in the affinity, K_2 $(=k_2/k_d)$, with an increase in the ratio of the fractal dimensions, D_{f2}/D_{fd}, for a dual-fractal analysis. For the data shown in Figure 12.10n, the ratio of the affinity, K_2, is given by:

$$K_2 = (8.79 \pm 7.52)[D_{f2}/D_{fd}]^{2.35 + 5.40} \tag{12.5n}$$

The fit is poor. Only three data points are available. The availability of more data points would lead to a more reliable fit. The poor fit is reflected in the data, and also in the order of dependence exhibited. Only the positive error in the order is given. The affinity, K_2, exhibits an order of dependence between two and two and a half (equal to 2.35) on the ratio of the fractal dimensions, D_{f2}/D_{fd}; the error in the order of dependence notwithstanding.

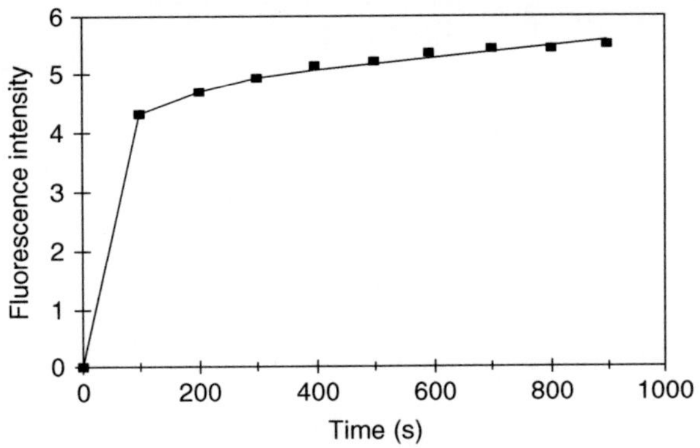

Figure 12.11
Binding of H9 avian influenza virus to cadmium quantum dots (Yun et al., 2007).

Yun et al. (2007) recently used cadmium telluride quantum dots as a proton flux sensor to detect the H9 avian influenza virus. Figure 12.11 shows the binding of the H9 avian influenza virus to the cadmium telluride quantum dot sensor. A single-fractal analysis is adequate to describe the binding kinetics. The values of the binding rate coefficient, k, and the fractal dimension, D_f, for a single-fractal analysis are given in Table 12.6.

Sauvage et al. (2007) recently pointed out the need for new sensor devices that can monitor ions, especially for fundamental studies in medicine. They report that, for example, sodium ions play a major role owing to their importance in the metabolism of the human body. They explain that sodium ion concentration plays a major role in blood pressure regulation, efficient muscle movement, and in the functioning of nerves. They also pointed out that various strategies are underway to help develop sensible, selective, and reproducible sodium ion sensors.

Kanoh et al. (1993) have reported the use of insertion materials for sensing purposes. Sauvage et al. (2004) used olivine-type $LiFePO_4$ insertion material for use as a good candidate for a Li-ion sensor. Tani and Umezawa (1998) used $Na_{0.44}MnO_2$ as a suitable material for a Na-type ion sensor. This material has the ability to reversibly accommodate lithium ions (Doeff et al., 1994, 1995). Sauvage et al. (2007) attempted to define the synthesis parameters for the formation of $Na_{0.44}MnO_2$, and then analyzed the potentiometric properties of this material towards sodium ions.

Figure 12.12 shows the open current voltage measurements (binding kinetics) using a $Na_{0.44}MnO_2$/C plastic composite electrode recorded in 1 M $NaNO_3$ concentration at 30° C. A dual-fractal analysis is required to adequately describe the "binding kinetics." The values of (a) the binding rate coefficient, k, and the fractal dimension, D_f, for a single-fractal analysis, and (b) the binding rate coefficients, k_1 and k_2, and the fractal dimensions, D_{f1} and D_{f2}, for a dual-fractal analysis are given in Table 12.6.

Table 12.6: Binding of H9 avian influenza virus to cadmium telluride quantum dots as a proton flux sensor (Yun et al., 2007) and sodium ions of $Na_{0.44-x}MnO_2$ to a sodium ion sensor (Sauvage et al., 2007).

Analyte in Solution/Sensor Type	k	k_1	k_2	D_f	D_{f1}	D_{f2}	References
H9 avian influenza virus/quantum dot	2.621 ± 0.023	na	na	2.7792 ± 0.0072	na	na	Yun et al. (2007)
Sodium ions of $Na_{0.44-x}MnO_2$	0.003 ± 0.00	0.0012 ± 0.00	0.01326 ± 0.0002	2.2708 ± 0.0746	2.0148 ± 0.1117	2.632 ± 0.05591	Sauvage et al. (2007)

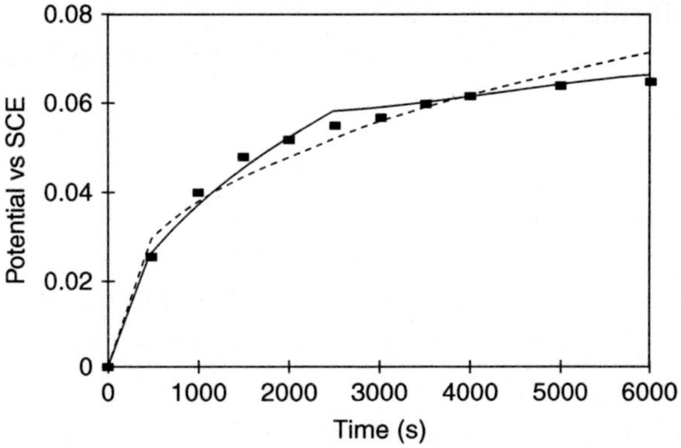

Figure 12.12
Binding of sodium ions of $Na_{0.44-x}MnO_2$ to a selective sodium ion sensor (Sauvage et al., 2007).

It is of interest to note that as the fractal dimension increases by a factor of 1.306 from a value of D_{f1} equal to 2.0148 to D_{f2} equal to 2.6318, the binding rate coefficient increases by a factor of 10.78 from a value of k_1 equal to 0.00123 to k_2 equal to 0.01326. The changes in the fractal dimension and in the binding rate coefficient are in the same direction.

The fractal analysis by linking the sensing surface with the "binding" rate coefficient should assist in understanding the insertion/de-insertion mechanism, and also help in facilitating the material into a practical Na^+ sensor (Sauvage et al., 2007). Also, more work is required to help turn this material into a practical Na^+ sensor.

12.4 Conclusions

A fractal analysis is presented for the binding and dissociation (if applicable) kinetics of different analytes in solution to different biosensing surfaces. The analysis includes: (a) the binding of different IgG species to a porous SiO_2 interferometric biosensor coated with protein A (Schwartz et al., 2007), (b) binding (hybridization) using differential surface plasmon resonance (Boecker et al., 2007), (c) binding of glucose to a One Touch II blood glucose meter, and a SERS sensor (Stuart et al., 2006), (d) binding of different concentrations (in nM) of Con A to a colloidal gold coated with a glycolipid/dodecanethiol bilayer (Guo et al., 2007), (e) binding of H9 avian influenza virus to calcium quantum dots (Yun et al., 2007), and (f) binding of sodium ions of $Na_{0.44-x}MnO_2$ to a selective sodium ion sensor (Sauvage et al., 2007). As in the previous chapters, the fractal analysis provides a quantitative measure of the degree of heterogeneity on the sensing surface, and links this degree of heterogeneity on the sensing surface to the binding and the dissociation (if applicable) rate coefficients. The versatility of the fractal analysis is, once again, demonstrated by its successful application to the kinetics of different analyte/receptor systems occurring on the different biosensing systems.

In some cases a single-fractal analysis is adequate to describe the binding and dissociation (if applicable) kinetics (Corel Quattro Pro 8.0, 1997). If the regression analysis indicates that the fit is not adequate (regression coefficient less than 0.95), only then is a dual-fractal analysis required.

Predictive relations are presented for the different analyte/receptor systems analyzed on the different biosensor surfaces. For example, (a) for the binding of 177 nM goat IgG in solution to a protein A-coated surface (Schwartz et al., 2007), the binding rate coefficient, k_1, exhibits a negative (−2.486) order of dependence on the fractal dimension, D_{f1}, or the degree of heterogeneity that exists on the protein A-coated porous SiO_2 surface. In this case, the binding rate coefficient, k_2, exhibits a 11.45 order of dependence on the fractal dimension, D_{f2}, or the degree of heterogeneity that exists on the protein A-coated porous SiO_2 surface, (b) for the binding of Con A in solution to a gold nanoparticle biosensor (Guo et al., 2007), and for a dual-fractal analysis, the binding rate coefficient, k_1, exhibits a slight (equal to 0.322) order of dependence on the Con A concentration in solution and the binding rate coefficient, k_2, also exhibits a slight (equal to 0.382) order of dependence on the Con A concentration in solution, in the 0.1-5.0 nM concentration range, (c) and the dissociation rate coefficient, k_d, for a single-fractal analysis exhibits a negligible (equal to 0.153) order of dependence on the Con A concentration in solution in the 0.1-5.0 nM range.

Similar predictive relations are developed for the fractal dimension, D_{f1}, and the fractal dimension, D_{fd}, as a function of the Con A concentration in solution. The binding rate coefficient, k_1, exhibits higher than second (equal to 2.204) order of dependence on the fractal dimension, D_{f1}. The binding rate coefficient, k_2, exhibits a higher than six and a half (equal to 6.607) order of dependence on the fractal dimension, D_{f2}. The binding rate coefficient, k_2, as noted, is more sensitive (by about three times) to the degree of heterogeneity on the surface than the binding rate coefficient, k_1.

The dissociation rate coefficient, k_d, is extremely sensitive to the fractal dimension in the dissociation phase, D_{fd}, as noted by the slightly higher than sixteen and a half (equal to 16.54) order of dependence on D_{fd}. Finally, predictive relations are presented for the affinities, K_1 and K_2, as a function of the Con A concentration in solution.

Only a few (six exactly) examples are presented where the fractal analysis is effective in describing the binding and dissociation (if applicable) kinetics of different analytes to receptors on different sensing surfaces. The analysis provides fresh physical insights into the reactions occurring on these sensing surfaces. The predictive relations are particularly useful since they provide a means by which the different parameters such as the binding and the dissociation rate coefficients, and thereby the corresponding affinities may be manipulated in desired directions.

More examples need to be analyzed for the binding of different analytes to different (and novel) sensing techniques by the fractal analysis method. This would further validate the fractal analysis method, especially if successfully applied to the novel biosensing techniques that are presently available in the literature or will, most probably be developed in the future. This is bound to happen owing to the increasing popularity and simplicity of measurement available in the biosensing techniques.

References

Aslan K, MJR Previte, Y Zhang, T Gallagher, L Baillie, and CD Geddes, Extraction and detection of DNA from *Bacillus anthracis* spores and the vegetative cells within one minute, *Analytical Chemistry*, **80**, 4125–4132 (2008).

Boecker D, A Zybin, V Horvatic, C Grunwald, and K Niemax, Differential surface plasmon resonance imaging for high-throughput bioanalysis, *Analytical Chemistry*, **79**, 702–709 (2007).

Chan S, PM Fauchet, Y Li, LJ Rothberg, and BL Miller, *Physics Status Solidi A*, **182**, 541–546 (2000).

Corel Quattro Pro 8.0, Corel Corporation Limited, Ottawa, Canada, 1997.

Dancil KPS, DP Grenier, and MJ Sailor, *Journal of the American Chemical Society*, **121**, 7925–7930 (1999).

Dieringer JA, AD Mcfarland, NC Shah, DA Stuart, AV Whitney, CR Yonzon, MA Young, XY Zhang, and RP Van Duyne, *Faraday Discussions*, **132**, 9–26 (2006).

Doeff MM, MY Peng, Y Ma, and LC De Jonghe, *Electrochemical Society*, **141**, 11 (1994).

Doeff MM, L Ding, and LC De Jonghe, *Materials Research Society Symposium*, **393** (1995).

Guo C, P Boullanger, L Jiang, and T Liu, Highly sensitive gold nanoparticles biosensor chips modified with a self-assembled bilayer for detection of con A, *Biosensors & Bioelectronics*, **22**, 1830–1834 (2007).

Ha TH, SO Jung, JM Lee, KY Lee, Y Lee, JS park, and BH Chung, Oriented immobilization of antibodies with the GST-fused multiple F_c-specific B-domains on a gold surface, *Analytical Chemistry*, **79**, 546–556 (2007).

Havlin S, Molecular diffusion and reactions in *The Fractal Approach to Heterogeneous Chemistry*: *Surfaces, Colloids, Polymers* (ed. D Avnir) Wiley, New York, 1989, pp. 251–269.

Kanoh H, Q Feng, Y Miyai, and K Oai, *Journal of the Electrochemical Society*, **140**(11) (1993).

Leidberg B, C Nylander, and I Lundstrom, *Biosensors & Bioelectronics*, **10**, I–IX (1995).

Lin VSY, K Moteisharei, KS Dancil, MJ Sailor, and MJ Ghadari, *Science*, **278**, 840–843 (1997).

Lyandres O, NC Shah, CR Yonzon, JT Walsh, MR Glucksberg, and RP Van Duyne, *Analytical Chemistry*, **77**, 6134–6139 (2005).

Ramakrishnan A and A Sadana, A single fractal analysis of cellular analyte-receptor binding kinetics using biosensors, *Biosystems*, **59**, 35–51 (2001).

Sadana A, A fractal analysis for the evaluation of hybridization kinetics in biosensors, *Journal of Colloid and Interface Science*, **151**(1), 166–177 (2001).

Sadana A, *Fractal Binding and Dissociation Kinetics for Different Biosensor Applications*, Elsevier, Amsterdam, 2005.

Sauvage F, E Baudrin, M Morcette, and JM Tarascon, *Electrochemistry and Solid State Letters*, **7**(1), A15 (2004).

Sauvage P, E Baudrin, and JM Tarason, Study of the potentiometric response towards sodium ions of $Na_{0.44-x}MnO_2$ for the development of sodium ion sensors, *Sensors & Actuators B*, **120**, 638–644 (2007).

Schwartz MP, AM Derfus, SD Alvarez, SN Bhatia, and MJ Sailor, *Langmuir*, **22**, 7084–7090 (2006).

Schwartz MP, SD Alvarez, and MJ Sailor, Porous SiO_2 interferometric biosensor for quantitative determination of protein interactions: Binding of protein A to immunoglobulins derived from different species, *Analytical Chemistry*, **79**, 327–334 (2007).

Shafer-Peltier KE, CL Haynes, MR Glucksberg, and RP Van Duyne, *Journal of the American Chemical Society*, **125**, 588–593 (2003).

Shumaker-Parry JS and CT Campbell, *Analytical Chemistry*, **76**, 907 (2004).

Starodub NF, LL Fedorenko, VM Starodub, SP Dikiju, and SV Sechnikov, *Sensors & Actuators B*, **79**, 44–47 (1996).

Stuart DA, CR Yonzon, XY Zhang, O Lyandres, NC Shah, MR Glucksberg, JT Walsh, and RP Van Duyne, *Analytical Chemistry*, **77**, 4013–4019 (2005).

Stuart DA, JM Yuen, N Shah, O Lyandres, CR Yonzon, MR Glucksberg, JT Walsh, and RP van Duyne, *In-vivo* glucose measurement by surface-enhanced Raman spectroscopy, *Analytical Chemistry*, **78**, 7211–7215 (2006).

Tani Y and Y Umezawa, *Mikrochim Acta*, **129**, 81–90 (1998).

Yonzon CR, CL Haynes, XY Zhang, JT Walsh, and RP Van Duyne, *Analytical Chemistry*, **76**, 78–85 (2004).

Yonzon CR, DA Stuart, XY Zhang, AD Mcfarland, CL Haynes, and RP Van Duyne, *Talanta*, **67**, 438–448 (2005).

Zhang J, H Qi, Y Li, J Yang, Q Gao, and C Zhang, Electrogenerated chemiluminescence DNA based on hairpin DNA probe labeled with ruthenium complex, *Analytical Chemistry*, **80**, 2888–2894 (2008).

Zhao Q, XF Li, Y Shao, and XC Le, Aptamer-based affinity chromatography assays for thrombin, *Analytical Chemistry*, (2008), in press.

Zybin A, C Grunwald, VM Mirsky, J Kuhlmann, OS Wolbeis, and K Neimax, *Analytical Chemistry*, **77**, 2393–2399 (2005).

CHAPTER 13

Binding of Different Analytes on Biosensor Surfaces

Chapter Outline

13.1 Introduction

Biosensor applications have expanded considerably over the last few years. Medical applications initially dominated the biosensor market and applications. However, over the years, the use of biosensors for the detection of analytes in other areas is gradually coming into prominence. This is primarily due to the ease and simplicity of the use of biosensors. In this chapter we analyze a few of the biosensor applications that have recently appeared in the literature. The fractal analysis method will be used, as in the previous chapters, to analyze the binding and dissociation (if applicable) kinetics of a wide variety of examples available in the literature. The systems analyzed were selected at random. The analyte-receptor systems analyzed include (a) the binding of perfectly matched oligodeoxynucleotide (ODN-P) and noncomplementary ODN during the hybridization assay with EST2-A34 reporter (Wang et al., 2007), (b) binding during the primer elongation reaction of DNA coupled directly to polyacrylic acid (PAC) and DNA coupled via biotin-streptavidin (Krieg et al., 2006), (c) binding and dissociation of trace amounts of trinitrotoluene (TNT) in μg/L to anti-TNT antibody immobilized on a prototype fluorescence-based detector system (KinExA Inline Biosensor, Sappidyne Instrument, Inc.; Bromage et al., 2007), (d) binding of EBP (estrogen binding protein) to *Saccharomyces cerevisiae* and the binding of 17β-estradiol (Baronian and Guruzada, 2007), (e) binding of different units of restriction endonuclease

Handbook of Biosensors and Biosensor Kinetics. DOI: 10.1016/B978-0-444-53262-6.00013-9

activity using a molecular beacon (MB; Ma et al., 2007), and (f) binding and dissociation of different rabbit IgG concentrations (in μM) in solution to A10B single-chain fragment variable (ScFv; Tang et al., 2006).

Some of the other analyte-receptor binding (and dissociation, if applicable) kinetics on biosensor surfaces that have appeared recently in the literature include (i) effects of geometry on transmission and sensing potential of tapered fiber sensors (Leung et al., 2006), (ii) screening antibody-antigen interactions in parallel using a Biacore A100 biosensor (Safsten et al., 2006), (iii) *in vivo* biosensors for organic pollutants (Tizzard and Lloyd-Jones, 2007), (iv) immunoassays based on microelectrodes arranged on a silicon chip for high-throughput screening of liver fibrosis markers in human serum (Shi et al., 2006), (v) detection of *Eschericihia coli* 0157:H7 with pure shear horizontal surface acoustic wave sensors (Berkenpas et al., 2006), (vi) nanoscale sensors for multi-modal detection of genotoxic agents (Heller et al., 2008), (vii) controlled assembly of biotinylated Au nanoparticles on metallic nanowires for nanobiosensing (Kizil, 2008), (viii) preparation of bioanalytical sensors by incorporating fluorophore in patternable poly(ethylene glycol) diacrylate-based membranes (Gao et al., 2008), and (ix) impedance biosensor for peanut protein allergens (Suni and Huang, 2008).

A fractal analysis method will be used to model the kinetics of analyte-receptor binding and dissociation (if applicable) on biosensor surfaces. As mentioned in previous chapters, this is one method of analyzing the binding and dissociation kinetics, with the added advantage that it permits a quantitative measure of the degree of heterogeneity on the biosensor surface. This quantitative measure of heterogeneity on the biosensor surface may then be used to enhance biosensor performance parameters such as selectivity, sensitivity, stability, detection time, etc.

13.2 Theory

13.2.1 Single-Fractal Analysis

Binding Rate Coefficient

Havlin (1989) notes that the diffusion of a particle (analyte [Ag]) from a homogeneous solution to a solid surface (e.g., receptor [Ab]-coated surface) on which it reacts to form a product (analyte-receptor complex; Ab·Ag) is given by:

$$(\mathrm{Ab \cdot Ag}) \approx \begin{cases} t^{(3-D_{\mathrm{f,bind}})/2} = t^p, & t < t_c \\ t^{1/2}, & t > t_c \end{cases} \tag{13.1}$$

Here $D_{\mathrm{f,bind}}$ or D_{f} is the fractal dimension of the surface during the binding step. t_c is the cross-over value. Havlin (1989) points out that the cross-over value may be determined by

$r_c^2 \sim t_c$. Above the characteristic length, r_c, the self-similarity of the surface is lost and the surface may be considered homogeneous. Above time, t_c, the surface may be considered homogeneous, since the self-similarity property disappears, and "regular" diffusion is now present. For a homogeneous surface where D_f is equal to 2, and when only diffusional limitations are present, $p = ½$ as it should be. Another way of looking at the $p = ½$ case (where $D_{f,bind}$ is equal to two) is that the analyte in solution views the fractal object, in this case, the receptor-coated biosensor surface, from a "large distance." In essence, in the association process, the diffusion of the analyte from the solution to the receptor surface creates a depletion layer of width $(Đt)^{½}$ where $Đ$ is the diffusion constant. This gives rise to the fractal power law, $(\text{Analyte}\cdot\text{Receptor}) \sim t^{(3-D_{f,\text{bind}})/2}$. For the present analysis, t_c is arbitrarily chosen and we assume that the value of t_c is not reached. One may consider the approach as an intermediate "heuristic" approach that may be used in future to develop an autonomous (and not time-dependent) model for diffusion-controlled kinetics.

Dissociation Rate Coefficient

The diffusion of the dissociated particle (receptor [Ab] or analyte [Ag]) from the solid surface (e.g., analyte [Ag]-receptor [Ab]) complex coated surface) into solution may be given, as a first approximation, by:

$$(\text{Ab}\cdot\text{Ag}) \approx -t^{(3-D_{f,\text{diss}})/2} = -t^p, \quad t > t_{\text{diss}} \tag{13.2}$$

Here $D_{f,diss}$ is the fractal dimension of the surface for the dissociation step. This corresponds to the highest concentration of the analyte-receptor complex on the surface. Henceforth, its concentration only decreases. The dissociation kinetics may be analyzed in a manner "similar" to the binding kinetics.

13.2.2 Dual-Fractal Analysis

Binding Rate Coefficient

Sometimes, the binding curve exhibits complexities and two parameters (k, D_f) are not sufficient to adequately describe the binding kinetics. This is further corroborated by low values of r^2 factor (goodness-of-fit). In that case, one resorts to a dual-fractal analysis (four parameters; k_1, k_2, D_{f1}, and D_{f2}) to adequately describe the binding kinetics. The single-fractal analysis presented above is thus extended to include two fractal dimensions. At present, the time ($t = t_1$) at which the "first" fractal dimension "changes" to the "second" fractal dimension is arbitrary and empirical. For the most part, it is dictated by the data analyzed and experience gained by handling a single-fractal analysis. A smoother curve is obtained in the "transition" region, if care is taken to select the correct number of points for the two regions.

In this case, the product (antibody-antigen; or analyte-receptor complex, Ab·Ag or analyte·receptor) is given by:

$$(\text{Ab}\cdot\text{Ag}) \approx \begin{cases} t^{(3-D_{f1,\text{bind}})/2} = t^{p1}, & t < t_1 \\ t^{(3-D_{f2,\text{bind}})/2} = t^{p2}, & t_1 < t < t_2 = t_c \\ t^{1/2}, \quad t > t_c \end{cases} \tag{13.3}$$

In some cases, as mentioned above, a triple-fractal analysis with six parameters (k_1, k_2, k_3, D_{f1}, D_{f2}, and D_{f3}) may be required to adequately model the binding kinetics. This is when the binding curve exhibits convolutions and complexities in its shape due perhaps to the very dilute nature of the analyte (in some of the cases to be presented) or for some other reasons. Also, in some cases, a dual-fractal analysis may be required to describe the dissociation kinetics.

13.3 Results

A fractal analysis is applied to the binding and dissociation (if applicable) kinetics of different analyte-receptor reactions occurring on different biosensor surfaces. Understandably, alternate expressions for fitting the data that include saturation, first-order reaction, and no diffusion limitations are available, but these expressions are apparently deficient in describing the heterogeneity that inherently exists on the surface. Another advantage of this technique is that the analyte-receptor binding (as well as the dissociation reaction) is a complex reaction, and the fractal analysis via the fractal dimension and the rate coefficient provide a useful lumped parameter(s) analysis of the diffusion-limited reaction occurring on a heterogeneous surface. In the classical situation to demonstrate fractality, one should make a log-log plot, and one should definitely have a large amount of data. It may be useful to compare the fit to some other forms, such as an exponential form or one involving saturation. At present, we do not present any independent proof or physical evidence of fractals in the examples presented. It is a convenient means (since it provides a lumped parameter) to make the degree of heterogeneity that exists on the surface more quantitative. Thus, there is some arbitrariness in the fractal model to be presented. One might justifiably argue that appropriate modeling may be achieved by using a Langmuirian or other approach. The Langmuirian approach has a major drawback because it does not allow for or accommodate the heterogeneity that exists on the surface.

Wang et al. (2007) have reported that the detection of DNA and RNA sequences is becoming important for the diagnosis of diseases (Nebling et al., 2004), and for the detection of pathogenic organisms in samples from the environment (Baeumner et al., 2003), food (Ko and Grant, 2006), and clinical (Mitterer et al., 2004) areas. Baeumner et al. (2004) have noted that nucleic acid sequences unique to a particular organism may be used to discriminate one self-replicating organism from another. Wang (2000) reports that as the demand for faster, simpler, and cheaper methods for obtaining sequence-specific information increases, increasing emphasis is being placed on nucleic acid-based biosensors.

Wang et al. (2007) have recently analyzed esterase 2-ODN conjugates as a sensitive reporter for the electrochemical detection of nucleic acid hybridization. They wanted to develop a reporter system consisting of a thermostable enzyme coupled with ODN conjugates for electrochemical microchip application. They report that the esterase 2 (EST 2) from *Allicyclobacillus acidocaldsrius* is a thermostable monomeric protein.

Figure 13.1a shows the binding of ODN-P during the hybridization assay with EST 2-A34 reporter (Wang et al., 2007). A dual-fractal analysis is required to adequately describe the binding kinetics. The values of (a) the binding rate coefficient, k and the fractal dimension, D_f for a single-fractal analysis, and (b) the binding rate coefficients, k_1 and k_2, and the fractal dimensions, D_{f1} and D_{f2} for a dual-fractal analysis and the dissociation rate coefficient, k_d and the fractal dimension, D_{fd} for the dissociation phase for a single-fractal analysis are given in Table 13.1(a) and 13.2(a). It is of interest to note that for a dual-fractal analysis, as the fractal dimension increases from a value of D_{f1} equal 0 to D_{f2} equal to 2.204 the binding rate coefficient increases by a factor of 56.06 from a value of k_1 equal to 0.949 to k_2 equal to 53.15. Note that changes in the fractal dimension or the degree of heterogeneity on the sensing surface and in the binding rate coefficient are in the same direction.

Figure 13.1b shows the binding of noncomplementary ODN during the hybridization assay with EST 2-A34 reporter (Wang et al., 2007). Once again, a dual-fractal analysis is required to adequately describe the binding kinetics. The values of (a) the binding rate coefficient, k and the fractal dimension, D_f for a single-fractal analysis, and (b) the binding rate coefficients, k_1 and k_2, and the fractal dimensions, D_{f1} and D_{f2} for a dual-fractal analysis are given in Table 13.1(b) and 13.2(b). It is of interest to note that for a dual-fractal analysis,

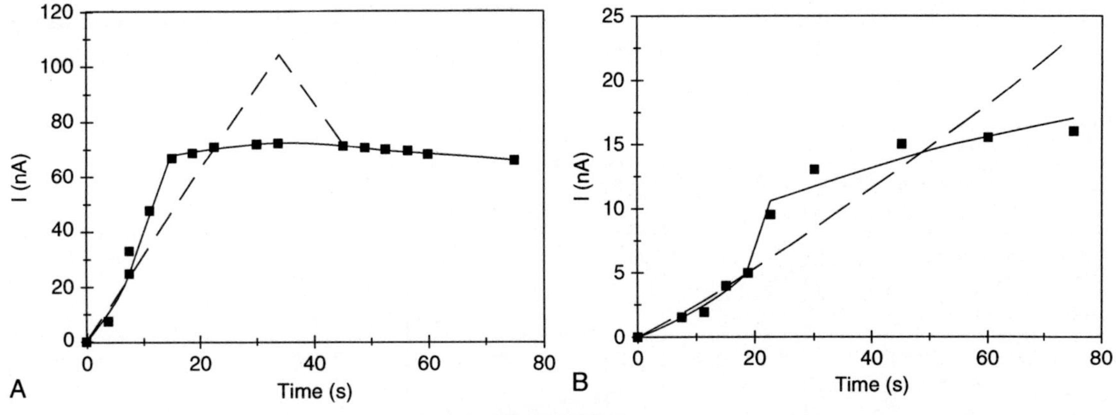

Figure 13.1

Binding of (a) perfectly matched ODN (ODN-P), and (b) the noncomplementary ODN during the hybridization assay with EST2-A34 reporter (Wang et al., 2007). When only a solid line is used then a single-fractal analysis applies. When both a solid line and a dashed line are used, then the dashed line is for a single-fractal analysis, and the solid line is for a dual-fractal analysis.

Table 13.1a,b Binding and dissociation rate coefficients for (a) DNA hybridization assay with EST2-A34 reporter (Wang et al., 2007), and (b) primer elongation of DNA coupled directly to PAC via biotin-streptavidin-biotin (Krieg et al., 2007)

Analyte/Receptor	k	k_1	k_2	k_d
(a) Perfectly matched ODN (ODN-P)/EST2-A34 reporter	3.256 ± 0.467	0.949 ± 0.267	2.204 ± 0.365	0.0608 ± 0.0069
Non-complementary ODN (ODN-N)/ES-T2-A34 reporter	0.1864 ± 0.0705	0.0981 ± 0.0832	3.365 ± 0.047	na
(b) Primer elongation reaction of DNA coupled directly to PAC	8.822 ± 0.424	na	na	na
Primer elongation reaction of DNA coupled to PAC via biotin-streptavidin-biotin reaction	6.074 ± 0.325	na	na	na

as the fractal dimension increases by a factor of 10.39 from a value of D_{f1} equal 0.212 to D_{f2} equal to 2.203, the binding rate coefficient increases by a factor of 34.3 from a value of k_1 equal to 0.0981 to k_2 equal to 3.365. Note, once again, that changes in the fractal dimension or the degree of heterogeneity on the sensing surface and in the binding rate coefficient are in the same direction. It is of interest to note that, and as may be expected, for a dual-fractal analysis, the binding rate coefficient k is higher for the ODN-P for the complementary

Table 13.2a,b Fractal dimensions for the binding and dissociation (a) DNA hybridization assay with EST2-A34 reporter (Wang et al., 2007), and (b) primer elongation of DNA coupled directly to PAC via biotin-streptavidin-biotin (Krieg et al., 2007)

Analyte/Receptor	D_f	D_{f1}	D_{f2}	D_{fd}
(a) Perfectly matched ODN (ODN-P)/EST2-A34 reporter	1.0296 ± 0.3798	$0 + 0.42434$	53.15 ± 1.16	0.0608 ± 0.0069
Non-complementary ODN (ODN-N)/ES-T2-A34 reporter	0.762 ± 0.293	$0.212 + 0.491$	2.2028 ± 0.3016	na
(b) Primer elongation reaction of DNA coupled directly to PAC	1.8418 ± 0.0398	na	na	na
Primer elongation reaction of DNA coupled to PAC via biotin-streptavidin-biotin reaction	1.7642 ± 0.0442	na	na	na

hybridization assay when compared with the binding of the noncomplementary ODN case. This is observed for both phases of binding, that is, for the binding rate coefficients k_1 and k_2. The binding rate coefficient for the ODN-P case is 9.65 times higher than that of the binding rate coefficient for the noncomplementary ODN during the hybridization assay with EST 2-A34 reporter.

Krieg et al. (2006) report that there is an increasing need for better sequencing methods (Shendure et al., 2004; Chan, 2005). A possible alternative is sequencing by hybridization (SBH) based on microarrays (Hacia, 1999). Krieg et al. (2006) note that new DNA sequencing techniques are being developed that use single-molecule fluorescence-based detection of enzymatic double-strand synthesis. They point out that such applications require surface structures on which single-stranded templates may be immobilized. Nonspecific adsorption should be a minimum, and Krieg et al. (2006) report that PAC coatings are a good choice to help minimize the nonspecific binding of the bases.

Figure 13.2a shows the primer elongation (binding) reaction of DNA coupled directly to PAC. A single-fractal analysis is adequate to describe the binding kinetics. The values of the binding rate coefficient k and the fractal dimension D_f for a single-fractal analysis are given in Table 13.1(b) and 13.2(b).

Figure 13.2b shows the primer elongation (binding) reaction of DNA coupled to PAC via biotin-streptavidin-biotin reaction. Once again, a single-fractal analysis is adequate to describe the binding kinetics. The values of the binding rate coefficient k and the fractal dimension D_f for a single-fractal analysis are given in Table 13.1(b) and 13.2(b). It is of interest to note that as one goes from the DNA coupled directly to the PAC via the biotin-streptavidn-biotin reaction the fractal dimension decreases by 4.22% from a value of

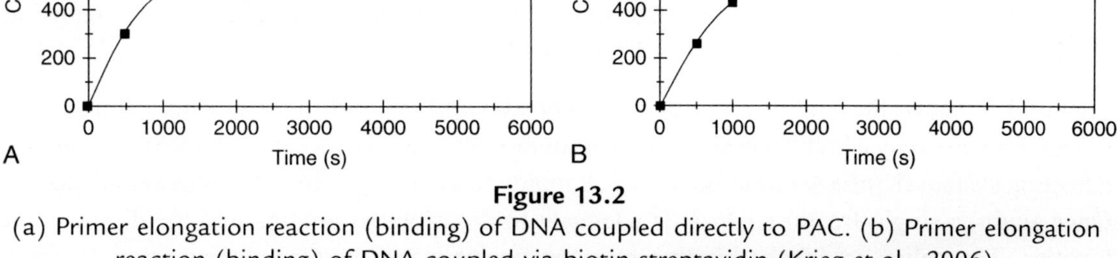

Figure 13.2

(a) Primer elongation reaction (binding) of DNA coupled directly to PAC. (b) Primer elongation reaction (binding) of DNA coupled via biotin-streptavidin (Krieg et al., 2006).

1.8418 to 1.7642, and the binding rate coefficient k decreases by 31.2% from a value of 8.822 to 6.074. Note that changes in the fractal dimension D_f or the degree of heterogeneity on the biosensor surface and in the binding rate coefficient are in the same direction.

Bromage et al. (2007) have recently developed a real-time biosensor for the detection of trace levels of TNT in aquatic environments. They report that most chemicals will have a biological impact on aquatic species at concentrations too low to be visually observed (Anderson et al., 1997). Hence the need to make rapid detection and quantitative trace concentrations of different chemicals. They focused on the detection of trace amounts of TNT in aqueous environments as environmental contamination occurs particularly from military sites due to manufacturing, storage, and handling of munitions. These authors further state that TNT is known to accumulate in plants and fish (Belden et al., 2005; Ownby et al., 2005), and have detrimental effects on fish (Ek et al., 2005), besides accumulating in aquatic sediments (Richter-Torres et al., 1995). They have developed a highly sensitive TNT immunosensor with a prototype fluorescence-based detector system (KinExA Inline Biosensor, Sapidyne Instrument Inc.). The antibody that they used had a high affinity for TNT, and a minimal cross reactivity for tetryl, 2,4-dinitrotoluene and 2-amino-4,6-dinitrotoluene. An additional advantage was that the sensor could be regenerated within 8 min, and permitted a minimum of 40 readings.

Figure 13.3a shows the binding and dissociation of 0 μg/L of TNT in the aqueous environment to anti-TNT mAb (anti-nitrophenol monoclonal antibody) by a highly sensitive TNT immunosensor with a prototype fluorescence-based detection system (KinExA Inline Biosensor, Sappidyne Instrument, Inc.; Bromage et al., 2007). A single-fractal analysis is adequate to describe the binding and the dissociation kinetics. The values of (a) the binding rate coefficient k and the fractal dimension D_f for a single-fractal analysis, and (b) the dissociation rate coefficient k_d and the fractal dimension for dissociation D_{fd} are given in Tables 13.3 and 13.4.

Figure 13.3b shows the binding and dissociation of 0.01 μg/L of TNT in the aqueous environment to anti-TNT mAb by a highly sensitive TNT immunosensor with a prototype fluorescence-based detection system (KinExA Inline Biosensor, Sappidyne Instrument, Inc.; Bromage et al., 2007). Once again, a single-fractal analysis is adequate to describe the binding and the dissociation kinetics. The values of (a) the binding rate coefficient k and the fractal dimension D_f for a single-fractal analysis, and (b) the dissociation rate coefficient k_d and the fractal dimension for dissociation D_{fd} are given in Tables 13.3 and 13.4.

Figure 13.3c shows the binding and dissociation of 0.05 μg/L of TNT in the aqueous environment to anti-TNT mAb by a highly sensitive TNT immunosensor with a prototype fluorescence-based detection system (KinExA Inline Biosensor, Sappidyne Instrument, Inc.; Bromage et al., 2007). Once again, a single-fractal analysis is adequate to describe the binding and the dissociation kinetics. The values of (a) the binding rate coefficient k and the fractal dimension D_f for a

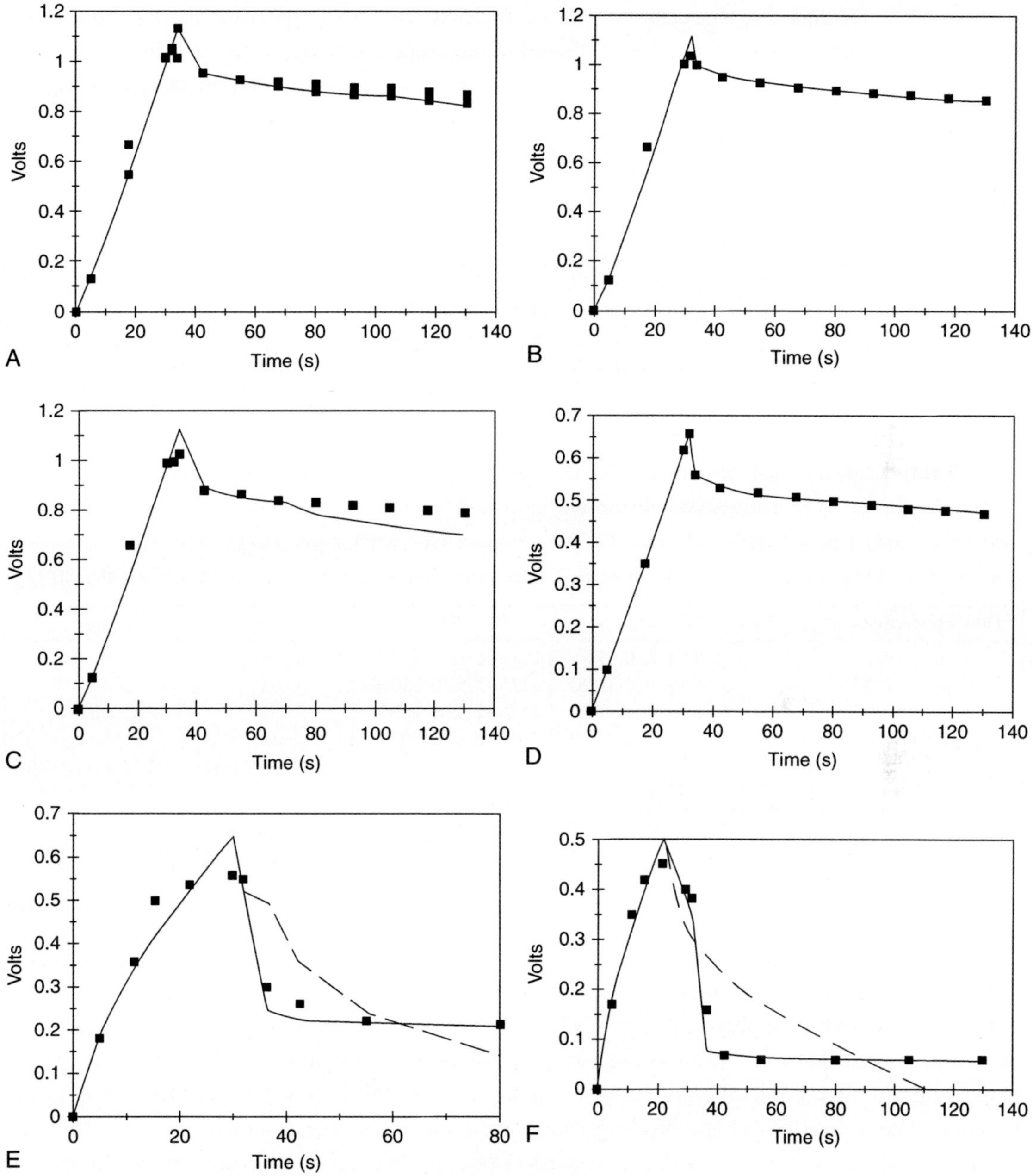

Figure 13.3

Binding of trace amounts of trinitrotoluene (TNT) (in μg/L) to anti-TNT antibody immobilized on a prototype fluorescence-based detector system (KinExA Inline Biosensor, Sapidyne Instrument, Inc.; Bromage et al., 2007): (a) 0, (b) 0.01, (c) 0.05, (d) 0.5, (e) 5, (f) 25. When only a solid line is used then a single-fractal analysis applies. When both a solid line and a dashed line are used, then the dashed line is for a single-fractal analysis, and the solid line is for a dual-fractal analysis.

Table 13.3: Binding and dissociation rate coefficients for TNT (trinitrotoluene) in aqueous environment to anti-TNT mAb (anti-nitorphenol monoclonal antibody, mAb) by a highly sensitive TNT immunosensor with a prototype fluorescence-based detection system (KinexA Inline Biosensor, Sappidyne Insrtrument, Inc.; Bromage et al., 2007).

TNT concentration (μg/L)	k	k_d	k_{d1}	k_{d2}
0	0.0239 ± 0.0037	0.03336 ± 0.00205	na	na
0.01	0.0206 ± 0.002	0.03347 ± 0.00202	na	na
0.05	0.02248 + 0.0337	0.0594 ± 0.0075	na	na
0.5	0.01935 ± 0.00027	0.08749 ± 0.0027	na	na
5	0.03409 ± 0.00547	0.0119 ± 0.0216	0.00546 + 0.0244	0.1471 ± 0.00042
25	0.02962 ± 0.00375	0.02437 + 0.02677	4.3E-05 + 5.7E-05	0.1843 ± 0.0016

Table 13.4: Fractal dimensions for the binding and the dissociation phase for TNT (trinitrotoluene) in aqueous environment to anti-TNT mAb (anti-nitorphenol monoclonal antibody, mAb) by a highly sensitive TNT immunosensor with a prototype fluorescence-based detection system (KinexA Inline Biosensor, Sappidyne Insrtrument, Inc.; Bromage et al., 2007).

TNT concentration (μg/L)	D_f	D_{fd}	D_{fd1}	D_{fd2}
0	0.812 ± 0.176	2.2506 ± 0.0340	na	na
0.01	0.6966 ± 0.1920	2.2582 ± 0.03302	na	na
0.05	0.7816 + 0.1727	2.2662 ± 0.1073	na	na
0.5	0.966 ± 0.0180	2.6742 ± 0.0017	na	na
5	1.6718 ± 0.2162	1.676 ± 0.560	0.594 + 1.864	2.9172 ± 0.00866
25	1.6162 ± 0.2184	2.0098 ± 0.5368	0 + 0	2.9740 ± 0.0130

single-fractal analysis, and (b) the dissociation rate coefficient k_d and the fractal dimension for dissociation D_{fd} are given in Tables 13.3 and 13.4.

Figure 13.3d shows the binding and dissociation of 0.5 μg/L of TNT in the aqueous environment to anti-TNT mAb by a highly sensitive TNT immunosensor with a prototype fluorescence-based detection system (KinExA Inline Biosensor, Sappidyne Instrument, Inc.; Bromage et al., 2007). Once again, a single-fractal analysis is adequate to describe the binding and the dissociation kinetics. The values of (a) the binding rate coefficient k and the fractal dimension D_f for a single-fractal analysis, and (b) the dissociation rate coefficient k_d and the fractal dimension for dissociation D_{fd} are given in Tables 13.3 and 13.4.

Figure 13.3e shows the binding and dissociation of 5.0 μg/L of TNT in the aqueous environment to anti-TNT mAb by a highly sensitive TNT immunosensor with a prototype fluorescence-based detection system (KinExA Inline Biosensor, Sappidyne Instrument, Inc.; Bromage et al., 2007). In this case, a single-fractal analysis is adequate to describe the binding kinetics.

However, a dual-fractal analysis is required to adequately describe the dissociation kinetics. The values of (a) the binding rate coefficient k and the fractal dimension D_f for a single-fractal analysis, (b) the dissociation rate coefficient k_d and the fractal dimension for dissociation D_{fd} for a single-fractal analysis, and (c) the dissociation rate coefficients k_{d1} and k_{d2} and the fractal dimensions D_{fd1} and D_{fd2} for a dual-fractal analysis are given in Tables 13.3 and 13.4.

It is of interest to note that as the fractal dimension increases by a factor of 4.91 from a value of D_{fd1} equal to 0.594 to D_{fd2} equal to 2.9172, the dissociation rate coefficient increases by a factor of 26.94 from a value of k_{d1} equal to 0.00546 to k_{d2} equal to 0.1471. Note that changes in the fractal dimension or the degree of heterogeneity on the sensing surface and in the dissociation rate coefficient are in the same direction.

Figure 13.3f shows the binding and dissociation of 25.0 μg/L of TNT in the aqueous environment to anti-TNT mAb by a highly sensitive TNT immunosensor with a prototype fluorescence-based detection system (KinExA Inline Biosensor, Sappidyne Instrument, Inc.; Bromage et al., 2007). In this case, a single-fractal analysis is adequate to describe the binding kinetics. Once again, a dual-fractal analysis is required to adequately describe the dissociation kinetics. The values of (a) the binding rate coefficient k and the fractal dimension D_f for a single-fractal analysis, (b) the dissociation rate coefficient k_d and the fractal dimension for dissociation D_{fd} for a single-fractal analysis, and (c) the dissociation rate coefficients k_{d1} and k_{d2} and the fractal dimensions D_{fd1} and D_{fd2} for a dual-fractal analysis are given in Tables 13.3 and 13.4.

Tables 13.3 and 13.4 and Figure 13.4a show the increase in the binding rate coefficient k with an increase in the fractal dimension D_f for a single-fractal analysis. For the data shown in Figure 13.4a, the binding rate coefficient k is given by:

$$k = (0.0242 \pm 0.0032) D_f^{0.492 \pm 0.147} \tag{13.4a}$$

The fit is reasonable. Only six data points are available. The availability of more data points would lead to a more reliable fit. The binding rate coefficient k is only mildly sensitive to the fractal dimension D_f or the degree of heterogeneity that exists on the sensing surface as noted by the slightly less than one-half (equal to 0.492) order of dependence exhibited.

Tables 13.3 and 13.4 and Figure 13.4b show the increase in the dissociation rate coefficient k_d with an increase in the fractal dimension D_{fd} for a single-fractal analysis. For the data shown in Figure 13.4b, the dissociation rate coefficient k_d is given by:

$$k_d = (0.000914 \pm 0.000343) D_{fd}^{4.65 \pm 2.17} \tag{13.4b}$$

The fit is reasonable. Only four data points are available. The availability of more data points would lead to a more reliable fit. The dissociation rate coefficient k_d is very sensitive to the

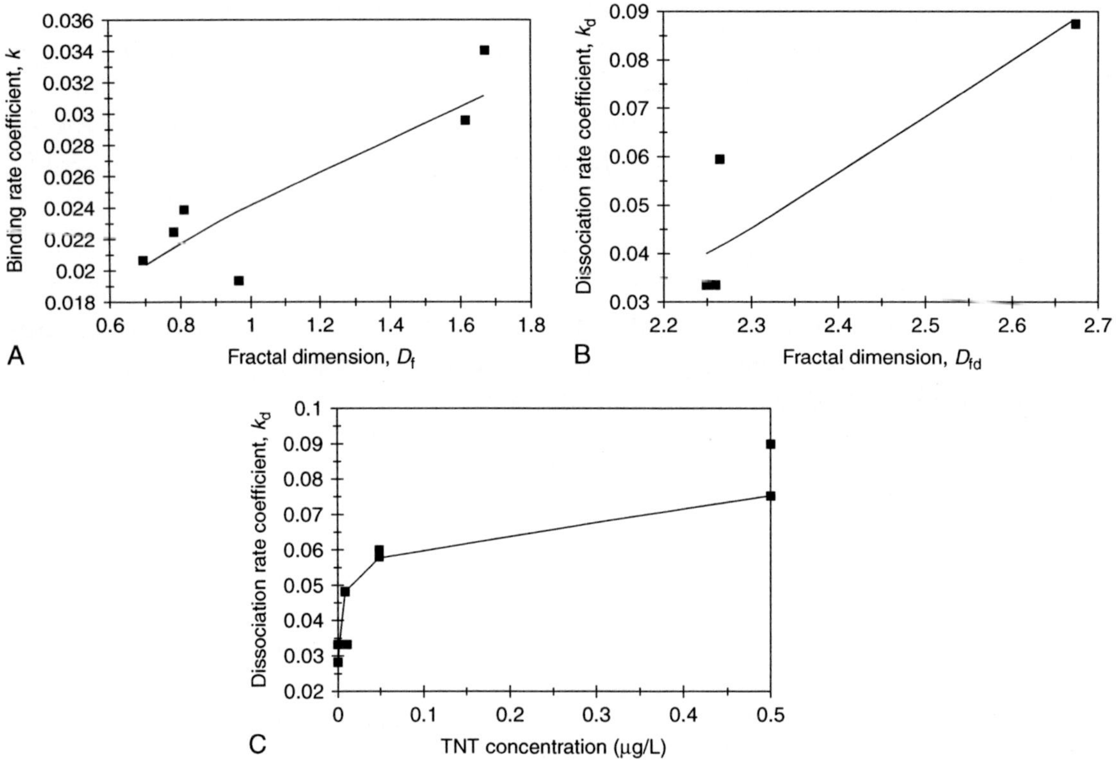

Figure 13.4

(a) Increase in the binding rate coefficient, k with an increase in the fractal dimension, D_f. (b) Increase in the dissociation rate coefficient, k_d with an increase in the fractal dimension, D_{fd}. (c) Increase in the dissociation rate coefficient, k_d with an increase in the TNT concentration (in μg/L) in the ground water.

fractal dimension D_{fd} or the degree of heterogeneity that exists on the sensing surface as noted by the order of dependence between four and one-half and five (equal to 4.65) exhibited.

Tables 13.3 and 13.4 and Figure 13.4c show the increase in the dissociation rate coefficient k_d with an increase in the TNT concentration in the 0 to 0.5 μg/L range for a single-fractal analysis. For the data shown in Figure 13.4c, the dissociation rate coefficient k_d is given by:

$$k_d = (0.0816 \pm 0.0295)[\mathrm{TNT}]^{0.115 \pm 0.0495} \tag{13.4c}$$

The fit is good. Only five data points are available. The availability of more data points would lead to a more reliable fit. The dissociation rate coefficient k_d exhibits a very low (almost negligible; equal to 0.115) order of dependence on the TNT concentration in solution.

Baronian and Gurazada (2007) have recently analyzed the electrochemical detection of wild-type *S. cerevisiae* responses to estrogens. They point out that the presence of an EBP and an endogeneous ligand in three yeast species was first reported in the early 1990s. This ligand was 17β-estradiol, and the binding affinity of this protein with a variety of estrogen and estrogen-like molecules has been noted for *S. cerevisiae* (Burshell et al., 1984), *Candida albicans* (Madani et al., 1994), and *Paracoccidioides brasiliensis*. Baronian and Gurazada (2007) indicate that the relative binding affinities of *S. cerevisiae* and *C. albicans* EBPs are similar for 17β-estradiol, estradiol, estrone, estriol, and 17α-estradiol.

Figure 13.5a shows the binding of EBP in *Saccharomyces cerevisiae*. A single-fractal analysis is adequate to describe the binding kinetics. The values of the binding rate coefficient k and the fractal dimension D_f for a single-fractal analysis are given in Table 13.5.

Figure 13.5b shows the binding of 1.1 nM 17β-estradiol. Once again, a single-fractal analysis is adequate to describe the binding kinetics. The values of the binding rate coefficient k and the fractal dimension D_f for a single-fractal analysis are given in Table 13.5.

It is of interest to note that as the fractal dimension increases by a factor of 1.29 from a value of 1.6944 for the wild-type case to 2.1904 for the 11 nM 17β-estradiol case, the binding rate coefficient increases by 1.37% (a very slight increase) from a value of k equal to 159.24 to k equal to 161.42. Note that change in the fractal dimension or the degree of heterogeneity on the electrochemical surface and in the binding rate coefficient are in the same direction.

Ma et al. (2007) have reported that restriction endonucleases are important enzymes in molecular biology. They are involved in recombinant technology, mapping, genotyping and

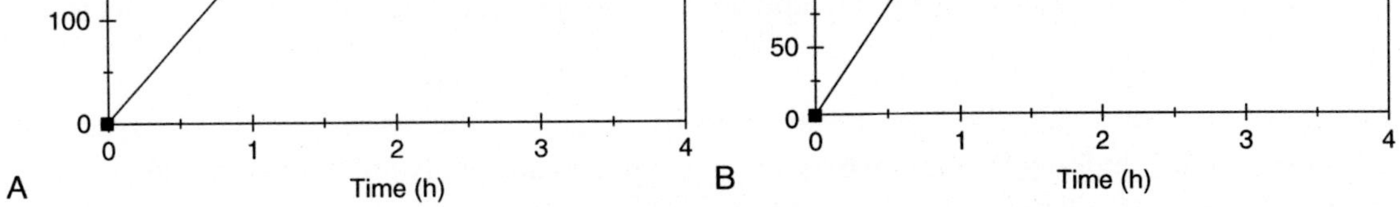

Figure 13.5
(a) Binding of EBP (estrogen binding protein) in *Saccharomyces cerevisiae*. (b) Binding of 1.1 nM 17β-estradiol (Baronian and Guruzada, 2007).

Table 13.5: Binding rate coefficient and fractal dimensions for (a) electrochemical detection of wild-type *Saccharomyces cerevisia* responses to estrogen (Baronian and Guruzada, 2007), and real-time monitoring of restriction endonuclease activity using a molecular beacon (Ma et al., 2007).

Analyte in Solution/Receptor on Surface	k	D_f
(a) EBP (estrogen binding protein) in *Saccaromyces cervisiae*	159.24 ± 2.91	1.6944 ± 0.0348
17β-estradiol	161.42 ± 4.81	2.1904 ± 0.05646
(b) 32 U endonuclease/molecular beacon	6.677 ± 0.344	1.7436 ± 0.0398
25 U endonuclease/molecular beacon	6.351 ± 0.379	1.7376 ± 0.0398
10 U endonuclease/molecular beacon	0.5446 ± 0.0048	0.972 ± 0.0104
4 U endonuclease/molecular beacon	0.34 ± 0.0	1.0 ± 0.0

in the sequencing of large strands of DNA (Bhagwat, 1992; Pingouid and Jeltsch, 2001). Due to their biological importance, assays for restriction endonucleases are essential (Ma et al., 2007). Ma et al. (2007) have recently used a MB for the real-time monitoring of restriction endonuclease activity. MBs have been used in biology, biotechnology, chemistry, and in medical science for biomolecular recognition due to their high selectivity and excellent specificity (Tyagi et al., 1998; Tan et al., 2004). MB may also be used to monitor DNA-protein interactions (Tang et al., 2003, 2005; Lin et al., 2005). Ma et al. (2007) developed a novel approach for the real-time monitoring of *Rsa*I endonuclease activity based on the primer extension reaction using MB.

Figure 13.6a shows the real-time monitoring of the DNA cleavage process catalyzed by 32 units of *Rsa*I endonuclease. A single-fractal analysis is adequate to describe the binding reaction. The values of the binding rate coefficient k and the fractal dimension D_f for a single-fractal analysis are given in Table 13.5.

Figure 13.6b shows the real-time monitoring of the DNA cleavage process catalyzed by 25 units of *Rsa*I endonuclease. Once again, a single-fractal analysis is adequate to describe the binding reaction. The values of the binding rate coefficient k and the fractal dimension D_f for a single-fractal analysis are given in Table 13.5. It is of interest to note that a decrease in the restriction endonuclease activity from 32 to 25 Units of activity decreases the binding rate coefficient k from a value of 6.677 to 6.351. Note that a change in the restriction endonuclease activity (U) and in the binding rate coefficient are in the same direction.

Figure 13.6c shows the real-time monitoring of the DNA cleavage process catalyzed by 10 units of *Rsa*I endonuclease. Once again, a single-fractal analysis is adequate to describe the binding reaction. The values of the binding rate coefficient, k and the fractal dimension, D_f for a single-fractal analysis are given in Table 13.5.

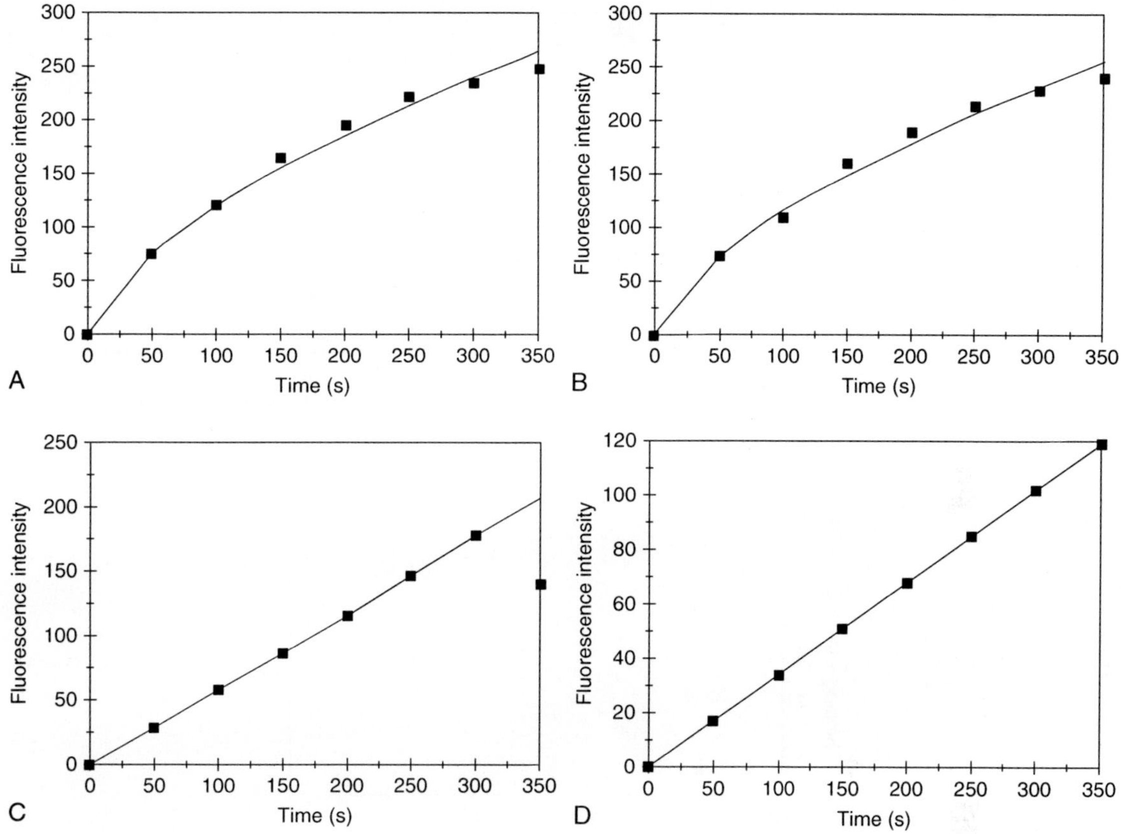

Figure 13.6
Binding of different units of restriction endonuclease activity using a molecular beacon (Ma et al., 2007): (a) 32, (b) 25, (c) 10, (d) 4.

Figure 13.6d shows the real-time monitoring of the DNA cleavage process catalyzed by 14 units of *Rsa*I endonuclease. Once again, a single-fractal analysis is adequate to describe the binding reaction. The values of the binding rate coefficient k and the fractal dimension D_f for a single-fractal analysis are given in Table 13.5.

Figure 13.7a and Table 13.5 show the increase in the binding rate coefficient k with an increase in the restriction endonuclease concentration in units, U. For the data shown in Figure 13.7a, the binding rate coefficient k is given by:

$$k = (0.0268 \pm 0.0209)[\text{endonuclease, unit}]^{1.593 \pm 0.349} \quad (13.5a)$$

The fit is good. Only three data points are available. The availability of more data points would lead to a more reliable fit. The binding rate coefficient k is sensitive to the

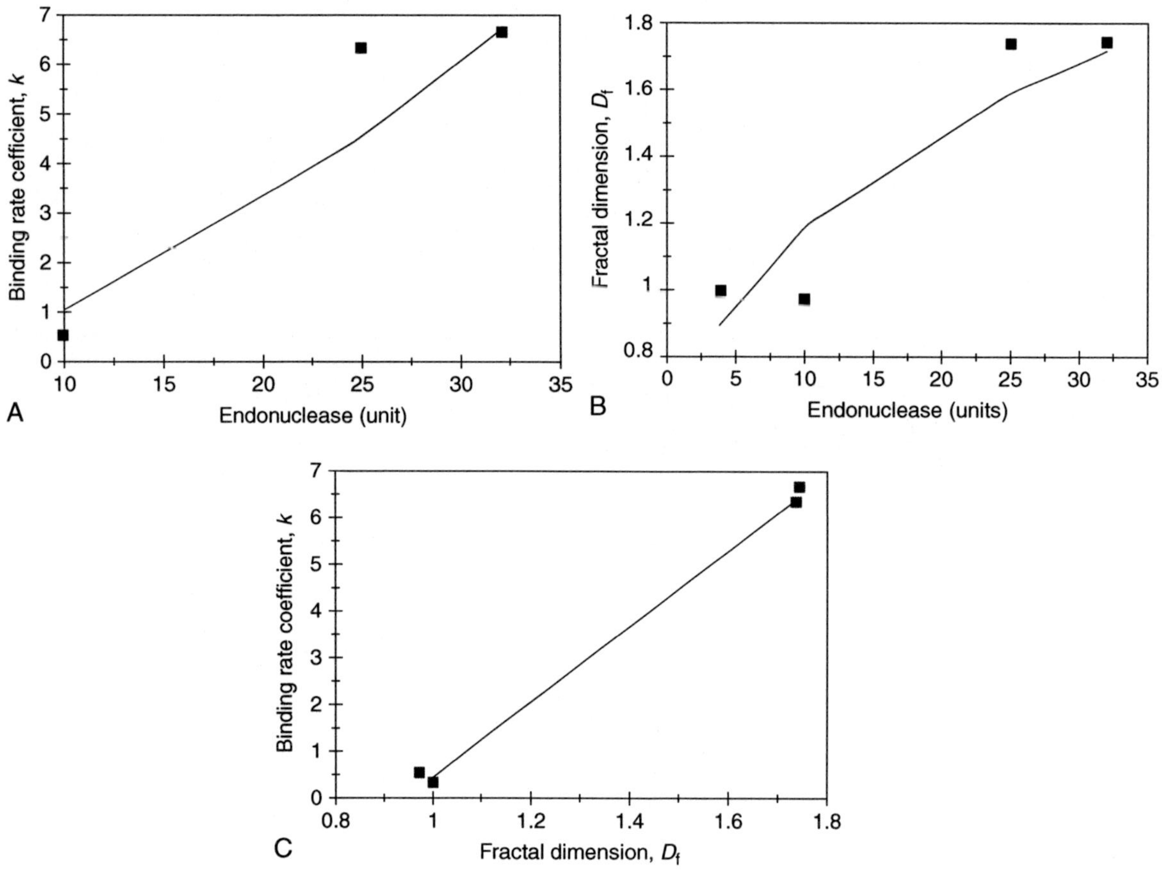

Figure 13.7

(a) Increase in the binding rate coefficient, *k* with an increase in the endonuclease concentration (in units). (b) Increase in the fractal dimension, D_f with an increase in the endonuclease concentration (in units). (c) Increase in the binding rate coefficient, *k* with an increase in the fractal dimension, D_f.

endonuclease concentration in units as noted by the greater than one and a half (equal to 1.593) order of dependence exhibited.

Figure 13.7b and Table 13.5 show the increase in the fractal dimension D_f with an increase in the restriction endonuclease concentration in units, U. For the data shown in Figure 13.7b, the fractal dimension D_f is given by:

$$D_f = (0.584 \pm 0.112)[\text{endonuclease, unit}]^{0.311 \pm 0.107} \quad (13.5b)$$

The fit is good. Only four data points are available. The availability of more data points would lead to a more reliable fit. The fractal dimension D_f is only mildly sensitive to the endonuclease concentration in units as noted by the low (equal to 0.311) order of dependence exhibited.

Figure 13.7c and Table 13.5 show the increase in the binding rate coefficient k with an increase in the fractal dimension D_f. For the data shown in Figure 13.7c, the binding rate coefficient k is given by:

$$k = (0.4638 \pm 0.1646)D_f^{4.753 \pm 0.5339} \tag{13.5c}$$

The fit is good. Only four data points are available. The availability of more data points would lead to a more reliable fit. The binding rate coefficient k is very sensitive to the fractal dimension D_f or the degree of heterogeneity that exists on the sensing surface as noted by the higher than four and a half (equal to 4.753) order exhibited.

Tang et al (2006) have developed and analyzed a nonregeneration protocol for surface plasmon resonance biosensors. They studied the high-affinity interaction with high-density biosensors. They report the "optimum" amount of regeneration involved in SPR biosensor applications, which has been around as a real-time, nonlabel technique for the analysis of biological interaction since the early 1990s (Karlsson et al., 1991; Jonsson and Malmquist, 1992). The essence of regeneration is to remove the bound analytes and still maintain the bioactivity of the ligand (receptor). Also, as Tang et al. (2006) report, the surface density of the ligand should be the same for each analyte injection to permit data analysis. The discovery or development of an acceptable regeneration protocol is not easy (Tang et al., 2006). To circumvent this, these authors have developed a nonregeneration protocol between successive analyte injections. This is especially useful for high-affinity antigen-antibody interactions. In essence, the nonregeneration protocol needs a relatively high ligand density on the biosensor surface so that more data points can be obtained before surface saturation. As a model system Tang et al. (2006) used rabbit IgG as the analyte and engineered recombinant antibody A10B ScFv as the ligand.

Figure 13.8a shows the binding and dissociation of 1.3 μM rabbit IgG in solution to engineered recombinant A10B ScFv immobilized on a high-density biosensor. A dual-fractal analysis is required to adequately describe the binding kinetics. A single-fractal analysis is adequate to describe the dissociation kinetics. The values of (a) the binding rate coefficient k and the fractal dimension D_f for a single-fractal analysis, (b) the binding rate coefficients k_1 and k_2 and the fractal dimensions D_{f1} and D_{f2} for a dual-fractal analysis, and (c) the dissociation rate coefficient k_d and the fractal dimension for dissociation D_{fd} for a single-fractal analysis are given in Tables 13.6 and 13.7.

Figure 13.8b shows the binding and dissociation of 0.53 μM rabbit IgG in solution to engineered recombinant A10B ScFv immobilized on a high-density biosensor. Once again, a dual-fractal analysis is required to adequately describe the binding kinetics. A single-fractal analysis is adequate to describe the dissociation kinetics. The values of (a) the binding rate coefficient k and the fractal dimension D_f for a single-fractal analysis, (b) the binding rate

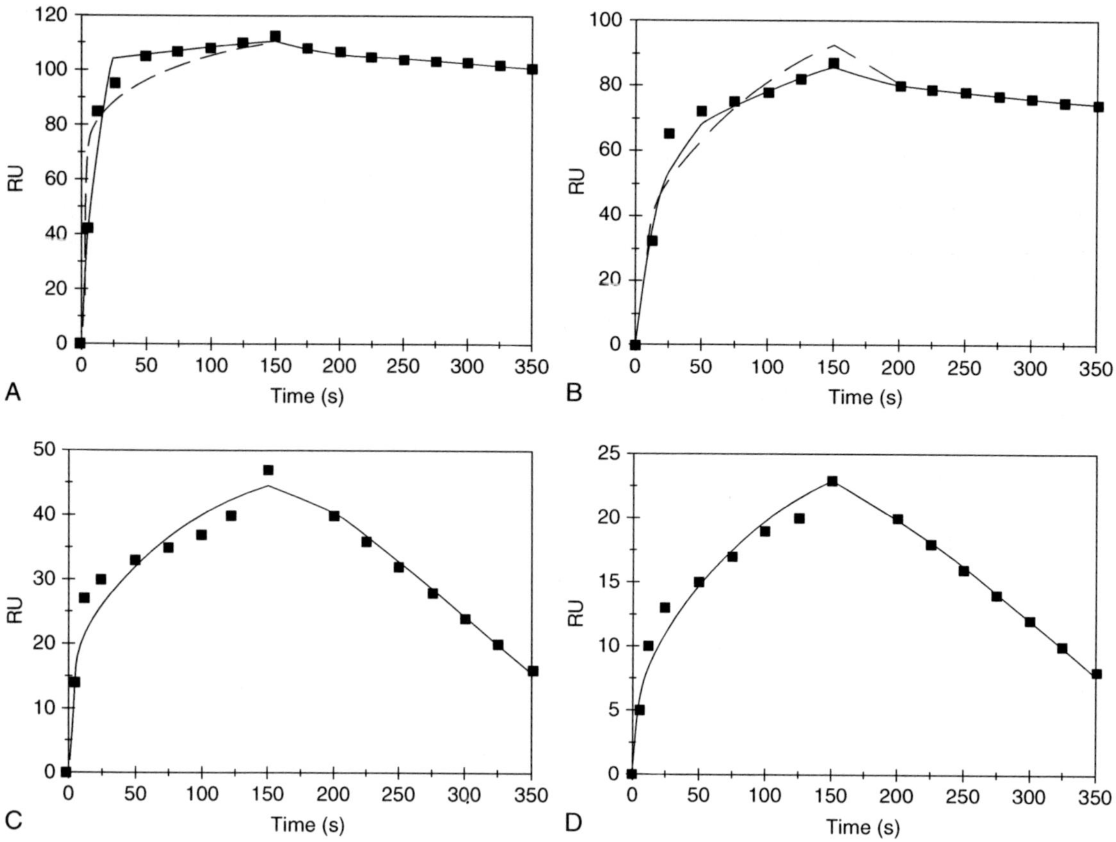

Figure 13.8

Binding of different rabbit IgG concentrations (in μM) in solution to A10B scFv (single-chain fragment variable; Tang et al., 2006): (a) 1.3, (b) 0.53, (c) 0.20, (d) 0.067. When only a solid line is used then a single-fractal analysis applies. When both a solid line and a dashed line are used, then the dashed line is for a single-fractal analysis, and the solid line is for a dual-fractal analysis.

Table 13.6: Binding and dissociation rate coefficients for different concentrations of rabbit IgG in solution engineered recombinant antibody A10B single chain variable (ScFv) (Tang et al., 2006).

Rabbit Concentration, Micromole/Antibody A10B	k	k_1	k_2	k_d
1.3	85.189 ± 60.167	21.714 ± 16.878	86.841 ± 0.358	1.237 ± 0.056
0.53	20.39 ± 17.35	10.767 ± 8.486	29.893 ± 0.452	1.172 ± 0.032
0.20	12.061 ± 10.406	na	na	0.0792 ± 0.0061
0.067	3.022 ± 0.449	na	na	0.0334 ± 0.0007

Table 13.7: Fractal dimensions for the binding and dissociation rate coefficients for different concentrations of rabbit IgG in solution engineered recombinant antibody A10B single chain variable (ScFv) (Tang et al., 2006).

Rabbit Concentration, Micromole/Antibody A10B	D_f	D_{f1}	D_{f2}	D_{fd}
1.3	2.758 ± 0.1992	1.8664 ± 0.4990	2.903 ± 0.0136	2.1522 ± 0.0734
0.53	2.3314 ± 0.1434	1.8558 ± 0.4716	2.5784 ± 0.0580	2.1032 ± 0.0442
0.20	2.4164 ± 0.0958	na	na	0.744 ± 0.1226
0.067	2.1882 ± 0.08976	na	na	0.688 ± 0.00326

coefficients k_1 and k_2 and the fractal dimensions D_{f1} and D_{f2} for a dual-fractal analysis, and (c) the dissociation rate coefficient k_d and the fractal dimension for dissociation D_{fd} for a single-fractal analysis are given in Tables 13.6 and 13.7.

Figure 13.8c shows the binding and dissociation of 0.20 μM rabbit IgG in solution to engineered recombinant A10B ScFv immobilized on a high-density biosensor. A single-fractal analysis is adequate to describe the binding and the dissociation kinetics. The values of (a) the binding rate coefficient k and the fractal dimension D_f for a single-fractal analysis, and (b) the dissociation rate coefficient k_d and the fractal dimension for dissociation D_{fd} for a single-fractal analysis are given in Tables 13.6 and 13.7.

Figure 13.8d shows the binding and dissociation of 0.067 μM rabbit IgG in solution to engineered recombinant A10B ScFv immobilized on a high-density biosensor. Once again, a single-fractal analysis is adequate to describe the binding and the dissociation kinetics. The values of (a) the binding rate coefficient k and the fractal dimension D_f for a single-fractal analysis, and (b) the dissociation rate coefficient k_d and the fractal dimension for dissociation D_{fd} for a single-fractal analysis are given in Tables 13.6 and 13.7.

Figure 13.9a and Table 13.6 show the increase in the dissociation rate coefficient k_d with an increase in the rabbit IgG concentration in solution in the μM range. For the data shown in Figure 13.9a, the dissociation rate coefficient k_d is given by:

$$k_d = (1.244 + 1.367)[\text{rabbit IgG concentration, } \mu\text{M}]^{1.374 \pm 0.336} \quad (13.6a)$$

The fit is reasonable. Only four data points are available. The availability of more data points would lead to a more reliable fit. The dissociation rate coefficient k_d is sensitive to the rabbit IgG concentration in solution as noted by the order of dependence between the first and one and a half (equal to 1.374) exhibited.

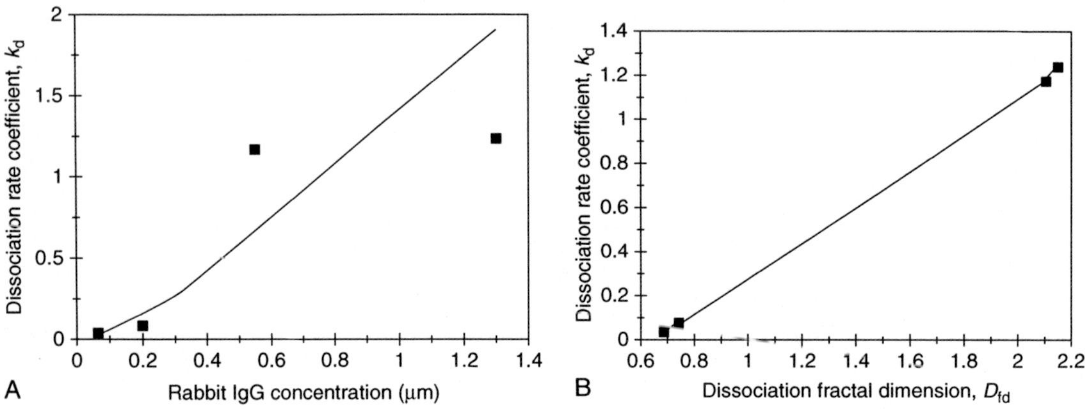

Figure 13.9
(a) Increase in the dissociation rate coefficient, k_d with an increase in the rabbit IgG concentration (in μm) (b) Increase in the dissociation rate coefficient, k_d with an increase in the fractal dimension, D_{fd}.

Figure 13.9b and Tables 13.6 and 13.7 show the increase in the dissociation rate coefficient k_d with an increase in the fractal dimension for dissociation D_{fd}. For the data shown in Figure 13.9b, the dissociation rate coefficient k_d is given by:

$$k_d = (0.295 + 1.038)D_{fd}^{1.94 \pm 1.38} \tag{13.6b}$$

The fit is not good. There is scatter in the data. The availability of more data points would lead to a more reliable fit. The dissociation rate coefficient k_d is sensitive to the degree of heterogeneity that exists on the sensing surface during the dissociation phase D_{fd} as noted by the close to second (equal to 1.94) order of dependence exhibited.

13.4 Conclusions

A fractal analysis is presented for the binding and dissociation (if required) of different analytes on different biosensor surfaces. Both a single- and a dual-fractal analysis were used. The dual-fractal analysis was used only when the single-fractal analysis did not provide an adequate fit. This was judged using Corel Quattro Pro 8.0 (Corel Quattro Pro, 1997) to see if the regression provided was adequate (regression coefficient greater than 0.95).

A wide variety of examples available in the literature were analyzed. The systems analyzed were selected at random. The analyte-receptor systems analyzed include (a) the binding of ODN-P and noncomplementary ODN during the hybridization assay with EST2-A34 reporter (Wang et al., 2007), (b) binding during the primer elongation reaction of DNA coupled directly to PAC and DNA coupled via biotin-streptavidin (Krieg et al., 2006), (c) binding and dissociation of trace amounts of TNT in μg/L to anti-TNT antibody immobilized on a prototype fluorescence-based detector system (KinExA Inline Biosensor, Sappidyne Instrument, Inc.; Bromage et al., 2007), (d) binding of EBP to *S. cerevisiae* and the binding of 17β-estradiol (Baronian and Guruzada, 2007), binding of different units of restriction endonuclease activity using a MB (Ma et al., 2007), and (f) binding and dissociation of different rabbit IgG concentrations (in μM) in solution to A10B scFv (Tang et al., 2006).

Predictive relations are developed for the binding and dissociation rate coefficients and for the fractal dimension in the binding phase. For example, for the binding of TNT in solution to anti-TNT antibody immobilized on a KinExA biosensor (Bromage et al., 2007), (a) the binding rate coefficient, k for a single-fractal analysis exhibits close to a one-half (equal to 0.492) order of dependence on the fractal dimension D_f, and the dissociation rate coefficient k_d exhibits a 4.65 order of dependence on the degree of heterogeneity that exists on the sensing surface. In this case, the dissociation rate coefficient is much more sensitive to the degree of heterogeneity that exists on the sensing surface than the binding rate coefficient. Also, (b) the binding rate coefficient k for a single-fractal analysis exhibits close to a one and a half (equal to 1.593) order of dependence on the endonuclease unit in solution for the real-time monitoring of the DNA cleavage process catalyzed by RsaI endonuclease (Ma et al., 2007), and (c) the binding rate coefficient k for a single-fractal analysis exhibits a 4.753 order of dependence on the fractal dimension D_f or the degree of heterogeneity on the sensing surface.

The predictive relationships developed and presented for the different analyte-receptor reactions occurring on the different biosensor surfaces are useful as they may be used to manipulate the different biosensor parameters (such as the binding and the dissociation rate coefficients) in required or desired directions. As the biosensor systems analyzed were selected at random, the fractal analysis technique may also be applied to other biosensor systems. A particular advantage of the fractal analysis method is that it provides a quantitative measure of the degree of heterogeneity that exists on the biosensor surface. This provides one with an extra variable that biosensorists may use to help enhance or modify the different and relevant biosensor parameters in desired directions. At times, this may require some ingenuity in the sense that changing one particular experimental variable may or may not affect the biosensor performance parameter(s) in different directions. For example, increasing the sensitivity may decrease the stability or increase the detection time. If one is fortuitous enough, then perhaps a change in an experimental variable may affect two (or more) biosensor performance parameters simultaneously in required or desired directions. For this to occur, it behooves one to know as much as one can about the biosensor system being analyzed. The fractal analysis presented is one step in that direction.

References

Anderson RS, LL Brubacher, LM Calvo, EM Burreson, and MA Unger, *Environmental Research*, **74**(1), 84–90 (1997).

Baeumner AJ, RN Cohen, V Miksic, and J Min, *Biosensors & Bioelectronics*, **18**(4), 405–413 (2003).

Baeumner AJ, J Pretz, and S Fang, *Analytical Chemistry*, **76**(4), 888–894 (2004).

Baronian KHR and S Guruzada, Electrochemical detection of wild-type *Saccharomyces cerevisiae* responses to estrogens, *Biosensors & Bioelectronics*, **22**, 2493–2499 (2007).

Belden JB, DR Ownby, GR Lotufo, and MJ Lydy, *Chemosphere*, **58**(9), 1161–1168 (2005).

Berkenpas E, P Millard, and M Pereira da Cunha, Detection of *Escherichia coli* 0157:H7 with Langasite pure shear horizontal surface acoustic wave sensors, *Biosensors & Bioelectronics*, **21**, 2255–2262 (2006).

Bhagwat AS, Restriction Enzymes: Properties and use, *Methods in Enzymology*, **216**, 199–224 (1992).

Bromage ES, T Lackie, MA Unger, J Ye, and SL Kattari, The development of a real-time biosensor for the detection of trace level of trinitrotoluene (TNT) in aquatic environments, *Biosensors & Bioelectronics*, **22**, 2532–2538 (2007).

Burshell A, DA Stathis, Y Do, SC Miller, and D Feldman, *Journal of Biological Chemistry*, **259**, 3450–3456 (1984).

Chan EY, Advances in sequencing technology, *Mutation Research*, **573**, 13–40 (2005).

Ek H, G Dave, J Sturve, BC Almroth, E Stephensen, L Forlin, and G Birgerson, *Aquatic Toxicology*, **72**(3), 221–230 (2005).

Gao Z, GS Kim, and DB Henthorn, Preparation of bioanalytical sensors by incorporating fluorophore in patternable poly(ethylene glycol) diacrylate-based membranes, paper 682e. In *American Institute of Chemical Engineers Meeting*, Philadelphia, Pennsylvania, November 16-21, 2008.

Hacia JG, Resequencing and mutational analysis using oligonucleotide microarrays, *Nature Genetics*, **21**, 42–47 (1999).

Havlin S, Molecular diffusion and reactions in *The Fractal Approach to Heterogeneous Chemistry: Surfaces, Colloids, Polymers* (ed. D Avnir), Wiley, New York, pp. 251–269 (1989).

Heller DA, H Jin, and MS Strano, Nanoscale sensors for multi-modal detection of genotoxic agents, paper 188ai. In *American Institute of Chemical Engineers Annual Meeting*, Philadelphia, Pennsylvania, November 16-21, 2008.

Jonsson U and M Malmquist, *Advances in Biosensors*, **2**, 291–336 (1992).

Karlsson R, A Michaelsson, and L Mattsson, *Journal of Immunological Methods*, **145**(1-2), 229–240 (1991).

Kizil R, Controlled assembly and disassembly of biotinylated Au nanoparticles on metallic nanowires for nanobiosensing, paper 655b. In *American Institute of Chemical Engineers Annual Meeting*, Philadelphia, Pennsylvania, November 16-21, 2008.

Ko WS and S Grant, *Biosensors & Bioelectronics*, **21**(7), 1283–1290 (2006).

Krieg A, T Ruckstuhl, and S Seeger, Towards single molecule DNA sequencing assays with low non-specific adsorption, *Analytical Biochemistry*, **349**, 181–185 (2006).

Leung A, K Rijal, PM Shankar, and R Mutharasan, Effects of geometry on transmission and sensing potential of tapered fiber sensors, *Biosensors & Bioelectronics*, **21**, 2202–2209 (2006).

Ma C, Z Tang, K Wang, W Tan, X Yang, W Li, Z Li, and X Lv, Real-time monitoring of restriction endonuclease activity using molecular beacon, *Analytical Biochemistry*, **363**, 294–296 (2007).

Madani ND, PJ Mlloy, P Rodriguez-Pombo, and AV Krishnan, *Proceedings of the National Academy of Sciences*, **91**, 922–926 (1994).

Mitterer G, M Huber, E Leidinger, C Kiristis, W Lubitz, MW Mueller, and WM Schmidt, *Journal of Clinical Microbiology*, **42**(3), 1048–1057 (2004).

Nebling E, T Grunwald, J Albers, P Schafer, and R Hintsche, *Analytical Chemistry*, **76**(3), 689–696 (2004).

Ownby DR, JB Belden, GR Lotufo, and MJ Lydy, *Chemosphere*, **58**(9), 1153–1159 (2005).

Pingouid A and A Jeltsch, Structure and function of type II restriction endonucleases, *Nucleic Acids Research*, **29**, 3705–3727 (2001).

Richter-Torres P, A Dorsey, and C Hodes, *Toxicological Profile for 2,4,6-Trinitrotoluene (TNT)*, US Department of Health and Human Services, Agency for Toxic Substances and Disease Registry, Atlanta, GA 1995.

Safsten P, SL Klakamp, AW Drake, R Karlsson, and DG Myszka, Screening antibody-antigen interactions in parallel using Biacore A100, *Analytical Biochemistry*, **35**, 181–190 (2006).

Shendure J, RD Mitra, C Varma, and GM Church, Advanced sequencing technologies: Methods and goals, *Nature Review of Genetics*, **5**, 335–344 (2004).

Shi M, Y Peng, J Zhou, B Liu, Y Huang, and J Kong, Immunoassays based on microelectrodes arrayed on a silica chip for high throughput screening of liver fibrosis markers in human serum, *Biosensors & Bioelectronics*, **21**, 2210–2216 (2006).

Suni II and Y Huang, *Impedance biosensor for peanut protein allergens,* paper 683b. In *Annual American Institute of Chemical Engineers Meeting*, Philadelphia, Pennsylvania, November 16-21, 2008.

Tan W, K Wang, and TJ Drake, Molecular beacons, *Current Opinions in Chemical Biology*, **8**, 547–553 (2004).

Tang D, R Yuan, Y Chai, Y Fu, J Dai, Y Liu, and Y Zhong, New amperometric and potentiometric immunosensors based on gold nanoparticles/tris (2,-2′-bipyridyl) cobalt (II) multilayer films for hepatitis B surface antigen determinations, *Biosensors and Bioelectronics*, **21**, 534–538 (2005).

Tang Y, R Mernaugh, and X Zeng, Nonregeneration protocol for surface plasmon resonance: Study of high-affinity interaction with high-density biosensors, *Analytical Chemistry*, **78**, 1841–1848 (2006).

Tizzard AC and G Lloyd-Jones, Bacterial oxygenases: *In vivo* enzyme biosensors for organic pollutants, *Biosensors & Bioelectronics*, **22**, 2400–2407 (2007).

Tyagi R, DP Bratu, and FR Kramer, Multicolor molecular beacons for allele discrimination, *Nature Biotechnology*, **16**, 49–53 (1998).

Wang J, *Nucleic Acids Research*, **28**, 3011–3016 (2000).

Wang Y, M Stanzel, W Gumbrecht, M Humenik, and M Spinzl, Esterase 2-oligodeoxynucleotide conjugates as sensitive reporter for electrochemical detection of nucleic acid hybridization, *Biosensors & Bioelectronics*, **22**, 1798–1806 (2007).

CHAPTER 14

Toxins and Pollutants Detection on Biosensor Surfaces

Chapter Outline

14.1 Introduction

The detection of toxins and pollutants is an important area of biosensor investigation. Food contamination by pathogenic bacteria is a cause for serious concern (Taylor et al., 2006). These authors point out that the following bacterial pathogens *Salmonella* spp., *Listeria monocytogenes*, *Campylobacter jejuni*, and *Escherichia coli* 0157:H7 are apparently responsible for about 67% (two-thirds) of food-related deaths (Mead et al., 1999). Buzby et al. (1996) report that the estimated economic impact of food-related illnesses is around $ 5.4 billion in the United States. Thus, Taylor et al. (2006) emphasize the need for a rapid, sensitive, and reliable biosensor to help detect these pathogenic bacteria that cause these serious illnesses. These authors have developed a multichannel surface plasmon resonance (SPR) biosensor for the quantitative, simultaneous detection of four food borne bacterial pathogens.

Walt and Franz (2000) report that *Staphylococcal enterotoxin B* (SEB) a 28.5-kDa protein may be used as a potential warfare agent and is a common source of food-related illnesses. Haes et al. (2006) recently developed a bead-assisted displacement immunoassay for SEB detection on a microchip using laser-induced fluorescence. Fluid delivery was electrokinetically controlled. Also, a monoclonal antibody that is specific for SEB was covalently attached to the silica beads. These authors explain that limited diffusion lengths and field-based enrichment similar to field-amplified stacking led to low detection limits.

Handbook of Biosensors and Biosensor Kinetics. DOI: 10.1016/B978-0-444-53262-6.00014-0

Dankbar et al. (2007) have recently developed a diagnostic low-density DNA microarray (B chip) for influenza B viruses. Their immunoassay is able to detect and identify the two currently circulating major lineages: the virus strains B/Victoria/2/87 and B/Yamagata/16/88. Their microarray procedure identifies the two virus strains in less than 8 h. The microarray procedure involves multiplex nucleic acid amplification and microarray hybridization of viral RNA.

Dankbar et al. (2007) point out that other microarray methods for the detection of influenza B virus are available, but these methods are unable to provide the lineage information on the virus (Li et al., 2001; Sengupta et al., 2003; Kessler et al., 2004). Previous reports are available on development of microarrays for the detection and identification of influenza A viruses (Mehlmann et al., 2006; Townsend et al., 2006; Dawson et al., 2006).

Doleman et al. (2007) have recently developed a bioluminescence DNA hybridization assay for *Plasmodium falciparum* based on the photoprotein aequorin. These authors report that *P. falciparum* is the most deadly and prevalent malaria species (Thayer, 2005). Mangold et al. (2005) point out that there are about 300-500 million cases of malaria each year. This disease kills about 3 million people every year, with about 90% of the deaths occurring in Africa. Doleman et al. (2007) explain that their bioluminescence hybridization assay has a detection limit of 3 pg/μL, and has been employed for the detection of target DNA in standard and spiked human samples. Furthermore, their hybridization assay has been developed in a microplate format without the need for sample PCR amplification.

Prabhakaran et al. (2007) have recently developed a naked-eye cadmium sensor which can detect cadmium (II) ion at the submicromolar levels using the Langmuir-Blodgett technique. These authors transferred molecular assemblies of 4-*n*-dodecyl-6-(2-thiazolylazo)resorcinol on precleaned microscopic glass slides. These assemblies served as the sensing probe. They point out that cadmium is a nonessential element; however, it is toxic to humans since it forms a strong bond with sulfur. Therefore, it can replace the essential metal ions such as Zn^{2+} and Ca^{2+} from the binding sites of certain enzymes (Arvidson, 1994; Wang et al., 1999; Lofts et al., 2005; Goel et al., 2006; Martelli et al., 2006; Prozialeck et al., 2006; Cerullia et al., 2006). Prabhakaran et al. (2007) point out that cadmium ions are released into the atmosphere via erosion, abrasion, and volcanic eruption. Kirk and Othmer (1982) indicate that cadmium may be found in water, soil, and food.

Phenols and their derivatives are persistent environmental pollutants in water streams, and some of them are carcinogenic. Anaissi and Toma (2005) have developed a bentonite-vanadium (V) oxide xerogel for the incorporation and detection of catechol. These authors attempted to confine the catechol species, by intercalation into a hybrid vanadium (V) oxide xerogel (VXG) + bentonite material (Anaissi et al., 1999, 2001). They point out that this

VXG + bentonite material is laminar, and is readily obtained by mixing polyvanadic acid and colloidal bentonite solutions, and forms a fluorescent green solid.

Vancura et al. (2007) have recently developed a novel resonant cantilever system for the detection of volatile organic compounds in liquid environments. These authors have coated their cantilevers with chemical sensitive polymers to detect low concentrations of toluene, xylene, and ethylbenzene in deionized water. They have been able to detect analytes in concentrations as low as single ppm. Their resonant sensor functions essentially as a balance. The resonance frequency depends on the cantilever mass loading. These authors explain that when the resonant cantilever is operated under liquid conditions, the liquid environment also needs to be taken into account besides the properties of the cantilever and the chemically sensitive coatings. The chemically sensitive coating consists of a polymer that adsorbs volatile organic carbons (VOCs) from the surrounding liquid. It is the adsorption of the analyte in the polymer matrix that leads to the increase in the mass loading on the cantilever surface.

Raguse et al. (2007) have developed a gold nanoparticle chemiresistor for the direct sensing of organics in an aqueous electrolyte solution. These authors point out that chemiresistors offer advantages for use as chemical sensors since they are easy to fabricate, easy to use, and are able to offer high sensitivity, good analyte discrimination, and have low operating cost. They point out that they are the first to demonstrate that high resistance (mega ohms) nanoparticle-based chemiresistors may be used to detect organic analytes in ionically conducive aqueous solutions.

Shen et al. (2007) have recently used fluorescence signaling DNA enzymes (OA-II, OA-III, and OA-IV) entrapped in sol-gel materials for the sensing of divalent metal ions such as Mg^{2+}, Ni^{2+}, Mn^{2+}, and Cd^{2+}. The OA-II, OA-III, and OA-IV DNA enzymes were entrapped in various silanes (diglycerylsilane (DGS), sodium silicate (SS), tetramethoxysilane (TMOS), and methyltrimethoxysilane (MTMOS) mixtures). These three DNAzymes were successfully immobilized within a series of sol-gel-derived materials. These authors explain that the DNAzymes are so designed that the binding of the appropriate metal ions induces the formation of a catalytic site. This catalytic site cleaves a ribonucleotide linkage within a DNA substrate. Upon catalytic cleavage at the RNA linkage there is a generation of fluorescence which may be measured. The authors report that the maximum sensitivity was obtained with composite mixtures of approximately 40% MTMS and approximately 60% TMOS.

Critterio et al. (2007) have recently developed a pH-independent fluorescent chemosensor for the selective detection of lithium ions. Even though this is not a toxin or pollutant application *per se*, it is worthwhile mentioning this here. These authors point out that lithium salts are used as pharmaceutically active compounds against manic-depressive psychosis. Their optode design is based on the synthesis of a novel lithium fluorionophore, KLI-1, and its polymer, KLI-2. These authors explain that lithium-containing medications may be taken by patients over a period of months or even years. They further point out that lithium ion concentrations have to be carefully monitored in the narrow range of 0.6-1.2 mM (Manji et al., 1995; Jope, 1999).

Ong et al. (2006) have recently developed a rapid and highly sensitive detection system for endotoxin. These authors explain that the early detection of endotoxin enables the administration of antibiotic therapy that prevents the onset of sepsis (Fink and Aranow, 1997). Sepsis, in common terms, is blood poisoning. It is a common blood stream infection, and involves the presence of bacteria, or other infectious organisms and other toxins in the blood. The National Vital Statistics Reports (1999) reports that sepsis is a common, expensive, and often fatal condition. Ong et al. (1999) explain that sepsis is the largest cause of mortality in ICUs (intensive care units). They point out that the early detection of sepsis significantly helps in improving patient survival.

Finally, Fonfria et al. (2007) have detected paralytic shellfish poisoning (PSP) toxins using a SPR-based biosensor. These authors emphasize that the early detection of these toxins is essential for human health preservation. Furthermore, PSP is a worldwide, algal-derived toxin that can cause serious food poisoning. They report that the neurotoxins responsible for PSP are saxitoxin (STX) and its derivatives. The authors point out that technical and ethical reasons have prompted the development of alternative assays to replace the original mouse bioassay. Fonfria et al. (2007) developed an inhibition assay that permitted making quantitative STX, decarbamoyl saxitoxin (dcSTX), gonyautoxin 2,3 (GTX2/3), gonyautoxin 5 (GTX5), and C 1,2 (C1/2) at concentrations of 2-50 ng/mL.

In this chapter we use fractal analysis to analyze the binding (and dissociation, if applicable) kinetics of (a) the binding (dose-response) of different concentrations (in mM) of phenol in solution to cells immobilized on a bio-MEMS based cell-chip (Yoo et al., 2007), (b) binding and dissociation of 0.88 mM hydrogen peroxide mixed with GC2 (*E. coli* strain) immobilized microcell chip (Yoo et al., 2007), (c) binding of catechol to bentonite-vanadium (V) oxide xerogels (Anaissi and Toma, 2005), (d) binding and dissociation of ethanol vapors in 40% RH (relative humidity) to a CTO (powdered sample of $Cr_{1.8}Ti_{0.2}O_3$; titanium substituted chromium oxide) thick film in a sol-gel-derived polycrystalline biosensor (Pokhrel et al., 2007), (e) and binding and dissociation of different concentrations of SEB in solution to the antibody-functionalized microbeads on a sensor chip (Haes et al., 2006). The fractal analysis may be considered as an alternative analysis for the kinetics observed in the different analyte-receptor reactions occurring on the biosensor surfaces presented above. In no way are we indicating that the present fractal analysis is better than any of the original kinetic analysis presented previously in the respective publications.

14.2 Theory

14.2.1 Single-Fractal Analysis

Binding Rate Coefficient

Havlin (1989) reports that the diffusion of a particle (analyte [Ag]) from a homogeneous solution to a solid surface (e.g., receptor [Ab]-coated surface) on which it reacts to form a product (analyte-receptor complex; (Ab·Ag)) is given by:

$$(\text{Ab}\cdot\text{Ag}) \approx \begin{cases} t^{(3-D_{\text{f,bind}})/2} = t^p, & t < t_c \\ t^{1/2}, & t > t_c \end{cases} \tag{14.1}$$

Here $D_{\text{f,bind}}$ or D_f is the fractal dimension of the surface during the binding step. t_c is the cross-over value. Havlin (1989) points out that the cross-over value may be determined by $r_c^2 \sim t_c$. Above the characteristic length, r_c, the self-similarity of the surface is lost and the surface may be considered homogeneous. Above time, t_c the surface may be considered homogeneous, since the self-similarity property disappears, and "regular" diffusion is now present. For a homogeneous surface where D_f is equal to 2, and when only diffusional limitations are present, $p = ½$ as it should be. Another way of looking at the $p = ½$ case (where $D_{\text{f,bind}}$ is equal to 2) is that the analyte in solution views the fractal object, in our case, the receptor-coated biosensor surface, from a "large distance." In essence, in the association process, the diffusion of the analyte from the solution to the receptor surface creates a depletion layer of width $(Đt)^{1/2}$ where $Đ$ is the diffusion constant. This gives rise to the fractal power law, (Analyte·Receptor) $\sim t^{(3-D_{\text{f,bind}})/2}$. For the present analysis, t_c is arbitrarily chosen and we assume that the value of the t_c is not reached. One may consider the approach as an intermediate "heuristic" approach that may be used in the future to develop an autonomous (and not time-dependent) model for diffusion-controlled kinetics.

Dissociation Rate Coefficient

The diffusion of the dissociated particle (receptor [Ab] or analyte [Ag]) from the solid surface (e.g., analyte [Ag]-receptor [Ab] complex coated surface) into solution may be given, as a first approximation by:

$$(\text{Ab}\cdot\text{Ag}) \approx -t^{(3-D_{\text{f,bind}})/2} = -t^p, \quad t > t_{\text{diss}} \tag{14.2}$$

Here $D_{\text{f,diss}}$ is the fractal dimension of the surface for the dissociation step. This corresponds to the highest concentration of the analyte-receptor complex on the surface. Henceforth, its concentration only decreases. The dissociation kinetics may be analyzed in a manner "similar" to the binding kinetics.

14.2.2 Dual-Fractal Analysis

Binding Rate Coefficient

Sometimes, the binding curve exhibits complexities and two parameters (k, D_f) are not sufficient to adequately describe the binding kinetics. This is further corroborated by low values of the r^2 factor (goodness-of-fit). In that case, one resorts to a dual-fractal analysis (four parameters; k_1, k_2, D_{f1}, and D_{f2}) to adequately describe the binding kinetics. The single-fractal analysis presented above is thus extended to include two fractal dimensions. At present, the time ($t = t_1$) at which the "first" fractal dimension "changes" to the "second" fractal dimension is arbitrary and empirical. For the most part, it is dictated by the data analyzed and experience gained by handling a single-fractal analysis. A smoother curve is obtained in the "transition" region, if care is taken to select the correct number of points for the two regions. In this case, the product (antibody-antigen; or analyte-receptor complex, Ab·Ag or analyte·receptor) is given by:

$$(\mathrm{Ab{\cdot}Ag}) \approx \begin{cases} t^{(3-D_{f1,\mathrm{bind}})/2} = t^{p1}, & t < t_1 \\ t^{(3-D_{f2,\mathrm{bind}})/2} = t^{p2}, & t_1 < t < t_2 = t_c \\ t^{1/2}, \quad t > t_c & \end{cases} \tag{14.3}$$

In some cases, as mentioned above, a triple-fractal analysis with six parameters (k_1, k_2, k_3, D_{f1}, D_{f2}, and D_{f3}) may be required to adequately model the binding kinetics. This is when the binding curve exhibits convolutions and complexities in its shape due perhaps to the very dilute nature of the analyte (in some of the cases to be presented) or for some other reasons. Also, in some cases, a dual-fractal analysis may be required to describe the dissociation kinetics.

14.3 Results

A fractal analysis is applied to the binding and dissociation (if applicable) kinetics of different analyte-receptor reactions occurring on different biosensor surfaces. Alternative expressions for fitting the data are available that include saturation, first-order reaction, and no diffusion limitations, but these expressions are apparently deficient in describing the heterogeneity that inherently exists on the surface. One might justifiably argue that the appropriate modeling may be achieved by using a Langmuirian or other approach. The Langmuirian approach may be used to model the data presented if one assumes the presence of discrete classes of sites (e.g., double exponential analysis as compared with a single-fractal analysis). Lee and Lee (1995) report that the fractal approach has been applied to surface science, for example, adsorption and reaction processes. These authors point out that the fractal approach provides a convenient means to represent the different structures and the morphology at the reaction surface. These authors also draw attention to the use of the fractal approach to develop optimal structures and as a predictive approach. Another advantage of the fractal technique is that the analyte-receptor association (as well as the dissociation reaction) is a complex reaction, and the fractal analysis via the fractal dimension and the rate coefficient provides a useful lumped parameter(s) analysis of the diffusion-limited reaction occurring on a heterogeneous surface.

In a classical situation, to demonstrate fractality, one should make a log-log plot, and one should definitely have a large amount of data. It may be useful to compare the fit to some other forms, such as exponential, or one involving saturation, etc. At present, no independent proof or physical evidence of fractals in the examples is presented. It is a convenient means (since it is a lumped parameter) to make the degree of heterogeneity that exists on the surface more quantitative. Thus, there is some arbitrariness in the fractal model to be presented. The fractal approach provides additional information about interactions that may not be obtained by conventional analysis of biosensor data.

There is no nonselective adsorption of the analyte. The present system (environmental pollutants in the aqueous or the gas phase) being analyzed may be typically very dilute. Nonselective adsorption would skew the results obtained very significantly. In these types of systems, it is imperative to minimize this nonselective adsorption. It is also recognized that, in some cases, this nonselective adsorption may not be a significant component of the adsorbed material and that this rate of association, which is of a temporal nature, would depend on surface availability. If the nonselective adsorption is to be accommodated into the model, there would be an increase in the heterogeneity on the surface, as, by its very nature, nonspecific adsorption is more heterogeneous than specific adsorption. This would lead to higher fractal dimension values since the fractal dimension is a direct measure of the degree of heterogeneity that exists on the surface.

Yoo et al. (2007) report that the use of living test organisms is a good and reliable method to analyze the toxicity of unknown samples. They point out that methods that incorporate bioluminescent bacteria are able to reliably detect unknown toxicity in water, soil, and sediment samples. For example, Kim and Gu (2003), and Lee et al. (2005) were able to achieve high throughput toxicity classification using different *E. coli* strains in a single 96-well plate or the 386-well plate using a Luria-Bertani (LB)-agar matrix. You et al. (2004) have developed a photolithography-based process of cell immobilization using polyvinyl alcohol-styrylpyridinium (PVA-SbQ) (a water-soluble and negative photosensitive polymer). They have used this PVA-SbQ for bioluminescent bacteria, and have applied this to fabricating cellular patterns in the microfluidic chip for toxicity monitoring. The emitted light intensity of bioluminescent bacteria changed in response to the presence of chemicals. Thus, the bacteria may be used as a toxicity indicator.

Yoo et al. (2007) obtained a dose-dependent bioluminescent response (binding) of different concentrations of phenol in the 0-21.28 mM range in solution to the immobilized cells on the bio-MEMS based cell-chip. Figure 14.1a shows the dose-dependent response of the immobilized cells to 0.67 mM phenol in solution. A dual-fractal analysis is required to adequately describe the binding kinetics. The values of (a) the binding rate coefficient, k, and the fractal dimension, D_f, for a single-fractal analysis, and (b) the binding rate coefficients, k_1 and k_2, and the fractal dimensions, D_{f1} and D_{f2}, for a dual-fractal analysis are given in Table 14.1. Corel Quattro Pro 8.0 (1997) was used to fit the data. The Corel Quattro Pro 8.0 program provided the parameter values presented in Table 14.1. The values of the parameters presented in Table 14.1 are within 95% confidence limits. For example, for a dual-fractal analysis, the value of the binding rate coefficient, k_1, for 0.67 mM phenol in solution is 16.64 ± 0.3246. The 95% confidence limit indicates that the k_1 value lies between 16.315 and 16.96. This indicates that the values are precise and significant.

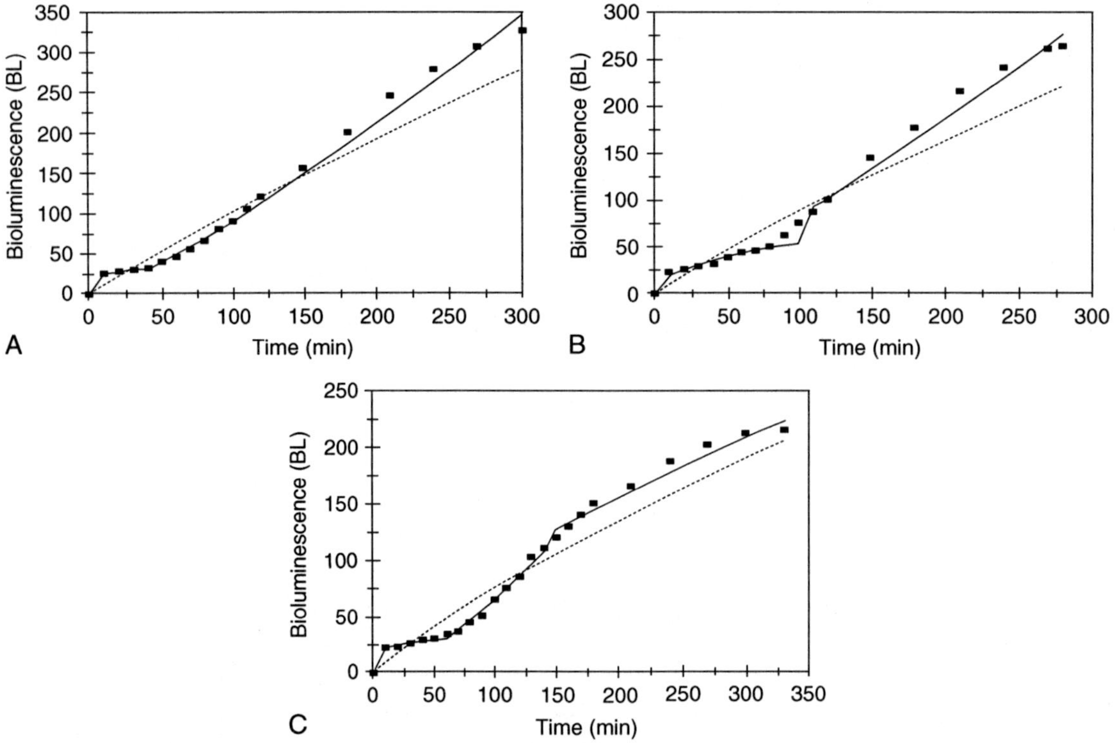

Figure 14.1

Binding (dose-response) of different concentrations (in mM) of phenol in solution to cells immobilized on a bio-MEMS based cell-chip (Yoo et al., 2007): (a) 0.67, (b) 1.33, (c) 2.66, (d) 5.32. When only a solid line (—) is used then a single-fractal analysis applies. When both a dashed (- - -) and a solid (—) line are used then the dashed line represents a single-fractal analysis and the solid line represents a dual-fractal analysis, and the dashed line represents the single-fractal analysis. In both cases, the solid line is the best fit line.

Table 14.1: Binding rate coefficients and fractal dimensions for different concentrations of phenol in solution to a bio-MEMS based cell chip (Yoo et al., 2007).

Analyte in Solution/ Receptor on Surface	k	k_1	k_2	k_3	D_f	D_{f1}	D_{f2}	D_{f3}
0.67 mM phenol/ cell chip	1.5102 ± 0.4616	16.64 ± 0.137	0.3246 ± 0.0231	na	1.1720 ± 0.1316	2.6494 ± 0.0157	0.5556 ± 0.0600	na
1.33 mM phenol/ cell chip	1.5379 ± 0.4928	6.7434 ± 0.917	0.3544 ± 0.0297	na	1.2390 ± 0.1388	2.0968 ± 0.1253	0.6386 ± 0.1310	na
2.66 mM phenol/ cell chip	1.5791 ± 0.4614	13.193 ± 0.635	0.06618 ± 0.0044	3.8005 ± 0.1800	1.3204 ± 0.1182	2.5884 ± 0.0739	0.0070 + 0.1583	1.5976 ± 0.0988
5.32 mM phenol/ cell chip	1.6189 ± 0.5180	12.135 ± 0.950	0.2060 ± 0.0230	1.4139 ± 0.1284	1.4800 ± 0.1280	2.6382 ± 0.0898	0.6702 ± 0.1154	1.3642 ± 0.2094

It is of interest to note that as the fractal dimension or the degree of heterogeneity on the biosensor chip surface decreases by a factor of 4.769 from a value of D_{f1} equal to 2.6494 to D_{f2} equal to 0.5556, the binding rate coefficient decreases by a factor of 51.26 from a value of k_1 equal to 16.64 to k_2 equal to 0.3246. The changes in the degree of heterogeneity or the fractal dimension on the biosensor chip surface and in the binding rate coefficient are in the same direction.

Figure 14.1b shows the dose-dependent response of the immobilized cells to 1.33 mM phenol in solution. Once again, a dual-fractal analysis is required to adequately describe the binding kinetics. The values of (a) the binding rate coefficient, k, and the fractal dimension, D_f, for a single-fractal analysis, and (b) the binding rate coefficients, k_1 and k_2, and the fractal dimensions, D_{f1} and D_{f2}, for a dual-fractal analysis are given in Table 14.1. It is of interest to note that as the fractal dimension or the degree of heterogeneity on the biosensor chip surface decreases by a factor of 3.283 from a value of D_{f1} equal to 2.0968 to D_{f2} equal to 0.6386, the binding rate coefficient decreases by a factor of 19.03 from a value of k_1 equal to 6.7434 to k_2 equal to 0.3544. Once again, changes in the degree of heterogeneity or the fractal dimension on the biosensor chip surface and in the binding rate coefficient are in the same direction.

Figure 14.1c shows the dose-dependent response of the immobilized cells to 2.66 mM phenol in solution. In this case, a triple-fractal analysis is required to adequately describe the binding kinetics. The values of (a) the binding rate coefficient, k, and the fractal dimension, D_f, for a single-fractal analysis, (b) the binding rate coefficients, k_1 and k_2, and the fractal dimensions, D_{f1} and D_{f2}, for a dual-fractal analysis, and (c) the binding rate coefficients, k_1, k_2, and k_3, and the fractal dimensions, D_{f1}, D_{f2}, and D_{f3}, for a triple-fractal analysis are given in Table 14.1. It is of interest to note that as the fractal dimension or the degree of heterogeneity on the biosensor chip surface changes, the binding rate coefficient too exhibits changes in the same direction. For example, as the fractal dimension decreases by a factor of 369.8 from a value of D_{f1} equal to 2.5884 to D_{f2} equal to 0.0072, the binding rate coefficient decreases by a factor of 199.35 from a value of k_1 equal to 13.193 to k_2 equal to 0.06618. Similarly, as the fractal dimension increases by a factor of 228.2 from a value of D_{f2} equal to 0.0070 to D_{f3} equal to 1.5976, the binding rate coefficient increases by factor of 57.43 from a value of k_2 equal to 0.06618 to k_3 equal to 3.8005.

Figure 14.1d shows the dose-dependent response of the immobilized cells to 5.32 mM phenol in solution. In this case, a triple-fractal analysis is required to adequately describe the binding kinetics. The values of (a) the binding rate coefficient, k, and the fractal dimension, D_f, for a single-fractal analysis, (b) the binding rate coefficients, k_1 and k_2, and the fractal dimensions, D_{f1} and D_{f2}, for a dual-fractal analysis, and (c) the binding rate coefficients, k_1, k_2, and k_3, and the fractal dimensions, D_{f1}, D_{f2}, and D_{f3}, for a triple-fractal analysis are given in Table 14.1. It is of interest to note that as the fractal dimension or the degree of heterogeneity on the biosensor chip surface changes, the binding rate coefficient too exhibits a change in the same direction. For example, as the fractal dimension decreases by a factor of 3.936 from a value of D_{f1} equal to 2.6382 to D_{f2} equal to 0.6702, the binding rate coefficient decreases by a factor of 58.91 from a value of k_1 equal to 12.135 to k_2 equal to 0.2060. Similarly, as the fractal dimension increases by a factor of 3.936 from a value of D_{f2} equal to 0.6702 to D_{f3} equal to 1.3642, the binding rate coefficient increases by a factor of 6.86 from a value of k_2 equal to 0.2060 to k_3 equal to 1.4139.

Figure 14.2a and Table 14.1 show the increase in the binding rate coefficient, k_1, with an increase in the fractal dimension, D_{f1}, for a dual-fractal analysis. For the data shown in Figure 14.2a, the binding rate coefficient, k_1, is given by:

$$k_1 = (0.6169 \pm 0.1007)D_{f1}^{3.225 \pm 0.7731} \qquad (14.4a)$$

The fit is reasonable. Only four data points are available. The availability of more data points would lead to a more reliable fit. The binding rate coefficient, k_1, is very sensitive to the

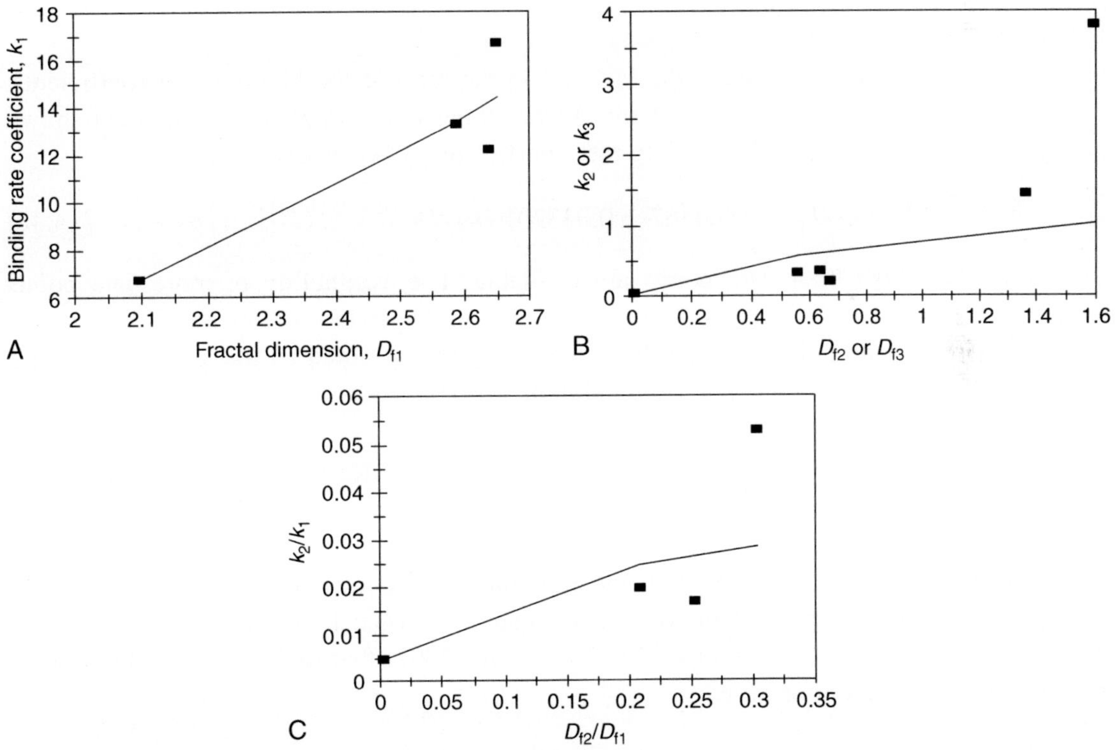

Figure 14.2

(a) Increase in the binding rate coefficient, k_1, with an increase in the fractal dimension, D_{f1}. (b) Increase in the binding rate coefficient, k_2 or k_3, with an increase in the fractal dimension, D_{f2} or D_{f3}. (c) Increase in the ratio k_2/k_1 with an increase in the ratio D_{f2}/D_{f1}.

fractal dimension, D_{f1}, or the degree of heterogeneity that exists on the biosensor surface as noted by the order of dependence between three and three and a half (equal to 3.225) exhibited.

Figure 14.2b and Table 14.1 show the increase in the binding rate coefficients, k_2 or k_3, with an increase in the fractal dimension, D_{f2} or D_{f3}, for a dual- and a triple-fractal analysis. For the data shown in Figure 14.2b, the binding rate coefficient, k_2 or k_3, is given by:

$$k_2 \text{ or } k_3 = (0.7649 + 1.2720)(D_{f2} \text{ or } D_{f3})^{0.5639 \pm 0.2167} \tag{14.4b}$$

The fit is poor. Seven data points are available. However, the data for k_2 and k_3 and D_{f2} and D_{f3} are plotted together. This may be the reason for the poor fit. The binding rate coefficients, k_2 and k_3, exhibit a low and close to one half (equal to 0.5638) order of dependence on the fractal dimension, D_{f2} and D_{f3}. The fractal dimensions, D_{f2} and D_{f3}, and the binding rate coefficients were plotted together in an attempt to see if there was some sort of continuity between these binding phases with regard to the surface heterogeneities on the biosensor surfaces.

Figure 14.2c and Table 14.1 show the increase in the ratio of the binding rate coefficients, k_2/k_1, with an increase in the ratio of the fractal dimensions, D_{f2}/D_{f1}. For the data shown in Figure 14.2c, the ratio of the binding rate coefficients, k_2/k_1, is given by:

$$k_2/k_1 = (0.04367 \pm 0.03320)(D_{f2}/D_{f1})^{0.3722 \pm 0.1435} \tag{14.4c}$$

The fit is poor. Only four data points are available. The availability of more data points would lead to a more reliable fit. The ratio of the binding rate coefficients, k_2/k_1, is only mildly sensitive to the ratio of fractal dimensions, D_{f2}/D_{f1}, as noted by the less than one half (equal to 0.3722) order of dependence exhibited.

Yoo et al. (2007) immobilized the bioluminescent bacteria, DK1 to a microcell chip to analyze the specificity of, for example, an oxidative toxicant. Figure 14.3 shows binding and dissociation of 0.88 μM hydrogen peroxide mixed with LB medium to the GC2 immobilized microcell chips. A 10-fold increase in the bioluminescence was observed when 0.88 μM hydrogen peroxide was present in solution compared to when it was absent. This 10-fold increase in the bioluminescence proved to Yoo et al. (2007) that their microbioluminescent bacteria chip could monitor toxicity.

A single-fractal analysis is adequate to describe the binding and the dissociation kinetics. The values of (a) the binding rate coefficient, k, and the fractal dimension, D_f, for a single-fractal analysis, and (b) the dissociation rate coefficient, k_d, and the fractal dimension, D_{fd}, for the dissociation phase for a single-fractal analysis are given in Table 14.2.

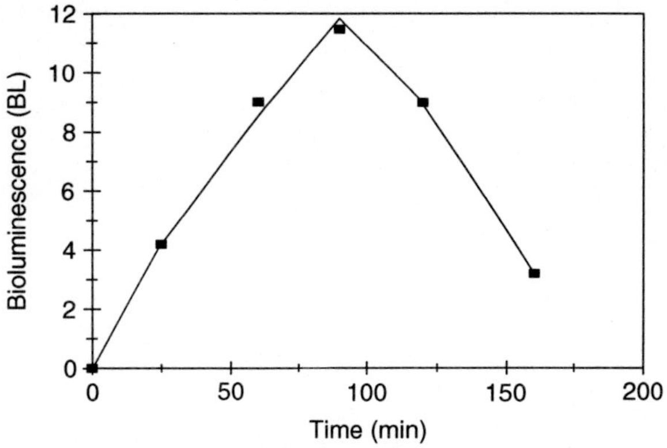

Figure 14.3

Binding and dissociation of 0.88 mM hydrogen peroxide mixed with LB (Luria-Bertani) medium to GC2 (*Escherichia coli* strain) immobilized microcell chips (Yoo et al., 2007).

Table 14.2: Binding and dissociation rate coefficients and fractal dimensions for the binding and the dissociation phase for 0.88 μM hydrogen peroxide mixed with LB (Luria-Bertani) medium in solution to the GC2 immobilized microcell chips (Yoo et al., 2007).

Analyte in Solution/Receptor on Surface	k	k_d	D_f	D_{fd}
0.88 μM hydrogen peroxide + LB medium	0.3248 ± 0.1097	0.0203 ± 0	1.400 ± 0.1272	0.168 ± 0

In this case, the affinity, K (= k/k_d) value is equal to 16.06.

Anaissi and Toma (2005) have developed a catechol incorporation and detection method using bentonite-vanadium (V) oxide xerogels. Catechol (a derivative of phenol) is a water pollutant, and Anaissi and Toma (2005) point out that in aqueous solution the electrochemistry is complicated by secondary reactions (Papouchado et al., 1972, 1975; Ryan et al., 1980). Anaissi and Toma (2005) report that due to the above complications biosensors are currently being used to detect catechols in aqueous solution (Cosnier et al., 1998, 2001; Rodriguez and Rivas, 2002; Kim and Lee, 2003; Timur et al., 2004). Anaissi and Toma (2005) have analyzed the intercalation of catechol into bentonite-vanadium (V) oxide xerogels (BV), and have characterized the resulting material electrochemically and spectroscopically.

Anaissi and Toma (2005) point out that for dilute solutions (e.g., less than 10^{-4} mol/dm^3) the intercalation process proceeds slowly. These authors monitored the kinetics (of binding of catechol in solution to the BV optode) spectrophotometrically by measuring the absorbance at 660 nm versus time. Figure 14.4 shows the incorporation of 7.8×10^{-5} mol/dm^3 catechol in solution and its detection using bentonite-vanadium (V) oxide xerogels. A dual-fractal analysis is required to adequately describe the binding kinetics. The values of (a) the binding

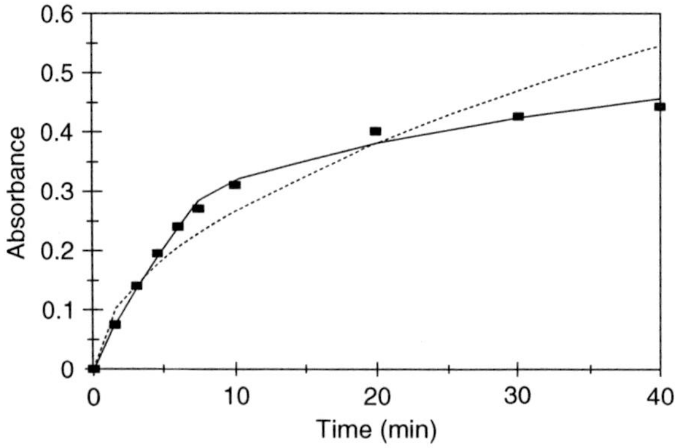

Figure 14.4

Binding of catechol to bentonite-vanadium (V) oxide xerogels (Anaissi and Toma, 2005). When only a solid line (—) is used then a single-fractal analysis applies. When both a dashed (- - -) and a solid (—) line are used then the dashed line represents a single-fractal analysis and the solid line represents a dual-fractal analysis, and the dashed line represents the single-fractal analysis. In both cases, the solid line is the best fit line.

Table 14.3: Binding and intercalation of dilute catechol solution (7.8 × 10^{-5} mol/dm^3) to the BV (bentonite-vanadium (V) oxide xerogel) (Anaissi and Toma, 2005).

Analyte in Solution /Receptor on Surface	k	k_1	k_2	D_f	D_{f1}	D_{f2}
7.8 × 10^{-5} mol/dm^3 catechol/ BV optode	0.8128 ± 0.0160	0.05583 ± 0.00258	0.1767 ± 0.0084	1.9692 ± 0.1174	1.3862 ± 0.0709	2.4884 ± 0.0894

rate coefficient, k, and the fractal dimension, D_f, for a single-fractal analysis, and (b) the binding rate coefficients, k_1 and k_2, and the fractal dimensions, D_{f1} and D_{f2}, for a dual-fractal analysis are given in Table 14.3.

It is of interest to note that as the fractal dimension or the degree of heterogeneity in the bentonite-vanadium (V) oxide xerogel increases by a factor of 1.795 from a value of D_{f1} equal to 1.3862 to D_{f2} equal to 2.4844 the binding rate coefficient increases by a factor of 3.165 from a value of k_1 equal to 0.05583 to k_2 equal to 0.1767. The changes in the degree of heterogeneity or the fractal dimension on the bentonite-vanadium (V) oxide xerogel surface and in the binding rate coefficient are in the same direction.

Pokhrel et al. (2007) indicate that solid-state sensors are very versatile and may be used to detect a wide variety of gases, besides which they may be used for other applications. These authors indicate that solid sate sensors are able to detect low ppm levels of gases, besides detecting high levels of combustible gases (Moos et al., 2003). Kanda and Maekawa (2005) point out that solid state sensors measure the change in the electrical resistance of an oxide film as a function of different gas concentrations. This is the functionality of these types of devices. Pokhrel et al. (2007) used sol-gel-derived polycrystalline $Cr_{1.8}Ti_{0.2}O_3$ thick films for alcohol sensing applications. These authors report that titanium-substituted chromium oxide, $Cr_{1.8}Ti_{0.2}O_3$ (CTO) is fabricated easily, demonstrates chemical stability to operating temperature, exhibits a resistance change that may be measured, and has a good gas response.

Pokhrel et al. (2007) obtained a powdered sample of CTO by the sol-gel method. The thick films were deposited (developed) on 4 mm length ceramic tubes. These ceramic tubes comprised of two Au-electrodes. An eight-layer film was prepared by mixing CTO with glass powder and α-terpinol.

Figure 14.5 shows the binding and dissociation (change in resistance) of ethanol at 40% RH to the TFM-850 [sintered at 850 °C; titanium substituted chromium oxide, CTO thick film] sensor. A dual-fractal analysis is required to adequately describe the binding kinetics. A single-fractal analysis is adequate to describe the dissociation kinetics. It is of interest to note that as the fractal dimension increases by a factor of 1.61 from a value of D_{f1} equal to 1.828 to D_{f2} equal to 2.948 the binding rate coefficient increases by a factor of 5.21 from a value of k_1 equal to 119.11 to k_2 equal to 620.27. Once again, changes in the degree of heterogeneity or the fractal dimension on the bentonite-vanadium (V) sol-gel surface and in the binding rate coefficient are in the same direction.

Pokhrel et al. (2007) analyzed the binding and dissociation kinetics of alcohol vapors to a TFE-850 ethanol sensor. Figure 14.6a shows the binding and dissociation kinetics observed when the TFE-850 ethanol sensor was exposed for 36 h to alcohol vapors. A dual-fractal

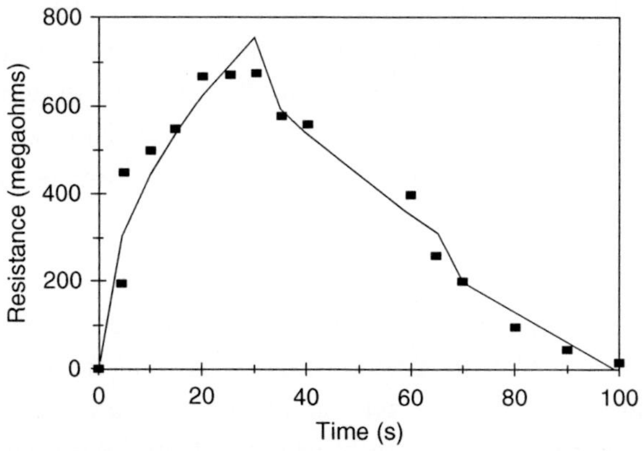

Figure 14.5
Binding and dissociation of ethanol vapors in 40% RH (relative humidity) to a CTO (powdered sample of $Cr_{1.8}Ti_{0.2}O_3$; titanium substituted chromium oxide) thick film in a sol-gel-derived polycrystalline biosensor (Pokhrel et al., 2007).

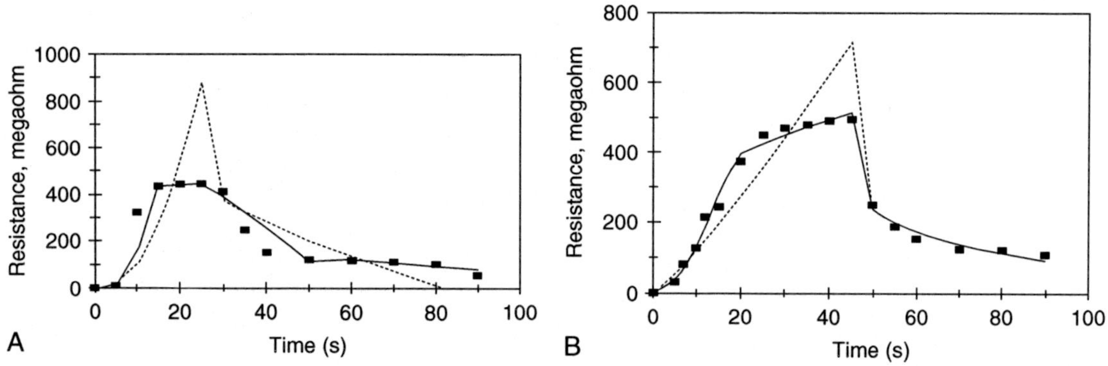

Figure 14.6

Binding and dissociation of alcohol vapors exposed for 36 and 24 h to an ethanol sensor (TFE-850) (Pokhrel et al., 2007): When only a solid line (—) is used then a single-fractal analysis applies. When both a dashed (- - -) and a solid (—) line are used then the dashed line represents a single-fractal analysis and the solid line represents a dual-fractal analysis, and the dashed line represents the single-fractal analysis. In both cases, the solid line is the best fit line.

analysis is required to describe the binding and the dissociation kinetics. The values of (a) the binding rate coefficient, k, and the fractal dimension, D_f, for a single-fractal analysis, and (b) the binding rate coefficients, k_1 and k_2, and the fractal dimensions, D_{f1} and D_{f2}, for a dual-fractal analysis, (c) the dissociation rate coefficient, k_d, and the fractal dimension, D_{fd}, for a single-fractal analysis, and (d) the dissociation rate coefficients, k_{d1} and k_{d2}, and the fractal dimensions D_{fd1} and D_{fd2}, for a dual-fractal analysis are given in Tables 14.4 and 14.5.

The estimated value of the fractal dimension in the binding phase is zero. This indicates that the TFE-850 sensor surface acts like a Cantor-like dust (Viscek, 1989). It is of interest to note that as the fractal dimension in the dissociation phase increases by a factor of 3.19 from a value of D_{fd1} equal to 0.8036 to D_{fd2} equal to 2.5608, the dissociation rate coefficient increases by factor of 14.56 from a value of k_{d1} equal to 9.8028 to k_{d2} equal to 142.27. Note that changes in the fractal dimension or the degree of heterogeneity on the TFE-850 ethanol sensor surface in the dissociation phase and in the dissociation rate coefficient are in the same direction.

Figure 14.6b shows the binding and dissociation kinetics observed when thee TFE-850 ethanol sensor was exposed for 24 h to alcohol vapors. A dual-fractal analysis is required to describe the binding kinetics. A single-fractal analysis is adequate to describe the dissociation kinetics. The values of (a) the binding rate coefficient, k, and the fractal dimension, D_f, for a single-fractal analysis, and (b) the binding rate coefficients, k_1 and k_2, and the fractal dimensions, D_{f1} and D_{f2}, for a dual-fractal analysis, and (c) the dissociation rate coefficient, k_d, and the fractal dimension, D_{fd}, for a single-fractal analysis are given in Tables 14.4 and 14.5.

Table 14.4: Binding and dissociation rate coefficients for (a) ethanol in 40% relative humidity (RH) to a CTO ($Cr_{1.8}Ti_{0.2}O_3$) thin film for sol-gel applications (TFM-850), and (b) alcohol vapors to an ethanol (TFE-850) sensor (Pokhrel et al., 2007).

Analyte in the Vapor Phase/Receptor	k	k_1	k_2	k_d	k_{d1}	k_{d2}
Ethanol in 40% RH/CTO thick film TFM-850 sensor	142.78 ± 42.88	119.10 ± 45.26	620.27 ± 0.867	24.922 ± 3.520	na	na
Alcohol vapors/ethanol (TFE-850) exposed to these vapors for 36 h	0.5277 ± 0.7966	0.03910 + 0.05578	374.01 ± 9.497	22.705 ± 13.699	9.8028 ± 6.726	142.72 ± 7.32
Alcohol vapors/ethanol (TFE-850) exposed to these vapors for 24 h	7.649 ± 2.789	1.5958 ± 0.3263	156.69 ± 7.79	186.57 ± 9.49	na	na

Table 14.5: Fractal dimensions for the binding and the dissociation phase for (a) ethanol in 40% relative humidity (RH) to a CTO ($Cr_{1.8}Ti_{0.2}O_3$) thin film for sol-gel applications (TFM-850), and (b) alcohol vapors to an ethanol (TFE-850) sensor (Pokhrel et al., 2007).

Analyte in the Vapor Phase/Receptor	D_f	D_{f1}	D_{f2}	D_{fd}	D_{fd1}	D_{fd2}
Ethanol in 40% RH/CTO thick film TFM-850 sensor	2.0174 ± 0.2834	1.8286 ± 0.4894	2.9482 ± 0.00974	1.4934 ± 0.1010	na	na
Alcohol vapors/ethanol (TFE-850) exposed to these vapors for 36 h	0 + 1.4478	0 + 2.2566	2.8847 ± 0.0073	1.5236 ± 0.3972	0.8036 ± 0.6822	2.5608 ± 0.2092
Alcohol vapors/ethanol (TFE-850) exposed to these vapors for 24 h	0.6126 ± 0.2680	0 + 0.4258	2.3772 ± 0.1434	2.5964 ± 0.05380	na	na

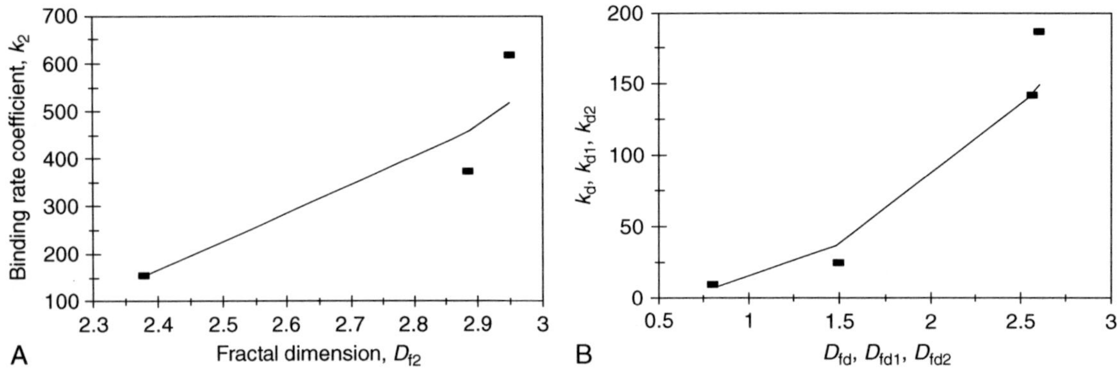

Figure 14.7

(a) Increase in the binding rate coefficient, k_2, with an increase in the fractal dimension, D_{f2}. (b) Increase in the dissociation rate coefficients, k_d, k_{d1}, and k_{d2}, with an increase in the fractal dimensions, D_{fd}, D_{fd1}, and D_{fd2}, respectively.

Once again, the estimated value of the fractal dimension in the binding phase is zero. This indicates, once again, that the TFE-850 sensor surface acts like a Cantor-like dust (Viscek, 1989).

Figure 14.7a and Table 14.3 show the increase in the binding rate coefficient, k_2, with an increase in the fractal dimension, D_{f2}, for a dual-fractal analysis. For the data shown in Figure 14.7a the binding rate coefficient, k_2, is given by:

$$k_2 = (1.1588 \pm 0.3618) D_{f2}^{5.643 \pm 1.621} \tag{14.5a}$$

The fit is reasonable. Only three data points are available. The availability of more data points would lead to a more reliable fit. The binding rate coefficient, k_2, is very sensitive to the fractal dimension or the degree of heterogeneity that exists on the bentonite-vanadium (V) sol-gel surface as noted by the greater than five and one-half (equal to 5.643) order of dependence exhibited.

Figure 14.7b and Tables 14.4 and 14.5 show the increase in the dissociation rate coefficient, k_d, k_{d1}, and k_{d2}, with an increase in the fractal dimension, D_{fd}, D_{fd1}, and D_{fd2}. For the data shown in Figure 14.7b the dissociation rate coefficient, k_d, k_{d1}, and k_{d2}, is given by:

$$k_d, k_{d1}, \text{ and } k_{d2} = (13.894 \pm 6.021)(D_{fd}, D_{fd1}, \text{ and } D_{fd2})^{2.486 \pm 0.374} \tag{14.5b}$$

The fit is reasonable. Only four three data points are available. The availability of more data points would lead to a more reliable fit. It is owing to the lack of points that k_d, k_{d1}, and k_{d2} are plotted together on the same plot. The dissociation rate coefficients are sensitive to the degree of heterogeneity present on the bentonite-vanadium (V) sol-gel surface as noted by the close to two and a half (equal to 2.486) order exhibited on the fractal dimensions present in the dissociation phase.

Haes et al. (2006) recently reported that detection techniques for target proteins are in demand; especially those that are fast, sensitive, selective, and without interference from background materials. Immunoassays are widely used bioanalytical methods (Elkins, 1989; Diamindis and Christpoulos, 1996). Haes et al. (2006) point out that all immunoassays use antibodies as capture molecules for strong and specific binding to target antigens. They further explain that immunoassay techniques have been miniaturized into microchip platforms (Sato et al., 2002, 2004; Gao et al., 2005; Rubina et al., 2005; Herr et al., 2005; Phillips and Cheng, 2005).

Haes et al. (2006) carefully delineate the parameters that are involved in immunoassay techniques. These include assay simplicity, convenience, cost, total assay time, assay sensitivity, and reagent stability. These authors also point out the limitations of immunoassay techniques which include limitations in the binding rate coefficient in the antigen-antibody interaction, the presence of non-specific binding, the diffusion distance between the antigen (e.g., in solution) and the antibody (e.g., immobilized on the sensor surface), detector sensitivity, and assay time. These authors have developed a displacement immunoassay for the detection of SEB on a glass microchip. They used laser-induced fluorescence detection and electrokinetically controlled fluidic delivery. The authors further point out that a monoclonal antibody that is specific for SEB is covalently attached to the microbeads. These beads were trapped on a microchip using narrow pathways on custom-made microchip devices. The authors emphasize that an apparent field enhancement enrichment phenomenon was obtained due to varying field strengths within the nonuniform channels.

Haes et al. (2006) analyzed the binding and dissociation of 100 aM (attamols) to 100 nM SEB in solution to the antibody-functionalized microbeads on a microchip. The size of the beads had two constraints: minimizing back pressure and the time required for the immunoassay. The back pressure was minimized by minimizing the bead length in the microbeads. The second constraint was satisfied by making sure that the antibody-coated beads attain saturation in a "short" time scale. These authors noted that a bead length of around 200 μm satisfied both of the above constraints. Also, excellent sensitivity and dynamic range was attained.

Figure 14.8a shows the binding and dissociation of 1 nM SEB in solution to the antibody-functionalized microbeads on a sensor chip. The monoclonal antibody that is specific for SEB is covalently attached to the silica beads. A dual-fractal analysis is required to describe the binding and the dissociation kinetics. The values of (a) the binding rate coefficient, k, and the fractal dimension, D_f, for a single-fractal analysis, (b) the binding rate coefficients, k_1 and k_2, and the fractal dimensions, D_{f1} and D_{f2}, for a dual-fractal analysis, (c) the dissociation rate coefficient, k_d, and the fractal dimension, D_{fd}, for a single-fractal analysis, and (d) the dissociation rate coefficients, k_{d1} and k_{d2}, and the fractal dimensions, D_{fd1} and D_{fd2}, for a dual-fractal analysis are given in Tables 14.6 and 14.7. The zero value obtained for the fractal dimension, D_{f1}, for a dual-fractal analysis indicates that for this initial phase of binding the

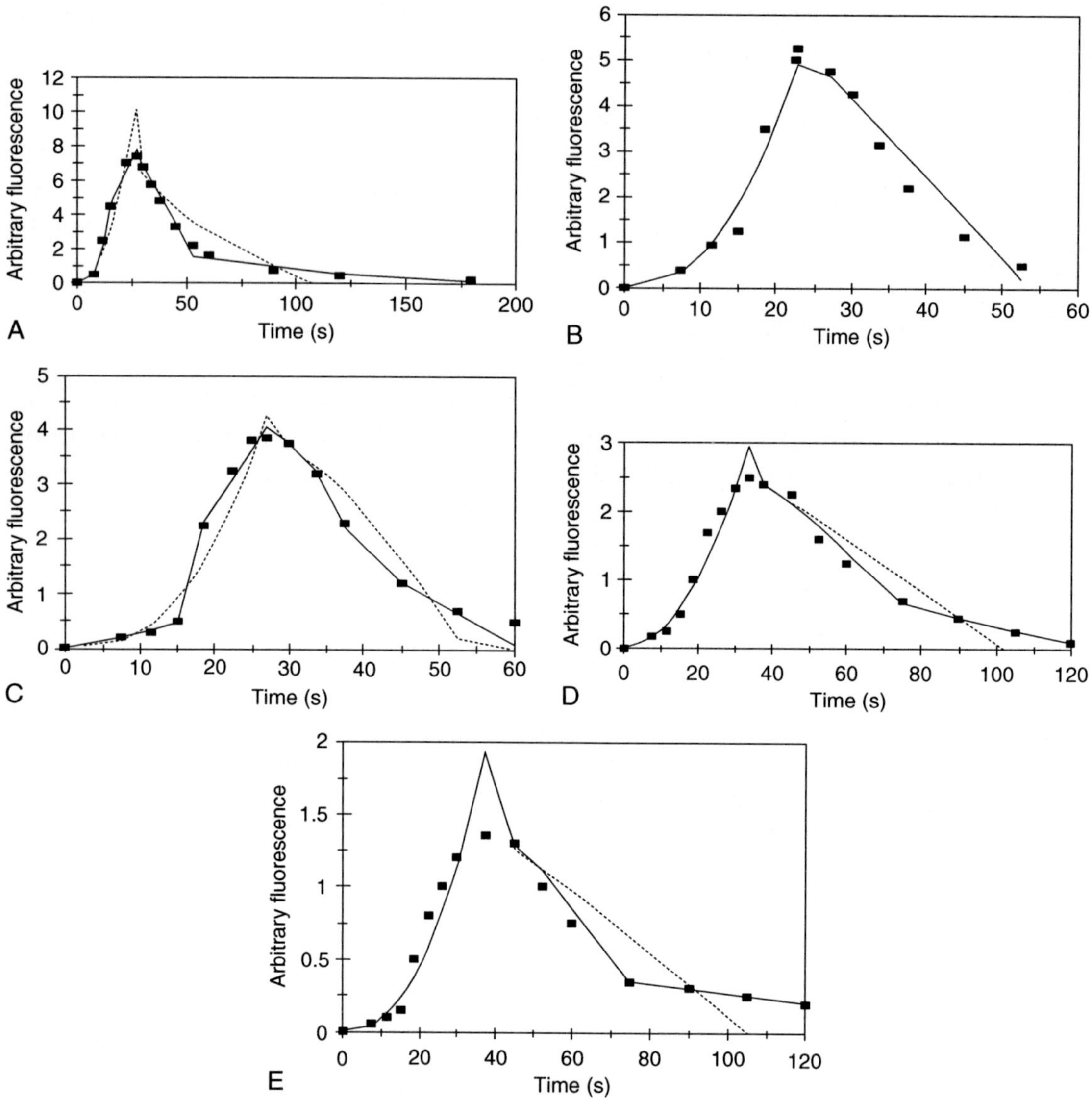

Figure 14.8

Binding and dissociation of different concentrations of *Staphylococcal enterotoxin B* (SEB) in solution to the antibody-functionalized microbeads on a sensor chip (Haes et al., 2006): (a) 1 nM, (b) 10 pM, (c) 100 fM, (d) 10 fM, (e) 1 fM. When only a solid line (—) is used then a single-fractal analysis applies. When both a dashed (- - -) and a solid (—) line are used then the dashed line represents a single-fractal analysis and the solid line represents a dual-fractal analysis, and the dashed line represents the single-fractal analysis. In both cases, the solid line is the best fit line.

Table 14.6: Binding and dissociation rate coefficients for different concentrations of *Staphylococcal enterotoxin B* (SEB) in solution to monoclonal antibody-functionalized microbeads on a sensor chip (Haes et al., 2006).

Analyte in Solution/ Receptor on Surface	k	k_1	k_2	k_d	k_{d1}	k_{d2}
1 nM SEB/monoclonal antibody-functionalized microbeads on sensor chip	0.0130 ± 0.00698	0.000827 ± 0.000250	0.4156 ± 0.0341	0.5201 ± 0.2078	0.2309 ± 0.0347	3.6949 ± 0.023
10 pM SEB/monoclonal antibody-functionalized microbeads on sensor chip	0.002975 ± 0.000725	na	na	0.1213 ± 0.0286	na	na
100 fM SEB/monoclonal antibody-functionalized microbeads on sensor chip	0.00638 ± 0.000314	0.01438 ± 0.00198	0.02969 ± 0.00198	0.02997 ± 0.01465	0.009114 ± 0.000876	0.4782 ± 0.0135
10 fM SEB/monoclonal antibody-functionalized microbeads on sensor chip	0.002653 ± 0.000655	na	na	0.02650 ± 0.00933	0.01598 ± 0.00524	0.4622 ± 0.002
1 fM SEB/monoclonal antibody-functionalized microbeads on sensor chip	0.000416 ± 0.00015	na	na	0.008844 ± 0.00577	0.001526 ± 0.000755	0.526 ± 0.0037

Table 14.7: Fractal dimensions for the binding and dissociation phase for different concentrations of *Staphylococcal enterotoxin B* (SEB) in solution to monoclonal antibody-functionalized microbeads on a sensor chip (Haes et al., 2006).

Analyte in Solution/Receptor on Surface	D_f	D_{f1}	D_{f2}	D_{fd}	D_{fd1}	D_{fd2}
1 nM SEB/monoclonal antibody-functionalized microbeads on sensor chip	0 + 0.8294	0 + 1.0696	1.2266 ± 0.3700	1.7860 ± 0.1866	1.0386 ± 0.1620	2.7294 ± 0.05224
10 pM SEB/monoclonal antibody-functionalized microbeads on sensor chip	0 + 0.4516	na	na	0.8058 ± 0.2232	na	na
100 fM SEB/monoclonal antibody-functionalized microbeads on sensor chip	0 + 0.7068	0.4265 + 0.5222	0.0126 + 0.4522	0.0346 + 0.3962	0 + 0.2044	1.8218 ± 0.1296
10 fM SEB/monoclonal antibody-functionalized microbeads on sensor chip	0 + 0.3260	na	na	0.8468 ± 0.2168	0.4136 ± 0.1549	2.2600 ± 0.0319
1 fM SEB/monoclonal antibody-functionalized microbeads on sensor chip	0 + 0.4426	na	na	2.6026 ± 0.4732	0 + 0.6832	2.6480 ± 0.02386

sensor chip surface acts like a Cantor-like dust (Viscek, 1989). As the fractal dimension in the dissociation phase increases by a factor of 2.63 from a value of D_{fd1} equal to 1.0386 to D_{fd2} equal to 2.7294, the dissociation rate coefficient increases by a factor of 16 from a value of k_{d1} equal to 0.2309 to k_{d2} equal to 3.6949. Changes in the fractal dimension or the degree of heterogeneity on the sensor chip surface and in the dissociation rate coefficient are in the same direction.

Figure 14.8b shows the binding and dissociation of 10 pM SEB in solution to the antibody-functionalized microbeads on a sensor chip. A single-fractal analysis is adequate to describe the binding and the dissociation kinetics. The values of (a) the binding rate coefficient, k, and the fractal dimension, D_f, for a single-fractal analysis, and (b) the dissociation rate coefficient, k_d, and the fractal dimension, D_{fd}, for a single-fractal analysis, are given in Tables 14.4 and 14.5. In this case, the affinity, K (= k/k_d), value is 0.0245. No reason is given here to indicate why at this lower SEB concentration (1 pM) a single-fractal analysis is adequate to describe the binding and dissociation kinetics, whereas at the higher SEB concentration (1 nM) a dual-fractal analysis is required to describe the binding and the dissociation kinetics. Once again, the zero value obtained for the fractal dimension, D_{f1}, for a dual-fractal analysis indicates that for this initial phase of binding the sensor chip surface acts like a Cantor-like dust (Viscek, 1989).

Figure 14.8c shows the binding and dissociation of 100 fM SEB in solution to the antibody-functionalized microbeads on a sensor chip. The monoclonal antibody that is specific for SEB is covalently attached to the silica beads. A dual-fractal analysis is required to describe the binding and the dissociation kinetics. The values of (a) the binding rate coefficient, k, and the fractal dimension, D_f, for a single-fractal analysis, (b) the binding rate coefficients, k_1 and k_2, and the fractal dimensions, D_{f1} and D_{f2}, for a dual-fractal analysis, (c) the dissociation rate coefficient, k_d, and the fractal dimension, D_{fd}, for a single-fractal analysis, and (d) the dissociation rate coefficients, k_{d1} and k_{d2}, and the fractal dimensions, D_{fd1} and D_{fd2}, for a dual-fractal analysis are given in Tables 14.6 and 14.7. In this case, during the initial phase of dissociation, the estimated value of the fractal dimension, D_{fd1}, is 0. This, once again, indicates that the sensor chip surface in this phase acts like a Cantor-like dust (Viscek, 1989). As the fractal dimension in the binding phase decreases by a factor of 33.85 from a value of D_{f1} equal to 0.4265 to D_{f2} equal to 0.0126, the binding rate coefficient increases by a factor of 2.06 from a value of k_1 equal to 0.0.01438 to k_2 equal to 0.02969. This is one of the few cases wherein the changes in the fractal dimension or the degree of heterogeneity on the sensor chip surface and in the binding rate coefficient are in the opposite directions. No reason is given, at present, to explain this.

Figure 14.8d shows the binding and dissociation of 10 fM SEB in solution to the antibody-functionalized microbeads on a sensor chip. A single-fractal analysis is adequate to describe the binding kinetics. A dual-fractal analysis is required to describe the dissociation kinetics. The values of (a) the binding rate coefficient, k, and the fractal dimension, D_f, for a single-fractal analysis, (b) the dissociation rate coefficient, k_d, and the fractal dimension, D_{fd}, for a single-fractal analysis, and (c) the dissociation rate coefficient, k_{d1} and k_{d2}, and the fractal dimensions D_{fd1} and D_{fd2}, for a dual-fractal analysis are given in Tables 14.6 and 14.7.

Figure 14.8e shows the binding and dissociation of 1 fM SEB in solution to the antibody-functionalized microbeads on a sensor chip. A single-fractal analysis is adequate to describe the binding kinetics. A dual-fractal analysis is required to describe the dissociation kinetics. The values of (a) the binding rate coefficient, k, and the fractal dimension, D_f, for a single-fractal analysis, (b) the dissociation rate coefficient, k_d, and the fractal dimension, D_{fd}, for a single-fractal analysis, and (c) the dissociation rate coefficient, k_{d1} and k_{d2}, and the fractal dimensions D_{fd1} and D_{fd2}, for a dual-fractal analysis are given in Tables 14.6 and 14.7.

Figure 14.9a and Tables 14.6 show the increase in the binding rate coefficient, k, with an increase in the SEB concentration (in fM) in solution for a single-fractal analysis. For the data shown in Figure 14.9a, the binding rate coefficient, k, is given by:

$$k = (0.00078 + 0.00149)[\text{SEB, in fM}]^{0.1681 \pm 0.1575} \quad (14.6a)$$

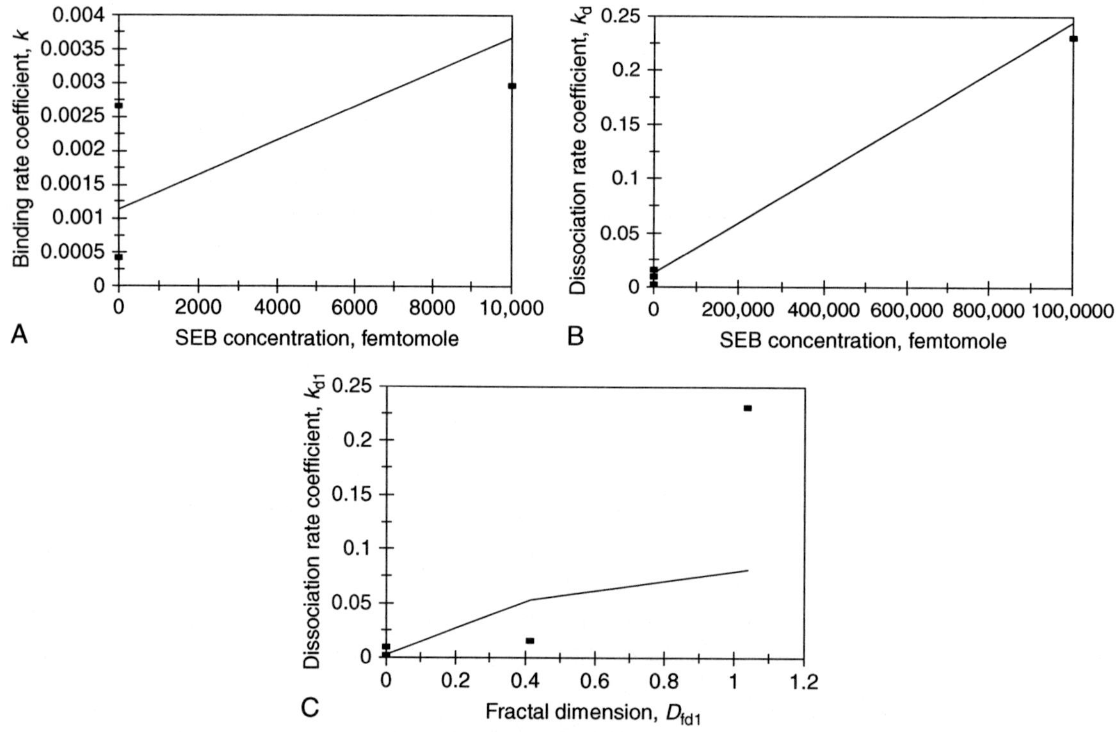

Figure 14.9

(a) Increase in the binding rate coefficient, k, with an increase in the *Staphylococcal enterotoxin B* (SEB) concentration in solution (in fM). (b) Increase in the dissociation rate coefficient, k_d, with an increase in the *Staphylococcal enterotoxin B* (SEB) concentration in solution (in fM). (c) Increase in the dissociation rate coefficient, k_{d1}, with an increase in the fractal dimension, D_{fd1}.

The fit is poor. Only three data points are available. The availability of more data points would lead to a more reliable fit. The poor fit is also reflected in the estimated value of the rate coefficient. The error is large, and thus only the positive value is given as the rate coefficient value cannot have a negative value. The binding rate coefficient, k, value exhibits only a very low (equal to 0.1681) order of dependence on the SEB concentration in solution in the range analyzed.

Figure 14.9b and Table 14.6 show the increase in the dissociation rate coefficient, k_{d1}, with an increase in the SEB concentration (in fM) in solution for a dual-fractal analysis. For the data shown in Figure 14.9b, the dissociation rate coefficient, k_{d1}, is given by:

$$k_d = (0.002823 + 0.00381)[\text{SEB, in fM}]^{0.3229 \pm 0.0815} \qquad (14.6b)$$

The fit is reasonable. The points are at extreme ends of the plot. Only four data points are available. The availability of more data points along with some points in the intermediate SEB concentration range would lead to a more reliable fit. The poor fit is also reflected in the estimated value of the rate coefficient. The error is large, and thus only the positive value is given as the rate coefficient value cannot have a negative value. The dissociation rate coefficient, k_d, value exhibits only a very low (equal to 0.3229) order of dependence on the SEB concentration in solution in the range analyzed.

Figure 14.9c and Tables 14.6 and 14.7 show the increase in the dissociation rate coefficient, k_{d1}, with an increase in the fractal dimension, D_{fd1}, for a dual-fractal analysis. For the data shown in Figure 14.9c, the dissociation rate coefficient, k_{d1}, is given by:

$$k_{d1} = (0.0797 + 0.258)D_{fd1}^{0.4547 \pm 0.2214} \qquad (14.6c)$$

The fit is very poor. Only three data points are available. The availability of more data may provide for a better fit. The very poor fit is also reflected in the estimated value of the rate coefficient. The error is large, and thus only the positive value is given as the rate coefficient value cannot have a negative value. The dissociation rate coefficient, k_{d1}, value exhibits close to one half (equal to 0.4547) order of dependence on the fractal dimension, D_{fd1}, present on the sensor chip surface in the initial dissociation phase.

Fonfria et al. (2007) have recently used a SPR-based biosensor in shellfish based matrixes to detect PSP toxins in contaminated shellfish. These authors point out that this is essential for human health conservation. Also, for ethical and technical reasons an alternative (detection method) to the mouse array is required. These authors used an inhibition assay using an anti-GTX2/3 antibody (GT-13-A) and a STX-CM5 chip. These authors were able to make quantitative the amounts of STX, dcSTX, GTX2/3, decarbamyl gonyautoxin 2,3 (dcGTX 2/3), GTX5, and C1,2 (C1/2) at concentrations ranging from 2 to 50 ng/mL. These authors explain that the technology that they have developed can be used as a PSP screening assay. Furthermore, they state that their biosensor is sensitive enough to quantify these PSP toxins in the European regulatory range limit of 80 μg/100 g of shellfish meat.

Fonfria et al. (2007) report that more than 24 compounds have been identified as PSP toxins (Schantz et al., 1975; Shimuzu et al., 1981; Yasumoto and Murata, 1993; Oshima et al., 1995; Negri et al., 2003). Fonfria et al. (2007) point out that all of these compounds differ in the combinations of hydroxyl and sulfate substitutions located at four sites of a 3,4,6-trialkyl tetrahydropurine backbone. The mechanism of action of all of these compounds is the same: the inhibition of the voltage-gated sodium channels in excitable cells (Kao, 1966; Catterall, 1980). Gessner and Middaugh (1995) and Rodriguez et al. (1990) report that neurological symptoms in humans are induced by the blockage of neuronal transmission. These symptoms include perioral paresthesia, dizziness, paralysis, and even respiratory illness and death. Paresthesia is an abnormal sensation of the skin. This may include numbness, tingling, burning, or creeping on the skin with no specific cause. Basically, it is an abnormal feeling.

Fonfria et al. (2007) have used a GT13-A-STX chip to detect PSP toxins in an inhibition assay using a SPR biosensor. These authors detected STX in solution by competition for binding to the GT13A-antibody also in solution with the STX immobilized to the chip surface. They report that the presence of STX in solution previously mixed with the GT13-A antibody inhibited the binding of the antibody to the STX chip surface.

Figure 14.10a shows the binding of the antibody control (0 ng/mL) to the STX sensor chip in a Biacore Q biosensor. A dual-fractal analysis is required to adequately describe the binding kinetics. The values of (a) the binding rate coefficient, k, and the fractal dimension, D_f, for a single-fractal analysis, and (b) the binding rate coefficients, k_1 and k_2, and the fractal dimensions, D_{f1} and D_{f2}, for a dual-fractal analysis are given in Tables 14.6 and 14.7. It is of interest to note that, once again, as the fractal dimension increases by a factor of 3.13 from a value of D_{f1} equal to 0.9576 to D_{f2} equal to 3.0 (maximum value), the binding rate coefficient increases by a factor of 63.34 from a value of k_1 equal to 9.473 to k_2 equal to 600. An increase in the fractal dimension or the degree of heterogeneity on the biosensor surface leads to an increase in the binding rate coefficient.

Figure 14.10b shows the binding of the antibody control (5 ng/mL) to the STX sensor chip in a Biacore Q biosensor. A dual-fractal analysis is, once again, required to adequately describe the binding kinetics. The values of (a) the binding rate coefficient, k, and the fractal dimension, D_f, for a single-fractal analysis, and (b) the binding rate coefficients, k_1 and k_2, and the fractal dimensions, D_{f1} and D_{f2}, for a dual-fractal analysis are given in Table 14.8. It is of interest to note that, once again, that as the fractal dimension increases by a factor of 2.87 from a value of D_{f1} equal to 1.0446 to D_{f2} equal to 3.0 (maximum value), the binding rate coefficient increases by a factor of 54.99 from a value of k_1 equal to 8.183 to k_2 equal to 450. Once again, an increase in the fractal dimension or the degree of heterogeneity on the biosensor surface leads to an increase in the binding rate coefficient.

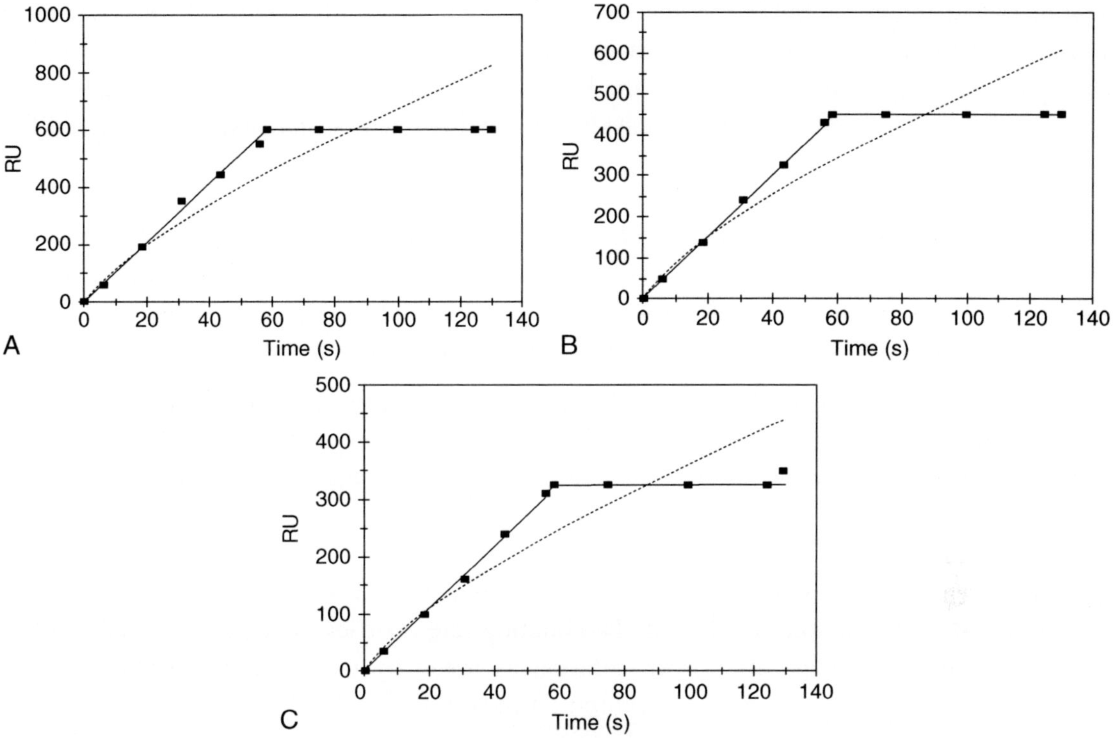

Figure 14.10

Binding of different concentrations of STX (in ng/mL) in solution premixed with GT13A-antibody to the STX immobilized on a chip surface in a Biacore Q SPR biosensor (Fonfria et al., 2007): (a) 0, (b) 5, (c) 10. When only a solid line (—) is used then a single-fractal analysis applies. When both a dashed (- - -) and a solid (—) line are used then the dashed line represents a single-fractal analysis and the solid line represents a dual-fractal analysis, and the dashed line represents the single-fractal analysis. In both cases, the solid line is the best fit line.

Table 14.8: Binding rate coefficients and fractal dimensions for paralytic shellfish poisoning (PSP) toxins in solution to surface plasmon resonance (SPR)-based biosensor in shellfish matrices (Fonfria et al., 2007).

STX Concentration (ng/mL)	k	k_1	k_2	D_f	D_{f1}	D_{f2}
0	20.626 ± 6.087	9.473 ± 0.642	600 ± 0.0	1.4850 ± 0.1835	0.9576 ± 0.0755	3.0 ± 7.5E-14
5	16.563 ± 4.489	8.183 ± 0.253	450 ± 0.0	1.520 ± 0.170	1.0446 ± 0.0351	3.0 ± 8.4E-14
10	11.336 ± 0.103	5.601 ± 0.224	325 ± 0.0	1.4966 ± 0.1718	1.1222 ± 0.0450	3.0 ± 9.0E-14

An anti-GTX/3 antibody immobilized on a saxitoxin (STX) CM5 chip.

Figure 14.10c shows the binding of the antibody control 10 ng/mL to the STX sensor chip in a Biacore Q biosensor. A dual-fractal analysis is, once again, required to adequately describe the binding kinetics. The values of (a) the binding rate coefficient, k, and the fractal dimension, D_f, for a single-fractal analysis, and (b) the binding rate coefficients, k_1 and k_2, and the fractal dimensions, D_{f1} and D_{f2}, for a dual-fractal analysis are given in Table 14.8. It is of interest to note that, once again, that as the fractal dimension increases by a factor of 2.67 from a value of D_{f1} equal to 1.1222 to D_{f2} equal to 3.0 (maximum value), the binding rate coefficient increases by a factor of 58.02 from a value of k_1 equal to 5.601 to k_2 equal to 325. Once again, an increase in the fractal dimension or the degree of heterogeneity on the biosensor surface leads to an increase in the binding rate coefficient.

Figure 14.11a and Table 14.8 show the decrease in the binding rate coefficient, k_1, with an increase in the STX concentration (in ng/mL) in solution for a dual-fractal analysis. For the data shown in Figure 14.11a, the binding rate coefficient, k_1, is given by:

$$k_1 = (7.476 \pm 2.046)[\text{STX, in ng/mL}]^{-0.0553 \pm 0.0449} \tag{14.7a}$$

The fit is reasonable. Only three data points are available. The availability of more data points would lead to amore reliable fit. The binding rate coefficient, k_1, is only very mildly sensitive to the STX concentration in solution as noted by the negative 0.0553 order of dependence exhibited on the STX concentration in solution in the 0-10 ng/mL range.

Figure 14.11b and Table 14.8 show the decrease in the binding rate coefficient, k_2, with an increase in the STX concentration (in ng/mL) in solution for a dual-fractal analysis. For the data shown in Figure 14.11b, the binding rate coefficient, k_2, is given by:

$$k_2 = (437.05 \pm 94.44)[\text{STX, in ng/mL}]^{-0.072 \pm 0.036} \tag{14.7b}$$

The fit is reasonable. Only three data points are available. The availability of more data points would lead to amore reliable fit. The binding rate coefficient, k_2, is only very mildly sensitive to the STX concentration in solution as noted by the negative 0.072 order of dependence exhibited on the STX concentration in solution in the 0-10 ng/mL range.

Figure 14.11c and Table 14.8 show the increase in the fractal dimension, D_{f1}, with an increase in the STX concentration (in ng/mL) in solution for a dual-fractal analysis. For the data shown in Figure 14.11c, the fractal dimension, D_{f1}, is given by:

$$D_{f1} = (1.0439 \pm 0.0441)[\text{STX, in ng/mL}]^{0.0194 \pm 0.00768} \tag{14.7c}$$

The fit is reasonable. Only three data points are available. The availability of more data points would lead to a more reliable fit. The fractal dimension, D_{f1}, exhibits close to a zero (equal to 0.0194) order of dependence on the STX concentration in solution in the 0-10 ng/mL range.

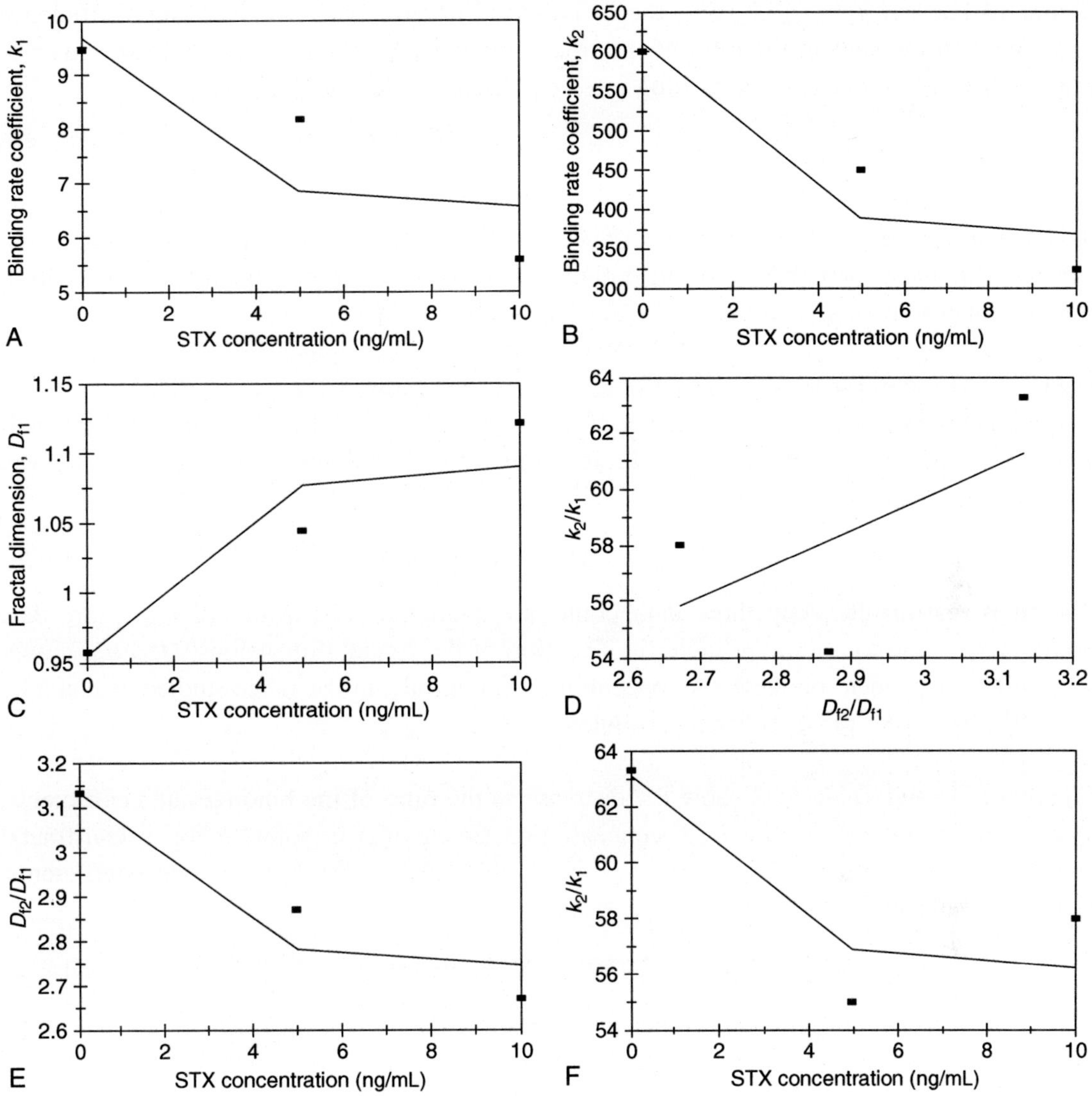

Figure 14.11

(a) Decrease in the binding rate coefficient, k_1, with an increase in the STX concentration (in ng/mL) in solution. (b) Decrease in the binding rate coefficient, k_2, with an increase in the STX concentration (in ng/mL) in solution. (c) Increase in the fractal dimension, D_{f1}, with an increase in the STX concentration (in ng/mL) in solution. (d) Increase in the binding rate coefficient ratio, k_2/k_1, with an increase in the ratio of the fractal dimensions, D_{f2}/D_{f1}. (e) Decrease in the ratio of the fractal dimension, D_{f2}/D_{f1}, with an increase in the STX concentration (in ng/mL) in solution. (f) Decrease in the ratio of the binding rate coefficients, k_2/k_1, with an increase in the STX concentration (in ng/mL) in solution.

Figure 14.11d and Table 14.8 show the increase in the ratio of the binding rate coefficients, k_2/k_1, with an increase in the ratio of the fractal dimensions, D_{f2}/D_{f1}. For the data shown in Figure 14.11d, the ratio of the binding rate coefficients, k_2/k_1, is given by:

$$k_2/k_1 = (1.092 \pm 0.349)(D_{f2}/D_{f1})^{0.5946 \pm 0.2834} \tag{14.7d}$$

There is scatter in the data. Only three data points are available. The availability of more data points would lead to a more reliable fit. The ratio of the binding rate coefficients, k_2/k_1, exhibits slightly greater than one half (equal to 0.5946) order of dependence on the ratio of fractal dimensions, D_{f2}/D_{f1}, present on the sensor chip surface.

Figure 14.11e and Table 14.8 show the decrease in the ratio of the fractal dimensions, D_{f2}/D_{f1}, with an increase in the STX concentration (in ng/mL) in solution for a dual-fractal analysis. For the data shown in Figure 14.11e, the ratio of the fractal dimensions, D_{f2}/D_{f1}, is given by:

$$D_{f2}/D_{f1} = (2.874 \pm 0.121)[\text{STX, in ng/mL}]^{-0.01943 \pm 0.0077} \tag{14.7e}$$

The fit is reasonable. Only three data points are available. The availability of more data points would lead to a more reliable fit. The ratio of the fractal dimensions, D_{f2}/D_{f1}, is only very mildly dependent on the STX concentration (in ng/mL) in the range studied as noted by the −0.01943 order of dependence exhibited.

Figure 14.11f and Table 14.8 show the decrease in the ratio of the binding rate coefficients, k_2/k_1, with an increase in the STX concentration (in ng/mL) in solution for a dual-fractal analysis. For the data shown in Figure 14.11f, the ratio of the binding rate coefficients, k_2/k_1, is given by:

$$k_2/k_1 = (58.461 \pm 2.779)[\text{STX, in ng/mL}]^{-0.0166 \pm 0.00863} \tag{14.7f}$$

The fit is reasonable. Only three data points are available. The availability of more data points would lead to a more reliable fit. The ratio of the fractal dimensions, k_2/k_1, is only very mildly dependent on the STX concentration (in ng/mL) in the range studied as noted by the −0.0166 order of dependence exhibited.

14.4 Conclusions

A fractal analysis is presented for the binding and dissociation (if applicable) of toxins and pollutants in solution to appropriate receptors immobilized on biosensor surfaces. Both single- and dual-fractal analysis are used to model the binding and dissociation kinetics. The dual-fractal analysis is used only when the single-fractal analysis does not provide an adequate fit. This was done by the regression analysis provided by Corel Quattro Pro 8.0 (1997).

The fractal analysis is used to analyze the binding (and dissociation, if applicable) kinetics of (a) the binding (dose-response) of different concentrations (in mM) of phenol in solution to cells immobilized on a bio-MEMS based cell-chip (Yoo et al., 2007), (b) binding and dissociation of 0.88 mM hydrogen peroxide mixed with GC2 (*E. coli* strain) immobilized microcell chip (Yoo et al., 2007), (c) binding of catechol to bentonite-vanadium (V) oxide xerogels (Anaissi and Toma, 2005), (d) binding and dissociation of ethanol vapors in 40% RH to a CTO (powdered sample of $Cr_{1.8}Ti_{0.2}O_3$; titanium substituted chromium oxide) thick film in a sol-gel-derived polycrystalline biosensor (Pokhrel et al., 2007), (e) binding and dissociation of different concentrations of SEB in solution to the antibody-functionalized microbeads on a sensor chip (Haes et al., 2006).

The fractal analysis is used to provide a better understanding of the kinetics of reactions (involving pollutants and toxins in solution), and to relate the binding and the dissociation rate coefficients with the fractal dimension or the degree of heterogeneity that exists on the sensor chip surface. The fractal analysis provides a quantitative indication of the state of disorder (fractal dimension) and the binding (and dissociation) rate coefficient values on the biosensor surface. In accord with the prefactor analysis for fractal aggregates (Sorenson and Roberts, 1997), quantitative (predictive) expressions are developed for (a) the binding rate coefficient, k_1, as a function of the fractal dimension, D_{f1}, for the dose-dependent response of immobilized cells to different phenol concentrations (in mM) in solution (Yoo et al., 2007), (b) the ratio of the binding rate coefficients, k_2/k_1, as a function of the ratio of the fractal dimensions, D_{f2}/D_{f1}, for the dose-dependent response of immobilized cells to different phenol concentrations (in mM) in solution (Yoo et al., 2007), (c) the binding rate coefficient, k_2, as a function of the fractal dimension, D_{f2}, for the binding and dissociation kinetics of alcohol vapors to a TFE-850 sensor (Pokhrel et al., 2007), (d) the dissociation rate coefficient, k_d, k_{d1}, and k_{d2}, as a function of the fractal dimensions, D_{fd}, D_{fd1}, and D_{fd2}, respectively, for the dissociation kinetics of alcohol vapors to a TFE-850 sensor (Pokhrel et al., 2007), (e) the binding rate coefficient, k, as a function of the SEB concentration (in fM) in solution (Haes et al., 2006), and (f) the dissociation rate coefficient, k_d, as a function of the SEB concentration (in fM) in solution (Haes et al., 2006).

The fractal dimension is not a classical independent variable such as analyte concentration in solution. Nevertheless, the expressions obtained for the binding (and the dissociation) rate coefficients for a single- and for a dual-fractal analysis as a function of the fractal dimension are valuable as they provide a means by which these rate coefficients may be manipulated by changing the degree of heterogeneity or the fractal dimension on the sensor chip surface.

The present analysis is applicable to other pollutant/toxin reactions occurring on different types of biosensor surfaces. Here it hoped that the kinetic analysis (albeit using fractals) would help shed some physical insights into these types of interactions occurring on biosensor surfaces. Pollution is a serious problem and the early detection of the different types of toxins and pollutants will go a long away to suppress the different types of diseases caused by them, along with other diseases that they may be indirectly involved in. Thus, there is a need to develop more and more sensitive sensors to help detect these harmful toxins and pollutants before they really become a major problem. It has to be recognized that the decontamination of pollutants and toxins from waste-water streams, or regular drinking sources is, in general, time-consuming, resource intensive, and very expensive. The sooner this problem is recognized, the sooner it can be arranged to have it rectified to meet the required levels set by the different governmental, state and other agencies.

References

Anaissi FJ and HE Toma, Catechol incorporation and detection using bentonite-vanadium (V) oxide xerogels, *Sensors & Actuators B*, **110**, 175–180 (2005).

Anaissi FJ, GJF Demets, HE Toma, and ACV Coehlo, Modified electrodes based on mixed bentonite vanadium (V) oxide xerogels, *Journal of Electroanalytical Chemistry*, **464**, 48–53 (1999).

Anaissi FJ, GJF Demets, HE Toma, S Dovidauskas, and ACV Coehlo, Characterization and properties of mixed bentonite-vanadium (V) oxide xerogels, *Materials Research Bulletin*, **36**, 289–306 (2001).

Arvidson B, A review of axonal transport of metals, *Toxicology*, **88**, 1–14 (1994).

Buzby JC, T Roberts, CT Lin, and J MacDonald, Agricultural Economic Report No. 741, Economic Research Service, U.S. Department of Agriculture, Washington, DC (1996).

Catterall WA, *Annual Reviews in Pharmacology and Toxicology*, **20**, 15–43 (1980).

Cerullia N, L Campanellab, R Grossib, L Politic, R Scandurrac, S Giovanni, F Galloe, S Damianie, A Alimontif, F Petruccif, and SJ Carolif, *Trace Elements in Medicine and Biology*, **20**, 171–179 (2006).

Corel Quattro Pro 8.0, Corel Corporation Limited, Ottawa, Canada, 1997.

Cosnier S, A Lepellec, I Guidetti, and I Rico-Lattes, Enhancement of biosensor sensitivity in aqueous and organic solvents using a combination of poly(pyrrole-ammonium) and poly(pyrrole-lactobionamide) films as host matrices, *Journal of Electroanalytical Chemistry*, **449**, 165–171 (1998).

Cosnier S, S Szunerits, RS Marks, JP Lellouche, and K Perie, Mediated electrochemical detection of catechol by tyropsinase-based poly(dicarbazole) electrodes, *Journal of Biochemical and Biophysical Methods*, **50**, 65–67 (2001).

Critterio D, J Takeda, M Kosugi, H Hisamoto, S Susaki, H Komatsu, and K Suzuki, pH-independent fluorescent chemosensor for highly sensitive selective lithium sensing, *Analytical Chemistry*, **79**, 1237–1242 (2007).

Dawson ED, CL Moore, JA Smagala, DM Dankbar, M Mehlmann, MB Townsend, CB Smith, NJ Cox, D Kuchta, and KL Rowlen, M Chip: A tool for influenza surveillance, *Analytical Chemistry*, **78**(22), 7610–7615 (2006).

Diamindis EP and TK Christpoulos, *Immunoassay*, Academic Press, San Diego, CA, 1996.

Doleman L, L Davies, L Rowe, EA Moschou, S Deo, and S Daunert, Bioluminescene DNA hybridization assay for *Plasmodium falciparum* based on the photporotein aequorin, *Analytical Chemistry*, **79**, 4149–4153 (2007).

Elkins RP, *Journal of Pharmaceutical and Biomedical Analysis*, **7**, 155–168 (1989).

Fink MP and JS Aranow, Gut barrier dysfunction and sepsis. In *Sepsis and Multiorgan Failure*, AM Fein, ED Abraham,, RA Balk,, GR Bernard,, RC Bone,, DR Dantzker,, MP Fink, (Eds.), Williams and Wilkins, Baltimore, MD, 1997, pp. 383–407.

Fonfria ES, N Vilarino, K Campbell, C Elliott, SA Haughey, B Ben-Gigirey, JM Vietes, K Kawatsu, and LM Botana, Paralytic shellfish poisoning detection by surface plasmon resonance-based biosensors in shellfish matrixes, *Analytical Chemistry*, **79**, 6303–6311 (2007).

Gao Y, FYH Lin, G Hu, PM Sherman, and D Li, Development of a novel electrokinetically driven microfluidic immunoassay for the detection of *Heliobacter pylori*, *Analytica Chimica Acta*, **543**, 109–116 (2005).

Goel J, K Kadirvelu, C Rajagopal, and VK Garg, *Industrial and Engineering Chemistry Research*, **45**, 6531–6537 (2006).

Haes AJ, A Terray, and GE Collins, Bead-assisted displacement immunoassay for *Staphylococcal enterotoxin B* on a microchip, *Analytical Chemistry*, **78**, 8412–8420 (2006).

Havlin S, Molecular diffusion and reactions in *The Fractal Approach to Heterogeneous Chemistry: Surfaces, Colloids, Polymers*, (ed. D Avnir), New York, pp. 251–269 (1989).

Herr AE, DJ Throckmorton, AA Davenport, and AK Singh, On-chip native gel electrophoresis-based immunoassays for tetanus antibody and toxin, *Analytical Chemistry*, **77**, 585–590 (2005).

Jope RS, *Psychiatry*, **4**, 117–128 (1999).

Kanda K and T Maekawa, Development of a WO_3 thick-film-based sensor for the detection of VOC, *Sensors & Actuators B*, **108**, 97–101 (2005).

Kessler N, O Ferraris, K Palmer, W Marsh, and AJ Steel, Use of the DNA flow thru chip, a three-dimensional biochip, for typing and subtyping of influenza viruses, *Journal of Clinical Microbiology*, **42**, 2173–2185 (2004).

Kim BC and MB Gu, *Biosensors and Bioelectronics*, **18**, 1015–1021 (2003).

Kim MA and WY Lee, Amperometric phenol biosensor based on sol-gel silicate/Nafion composite film, *Analytica Chimia Acta*, **479**, 143–150 (2003).

Kirk RE and DF Othmer, *Encyclopedia of Chemical Technology*, Interscience, New York, NY, 1982.

Lee CK and SL Lee, Multi-fractal scaling analysis of reactions over fractal surfaces, *Surface Science*, **325**, 294–310 (1995).

Lee JH, RJ Mitchell, BC Kim, DC Cullen, and MB Gu, A cell array biosensor for environmental toxicity analysis, *Biosensors & Bioelectronics*, **21**, 500–507 (2005).

Li J, S Chen, and DH Evans, Typing and subtyping in influenza virus using DNA microarrays and multiplex reverse transcriptase PCR, *Journal of Clinical Microbiology*, **39**, 696–704 (2001).

Lofts S, D Spurgeon, and C Svendsen, *Environmental Science & Technology*, **39**, 8533–8540 (2005).

Mangold KA, RU Manson, ESC Koay, L Stephens, M Regner, RB Thomson Jr, LR Peterson, and KL Kaul, Real-time PCR for detection and identification of *Plasmodium* spp, *Journal of Clinical Microbiology*, **43**, 2435–2440 (2005).

Manji HK, WZ Potter, and RH Lenox, Signal transduction pathways: Molecular targets for lithium's actions, *Archives of General Psychiatry*, **52**, 531–543 (1995).

Martelli A, E Rousselet, C Dycke, A Bouron, and JM Moulis, *Biochimie*, **88**, 1707–1719 (2006).

Mead PS, L Slutsker, V Dietz, LF McCaig, JS Bresee, C Shapiro, PM Griffin, and RV Tauxe, *Emerging Infectious Diseases*, **5**, 607–625 (1999).

Mehlmann M, ED Dawson, MB Townsend, JA Smagala, CL Moore, CB Smith, NJ Cox, RD Kuchta, and KW Rowlen, Robust sequence selection method used to develop the fluchip diagnostic microarray for influenza virus, *Journal of Clinical Microbiology*, **44**, 2857–2862 (2006).

Moos R, F Rettig, A Hurland, and C Plog, Temperature independent resistive oxygen exhaust gas sensor for lean-burn engines in thick-film technology, *Sensors & Actuators B*, **93**, 43–50 (2003).

National Vital Statistics Reports, June **30**, 47, 1999.

Negri A, D Stirling, M Quilliam, S Blackburn, C Bolch, I Burton, G Eaglesham, K Thomas, J Walter, and R Willis, Three novel hydroxybenzoate saxitoxin analogues isolated form dinoflagellate *Gymnodium catenatum, Chemical Research and Toxicology*, **16**, 1029–1033 (2003).

Ong KG, JM Leland, K Zeng, G Barret, and M Zourob, A rapid highly-sensitive endotoxin detection system, *Biosensors & Bioelectronics*, **21**, 2270–2274 (2006).

Oshima Y, Chemical and enzymatic transformation of paralytic shellfish toxins in marine organisms. In P Lassus, G Arzul, P Gentien, and C Marcaillou, (Eds.), Lavoisier Publishers, Paris, France, 1995, pp. 475–480.

Papouchado L, G Petrie, and RN Adams, Anodic-oxidation pathways of phenolic compounds, 1, *Journal of Electroanalytical Chemistry*, **38**, 389–394 (1972).

Papouchado L, RW Sandford, G Petrie, and RN Adams, Anodic-oxidation pathways of phenolic compounds, 1, *Journal of Electroanalytical Chemistry*, **65**, 275–284 (1975).

Phillips KS and Q Cheng, Microfluidic immunoassay for bacterial toxins with supported phosphplipid bilayer membranes on poly(dimethylsiloxane), *Analytical Chemistry*, **77**, 327–334 (2005).

Pokhrel S, L Huo, H Zhao, and S Gao, Sol-gel derived polycrystalline $Cr_{1.8}Ti_{0.2}O_3$ thick film for alcohol sensing applications, *Sensors & Actuators B*, **120**, 560–567 (2007).

Prabhakaran D, M Yuehong, H Nanjo, and H Matsunaga, Naked-eye cadmium sensor: Using chromoionophore arrays of Langmuir-Blodgett molecular assemblies, *Analytical Chemistry*, **79**, 4056–4065 (2007).

Prozialeck WC, JR Edwards, and JM Woods, *Life Science*, **79**, 1493–1506 (2006).

Raguse B, E Chow, CS Barton, and L Wieczorek, Gold nanoparticle chemiresistor sensors: Direct sensing of organics in aqueous electrolyte solution, *Analytical Chemistry* (2007), published on the Web August 28, 2007.

Rodriguez DC, RA Etzel Sttall, E de Porras, OH Velasquez, RV Tauxe, EM Kilbourne, and PA Blake, *American Journal of Tropical Medicine and Hygiene*, **42**, 267–274 (1990).

Rodriguez MC and GA Rivas, Glassy carbon paste electrodes modified with polyphenol oxidase, *Analytica Chimica Acta*, **459**, 43–51 (2002).

Rubina AY, VI Dyukova, EI Dementiieva, AA Stomakhin, VA Nesmeyanov, EV Grishin, and AS Zasedatelev, Quantitative immunoassay of biotoxins on hydrogel-based protein microchips, *Analytical Biochemistry*, **340**, 317–329 (2005).

Ryan MD, A Yueh, and WY Chen, The electrochemical oxidation of substituted catechols, *Journal of the Electrochemical Society*, **127**, 1489–1495 (1980).

Sato K, M Yamanaka, H Takahashi, M Tokeshi, H Kimura, and T Kitamori, *Electrophoresis*, **23**, 734–739 (2002).

Sato K, M Yamanaka, T Hagino, M Tokeshi, H Kimura, and T Kitamori, *Lab Chip*, **4**, 570–575 (2004).

Schantz EJ, VE Ghazarossian, HK Schnoes, FM Strong, JP Springer, JO Pezzanite, and J Clardy, Structure of saxitoxins, *Journal of the American Chemical Society*, **97**, 1238–1239 (1975).

Sengupta S, K Onodera, A Lai, and U Melcher, Molecular detection and identification of influenza viruses by oligonucleotide microarray hybridization, *Journal of Clinical Microbiology*, **41**, 4542–4550 (2003).

Shen Y, G Mackey, N Rupcich, D Gloster, W Chiuman, Y Li, and JD Brennan, Entrapment of fluorescence signaling enzymes in sol-gel derived materials for metal-ion sensing, *Analytical Chemistry*, **79**, 3494–3503 (2007).

Shimuzu Y, H Chien-Ping, and AA Genenah, Structure of saxitoxins in solution and stereochemistry of dihydrosaxitoxins, *Journal of the American Chemical Society*, **103**, 605–609 (1981).

Sorenson CM and GC Roberts, The prefactor of fractal aggregates, *Journal of Colloid and Interface Science*, **186**, 447–452 (1997).

Taylor AD, J Ladd, Q Yu, S Chen, J Homola, and S Jiang, Quantitative and simultaneous detection of our food borne bacterial pathogens with a multi-channel SPR sensor, *Biosensors & Bioelectronics*, (2006).

Thayer AM, *Chemical and Engineering News*, **83**(43), 69–82 (2005).

Timur S, N Pazarliogly, R Pilloton, and A Telefoncu, Thick film sensors based on lactases from different sources immobilized in polyaniline matrix, *Sensors & Actuators B*, **97**, 132–136 (2004).

Townsend MB, ED Dawson, M Mehlmann, JA Smagala, DM Dankbar, CL Moore, CB Smith, NJ Cox, RD Kuchta, and KL Rowlen, Experimental evaluation of the fluchip diagnostic microarray for influenza virus surveillance, *Journal of Clinical Microbiology*, **44**, 2863–2871 (2006).

Vancura C, Y Li, J Lichtenberg, KU Kirsten, A Hierlemann, and F Josse, *Analytical Chemistry*, **79**, 1646–1654 (2007).

Viscek T, *Fractal Growth Phenomena*, World Scientific Publishing Company, Singapore, 1989.

Walt DR and DR Franz, Biological warfare detection, *Analytical Chemistry*, **5**, 738A–746A (2000).

Wang C, Y Fang, S Peng, D Ma, and J Zhao, *Chemical Research and Toxicology*, **12**, 331–334 (1999).

Yasumoto T and M Murata, Marine toxins, *Chemical Reviews*, **93**, 1897–1909 (1993).

You SK, SS Yun, JH Lee, MB Gu, and JH Lee, *International Conference in Electrical Engineering*, **3**(2), 1048–1052 (2004).

Yoo SK, JH Lee, MB Gu, and JH Lee, Fabrication of a bio-MEMS cell-chip for toxicity monitoring, *Biosensors & Bioelectronics*, **22**, 1586–1592 (2007).

CHAPTER 15

Fractal Analysis of the Binding and Dissociation Kinetics of Protein Biomarkers and Other Medically Oriented Analytes on Biosensor Surfaces

Chapter Outline

15.1 Introduction

The detection of the biomarkers of different diseases is an important area of biosensor development, as the early detection of these biomarkers generally leads to a better prognosis of the disease. College Hill (2009) reports that "Biomarkers are valued tools used across the biological spectrum from research to diagnostics, as indicators of normal or disease processes or to assess pharmacological response." Recently this area of biosensor investigation has expanded considerably. Biosensors have also been used to detect medically-related analytes which, above a certain threshold level, indicate the early incidence or onset of certain diseases. In this chapter we analyze the kinetics of binding and dissociation (if applicable) of biomarkers of diseases and medically-related analytes, which indicate the early onset of certain diseases.

Some of the more recent studies that have appeared in the current literature include: (a) sensitive immunoassay of a biomarker tumor necrosis factor (TNF)-α based on poly(guanine)-functionalized silica nanoparticle label (Wang et al., 2006), (b) development of a screen-printed

Handbook of Biosensors and Biosensor Kinetics. DOI: 10.1016/B978-0-444-53262-6.00015-2

cholesterol biosensor (Shin and Liu, 2007), (c) a doubly amplified electrochemical assay for carcinoembryonic antigen (CEA) (Gao et al., 2009), (d) a cholesterol biosensor based on poly-(3-hexylthiophene) self-assembled monolayer using surface plasmon resonance technique (Arya et al., 2007), (e) protein kinase assay using peptide-conjugated gold nanoparticles (Kim et al., 2008a,b), (f) aptamer evolution for assay-based diagnostics for thrombin in solution (Platt et al., 2009a,b), (g) detection of thrombin by an electrochemical aptamer-based assay coupled to magnetic beads (Centi et al., 2008), (h) gold nanoparticles for quantification of prostate specific antigen (PSA) protein biomarker (Cao et al., 2009), (i) label-free analysis of transcription factors using microcantilever arrays (Huber et al., 2006), (j) carp vitellogenin (a potent fish biomarker for estrogenic activity; Kim et al., 2008a,b), and (k) point-of-care (POC) biosensor systems for cancer diagnostics/prognostics (Soper et al., 2006).

Commercial reports on biomarkers (available at a price; generally expensive, though) have also recently appeared. These include, "In-vitro diagnostics: market analysis 2009-2024" (email from ewa-malkowska@vg-sales.com, Malkowska, 2009), and "Cancer biomarkers: adoption is driving growth" (email from jimp@healthtech.com, Jimp, 2009). The first report highlights the detection of biomarkers, and points out that this constitutes a major advance and an expanding market opportunity. The second report reviews emerging cancer biomarker types. It presents business models behind cancer biomarker products and a SWOT profile analysis associated with specific strategies. Projection for growth areas within the cancer biomarker products are also given.

15.2 Theory

Havlin (1989) has reviewed and analyzed the diffusion of reactants towards fractal surfaces. The details of the theory and the equations involved for the binding and the dissociation phases for analyte-receptor binding are available (Sadana, 2001). The details are not repeated here; only the equations are given to permit an easier understanding. These equations have been applied to other biosensor systems (Sadana, 2001, 2005; Ramakrishnan and Sadana, 2001). For most applications, a single- or a dual-fractal analysis is often adequate to describe the binding and the dissociation kinetics. Peculiarities in the values of the binding and the dissociation rate coefficients in the systems being analyzed will be carefully noted, if applicable.

15.2.1 Single-Fractal Analysis

Binding Rate Coefficient

Havlin (1989) reports that the diffusion of a particle (analyte [Ag]) from a homogeneous solution to a solid surface (e.g., receptor [Ab]-coated surface) on which it reacts to form a product (analyte-receptor complex; (Ab·Ag)) is given by:

$$(\mathrm{Ab{\cdot}Ag}) \approx \begin{cases} t^{(3-D_{\mathrm{f,bind}})/2}, & t < t_{\mathrm{c}} \\ t^{1/2}, & t > t_{\mathrm{c}} \end{cases} \tag{15.1}$$

Here $D_{f,bind}$ or D_f (used later on in the book) is the fractal dimension of the surface during the binding step. t_c is the cross-over value. Havlin (1989) reports that the cross-over value may be determined by $r_c^2 \sim t_c$. Above the characteristic length, r_c, the self-similarity of the surface is lost and the surface may be considered homogeneous. Above time, t_c, the surface may be considered homogeneous, as the self-similarity property disappears, and "regular" diffusion is now present. For a homogeneous surface where D_f is equal to 2, and when only diffusional limitations are present, $p = ½$ as it should be. Another way of looking at the $p = ½$ case (where $D_{f,bind}$ is equal to 2) is that the analyte in solution views the fractal object, in our case, the receptor-coated biosensor surface, from a "large distance." In essence, in the association process, the diffusion of the analyte from the solution to the receptor surface creates a depletion layer of width $(Đt)^{½}$ where $Đ$ is the diffusion constant. This gives rise to the fractal power law, (Analyte·Receptor) $\sim t^{(3-D_{f,bind})/2}$. For the present analysis, t_c is arbitrarily chosen and we assume that the value of the t_c is not reached. One may consider the approach as an intermediate "heuristic" approach that may be used in future to develop an autonomous (and not time-dependent) model for diffusion-controlled kinetics.

Dissociation Rate Coefficient

The diffusion of the dissociated particle (receptor [Ab] or analyte [Ag]) from the solid surface (e.g., analyte [Ag]-receptor [Ab] complex coated surface) into the solution may be given, as a first approximation by:

$$(\text{Ab}\cdot\text{Ag}) \approx -t^{(3-D_{f,bind})/2} = -t^p, \quad t > t_{diss} \tag{15.2}$$

Here $D_{f,diss}$ is the fractal dimension of the surface for the dissociation step. This corresponds to the highest concentration of the analyte-receptor complex on the surface. Henceforth, its concentration only decreases. The dissociation kinetics may be analyzed in a manner "similar" to the binding kinetics.

15.2.2 Dual-Fractal Analysis

Binding Rate Coefficient

Sometimes, the binding curve exhibits complexities and two parameters (k, D_f) are not sufficient to adequately describe the binding kinetics. This is further corroborated by low values of the r^2 factor (goodness-of-fit). In that case, one resorts to a dual-fractal analysis (four parameters; k_1, k_2, D_{f1}, and D_{f2}) to adequately describe the binding kinetics. The single-fractal analysis presented above is thus extended to include two fractal dimensions. At present, the time ($t = t_1$) at which the "first" fractal dimension "changes" to the "second" fractal dimension is arbitrary and empirical. For the most part, it is dictated by the data analyzed and experience gained by handling a single-fractal analysis. A smoother curve is obtained in the "transition" region, if care is taken to select the correct number of points for the two regions. In this case, the product (antibody-antigen; or analyte-receptor complex, Ab·Ag or analyte·receptor) is given by:

$$(\text{Ab}\cdot\text{Ag}) \approx \begin{cases} t^{(3-D_{f1,\,\text{bind}})/2} = t^{p1}, & t < t_1 \\ t^{(3-D_{f2,\,\text{bind}})/2} = t^{p2}, & t_1 < t < t_2 = t_c \\ t^{1/2}, & t > t_c \end{cases} \tag{15.3}$$

In some cases, as mentioned above, a triple-fractal analysis with six parameters (k_1, k_2, k_3, D_{f1}, D_{f2}, and D_{f3}) may be required to adequately model the binding kinetics. This is when the binding curve exhibits convolutions and complexities in its shape due perhaps to the very dilute nature of the analyte (in some of the cases to be presented) or for some other reasons. Also, in some cases, a dual-fractal analysis may be required to describe the dissociation kinetics.

15.3 Results

We now present a fractal analysis of the binding and dissociation (if applicable) of the kinetics of (a) CEA in glucose using enzyme-catalyzed deposition of a redox polymer and electrolytic oxidation of ascorbic acid (AA) (Gao et al., 2009), (b) TNF-α using poly(guanine) functionalized silica nanoparticles (NPs) (Wang et al., 2006), (c) binding of IgG-antithrombin in solution to immobilized biotinylated thrombin, and binding of thrombin in solution to immobilized biotinylated aptamer (Centi et al., 2008), (d) binding of cholesterol to a cholesterol biosensor (Shin and Liu, 2007), (e) binding and dissociation of the protein biomarker, protein specific antigen (PSA) to different gold nanocrystals for different immunoprobe concentrations in solution (Cao et al., 2009), (f) binding of the transcription factors (rhSP1 and rhNF-κB) to oligonucleotides on a microcantilever array (Huber et al., 2006), and (g) binding and dissociation of different concentrations of thrombin in solution to the best aptamer in generation 4 (G4.0422) immobilized on a SA chip (Platt et al., 2009a,b).

Gao et al. (2009) have recently developed a doubly amplified electrochemical immunoassay (EIA) for the detection of CEA. Their assay uses enzyme-catalyzed deposition of a redox polymer and electrolytic oxidation of AA by the deposited redox polymer. This is a dual-amplification scheme that enhances analytical signals.

Gao et al. (2009) have expressed the need to detect a few key proteins at the POC such as in clinics and in doctors' offices (Emili and Cagney, 2000; Mcbeath, 2002; Mitchell, 2002). They further propose that electrochemical biosensors may be used to measure different analytes in blood without interference from blood cells, proteins, and fat globules (Yao et al., 1995; Hirsch et al., 2003). They have developed a dual amplification scheme in EIA to detect a colon cancer biomarker, CEA and claim that the detection limit and sensitivity are far superior to the previous biosensor detection devices.

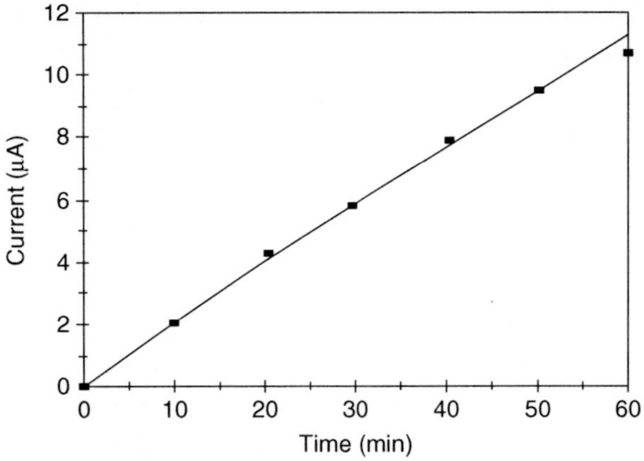

Figure 15.1

Binding of 5.0 ng/ml carcinoembryonic antigen (CEA) in glucose using enzyme-catalyzed deposition of a redox polymer and electrolytic oxidation of ascorbic acid (AA; Gao et al., 2009).

Figure 15.1 shows the binding of 5.0 ng/ml CEA in glucose/PVIA-Os solution (Gao et al., 2009). The authors report that PVIA gave the highest redox polymer loading amongst the polymer backbones. PVIA-Os is PVIA complexed with $[Os(bpy)(tpy)]^{2+/3+}$. A single-fractal analysis is adequate to describe the binding kinetics. The values of the binding rate coefficient k and the fractal dimension D_f for a single-fractal analysis are given in Table 15.1.

Wang et al. (2006) have recently defined a biomarker as, "a characteristic that is measured and evaluated objectively as an indicator of normal biological processes, pathogenic processes, or pharmacologic responses to a therapeutic intervention." TNF-α is a biomarker that

Table 15.1: Binding rate coefficient and fractal dimension for the binding phase for the detection of different biomarkers by an electrochemical immunoassay (EIA): (a) carcinoembryonic antigen (CEA) using enzyme-catalyzed deposition of a redox polymer and electrolytic oxidation of ascorbic acid (AA; Gao et al., 2009), (b) tumor necrosis factor-α (TNF-α) using poly(guanine) functionalized silica nanoparticles (NPs; Wang et al., 2006), and (c) thrombin using aptamer-based assay coupled to magnetic beads and anti-thrombin to immobilized biotinylated thrombin (Centi et al., 2008).

Analyte in Solution	k	D_f	References
5.0 ng/ml CEA in glucose/PVIA-OS	0.2451 ± 0.0158	1.132 ± 0.0840	Gao et al. (2009)
1.0 ng/ml TNF-α in 0.1M PBS buffer containing $Ru(bpy)_3Cl_2$	2.360 ± 0.325	1.8350 ± 0.3268	Wang et al. (2006)
400 ppm IgG antithrombin to immobilized biotinylated thrombin	81.24 ± 1.40	2.3794 ± 0.0137	Centi et al. (2008)
20 nM thrombin to immobilized biotinylated aptamer	9.879 ± 0.474	1.5386 ± 0.0364	Centi et al. (2008)

is an extremely potent peptide cytokine that serves as an endogeneous mediator of inflammatory, immunodefense, and host-defense functions (Old, 1987). DeKossodo et al. (1995) report that elevated concentrations of TNF in serum indicate a broad series of pathogenic states such as neonatal listeriosis, severe meningococcemia, HIV infection, systemic erythema nodosum leprosum, endotoxic shock, graft rejection, and rheumatoid arthritis. Wang et al. (2006) emphasize that sensitive detection methods are required to help detect this trace biomarker to promote the understanding of tumor biological processes and inherent mechanisms, and discover drugs that have a therapeutic potential for the treatment of diseases.

College Hill (2009) in the journal Nature defines biomarkers, and as stated in the beginning of the chapter, "as valuable tools used across the biological science spectrum. These tools are used both in research and in diagnostics as indicators of normal and disease processes or to assess pharmacologic response." Some of the more recent methods used to detect TNF-α include ELISA (enzyme-linked immunosorbent assay; Yates et al., 1999; Rossomando and White, 1993), radioimmunoassay (Teppo et al., 1987), and fluorescence immunoassay (Okubo et al., 1998; Cesaro-Tadic et al., 2004) and by reverse transcriptase-polymerase chain reaction (Takahashi et al., 2001).

Wang et al. (2006) report that the electrochemical technique is useful in detecting biomarkers, because of its high sensitivity, inherent simplicity, miniaturization, and cost. The detection techniques mentioned in the previous paragraph, (Wang et al., 2006) are either hazardous to health, time-consuming and labor intensive or require highly qualified and sophisticated personnel. These authors have developed poly(guanine) (poly[G])-functionalized silica NPs and mediator-induced catalytic oxidation of guanine for an amplified EIA of TNF-α.

Figure 15.2 shows the binding of 1.0 ng/ml TNF-α in 0.1 M PBS buffer containing $Ru(bpy)_3Cl_2$ to poly(guanine) functionalized nanoparticles (NPs; Wang et al., 2006). A single-fractal

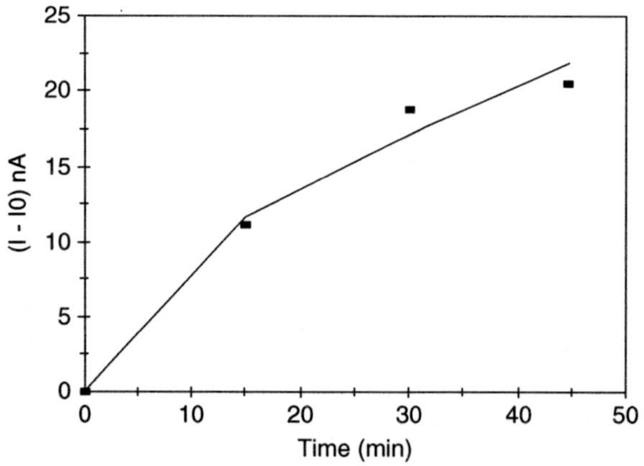

Figure 15.2
Binding of tumor necrosis factor (TNF-α) using poly(guanine) functionalized silica nanoparticles (NPs; Wang et al., 2006).

analysis is adequate to describe the binding kinetics. The values of the binding rate coefficient k and the fractal dimension D_f for a single-fractal analysis is given in Table 15.1.

Centi et al. (2008) have recently developed an electrochemical aptamer-based assay coupled to magnetic beads for the detection of thrombin. These authors indicate that aptamers are nucleic acid ligands that can be generated against amino acids, drugs, proteins, and other molecules.

SELEX (Systematic Evolution of Ligands by Exponential Enrichment) is used to isolate aptamers from a random library of synthetic nucleic acids by an iterative process of binding, separation, and amplification. Aptamers have been established recently as biorecognition elements (Ellington and Szoztak, 1990; Tuerk and Gold, 1990; Tombelli et al., 2007).

Centi et al. (2007) have reported the application of an aptamer-based electrochemical sandwich assay coupled with magnetic beads. They affirm that their assay demonstrated good reproducibility. They also point out that the aptamer-based assay may be used in the following formats: sandwich or competitive assay, and direct or indirect assay. The choice of the format is primarily dependent on the analyte's molecular size and cost.

Figure 15.3a shows the binding of 400 ppm IgG antithrombin in solution to immobilized biotinylated thrombin (Centi et al., 2008). A single-fractal analysis is adequate to describe the binding kinetics. The values of the binding rate coefficient k and the fractal dimension D_f for a single-fractal analysis are given in Table 15.1.

Figure 15.3b shows the binding of 200 nM-thrombin in solution to immobilized biotinylated aptamer (Centi et al., 2008). A single-fractal analysis is, once again, adequate to describe the

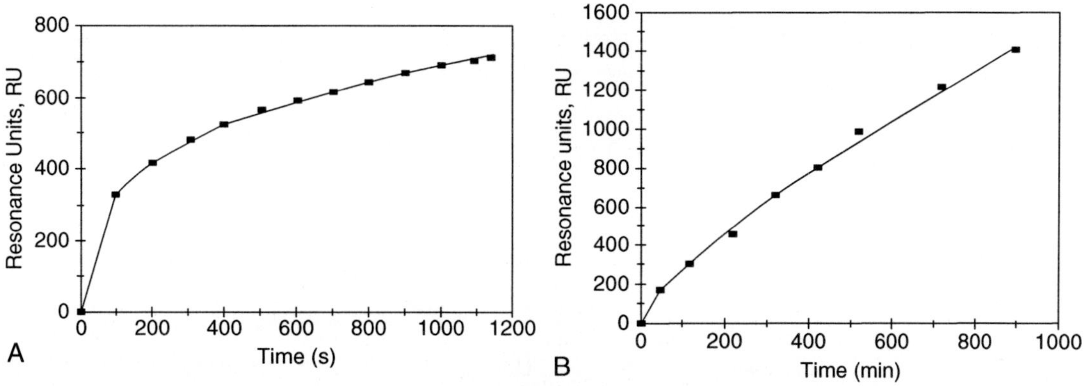

Figure 15.3

(a) Binding of 400 ppm IgG antithrombin in solution to immobilized biotinylated thrombin (Centi et al., 2008). (b) Binding of 200 nM thrombin in solution to immobilized biotinylated aptamer (Centi et al., 2008).

binding kinetics. The values of the binding rate coefficient k and the fractal dimension D_f for a single-fractal analysis are given in Table 15.1.

Shin and Liu (2007) have recently developed a screen-printed cholesterol biosensor. These authors compared the performance of gold (Au) and platinum (Pt) as the working electrode material. A self-assembly approach was used to fabricate the biosensor. These authors indicate that both the gold and the platinum would detect the cholesterol in solution through the electrochemical oxidation of H_2O_2. However, the authors selected gold as the working electrode material as it exhibited a higher response current and better sensitivity. The enzyme cholesterol oxidase (Chox, E.C. 1.1.3.6) was immobilized on the Au working electrode by using a self-assembly approach. The authors report that their thick-film screen-printed cholesterol biosensor comprises three electrodes on an alumina-substrate. 3-mercaptopropionic acid (MPA) was self-assembled onto the gold working electrode and formed a thin organic layer. This thin layer of MPA served as an anchor for enzyme immobilization.

Figure 15.4a shows the binding of 20 mg/dl cholesterol in solution to the cholesterol biosensor. A single-fractal analysis is adequate to describe the binding kinetics. The values of the binding rate coefficient k and the fractal dimension D_f are given in Table 15.2.

Figure 15.4b shows the binding of 100 mg/dl cholesterol in solution to the cholesterol biosensor. A dual-fractal analysis is required to adequately describe the binding kinetics. The values of (a) the binding rate coefficient k and the fractal dimension D_f for a single-fractal analysis, and (b) the binding rate coefficients k_1 and k_2 and the fractal dimensions D_{f1} and D_{f2} for a dual-fractal analysis are given in Table 15.2.

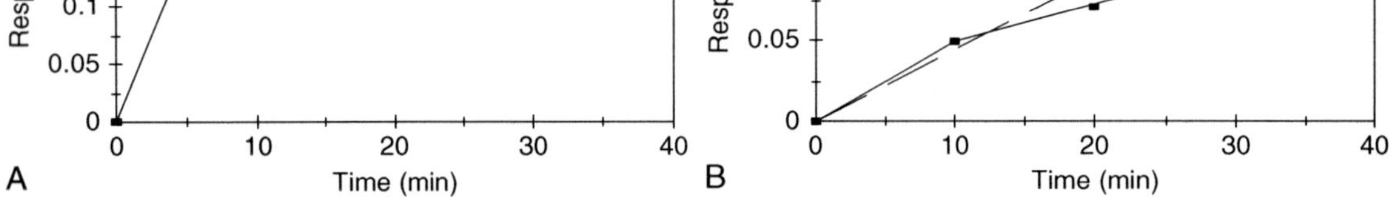

Figure 15.4
Binding of different concentrations of cholesterol (in mg/dl) to a cholesterol biosensor (Shin and Liu, 2007): (a) 20

Continued

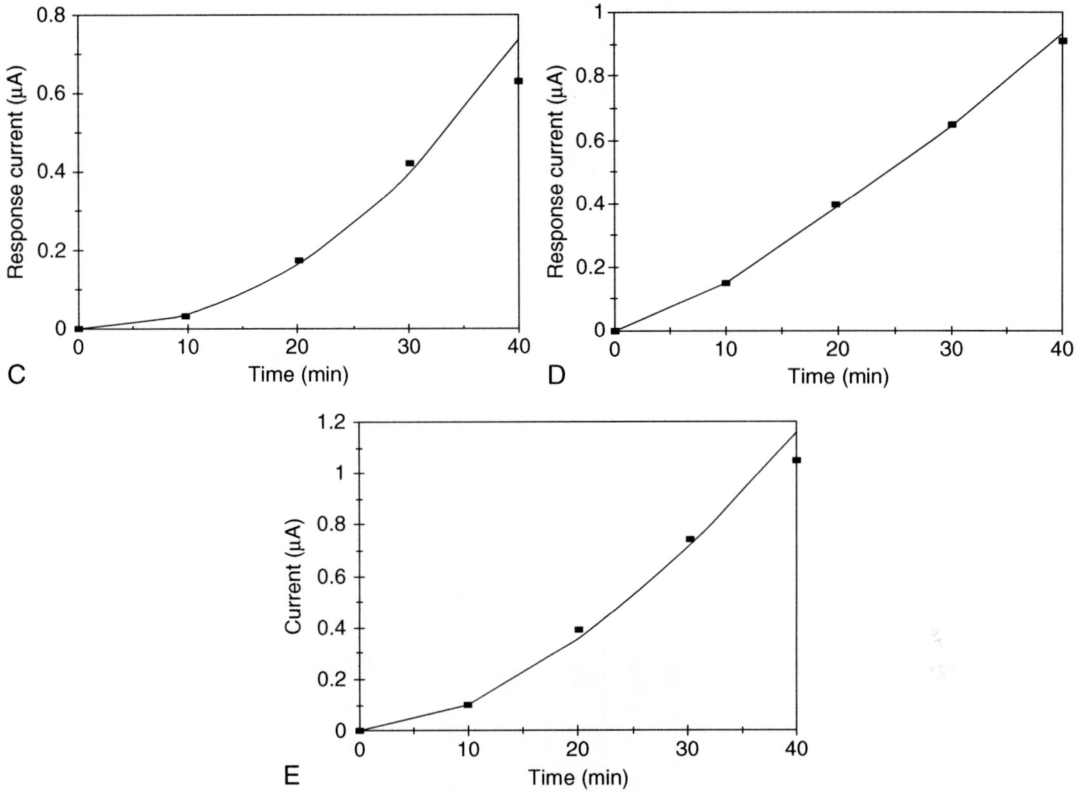

Figure 15.4—cont'd

(b) 100 (c) 200 (d) 250 (e) 300. When only a solid line (—) is used then a single-fractal analysis applies. When both a dashed (- - -) and a solid (—) line are used then the dashed line represents a single-fractal analysis and the solid line represents a dual-fractal analysis.

Figure 15.4c shows the binding of 200 mg/dl cholesterol in solution to the cholesterol biosensor. A single-fractal analysis is adequate to describe the binding kinetics. The values of the binding rate coefficient k and the fractal dimension D_f are given in Table 15.2.

Figure 15.4d shows the binding of 250 mg/dl cholesterol in solution to the cholesterol biosensor. A single-fractal analysis is adequate to describe the binding kinetics. The values of the binding rate coefficient k and the fractal dimension D_f are given in Table 15.2.

Figure 15.4e shows the binding of 300 mg/dl cholesterol in solution to the cholesterol biosensor. A single-fractal analysis is adequate to describe the binding kinetics. The values of the binding rate coefficient, k and the fractal dimension, D_f are given in Table 15.2.

Figure 15.5 shows the increase in the binding rate coefficients k or k_1 with an increase in the fractal dimensions D_f or D_{f1}. Not enough data points are available; therefore these two

Table 15.2: Binding rate coefficients and fractal dimensions for binding of different concentrations of cholesterol (in mg/dl) to the cholesterol biosensor (Shin and Liu, 2007).

Cholesterol Concentration (mg/dl)	k	k_1	k_2	D_f	D_{f1}	D_{fd2}
20	0.2509 ± 0.0042	na	na	2.8486 ± 0.03196	na	na
100	0.00384 ± 0.00169	0.01183 ± 0.00101	0.00031 ± 0.000115	0.9172 ± 0.7016	1.7694 ± 0.2072	0 + 1.2802
200	0.000189 ± 0.000038	na	na	0. + 0.3474	na	na
250	0.0007781 ± 0.00040	na	na	0.406 ± 0.0963	na	na
300	0.002053 ± 0.000286	na	na	na	na	na

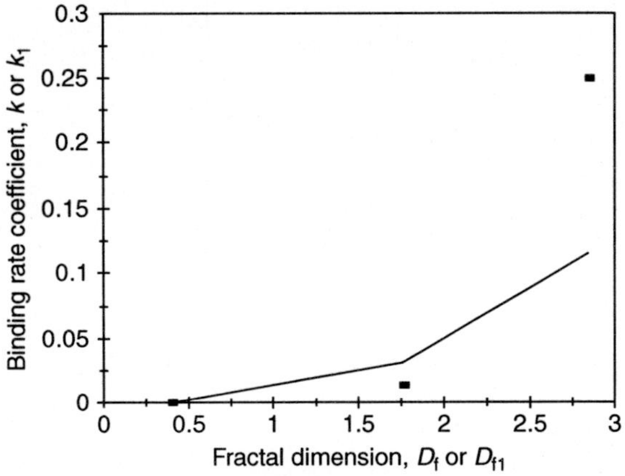

Figure 15.5
Increase in the binding rate coefficient, k or k_1 with an increase in the fractal dimension, D_f or D_{f1}.

binding rate coefficients are plotted together on the same graph. For the data presented in Figure 15.5, the binding rate coefficients k or k_1 are given by:

$$k \text{ or } k_1 = (0.0069 + 0.0181)(D_f \text{ or } D_{f1})^{2.7} \quad (15.4)$$

The fit is not good, and this is not unexpected as two different binding rate coefficients are plotted together due to the lack of data points for each of them. The binding rate coefficients exhibit an order of dependence between 2.5 and 3.0 (equal to 2.7) on the fractal dimension or the degree of heterogeneity that exists on the biosensor surface.

Cao et al. (2009) have recently developed an alternative sensitive platform for the detection of protein biomarkers, PSA, and PSA-ACT complex. These authors indicate that their biosensor translates the immunosensing event to a gold nanoparticle growth signal. This signal, the authors claim, can either be recognized by the unaided eye or by an UV-vis spectrophotometer.

Cao et al. (2009) report that the use of protein biomarkers for disease detection is still a major driving force for biosensor research. Ambrosi et al. (2007) had indicated that gold nanoparticles (NPs) coupled with antigens or antibodies were used as optical labels. Das et al. (2006) and Zhou et al. (2006) have used these AuNPs as electrochemical markers. Cao et al. (2009) have recently used a homogeneous detection assay that incorporates Au nanocrystalline growth with the use of MMPs (magnetic microbeads). They used a sandwich type assay to detect protein biomarkers, and used prostate specific antigen (PSA-ACT) as a model reaction. This, the authors report, is a valuable signature in the diagnosis of prostate cancer (Linstedt et al., 1990; Curry et al., 1996). Cao et al. (2009) affirm that their highly sensitive detection technique is much better for PSA as far as detection limits are concerned, when compared with presently available commercial detection techniques.

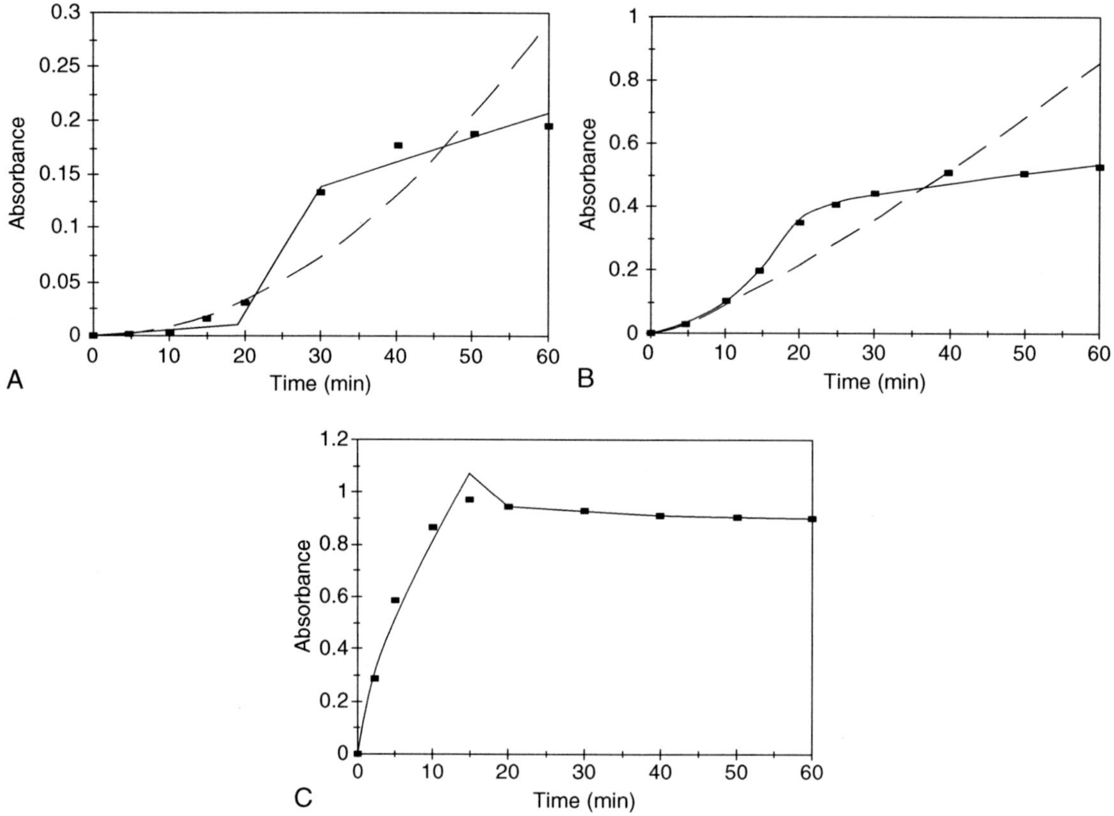

Figure 15.6

Binding and dissociation of PSA (protein specific antigen) to different gold nanocrystals for immunoprobe concentrations (in M) in solution (Cao et al., 2009): (a) 0.65×10^{-7}, (b) 1.3×10^{-7}, (c) 5.2×10^{-7}. When only a solid line (—) is used then a single-fractal analysis applies. When both a dashed (- - -) and a solid (—) line are used then the dashed line represents a single-fractal analysis and the solid line represents a dual-fractal analysis.

Figure 15.6a shows the binding of the protein biomarker PSA-ACT to 0.65×10^{-7} M immunoprobe concentration in solution (Cao et al., 2009). As mentioned, the detection of PSA-ACT is based on the homogeneous growth of nanocrystals in the solution phase. For the 0.65×10^{-7} M immunoprobe concentration a dual-fractal analysis is required to adequately describe the binding kinetics. The values of (a) the binding rate coefficient k and the fractal dimension D_f for a single-fractal analysis, and (b) the binding rate coefficients k_1 and k_2 and the fractal dimensions D_{f1} and D_{f2} for a dual-fractal analysis are given in Tables 15.3 and 15.4. Figure 15.6b shows the binding of the protein biomarker, PSA-ACT to 1.3×10^{-7} M immunoprobe concentration in solution (Cao et al., 2009). As before, the detection of PSA-ACT is based on the homogeneous growth of nanocrystals in the solution phase. For the 0.65×10^{-7} M immunoprobe concentration a dual-fractal analysis is required to adequately describe the binding kinetics. The values of (a) the binding rate coefficient k and the fractal dimension D_f for a single-fractal analysis, and (b) the binding rate coefficients

Table 15.3: Binding and dissociation rate coefficients for the PSA protein biomarker on different gold nanocrystals for immunoprobe concentrations in solution (Cao et al., 2009).

PSA Concentration (M)	k	k_1	k_2	k_d
0.65×10^{-7}	0.000077 ± 0.000037	0.000182 ± 0.000077	0.01948 ± 0.00188	na
1.3×10^{-7}	0.005087 ± 0.002666	0.000834 ± 0.000153	0.1692 ± 0.0077	na
5.2×10^{-7}	0.1825 ± 0.030	na	na	0.004847 ± 0.000380

Table 15.4: Fractal dimensions for the binding and dissociation phase for PSA protein biomarker on different gold nanocrystals immunoprobe concentrations in solution (Cao et al., 2009).

PSA Concentration (M)	D_f	D_{f1}	D_{f2}	D_{fd}
0.65×10^{-7}	0 + 0.3478	0 + 0.8996	1.8470 ± 0.3568	na
1.3×10^{-7}	0.496 ± 0.03724	0 + 0.3224	2.4488 ± 0.1253	na
5.2×10^{-7}	1.7042 ± 0.2224	na	na	1.7042 ± 0.8666

k_1 and k_2 and the fractal dimensions D_{f1} and D_{f2} for a dual-fractal analysis are given in Tables 15.3 and 15.4.

Huber et al. (2006) have recently used cantilever microarray technology to analyze the binding of two different DNA-binding proteins, the transcription factors SP1 and NF-κB. Their use of cantilever arrays permits them to use label-free measurement of different biomolecular interactions in parallel. They took double-stranded DNA oligonucleotides containing a specific binding site for a transcription factor and these were sensitized on gold-coated cantilevers. The binding of the transcription factor created a surface stress. This surface stress resulted in a bending of the cantilevers.

Huber et al. (2006) report that cells respond to changes in the environment by regulating gene expression. Honore et al. (2004) have specified the different technologies that have been used to study this, such as DNA microarrays, 2D electrophoresis of proteins, and ELISA. Huber et al. (2006) have proposed microcantilever array technology to analyze gene regulation not only at the mRNA or protein level but also at the level of transcription factors. They report that microcantilver arrays have been used to analyze DNA hybridization (Fritz et al., 2000; McEndry et al., 2004), and antigen-antibody interaction (Wu et al., 2001; Arntz et al., 2003). They coated one side of the microfabricated silicon cantilevers with gold. This permitted the sensitization of the thiol-modified receptor molecules (DNA oligonucleotides or antibodies).

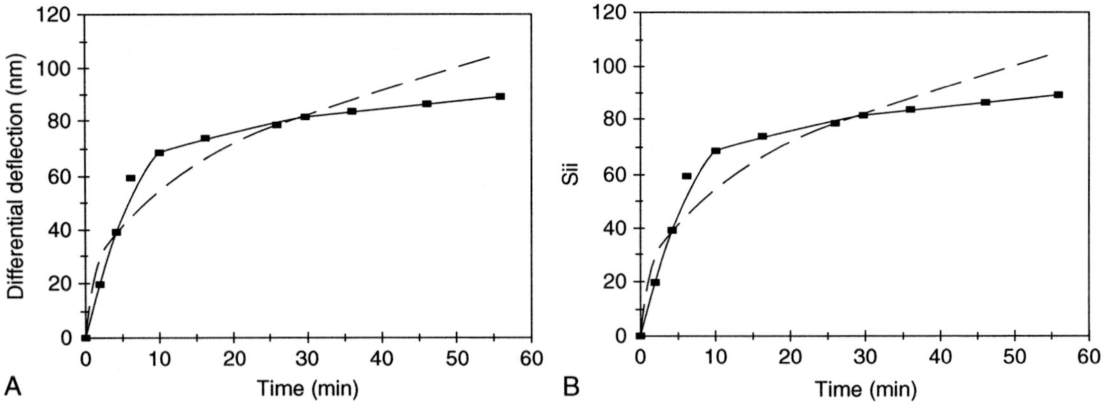

Figure 15.7

(a) Binding of 80 nM transcription factor (rhSP1) to rhNF-κB binding oligonucleotide on a microcantilever array (Huber et al., 2006). (b) Binding of 100 nM transcription factor (rhNF-κB) to SP1 binding oligonucleotide on a microcantilever array (Huber et al., 2006). When only a solid line (—) is used then a single-fractal analysis applies. When both a dashed (- - -) and a solid (—) line are used then the dashed line represents a single-fractal analysis and the solid line represents a dual-fractal analysis.

The biomoelcular recognition event of the binding of the ligand to the receptor on the surface resulted in a bending of the microcantilever. This bending is detected by laser beam deflection.

Figure 15.7a shows the binding of 80 nM transcription factor rhSP1 in solution to the cantilever array functionalized with the binding oligonucleotide. A dual-fractal analysis is required to adequately describe the binding kinetics. The values of (a) the binding rate coefficient k and the fractal dimension D_f for a single-fractal analysis, and (b) the binding rate coefficients k_1 and k_2 and the fractal dimensions D_{f1} and D_{f2} for a dual-fractal analysis are given in Table 15.5.

It is of interest to note that as the degree of heterogeneity or the fractal dimension increases by a factor of 1.58 from a value of D_{f1} equal to 1.3908 to D_{f2} equal to 2.2030, the binding rate coefficient increases by a factor of 3.91 from a value of k_1 equal to 12.367 to k_2 equal to 49.233. Once again, an increase in the degree of heterogeneity or the fractal dimension on the cantilever array (biosensor) surface leads to an increase in the binding rate coefficient.

Figure 15.7b shows the binding of 100 nM rhNF-κB transcription factor, in solution to the cantilever array functionalized with the SP1 binding oligonucleotide. A dual-fractal analysis is, once again, required to adequately describe the binding kinetics. The values of (a) the binding rate coefficient, k and the fractal dimension, D_f for a single-fractal analysis, and (b) the binding rate coefficients, k_1 and k_2, and the fractal dimensions, D_{f1} and D_{f2} for a dual-fractal analysis are given in Table 15.5.

It is of interest to note that as the degree of heterogeneity or the fractal dimension increases by a factor of 1.27 from a value of D_{f1} equal to 2.3742 to D_{f2} equal to 2.8875, the binding rate coefficient increases by a factor of 2.42 from a value of k_1 equal to 13.708 to k_2 equal

Table 15.5: Binding rate coefficients and fractal dimensions for transcription factors to microcantilever arrays (Huber et al., 2006).

Analyte in Solution/Receptor on Surface	k	k_1	k_2	D_f	D_{f1}	D_{f2}
80 nM rhSPl/rhNF-κB binding oligonucleotide	22.750 ± 3.380	12.367 ± 1.941	49.233 ± 0.322	2.82412 ± 1.2598	1.3908 ± 0.2476	2.2030 ± 0.01719
100 nM rhNF-κB/SP1-binding oligonucleotide	16.941 ± 1.773	13.708 ± 1.620	33.166 ± 0.326	2.4926 ± 0.07238	2.2742 ± 0.1797	2.88748 ± 0.02348
200 nM rhSPlin SP1 binding buffer	18.843 ± 0.621	na	na	2.3718 ± 0.4756	na	na
400 nM rhNF-κB in NF-κB binding buffer	5.786 ± 0.115	na	na	1.6270 ± 0.01876	na	na

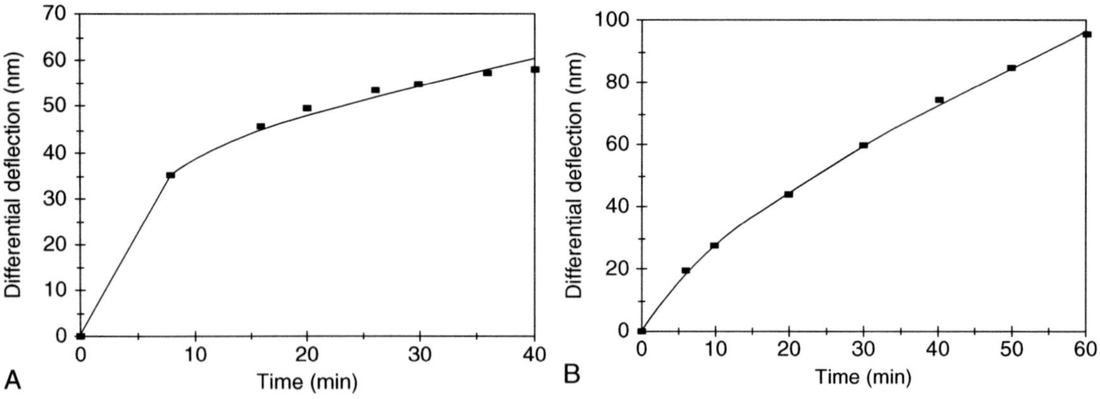

Figure 15.8
(a) Binding of 280 nM transcription factor (rhSP1) in SP1 binding buffer on a microcantilever array (Huber et al., 2006). (b) Binding of 400 nM transcription factor rhNF-κB in N-κB binding buffer to SP1 binding on a microcantilever array (Huber et al., 2006).

to 33.166. Once again, an increase in the degree of heterogeneity or the fractal dimension on the cantilever array (biosensor) surface leads to an increase in the binding rate coefficient.

Figure 15.8a shows the binding of 280 nM rhSP1 in SP1 binding buffer to the cantilever array (Huber et al., 2006). A single-fractal analysis is adequate to describe the binding kinetics. The values of the binding rate coefficient k and the fractal dimension D_f for a single-fractal analysis are given in Table 15.5.

Figure 15.8b shows the binding of 400 NF-κB in NF-κB binding buffer to the cantilever array (Huber et al., 2006). A single-fractal analysis is adequate to describe the binding kinetics. The values of the binding rate coefficient k and the fractal dimension D_f for a single-fractal analysis are given in Table 15.5.

If one were permitted to compare the results in Figures 15.8a and b (same cantilever, different analytes), then a decrease in the fractal dimension by a factor of 1.46 from a value of D_f equal to 2.3718 (200 nM rhSP1) to D_f equal to 1.6270 (400 nM rhNF-κB) leads to a decrease in the binding rate coefficient by a factor of 3.26 from a value of k equal to 18.843 to k equal to 5.786. Changes in the binding rate coefficient and in the fractal dimension are in the same direction.

Platt et al. (2009a,b) have recently analyzed aptamer evolution for array-based diagnostics. These authors report that aptamers are oligonucleotides that may bind to target ligands. Their affinities are comparable to those of antibodies. Bunka and Stockley (2006) indicate that aptamers have found applications ranging from biosensors to therapeutics. Balamurugan et al., 2008) indicate that an “on-chip” optimization procedure is often required that involves immobilization and tethering strategies.

Platt et al. (2009a,b) have recently demonstrated the "on-chip" evolution for fluorescently tagged protein targets. They report that thrombin has been widely used in array experiments (Bock et al., 1992; Cho et al., 2006). Thrombin was selected as a target due to its binding to its motif,-$GGN_{2\text{-}5}$GGT(A/T)GG. Platt et al. (2009a,b) reiterate that one should be able to demonstrate or differentiate the actual protein binding events from nonspecific interactions. That is the key (Warren et al., 2006). Platt et al. (2006) have demonstrated that biotin labeling followed by a posthybridization stage with a streptavidin-fluorophore conjugate has lower nonspecific and background interactions.

Figure 15.9a shows the binding of 16 nM thrombin in solution to aptamer (G4.04422; best aptamer in generation 4) immobilized on a SA chip. A single-fractal analysis is adequate to describe the binding and the dissociation kinetics. The values of (a) the binding rate coefficient k and the fractal dimension D_f for a single-fractal analysis, and (b) the dissociation

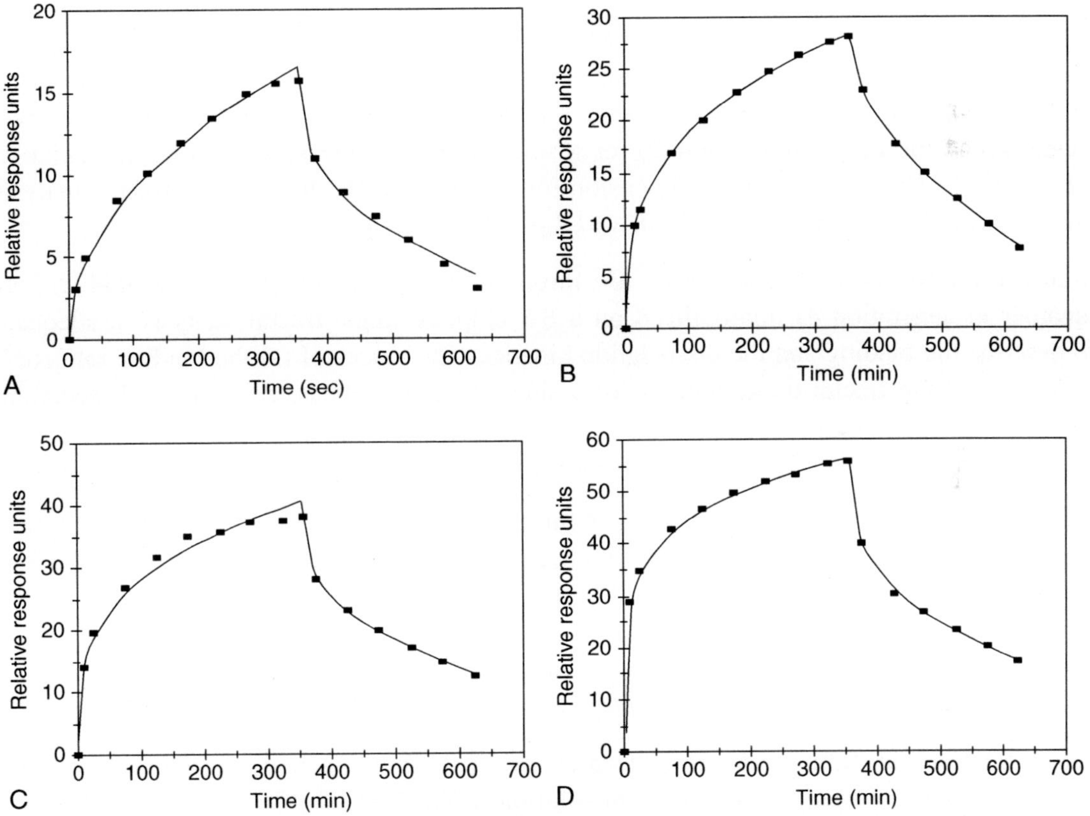

Figure 15.9

Binding and dissociation of different concentrations (in nM) of thrombin in solution to best aptamer in generation 4 (G4.0422) immobilized on a SA chip (Platt et al., 2009a,b): (a) 16, (b) 32, (c) 65, (d) 130.

Table 15.6: Binding and dissociation rate coefficients for the binding and dissociation phase for different concentrations (in nM) of thrombin to best aptamer in generation 4 (G4.04422) immobilized on a SA chip (Platt et al., 2009a,b).

Thrombin Concentration in Solution (nM)	k	k_d	D_f	D_{fd}
16	0.9997 ± 0.00591	1.4477 ± 0.0967	2.0416 ± 0.0341	2.2498 ± 0.0604
32	4.314 ± 0.128	1.0428 ± 0.0364	2.3586 ± 0.01748	1.9400 ± 0.0320
65	7.519 ± 0.486	3.2895 ± 0.107	2.4362 ± 0.0371	2.2744 ± 0.02982
130	18.714 ± 0.348	6.0088 ± 0.1082	2.6236 ± 0.0109	2.3398 ± 0.02874

rate coefficient k_d and the fractal dimension for dissociation D_{fd} for a single-fractal analysis are given in Table 15.6. The affinity K $(=k/k_d)$ value is 0.691.

Figure 15.9b shows the binding of 32 nM thrombin in solution to aptamer (G4.04422; best aptamer in generation 4) immobilized on a SA chip. A single-fractal analysis is adequate to describe the binding and the dissociation kinetics. The values of (a) the binding rate coefficient k and the fractal dimension D_f for a single-fractal analysis, and (b) the dissociation rate coefficient k_d and the fractal dimension for dissociation D_{fd} for a single-fractal analysis are given in Table 15.6. The affinity K $(=k/k_d)$ value is 4.14.

Figure 15.9c shows the binding of 65 nM thrombin in solution to aptamer (G4.04422; best aptamer in generation 4) immobilized on a SA chip. A single-fractal analysis is adequate to describe the binding and the dissociation kinetics. The values of (a) the binding rate coefficient k and the fractal dimension D_f for a single-fractal analysis, and (b) the dissociation rate coefficient k_d and the fractal dimension for dissociation D_{fd} for a single-fractal analysis are given in Table 15.6. The affinity K $(=k/k_d)$ value is 2.29.

Figure 15.9d shows the binding of 130 nM thrombin in solution to aptamer (G4.04422; best aptamer in generation 4) immobilized on a SA chip. A single-fractal analysis is adequate to describe the binding and the dissociation kinetics. The values of (a) the binding rate coefficient k and the fractal dimension D_f for a single-fractal analysis, and (b) the dissociation rate coefficient k_d and the fractal dimension for dissociation D_{fd} for a single-fractal analysis are given in Table 15.6. The affinity K $(=k/k_d)$ value is 3.11.

Figure 15.10a and Table 15.6 show the increase in the binding rate coefficient k with an increase in the thrombin concentration in solution in the 16-130 nM range for binding to a best aptamer in generation 4(G4.04422). For the data shown in Figure 15.10a, the binding rate coefficient k is given by:

$$k = (0.030 \pm 0.010)\,[\text{thrombin concentration, nM}]^{1.34 \pm 0.181} \qquad (15.5a)$$

The fit is very good. Only four data points are available. The availability of more data points would lead to a more reliable and better fit. The binding rate coefficient k is sensitive to the thrombin concentration (16-130 nM) in solution as it exhibits an order of dependence between first and one and a half (equal to 1.34) on the thrombin concentration in solution.

Figure 15.10b and Table 15.6 show the increase in the dissociation rate coefficient k_d with an increase in the thrombin concentration in solution in the 16-130 nM range for binding to a best aptamer in generation 4(G4.04422). For the data shown in Figure 15.10b, the dissociation rate coefficient k_d is given by:

$$k_d = (0.120 + 0.190)\left[\text{thrombin concentration, nM}\right]^{0.726 \pm 0.292} \tag{15.5b}$$

The fit is reasonable. There is scatter in the data, and this is reflected in the estimate for the dissociation rate coefficient k_d. Only four data points are available. The availability of more data points would lead to a more reliable and better fit. Only the positive value of the error is given as the dissociation rate coefficient cannot have a negative value. The dissociation rate coefficient k_d for a single-fractal analysis exhibits less than first- (equal to 0.726) order of dependence on the thrombin concentration (16-130 nM) in solution. Figure 15.10c and Table 15.6 show the increase in the binding rate coefficient k with an increase in the fractal dimension D_f. For the data shown in Figure 15.10c, the binding rate coefficient k is given by:

$$k = (0.000235 \pm 0.0000033)D_f^{11.62 \pm 0.714} \tag{15.5c}$$

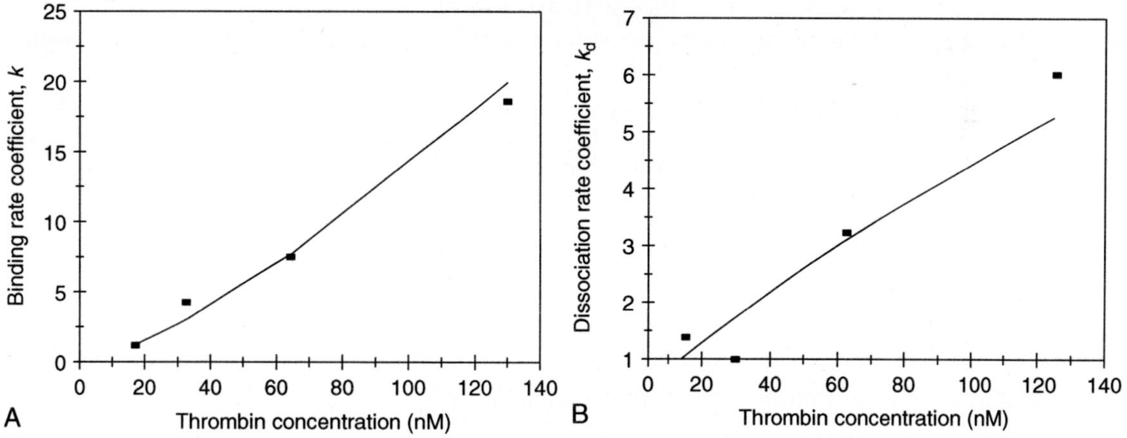

Figure 15.10

(a) Increase in the binding rate coefficient, k for a single-fractal analysis with an increase in the thrombin concentration (in nM) in solution. (b) Increase in the dissociation rate coefficient, k_d for a single-fractal analysis with an increase in the thrombin concentration (in nM) in solution.

Continued

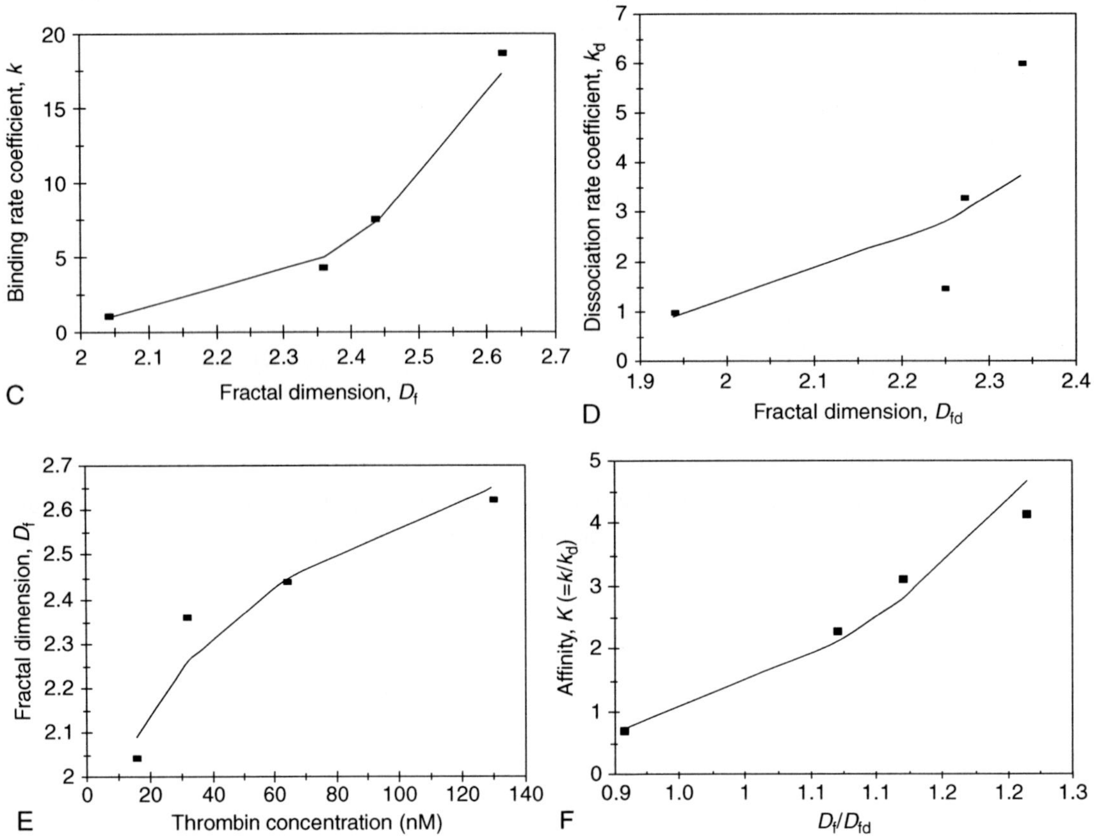

Figure 15.10—cont'd

(c) Increase in the binding rate coefficient, k for a single-fractal analysis with an increase in the fractal dimension, D_f. (d) Increase in the dissociation rate coefficient, k_d for a single-fractal analysis with an increase in the fractal dimension, D_{fd}. (e) Increase in the fractal dimension, D_f with an increase in the thrombin concentration (in nM) in solution. (f) Increase in the affinity, K (=k/k_d) with an increase in the fractal dimension ratio (D_f/D_{fd}).

The fit is very good. Only four data points are available. The availability of more data points would lead to a more reliable and better fit. The binding rate coefficient k for a single-fractal analysis is extremely sensitive to the fractal dimension D_f or the degree of heterogeneity that exists on the SA chip surface as it exhibits an order of dependence between eleven and twelve (equal to 11.62) on the fractal dimension D_f.

Figure 15.10d and Table 15.6 show the increase in the dissociation rate coefficient k_d with an increase in the fractal dimension D_{fd}. For the data shown in Figure 15.10d, the dissociation rate coefficient k_d is given by:

$$k_d = (0.0061 \pm 0.0048) D_{fd}^{7.56 \pm 3.99} \quad (15.5d)$$

The fit is poor. This is reflected in the estimates of the error provided. There is scatter in the data. Only four data points are available. The availability of more data points would lead to a more reliable and better fit. The dissociation rate coefficient k_d for a single-fractal analysis is extremely sensitive (just like the binding rate coefficient k) to the fractal dimension D_{fd} or the degree of heterogeneity that exists on the aptamer-SA chip surface as it exhibits an order of dependence slightly higher than seven and a half (equal to 7.56) on the fractal dimension D_{fd}.

Figure 15.10e and Table 15.6 show the increase in the fractal dimension D_f for a single-fractal analysis with an increase in the thrombin concentration in solution in the 16-130 nM concentration range. For the data shown in Figure 15.10e, the fractal dimension D_f is given by:

$$D_f = (1.589 \pm 0.064)[\text{thrombin concentration, in nM}]^{0.112 \pm 0.022} \quad (15.5e)$$

The fit is good. Only four data points are available. The availability of more data points would lead to a more reliable and better fit. The fractal dimension D_f exhibits a very slight dependence on the thrombin concentration in solution in the 16-130 nM concentration range as it exhibits a close to zero (equal to 0.112) order of dependence on the thrombin concentration in solution.

Figure 15.10f and Table 15.6 show the increase in the affinity, K ($=k/k_d$) with an increase in the ratio of the fractal dimensions, D_f/D_{fd}. For the data shown in Figure 15.10f, the affinity K is given by:

$$K = (1.367 \pm 0.205)(D_f/D_{fd})^{6.33 \pm 0.658} \quad (15.5f)$$

The fit is very good. Only four data points are available. The availability of more data points would lead to a more reliable and better fit. The affinity K is extremely sensitive to the ratio of the fractal dimensions as noted by the order of dependence between six and six and a half (equal to 6.33) exhibited.

15.4 Conclusions

A fractal analysis is presented for the binding and the dissociation (if applicable) kinetics of protein biomarkers and other medically-oriented analytes on biosensor surfaces. Both a single- and a dual-fractal analysis were utilized to model the binding and the dissociation kinetics. The dual-fractal analysis was used only when the single-fractal analysis did not provide an adequate fit. Corel Quattro Pro 8.0 was used to provide the regression analysis.

The examples analyzed include the following: the binding of CEA (Gao et al., 2009), (b) the binding of TNF-α using poly(guanine) functionalized silica nanoparticles (Wang et al., 2006), (c) binding of 400 ppm IgG antithrombin in solution to immobilized biotinylated thrombin, and binding of 200 nM thrombin in solution to immobilized biotinylated aptamer

(Centi et al., 2008), (d) binding of different concentrations of cholesterol (in mg/dl) to a cholesterol biosensor (Shin and Liu, 2007), (e) binding and dissociation of PSA to different gold nanocrystals for different immunoprobe concentrations (10^{-7} M) in solution (Cao et al., 2009), (f) binding of the transcription factors rhSP1 and rhNF-κB to their corresponding oligonucleotides on a microcantilever array (Huber et al., 2006), and (g) binding and dissociation of different concentrations (in nM) of thrombin in solution to the best aptamer in generation 4(G4.0422) immobilized on a SA chip (Platt et al., 2009a,b). The fractal dimension values provide a quantitative indicator of the degree of heterogeneity present on the sensor chip surface. Binding and dissociation values, and affinity values are provided whereever possible. The fractal dimension for the binding and the dissociation phase is not a typical independent variable that may be directly manipulated. It is estimated from Equations (15.1-15.3), and one may consider it as a derived variable.

In a general sense, fractal models are fascinating. Newer avenues are required to analyze and help detect protein biomarkers for disease and medically-oriented analytes at very dilute levels. The sooner one is able to help detect these protein biomarkers for certain diseases accurately, the better is the prognosis for that disease. Of course, the detection of these protein biomarkers (and subsequently the onset of the disease) is more and more difficult during the early stages. If one may take the liberty of mentioning cancer, it generally goes through roughly three stages: the initial, intermediate, and "blast" stage. The initial stage is rather lengthy (perhaps a time period of years) and is very difficult to detect. During the blast stage (time period of months or even weeks) the cancer is easier to detect, but by then perhaps it is too late. This example lays emphasis on the ability to be able to detect the different protein biomarkers for the different diseases at continuously lower and lower levels.

An increase in the fractal dimension value or the degree of heterogeneity on the biosensor surface leads, in general, to an increase in the binding and in the dissociation rate coefficient. It is suggested that the fractal surface (roughness) leads to turbulence, which enhances mixing, decreases diffusional limitations, and leads to an increase in the binding rate coefficient (Martin et al., 1991). For this to occur, the characteristic length of the turbulent boundary layer may have to extend a few monolayers above the senor chip surface to affect bulk diffusion to and from the surface. However, given the extremely laminar regimes in most biosensors, this may not actually take place. The sensor chip surface is characterized by grooves and ridges, and this surface morphology may lead to eddy diffusion. This eddy diffusion can then help to enhance the mixing and extend the characteristic length of the boundary layer to affect the bulk diffusion to and from the surface.

The predictive relationships are developed for the binding rate coefficient as (a) a function of the fractal dimension for the screen-printed cholesterol biosensor (Shin and Liu, 2007), and (b) for the binding rate coefficient k as a function of the thrombin concentration in the 16-130 nM range and (c) for the binding to the best aptamer in generation 4 (G4.04422) are of considerable value because they directly link the binding rate coefficient either to the degree of heterogeneity that exists on the sensor surface or to the analyte concentration in solution. This provides a means by which the binding rate coefficient may be manipulated by either changing the degree of heterogeneity on the sensor chip surface or by changing the analyte concentration in solution as the case may be. The really interesting test of the fractal model would be its ability to make a prediction that turns out to be correct. This would prove to be extremely valuable, especially in the detection of protein biomarker(s) and medically-related analytes.

References

Ambrosi MT, AJ Castenada, AJ Killard, MR Smyth, S Alegret, and Amarkoci, *Analytical Chemistry*, **79**, 5232–5240 (2007).

Arntz Y, JD Seelig, HP Lang, J Zhang, P Hunziker, JP Ramseyer, E Meyer, M Hegner, and CH Gerber, Label free protein assay based on a nanomechanical cantilever array, *Nanotechnology*, **14**(1), 86–90 (2003).

Arya SK, PR Solanki, SP Singh, K Kaneto, MK Pandey, M Dutta, and BD Malhotra, Poly-(3-hexylthiphene) self-assembled monolayer based cholesterol biosensor using surface plasmon resonance technique, *Biosensors & Bioelectronics*, **22**, 2516–2524 (2007).

Balamuragan S, A Obubuato, S Sofer, and D Spivak, Surface immobilization methods for aptamer diagnostic applications, *Analytical and Bioanalytical Chemistry*, **390**, 1009–1021 (2008).

Bock LC, LC Griffin, JA Latham, EH Vermass, and JJ Toole, Selection of single-stranded DNA molecules that bind and inhibit human thrombin, *Nature*, **355**, 564–566 (1992).

Bunka DHJ and PG Stockley, Aptamers come of age—At last, *Natural Reviews of Microbiology*, **4**, 588–596 (2006).

Cao C, X Li, J Lee, and SJ Sim, Homogeneous growth of gold nanocrystals for quantification of PSA protein biomarkers, *Biosensors & Bioelectronics*, **24**(5), 1292–1297 (2009).

Centi S, S Tombelli, M Minunni, and M Mascini, Aptamer-based detection of plasma proteins by an electrochemical assay coupled to magnetic beads, *Analytical Chemistry*, **79**, 1466–1473 (2007).

Centi S, G Messina, S Tombelli, I Palchetti, and M Mascini, Different approaches for the detection of thrombin by an electrochemical aptamer-based assay coupled to magnetic beads, *Biosensors & Bioelectronics*, **23**(11), 1602–1609 (2008).

Cesaro-Tadic S, G Dernick, JD Juncker, G Buurman, H Kropshofer, H Michel, C Fattinger, and E Delamarche, *Lab Chip*, **4**, 563–569 (2004).

Cho EJ, JR Collett, AE Szalrauska, and ED Ellington, Optimization of aptamer microarray technology for multiple protein targets, *Analytica Chimica Acta*, **564**, 82–90 (2006).

College Hill , Valued indicators, *Nature Methods*, **6**(6), (2009) advertising feature.

Curry C, L Pham, W Kramp, T Sharp, and TF Soriano, *Clinical Chemistry*, **42**, 143–149 (1996).

Das J, MA Aziz, and H Yang, *Journal of the American Chemical Society*, **128**, 16022–16023 (2006).

DeKossodo S, V Houba, GE Grau, and WCS Group: *Journal of Immunological Methods*, **182**, 1078–1114 (1995).

Ellington AD and JW Szoztak, *In vitro* selection of RNA molecules that bind specific ligands, *Nature*, **346**, 818 (1990).

Emili AG and G Cagney, Large-scale functional analysis using peptide or protein microarrays, *Nature Biotechnology*, **18**, 393–397 (2000).

Fritz J, MK Baller, HP Lang, H Rothuizen, P Vettiger, E Meyer, HJ Guntherhodt, Ch Gerber, and JK Gimewski, Translating biomolecular recognition into nanomechanics, *Science*, **288**(5464), 316–318 (2000).

Gao Z, J Zhang, and BP Ting, A doubly amplified electrochemical immunoassay for carcinoembryonic antigen, *Biosensors & Bioelectronics*, **24**(7), 1825–1830 (2009).

Havlin S, Molecular diffusion and reactions in *The Fractal Approach to Heterogeneous Chemistry: Surfaces, Colloids, Polymers* (ed. D Avnir), Wiley, New York, 1989, pp. 251–269.

Hirsch LR, JB Jackson, A Lee, NJ Halos, and J West, A whole blood immunoassay using fold nanoshells, *Analytical Chemistry*, **75**, 2377–2381 (2003).

Honore B, M Ostergaard, and H Vorum, Functional genomics studied by proteomics, *Bioassays*, **26**(8), 901–915 (2004).

Huber F, M Hegner, C Gerber, H-J Guntherodt, and HP Lang, Label-free analysis of transcription factors using microcantilever arrays, *Biosensors & Bioelectronics*, **21**(8), 1599–1605 (2006).

Jimp, Cancer biomarkers: Adoption is driving growth report, email, personal communication, jimp@healthtech.com, June 08, 2009.

Kim N, DK Kim, YJ Cho, DK Moon, and WY Kim, Carp vitellogenin detection by an optical waveguide lightmode spectroscopic biosensor, *Biosensors & Bioelectronics*, **24**(3), 391–396 (2008a).

Kim YP, YH Oh, and NS Kim, Protein kinase assay on peptide-conjugated gold nanoparticles, *Biosensors & Bioelectronics*, **23**(7), 980–986 (2008b).

Linstedt G, A Jacobsson, PA Lundberg, H Hedelin, B Pettersson, and B Linsgaard, *Clinical Chemistry*, **36**, 53–58 (1990).

Malkowska E, *In-vitro* diagnostics: Market analysis 2009-2024 report, publication date, 29 May, 2009, email, personal communication, June 2, 2009.

Martin SJ, VE Granstaff, and GC Frye, Effect of surface roughness on the response of thickness-shear mode resonators in liquids, *Analytical Chemistry*, **65**, 2910–2922 (1991).

Mcbeath G, Protein microarrays and proteomics, *Nature Genetics*, **32**, 526–532 (2002).

McEndry R, J Zhang, Y Arntz, T Strunz, M Hegner, HP Lang, MK Baller, U Cesta, E Meyer, H-J Gunterherodt, and Ch Gerber, Multiple label-free biodetection and quantitative DNA binding assays on a nanobiomechanical cantilever array, *Proceedings of the National Academy of Sciences*, **26**(8), 901–915 (2004).

Mitchell PA, A perspective on protein microarrays, *Nature Biotechnology*, **20**, 225–229 (2002).

Okubo I, MP Brown, K Chiba, R Kasukawa, and T Nishimaki, *Molecular Biology*, **25**, 217–224 (1998).

Old L, Polypeptide mediator network, *Nature*, **326**, 330–331 (1987).

Platt M, W Rowe, L Knowles, PJ Day, and DB Kell, Analysis of aptamer sequence activity relationships, *Integrative Biology*, **1**, 116–122 (2009a).

Platt M, W Rowe, DC Wedge, DR Kell, J Knowles, and PJR Day, Aptamer evolution for array-based diagnostics, *Analytical Biochemistry*, (2009b). In press.

Ramakrishnan A and A Sadana, A single fractal analysis of cellular analyte-receptor binding kinetics using biosensors, *BioSystems*, **59**, 35–51 (2001).

Rossomando EF and I White, *Journal of Peridontology*, **64**, 445–449 (1993).

Sadana A, A fractal analysis for the evaluation of hybridization kinetics in biosensors, *Journal of Colloid and Interface Science*, **151**(1), 166–177 (2001).

Sadana A, *Fractal Binding and Dissociation Kinetics for Different Biosensor Applications*, Elsevier, Amsterdam, 2005.

Shin J and C-C Liu, Development of a screen-printed cholesterol biosensor: Comparing the performance of gold and platinum as the working electrode material and fabrication using a self-assembly approach, *Sensors & Actuators*, **120**(2), 417–425 (2007).

Soper SA, K Brown, A Ellington, B Frazier, GG Manero, V Gau, SI Gutman, DF Hayes, B Korte, JL Landers, D larson, F Ligler, A Majumdar, M Mascini, D Nolte, Z Rosenzweig, J Wang, and D Wilson, Point-of-care biosensor systems for cancer diagnostics/prognostics, *Biosensors & Bioelectronics*, (2006).

Takahashi M, T Funato, KK Ishii, M Kaku, and T Sasaki, Measurement of tumor necrosis factor-α messenger RNA in synovial fibroblasts by real-time quantitative reverse transcriptase-polymerase chain reaction, *Journal of Laboratory and Clinical Medicine*, **137**, 101–106 (2001).

Teppo AM and CPJ Maury, Radioimmunoassay of tumor necrosis factor in serum, *Clinical Chemistry*, **33**, 2024–2027 (1987).

Tombelli S, M Minuuni, and M Mascini, *Biomolecular Engineering*, **24**, 191–200 (2007).
Tuerk C and L Gold, Systematic evolution of ligands by exponential enrichment: RNA ligands to bacteriophage T4 DNA polymerase, *Science*, **249**, 505 (1990).
Wang J, G Liu, MH Engelhard, and Y Lin, Sensitive immunoassay of a biomarker tumor necrosis-α based on poly (guanine)-functionalized silicon nanoparticle label, *Analytical Chemistry*, **78**, 6974–6979 (2006).
Warren CL, NCS Kratochvil, KE Hauschild, S Folster, ML Brezinski, PB Dervan, GN Phillips, and AZ Ansari, Defining the sequence-recognition profile of DNA-binding molecules, *Proceedings of the National Academy of Sciences*, **103**, 867–872 (2006).
Yao H, HB Halsall, WR Heineman, and SH Jenkins, *Clinical Chemistry*, **41**, 591–593 (1995).
Yates AM, SJ Elvin, and DEJ Williamson, *Journal of Immunoassay*, **20**, 31–44 (1999).
Zhou N, J Wang, T Chen, Z Yu, and G Li, *Analytical Chemistry*, **78**, 5227–5230 (2006).

CHAPTER 16

Binding and Dissociation Kinetics During Hybridization on Biosensor Surfaces

Chapter Outline

16.1 Introduction

Mao et al. (2009) recently reported that the nucleic acid test is important in the diagnosis and treatment of genetic disorders, for the detection of infectious agents, drug discovery, and in the warnings against biowarfare agents (Wang, 1999; Palacek and Fojta, 2001; Gooding, 2002; Drummond et al., 2003). Although new methods such as real-time polymerase chain reaction (RT-PCR) (Kaltenbeck and Wang, 2005), DNA microarrays (gene chip) (Brown and Botstein, 1999), SPR BIAcore instrument (http://wwwbiacore.com), and GeneXpert system (Petersen et al., 1999) are fast and sensitive tools to detect nucleic acid sequences, there is a high equipment cost and a need for highly-trained personnel. An ideal tool is required for fast, sensitive, low-cost, and easy-to-use detection of nucleic acids (Piunno and Krull, 2005; Hahns et al., 2005). Mao et al. (2009) point out that nucleic acid biosensors may be good candidates to meet these standards. Odenthal and Gooding (2007) and Leung et al. (2007) have reported on the use of different nucleic acid biosensors with different transducers. Nanomaterial labels and novel signal amplification strategies labels have increased sensitivity (Wang, 2005). Mao et al. (2009) have recently reported on disposable nucleic acid biosensors based on gold nanoparticle probes and a lateral flow strip.

Handbook of Biosensors and Biosensor Kinetics. DOI: 10.1016/B978-0-444-53262-6.00016-4

Some of the more recent studies that have appeared in the literature include:

(a) Locked nucleic acid based biosensors for surface interaction studies and biosensor development (Martinez et al., 2009),
(b) Real-time monitoring of the activity and kinetics of T4 polynucleotide kinase (PNK) by a singly labeled DNA-hairpin smart probe coupled with λ exonuclease cleavage (Song and Zhao, 2009),
(c) A competitive kinetic model of nucleic acid surface hybridization in the presence of point mutants (Bishop et al., 2006),
(d) Enhancement of DNA immobilization and hybridization on gold electrode modified by nanogold electrodes (Liu et al., 2005),
(e) Sequential injection analysis system for the sandwich hybridization-based detection of nucleic acids (Edwards and Baeumner, 2006),
(f) A novel fluorescence-based array biosensor:principle and application to DNA hybridization arrays (Schultz et al., 2008),
(g) Electrochemical detection of 17-β estradiol using DNA aptamer immobilized gold electrode chip (Kim et al., 2007),
(h) Surface plasmon resonance study of cooperative interactions of estrogen receptor α and transcriptional factor Sp1 with composite DNA elements (Neo et al., 2009), and Ultrasensitive optical DNA biosensor based on surface immobilization of molecular beacon (MB) by a bridge structure (Li et al., 2001).

In this chapter we use fractal analysis to analyze the binding and dissociation (if applicable) kinetics of some of the examples presented above. These examples were randomly selected from the literature. It should be indicated, as done elsewhere in book, that the fractal analysis method is just one method of analyzing the (external) diffusion-limited kinetics of binding and dissociation occurring on the biosensor surface. Other possible methods of analyzing the kinetics are also available. One particular advantage of the fractal analysis method is that the fractal dimension provides a quantitative measure of the degree of heterogeneity on the biosensor surface. This method is particularly advantageous in that it helps link the degree of heterogeneity on the surface with the binding and the dissociation rate coefficients.

16.2 Theory

Havlin (1989) has reviewed and analyzed the diffusion of reactants towards fractal surfaces. The details of the theory and the equations involved for the binding and the dissociation phases for analyte-receptor binding are available in the literature (Sadana, 2001). The details are not repeated here except that the equations are given to permit an easier reading. These equations have been applied to other biosensor systems (Sadana, 2001; Ramakrishnan and Sadana, 2001; Sadana, 2005). For most applications, a single- or a dual-fractal analysis is often adequate to describe the binding and the dissociation kinetics. Peculiarities in the values of the binding and the dissociation rate coefficients, in the systems being analyzed will be carefully noted, if applicable.

16.2.1 Single-Fractal Analysis

Binding Rate Coefficient

Havlin (1989) reports that the diffusion of a particle (analyte [Ag]) from a homogeneous solution to a solid surface (e.g., receptor [Ab]-coated surface) on which it reacts to form a product (analyte-receptor complex; (Ab·Ag)) is given by:

$$(\mathrm{Ab\cdot Ag}) \approx \begin{cases} t^{(3-D_{\mathrm{f,bind}})/2} = t^p, & t < t_c \\ t^{1/2}, & t > t_c \end{cases} \tag{16.1}$$

Here $D_{\mathrm{f,bind}}$ or D_{f} is the fractal dimension of the surface during the binding step. t_c is the cross-over value. Havlin (1989) points out that the cross-over value may be determined by $r_c^2 \sim t_c$. Above the characteristic length, r_c, the self-similarity of the surface is lost and the surface may be considered homogeneous. Above time, t_c the surface may be considered homogeneous, since the self-similarity property disappears, and "regular" diffusion is now present. For a homogeneous surface where D_{f} is equal to 2, and when only diffusional limitations are present, $p = ½$ as it should be. Another way of looking at the $p = ½$ case (where $D_{\mathrm{f,bind}}$ is equal to 2) is that the analyte in solution views the fractal object, in our case, the receptor-coated biosensor surface, from a "large distance." In essence, in the association process, the diffusion of the analyte from the solution to the receptor surface creates a depletion layer of width $(Đt)^{½}$ where $Đ$ is the diffusion constant. This gives rise to the fractal power law, (Analyte·Receptor) $\sim t^{(3-D_{\mathrm{f,bind}})/2}$. For the present analysis, t_c is arbitrarily chosen and we assume that the value of the t_c is not reached. One may consider the approach as an intermediate "heuristic" approach that may be used in the future to develop an autonomous (and not time-dependent) model for diffusion-controlled kinetics.

Dissociation Rate Coefficient

The diffusion of the dissociated particle (receptor [Ab] or analyte [Ag]) from the solid surface (e.g., analyte [Ag]-receptor [Ab] complex coated surface) into solution may be given, as a first approximation by:

$$(\mathrm{Ab\cdot Ag}) \approx -t^{(3-D_{\mathrm{f,bind}})/2} = -t^p, \quad t > t_{\mathrm{diss}} \tag{16.2}$$

Here $D_{\mathrm{f,diss}}$ is the fractal dimension of the surface for the dissociation step. This corresponds to the highest concentration of the analyte-receptor complex on the surface. Henceforth, its concentration only decreases. The dissociation kinetics may be analyzed in a manner "similar" to the binding kinetics.

16.2.2 Dual-Fractal Analysis

Binding Rate Coefficient

Sometimes, the binding curve exhibits complexities and two parameters (k, D_f) are not sufficient to adequately describe the binding kinetics. This is further corroborated by low values of the r^2 factor (goodness-of-fit). In that case, one resorts to a dual-fractal analysis (four parameters; k_1, k_2, D_{f1}, and D_{f2}) to adequately describe the binding kinetics. The single-fractal analysis presented above is thus extended to include two fractal dimensions. At present, the time ($t = t_1$) at which the "first" fractal dimension "changes" to the "second" fractal dimension is arbitrary and empirical. For the most part, it is dictated by the data analyzed and experience gained by handling a single-fractal analysis. A smoother curve is obtained in the "transition" region, if care is taken to select the correct number of points for the two regions. In this case, the product (antibody-antigen; or analyte-receptor complex, Ab·Ag or analyte·receptor) is given by:

$$(\text{Ab}\cdot\text{Ag}) \approx \begin{cases} t^{(3-D_{f1,\text{bind}})/2} = t^{p1}, & t < t_1 \\ t^{(3-D_{f2,\text{bind}})/2} = t^{p2}, & t_1 < t < t_2 = t_c \\ t^{1/2}, \quad t > t_c & \end{cases} \tag{16.3}$$

In some cases, as mentioned above, a triple-fractal analysis with six parameters (k_1, k_2, k_3, D_{f1}, D_{f2}, and D_{f3}) may be required to adequately model the binding kinetics. This is when the binding curve exhibits convolutions and complexities in its shape due perhaps to the very dilute nature of the analyte in some of the cases to be presented or for some other reasons. Also, in some cases, a dual-fractal analysis may be required to describe the dissociation kinetics.

16.3 Results

In this chapter we use fractal analysis to analyze the binding and dissociation (if applicable) kinetics of 9a0 T4 PNK by a singly labeled DNA-hairpin smart probe coupled with λ exonuclease cleavage (Song and Zhao, 2009), (b) binding of 10 mM target to LNA (locked nucleic acid bases) molecular beacons (LNB) (Martinez et al., 2009), (c) binding of 500 pM cy3-labelled targets (both raw and corrected data) to different areas on a 20-mer capture probe spotted on glass (Schultz et al., 2008), and (d) the electrochemical detection of 17β-estradiol using a DNA aptamer immobilized gold electrode chip (Kim et al., 2007).

Song and Zhao (2009) have recently analyzed the real-time monitoring of the activity and kinetics of T4 PNK by a single labeled DNA-hairpin smart probe coupled with λ exonuclease cleavage. These authors report that T4 PNK catalyzes the transfer of gamma-phosphate residue of ATP to the 5′-hydroxyl group of nucleic acids and oligonucleotides (Richardson, 1965). This enzyme has been widely used in the detection of DNA adducts (Lee et al., 1995;

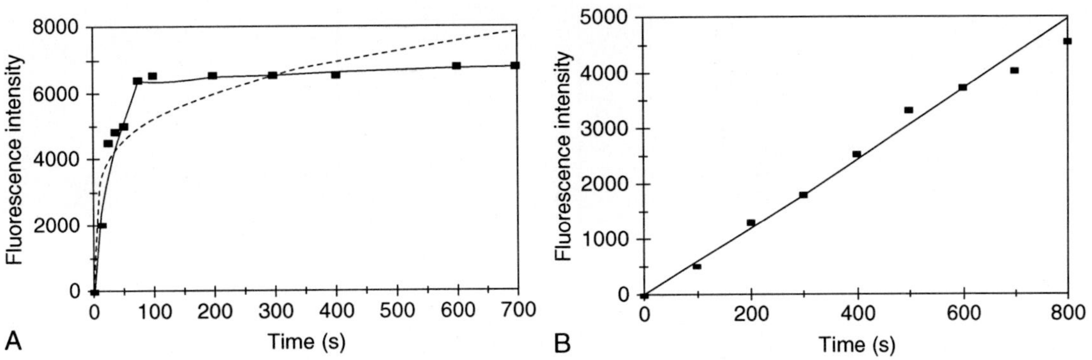

Figure 16.1

(a) Binding of 40 nM FAM-SP-2p + 10 units of λ exonuclease in solution to the singly labeled DNA hairpin smart probe (Song and Zhao, 2009). (b) Binding of 40 nM FAM-SP-2p + 10 units λ exonuclease +5.6 nM/s T4PNK (Song and Zhao, 2009). When only a solid line (—) is used then a single-fractal analysis applies. When both a dashed (- - -) and a solid (—) line are used then the dashed line represents a single-fractal analysis and the solid line represents a dual-fractal analysis.

Phillips and Arit, 2007), DNA oligonucleotides (El Atifi et al., 2003; Frauendorf et al., 2003), and in the repair of DNA lesions (Chappell et al., 2002; Rasouli-Nia et al., 2004).

Figure 16.1a shows the binding of FAM (fluorescein)SP-2p+10 units of λ exonuclease in solution to the singly labeled DNA hairpin smart probe (Song and Zhao, 2009). FAM-SP-2 is the oligonucleotide sequence (5′-3′) *GGGCC(AG*$_{10}$*)GGCCC*-FAM, SP: DNA hairpin quenched solely by the DNA base guanine were introduced as smart probes (Knemeyer et al., 2000). A dual-fractal analysis is required to adequately describe the binding kinetics. The values of (a) the binding rate coefficient, k, and the fractal dimension, D_f, for a single-fractal analysis, and (b) the binding rate coefficients, k_1 and k_2, and the fractal dimensions, D_{f1} and D_{f2}, for a dual-fractal analysis are given in Table 16.1.

It is of interest to note that for the dual-fractal analysis, as the fractal dimension increases by a factor of 1.51 from a value of D_{f1} equal to 1.9390 to D_{f2} equal to 2.9296, the binding rate coefficient increases by a factor of 8.34 from a value of k_1 equal to 640.96 to k_2 equal to 5243.91. The changes in the degree of heterogeneity or the fractal dimension on the biosensor surface and in the binding rate coefficient are in the same direction.

Figure 16.1b shows the binding of 40 nM FAM-SP-2p + 10 units λ exonuclease + 5.6 nM/s T4 PNK (T4 PNK) in solution to the singly labeled DNA hairpin smart probe (Song and Zhao, 2009). In this case, a single-fractal analysis is adequate to describe the binding kinetics. The values of the binding rate coefficient, k, and the fractal dimension, D_f, for a single-fractal analysis are given in Table 16.1.

Figure 16.2a shows the binding of 40 nM FAM-SP-2 + 5.6 nM/s 4T PNK + 1 mM ATP and 0.5 units of λ exonuclease at pH 8.0 to the singly labeled DNA hairpin smart probe (Song and Zhao, 2009). A dual-fractal analysis is required to adequately describe the binding kinetics.

Table 16.1: Binding rate coefficients and fractal dimensions for the binding of FAM-SP-2 + $lambda_{exo}$ + T4PNK and FAB-SP-2p + $lambda_{exo}$ in solution to the G-quenched singly-labeled DNA-hairpin smart probe coupled with λ exonuclease (Song and Zhao, 2009).

Analyte in Solution/ Receptor on Surface	k	k_1	k_2	D_f	D_{f1}	D_{f2}
40 nM FAM + SP2 + $lambda_{exo}$(10 units)/G quenched smart probe	1972.38 ± 512.55	640.96 ± 124.0	5343.91 ± 63.44	2.5754 ± 0.1088	1.9390 ± 0.2090	2.9296 ± 0.02314
40 nM FAM + SP2 + $lambda_{exo}$(10 units) + 40 nM 4PNK//G quenched smart probe	4.576 ± 0.439	na	na	0.9110 ± 0.0986	na	na

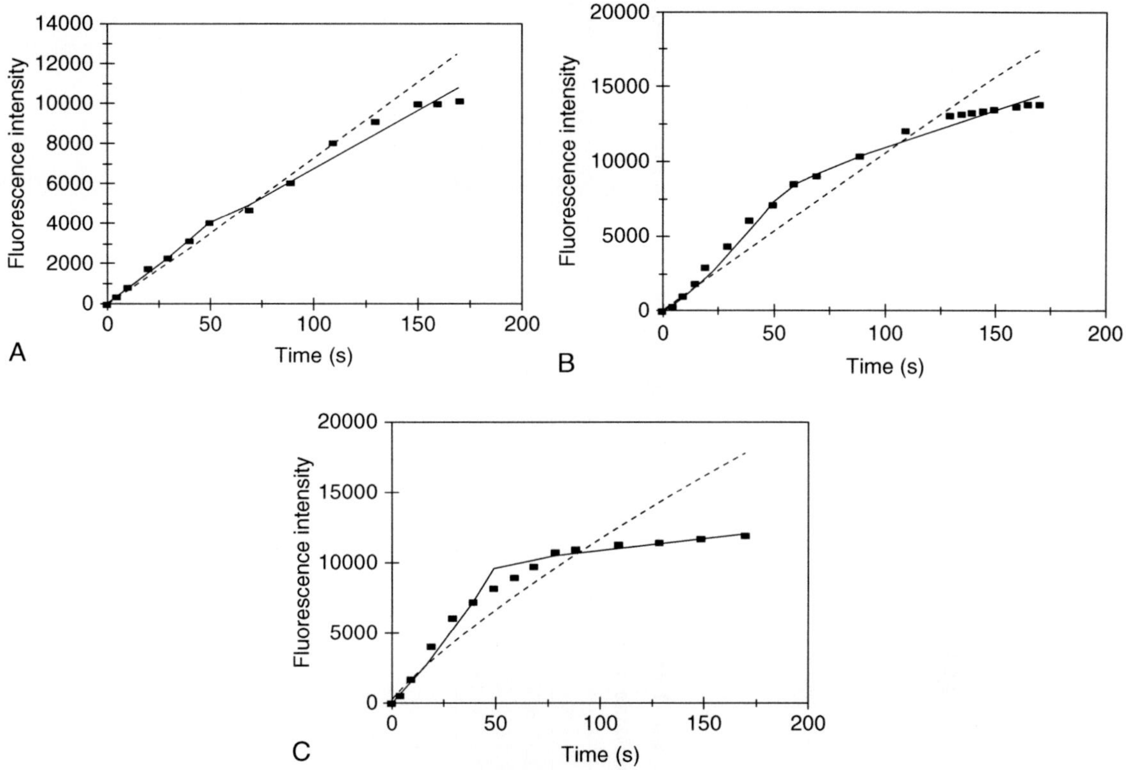

Figure 16.2

Binding of 40 nM FAM-SP-2 + 5.6 nM/s 4TPNK + 1 mM ATP, pH 8.0, to the singly labeled DNA hairpin smart probe (Song and Zhao, 2009). Influence of λ exonuclease units: (a) 0.5, (b) 1.0, (c) 2.5. When only a solid line (—) is used then a single-fractal analysis applies. When both a dashed (- - -) and a solid (—) line are used then the dashed line represents a single-fractal analysis and the solid line represents a dual-fractal analysis.

The values of (a) the binding rate coefficient, k, and the fractal dimension, D_f, for a single-fractal analysis, and (b) the binding rate coefficients, $k_1$and k_2, and the fractal dimensions, D_{f1} and D_{f2}, for a dual-fractal analysis are given in Table 16.2.

It is of interest to note that as the fractal dimension increases by a factor of 1.475 from a value of D_{f1} equal to 0.822 to D_{f2} equal to 1.213, the binding rate coefficient increases by a factor of 1.89 from a value of k_1 equal to 58.33 to k_2 equal to 110.31. The changes in the binding rate coefficient and in the degree of heterogeneity or the fractal dimension on the biosensor surface are in the same direction.

Figure 16.2b shows the binding of 40 nM FAM-SP-2 + 5.6 nM/s 4T PNK + 1 mM ATP and 1.0 units of λ exonuclease at pH 8.0 to the singly labeled DNA hairpin smart probe (Song and Zhao, 2009). Once again, a dual-fractal analysis is required to adequately describe the binding kinetics. The values of (a) the binding rate coefficient, k, and the fractal dimension, D_f, for a single-fractal analysis, and (b) the binding rate coefficients, k_1 and k_2, and the fractal dimensions, D_{f1} and D_{f2}, for a dual-fractal analysis are given in Table 16.2.

Table 16.2: Binding of 40 nM FAM-SP-2, 5.6 nM/s T4PNK, and 1 mM ATP to the singly labeled DNA-hairpin probe coupled with λ exonuclease at pH 8.0.

λ exonuclease	k	k_1	k_2	D_f	D_{f1}	D_{f2}
0.5	69.96 ± 65.16	58.33 ± 6.93	110.31 ± 7.19	0.906 ± 0.338	0.822 ± 0.114	1.213 ± 0.1578
1.0	118.57 ± 34.29	50.75 ± 9.98	1109.57 ± 29.72	1.0532 ± 0.1124	0.458 ± 0.143	2.004 ± 0.0464
2.5	241.15 ± 87.74	98.96 ± 23.1	4750.65 ± 133.47	13224 ± 0.1658	0.690 ± 0.189	2.638 ± 0.0682

Influence of λ exonuclease (Song and Zhao, 2009).

It is of interest to note, once again, that as the fractal dimension increases by a factor of 4.375 from a value of D_{f1} equal to 0.458 to D_{f2} equal to 2.004, the binding rate coefficient increases by a factor of 21.86 from a value of k_1 equal to 50.75 to k_2 equal to 1109.57. The changes in the binding rate coefficient and in the degree of heterogeneity or the fractal dimension on the biosensor surface are, once again, in the same direction.

Figure 16.2c shows the binding of 40 nM FAM-SP-2 + 5.6 nM/s 4T PNK + 1 mM ATP and 2.50 units of λ exonuclease at pH 8.0 to the singly labeled DNA hairpin smart probe (Song and Zhao, 2009). Once again, a dual-fractal analysis is required to adequately describe the binding kinetics. The values of (a) the binding rate coefficient, k, and the fractal dimension, D_f, for a single-fractal analysis, and (b) the binding rate coefficients, k_1 and k_2, and the fractal dimensions, D_{f1} and D_{f2}, for a dual-fractal analysis are given in Table 16.2.

It is of interest to note, once again, that as the fractal dimension increases by a factor of 3.823 from a value of D_{f1} equal to 0.690 to D_{f2} equal to 2.638, the binding rate coefficient increases by a factor of 48 from a value of k_1 equal to 98.96 to k_2 equal to 4750.65. The changes in the binding rate coefficient and in the degree of heterogeneity or the fractal dimension on the biosensor surface are, once again, in the same direction.

Figure 16.3a and Table 16.2 show the increase in the binding rate coefficient, k_1, with an increase in the λ nuclease units in solution for a dual-fractal analysis. For the data shown in Figure 16.3a the binding rate coefficient, k_1, is given by:

$$k_1 = (64.71 \pm 12.52)(\lambda \text{ exonuclease units})^{0.3494 \pm 0.2613} \quad (16.4a)$$

The fit is reasonable. Only three data points are available. The availability of more data points would lead to a more reliable fit. The binding rate coefficient, k_1, exhibits only a mild dependence on the λ exonuclease units in solution as shown by the close to 0.35 (equal to 0.3494) order of dependence exhibited.

Figure 16.3b and Table 16.2 show the increase in the binding rate coefficient, k_2, with an increase in the λ exonuclease units in solution for a dual-fractal analysis. For the data shown in Figure 16.3b the binding rate coefficient, k_2, is given by:

$$k_2 = (703.48 \pm 127.94)(\lambda \text{ exonuclease units})^{2.299 \pm 0.2613} \quad (16.4b)$$

The fit is good. Only three data points are available. The availability of more data points would lead to a more reliable fit. The binding rate coefficient k_2, exhibits an order of dependence between two and two and a half (equal to 2.299) on the λ exonuclease units in solution.

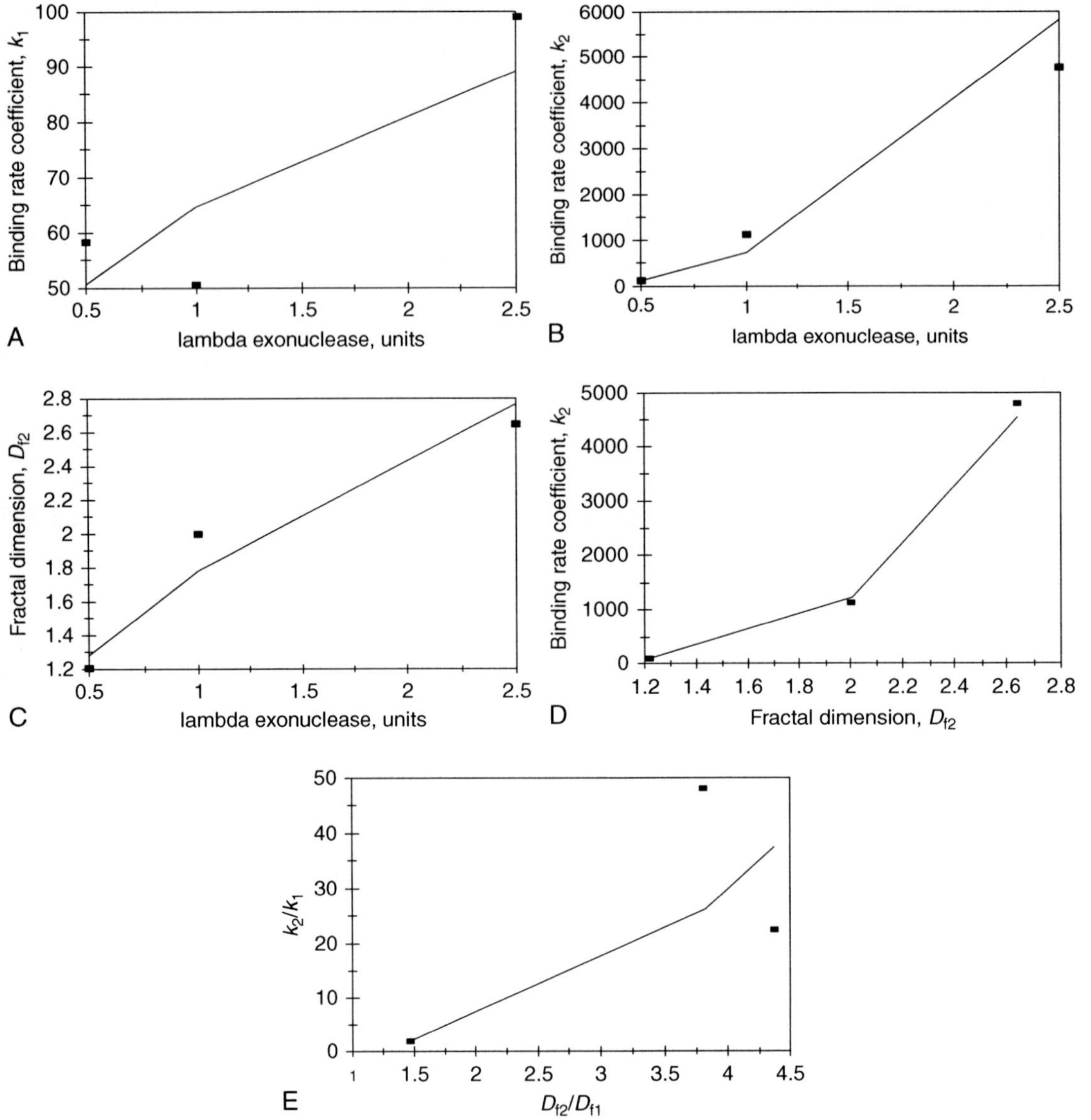

Figure 16.3

(a) Increase in the binding rate coefficient, k_1 for a dual-fractal analysis with an increase in the λ exonuclease units. (b) Increase in the binding rate coefficient, k_2, for a dual-fractal analysis with an increase in the λ exonuclease units. (c) Increase in the fractal dimension, D_{f2}, for a dual-fractal analysis with an increase in the λ exonuclease units. (d) Increase in the binding rate coefficient k_2 with an increase in the fractal dimension D_{f2}, and (e) increase in the binding rate coefficient ratio k_2/k_1 with an increase in the fractal dimension ratio D_{f2}/D_{f1}.

Figure 16.3c and Table 16.2 show the increase in the fractal dimension, D_{f2}, with an increase in the λ nuclease units in solution for a dual-fractal analysis. For the data shown in Figure 16.3c the fractal dimension, D_{f2}, is given by:

$$D_{f2} = (1.791 \pm 0.260)(\lambda \text{ exonuclease units})^{0.475 \pm 0.119} \quad (16.4c)$$

The fit is good. Only three data points are available. The availability of more data points would lead to a more reliable fit. The fractal dimension, D_{f2}, exhibits only a mild dependence on the λ exonuclease units in solution as shown by the less than one-half (equal to 0.475) order of dependence exhibited.

Figure 16.3d and Table 16.2 show the increase in the binding rate coefficient, k_2, with an increase in the fractal dimension, D_{f2}, for a dual-fractal analysis. For the data shown in Figure 16.3d the binding rate coefficient, k_2, is given by:

$$k_2 = (42.399 \pm 4.055)(D_{f2})^{4.816 \pm 0.164} \quad (16.4d)$$

The fit is good. Only three data points are available. The availability of more data points would lead to a more reliable fit. The binding rate coefficient, k_2, is very sensitive to the fractal dimension, D_{f2}, as noted by the close to fifth (equal to 4.816) order of dependence exhibited.

Figure 16.3e and Table 16.2 show the increase in the binding rate coefficient ratio, k_2/k_1, with an increase in the fractal dimension ratio, D_{f2}/D_{f1}, for a dual-fractal analysis. For the data shown in Figure 16.3e the binding rate coefficient ratio, k_2/k_1, is given by:

$$k_2/k_1 = (0.729 + 0.918)(D_{f2}/D_{f1})^{2.668 \pm 0.970} \quad (16.4e)$$

There is scatter in the data. This is reflected in the error for the ratio of the binding rate coefficients, k_2/k_1. Only three data points are available. The availability of more data points would lead to a more reliable fit. The ratio of the binding rate coefficients, k_2/k_1, is sensitive to the fractal dimension ratio, (D_{f2}/D_{f1}) as noted by the order of dependence between two and a half and third (equal to 2.668) exhibited.

Figure 16.4a shows the binding of 20 nM FAM-SP-2 to T4 PNK immobilized on the sensor surface. A single-fractal analysis is adequate to describe the binding kinetics. The values of the binding rate coefficient, k, and the fractal dimension, D_f, for a single-fractal analysis is given in Table 16.3.

Figure 16.4b shows the binding of 30 nM FAM-SP-2 to T4 PNK immobilized on the sensor surface. A single-fractal analysis is adequate to describe the binding kinetics. The values of the binding rate coefficient, k, and the fractal dimension, D_f, for a single-fractal analysis is given in Table 16.3.

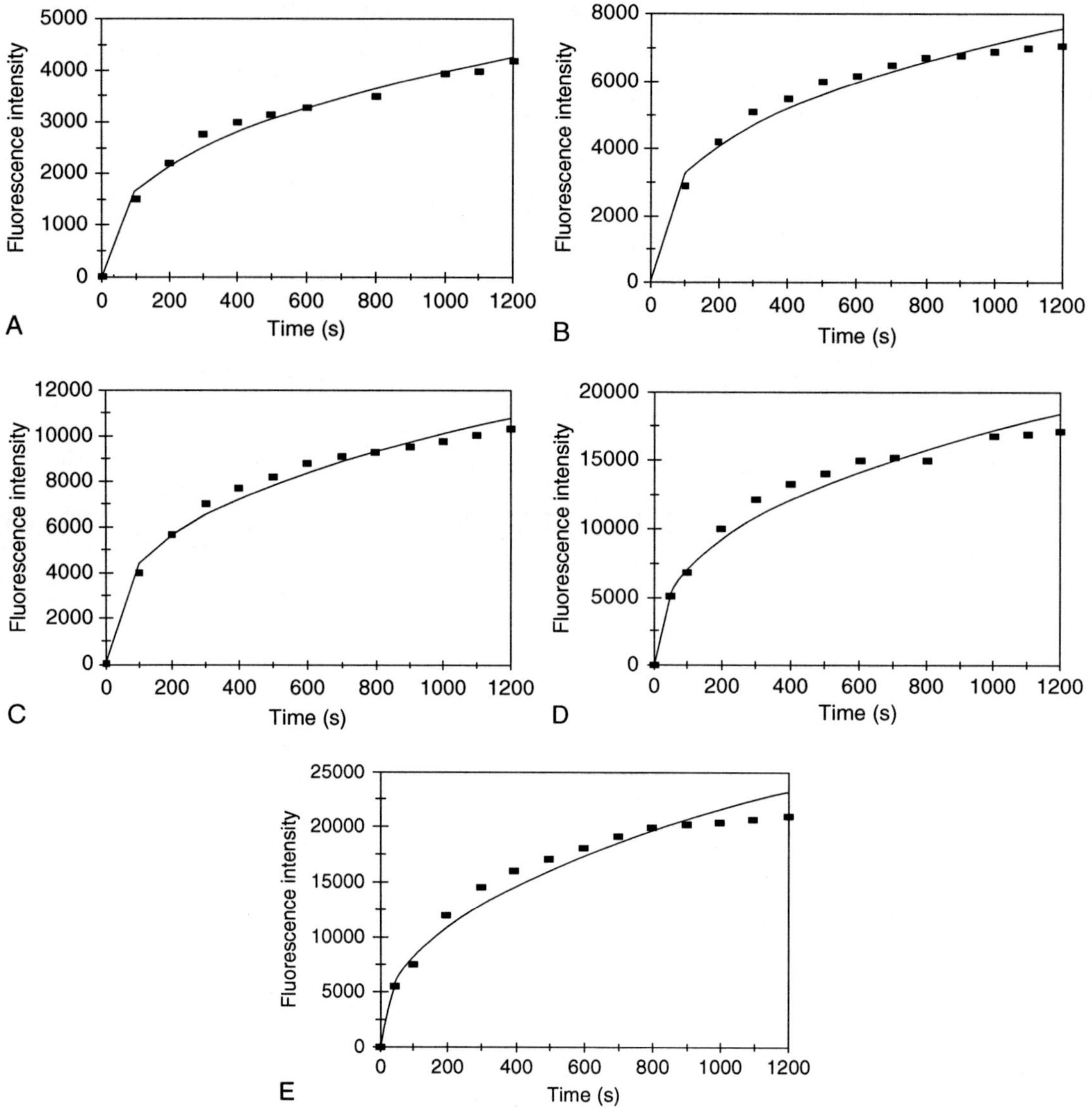

Figure 16.4

Binding curves for different concentrations (in nM) of FAM-SP-2 to T4PNK Immobilized on the sensor surface (Song and Zhao, 2009): (a) 20, (b) 30, (c) 40, (d) 50, (e) 60.

Figure 16.4c shows the binding of 40 nM FAM-SP-2 to T4 PNK immobilized on the sensor surface. A single-fractal analysis is adequate to describe the binding kinetics. The values of the binding rate coefficient, *k*, and the fractal dimension, D_f, for a single-fractal analysis is given in Table 16.3.

Figure 16.4d shows the binding of 50 nM FAM-SP-2 to T4 PNK immobilized on the sensor surface. A single-fractal analysis is adequate to describe the binding kinetics. The values of

Table 16.3: Binding (phosphorylation) of FAM-SP-2 by T4 PNK by a single-labeled DNA-hairpin smart probe with λ exonuclease.

FAM-SP-2 concentration (nM)	k	D_f
20	281.63 ± 16.70	2. 2322 ± 0.0474
30	670.97 ± 41.71	2.3148 ± 0.0482
40	829.22 ± 44.39	2. 2754 ± 0.0416
50	1213.94 ± 91.79	2.2300 ± 0.0442
60	1145.14 ± 97.09	2.1492 ± 0.0480

Influence of substrate (FAM-SP-2) concentration (Song and Zhao, 2009). T4 PNK, λ exonuclease and ATP concentration are 5.6 nM/s, 10 units, and 1.0 nM, respectively.

the binding rate coefficient, k, and the fractal dimension, D_f, for a single-fractal analysis is given in Table 16.3.

Figure 16.4e shows the binding of 60 nM FAM-SP-2 to T4 PNK immobilized on the sensor surface. A single-fractal analysis is adequate to describe the binding kinetics. The values of the binding rate coefficient, k, and the fractal dimension, D_f, for a single-fractal analysis is given in Table 16.3.

Figure 16.5 and Table 16.3 show for a single-fractal analysis the increase in the binding rate coefficient, k, with an increase in the FAM-SP-2 concentration in solution in the 20-60 nM range in solution. For the data shown in Figure 16.5, the binding rate coefficient, k, is given by:

$$k = (6.502 \pm 1.336)[\text{FAM-SP-2, nM}]^{1.307 \pm 0.126} \tag{16.5}$$

The fit is good. Only five data points are available. The availability of more data points would lead to a more reliable fit. The binding rate coefficient, k, is sensitive to the FAM-SP-2

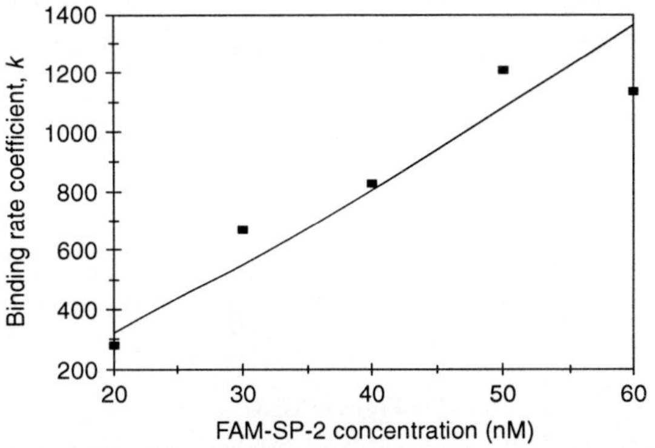

Figure 16.5
Increase in the binding rate coefficient, k, for a single-fractal analysis with an increase in the FAM-SP-2 concentration (in nM) in solution.

concentration in solution in the 20-60 nM range as noted by the dependence between one and one and a half (equal to 1.307) order exhibited.

Martinez et al. (2009) recently pointed out that DNA/RNA-based analytical methods are important in molecular biology, disease diagnosis, and gene expression. They emphasize that biosensors are popular as they permit the detection of a large number of samples using a fast and simple setup. Biosensors permit the real-time detection of nucleic acid molecules and gene expression changes (Feng et al., 1999; Steel et al., 2000; Steemers et al., 2000; Broude et al., 2001; Preininger and Chiarelli, 2001; Culha et al., 2004; Gang et al., 1999). Martinez et al. (2009) explain the importance of efficiency, reproducibility, and the stable immobilization of the DNA probes onto specific surfaces.

Martinez et al. (2009) report that their design of a new MB biosensor overcomes the previous limitations of MBs for surface immobilization. In essence, their new design adds LNAs to the beacon structure, resulting in a LNB that exhibits robust stability after surface immobilization. They indicate that their LNB-based biosensor exhibited better stability, reproducibility, selectivity, and robustness when compared with the regular molecular beacons (RMBs).

Figure 16.6 shows the binding of 10 μM of the target to the LNB-based biosensor. A dual-fractal analysis is required to adequately describe the binding kinetics. The values of (a) the binding rate coefficient, k, and the fractal dimension, D_f, for a single-fractal analysis, and (b) the binding rate coefficients, k_1 and k_2, and the fractal dimensions, D_{f1} and D_{f2}, for a dual-fractal analysis are given in Table 16.4.

It is of interest to note that as the fractal dimension increases by a factor of 1.67 from a value of D_{f1} equal to 1.406 to D_{f2} equal to 2.3538, the binding rate coefficient increases by a factor

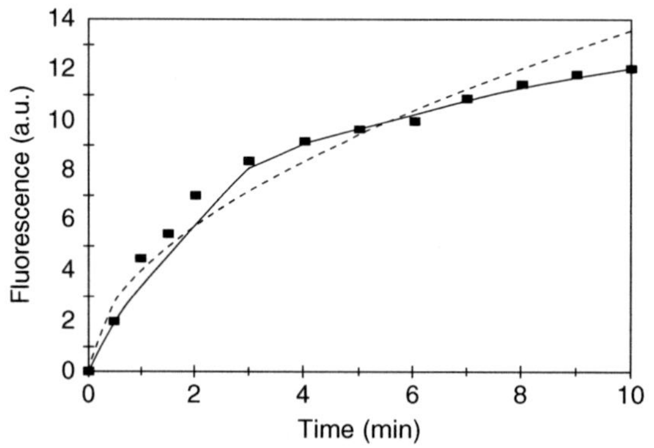

Figure 16.6
Binding of 10 mM target to LNA (locked nucleic acid bases) molecular beacon (Martinez et al., 2009). When only a solid line (—) is used then a single-fractal analysis applies. When both a dashed (- - -) and a solid (—) line are used then the dashed line represents a single-fractal analysis and the solid line represents a dual-fractal analysis.

Table 16.4: Binding rate coefficients and fractal dimensions for the binding of the target in solution to the LNB-based biosensor (Martinez et al., 2009).

Analyte/Receptor	k	k_1	k_2	D_f	D_{f1}	D_{f2}
Target/LNB-based biosensor	4.6647 ± 0.6508	3.878 ± 0.517	5.791 ± 0.107	1.9346 ± 0.095	1.406 ± 0.1824	2.3538 ± 0.0456

of 1.49 from a value of k_1 equal to 3.878 to k_2 equal to 5.791. This is almost a linear increase or close to it.

Schultz et al. (2008) recently developed a novel fluorescence-based array biosensor for field applications, such as environmental monitoring. It is based on DNA hybridization assays. The intent was to meet the demand for a portable, automated, and easy-to-maintain biosensor that permitted the continuous monitoring of surface reactions. Their sensor is based on a microscopic slide that serves as a transducer as well as a biological array sensor.

Schultz et al. (2008) have pointed out the need for integrated microdevices that combine integrated circuit elements, electro-optic excitations/detection systems, and bioreceptor probes (Malakios et al., 2004; Ito et al., 1996). Schultz et al. (2005) explain that fluorescence based assays predominate the microarray field owing to their better sensitivity and specificity, and decreased background signals when compared with free-label assays (Baldini and Giannetti, 2005).

Karpf et al. (1988) first demonstrated the use of hybridization assays. Later Eggers et al. (1994) applied this to a CCD chip with radio-active fluorescent labeling of target molecules. Vo-Dinh et al. (2004) showed the usefulness and feasibility of a DNA biochip using a phototransistor integrated circuit. Later there was an effort to reduce the size of the instrument and increase the number of analytes that could be screened. Anderson et al. (2000) developed the fiber-optic immunosensor, RAPTOR that permitted the detection of four bacterial agents, viruses, and toxins. Rodriguez-Mozaz et al. (2004) developed the RIANA (based on total internal reflection fluorescence) biosensor that was able to detect three contaminants (atrazene, isoprofuron, and estrone) in natural waters. Schultz et al. (2008) have developed a novel array biosensor method that uses the FCFD (fluorescent capillary filler device) on glass slides, and permits an automated continuous flow assay. DNA hybridization assays are used.

Figure 16.7a shows the binding of 500 pM cy3-labelled target (raw data) to a 20-mer capture probe immobilized on Area 1 on the microarray biosensor (Schultz et al., 2008). These authors showed the vignetting effect, that is the distance of the Area (1, 2, or 3) from the detector. Figure 16.7a shows that a dual-fractal analysis is required to adequately describe the binding kinetics. The values of (a) the binding rate coefficient, k, and the fractal

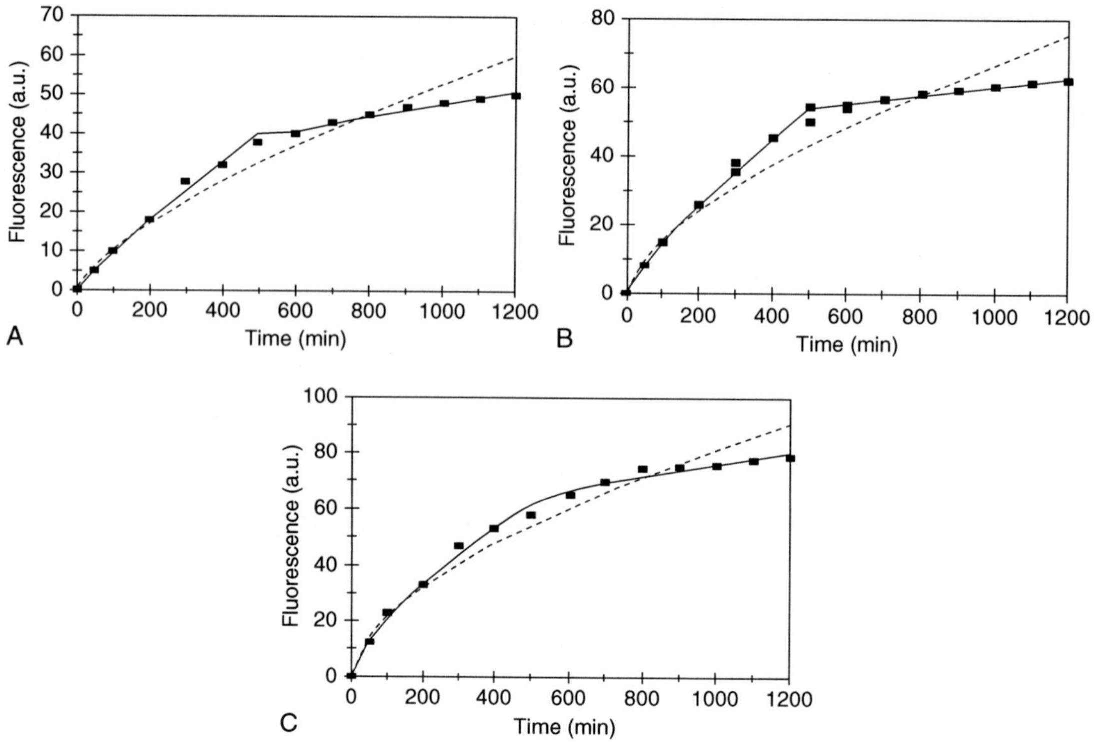

Figure 16.7

Binding (raw data) of 500 pM cy3-labelled targets to different areas on a 20-mer capture probe spotted on glass (Schultz et al., 2008): (a) Area 1, (b) Area 2, (c) Area 3. When only a solid line (—) is used then a single-fractal analysis applies. When both a dashed (- - -) and a solid (—) line are used then the dashed line represents a single-fractal analysis and the solid line represents a dual-fractal analysis.

dimension, D_f, for a single-fractal analysis, and (b) the binding rate coefficients, k_1 and k_2, and the fractal dimensions, D_{f1} and D_{f2}, for a dual-fractal analysis are given in Table 16.5.

For a dual-fractal analysis, it is of interest to note that as the fractal dimension increases by a factor of 1.93 from a value of D_{f1} equal to 1.2272 to D_{f2} equal to 2.3698, the binding rate coefficient increases by a factor of 33.29 from a value of k_1 equal to 0.1629 to k_2 equal to 5.4224. The changes in the degree of heterogeneity or the fractal dimension on the sensor surface and in the binding rate coefficient are in the same direction.

Figure 16.7b shows the binding of 500 pM cy3-labelled target (raw data) to a 20-mer capture probe immobilized on Area 2 (control) on the microarray biosensor (Schultz et al., 2008). It shows, once again, that a dual-fractal analysis is required to adequately describe the binding kinetics. The values of (a) the binding rate coefficient, k, and the fractal dimension, D_f, for a single-fractal analysis, and (b) the binding rate coefficients, k_1 and k_2, and the fractal dimensions, D_{f1} and D_{f2}, for a dual-fractal analysis are given in Table 16.5.

Table 16.5: Binding of 500 pM cy3-labelled targets to Area 1, Area 2, and Area 3 on a 20-mer capture probe spotted on glass (Schultz et al., 2008).

Area Number	k	k_1	k_2	D_f	D_{f1}	D_{f2}
One	0.3838 ± 0.0521	0.1629 ± 0.0105	5.4224 ± 0.0734	1.5734 ± 0.0856	1.2272 ± 0.0634	2.3698 ± 0.0442
Two (control)	0.8203 ± 0.1327	0.3445 ± 0.020	17.362 ± 0.287	1.7224 ± 0.08868	1.3724 ± 0.0574	2.6376 ± 0.0538
Three	1.4986 ± 0.1611	0.9099 ± 0.0762	12.179 ± 0.305	1.8408 ± 0.0604	1.6406 ± 0.0816	2.4670 ± 0.0812

Influence of the vignetting effect (distance of area from the detector).

For a dual-fractal analysis, it is of interest to note that as the fractal dimension increases by a factor of 1.92 from a value of D_{f1} equal to 1.3724 to D_{f2} equal to 2.6376, the binding rate coefficient increases by a factor of 50.40 from a value of k_1 equal to 0.3445 to k_2 equal to 17.362. Once again, the changes in the degree of heterogeneity or the fractal dimension on the sensor surface and in the binding rate coefficient are in the same direction.

Figure 16.7c shows the binding of 500 pM cy3-labelled target (raw data) to a 20-mer capture probe immobilized on Area 3 on the microarray biosensor (Schultz et al., 2008). It shows, once again, that a dual-fractal analysis is required to adequately describe the binding kinetics. The values of (a) the binding rate coefficient, k, and the fractal dimension, D_f, for a single-fractal analysis, and (b) the binding rate coefficients, k_1 and k_2, and the fractal dimensions, D_{f1} and D_{f2}, for a dual-fractal analysis are given in Table 16.5.

For a dual-fractal analysis, it is of interest to note that as the fractal dimension increases by a factor of 1.50 from a value of D_{f1} equal to 1.6406 to D_{f2} equal to 2.4670, the binding rate coefficient increases by a factor of 13.38 from a value of k_1 equal to 0.9099 to k_2 equal to 12.179. Once again, the changes in the degree of heterogeneity or the fractal dimension on the sensor surface and in the binding rate coefficient are in the same direction.

Figure 16.8a and Table 16.6 show the increase in the binding rate coefficient, k_1, with an increase in the fractal dimension, D_{f1}, for a dual-fractal analysis. For the data shown in Figure 16.8a, the binding rate coefficient, k_1, is given by:

$$k_1 = (0.0506 \pm 0.0037)D_{f1}^{5.880 \pm 0.357} \tag{16.6}$$

The fit is good. Only three data points are available. The availability of more data points would lead to better fit. The binding rate coefficient, k_1, is very sensitive to the fractal dimension, D_{f1}, or the degree of heterogeneity that exists on the microarray biosensor surface as noted by the order of dependence between five and a half and six (equal to 5.880) exhibited.

Figure 16.8b and Table 16.6 show the increase in the binding rate coefficient, k_2, with an increase in the fractal dimension, D_{f2}, for a dual-fractal analysis. For the data shown in Figure 16.8b, the binding rate coefficient, k_2, is given by:

$$k_2 = (0.000869 \pm 0.000305)D_{f2}^{10.303 \pm 3.933} \tag{16.7}$$

The fit is good. Only three data points are available. The availability of more data points would lead to better fit. The binding rate coefficient, k_2, is extremely sensitive to the fractal dimension, D_{f2}, or the degree of heterogeneity that exists on the microarray biosensor surface as noted by the order of dependence between 10 and 10 and one-half (equal to 10.303) exhibited.

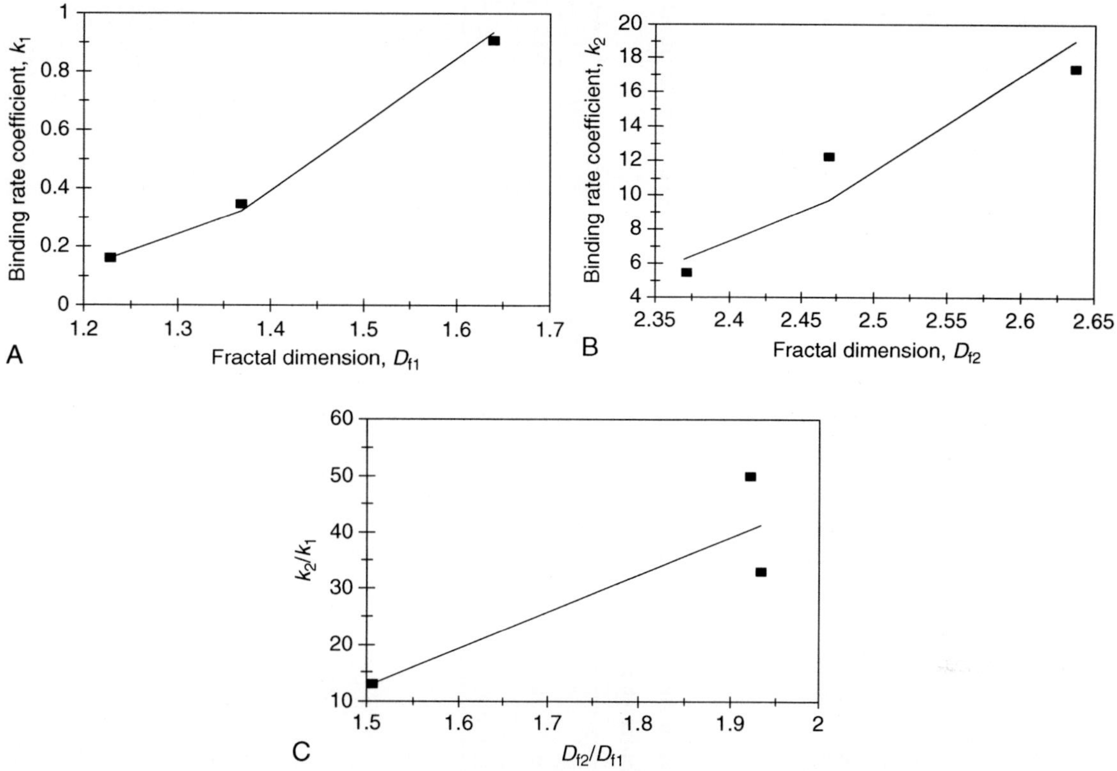

Figure 16.8

(a) Increase in the binding rate coefficient, k_1, for a dual-fractal analysis with an increase in the fractal dimension, D_{f1}. (b). Increase in the binding rate coefficient, k_2, for a dual-fractal analysis with an increase in the fractal dimension, D_{f2}. (c) Increase in the binding rate coefficient ratio, k_2/k_1, for a dual-fractal analysis with an increase in the fractal dimension ratio, D_{f2}/D_{f1}.

Figure 16.8c and Table 16.6 show the increase in the binding rate coefficient ratio, k_2/k_1, with an increase in the fractal dimension ratio, D_{f2}/D_{f1}, for a dual-fractal analysis. For the data shown in Figure 16.8c, the binding rate coefficient ratio, k_2/k_1, is given by:

$$k_2/k_1 = (1.0494 \pm 0.133)\left(\frac{D_{f2}}{D_{f1}}\right)^{4.884 \pm 2.767} \tag{16.8}$$

The fit is reasonable. There is scatter in the data. Only three data points are available. The availability of more data points would lead to better fit. The ratio of the binding rate coefficients, k_2/k_1, is very sensitive to the ratio of the fractal dimensions, D_{f2}/D_{f1}, as noted by the close to five (equal to 4.884) order of dependence exhibited for a dual-fractal analysis.

Figure 16.9a shows the binding of 500 pM cy3-labelled target (raw data) to a 20-mer capture probe immobilized on Area 1 on the microarray biosensor (Schultz et al., 2008).

Table 16.6: Binding (corrected) of 500 pM cy3-labelled targets to Area 1, Area 2, and Area 3 on a 20-mer capture probe spotted on glass (Schultz et al., 2008).

Area Number	k	k_1	k_2	D_f	D_{f1}	D_{f2}
One	0.5946 ± 0.1037	0.2108 ± 0.0207	7.908 ± 0.100	1.5288 ± 0.0998	1.1008 ± 0.1100	2.3186 ± 0.03638
Two (control)	0.3441 ± 0.0644	0.1378 ± 0.0145	5.680 ± 0.122	1.5068 ± 0.1066	1.1342 ± 0.1018	2.3628 ± 0.0614
Three	0.5751 ± 0.0990	0.1923 ± 0.0213	5.684 ± 0.060	1.6018 ± 0.0940	1.1520 ± 0.1244	2.2946 ± 0.02624

Influence of the vignetting effect (distance of area from the detector).

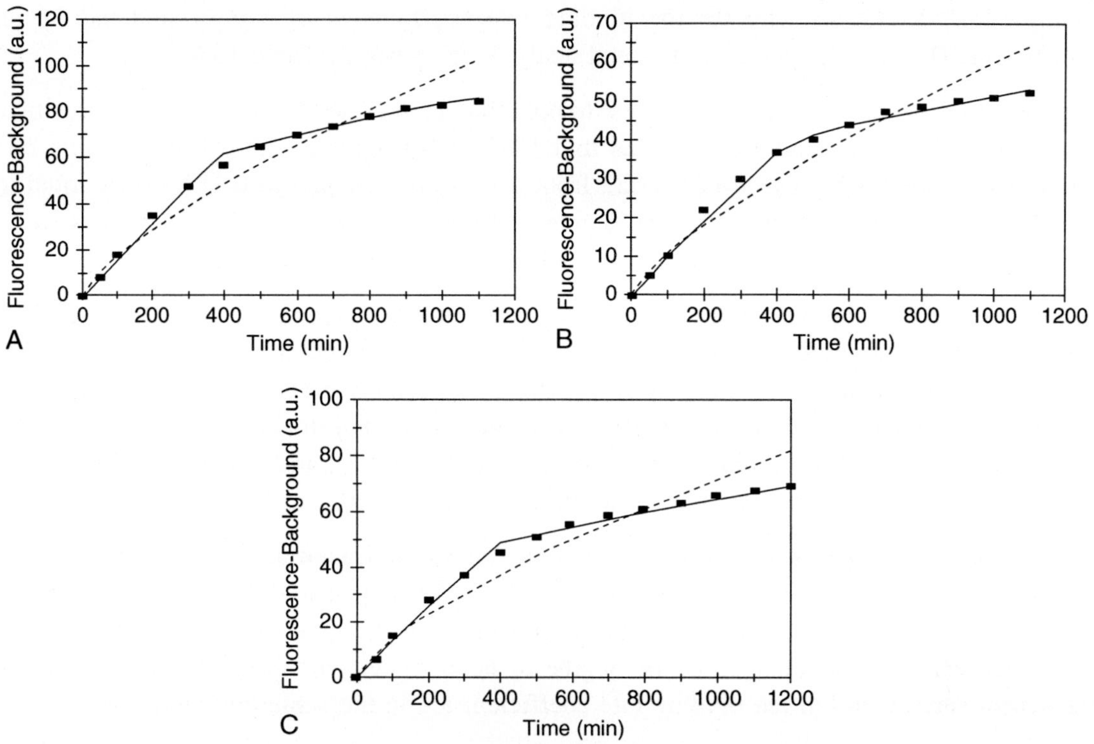

Figure 16.9

Binding (corrected data) of 500 pM cy3-labelled targets to different Areas on a 20-mer capture probe spotted on glass (Schultz et al., 2008): (a) Area 1, (b) Area 2, (c) Area 3. When only a solid line (—) is used then a single-fractal analysis applies. When both a dashed (- - -) and a solid (—) line are used then the dashed line represents a single-fractal analysis and the solid line represents a dual-fractal analysis.

These authors showed the vignetting effect, that is the distance of the Area (1, 2, or 3) from the detector. Figure 16.9a shows that a dual-fractal analysis is required to adequately describe the binding kinetics. The values of (a) the binding rate coefficient, k, and the fractal dimension, D_f, for a single-fractal analysis, and (b) the binding rate coefficients, k_1 and k_2, and the fractal dimensions, D_{f1} and D_{f2}, for a dual-fractal analysis are given in Table 16.6.

For a dual-fractal analysis, it is of interest to note that as the fractal dimension increases by a factor of 2.11 from a value of D_{f1} equal to 1.1008 to D_{f2} equal to 2.3186, the binding rate coefficient increases by a factor of 37.51 from a value of k_1 equal to 0.2108 to k_2 equal to 7.908. The changes in the degree of heterogeneity or the fractal dimension on the sensor surface and in the binding rate coefficient are in the same direction.

Figure 16.9b shows the binding of 500 pM cy3-labelled target (raw data) to a 20-mer capture probe immobilized on Area 2 (control) on the microarray biosensor (Schultz et al., 2008). It shows, once again, that a dual-fractal analysis is required to adequately describe the binding kinetics. The values of (a) the binding rate coefficient, k, and the fractal dimension, D_f, for a single-fractal analysis, and (b) the binding rate coefficients, k_1 and k_2, and the fractal dimensions, D_{f1} and D_{f2}, for a dual-fractal analysis are given in Table 16.6.

For a dual-fractal analysis, it is of interest to note that as the fractal dimension increases by a factor of 2.08 from a value of D_{f1} equal to 1.1342 to D_{f2} equal to 2.3628, the binding rate coefficient increases by a factor of 41.22 from a value of k_1 equal to 0.1378 to k_2 equal to 5.680. Once again, the changes in the degree of heterogeneity or the fractal dimension on the sensor surface and in the binding rate coefficient are in the same direction.

Figure 16.9c shows the binding of 500 pM cy3-labelled target (raw data) to a 20-mer capture probe immobilized on Area 3 on the microarray biosensor (Schultz et al., 2008). It shows, once again, that a dual-fractal analysis is required to adequately describe the binding kinetics. The values of (a) the binding rate coefficient, k, and the fractal dimension, D_f, for a single-fractal analysis, and (b) the binding rate coefficients, k_1 and k_2, and the fractal dimensions, D_{f1} and D_{f2}, for a dual-fractal analysis are given in Table 16.6.

For a dual-fractal analysis, it is of interest to note that as the fractal dimension increases by a factor of 1.99 from a value of D_{f1} equal to 1.1520 to D_{f2} equal to 2.2946, the binding rate coefficient increases by a factor of 29.56 from a value of k_1 equal to 0.1923 to k_2 equal to 5.684. Once again, the changes in the degree of heterogeneity or the fractal dimension on the sensor surface and in the binding rate coefficient are in the same direction.

Figure 16.10 and Table 16.6 show the decrease in the binding rate coefficient, k_1, with an increase in the fractal dimension, D_{f1}, for a dual-fractal analysis. For the data shown in Figure 16.10, the binding rate coefficient, k_1, is given by:

$$k_1 = (0.276 \pm 0.094) D_{f1}^{-3.65 \pm 8.96} \tag{16.9}$$

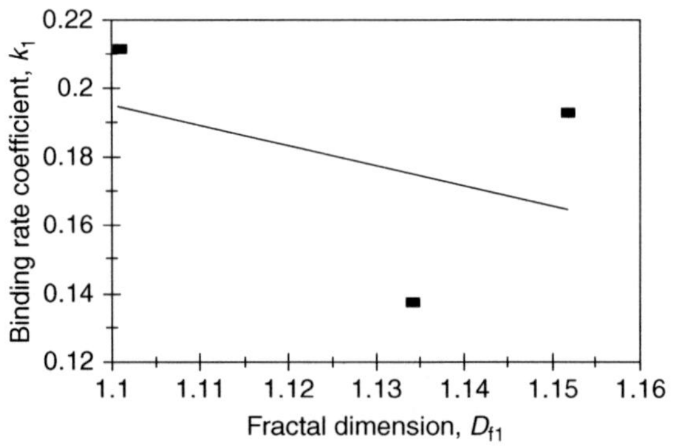

Figure 16.10
Decrease in the binding rate coefficient, k_1, for a dual-fractal analysis with an increase in the fractal dimension, D_{f1}.

The fit is not good. There is scatter in the data. Only three data points are available. The availability of more data points would lead to a more reliable fit. The poor fit is also noticed by the large error in the power to which the fractal dimension D_{f1}, is raised to.

The binding rate coefficient, k_1, is very sensitive to the fractal dimension, D_{f1}, or the degree of heterogeneity that exists on the microarray biosensor surface as noted by the negative order of dependence between three and a half and four (equal to −3.65) exhibited.

Feng et al. (1999) have developed a biotinylated ssDNA MB for DNA hybridization studies at a solid interface. These authors point out that DNA hybridization and molecular interaction studies assist in the diagnosis of genetic disease. The clinical symptoms are linked to the alterations in DNA. MBs were first developed by Tyagi and Kramer (1998). These are single-stranded oligonucleotides that possess a stem-and-loop structure. Feng et al. (1999) explain that the loop portion of the molecule can report the presence of a specific complementary nucleic acid (Kostrikis et al., 1998a,b; Piatek et al., 1998; Tyagi et al., 1998). The five bases at the two ends of the MB are complementary to each other forming the stem.

Figure 16.11 shows the binding of the target complementary DNA to the biotinylated and immobilized MB (5′-(_TMR) GCA CGT CCA TGC CCA GGA AGG AAC G(Biotin dT) GC(-DABCYL)-3′. TMR is tetramethylrhodmine is the fluorophore, and DABCYL (dimethylaminoazobenzeaminoexal-3-acryinido) is the quencher.

A dual-fractal analysis is required to adequately describe the binding kinetics. The values of (a) the binding rate coefficient, k, and the fractal dimension, D_f, for a single-fractal analysis, and (b) the binding rate coefficients, k_1 and k_2, and the fractal dimensions, D_{f1} and D_{f2}, for a dual-fractal analysis are given in Table 16.7.

It is of interest to note that for a dual-fractal analysis, as the fractal dimension increases by a factor of 1.49 from a value of D_{f1} equal to 1.8126 to D_{f2} equal to 2.7108, the binding rate

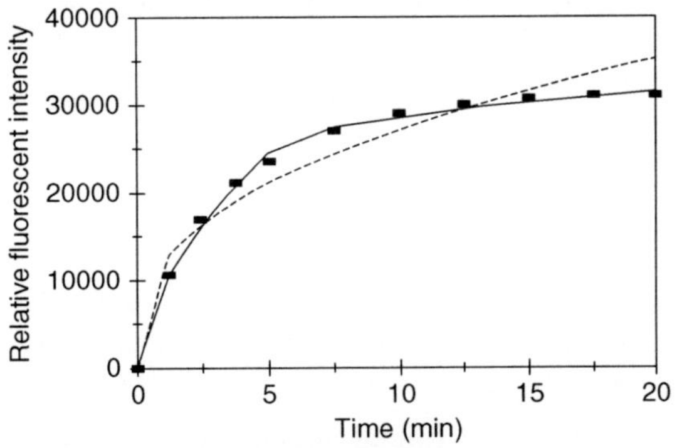

Figure 16.11

Binding of complementary ss DNA to a biotinylated molecular beacon immobilized on a biosensor surface (Feng et al., 1999). When only a solid line (—) is used then a single-fractal analysis applies. When both a dashed (- - -) and a solid (—) line are used then the dashed line represents a single-fractal analysis and the solid line represents a dual-fractal analysis.

Table 16.7 Binding rate coefficient and the fractal dimension for a novel molecular beacon (MB) for surface immobilized DNA hybridization studies (Feng et al., 1999).

Analyte in Solution/Receptor (MB) on micro-array surface	k	k_1	k_2	D_f	D_{f1}	D_{f2}
Complementary target DNA at pH 8.0/Biotinylated molecular beacon (MB)	11686 ± 1348	9435 ± 436	20554 369	2.2632 ± 0.0796	1.8126 ± 0.0868	2.702 ± 0.0436

coefficient increases by a factor of 2.18 from a value of k_1 equal to 9435 to k_2 equal to 20554. The changes in the fractal dimension or in the degree of heterogeneity on the microarray surface and in the binding rate coefficient are in the same direction.

Li et al. (2001) have developed an optical DNA biosensor based on surface immobilization of an MB by a bridge structure. The use of molecular beacons has already been presented in a previous example. Another example on MBs is now presented. These authors report that DNA biosensors exhibit potential for obtaining sequence-specific information when compared to traditional hybridization assays in a rapid, simpler, and less-expensive manner (Okahata et al., 1988). Li et al. (2001) provide examples wherein these DNA biosensors have been developed which include the surface of a modified electrode (Steel et al., 1998), a piezoelectrode quartz crystal microbalance (Okahata et al., 2000), a surface plasmon resonance biosensor (Nilsson et al., 1995), an optical fiber (Mehrvar et al., 2000), and a planar waveguide (Derisi et al., 1997). Li et al. (2001) emphasize the two customary fabrication steps involved in these types of biosensors that include: (a) a chemical modification of the polymer or the activation of the matrix, and (b) the immobilization of the single-stranded DNA on the sensor surface (Mehrvar et al., 2000).

Li et al. (2001) have developed an optical DNA biosensor based on a streptavidin-biotin interaction and is based on a bridge immobilization method. This according to them enhanced the freedom of the immobilized MB. The MB is a new DNA fluorescence probe which is a single-stranded oligonucleotide. This oligonucleotide consists of a probe sequence embedded within complementary sequences that form a hairpin stem (Kostrikis et al., 1998a,b; Tyagi and Kramer, 1998).

Figure 16.12a shows the binding of 50 nM complementary target 5′-GCG ACC ATA GCG ATT TAG (A-3′) in solution to the MB (5′-TMR-CCT AGC TCT AAA TCG CTA TGG TCG CGC (biotin dT)AG G-DABCYL-3′) immobilized using streptavidin-biotin immobilized on a biosensor surface (Li et al., 2001) A dual-fractal analysis is required to adequately describe the binding kinetics. The values of (a) the binding rate coefficient, k, and the fractal dimension, D_f, for a single-fractal analysis, and (b) the binding rate coefficients, k_1 and k_2 and the fractal dimensions, D_{f1} and D_{f2}, for a dual-fractal analysis are given in Table 16.8.

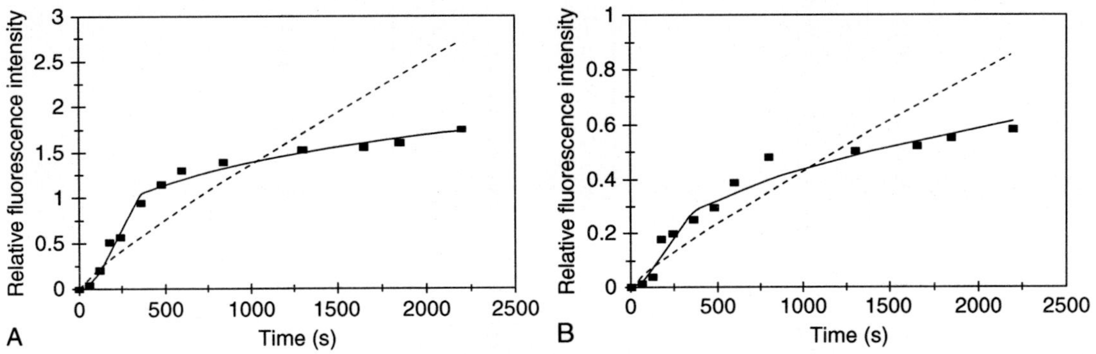

Figure 16.12

Binding of complementary ss DNA to a molecular beacon. Influence of different immobilization techniques (Li et al., 2001): (a) Streptavidin-biotin, (b) BSA-streptavidin-biotin. When only a solid line (—) is used then a single-fractal analysis applies. When both a dashed (- - -) and a solid (—) line are used then the dashed line represents a single-fractal analysis and the solid line represents a dual-fractal analysis.

Figure 16.12b shows the binding of 50 nM complementary target 5′-GCG ACC ATA GCG ATT TAG (A-3′) in solution to the MB (5′-TMR-CCT AGC TCT AAA TCG CTA TGG TCG CGC (biotin dT)AG G-DABCYL-3′) immobilized using BSA-streptavidin-biotin immobilized on a biosensor surface (Li et al., 2001). Once again, a dual-fractal analysis is required to adequately describe the binding kinetics. The values of (a) the binding rate coefficient, k, and the fractal dimension, D_f, for a single-fractal analysis, and (b) the binding rate coefficients, k_1 and k_2, and the fractal dimensions, D_{f1} and D_{f2}, for a dual-fractal analysis are given in Table 16.8 and Table 16.9.

It is of interest to note that when one compares the binding rate coefficients, k_1 and k_2, for a dual-fractal analysis when streptavidin-biotin is used with when BSA-streptavidin-biotin is used, both of the binding rate coefficients, k_1 and k_2 are higher when BSA-streptavidin-biotin is used.

Figure 16.13a shows the binding of 50 nM complementary oligonucleotide target (5′-GCG ACC ATA GCG ATT TAG(A-3′) in solution to the MB immobilized on the biosensor surface by BSA-streptavidin-biotin (Li et al., 2001). A dual-fractal analysis is required to adequately describe the binding kinetics. The values of (a) the binding rate coefficient, k, and the fractal dimension, D_f, for a single-fractal analysis, and (b) the binding rate coefficients, k_1 and k_2, and the fractal dimensions, D_{f1} and D_{f2}, for a dual-fractal analysis are given in Tables 16.10 and 16.11.

Figure 16.13b shows the binding of 50 nM one base mismatch oligonucleotide target (5′-GCG ACC ATA TCG ATT TAG(A-3′) in solution to the MB immobilized on the biosensor surface by BSA-streptavidin-biotin (Li et al., 2001). Once again, a dual-fractal analysis is required to adequately describe the binding kinetics. The values of (a) the binding rate coefficient, k, and the fractal dimension, D_f, for a single-fractal analysis, and (b) the binding rate

Table 16.8: Binding rate coefficients of complementary ss DNA target to molecular beacon (MB) at pH 8.0 (Li et al., 2001).

Complementary Target in Solution/Molecular Beacon (MB) on Surface	Immobilization Method	k	k_1	k_2
50 nM 5′-GCG ACC ATA GCG ATT TAG(A-3′)/MB (5′-TMR-CCT AGC TCT AAA TCG CTA TGG TCG CGC (biotin dT)AG G-DABCYL-3′)	Streptavidin-biotin	0.00114 ± 0.00064	$3.3\times10^{-5} \pm 1.6\times10^{-5}$	0.02271 ± 0.00261
50 nM 5′-GCG ACC ATA GCG ATT TAG(A-3′)/MB (5′-TMR-CCT AGC TCT AAA TCG CTA TGG TCG CGC (biotin dT)AG G-DABCYL-3′)	BSA-streptavidin-biotin	0.003416 ± 0.002325	$3.6\times10^{-5} \pm 1.0\times10^{-5}$	0.1956 ± 0.0117

Influence of different immobilization techniques.

Table 16.9: Fractal dimensions of complementary ss DNA target to molecular beacon (MB) at pH 8.0 (Li et al., 2001).

Complementary Target in Solution/Molecular Beacon (MB) on Surface	Immobilization Method	D_f	D_{f1}	D_{f2}
50 nM 5′-GCG ACC ATA GCG ATT TAG(A-3′)/MB (5′-TMR-CCT AGC TCT AAA TCG CTA TGG TCG CGC (biotin dT)AG G-DABCYL-3′)	Streptavidin-biotin	1.2786 ± 0.2318	$0 + 0.5812$	2.1432 ± 0.1216
50 nM 5′-GCG ACC ATA GCG ATT TAG(A-3′)/MB (5′-TMR-CCT AGC TCT AAA TCG CTA TGG TCG CGC (biotin dT)AG G-DABCYL-3′)	BSA-streptavidin-biotin	1.2650 ± 0.2702	$0 + 0.4366$	2.4316 ± 0.03263

Influence of different immobilization techniques.

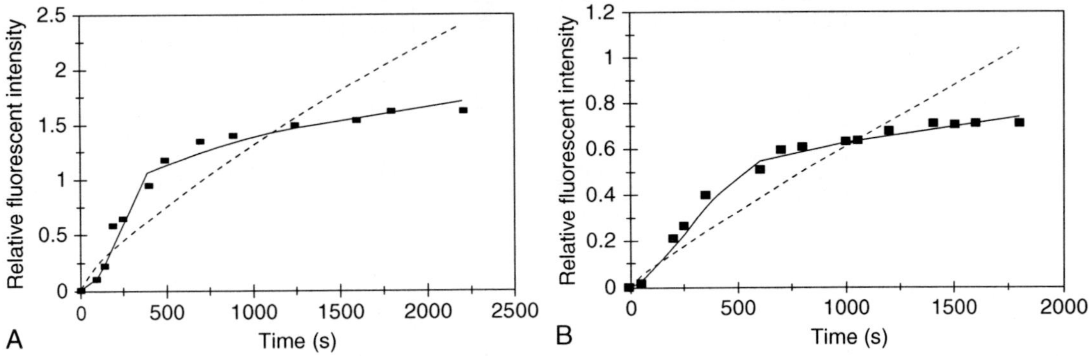

Figure 16.13

Binding of 50 nM complementary oligonucleotide and one base mismatch to the molecular beacon (MB) (Li et al., 2001): (a) Complementary oligonucleotide, (b) one base mismatch oligonucleotide. When only a solid line (—) is used then a single-fractal analysis applies. When both a dashed (- - -) and a solid (—) line are used then the dashed line represents a single-fractal analysis and the solid line represents a dual-fractal analysis.

coefficients, k_1 and k_2, and the fractal dimensions, D_{f1} and D_{f2}, for a dual-fractal analysis are given in Table 16.10 and Table 16.11.

It is of interest to note that as the fractal dimension increases by a factor of 9.08 from a value of D_{f1} equal to 0.268 to D_{f2} equal to 2.4330, the binding rate coefficient increases by a factor of 524.4 from a value of k_1 equal to 0.00017 to k_2 equal to 0.08915. The changes in the degree of heterogeneity or the fractal dimension on the biosensor surface and in the binding rate coefficient are in the same direction.

Figure 16.14 and Tables 16.10 and 16.11 show the increase in the binding rate coefficient, k_2, with an increase in the fractal dimension, D_{f2}, for a dual-fractal analysis. For the data shown in Figure 16.14, the binding rate coefficient, k_2, is given by:

$$k_2 = (2.6 \times 10^{-7} \pm 1.5 \times 10^{-7}) D_{f2}^{14.92 \pm 4.08} \qquad (16.10)$$

The fit is not good. There is scatter in the data. Only four data points are available. The availability of more data points would lead to a more reliable fit. Note that owing to the lack of data points, all the data presented until now are presented together:(a) complementary target binding to the MB immobilized using streptavidin-biotin on a biosensor, (b) complementary target binding to immobilized MB using BSA-streptavidin-biotin on a biosensor, (c) complementary target binding to immobilized MB using BSA-streptavidin-biotin on a biosensor (repeat), and (d) one base mismatch target binding to immobilized MB using BSA-streptavidin-biotin on a biosensor. Nevertheless, it is of interest to note that the binding rate coefficient, k_2, is extremely sensitive to the fractal dimension, D_{f2}, or the degree of heterogeneity that exists on the biosensor surface as noted by the close to fifteenth (equal to 14.92) order of dependence exhibited.

Table 16.10: Binding rate coefficients for 50 nM complementary oligonucleotide and one base mismatch to the molecular beacon (MB) (Li et al., 2001).

Target in Solution/Molecular Beacon (MB) on Surface	Immobilization Method	k	k_1	k_2
50 nM 5′-GCG ACC ATA GCG ATT TAG(A-3′)/MB (5′-TMR-CCT AGC TCT AAA TCG CTA TGG TCG CGC (biotin dT)AG G-DABCYL-3′)	BSA-streptavidin-biotin	0.006209 ± 0.003251	$5\times10^{-5} \pm 1.8\times10^{-5}$	20.1984 ± 0.0137
50 nM 5′-GCG ACC ATA GCG ATT TAG(A-3′)/MB (5′-TMR-CCT AGC TCT AAA TCG CTA TGG TCG CGC (biotin dT)AG G-DABCYL-3′) *(one base mismatch)*	BSA-streptavidin-biotin	0.00136 ± 0.00059	0.000117 ± 0.000042	0.08915 ± 0.00328

Table 16.11: Fractal dimensions for 50 nM complementary oligonucleotide and one base mismatch to the molecular beacon (MB) (Li et al., 2001).

Complementary Target in Solution/Molecular Beacon (MB) on Surface	Immobilization Method	D_f	D_{f1}	D_{f2}
50 nM 5′-GCG ACC ATA GCG ATT TAG(A-3′)/MB (5′-TMR-CCT AGC TCT AAA TCG CTA TGG TCG CGC (biotin dT)AG G-DABCYL-3′)	BSA-Streptavidin-biotin	1.4492 ± 0.2452	$0 + 0.6076$	2.4386 ± 0.0816
50 nM 5′-GCG ACC ATA GCG ATT TAG(A-3′)/MB (5′-TMR-CCT AGC TCT AAA TCG CTA TGG TCG CGC (biotin dT)AG G-DABCYL-3′) *(one base mismatch)*	BSA-streptavidin-biotin	1.2264 ± 0.1986	$0.268 + 0.3290$	2.4330 ± 0.0652

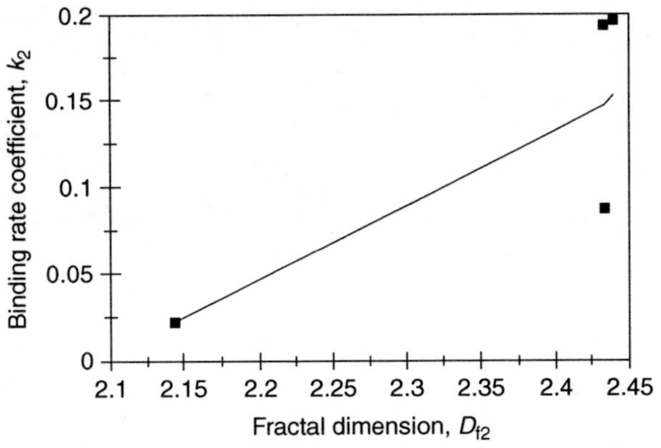

Figure 16.14
Increase in the binding rate coefficient, k_2, with an increase in the fractal dimension, D_{f2}.

Li et al. (2001) report that there are phosphate groups on the oligonucleotide. Single strand DNA will dissociate in solution and process a negative charge. This will hinder the formation of a duplex for the electrostatic repulsion. They further point out that the presence of cations in solution could counteract the negative charge on the oligonucleotide and accelerate the hybridization reaction. Thus, they analyzed the influence of Mg^{++} ($MgCl_2$) concentration (0.1 to 500 mM) on the hybridization of the immobilized MB with its target DNA. They noted that the hybridization efficiency improved when the ion concentration in solution was increased.

Figure 16.15a shows the binding of cDNA to the MB immobilized on the biosensor surface in the presence of 0.1 mM $MgCl_2$. A dual-fractal analysis is required to adequately describe the binding kinetics. The values of (a) the binding rate coefficient, k, and the fractal dimension, D_f, for a single-fractal analysis, and (b) the binding rate coefficients, k_1 and k_2, and the fractal dimensions, D_{f1} and D_{f2}, for a dual-fractal analysis are given in Table 16.12. It is of interest to note that as the fractal dimension increases by a factor of 2.68 from a value of D_{f1} equal to 1.0482 to D_{f2} equal to 2.9105, the binding rate coefficient increases by a factor of 802.2 from a value of k_1 equal to 0.000226 to k_2 equal to 0.1813. The changes in the degree of heterogeneity or the fractal dimension on the biosensor surface and in the binding rate coefficient are in the same direction.

Figure 16.15b shows the binding of cDNA to the MB immobilized on the biosensor surface in the presence of 1 mM $MgCl_2$. Once again, a dual-fractal analysis is required to adequately describe the binding kinetics. The values of (a) the binding rate coefficient, k, and the fractal dimension, D_f, for a single-fractal analysis, and (b) the binding rate coefficients, k_1 and k_2, and the fractal dimensions, D_{f1} and D_{f2}, for a dual-fractal analysis are given in Table 16.12.

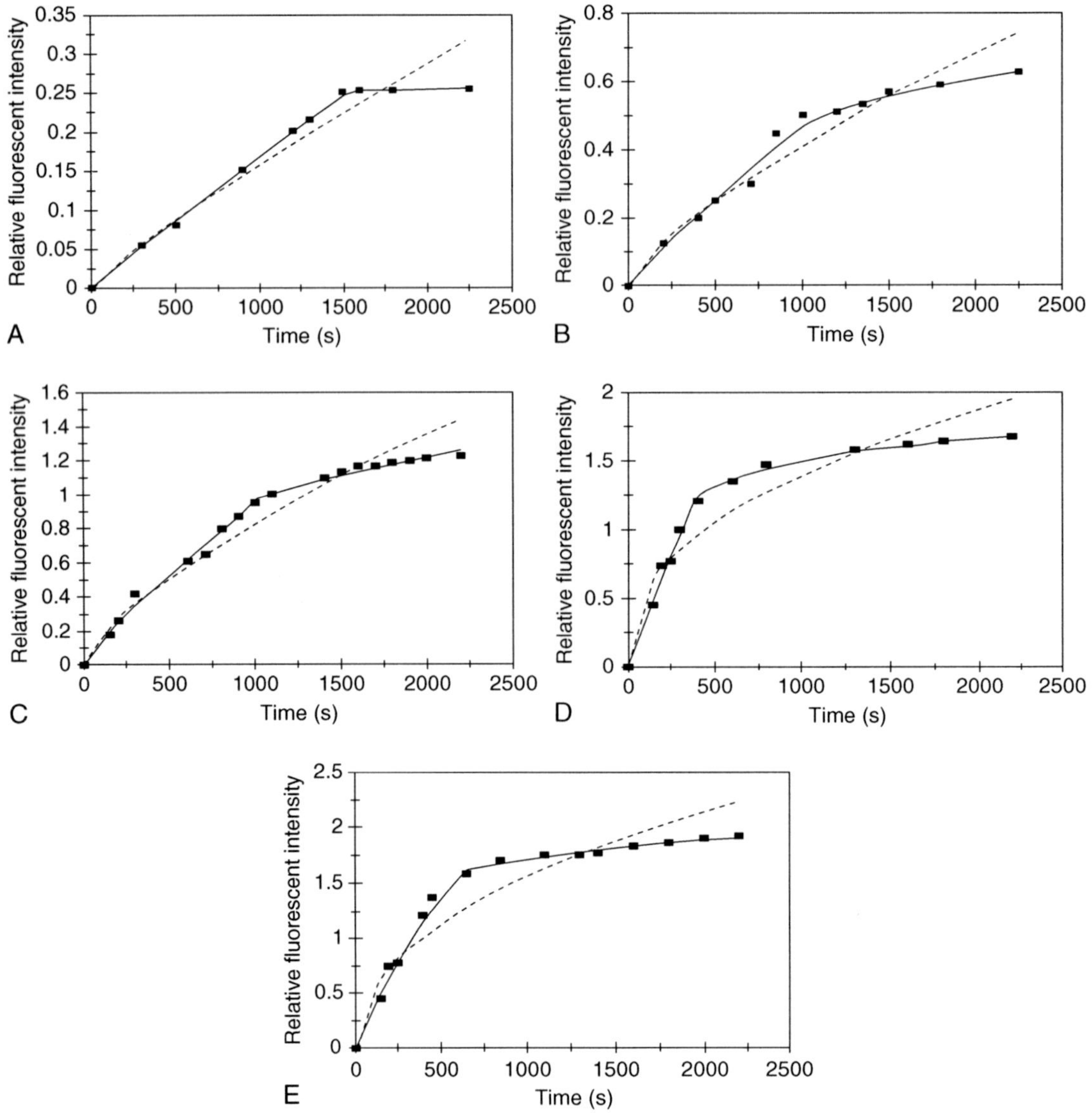

Figure 16.15

Binding (hybridization) of the complementary DNA (cDNA) to the molecular beacon (MB) immobilized on the biosensor surface (Li et al., 2001). Influence of ion ($MgCl_2$) concentration (in mM) in solution: (a) 0.1, (b) 1, (c) 10, (d) 100, (e) 500. When only a solid line (—) is used then a single-fractal analysis applies. When both a dashed (- - -) and a solid (—) line are used then the dashed line represents a single-fractal analysis and the solid line represents a dual-fractal analysis.

It is of interest to note that as the fractal dimension increases by a factor of 1.852 from a value of D_{f1} equal to 1.2708 to D_{f2} equal to 2.3546, the binding rate coefficient increases by a factor of 43.77 from a value of k_1 equal to 0.001195 to k_2 equal to 0.05231. The changes in the degree of heterogeneity or the fractal dimension on the biosensor surface and in the binding rate coefficient are, once again, in the same direction.

Table 16.12: Binding rate coefficients and fractal dimensions for the hybridization of 50 mM cDNA in solution at pH 8.0 (Li et al., 2001).

Ion ($MgCl_2$) concentration (mM)	k	k_1	k_2	D_f	D_{f1}	D_{f2}
0.1	0.00043 ± 0.00005	0.000226 ± 0.000009	0.1813 ± 0.00036	1.2886 ± 0.1191	1.0842 ± 0.0536	2.9105 ± 0.00162
1	0.00234 ± 0.00097	0.001195 ± 0.00013	0.05231 ± 0.00104	1.5648 ± 0.1046	1.2708 ± 0.1563	2.3546 ± 0.07922
10	0.006961 ± 0.000824	0.003359 ± 0.000352	0.09877 ± 0.0017	1.6138 ± 0.06710	1.3672 ± 0.1132	2.3378 ± 0.04482
100	0.0802 ± 0.01555	0.003981 ± 0.000438	0.4834 ± 0.0079	2.1698 ± 0.1172	1.0788 ± 0.2786	2.6728 ± 0.02882
500	0.06276 ± 0.01168	0.007705 ± 0.000896	0.6786 ± 0.0066	2.0704 ± 0.1051	1.3266 ± 0.1786	2.7308 ± 0.02306

Effect of ion concentration.

Figure 16.15c shows the binding of cDNA to the MB immobilized on the biosensor surface in the presence of 10 mM $MgCl_2$. Once again, a dual-fractal analysis is required to adequately describe the binding kinetics. The values of (a) the binding rate coefficient, k, and the fractal dimension, D_f, for a single-fractal analysis, and (b) the binding rate coefficients, k_1 and k_2, and the fractal dimensions, D_{f1} and D_{f2}, for a dual-fractal analysis are given in Table 16.12. It is of interest to note that as the fractal dimension increases by a factor of 1.71 from a value of D_{f1} equal to 1.3672 to D_{f2} equal to 2.3378, the binding rate coefficient increases by a factor of 29.4 from a value of k_1 equal to 0.003359 to k_2 equal to 0.09877. Note that changes in the degree of heterogeneity or the fractal dimension on the biosensor surface and in the binding rate coefficient are, once again, in the same direction.

Figure 16.15d shows the binding of cDNA to the MB immobilized on the biosensor surface in the presence of 100 mM $MgCl_2$. Once again, a dual-fractal analysis is required to adequately describe the binding kinetics. The values of (a) the binding rate coefficient, k, and the fractal dimension, D_f, for a single-fractal analysis, and (b) the binding rate coefficients, k_1 and k_2, and the fractal dimensions, D_{f1} and D_{f2}, for a dual-fractal analysis are given in Table 16.12. It is of interest to note that as the fractal dimension increases by a factor of 2.48 from a value of D_{f1} equal to 1.0788 to D_{f2} equal to 2.6728, the binding rate coefficient increases by a factor of 110.1 from a value of k_1 equal to 0.003981 to k_2 equal to 0.4834. Note that changes in the degree of heterogeneity or the fractal dimension on the biosensor surface and in the binding rate coefficient are, once again, in the same direction.

Figure 16.15e shows the binding of cDNA to the MB immobilized on the biosensor surface in the presence of 500 mM $MgCl_2$. Once again, a dual-fractal analysis is required to adequately describe the binding kinetics. The values of (a) the binding rate coefficient, k, and the fractal dimension, D_f, for a single-fractal analysis, and (b) the binding rate coefficients, k_1 and k_2, and the fractal dimensions, D_{f1} and D_{f2}, for a dual-fractal analysis are given in Table 16.12. It is of interest to note that as the fractal dimension increases by a factor of 2.058 from a value of D_{f1} equal to 1.3266 to D_{f2} equal to 2.7308, the binding rate coefficient increases by a factor of 88.07 from a value of k_1 equal to 0.007705 to k_2 equal to 0.46786. The changes in the degree of heterogeneity or the fractal dimension on the biosensor surface and in the binding rate coefficient are, once again, in the same direction.

Figure 16.16a and Table 16.12 show the increase in the binding rate coefficient, k_1, with an increase in the ion concentration ($MgCl_2$) in solution in the 0.1-500 mM range for a dual-fractal analysis. For the data shown in Figure 16.16a, the binding rate coefficient, k_1, is given by:

$$k_1 = (0.000841 \pm 0.000541)[\text{ion concentration, mM}]^{0.378 \pm 0.0694} \tag{16.10a}$$

The fit is good. Only five data points are available. The availability of more data points would lead to a more reliable fit. The binding rate coefficient, k_1, for a dual-fractal analysis

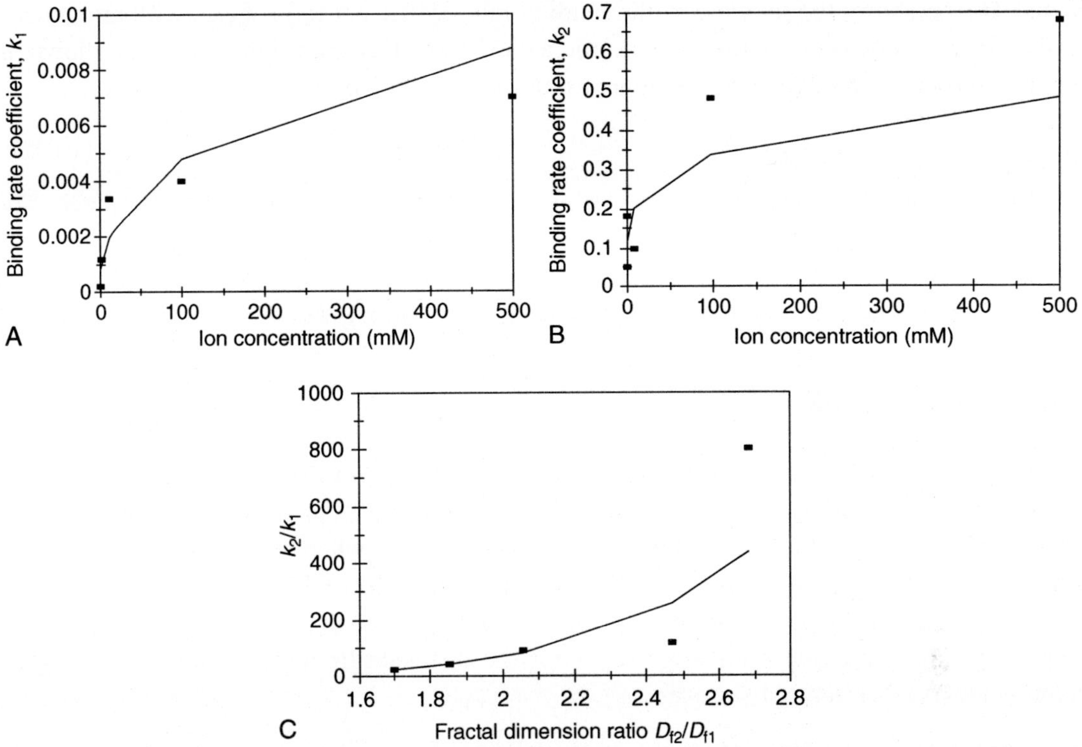

Figure 16.16

(a) Increase in the binding rate coefficient, k_1, for a dual-fractal analysis with an increase in the ion concentration (in mM) in solution. (b) Increase in the binding rate coefficient, k_2, for a dual-fractal analysis with an Increase in the ion concentration (in mM) in solution. (c) Increase in the binding rate coefficient ratio, k_2/k_1, with an increase in the fractal dimension ratio, D_{f2}/D_{f1}.

exhibits only a mild (less than one-half (equal to 0.378)) order of dependence on the ion concentration in solution in the 0.1-500 mM range.

Figure 16.16b and Table 16.12 show the increase in the binding rate coefficient, k_2, with an increase in the ion concentration ($MgCl_2$) in solution in the 0.1-500 mM range for a dual-fractal analysis. For the data shown in Figure 16.16b, the binding rate coefficient, k_2, is given by:

$$k_2 = (01232 + 0.1726)[\text{ion concentration, mM}]^{0.123+0.173} \qquad (16.10b)$$

The fit is not good. There is scatter in the data, and this is reflected in the error of the estimated rate coefficient values. Only the positive value of the errors is given. Only five data points are available. The availability of more data points would lead to a more reliable fit. The binding rate coefficient, k_2, for a dual-fractal analysis exhibits close to a zero (equal to 0.123) order of dependence on the ion concentration in solution in the 0.1-500 mM range.

Figure 16.16c shows the increase in the binding rate coefficient ratio, k_2/k_1, with an increase in the fractal dimension ratio, D_{f2}/D_{f1}, for a dual-fractal analysis. For the data shown in Figure 17.16c, the binding rate coefficient ratio, k_2/k_1, is given by:

$$k_2/k_1 = (1.0127 \pm 0.7851)(D_{f2}/D_{f1})^{6.156 \pm 1.498} \tag{16.10c}$$

The fit is reasonable. There is some scatter at the higher values of the D_{f2}/D_{f1} presented. The availability of more data points at the higher D_{f2}/D_{f1} ratios would lead to a better fit. The binding rate coefficient ratio, k_2/k_1, exhibits a very high and close to sixth (equal to 6.156) order of dependence on the ratio of the fractal dimensions, D_{f2}/D_{f1}.

16.4 Conclusions

A fractal analysis is presented for the binding of different hybridization reactions occurring on different biosensor surfaces. Both a single- and a dual-fractal analysis were used to model the binding kinetics. The dual-fractal analysis was used only when the single-fractal analysis did not provide an adequate fit (sum of least squares less than 0.97). Corel Quattro Pro 8.0 (1997) was used to model the binding kinetics. The fractal analysis is an alternate and convenient means to provide a kinetic analysis of the diffusion-limited reactions occurring on heterogeneous or structured surfaces.

In accord with the prefactor analysis of fractal aggregates (Sorenson and Roberts, 1997), quantitative (predictive) expressions are developed for (a) the increase in the binding rate coefficient, k_1 and k_2, for a dual-fractal analysis with an increase in the λ exonuclease units in solution during the real-time monitoring of the activity and kinetics of T4 PNK by a single labeled DNA-hairpin smart probe coupled with λ exonuclease cleavage (Song and Zhao, 2009), (b) the increase in the fractal dimension, D_{f2}, for a dual-fractal analysis with an increase in the λ exonuclease units in solution during the real-time monitoring of the activity and kinetics of the T4 PNK by a single labeled DNA-hairpin smart probe coupled with λ exonuclease cleavage (Song and Zhao, 2009), (c) the increase in the binding rate coefficient, k_2, and in the ratio of the binding rate coefficients, k_2/k_1, with a increase in the fractal dimension, D_{f2}, and in the ratio D_{f2}/D_{f1}, respectively (d) an increase in the binding rate coefficients, k_2 and k_1, with an increase in the fractal dimensions, D_{f2} and D_{f1}, respectively, during the binding of 500 pM cy3-labeled target (raw data) to a 20-mer capture probe immobilized on different areas (the vignetting effect-distance from the detector) on a microarray biosensor (Schultz et al., 2008). In some cases these predictive relations indicate that the binding rate coefficients are very sensitive to the fractal dimension or the degree of heterogeneity that exists on the biosensor surface. This is explained for other reactions occurring on biosensor surfaces in chapters elsewhere in this book. These predictive relationships presented provide a means by which these binding rate coefficients may be manipulated by changing either the analyte concentration in solution or the degree of heterogeneity that exists on the biosensor surface. A change in the degree of heterogeneity on the surface generally leads to a change in the binding rate coefficient in the same direction.

It is suggested that the fractal surface (roughness) leads to turbulence, which enhances mixing, decreases diffusional limitations, and leads to an increase in the binding rate coefficient (Martin et al., 1991). For this to occur, the characteristic length of the turbulent boundary layer may have to extend a few monolayers above the sensor chip or affect bulk diffusion to and from the surface. However, given the extremely laminar flow regimes in most biosensors this may not actually take place. The sensor chip surface is characterized by grooves and ridges, and this surface morphology may lead to eddy diffusion. This eddy diffusion can then help to enhance the mixing and extend the characteristic length of the boundary layer to affect the bulk diffusion to and from the surface.

The analysis of the different hybridization examples especially of biomedical interest presented in this chapter (and in other chapters of the book) should encourage experimentalists to pay more attention to the nature of the surface, and how it may be manipulated to advantage in desired directions. This is of particular value for biomedical cases (or the community) wherein, with more sensitive and better hybridization biosensors, the earlier one may detect and diagnose the probable onset of diseases, the earlier one can begin the medical protocols necessary to help prevent, alleviate, or correct the onset of especially debilitating and intractable diseases. The importance of a better understanding of hybridization reactions with regard to disease prevention and management cannot be over-emphasized.

References

Baldini F and A Giannetti, *Proceedings of SPIE*, **5826**, 485–1400 (2005).

Bishop J, S Blair, and AM Chagovetz, A comparative kinetic model of nucleic acid surface hybridization in the presence of point mutants, *Biophysical Journal*, **90**, 831–840 (2006).

Broude NE, K Woodward, R Cavallo, CR Cantor, and D Englert, DNA microarrays with stem-loop DNA probes; preparation and applications, *Nucleic Acids Research*, **29**, e91/1–e91/11 (2001).

Brown PO and D Botstein, Exploring the new world of genome with DNA microarrays, *Nature Genetics*, **21**, 33–37 (1999).

Chappell C, LA Hankahi, F Karimi-Bucheri, M Weinfeld, and SC West, Involvement of human polynucleotide kinase in double-strand break repair by non-homogeneous end joining, *EMBO Journal*, **21**, 2827–2832 (2002).

Corel Quattro Pro 8.0 , *Corel Corporation Limited*, Ottawa, Canada, 1997.

Culha H, M Stokes, DL Griffin, and T Vo-Dinh, *Biosensors & Bioelectronics*, **19**, 1007–1012 (2004).

Derisi JL, VR Iyer, and PO Borwn, Exploring the metabolic and genetic control of gene expression on a genomic scale, *Science*, **278**, 680 (1997).

Drummond TG, MG Hill, and JK Barton, Electrochemical DNA sensors, *Nature Biotechnology*, **21**, 1192–1199 (2003).

Edwards KA and AJ Baeumner, Sequential injection analysis system for the sandwich hybridization-based detection of nucleic acids, *Analytical Chemistry*, **78**, 1958–1966 (2006).

Eggers M, M Hogan, RK Reich, J Lamlure, D Ehrlich, M Hollis, B Kosicki, T Powdrill, K Beattie, and S Smith, *Biotechniques*, **17**, 516–525 (1994).

El Atifi M, I Dupre, B Rostaining, AL Benabidi, and F Berger, Single nucleotide polymorphism discrimination assisted by imp, *Biotechniques*, **35**, 262 (2003).
Feng X, X Liu, S Schuster, and W Tan, Designing a novel molecular beacon for surface immobilized DNA hybridization studies, *Journal of the American Chemical Society*, **121**, 2921–2922 (1999).
Frauendorf C, F Hausch, I Rohl, A Lichte, S Vonhoff, and S Klussman, *Nucleic Acids Research*, **31**, e34 (2003).
Gang X, X Liu, S Schuster, and W Tan, Designing a novel new molecular beacon for surface-immobilized DNA hybridization studies, *Journal of the American Chemical Society*, **121**, 2921–2922 (1999).
Gooding JJ, Electrochemical DNA hybridization biosensors, *Electroanalysis*, **14**, 1149–1156 (2002).
Hahns S, S Mergenthaler, B Zimmermann, and W Holzgreve, *Bioelectrochemistry*, **67**, 151–154 (2005).
Havlin S, Molecular diffusion and reaction in *The Fractal Approach to Heterogeneous Chemistry: Surfaces, Colloids, Polymers* (ed. D Avnir), Wiley, New York, pp. 251–269 (1989).
Ito K, K Hashimoto, and Y Ishimori, *Analytica Chimica Acta*, **327**, 29 (1996).
Kaltenbeck B and C Wang, *Advances in Clinical Chemistry*, **40**, 219–259 (2005).
Karpf H, A Leitner, and H Marsoner, US patent 475 5667 (1988).
Kim YS, HS Jung, T Matsuura, HY Lee, T Kawai, and MB Gu, Electrochemical detection of 17β-estradiol using DNA aptamer immobilized gold electrode chip, *Biosensors & Bioelectronics*, **22**, 2525–2531 (2007).
Knemeyer JP, N Marine, and M Sauer, Probes for detection of specific DNA sequences at the single-molecule level, *Analytical Chemistry*, **72**, 3717–3724 (2000).
Kostrikis LG, S Tyagi, MM Mhlanga, DD Ho, and FR Kramer, Special genotyping of human alleles, *Science*, **279**, 1228–1229 (1998a).
Kostrikis LG, Y Huang, JP Moore, SM Wolinsky, L Q Zhang, Y Guo, L Deutsch, J Phair, AU Neumann, and DD Ho, *Nature Medicine*, **4**, 350–353 (1998b).
Lee CK, BG Brown, BR Bombisk, and DJ Doolittle, *Analytical Biochemistry*, **227**, 156–161 (1995).
Leung A, PM Shankar, and R Mutharasan, *Sensors & Actuators B*, **125**, 688–703 (2007).
Li J, W Tan, K Wang, D Xiao, X Yang, and Z Tang, Ultrasensitive optical DNA biosensor based on surface immobilization of molecular beacon by a bridge structure, *Analytical Sciences*, **17**, 1149–1153 (2001).
Liu CK, YF Li, JR Li, and L Jiang, Enhancement of DNA immobilization and hybridization on gold electrode modified by nanogold electrodes, *Biosensors & Bioelectronics*, **21**, 787–795 (2005).
Mao X, Y Ma, A Zhang, L Zhang, L Zeng, and G Lu, Disposable nucleic acid biosensors based on gold nanoparticle probes and lateral flow strip, *Analytical Chemistry*, **81**, 1660–1668 (2009).
Martin JS, GC Frye, AJ Ricco, and AD Sentoria, Effect of surface roughness on the response of thickness-shear mode resonators, *Analytical Chemistry*, **65**, 2910–2922 (1991).
Martinez K, MC Estevez, Y Wu, JA Phillips, CD Medley, and W Tan, Locked nucleic acid based beacons for surface interaction studies and biosensor development, *Analytical Chemistry*, **81**, 3448–3454 (2009).
Mehrvar M, C Bis, JM Scharer, MM Young, and JH Luong, *Analytical Sciences*, **16**, 667 (2000).
Neo SJ, X Su, and JS Thomsen, Surface plasmon resonance study of cooperative interactions of estrogen receptor α and transcriptional factor Sp1 with composite DNA elements, *Analytical Chemistry*, **81**, 3344–3349 (2009).
Nilsson P, B Persson, M Uhlen, and P Nygren, *Analytical Biochemistry*, **224**, 400 (1995).
Odenthal KJ and JJ Gooding, *Analyst*, **132**, 603–610 (2007).
Okahata Y, M Kawase, K Niikura, F Ohtake, H Furusawa, and Y Ehara, Kinetic measurement of DNA hybridization on an oligonucleotide-immobilized 27-MHz quartz crystal microbalance, *Analytical Chemistry*, **70**, 1288 (1988).
Okahata Y, K Niikura, H Furusawa, and H Matauno, *Analytical Sciences*, **16**, 1113 (2000).
Palacek E and M Fojta, Detecting DNA hybridization and damage, *Analytical Chemistry*, **73**, 75A–83A (2001).
Petersen KE, WA Kovacs, and SJ Young, U.S. Patent 5958349 (1999) .
Phillips DH and VM Arit, *Natural Protocol*, **2**, 2772 (2007).
Piatek AS, S Tyagi, AC Pol, A Telenti, LP Miller, FR Kramer, and D Alland, Nature *Biotechnology*, **16**, 359–363 (1998).
Piunno PAE and UJ Krull, *Analytical and Bioanalytical Chemistry*, **381**, 1004–1011 (2005).
Preininger C and P Chiarelli, *Talanta*, **55**, 973–980 (2001).

Ramakrishnan A and A Sadana, A single fractal analysis of cellular analyte-receptor binding kinetics using biosensors, *BioSystems*, **59**, 35–51 (2001).

Rasouli-Nia A, F Karimi-Busheri, and M Weinfeld, Phosphorylatin of nucleic acid by an enzyme rom T4 bacteriophage-infected *Escherichia coli*, *Proceedings of the National Academy of Sciences USA*, **101**, 6905–6910 (2004).

Richardson CC, Stable down-regulation of human polynucleotie kinase enhances spontaneous mutation frequency and sensitizes cells to genomic agents, *Proceedings of the National Academy of Sciences USA*, **54**, 158–165 (1965).

Rodriguez-Mozaz S, S Reder, M Lpez de Alda, G Gauglitz, and D Barccelo, Simultaneous multi-analyte determination of estrone, isoproturon, and atrazine in natural waters by the River ANAlyser, *Biosensors &* Bioelectronics, **19**, 633–640 (2004).

Sadana A, A fractal analysis for the evaluation of hybridization kinetics in biosensors, *Journal of Colloid and Interface Science*, **151**(1), 166–177 (2001).

Sadana A, *Fractal Binding and Dissociation Kinetics for Different Biosensor Applications*, Elsevier, Amsterdam, 2005.

Schultz E, R Galland, D Du Bouetiez, T Flahaut, A Planat-Chretien, F lesbre, H Volland, and F Perraut, A novel fluorescence-based array biosensor: Principle and application to DNA hybridization, *Biosensors & Bioelectronics*, **23**(7), 9977–9984 (2008).

Song C and M Zhao, Real-time monitoring of the activity and kinetics of T4 polynucleotide kinase by a singly labeled DNA-hairpin smart probe coupled with exonuclease cleavage, *Analytical Chemistry*, **81**, 1383–1388 (2009).

Sorenson CM and GC Roberts, The prefactor of fractal aggregates, *Journal of Colloid & Interface Science*, **186**, 447–452 (1997).

Steel AB, TM Herne, and MJ Tarlov, Electrochemical quantitation of DNA immobilized on gold, *Analytical Chemistry*, **70**, 46670 (1998).

Steel AB, RL Levicky, TM Herne, and MJ Tarlov, Immobilization of nucleic acids at solid surfaces: Effect of oligonucleotide length on layer assembly, *Biophysical Journal*, **79**, 975–981 (2000).

Steemers FJ, JA Ferguson, and DR Walt, Screening unlabeled DNA targets with randomly ordered fiber-optic gene arrays, *Nature Biotechnology*, **18**, 91–94 (2000).

Tyagi S and FR Kramer, Molecular beacons: Probes that fluoresce upon hybridization, *Nature Biotechnology*, **14**, 303–308 (1998).

Tyagi S, D Bratu, and FR Kramer, Multicolor molecular beacons for allele discrimination, *Nature Biotechnology*, **16**, 49–53 (1998).

Vo-Dinh T, JP Alarie, N Isola, DI Landis, AL Wintenberg, and MN Ericson, DNA biochip using a phototransistor integrated circuit, *Analytical Chemistry*, **71**, 63–640 (2004).

Wang J, *Chemistry- European Journal*, **5**, 1681–1685 (1999).

Wang J, *Small*, **1**(11), 1036–1043 (2005).

CHAPTER 17

Economics of Biosensors

Chapter Outline

17.1 Introduction

Biosensor applications have expanded substantially from the initial detection of glucose for diabetes management. This is owing to the simplicity of and also the simplicity in the use of the biosensor in different areas. One use that presumably constrains the use of biosensors in other areas is the lack of economic information available in the open literature on setting up a biosensor industry and on the pitfalls the potential companies will face, as they attempt to get into this market. Another factor that, not unexpectedly, plays a very significant role is the biosensor market that exists, and future trends and predictions in the different areas of biosensor market growth such as, besides medical applications, the environment, security, drugs, pathogen detection in food and in air, besides other areas. Until a few years ago, medical applications dominated the biosensor market, but slowly and surely, other areas, especially the security area is picking up relative to the medical market. It is probably safe to suggest that the medical area will dominate the biosensor market for the next decade if not more; however, one should pay attention to other areas too.

This chapter attempts to shed some light and perspective on the biosensor market. As expected predictions are difficult to make, but this information is useful. There are reports that are available, but they are expensive (generally cost thousands of dollars), and thanks to the dynamics of the biosensor market, they lose their value rather quickly. This author has attempted to provide the economics of biosensors in the last chapter in his previous books. In the same vein, this last, perhaps capstone chapter, the economics of biosensors is again emphasized.

Most, if not all, of the information is gathered from the open literature, and quite reasonably, one may question the reliability of the information presented. Keeping this caveat in mind, the next section that follows analyzes market reports and trends. The section after that

Handbook of Biosensors and Biosensor Kinetics. DOI: 10.1016/B978-0-444-53262-6.00017-6

provides examples of investment/financing in biosensor companies. It is the intention of this chapter to provide some perspective on the biosensor market, at least to the novice reader. Finally, the author admits that most of his life has been spent in the academic area, apart from a 5 year stint at a National Chemical Laboratory in India, and about a year at different laboratories here in the United States. Besides some minor consulting in the biosensor and catalyst deactivation area most of the working life of the author has been spent at a university. Nevertheless, this author recognizes the importance of economics, especially in the biotechnology area; a similar chapter on the economics of bioseparations has been written by this same author. Finally, one may consider this chapter as an initial and iterative solution to a nonlinear problem. This chapter can be taken as a starter to keep improving the solution using some effective algorithm or logical reasoning to arrive at the optimum solution. The question may eventually be, "should we invest in a biosensor industry?"

17.2 Market Reports and Trends

A recent report has analyzed the biosensor market in medical diagnostics between the years 2000-2015 (Global Strategic Biosensor Report, 2009). The report examines the biosensor market for medical diagnostics, medical biosensors, glucose, and other medical biosensors, environmental biosensors, and other biosensors. There has been a considerable increase in medical applications, besides the original glucose market owing to the relative ease of applications of biosensors. The report was careful to analyze each area geographically such as the United States, Canada, Europe, the Asia-Pacific region, Japan, and the rest of the world. The report also analyzed the major companies throughout the world that were major players in the biosensor market.

Newsguide US (2009) recently pointed out that the market for biosensors will reach $6.1 billion by the year 2012 on the basis of a new report by Global Industry Analysts, Inc. According to this report this will happen because of the increasing opportunities for use of biosensors in industrial, environmental, and especially medical diagnostic applications. Factors that will lead to this growth are a growing population, rising incidence of chronic diseases (such as diabetes), and the growing need for environmental monitoring.

The advent of microfabrication technology has led to the miniaturization of biosensors. Nanosize electronic components under development will be incorporated in future commercial biosensors. The authors point out the slow pace of commercialization of biosensors because of the high cost and the availability of effective alternative technologies. Also, issues such as stability, sensitivity, and quality assurance have still to be worked out for complete satisfaction. The authors add that medical applications of biosensors is still an attractive proposition, and many players are putting in significant resources to develop this, as yet largely untapped, market section. According to this report, the United States and Europe are still the major players in the world capturing approximately 69.7% of the world's share in the year 2012. However, by the year 2012, the Asia-Pacific biosensor market is expected to grow to equal the U.S. $794 million market. The major market for biosensors is still the medical diagnostic market with biosensors finding applications in blood gas analyzers, electrolyte analyzers, metabolite analyzers, and glucose meters.

The report points out that glucose meters, as a management tool for diabetes have undergone a dramatic change in the past few years. Some of the changes include wireless and sensor technologies, and noninvasive glucose meters. By the year 2012, the U.S. market (the largest one for glucose meters) is expected to grow to $1.28 billion Market Research, 2008. The report further adds that in the year 2008, France and Germany together constituted 55.3% of the biosensor market in Europe. By the year 2015, the environmental biosensor market is expected to grow to U.S. $32.7 million in Germany.

The report also mentions that biosensor technology is being developed, amongst other institutions at (in alphabetical order): AgMatrix, Inc., Cranfield Biotechnolgy Center, LfeSensors, Inc., M-Biotech, and Nova Biomedical. The report also mentions that lead manufacturers of biosensors include, once again, in alphabetical order: Abbott Point of Care, Affinity Sensors, Animas Corporation, LifeScan, Inc., Medtronic Diabetes, Neo Sensors Limited, Roche Diagnostics, and Science Healthcare Diagnostics.

Wei (2009) in a personal communication (e-mail) has pointed out that Visiongain has recently come out with a report entitled, "In-vitro diagnostics market report and analysis 2008-2023." The report examines the commercial prospects for *in vitro* diagnostics (IVD) in the global market. The report emphasizes that molecular diagnostics is increasing the prospects of personalized medicine which should assist patients, companies, and healthcare providers. The FDA (Food and Drug Administration) is also supporting IVD development.

Visiongain believes that the timing is right for the fostering of mutually beneficial partnerships. In its report Visiongain also used in-house forecasting techniques, and analysis of drivers and constraints. The report also identified emerging and established key industrial players. China and India were also included in their analysis. Visiongain predicts that the market for IVD will exhibit strong growth during the period 2008-2023. Besides, far-sighted companies and health-care providers will make significant investments in these IVD devices. Furthermore, Visiongain predicts that molecular diagnostics, though still in its infancy, will continue to make an increasing impact in the future of IVDs.

This is an expensive 240 page report with a three tier pricing: (a) individual copies £1499 (U.S. $2237, exchange rate 1 GBP (pound) = 1.492 U.S. dollars), departmental copy 2999 pounds (U.S. $4475), and whole company 4999 pounds (U.S. $7459).

Reports-research.com have recently come out with a SWOT (Strengths, Weaknesses, Opportunities, and Threats) analysis on Universal Biosensors, Inc. Healthcare-Medical Equipment. It is a market study (Marketstudie) and is published by Global Markets Direct.

The SWOT analysis involves specifying the objective of the business venture or project and identifying the internal and external factors that are favorable and unfavorable to achieving that objective. The technique is credited to Albert Humphrey of Stanford University who used data from Fortune 500 companies in the 1960s and 1970s.

This is a very expensive 30 page report that costs U.S. $12,500 (PDF; e-mail). The comprehensive report of this company includes both qualitative and quantitative research. The report carefully examines the company's business structure and operations, history and products. The SWOT analysis studies the major factors which will affect the company's performance, for example, strengths, costs, and revenues. Also considered are external factors such as competitive positioning and industry trends. Finally, the report does provide a balanced presentation of the opportunities that the company may exploit, and the threats facing it.

In a recent report titled, "World Biosensors Market 2007" the Frost and Sullivan Service (2009) also provided an overview of the world biosensor market. The report examined different product and end-user segments in terms of revenues. The report also provides the key market drivers as well as the constraints and trends for the future growth of the biosensor market. This is almost like a theoretical optimization problem wherein one optimizes an objective (theoretical) function within constraints, rather than the bottom line, return on investment (ROI) or any other suitable economic parameter in our case. The report provides predictions or projections up until the year 2013, and emphasizes that research is continuously providing newer concepts. These newer concepts, at least some of them, have the potential to not only make the prevalent biosensor better, but also to help increase the range of biosensor applications. The author of this report concludes that one of the factors that would lead the biosensor industry into the next generation is a biosensor that is based on a single-platform that one is able to adapt to multiple test/detection procedures. This author suggests that to successfully survive, at least the following criteria should be followed: user-friendly instruments in critical applications, monitoring continuously the evolving and ever-changing requirements for the different end-users, and also catering to their needs. Besides, one needs to keep in touch with the rapidly changing technological changes, for example, the recent changes in nanotechnology/nanobiotechnology, and as it applies to the growth of the biosensor markets.

In another report entitled "Biosensors open umpteen opportunities for security and biodefense detection," Frost and Sullivan (2007) point out that the demand for multiple applications is increasing. The report emphasizes the need for efficient, compatible, and user-friendly detection for the diagnosis and monitoring of chronic diseases such as cancer and diabetes. The report indicates that though biosensors are exhibiting a double-digit growth, there are still a few challenges and problems that need to addressed. The area of each biosensor application, besides glucose monitoring in diabetic patients, is narrow, but the cost of development for each biosensor application according to this report is around $30 million. Thus, according to Frost and Sullivan (2008) this high development cost constrains the development of biosensors for other applications.

This report indicates that the demand for biodefense biosensors is expected o grow at a double-digit annual rate. Precision and the time taken for detection still remains a major handicap. By the year 2013, this report estimates that the annual growth rate will be around 16.9%. The CAGR (compound annual growth rate) of 14.6% is projected for the years 2007 till the year 2013.

The report also provides some market share for biosensor applications (a) West Nile virus 12.9%, (b) SARS 12.7%, and (c) *Escherichia coli* 12.1%. Though no numbers are provided, at least some sort of perspective has been provided. It should be noted that these numbers were given for the year 2006. The report emphasizes that the biosensor market for the detection of West Nile virus is estimated to increase till the year 2013. Similarly, the report estimates that the biosensor market for anthrax detection will also increase till the year 2013. The report also provides biosensor market estimates for drugs such as cocaine and ecstasy. Finally, as expected, with the general continuous deterioration of geopolitical conditions, the report concludes that the highest growth market will be in the security and biodefense areas.

In a recent theoretical analysis with regard to a biosensor commercialization strategy the Graduate Institute of Management Science in Taiwan (2009), has estimated the total original biosensor market in the year 2007 at $10.8 billon. This analysis was originally published on January 1, 2005. These authors emphasize that the emerging biosensor market presents both opportunities as well as obstacles to biosensor entrepreneurs who are willing to start new biosensor companies. Technology is the key issue according to these authors: how does one effectively (a) predict the emerging biosensor market, and (b) apply new technology to a simple biosensor commercialization network. These authors were careful enough to present alternative commercialization strategies, including how to select the optimum strategy for one's particular environment and application. The authors point out that their approach is, in general, a good starting point for start-up biosensor companies. It is presumed that, as in an iterative solution, using suitable algorithms (or relevant biosensor information, in this case), helps converge to the best solution. In the biosensor case, presumably, the "optimum" solution is a dynamic one and by its very nature, continuously changing.

In a 2006 article entitled, "BioMarket trends: Market expectations outperform expectations," Genetic and Engineering News (2006) reports that biosensors have significantly penetrated the nonmedical applications market. Medical applications traditionally include glucose, cholesterol, and coagulation monitoring. Rapidly advancing technologies have resulted in a wider variety of applications, besides changing the definition of what a biosensor is. According to these authors biosensors have evolved to become more optics-based and less electrochemical-based than projected. Advances in surface characterization, molecular markers, and nanotechnology have further impacted biosensor R&D.

These authors also point out that microelectronics, smart sensors, and MEMS (microelectromechanical systems) have apparently solved most sensitivity drawbacks. However, there are drawbacks in the sense that biosensor commercialization has continued to lag behind research by several years. Besides, biosensors have to compete with existing technologies. These authors estimate that it takes about five years to get a medial biosensor to the market, and at a cost of more than $40 million. Thus, often, these authors are of the opinion that unless well capitalized, sensor developers may go out of business before attaining commercialization.

Morrow (2008) in a recent article discussing innovations that have recently been made in biosensor development points out that advances in biosensor development have recently impacted pharmacology and molecular biology significantly. The author emphasizes that label-free biosensors exhibit significant potential in the analysis of critical disease-related molecules. Furthermore, the kinetic information provided by biosensors facilitate the design of new pharmaceutical agents. The author emphasizes that the biosensor industry is driven largely by large-scale drug discovery programs that are carried out by big pharmaceuticals.

Morrow (2008) further reports that biosensors may be effectively used in a clinical setting; however, there are challenges with regard to accepting label-free technology. This is owing to the poor selectivity in the presence of crude samples, and a lack of workable instrumentation for hand-held, point-of-care (POC) applications. Furthermore, high-levels of nonspecific binding severely constrains the applications of label-free biosensors at least for clinical diagnostic tests, wherein tissue, serum, and feces may be involved. It is hoped that the gap between academic research and commercial application of biosensors will reduce, rather than widen, in the future with effective research and application, as well as appropriate feedback.

Research and Insights (2009) points out, not surprisingly, that most of the biosensor research is being done in the universities. Private companies, however, do commercialize these biosensors. There is a need to minimize the cost of these biosensors, as they are, generally, expensive devices, and there is an ongoing effort to make them more economical. This report points out that customized biosensors have the potential to open up markets for themselves in high value-added products. For example, a biosensor can rapidly and accurately, continuously measure in real time the absence or presence of different concentrations of different analytes. This has been used to help develop a handheld alcohol sensor. Also, with illicit drug use, time is of the essence to help an overdosed victim by identifying the cocktail of drugs that an individual has ingested. This can help save lives that may otherwise be unnecessarily lost. An important piece of information provided by this report is the list of key companies to watch and the developers, as well as the timelines required for commercializing the different biosensor technologies. Furthermore, it provides information on the key technological advances that are taking place in the different laboratories that are involved in biosensor R&D.

One of the distinct advantages of biosensors is the continuous real-time monitoring at close proximity. Thus, this is very suitable for health-care applications such as patient monitoring. Other applications could also include monitoring of industrial operations and of exposure of humans to chemical and biological hazards, especially under warfare conditions. Biosensors are successful in economically monitoring health care due to the large number of patients, for example, blood glucose monitoring for diabetics or prediabetics. However, the volume of "target applications" severely constrains the applications of biosensors especially for non-health applications, where the volume of targets is relatively small. One is thus unable to offset the large expense and time required (to bring the biosensor from the bench to the market) to develop the biosensor for a specific analyte. In general, there should be either an order of magnitude increase in the number of targets or an order of magnitude decrease in the development cost and time to market. This time to market is also of importance, owing to intense competition. Another way, for biosensors to provide a reasonable ROI, is if one is able to find a high-end market application, as indicated earlier, for the biosensor with a reasonable number of possible targets.

Because a biosensor traditionally involves converting a biological or chemical signal to an electrical signal (output), in other words, a transduction, expertise in both areas is essential, as otherwise this can very significantly affect the efficiency of the biosensor. This increases the expense of the biosensor development. Other cost sensitive parts in biosensor development is the fabrication of biosensors in bulk, and the constraints of single-use or at best using a biosensor for a few times only. This is owing to contamination and reproducibility issues. Hence, the emphasis on "disposable" type biosensors; but again these, by their very nature, need to be inexpensive.

Biosensors have different inherent performance parameters such as sensitivity, selectivity, response time, simplicity (ease of use), reproducibility, robustness, detection time, and detection limits. To economically justify the use of biosensors for different and varying applications, it is necessary to clearly justify the substantial improvement or enhancement of one (or preferably more) of the above mentioned performance parameters over other conventional means of detection. This is especially true if it is not possible to justify the use of one method of detection over the other strictly on an economic basis. All of this points to the formation of carefully-thought out partnerships wherein different and synergistic strengths may be brought to the table, keeping in mind at all times, that the market is not very large, and any failure should ensure that the loss is not overly unbearable. More often than not reasonable investment sources are required, keeping in mind that it will be close to 10 years before any sort of reasonable profit may be made from these biosensor devices. It is therefore, not surprising that big companies with deep pockets have dominated the biosensor market. This is amply demonstrated by the acquisition of Biacore by General Electric a few years ago. It must be remembered that Biacore in Sweden developed and marketed the first operating surface plasmon resonance (SPR) biosensor.

Biacore was the first company to manufacture and market the SPR biosensor. However, other companies have also joined in marketing the SPR biosensor because of its inherent advantages. For example, Sierra Sensors GmbH in Hamburg, Germany makes a label-free detector that is extremely powerful for the characterization of molecular interactions. This, the company claims becomes a powerful research tool when combined with correct supporting technologies.

This company combines label-free detection with modern microfluidic sample delivery, along with sensor design and automation. These biosensors are practical to operate. They have just launched a new SPR-2 analytical system. Some of the parameters that their biosensor can determine includes binding specificity (which molecules bind to the target), kinetic rates (the speed of the interactions), affinity constants (how strong are the interactions, what are the on-off rates, that is the binding and the dissociation rate coefficients), concentration (how much target is present), and structural change (does the binding or dissociation, for example, result in any structural change). A further distinct advantage of their system is quick results, direct and even multistage analysis, and minimal sample preparation. Their high-performance biosensor systems are well-suited for both research and industrial environments.

The Economic Times (2008) recently reported that the Indian Institute of Technology (IIT) in Mumbai has developed a biosensor to detect heart attacks, and that the biosensor will be in the market soon. The low-cost biosensor "i-Sens" was developed by the Center for Excellence on Nanoelectronics at IIT and is expected to detect heart attacks well in advance before they occur. The "i-Sens" biosensor will detect acute myocardial infarctions and should have gone into commercial production at the end of the year 2008. The "i-Sens" biosensor consists of an indigenous and sensitive detector that works with a table-top box that costs Rupees 5000-10,000, which is equivalent to $98.9-$197.8 (Exchange rate $1 = 50.57 Rupees, April 2, 2009). The Director of the project, Dr. V. Ramgopal Rao indicates that a disposal card costs only a few hundred rupees each time a test is conducted. The cost per patient is less than 1000 Rupees. This is an order of magnitude less costly than Rupees 20,000 which is the cost of the existing test. Dr. Rao further adds that as this is only a diagnostic tool, the field trials are not expected to take time. A particular advantage of the "i-Sens" is that it warns a possible patient 6 months in advance. This makes the prognosis definitely better, as any advance warning of an impending heart attack is extremely beneficial to both the patient as well as the attending physician.

In a report on medical and biological sensors and sensor systems, Kalorama Information (2002) estimates that it takes 5 years and $40 million to get a medical sensor to the market. This is data for the year 2000. A 7-8 year inflation index needs to be incorporated in the cost numbers. The time to take a biosensor from the bench to the market may have also, in all probability, increased say from 7 to 9 years. This report examines and analyzes the strategies and trends (note 7-8 years ago) in the medical sensor market. It also provides forecasts for the medical and medical sensor systems. Once again, it has to be borne in mind that this is a relatively old report. The biosensor market is such a dynamic and changing field, that even a report that is 1 or 2 years old, may be out of date, especially in certain areas. This is primarily because of the rapidly expanding research in areas such as nanotechnology, and microfabrication techniques, which have and will continue to have a very direct impact on biosensor development, and subsequent commercialization.

Takeda Pacific (2005) in a 81-page, $1995 report on medical biosensor applications predicts the condition of the market up to the year 2008. The report is relatively old. Nevertheless, the report presents key information, analysis, and developments that were driving the biosensor market. The report points out the key medical applications that biosensors are involved in. This includes home blood glucose monitoring used for diabetes management as well as clinical applications for POC devices.

This report also points out that the increase in obesity in the developed world is exacerbating the increasing number of prediabetics and diabetics, and is driving the need for better diabetes management devices. In hospitals, this report notes that biosensor based devices are required for monitoring blood gases, other vital signs, and related clinical applications. The report estimated the biosensor market growth rate at 9.7% from the year 2004 to the year 2008. In the year 2004, the estimated market size was $7.1 billon. For the years 2005, 2006, 2007, and 2008 the global biosensor market was estimated to be $7.8, $8.5, $9.3, and $10.2 billion, respectively.

With the advent of modern technology, such as nanotechnology and nanobiotechnology, and its direct impact on biosensor development, and the increasing obesity worldwide, the assumption that the biosensor market will exhibit at least a 10% growth rate for the next 5 years, from the years 2009 through 2013 is probably a conservative estimate. Thus, for the years 2009, 2010, 2011, 2012, and 2013, the global medical biosensor market may be estimated to be $11.2, $12.3, $13.5, $14.9, and $16.4 billion, respectively.

The report also provides insights into products and trends. Even though the report is old, nevertheless, it is useful in providing some general trends which are presumably inherent to the biosensor market and its development. Finally, the report should be useful for CEOs, vice-presidents, directors of business development, research directors, entrepreneurs, etc.

In an earlier report by Business Communications Company, Inc. (2005) (bccreseach.com), RB-159R titled, "Biosensors and Bioelectronics," the estimated global market was expected to grow from $6.1 billion to $8.2 billion in the AAGR (average annual growth rate) of 6.3%. This is lower than the expected growth rate mentioned in the above report. This is not entirely unexpected, as these growth rates are based on the assumptions and the biosensor considered by each author Nevertheless, these sets of numbers provide one with a reasonable estimate of the market size. Surely, if more predictions are available of the global market size up until the year 2009, and perhaps for later years, it would help increase the reliability of these numbers. Inherently, these predictive numbers for growth rates and market sizes have rather low reliability factors. The further into the future these predictions are made, the lower will be the reliability of these numbers, thanks to the rapidly changing dynamics in the biosensor field.

A recent report by reportlinker.com (2008) titled, "Blood glucose testing and diabetes management," published in July 2007 indicates that the global market for glucose testing products is undergoing a significant change due to the introduction of new technologies and developments. This is an expensive ($3400), 286 page report, and explains that even though the IVD industry is mature, there is considerable room for growth in the home testing devices for diabetes management, especially in the area of noninvasive management. The report indicates that consumers prefer to test at home (without visiting a doctor), and the less invasive the procedure the more comfortable consumers are; besides there is considerably better testing compliance.

Furthermore, a whole wide range of self-testing devices are available now; and some are even given away for free. In these cases, it is important to recognize that the money is made through the testing strips, which can range from 60-80 cents per strip. This is based on the business model of the printer-ink cartridge, or the shaving razor-blade model, wherein, for example, the printer is relatively cheap, but the ink cartridges are expensive; or alternatively the shaving razor is relatively inexpensive, but the company makes its money via the razor blades.

This is an expensive report, but rather exhaustive. Some of the chapters address the following topics (in chronological order): economics of biosensors, business trends in glucose testing, confluence of new technology, important technology trends, biosensor technology, the market players, increased market penetration, costs of doing business in Europe, cost containment in Europe, price competition, government regulation of medical devices in different countries such as US, UK, France, Japan, FDA regulations, legal liabilities of glucose meters, etc. Also, corporate profiles of different companies are given.

SRI, Consulting Business Intelligence (2009) points out the advantages and disadvantages of using biosensors. The advantages include the ability to measure the different analytes accurately, quickly, and with a high degree of specificity. However, the disadvantages include the instability of the biological molecule as it is removed from its native environment. The report emphasizes that in nonhealth care applications biosensors will find their mark or niche only when inexpensive and reliable biosensors become available.

Various electronic and life-science companies are becoming interested in biosensors owing to the significant technologies that have become available. However, this report explains that varying skills are required in this interdisciplinary area. More often than not, to be successful, strategic partnerships will be required.

Fabrication of these biosensor devices in bulk will always be a constraint. Furthermore, and as indicated previously, this report also points out that the transduction of the biological signal to an electrical signal will continue to challenge the best of minds involved in the development of miniaturized next generation biosensors. Clearly, biosensor companies need to exploit niches, especially as quite a few competitive analytical technologies exist at present.

The Fraunhofer Institute (IBMT, Institute Biomedizinische Technik) (2009) indicates that biosensors and immunoassays are cost-efficient techniques to detect analytes in complex samples. The authors add that there is a growing need for multiplexed tests; for example, hormone profiling in menopausic women. These authors emphasize that a flexible technological platform which permits the detection of several analytes in parallel will significantly reduce the time-to-market. They point out that miniaturized multiplex biochips and biosensors are the key to economical and maketable applications. They further report that if minimum system costs are of extreme importance then electrochemical schemes can be very beneficial. They have developed regenerable immunoassay chips for the parallel detection of several steroid hormones (e.g., progesterone, estradiol, and testosterone).

In a recent report titled "Emerging Healthcare Applications, Market Trends Analysis," Infoshop (2009) has written a report that is beneficial to strategic planners, marketing managers at medical device companies in the life science and biotechnology companies who may be interested in nanotechnology and its impact on their devices. A set of companies of current interest that were targeted were diagnostic companies interested in nanosensors and nanoarrays that are capable of rapid and real-time detection of single molecules. Also, microarrays and biochip companies would find the analysis of interest. The authors of the report interviewed key business development personnel, CEOs, and marketing executives, amongst others. In these expert interviews the questions addressed included strategic focus, technologies under development, diseases targeted, strategic alliances, barriers to expansion, time to technology introduction, sales data, and revenue projections. Technological and market limiters were also addressed. Apparently, quite a few biosensor based companies may find the information in this report useful in some form or the other. Of course, this is with the implicit understanding that these types of reports lose their significant impact or usefulness rather quickly owing to the rapidly changing biosensor field.

Harrop (2009) of IDTechEx reports that Faunhofer, IZM, Germany in an article on manufacturing microfluidic biosensors explains that high throughput manufacturing of biosensors is still a challenge. This, according to them, is owing to the survival of the biological component on the one hand, and current MEMS technologies on the other. This they emphasize has been directly responsible (hindrance) for the commercialization of biosensors for a wide range of applications. The manufacturing challenges lie in the lack of compatibility of MEMS processes and biomolecular survival. They emphasize that, for example, if the biosensor reaction system requires more than a single reagent for the assay, then only closed microfluidic systems will provide the required functionality. One of the keys is to stabilize the biomolecules by immobilization and by protective measures. Also, novel assembly methods need to be developed in order to increase the survival of the biomolecular compounds. Emphasis needs to be placed at the interface where the chemical or biological signal is transduced to the electrical or other measurable signal.

Ingenta (2009) recently pointed out that there has been an exponential rise in the development of biosensors during the last 40 years or so. This is primarily owing to the high selectivity and sensitivity offered by the biosensor in detecting the analyte. The key to biosensor development lies in the transduction step and the biological receptor. However, Ingenta (2009) explains the importance of the immobilization step which significantly influences shelf-life and surface regeneration. Thus, both random as well as oriented immobilization procedures are being explored and analyzed. For example, Ingenta (2009) reports that among the random procedures being explored are adsorption, entrapment, cross-linking, and electrostatic interactions. Among the oriented immobilization procedures being explored are covalent binding and affinity interactions. These provide controlled and reproducible receptor surfaces.

The Ministry of Foreign Affairs of Denmark (2007) reports that the Danish biotech firm Areas A/S has been selected by the World Economic Forum (WEF) as a Technology Pioneer. This is because of their development of a biosensor for the detection of landmines and unexploded ordnance devices. This Areas biosensor can sense nitrogen dioxide in the soil and exhibits a change in color from green to red near the vicinity of the landmine. The importance of this detection device cannot be overstated, as landmines kill or maim around 27,000 persons each year. Landmines continue to be a hazard even where original conflicts have subsided. This recognition technique should assist in raising more capital for its biosensor development. Areas has already raised 55 million Danish Kroner (around U.S. $9.4 million) from its bourse listing on First North. This is an alternate marketplace for small growth companies, and is part of OMX Nordic Exchange (The Ministry of Foreign Affairs, Denmark, 2007).

In an interesting communication Sullivan (2009) explains the commercial opportunities in diagnostics and in drug development for cancer biomarkers from early discovery through to the clinic. The author does caution, however, that the discovery of cancer biomarkers, and the discovery and validation of assays is not only complex, but is also expensive. Besides, there is no guarantee that an investment will provide a commercial return.

This author provides an example of pairing a diagnostic test with a specific therapeutic agent. The author cites the example of the immunoassay for HER2 protein (Hercep Test; Dako) and the monoclonal antibody trastuzumab (Herceptin; Genentech). This author points out that this is an example of an incentive to develop such tests. These tests support the use of Hereptin in patients who are shown by the assays to be HER2 positive. Sullivan (2009) points out that these tests therefore drive commercial development.

The Freedonia Group (2008) recently published a report on the likely demand for sensors till the year 2012. This is an expensive 328-page report that costs U.S. $4600. It even has a U.S. $30 per page charge. The report indicates that U.S. sensor demand will grow 4.3% annually through the year 2012. Motor vehicles will exhibit an increasing major-sensor containing products. The report further states that new technologies, such as MEMS-based imaging will exhibit one of the fastest growth gains. According to this report the automotive industry will remain the biggest market. This may be tempered by the current turmoil in the financial and housing markets, and in the automotive industry. This report was written and completed before the current downward trend in the market as a whole. The report adds, however, that growth in the military and in the aerospace industry will be strong.

The study analyzes the $10.3 billion U.S. sensor industry. The report is careful to breakup its analysis into:

(a) Sensor type—Process variable, chemical property, proximity and positioning, physical property, and imaging.
(b) Market—motor vehicles, industrial, military/aerospace, and electronic security.

Apparently, this report does not consider the medical IVD (biosensor) industry which is a major player in the biosensor market. In any case, the analysis is worth mentioning as it places in perspective, at least in some sense, the biosensor and sensor markets.

Over the years the biosensor market and its economics have begun to develop and take shape, and Cranfield University in the United Kingdom (2009) has introduced a short course titled, "Introduction to Biosensor Technology". In the year 2009, the course was offered on May 07 and 08, at Cranfield Health on the Cranfield campus. This two-day residential course provides an introduction to the biosensor market. It enumerates the key drivers for future research and development. The course also provides an overview of the principal manufacturing techniques, and some comments on the future technological and market trends. It is of interest to note that though quite a few courses on biosensors are taught throughout the world at different universities, and courses are offered at industrial organizations specific to the need of these organizations, this is presumably the first course that discusses the biosensor market and where it may be heading in the future. This type of information, as expected, is difficult to obtain in the open literature. Thus, Cranfield University is clearly doing the biosensor manufacturers a favor by providing some sort of science in the development of the biosensor market.

As expected there is emphasis on diabetes and its role as a medical and economic driver for biosensor development. POC tests are important in the health area, and as expected the course analyzes the recent advances in the POC market.

17.3 Examples of Investment/Financing in Biosensor Companies

This section provides some examples of investment/financing in biosensor companies. The examples were selected at random from what is available on the internet. They are provided here together for instructive purposes. There is no hidden agenda behind the presentation of these examples. The readers may be advised to accept them as is, and reach/arrive at their own conclusion(s) based on their readings of just a particular example, or a set of examples. At the very least, the examples should indicate the amount of financing/investment that is required for a biosensor start-up, to keep it going, and hopefully reach a profitable state.

It would be helpful to have information on how many start-ups begin, and manage to go through the different stages of financing and development (from bench scale to bedside) before the biosensor is ready to be introduced into the market. This sort of information is not easily available. Perhaps, it exists with the more successful and presumably larger biosensor companies or with venture capitalist companies who finance these types of start-up companies or support such types of entities during the different stages of their development.

Analytica-World (2009) indicates that MedMira, Inc. located in Halifax, Nova Scotia, Canada has received $3 million from ACOA's Atlantic Innovation Fund for its biosensor division for the development of revolutionary diagnostics. The company reports that this infusion of capital from the Atlantic Innovation Fund (a government of Canada entity) will support the development of the Maple Bioscience biosensor technology. This new diagnostic instrumentation proposed to be developed will replace the 30 year old technology. Furthermore, it will place the company in a favorable position to compete in the U.S. diagnostic market valued at U.S. $22 billion.

Bioportfolio (2008) recently indicated that U.S. genomics has been awarded a $9.1 million contract for the development of an advanced biosensor by the U.S. DHS (United States Department of Homeland Security). This Phase IIIX contract under the Bioagent Autonomous Networked Detectors (BAND) program will permit the company to continue its development, testing, and optimization of its sophisticated biological sensor for the detection of airborne pathogens using single molecule DNA mapping technology. U.S. genomics CEO, John J. Caneppa indicates that the resources provided under the new contract will permit the company to carry out extensive operational testing, and to advance the capabilities of its prototype systems. The company indicates that their prototypes demonstrate the capabilities of their unique proprietary approach to rapidly detect multiple bacterial pathogens, toxins, and viruses simultaneously in an environmental sample using a single reagent set. The company points out that this should help in improving the security of U.S. citizens. Furthermore, their detection technique is flexible enough to be applicable to forensic, human diagnostic, and military applications.

The Oxford Investment Opportunity Network (OION) (2009) has provided Cybersense Biosystems Ltd. in Oxford's Center for Ecology and Hydrology with £225,000 (equivalent to $330,075; exchange rate £I = $1.467 April 11, 2009). This infusion of capital will permit the company to develop a prototype and conduct nationwide trials on its ROTAS™ portable instrument. This portable instrument permits the rapid detection of toxicity in contaminated land. Legislation aimed at cleaning up contaminated land and water resources has apparently spurred the market in this area, which is estimated to be £440-450 million (equivalent to U.S. $587-660 million). The ROTAS™ instrument is meant for day-to-day testing at work sites where workers may be exposed to toxic chemicals. OION explained that Cybersense presented a sound business proposition, which was based on innovative technology. Besides, there is a growing market for their proposed instrumentation.

Innovative Biosensors, Inc. (IBI) is a company in College Park, Maryland that is developing rapid, ultrasensitive tests to detect harmful pathogens. It has recently expanded its round table financing to $6.25 million that includes additional investors, which are Chart Venture Partners and CNF Investments, LLC, an affiliate of Clark Enterprises, Inc. (CEI). IBI reports that it will use the cash infusion to develop human clinical assays and the commercialization of biodefense testing products. Joe Hernandez, President, IBI explains that this additional financing will hasten the availability of these rapid tests to markets that require quick and accurate results with critical, time-sensitive decision making.

Information is required instantaneously in the biodefense and human clinical markets, and IBI's CANARY™ technology (developed at MIT) does have the capacity to provide this information. One of the investors, CNF adds that they are always looking for emerging industry leaders, and they are looking forward to supporting IBI as it helps guide and support the future of pathogen detection. Besides, this is dual-purpose technology, in that it is applicable to both the defense and the human clinical (commercial) sectors. Besides, the overall intention of IBI is to develop detection devices for broad applications in food testing, animal health, human health care, drug development, and disease diagnosis.

Vargas (2008) reports that investors are collaborating with Cambridge Medial Innovation (CMI) to help develop its handheld, low power device that can detect biological agents rapidly and accurately. Some of the organizations include the U.S. military, the UK government, Unipath, and a leading technology consultancy company. Inverness Medical Innovations (IMI) set up CMI and the U.S. Army Medial Research Institute of Infectious Diseases (USAAMRIID) has given CMI $3 million to help develop its detection device. In return, CMI will give USAAMRIID access to its breakthrough acoustic detection technologies including proprietary Resonant Acoustic Profiling (RAP) system.

The USAAMRIID will use this technology to optimize and to develop assays for the detection of select biothreat-viruses, bacteria, and toxins. The U.S. government has given CMI £826,000 (equivalent to U.S. $1.1 million) to help develop the hand held technology that would assist medical doctors to make instant and accurate diagnosis of diseases such as malaria and meningitis.

THE CMI device uses a simple quartz crystal element and can be powered by standard batteries. The device permits the medical doctors to make instant, accurate, at-the-bedside or in-the-field medical diagnoses from blood or other samples.

CMI emphasizes the advantages of being able to diagnose one's own viruses which lead to benefits for the consumer under the healthcare system. Their founder and managing director Dr. Matthew Cooper insists that CMI's focus on low cost, high performance tools involves slightly high risk, but likewise the rewards are high. He adds that the Life Sciences tools market is £2 billion (equivalent to U.S. $2.93 billion) and the diagnostic market that they are aiming at is £45 billion (equivalent to U.S. $65.92 billion).

CMI will focus on the home diagnostic OTC (over-the-counter) devices. Emphasis will be on infection detection. Dr. Cooper adds that it will take around 12 months to obtain regulatory approval for both Europe and the United States once the necessary R&D is finished. Incidentally, the technology which is the basis for CMI was developed in the departments of Chemistry and Pathology of the University of Cambridge.

Cambridge Healthtech Institute (CHI) (2009) in a personal communication about their conference titled, "Next generation Dx summit: development, commercialization, clinical adoption of novel assays," reports that molecular diagnostics is the fastest growing segment of the IVD market. They point out that the clinical application of molecular diagnostics to identify, diagnose, and monitor infections, cancer, and other diseases has made molecular diagnostics a strong player in healthcare. Furthermore, the range of utility of POC testing is broad, and includes influenza, HIV, emerging pathogens, cardiology, stroke, and even cancer applications. They further explain that the demand for automated and integrated systems and the number of players is growing exponentially. Also, gene expression panels from microarray data and the next generation sequencing are being used for the early detection and prognosis of cancer. The importance of sequencing has been mentioned earlier. CHI further explains that coupling molecular methods with advances in nanotechnology, microfluidics, information technology, and fabrication will all lead to newer platforms that continue to emerge in the detection and diagnosis of infectious diseases. They further point out that to make these tests reach the market clinically useful tests that are well validated and standardized need to be developed.

Finally, the medical community needs to clinically adopt these novel types of tests based on molecular diagnostics. The process by which these tests may be adopted were to be explored at the conference which was to include comments and lectures by the medical and regulatory community. Regulatory requirements are continuously changing, and the earlier one is aware of the major changes that are involved, the sooner one may modify one's development process for the diagnostic device, thereby helping to save precious resources, such as time and money. One may also help save time-to-market also.

Pisano (2009) describes a new method to develop protein assays called peptide MRM. This method, the author claims is quickly being realized as a solution to the current bottlenecks in protein biomarker development. The author explains that the new method does not use antibodies for making the proteins quantitative. The high specificity is determined prior to the assay's development. The author points out that the assay development time ranges from a week to several months (not years). The development cost for the assays can be less than $2000 per protein.

Lochhead (2009) recently analyzed a low-cost system for multiple pathogen, POC infectious disease diagnosis. The author emphasizes that cost effective POC diagnoses are a critical need for infectious diseases management, especially in resource-limited surroundings. The author points out that there is a significant gap between the low-cost, single analyte rapid tests that are in the market, and the multiplexed systems found in clinical laboratories. The author is presently a vice-president at mBio Diagnostics, and reports that mBio Diagnostics has developed a robust, low-cost fluidic cartridge and fluorescence imaging system, for POC, multiplexed protein, nucleic and cellular assays. Lochhead (2009) further states that their system leverages advances in volume-manufactured electronic components and microarray technology.

Olano (2009) recently analyzed emerging and reemerging diseases. He points out that the application of molecular diagnostic tools opens up a new era of clinical diagnostics for traditionally neglected pathogens in the clinical setting. He reports that many diseases remain alarmingly underdiagnosed due to the lack of commercially available assays. Perkins (2009) recently addressed the issue of bringing molecular diagnostics toward POC in the developing countries. The author explains that in spite of the advantages offered by molecular testing there has been very little impact on clinical care in the developing countries. This is primarily owing to the cost and the complexity of the test methods. Furthermore, the health systems in these countries are rather frail. The author examines methods by which infectious disease diagnosis may be brought to the POC in disease endemic countries.

Finally, Krieswirth (2009) recently analyzed developing rapid diagnostics for bacterial pathogens. The author explains the challenges in developing rapid diagnostics for the identification and sub-speciation of bacterial pathogens. This could arise from processing diverse primary samples to providing epidemiological information for infection control to practitioners. The author points out that the following issues are important in assay development: specificity, sensitivity, speed, cost, stability of reagents, reproducibility, scalability, platform, and ease of interpretation.

References

Analytica-world, Medmira's biosensor division to receive $3 million from ACOA's Atlantic Innovation Fund, http://www.analytica-world-.com/news/e/60934, downloaded April 9, 2009.

Bioportfolio, Genomics awarded $ 9.1 million contract, 2008, Business Communication Company (2005), RB-159R, Biosensors & bioelectronics, 2008, downloaded April 7, 2009, http://www.bcc.

Cambridge Healthtech Institute, Next generation Dx summit: Development, commercialization, clinical adoption of novel assays, Personal communication (e-mail), March 30, 2009.

Cranfield University, Introduction to biosensor technology, Short course, May 07-09, 2009, http://www.cranfiel.ac.uk/health/shortcpurses/page 26524.jsp, downloaded April 22, 2009.

Economic Times, IIT Mumbai biosensor to detect heart attack, soon in market, May 5, 2008, http://www.economictimes.indiatimes.com/News?news_by_industry/healthcae_Biotech/Health, downloaded April 2, 2009.

Franhoufer Institute, Biosensor and immunoassay chip development, www.ibmt.frauenhofer.de, downloaded April 2, 2009.

Frost and Sullivan Research Service, World Biosensors Markets, 2007, http://www.researchand markets.com/reports/c74394, downloaded March 19, 2009.

Genetic & Engineering News, Biomarket trends: Biosensor markets outperform expectations, **26**(16), September 2006, http://www.genengnews.com/articles/chitm.aspx?aid=1864&chid=0, downloaded March 31, 2009.

Global Strategic Report, Biosensors in Medical Diagnostics, downloaded March 19, 2009, http://www.earthturns.org/articles/biosensorsinmedicaldiagnostics.

Graduate Institute of Management Science, Taiwan, ROC, Biosensor commercialization strategy—A theoretical approach, originally published January 1, 2005, journal unknown, http://www.ncbi.nlm.nih.gov/pubmed/15574353, downloaded March 18, 2009.

Harrop P, Manufacturing microfluidic biosensors, downloaded April 2, 2009, http://www.printedelectronicsworld.com/articles/manufacturing_microfluidic-biosensors.

Ingenta, Biomolecule immobilization in biosensor development: tailored strategies, http://www.ingentsconnect.com/content/ben/ppl/2008/0000001/art00002?craw, downloaded April 2, 2009.

Innovative Biosensors Inc., Innovative Biosensors Inc. (IBI) expands A round financing to total $ 6.25 million, info@innovativebiosensors.com, October 16, 2006, College Park, Maryland, USA.

Kalorama Information, Medical and biological sensors and sensor systems: markets, applications and competitors worldwide, 2002, publication ID: KLI751025, July 2002, c/o Marketresarch.com, 641 Avenue of the Americas, 5th Floor, New York, NY 10011-2002.

Kriesworth BN, Developing rapid diagnostics for bacterial pathogens. In *Next Generation Dx Summit: Moving Assays to the Clinic*, Washington, DC, August 10-12, 2009.

Lochhead M, Low-cost system for multiple pathogen, point-of-care infectious disease diagnosis. In *Next Generation Dx Summit: Moving Assays To The Clinic*, Washington, DC, August 10-12, 2009.

Market Research, Medical biosensor applications and market to 2008—Diabetes management, point-of-care and related applications, http://www.marketresearch.com/product/display.asp?productid=1100483&g=1, downloaded April 2, 2009.

Morrow KJ, Innovations mark biosensor development: advances improve quality of information for component interactions and screening, *Genetic and Engineering News*, **28**(5), 2008, http://www.genengnews.com/articles/chitem.aspxaid=2393&chid=1, downloaded March 19, 2009.

Newsguide US, Diabetes to reach $6.1 billon by 2012, http://www.newsguied.us/halth-medical/diabetes/Market-for-Biosensors-to-reach 6-1 billion, downloaded April 15, 2009.

Olano JP, Emerging and re-emerging infectious diseases: Molecular diagnostic approaches. In *Next Generation Dx Summit: Moving Assays To The Clinic*, Washington, DC, August 10-12, 2009.

Oxford Investment Opportunity Network, Cybersense biosystems, Ltd., http://www.oion.co.uk/success.cybersense.html, 2009.

Pisano M, New method provides the solution for the bottleneck in protein assay development. In *Next Generation Dx Summit: Moving Assays To The Clinic*, Washington, DC, August 10-12, 2009.

Reportlinker.com Blood glucose testing and diabetes management, http://www.reportlinker.com/p050718/bood-glucose-testing.html, July 2007.

Research and markets, Biosensors: Emerging technologies and growth opportunities, (technical insights), markets.com/reportinfo.asp?report_id364432, downloaded April 14, 2009.

SRI Consulting Business Intelligence, Bringing futures into focus, Biosensors, http://www.sric-bi.com/explorer/BS.shtml, downloaded April 9, 2009.

Sullivan L, Personal communication, Commercial opportunities for cancer biomarkers, email received April 8, 2009.

Takeda Pacific, Medical biosensor applications and market to 2008—Diabetes management, point-of-care and related applications, downloaded April 5, 2009, http://www.mindbranch.com/products/R183-013.html.

The Freedonia Group, Sensors to 2012—Market research, market share, market size, sales, demand forecast, market leaders, company profiles, industry trends, Study 2377, 328pp, http://www.freedoniagroup.com/Sensors.html, 2008.

The Infoshop.com, Emergimg healthcare applications, market research trends, analysis, http:www. the-infoshop.com/study/tv19465_emergimg_healthcare_.html, downloaded April 7, 2009.

The Ministry of Foreign Affairs of Denmark, WEF honors Danish biotech firm for landmine biosensor, http://www.invstindk.com/visNhed.asp?artikelID=16814, downloaded April 15, 2009, 2007.

Vargas L, Recycled Cambridge biosensor technology attracts millions in investment, http://www.businessweekly.co.uk/2008082732395/life_science/recycled-cambridge-biosensor, downloaded April 16, 2009, August 27, 2008.

Wei N, In-vitro diagnostics market report and analysis, 2008-2023, 2009, Personal communication, April 16, 2009, Visiongain Ltd. BSG House, 26-236 City Road, London, ECIV 2QY.

Index

Note: Page numbers followed by *f* indicate figures *t* indicate tables.

G

O

T